MATHEMATICS FOR THE MANAGEMENT, LIFE, AND SOCIAL SCIENCES

Larry J. Goldstein
David C. Lay
David I. Schneider

Department of Mathematics, University of Maryland

Prentice-Hall, Inc., Englewood Cliffs, N.J. 07632

MATHEMATICS FOR THE MANAGEMENT, LIFE, AND SOCIAL SCIENCES

second edition

Library of Congress Cataloging in Publication Data

GOLDSTEIN, LARRY JOEL.
    Mathematics for the management, life, and social
sciences.

    Rev. ed. of: Modern mathematics and its applications.
c1980.
    Includes index.
    1. Mathematics — 1961 –   . I. Lay, David C.
II. Schneider, David I.   III. Title.
QA37.2.G64 1984   510   83-3081
ISBN 0-13-562512-2

Editorial/production supervision: Kathryn Gollin Marshak
Interior design: Lee Cohen and Walter A. Behnke
Cover design: Lee Cohen
Cover art: Paul Silverman
Manufacturing buyer: John Hall

MATHEMATICS FOR THE MANAGEMENT, LIFE, AND SOCIAL SCIENCES, second edition
LARRY J. GOLDSTEIN / DAVID C. LAY / DAVID I. SCHNEIDER

© 1984, 1980 by Prentice-Hall, Inc., Englewood Cliffs, New Jersey 07632

Printed in the United States of America

10 9 8 7 6 5 4 3 2 1

Previously published under the title MODERN MATHEMATICS AND ITS APPLICATIONS

ISBN 0-13-562512-2

PRENTICE-HALL INTERNATIONAL, INC., *London*
PRENTICE-HALL OF AUSTRALIA PTY. LIMITED, *Sydney*
EDITORA PRENTICE-HALL DO BRASIL, LTDA., *Rio de Janeiro*
PRENTICE-HALL OF CANADA INC., *Toronto*
PRENTICE-HALL OF INDIA PRIVATE LIMITED, *New Delhi*
PRENTICE-HALL OF JAPAN, INC., *Tokyo*
PRENTICE-HALL OF SOUTHEAST ASIA PTE. LTD., *Singapore*
WHITEHALL BOOKS LIMITED, *Wellington, New Zealand*

Chapter Opening Photo Credits

Mathematics and Its Applications

This volume is one of a collection of texts for freshman and sophomore college mathematics courses. Included in this collection are the following.

Calculus and Its Applications, third edition by L. Goldstein, D. Lay, and D. Schneider. A text designed for a two-semester course in calculus for students of business and the social and life sciences. Emphasizes an intuitive approach and integrates applications into the development. Much expanded from the highly successful second edition.

Brief Calculus and Its Applications, third edition by L. Goldstein, D. Lay, and D. Schneider. Consists of chapters 0–7 of the above book. Suitable for shorter courses.

Finite Mathematics and Its Applications, second edition by L. Goldstein and D. Schneider. A traditional finite mathematics text for students of business and the social and life sciences. Allows courses to begin with either linear mathematics (linear programming, matrices) or probability and statistics.

Mathematics for the Management, Life, and Social Sciences, second edition by L. Goldstein, D. Lay, and D. Schneider. A text for a two-semester course covering finite mathematics, precalculus, and calculus.

Contents

Algebra and Its Applications

* Sections preceded by a * are optional in the sense that they are not prerequisites for later material.

II   Linear Mathematics and Its Applications

Preface

We have been very pleased with the enthusiastic response accorded the Mathematics and Its Applications Series, of which this volume is a part. We have been privileged to hear from a number of teachers and students. The present revision incorporates many of their suggestions.

The current edition, like its predecessor, is designed for a one-year course introducing various topics in algebra, linear mathematics, probability and statistics, and calculus to students of business and the biological and social sciences. Although there are many changes in this edition, we have preserved the approach and the flavor. Our goals remain the same: to present mathematics in an intuitive yet intellectually satisfying way, and to illustrate the many applications of mathematics to the biological, social, and management sciences. We have tried to achieve these goals while paying close attention to students' real and potential problems in learning calculus. Our main concern, as always, is: Will it work for the students? Listed on the following pages are some of the features that illustrate various aspects of this student-oriented approach.

Applications We provide realistic applications that illustrate the uses of calculus in other disciplines. The reader may survey the variety of applications by turning to the Index of Applications on page xvii. Wherever possible, we have attempted to use applications to motivate the mathematics. For example, the integral is introduced in Chapter 16 via a discussion of world oil consumption.

Examples We have included many more worked examples than is customary (656). Furthermore, we have included computational details to enhance readability by students whose basic skills are weak.

Exercises The 3283 exercises comprise about one-quarter of the text — the most important part of the text in our opinion. The exercises at the end of the sections are usually arranged in the order in which the text proceeds, so that the homework assignments may easily be made after only part of a section is discussed. Interesting applications and more challenging problems tend to be located near the ends of the exercise sets. Supplementary exercises at the end of each chapter expand the other exercise sets and provide cumulative exercises that require skills from earlier chapters.

Practice Problems The practice problems introduced in the previous edition have proved to be a popular and useful feature and are included in the present edition. The practice problems are carefully selected exercises that are located at the end of each section, just before the exercise set. Complete solutions are given following the exercise set. The practice problems often focus on points that are potentially confusing or are likely to be overlooked. We recommend that the reader seriously attempt the practice problems and study their solutions before moving on to the exercises. In effect, the practice problems constitute a built-in workbook.

New in This Edition

Among the many changes in this edition, the following are the most significant.

1. *Additional Examples and Exercises.*   The already ample stock of examples and exercises has been expanded, the examples by 15% and the exercises by 34%. The new examples introduce new topics and applications, and provide further explanation where needed. Many of the exercise sets have been revised with some "drill" problems added. Among the new exercises are some that will challenge the better students.

2. *Algebra Review.*   The first part of the book is devoted to algebra review. This section is significantly expanded and now contains enough material for a full quarter course. This material may, however, be omitted without loss of continuity. The remaining material is still sufficient for a full-year or three-quarter course.

3. *Mathematics of Finance.*   Our chapter on the mathematics of finance now treats the subject from a traditional viewpoint. This treatment relies on the use of either a calculator or the financial tables included in the Appendix.

4. *Duality Theory.*   We have added a section on the duality theorem for the simplex method. This section may be used to provide an independent treatment of minimum problems or it may be used as a supplement to the discussion of minimum problems in Section 8.3.

5. *Probability.*   We have rewritten certain portions of the discussion on probability, especially the discussion of conditional probability and independence of events.

6. *The Exponential and Natural Logarithm Functions.*   The two basic chapters on these functions have been revised somewhat and combined into one chapter. The applications chapter has also been revised and enlarged to include a section on percentage rates of change and elasticity of demand.

7. *Limits.*   We have rewritten Section 11.3 for greater clarity. However, as in the previous edition, the instructor need cover only the first two pages of this section in order to introduce the limit notation. This is the only prerequisite from Section 11.3 for the remainder of the book. In addition, Section 11.4 on Differentiability and Continuity is optional.

8. *Additional Material.*   In addition to the new topics mentioned above, we have added a section on related rates and implicit differentiation. We have also added a discussion of level curves to the chapter on calculus of several variables.

The additional topics provide the instructor with considerable flexibility in curriculum design.

Answers to the odd-numbered exercises are included at the back of the book. Answers to the even-numbered exercises are contained in the Instructor's Manual.

We welcome any comments or suggestions you may have and hope that you enjoy using this text as much as we have enjoyed writing it.

Acknowledgments

While writing this book, we have received assistance from many persons. And our heartfelt thanks goes out to them all. Especially, we should like to thank the following reviewers, who took the time and energy to share their ideas, preferences, and often their enthusiasm, with us.

Reviewers of the first edition: Evan G. Houston, University of North Carolina, Charlotte; Richard Bouldin, University of Georgia; David H. Carlson, University of Missouri, Columbia; Marc Konvisser, Wayne State University; Kenneth N. Berk, Illinois State University; Rebecca Klemm, Georgetown University; Jack P. Tull, Ohio State University; James F. Hurley, University of Connecticut.

Reviewers of the second edition: Fred Wright, Iowa State University; Carroll Wells, Western Kentucky University; David Pentico, Virginia Commonwealth University; Perrin Wright, Florida State University; Gordon Brown, University of Colorado; James Buckley, University of Alabama; Faye Hendrix Thames, Lamar University.

The authors would like to thank the many people at Prentice-Hall who have contributed to the success of our books. We appreciate the tremendous efforts of the production, art, manufacturing, marketing, and sales departments. Our sincere thanks go to Kathryn Marshak for courageously and effectively undertaking the mammoth editorial task posed by our series of books, and to Lee Cohen, for her beautiful design work. An extra special thanks to Robert Sickles, Executive Editor of Prentice-Hall, for his help in planning and executing these new editions. His partnership and friendship have added a warm personal dimension to the writing process.

Larry J. Goldstein
David C. Lay
David I. Schneider

Index
of Applications

MATHEMATICS
FOR THE
MANAGEMENT,
LIFE, AND
SOCIAL SCIENCES

Basic Algebraic Techniques

Chapter 1

1.1. The Real Numbers

In this section we classify numbers and discuss some basic operations with numbers.

Each number corresponds to a point on the number line. In Fig. 1 we have labeled several points with their corresponding numbers. Numbers to the right of 0 are called *positive numbers* and numbers to the left of 0 are called *negative numbers*.

Some important types of numbers are integers, rational numbers, and irrational numbers.

☐ Integers The positive and negative whole numbers and zero are called *integers*. That is, the integers are 0, 1, -1, 2, -2, 3, -3, and so on. The sum or product of two integers is another integer.

☐ Rational Numbers A number that can be written as a quotient of two integers is called a *rational number*. Some examples are $\frac{3}{2}, \frac{17}{100}, \frac{-7}{5},$ and $\frac{4}{-8}$. Every integer is a rational number $\left(\text{for instance, } 5 = \frac{5}{1}\right)$, and so are certain mixed numbers $\left(\text{for instance, } 5\frac{1}{4} = \frac{21}{4}\right)$. All finite decimals are rational numbers $\Big(\text{for instance, } .39 = \frac{39}{100}\Big)$, and so are certain infinite decimals $\left(\text{for instance, } .333\ldots = \frac{1}{3}\right)$.

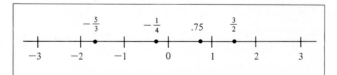

FIGURE 1

The sum, product, and quotient (provided that the denominator is not 0) of two rational numbers is also a rational number.

☐ Irrational Numbers   The totality of all numbers corresponding to points on the number line is called the *real numbers*. Those real numbers which are not rational numbers are called *irrational numbers*. Some examples are π, $\sqrt{2}$, and $-\sqrt{5}$. Irrational numbers can be thought of as corresponding to certain infinite decimals. However, when computing with irrational numbers we usually approximate the numbers by finite decimals, for instance, $\pi \approx 3.1416$ and $\sqrt{2} \approx 1.4142$.

For any number a, $-a$ is the number which when added to a gives 0. On the number line, a and $-a$ have the same distance from 0 and lie on opposite sides of 0. For instance, if $a = 5$, $-a = -5$ and if $a = -3$, $-a = 3$.

☐ Absolute Value   The absolute value of the number a, written $|a|$, is the distance from a to 0. For instance, $|5| = 5$ and $|-3| = 3$. The absolute value of a positive number is always the number itself and the absolute value of a negative number is the positive number that results from deleting the minus sign.

☐ Laws of Signs   The product of two positive numbers or two negative numbers is positive. The product of a positive number and a negative number is negative. For instance, $2 \cdot 3 = 6$, $-2 \cdot 3 = -6$, and $-2 \cdot (-3) = 6$. Using these laws, we see that for any number a, $(-1) \cdot a = -a$. The laws of signs for division are identical to the laws for multiplication.

The sum of two positive numbers is always positive and the sum of two negative numbers is always negative. The sum of a positive and a negative number has the same sign as the number of largest absolute value. Questions concerning subtraction are reduced to questions of addition by regarding $a - b$ as $a + (-b)$.

EXAMPLE 1   Compute.

(a) $12 \div (-8)$ (b) $(-3) \cdot (-1)$ (c) $3 + (-5)$ (d) $3 - 5$

Solution   (a) Since the two numbers have different signs, the quotient is a negative number.

$$\frac{12}{-8} = -\frac{12}{8} = -\frac{3}{2}.$$

(b) Since the two numbers have the same sign, the product is a positive number.

$$(-3)(-1) = 3 \cdot 1 = 3.$$

(c) The number -5 has a larger absolute value than the number 3. Hence, the sum is a negative number.

$$3 + (-5) = -2.$$

(d)
$$3 - 5 = 3 + (-5) = -2.$$

Inequalities We say that *a is less than b* (or *b is greater than a*) if *a* is to the left of *b* on the number line. We write $a < b$ ($b > a$). For instance, $3 < 7$, $2 > 1$, and $-5 < -4$. The symbol "$\leq$" means "less than or equal to." For instance, $2 \leq 4$ and $2 \leq 2$. Similarly, the symbol "$\geq$" means "greater than or equal to." (*Note:* Inequality symbols point to the smaller number.)

The notation for inequalities provides a convenient shorthand in many situations. Instead of saying that *a* is positive, we write $a > 0$. We write $a < x < b$ to mean that *x* is strictly between *a* and *b*.

Arithmetic operations on inequalities sometimes reverse the direction of the inequality. If both sides of an inequality are multiplied by the same positive number, the direction of the inequality is unchanged. However, multiplication by a negative number reverses the direction of the inequality. For instance, consider the inequality $-2 \leq 4$. Multiply both sides by 3 and we obtain $-6 \leq 12$, whereas multiplication by -3 yields $6 \geq -12$.

Adding or subtracting the same number from both sides of an inequality does not change the direction of the inequality. For instance, consider the inequality $3 \leq 5$. Adding 4 to both sides yields $7 \leq 9$ and subtracting 4 from both sides yields $-1 \leq 1$.

PRACTICE PROBLEMS 1

1. Explain why $\dfrac{3}{-4}$ is a negative number.

2. Is $-2 \geq -3$?

3. Express the statement "*a* is not smaller than *b*" using inequality signs.

EXERCISES 1

In Exercises 1–12, express the given number as a quotient of two integers.

1. $3\frac{1}{2}$	2. -5	3. $.7$	4. $1\frac{2}{3}$
5. -1	6. $-5\frac{1}{4}$	7. 21.3	8. $.009$
9. 5%	10. 1.3%	11. $\frac{1}{2}\%$	12. $13\frac{1}{4}\%$

Calculate.

13. $|-8|$	14. $|0|$	15. $|3.6|$	16. $\left|\dfrac{-2}{5}\right|$
17. $5(-3)$	18. $(-4)(-1)$	19. $\dfrac{-5}{3}(-6)$	20. $(-.1)(2.2)$
21. $5 + (-2)$	22. $-6 + 4$	23. $3 - 8$	24. $2 + (-7)$
25. $4 - (-1)$	26. $-\left(\dfrac{1}{-2}\right)$	27. $.5 - 1.1$	28. $6 - (-1.4)$

29. $|1 - 2|$ 30. $|2 - 1|$ 31. $-|-3|$ 32. $|-2| - |3|$

33. $-|2| + |3|$ 34. $-|3| - |-5|$ 35. $-4 \cdot |-4|$ 36. $|.2 - 1|$

37. $|3 - 3|$ 38. $|5| \cdot |-5|$ 39. $|-(-2)|$ 40. $|0 - 5|$

41. $\left(\dfrac{-1}{4}\right)(.3)$ 42. $.6 - \dfrac{1}{5}$ 43. $|(-3) \cdot (-4)|$ 44. $\left(\dfrac{4}{-5}\right) \cdot \left(\dfrac{-1}{2}\right)$

In Exercises 45–56, fill in the blank with $<$, $>$, or $=$.

45. $3 \_\_ 5$ 46. $7 \_\_ 4$ 47. $2 \_\_ -5$ 48. $-3 \_\_ 1$

49. $-6 \_\_ -4$ 50. $-5 \_\_ -7$ 51. $\dfrac{3}{4} \_\_ \dfrac{5}{7}$ 52. $\dfrac{-4}{5} \_\_ -.8$

53. $|5| - |4| \_\_ |5 - 4|$ 54. $\dfrac{2}{5} \_\_ \dfrac{5}{13}$

55. $|3 - 2| \_\_ |2 - 3|$ 56. $0 \_\_ -.1$

In Exercises 57–64, express the given statement using inequality signs. (See Practice Problem 3.)

57. a is nonnegative   58. b is not more than a

59. b is at least as large as a 60. a is positive

61. a is not less than b 62. a is not larger than b

63. a is not positive   64. b is at most as large as a

In Exercises 65–74, begin with the given inequality and alter it by performing the given operation to both sides of the inequality.

65. $2 < 3$, multiply by 4   66. $0 > -5$, multiply by 2

67. $-5 > -6$, multiply by -1 68. $2 \le 3$, multiply by -4

69. $-2 < 3$, add 4   70. $5 > 2$, subtract 8

71. $3 > -1$, subtract 6   72. $-7 < -5$, add 9

73. $-1 \le 3$, multiply by -2 74. $5 \ge -2$, multiply by -3

SOLUTIONS TO PRACTICE PROBLEMS 1

1. The number $\dfrac{3}{-4}$ can be thought of as $3 \div (-4)$. The quotient of a positive and a negative number is negative. Hence, $\dfrac{3}{-4} = -\dfrac{3}{4}$.

2. Yes. We have $-2 > -3$ since -2 is to the right of -3 on the number line. It is also correct to use $\ge$ since $\ge$ means "greater than *or* equal to."

3. $a \ge b$.

1.2. Linear Equations and Inequalities in One Variable

A *linear equation* (in one variable) is an equation that can be written in the form $ax + b = c$, where a, b, and c are specific numbers and $a \neq 0$. Some examples are $2x + 4 = 10$ and $-x + 4 = 7$. A solution of an equation is a number that yields a correct statement when substituted for the variable. For instance, 3 is a solution to $2x + 4 = 10$ since $2(3) + 4 = 10$ is correct.

In Example 1, we present a systematic technique for solving a linear equation.

EXAMPLE 1 Solve the linear equation $5x + 3 = 9$.

Solution Our goal is to transform the equation into the form $x = a$, where a is the solution. That is, we would like to eventually have x alone on the left side of the equation and a single number on the right. So we begin by taking all terms not involving x to the right.

To take "3" to the right side of the equation, subtract 3 from both sides of the equation.

$$5x + 3 = 9$$
$$5x + 3 - 3 = 9 - 3$$
$$5x = 6.$$

Since we want "x" on the left, not "$5x$," we multiply both sides of the equation by $\frac{1}{5}$.

$$5x = 6$$
$$\tfrac{1}{5} \cdot 5x = \tfrac{1}{5} \cdot 6$$
$$x = \tfrac{6}{5}.$$

To check that we have correctly solved the equation, we substitute $\frac{6}{5}$ for x in the original equation.

$$5x + 3 = 9$$
$$5(\tfrac{6}{5}) + 3 = 9$$
$$6 + 3 = 9$$
$$9 = 9.$$

The following operations are useful when solving any equation.

1. Add or subtract the same quantity from both sides of the equation.
2. Multiply both sides of the equation by the same nonzero quantity.

EXAMPLE 2   Solve the equation $5x + 4 = 3 - 2x$.

Solution

$$5x + 4 = 3 - 2x$$

Add $2x$ to both sides $7x + 4 = 3$

Subtract 4 from both sides $7x = -1$

Multiply both sides by $\frac{1}{7}$ $x = -\frac{1}{7}$.

The ability to solve linear equations enables us to solve certain other types of equations.

EXAMPLE 3   Solve $5 - \dfrac{2}{x} = 1$.

Solution

$$5 - \frac{2}{x} = 1$$

Multiply both sides by x $5x - 2 = x$

Now solve the linear equation $4x = 2$

$$x = \tfrac{1}{2}.$$

Linear Inequalities   The technique for solving linear equations can also be used to solve linear inequalities. However, when multiplying both sides of an inequality by a negative number we must remember to reverse the direction of the inequality.

EXAMPLE 4   Solve the linear inequality $5 - \frac{1}{3}x < 4$.

Solution

$$5 - \tfrac{1}{3}x < 4$$

Subtract 5 from both sides $-\frac{1}{3}x < -1$

Multiply both sides by -3 $x > 3$.

The solution to Example 4 was not a single number but rather the collection of all numbers greater than 3. All linear inequalities have infinitely many solutions.

EXAMPLE 5   Solve the linear inequality $3(x + 2) \geq 4 - x$.

Solution

$$3(x + 2) \geq 4 - x$$

Multiply out the left side $3x + 6 \geq 4 - x$

Subtract 6 from both sides $3x \geq -2 - x$

Add x to both sides $4x \geq -2$

Multiply both sides by $\frac{1}{4}$ $x \geq -\frac{1}{2}$.

PRACTICE PROBLEMS 2

1. Solve the equation $5 - \dfrac{x}{1 + x} = 6$.

2. Solve the inequality $4 - .3x \geq 7$.

EXERCISES 2

Solve the equations in Exercises 1–24.

1. $3x - 5 = 11$
2. $2 + 5x = 17$
3. $4 - 2x = 3$

4. $3 = 5x - 4$
5. $\dfrac{1}{3}x + 5 = 6$
6. $3 - \dfrac{x}{2} = 5$

7. $5x - 4 = 2 + 3x$
8. $\dfrac{x + 1}{2} = 3 - x$
9. $\dfrac{x}{3} + 1 = 4 - x$

10. $2(x + 3) = 4 - 5x$
11. $.3x + 2 = 8$
12. $.2x - .05 = 1$

13. $\dfrac{3x - 4}{5} = 2$
14. $5 - .4x = 2$
15. $.7(x + 3) = 4.9$

16. $\dfrac{1}{3}x - 5 = 2(x + 1)$
17. $8 + \dfrac{3}{x} = 6$
18. $\dfrac{1}{x} - 4 = 7$

19. $\dfrac{x}{1 + x} - 2 = 3$
20. $1 - \dfrac{2}{x - 2} = 3$
21. $\dfrac{3}{2x - 1} + 2 = 5$

22. $\dfrac{4x + 5}{x} = 2$
23. $\dfrac{.2x + 1}{3x} = .4$
24. $6 + \dfrac{.3}{5 - 3x} = 1$

Solve the inequalities in Exercises 25–38.

25. $3x - 2 > 10$
26. $4 - 5x < 9$
27. $-2x + 3 \leq 4$

28. $5x + 1 \geq 4$
29. $\frac{1}{2}x - 1 > 5$
30. $3 + \frac{1}{4}x < 11$

31. $5 - \dfrac{x}{5} \geq 2$
32. $.3x - 4 \geq 8$
33. $.5 + .3x \leq 2$

34. $1.1x - 3 \leq 19$
35. $4(x - 3) \geq 5 + x$
36. $3(4 - 2x) \leq x + 1$

37. $-2(x - 1) \leq 3x + 1$
38. $2(3 - 4x) \leq 3$

SOLUTIONS TO PRACTICE PROBLEMS 2

1. When a term of an equation contains the variable in its denominator, it is usually a good idea to begin by multiplying both sides of the equation by this denominator.

$$5 - \frac{x}{1 + x} = 6$$

Multiply both sides by $1 + x$ $\quad 5(1 + x) - x = 6(1 + x)$

Multiply out on both sides $\quad 5 + 5x - x = 6 + 6x$

Solve $\qquad\qquad -2x = 1$

$$x = -\tfrac{1}{2}.$$

2. $\qquad\qquad\qquad\qquad\qquad 4 - .3x \geq 7$

Subtract 4 from both sides $\qquad -.3x \geq 3$

Multiply both sides by $\dfrac{1}{-.3}$ $\qquad x \leq \dfrac{3}{-.3}$

$$x \leq -10.$$

1.3. Exponents

In this section we review the operations with exponents that occur frequently throughout the text. We begin with the definition of b^r for various types of numbers b and r.

For any nonzero number b and any positive integer n, we have by definition that

$$b^n = \underbrace{b \cdot b \cdot \ldots \cdot b}_{n \text{ times}},$$

$$b^{-n} = \frac{1}{b^n},$$

and

$$b^0 = 1.$$

For example, $2^4 = 2 \cdot 2 \cdot 2 \cdot 2 = 16$, $2^{-4} = \dfrac{1}{2^4} = \tfrac{1}{16}$, and $2^0 = 1$.

Next, we consider numbers of the form $b^{1/n}$, where n is a positive integer. For instance,

$2^{1/2}$ is the positive number whose square is 2: $\qquad 2^{1/2} = \sqrt{2}$;

$2^{1/3}$ is the positive number whose cube is 2: $\qquad 2^{1/3} = \sqrt[3]{2}$;

$2^{1/4}$ is the positive number whose fourth power is 2: $2^{1/4} = \sqrt[4]{2}$;

and so on. In general, when b is zero or positive, $b^{1/n}$ is zero or the positive number whose nth power is b.

In the special case when n is odd, we may permit b to be negative as well as positive. For example, $(-8)^{1/3}$ is the number whose cube is -8; that is,

$$(-8)^{1/3} = -2.$$

Thus, when b is negative and n is odd, we again define $b^{1/n}$ to be the number whose nth power is b.

Finally, let us consider numbers of the form $b^{m/n}$ and $b^{-m/n}$, where m and n are positive integers. We may assume that the fraction m/n is in lowest terms (so that m and n have no common factor). Then we define

$$b^{m/n} = (b^{1/n})^m$$

whenever $b^{1/n}$ is defined, and

$$b^{-m/n} = \frac{1}{b^{m/n}}$$

whenever $b^{m/n}$ is defined and is not zero. For example,

$$8^{5/3} = (8^{1/3})^5 = (2)^5 = 32,$$

$$8^{-5/3} = \frac{1}{8^{5/3}} = \frac{1}{32},$$

$$(-8)^{5/3} = [(-8)^{1/3}]^5 = [-2]^5 = -32.$$

Exponents may be manipulated algebraically according to the following rules:

Laws of Exponents

1. $b^r b^s = b^{r+s}$	4. $(b^r)^s = b^{rs}$
2. $b^{-r} = \dfrac{1}{b^r}$	5. $(ab)^r = a^r b^r$
3. $\dfrac{b^r}{b^s} = b^r \cdot b^{-s} = b^{r-s}$	6. $\left(\dfrac{a}{b}\right)^r = \dfrac{a^r}{b^r}$

EXAMPLE 1 Use the laws of exponents to calculate the following quantities.

(a) $2^{1/2} 50^{1/2}$ (b) $(2^{1/2} 2^{1/3})^6$ (c) $\dfrac{5^{3/2}}{\sqrt{5}}$

Solution (a) $2^{1/2} 50^{1/2} = (2 \cdot 50)^{1/2}$ (Law 5)

$$= \sqrt{100}$$

$$= 10.$$

(b) $(2^{1/2}2^{1/3})^6 = (2^{(1/2)+(1/3)})^6$ (Law 1)

$= (2^{5/6})^6$

$= 2^{(5/6)6}$ (Law 4)

$= 2^5$

$= 32.$

(c) $\dfrac{5^{3/2}}{\sqrt{5}} = \dfrac{5^{3/2}}{5^{1/2}}$

$= 5^{(3/2)-(1/2)}$ (Law 3)

$= 5^1$

$= 5.$

EXAMPLE 2 Simplify the following expressions.

(a) $\dfrac{1}{x^{-4}}$ (b) $\dfrac{x^2}{x^5}$ (c) $\sqrt{x}(x^{3/2} + 3\sqrt{x})$

Solution (a) $\dfrac{1}{x^{-4}} = x^{-(-4)}$ (Law 2 with $r = -4$)

$= x^4.$

(b) $\dfrac{x^2}{x^5} = x^{2-5}$ (Law 3)

$= x^{-3}.$

It is also correct to write this answer as $\dfrac{1}{x^3}$.

(c) $\sqrt{x}(x^{3/2} + 3\sqrt{x}) = x^{1/2}(x^{3/2} + 3x^{1/2})$

$= x^{1/2}x^{3/2} + 3x^{1/2}x^{1/2}$

$= x^{(1/2)+(3/2)} + 3x^{(1/2)+(1/2)}$ (Law 1)

$= x^2 + 3x.$

A *power function* is a function of the form

$$f(x) = x^r,$$

for some number r.

EXAMPLE 3 Let $f(x)$ and $g(x)$ be the power functions

$$f(x) = x^{-1} \qquad \text{and} \qquad g(x) = x^{1/2}.$$

Determine the following functions.

(a) $\dfrac{f(x)}{g(x)}$ (b) $f(x)g(x)$ (c) $\dfrac{g(x)}{f(x)}$

Solution (a) $\dfrac{f(x)}{g(x)} = \dfrac{x^{-1}}{x^{1/2}}$ (b) $f(x)g(x) = x^{-1}x^{1/2}$ (c) $\dfrac{g(x)}{f(x)} = \dfrac{x^{1/2}}{x^{-1}}$

$= x^{-1-(1/2)}$ $= x^{-1+(1/2)}$ $= x^{(1/2)-(-1)}$

$= x^{-3/2}$ $= x^{-1/2}$ $= x^{3/2}.$

$= \dfrac{1}{x^{3/2}}.$ $= \dfrac{1}{x^{1/2}}$

$= \dfrac{1}{\sqrt{x}}.$

PRACTICE PROBLEMS 3

1. Compute. (a) -5^2 (b) $16^{.75}$

2. Simplify. (a) $(4x^3)^2$ (b) $\dfrac{\sqrt[3]{x}}{x^3}$ (c) $\dfrac{2 \cdot (x+5)^6}{x^2 + 10x + 25}$

EXERCISES 3

In Exercises 1–28, compute the numbers.

1. 3^3 2. $(-2)^3$ 3. 1^{100} 4. 0^{25}

5. $(.1)^4$ 6. $(100)^4$ 7. -4^2 8. $(.01)^3$

9. $(16)^{1/2}$ 10. $(27)^{1/3}$ 11. $(.000001)^{1/3}$ 12. $\left(\dfrac{1}{125}\right)^{1/3}$

13. 6^{-1} 14. $\left(\dfrac{1}{2}\right)^{-1}$ 15. $(.01)^{-1}$ 16. $(-5)^{-1}$

17. $8^{4/3}$ 18. $16^{3/4}$ 19. $(25)^{3/2}$ 20. $(27)^{2/3}$

21. $(1.8)^0$ 22. $9^{1.5}$ 23. $16^{.5}$ 24. $(81)^{.75}$

25. $4^{-1/2}$ 26. $\left(\dfrac{1}{8}\right)^{-2/3}$ 27. $(.01)^{-1.5}$ 28. $1^{-1.2}$

In Exercises 29–40, use the laws of exponents to compute the numbers.

29. $5^{1/3} \cdot 200^{1/3}$ 30. $(3^{1/3} \cdot 3^{1/6})^6$ 31. $6^{1/3} \cdot 6^{2/3}$ 32. $(9^{4/5})^{5/8}$

33. $\dfrac{10^4}{5^4}$ 34. $\dfrac{3^{5/2}}{3^{1/2}}$ 35. $(2^{1/3} \cdot 3^{2/3})^3$ 36. $20^{.5} \cdot 5^{.5}$

37. $\left(\dfrac{8}{27}\right)^{2/3}$ **38.** $(125 \cdot 27)^{1/3}$ **39.** $\dfrac{7^{4/3}}{7^{1/3}}$ **40.** $(6^{1/2})^0$

In Exercises 41–70, use the laws of exponents to simplify the following algebraic expressions. Your answer should not involve parentheses or negative exponents.

41. $(xy)^6$ **42.** $(x^{1/3})^6$ **43.** $\dfrac{x^4 \cdot y^5}{xy^2}$ **44.** $\dfrac{1}{x^{-3}}$

45. $x^{-1/2}$ **46.** $(x^3 \cdot y^6)^{1/3}$ **47.** $\left(\dfrac{x^4}{y^2}\right)^3$ **48.** $\left(\dfrac{x}{y}\right)^{-2}$

49. $(x^3y^5)^4$ **50.** $\sqrt{1+x}(1+x)^{3/2}$ **51.** $x^5 \cdot \left(\dfrac{y^2}{x}\right)^3$ **52.** $x^{-3} \cdot x^7$

53. $(2x)^4$ **54.** $\dfrac{-3x}{15x^4}$ **55.** $\dfrac{-x^3y}{-xy}$ **56.** $\dfrac{x^3}{y^{-2}}$

57. $\dfrac{x^{-4}}{x^3}$ **58.** $(-3x)^3$ **59.** $\sqrt[3]{x} \cdot \sqrt[3]{x^2}$ **60.** $(9x)^{-1/2}$

61. $\left(\dfrac{3x^2}{2y}\right)^3$ **62.** $\dfrac{x^2}{x^5y}$ **63.** $\dfrac{2x}{\sqrt{x}}$ **64.** $\dfrac{1}{yx^{-5}}$

65. $(16x^8)^{-3/4}$ **66.** $(-8y^9)^{2/3}$ **67.** $\sqrt{x}\left(\dfrac{1}{4x}\right)^{5/2}$ **68.** $\dfrac{(25xy)^{3/2}}{x^2y}$

69. $\dfrac{(-27x^5)^{2/3}}{\sqrt[3]{x}}$ **70.** $(-32y^{-5})^{3/5}$

The expressions in Exercises 71–74 may be factored as shown. Find the missing factors.

71. $\sqrt{x} - \dfrac{1}{\sqrt{x}} = \dfrac{1}{\sqrt{x}}(\quad)$ **72.** $2x^{2/3} - x^{-1/3} = x^{-1/3}(\quad)$

73. $x^{-1/4} + 6x^{1/4} = x^{-1/4}(\quad)$ **74.** $\sqrt{\dfrac{x}{y}} - \sqrt{\dfrac{y}{x}} = \sqrt{xy}(\quad)$

75. Explain why $\sqrt{a} \cdot \sqrt{b} = \sqrt{ab}$. **76.** Explain why $\sqrt{a}/\sqrt{b} = \sqrt{a/b}$.

In Exercises 77–84, evaluate $f(4)$.

77. $f(x) = x^2$ **78.** $f(x) = x^3$ **79.** $f(x) = x^{-1}$ **80.** $f(x) = x^{1/2}$

81. $f(x) = x^{3/2}$ **82.** $f(x) = x^{-1/2}$ **83.** $f(x) = x^{-5/2}$ **84.** $f(x) = x^0$

SOLUTIONS TO PRACTICE PROBLEMS 3

1. (a) $-5^2 = -25$. [Note that -5^2 is the same as $-(5^2)$. This number is different from $(-5)^2$, which equals 25. Whenever there are no parentheses, apply the exponent first and then apply the other operations.]

   (b) Since $.75 = \frac{3}{4}$, $16^{.75} = 16^{3/4} = (\sqrt[4]{16})^3 = 2^3 = 8$.

2. (a) Apply Law 5 with $a = 4$ and $b = x^3$. Then use Law 4.

$$(4x^3)^2 = 4^2 \cdot (x^3)^2 = 16 \cdot x^6.$$

[A common error is to forget to square the 4. If that had been our intent, we would have asked for $4(x^3)^2$.]

(b) $\dfrac{\sqrt[3]{x}}{x^3} = \dfrac{x^{1/3}}{x^3} = x^{(1/3)-3} = x^{-8/3}$. [The answer can also be given as $1/x^{8/3}$.] When simplifying expressions involving radicals, it is usually a good idea to first convert the radical to an exponent.

(c) $\dfrac{2(x+5)^6}{x^2 + 10x + 25} = \dfrac{2 \cdot (x+5)^6}{(x+5)^2} = 2(x+5)^{6-2} = 2(x+5)^4$. [Here the third law of exponents was applied to $(x+5)$. The laws of exponents apply to any algebraic expression.]

1.4. Polynomials and Factoring

When solving applied problems, we often use letters (such as x, y, z, r, and t) to stand for numbers that may be unknown or may vary. These letters are called *variables*, and they may be combined in various ways with numerals, called *constants*, to form algebraic expressions. Since the variables represent real numbers, we can use the standard properties of real numbers to combine or simplify algebraic expressions.

The simplest sort of algebraic expression is a *monomial*, which consists of a numeral times one or more variables. Typical monomials are $3x$, $\frac{1}{2}y$, and $-.5xt$. The numeral preceding the variable is called the *coefficient*. Monomials can be added, subtracted, multiplied, and divided.

EXAMPLE 1   Determine.

(a) $3x + 4x$ (b) $y - \frac{1}{3}y$ (c) $(3x)(4x)$

(d) $3x \cdot 4y$ (e) $\dfrac{3x}{4x}$ (f) $\dfrac{6x}{3y}$

Solution   (a) $3x + 4x = 7x$. [In general, if c and d are numbers, then $cx + dx = (c + d)x$.]

(b) $y - \frac{1}{3}y = \frac{2}{3}y$. [The variable y is the same as $1 \cdot y$. In general, if c and d are numbers, then $cy - dy = (c - d)y$.]

(c) $3x \cdot 4x = 12x^2$. [The expression x^2 means $x \cdot x$. When multiplying two monomials, multiply the coefficients and then multiply the variables.]

(d) $3x \cdot 4y = 12xy$.

(e) $\dfrac{3x}{4x} = \dfrac{3}{4}$. $\Big[$ We can cancel like factors in the numerator and denominator. The

justification is: $\dfrac{3x}{4x} = \dfrac{3}{4} \cdot \dfrac{x}{x} = \dfrac{3}{4} \cdot 1 = \dfrac{3}{4}. \Big]$

(f) $\dfrac{6x}{3y} = 2 \cdot \dfrac{x}{y}$. $\Big[$ In detail, $\dfrac{6x}{3y} = \dfrac{6}{3} \cdot \dfrac{x}{y} = 2 \cdot \dfrac{x}{y}. \Big]$

The next example shows how to combine monomials that involve products of variables.

EXAMPLE 2 Determine.

(a) $4xy + 5xy$ (b) $3y^2 - 5y^2$ (c) $3x \cdot 4x^2$ (d) $3x \cdot 4y^2$

Solution (a) $4xy + 5xy = 9xy$. [Two monomials having the same variables to the same powers are added by adding their coefficients.]

(b) $3y^2 - 5y^2 = -2y^2$. [Two monomials having the same variables to the same powers are subtracted by subtracting their coefficients.]

(c) $3x \cdot 4x^2 = 12x^3$. [Here the coefficients are multiplied together and the variables are multiplied together.]

(d) $3x \cdot 4y^2 = 12xy^2$.

Remark: Constants standing alone without a variable are also regarded as monomials. For instance, 2 is a monomial and $2 \cdot 3x = 6x$.

Polynomials Polynomials are sums of monomials, such as $x^2 + 2x + 1$, $5x + 4y$, and $3x^2y + 2x$. The monomials are often referred to as the *terms* of the polynomial.

EXAMPLE 3 Determine.

(a) $(x^2 + 2x + 1) + (3x^2 + 5x - 4)$ (b) $(5x + 4y) + (3x^2y + 2x)$

(c) $(5y^2 - 3y) - (2y^2 - 4)$ (d) $5x(2x + 3)$

(e) $(5x + 4y)(2x + 3)$

Solution (a) $(x^2 + 2x + 1) + (3x^2 + 5x - 4) = x^2 + 3x^2 + 2x + 5x + 1 - 4 = 4x^2 + 7x - 3$. [To add two polynomials, add together the terms having the same variables *to the same powers.*]

(b) $(5x + 4y) + (3x^2y + 2x) = 5x + 2x + 4y + 3x^2y = 7x + 4y + 3x^2y$. [This sum cannot be simplified any further since only terms having the same variables *to the same powers* can be combined under addition.]

(c) $(5y^2 - 3y) - (2y^2 - 4) = 5y^2 - 3y - 2y^2 + 4 = 5y^2 - 2y^2 - 3y + 4 = 3y^2 - 3y + 4.$

(d) $5x(2x + 3) = (5x)(2x) + (5x) \cdot 3 = 10x^2 + 15x$.

(e) $(5x + 4y)(2x + 3) = 5x(2x + 3) + 4y(2x + 3) = 10x^2 + 15x + 8xy + 12y$.
[We have multiplied each term of the left polynomial by each term of the right polynomial.]

Remark The product in part (e) is of the type $(a + b)(c + d)$, where a, b, c, and d are monomials. Since this situation occurs so frequently, a memory technique has been devised to form the product. Think of the word "FOIL."

F (first):   multiply the first terms of each polynomial (a and c)

O (outer):   multiply the two outer terms (a and d)

I (inner):   multiply the two inner terms (b and c)

L (last):   multiply the last terms of each polynomial (b and d)

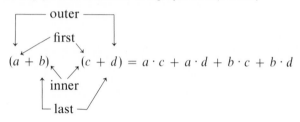

EXAMPLE 4   Use the FOIL method to determine:

(a) $(x + y)^2$ (b) $(x + y)(x - y)$

Solution   (a) $(x + y)^2 = (x + y)(x + y) = x^2 + xy + yx + y^2 = x^2 + 2xy + y^2$.

(b) $(x + y)(x - y) = x^2 - xy + yx - y^2 = x^2 - y^2$.

These two formulas should be memorized:

$$(x + y)^2 = x^2 + 2xy + y^2 \qquad (1)$$
$$x^2 - y^2 = (x + y)(x - y) \qquad (2)$$

Factoring   Often it is useful to do the reverse of the procedures discussed above. To factor a polynomial means to write it as a product of polynomials involving smaller exponents.

EXAMPLE 5   Factor the polynomials.

(a) $x^2 + 7x + 12$ (b) $x^2 - 13x + 12$ (c) $x^2 - 4x - 12$ (d) $x^2 + 4x - 12$

Solution   Note first that for any numbers c and d,

$$(x + c)(x + d) = x^2 + (c + d)x + cd.$$

In the expression on the right, the constant term is the product cd, while the coefficient of x is the sum $c + d$.

(a) Think of all integers c and d such that $cd = 12$. Then choose the pair that satisfies $c + d = 7$; that is, take $c = 3, d = 4$. Thus

$$x^2 + 7x + 12 = (x + 3)(x + 4).$$

(b) We want $cd = 12$. Since 12 is positive, c and d must be both positive or both negative. We must also have $c + d = -13$. These facts lead us to

$$x^2 - 13x + 12 = (x - 12)(x - 1).$$

(c) We want $cd = -12$. Since -12 is negative, c and d must have opposite signs. Also, they must sum to give -4. We find that

$$x^2 - 4x - 12 = (x - 6)(x + 2).$$

(d) This is almost the same as (c).

$$x^2 + 4x - 12 = (x + 6)(x - 2).$$

EXAMPLE 6   Factor the polynomials.

(a) $x^2 - 6x + 9$ (b) $x^2 - 25$ (c) $3x^2 - 21x + 30$ (d) $20 + 8x - x^2$

Solution   (a) We look for $cd = 9$ and $c + d = -6$. The solution is $c = d = -3$, and

$$x^2 - 6x + 9 = (x - 3)(x - 3) = (x - 3)^2.$$

In general,

$$x^2 - 2cx + c^2 = (x - c)(x - c) = (x - c)^2.$$

(b) We use the identity

$$x^2 - c^2 = (x + c)(x - c).$$

Hence

$$x^2 - 25 = (x + 5)(x - 5).$$

(c) We first factor out a common factor of 3, and then use the method of Example 5.

$$3x^2 - 21x + 30 = 3(x^2 - 7x + 10)$$
$$= 3(x - 5)(x - 2).$$

(d) We first factor out a -1 in order to make the coefficient of x^2 equal to $+1$.

$$20 + 8x - x^2 = (-1)(x^2 - 8x - 20)$$
$$= (-1)(x - 10)(x + 2).$$

EXAMPLE 7   Factor the polynomials.

(a) $x^2 - 8x$ (b) $x^3 + 3x^2 - 18x$ (c) $x^3 - 10x$

Solution   In each case we first factor out a common factor of x.

(a) $x^2 - 8x = x(x - 8).$

(b) $x^3 + 3x^2 - 18x = x(x^2 + 3x - 18) = x(x + 6)(x - 3)$.

(c) $x^3 - 10x = x(x^2 - 10)$. To factor $x^2 - 10$, we use the identity $x^2 - c^2 = (x + c)(x - c)$, where $c^2 = 10$ and $c = \sqrt{10}$. Thus $x^3 - 10x = x(x^2 - 10) = x(x + \sqrt{10})(x - \sqrt{10})$.

PRACTICE PROBLEMS 4

In Problems 1–3, perform the indicated operations.

1. $\dfrac{x}{3} + 2x$

2. $(4x)^2 - 4x^2$

3. $3x^2 - x(1 + x)$

4. Factor the polynomial $12 + 2x - 2x^2$.

EXERCISES 4

Specify the coefficient of each of the monomials in Exercises 1–4.

1. $5x^2$

2. $\dfrac{x}{3}$

3. $\dfrac{2xy}{-3}$

4. $-x^2$

Perform the indicated arithmetic operations in Exercises 5–44.

5. $3y + 5y$

6. $2x^2 + .5x^2$

7. $(\frac{1}{2}x)(\frac{2}{3}y)$

8. $\dfrac{8x^2}{-2x^2}$

9. $2xy - 5xy$

10. $(2x) \cdot x$

11. $\dfrac{8x}{4xy}$

12. $\dfrac{y}{\frac{1}{3}}$

13. $(2x)(3xy)$

14. $\frac{1}{2}x + \frac{1}{3}x$

15. $(3x)^2$

16. $-x^2 + 4x^2$

17. $(3x^2 + 2x + 1) + (5x^2 + 4x + 6)$

18. $(x^2 + 4x - 5) + (3x - 2)$

19. $(2x + 1) - (x^2 + 2)$

20. $(.02x^2 + 3) - (.005x^2 - 2)$

21. $(3x + 1) + (5y + 2)$

22. $(5x + 3xy) + (2xy + 2x - 3)$

23. $(3xy + 2x - y) - (x + 2xy)$

24. $(x^2 - y^2) + (x^2 + y^2)$

25. $2x(3x + 4)$

26. $4y(5 - 3y)$

27. $(\frac{3}{5}x + 2)(5x)$

28. $(x - 2y)(-3x)$

29. $(x + 3)(y + 2)$

30. $(3 - x)(y + 5)$

31. $(x - 4)(x + 5)$

32. $(2x + 1)(x + 3)$

33. $(x + 4)^2$

34. $(x - 2)(x + 2)$

35. $(2x + 3)(4x - 5)$

36. $(3x - 4)^2$

37. $(3x + 1)^2$

38. $(x - y)^2 + (x + y)^2$

39. $-x(2x + 3) + 6x^2$

40. $(5x)^2 - 5x^2$

41. $4x - x(1 - x)$

42. $5x^2 - 2(x - 1)^2$

43. $r(3r - 2)$

44. $2r + \pi r$

Factor the polynomials in Exercises 45–62.

45. $x^2 + 8x + 15$

46. $x^2 - 10x + 16$

47. $x^2 - 16$

48. $x^2 - 1$

49. $3x^2 + 12x + 12$

50. $2x^2 - 12x + 18$

51. $30 - 4x - 2x^2$

52. $15 + 12x - 3x^2$

53. $3x - x^2$

54. $4x^2 - 1$

55. $6x - 2x^3$

56. $16x + 6x^2 - x^3$

57. $x^3 + 6x^2 + 9x$

58. $x^2 + 2x + 1$

59. $\frac{1}{2}x^2 + \frac{1}{2}x - 1$

60. $\frac{1}{3}x^4 - 3x^2$

61. $3x + x^2 + 2$

62. $-x^2 + x$

SOLUTIONS TO PRACTICE PROBLEMS 4

1. $\dfrac{x}{3}$ is the same as $\frac{1}{3}x$. Therefore, $\dfrac{x}{3} + 2x = \frac{1}{3}x + 2x = (\frac{1}{3} + 2)x = \frac{7}{3}x$.

2. $(4x)^2 - 4x^2 = (4x)(4x) - 4x^2 = 16x^2 - 4x^2 = 12x^2$. [A common mistake is to compute $(4x)^2$ as $4x^2$ and end up with the answer 0. The parentheses tell us to square the entire term $4x$.]

3. $3x^2 - x(1 + x) = 3x^2 - (x + x^2) = 3x^2 - x - x^2 = 2x^2 - x$.

4. Factor out -2. Then $12 + 2x - 2x^2 = -2(x^2 - x - 6)$ and the problem is reduced to factoring $x^2 - x - 6$. We would like to write $x^2 - x - 6$ as $(x + c)(x + d)$, where c and d are integers with $cd = -6$ and $c + d = -1$. Looking at the condition $cd = -6$ we have the following possible pairs:

$$-6, 1 \qquad 6, -1 \qquad -3, 2 \qquad 3, -2.$$

The correct choice is $-3, 2$, because these numbers add up to -1. So $x^2 - x - 6 = (x - 3)(x + 2)$. Therefore,

$$12 + 2x - 2x^2 = -2(x - 3)(x + 2).$$

1.5. Algebraic Fractions

In this section we extend the rules for the arithmetic of rational numbers to rules for manipulating algebraic expressions. We begin by reviewing the arithmetic of fractions.

Arithmetic of Fractions

I. $\dfrac{P}{Q} \cdot \dfrac{R}{S} = \dfrac{PR}{QS}$

II. $\dfrac{P}{Q} \div \dfrac{R}{S} = \dfrac{P}{Q} \cdot \dfrac{S}{R}$

III. $\dfrac{P}{Q} + \dfrac{R}{Q} = \dfrac{P + R}{Q}$

IV. $\dfrac{P}{Q} - \dfrac{R}{Q} = \dfrac{P - R}{Q}$

V. $\dfrac{P}{Q} = \dfrac{PR}{QR}$

In the rules above, all denominators are assumed to be nonzero. Rules I to IV should be quite familiar. Rule V, however, calls for some comment.

Rule V is used in two ways to convert fractions to equivalent fractions. When read from left to right it is used to convert a fraction to an equivalent fraction having a specified denominator. This is most often utilized in the addition of fractions to convert the fractions to fractions having common denominators.

When read from right to left, Rule V points out that we can cancel common factors from the numerator and denominator of a fraction. When a fraction has been simplified to an equivalent fraction in which the numerator and denominator have no common factors, we say that the fraction has been *reduced to lowest terms*.

Whenever we use Rule V, we are in essence just multiplying numbers by 1.

$$\frac{P}{Q} = \frac{P}{Q} \cdot 1 = \frac{P}{Q} \cdot \frac{R}{R} = \frac{PR}{QR}.$$

Here we wrote 1 in the form R/R.

EXAMPLE 1   (a) Write $\frac{3}{5}$ as a fraction with 20 as denominator.

(b) Reduce $\frac{12}{14}$ to lowest terms.

Solution   Rule V applies to both of these situations.

(a) $\dfrac{3}{5} = \dfrac{3 \cdot 4}{5 \cdot 4} = \dfrac{12}{20}$. Here $P = 3$, $Q = 5$, and $R = 4$.

(b) $\dfrac{12}{14} = \dfrac{6 \cdot 2}{7 \cdot 2} = \dfrac{6}{7}$. Here $P = 6$, $Q = 7$, and $R = 2$.

EXAMPLE 2   Calculate $\frac{3}{10} + \frac{5}{4}$.

Solution   We cannot apply Rule III directly because the two fractions have different denominators. A "common denominator" is $10 \cdot 4 = 40$. We can use Rule V to express both fractions with denominator 40.

$$\frac{3}{10} = \frac{3 \cdot 4}{10 \cdot 4} = \frac{12}{40}, \qquad \frac{5}{4} = \frac{5 \cdot 10}{4 \cdot 10} = \frac{50}{40}.$$

Then

$$\frac{3}{10} + \frac{5}{4} = \frac{12}{40} + \frac{50}{40} = \frac{62}{40}$$

$$= \frac{31 \cdot 2}{20 \cdot 2} = \frac{31}{20}.$$

(In the last step we used Rule V.)

Another way to compute $\frac{3}{10} + \frac{5}{4}$ is to notice first that 20 is a common multiple of the denominators 10 and 4. So we may convert the given fractions into fractions with the common denominator 20:

$$\frac{3}{10} + \frac{5}{4} = \frac{3 \cdot 2}{10 \cdot 2} + \frac{5 \cdot 5}{4 \cdot 5} = \frac{6}{20} + \frac{25}{20} = \frac{31}{20}.$$

Rules I to V can be used to define arithmetic operations for any algebraic fractions, such as quotients of polynomials.

EXAMPLE 3   Calculate.

(a) $\dfrac{x}{1-x} \cdot \dfrac{3}{1+x}$

(b) $\dfrac{6}{x^2+1} \div \dfrac{3}{x}$

(c) $\dfrac{x}{5x-1} + \dfrac{1+x}{5x-1}$

(d) $\dfrac{3}{x} - \dfrac{4}{1+x}$

Solution   (a) Apply Rule I.

$$\frac{x}{1-x} \cdot \frac{3}{1+x} = \frac{3x}{(1-x)(1+x)} = \frac{3x}{1-x^2}.$$

(b) Apply Rule II.

$$\frac{6}{x^2+1} \div \frac{3}{x} = \frac{6}{x^2+1} \cdot \frac{x}{3} = \frac{6x}{3(x^2+1)}.$$

Since both numerator and denominator are divisible by 3, the result can be further simplified. By Rule V,

$$\frac{6x}{3(x^2+1)} = \frac{3 \cdot 2 \cdot x}{3(x^2+1)} = \frac{2x}{x^2+1}.$$

(c) By Rule III,

$$\frac{x}{5x-1} + \frac{1+x}{5x-1} = \frac{x+(1+x)}{5x-1} = \frac{1+2x}{5x-1}.$$

(d) In order to apply Rule IV, the two fractions must have a common denominator. The simplest common denominator is $x(1+x)$. Using Rule V twice, we obtain

$$\frac{3}{x} - \frac{4}{1+x} = \frac{3(1+x)}{x(1+x)} - \frac{4x}{x(1+x)}$$

$$= \frac{3(1+x) - 4x}{x(1+x)}$$

$$= \frac{3-x}{x+x^2}.$$

EXAMPLE 4   Reduce $\dfrac{x^2+5x+6}{x^3-4x}$ to lowest terms.

Solution Factor the numerator and denominator and use Rule V to cancel common factors.

$$\frac{x^2 + 5x + 6}{x^3 - 4x} = \frac{(x + 3)(x + 2)}{x(x^2 - 4)} = \frac{(x + 3)(x - 2)}{x(x + 2)(x - 2)} = \frac{x + 3}{x(x + 2)}.$$

PRACTICE PROBLEMS 5

Perform the indicated algebraic operation and reduce the answer to lowest terms.

1. $\dfrac{x}{3} \div \dfrac{x^2 + x}{x + 2}$

2. $\dfrac{x}{x - 2} - \dfrac{x - 1}{x}$

EXERCISES 5

In Exercises 1–20, reduce each fraction to lowest terms.

1. $\dfrac{6}{8}$

2. $\dfrac{15}{10}$

3. $\dfrac{34}{17}$

4. $-\dfrac{20}{45}$

5. $\dfrac{25x^2}{30x}$

6. $\dfrac{39x}{3x^2}$

7. $\dfrac{5x^3}{x}$

8. $\dfrac{6xy}{36y^2}$

9. $\dfrac{x^2 - 1}{x + 1}$

10. $\dfrac{x^3 + x^2}{5x^4}$

11. $\dfrac{3x}{3x^2 + x}$

12. $\dfrac{9r^2}{12rh}$

13. $\dfrac{15y}{18xy^2}$

14. $\dfrac{3x^2 - 12}{9x - 18}$

15. $\dfrac{-x - 3}{x^2 + 6x + 9}$

16. $\dfrac{3x + 6}{2x^2 + 2x - 4}$

17. $\dfrac{14x^3}{21x^2 + 42x}$

18. $\dfrac{x^2 + 5x + 6}{x^2 + 4x + 4}$

19. $\dfrac{x^2 - 25}{x^2 - 10x + 25}$

20. $\dfrac{15x^5}{x^3 + 2x^2}$

In Exercises 21–50, perform the indicated algebraic operation. Reduce all answers to lowest terms.

21. $\dfrac{2}{3} + \dfrac{1}{5}$

22. $\dfrac{3}{4} - \dfrac{2}{10}$

23. $\dfrac{4}{5} \div \dfrac{2}{7}$

24. $\dfrac{5}{8} \cdot \dfrac{4}{25}$

25. $\dfrac{5}{27} - \dfrac{5}{12}$

26. $\dfrac{2}{3} \div \dfrac{6}{11}$

27. $\dfrac{x}{x + 2} \cdot \dfrac{x + 1}{x^2}$

28. $\dfrac{3x + 1}{x + 2} \cdot \dfrac{x}{3}$

29. $\dfrac{4x^2}{x^2 - 9} + \dfrac{12x}{x^2 - 9}$

30. $\dfrac{x}{x + 5} - \dfrac{x^2 + x}{x + 5}$

31. $\dfrac{3}{x} + \dfrac{x}{x + 2}$

32. $\dfrac{1 + x}{x} + \dfrac{2}{x^2}$

33. $\dfrac{3}{x^2 + x} - \dfrac{2}{x}$

34. $\dfrac{2}{x - 1} + \dfrac{3}{x + 1}$

35. $\dfrac{1}{x^2 - 1} - \dfrac{x}{x^2 + 3x + 2}$

36. $\dfrac{3}{x + 2} - \dfrac{x}{x^2 + 5x + 6}$

37. $\dfrac{x}{x + 2} \cdot \dfrac{3x + 6}{x^2 + x}$

38. $\dfrac{x - 1}{x - 2} \cdot \dfrac{x - 3}{x - 4}$

39. $\dfrac{x^2 + 1}{x} \div \dfrac{3}{x}$

40. $\dfrac{x^2 - 16}{5} \div \dfrac{x + 4}{3}$

41. $\dfrac{6}{x^2 - 25} \div \dfrac{9}{x + 5}$

42. $\dfrac{4x}{x - 1} \div \dfrac{x}{x^2 + 7x - 5}$

43. $\dfrac{1 + \dfrac{1}{x}}{x^2 - 1}$

44. $\dfrac{\dfrac{3}{1 + x} - \dfrac{2}{x}}{x^2 - 2x + 2}$

45. $\dfrac{3 + \dfrac{x}{1 + x}}{\dfrac{2}{x}}$

46. $\dfrac{x^2 + 2}{\dfrac{1}{x} - \dfrac{x}{3}}$

47. $\dfrac{5 + \dfrac{2}{x}}{x - \dfrac{1}{1 + x}}$

48. $\dfrac{\dfrac{4}{x - 3} - \dfrac{2}{x}}{\dfrac{x}{5} + \dfrac{1}{1 + x}}$

49. $\left(1 - \dfrac{1}{x}\right)\left(2 + \dfrac{x}{x + 5}\right)$

50. $\left(\dfrac{x}{x + 3} - \dfrac{1}{x}\right)\left(x + \dfrac{2}{x}\right)$

SOLUTIONS TO PRACTICE PROBLEMS 5

1. By Rule II,

$$\frac{x}{3} \div \frac{x^2 + x}{x + 2} = \frac{x}{3} \cdot \frac{x + 2}{x^2 + x}$$

$$= \frac{x(x + 2)}{3(x^2 + x)}.$$

Since we want all final answers reduced to lowest terms, we do not multiply out just yet, but rather we try to factor further and perhaps cancel some common factors.

$$= \frac{x(x + 2)}{3x(x + 1)}$$

$$= \frac{x + 2}{3(x + 1)}$$

$$= \frac{x + 2}{3x + 3}.$$

2. Before subtracting (or adding) two fractions, we must first convert them into fractions having the same denominator. This can always be accomplished by taking the common denominator to be the product of the two denominators.

$$\frac{x}{x-2} - \frac{x-1}{x} = \frac{x}{x-2} \cdot \frac{x}{x} - \frac{x-1}{x} \cdot \frac{x-2}{x-2}$$

$$= \frac{x^2}{x(x-2)} - \frac{x^2 - 3x + 2}{x(x-2)}$$

$$= \frac{x^2 - x^2 + 3x - 2}{x(x-2)}$$

$$= \frac{3x - 2}{x(x-2)}.$$

There is no opportunity to cancel common factors here, so we just multiply out the denominator.

$$= \frac{3x - 2}{x^2 - 2x}.$$

1.6. Solving Quadratic Equations

A quadratic equation is an equation of the form $ax^2 + bx + c = 0$ where a, b, and c are given numbers and $a \neq 0$. In this section we present two methods for solving quadratic equations—factoring and the quadratic formula.

When nonzero numbers are multiplied together the product must be another nonzero number. Hence if a product of numbers turns out to be zero, then at least one of the numbers must be zero. This observation is the key to solving quadratic equations by the factoring method.

EXAMPLE 1   Solve the equation $(x - 2)(3x - 4) = 0$.

Solution   Since the product of $x - 2$ and $3x - 4$ is zero, we must have $x - 2 = 0$ or $3x - 4 = 0$. These two linear equations are easily solved.

$$x - 2 = 0 \qquad \text{or} \qquad 3x - 4 = 0$$
$$x = 2 \qquad\qquad\qquad 3x = 4$$
$$x = \tfrac{4}{3}.$$

So, the equation has two solutions, $x = 2$ and $x = \tfrac{4}{3}$.

EXAMPLE 2   Solve the quadratic equation $x^2 - 2x - 3 = 0$.

Solution   Factor the polynomial $x^2 - 2x - 3$ and then solve as in Example 1.

$$x^2 - 2x - 3 = 0$$
$$(x - 3)(x + 1) = 0$$
$$x - 3 = 0 \qquad \text{or} \qquad x + 1 = 0$$
$$x = 3 \qquad\qquad\qquad x = -1.$$

The two solutions are $x = 3$ and $x = -1$.

EXAMPLE 3   Solve the quadratic equation $5x^2 - 4x = 0$.

Solution

$$5x^2 - 4x = 0$$
$$(5x - 4)x = 0$$
$$5x - 4 = 0 \quad \text{or} \quad x = 0$$
$$5x = 4$$
$$x = \tfrac{4}{5}.$$

The two solutions are $x = \tfrac{4}{5}$ and $x = 0$.

EXAMPLE 4   Solve the equation $3x^2 = 30x - 75$.

Solution   The first step is to transform the equation into a quadratic equation of the form $ax^2 + bx + c = 0$. This is accomplished here by moving all nonzero terms to the left side of the equation, arriving at

$$3x^2 - 30x + 75 = 0.$$

Each of the coefficients $3, -30, 75$ is divisible by 3. Hence we can simplify the equation by multiplying each side by $\tfrac{1}{3}$. We obtain

$$x^2 - 10x + 25 = 0.$$

Now factor and solve:

$$(x - 5)(x - 5) = 0.$$

In this case the equation has just one solution, $x = 5$.

Some quadratic equations are difficult to solve by factoring; other quadratic equations have no solution and hence are impossible to solve by any method. The quadratic formula, given below, takes care of both situations. It gives the solution or solutions if a solution exists, and it shows when no solution exists. A derivation of the quadratic formula appears at the end of this section.

The Quadratic Formula for Solving $ax^2 + bx + c = 0$

1. If $b^2 - 4ac < 0$, the equation has no solution.
2. If $b^2 - 4ac \geq 0$, all solutions are given by

$$x = \frac{-b \pm \sqrt{b^2 - 4ac}}{2a}.$$

EXAMPLE 5   Use the quadratic formula to solve the following quadratic equations.

(a) $x^2 - 2x - 15 = 0$ (b) $4x^2 - 4x + 1 = 0$ (c) $x^2 + 2x + 3 = 0$

Solution (a) Here $a = 1$, $b = -2$, and $c = -15$.

$$b^2 - 4ac = (-2)^2 - 4(1)(-15)$$
$$= 4 + 60 = 64$$
$$x = \frac{-(-2) \pm \sqrt{64}}{2(1)}$$
$$= \frac{2 \pm 8}{2}.$$

$$\frac{2 + 8}{2} = 5 \quad \text{and} \quad \frac{2 - 8}{2} = -3.$$

Thus the quadratic equation has the two solutions $x = 5$ and $x = -3$.

(b) Here $a = 4$, $b = -4$, and $c = 1$.

$$b^2 - 4ac = (-4)^2 - 4(4)(1)$$
$$= 16 - 16 = 0$$
$$x = \frac{-(-4) \pm \sqrt{0}}{2(4)}$$
$$= \frac{4}{8} = \frac{1}{2}.$$

Thus the quadratic equation has just one solution, $x = \frac{1}{2}$.

(c) Here $a = 1$, $b = 2$, and $c = 3$.

$$b^2 - 4ac = (2)^2 - 4(1)(3)$$
$$= 4 - 12 = -8.$$

Since $-8 < 0$, we conclude that this quadratic equation has no solution.

In Example 5 we saw the three different possibilities for the solution of a quadratic equation.

Two solutions   if $b^2 - 4ac > 0$.
One solution   if $b^2 - 4ac = 0$.
No solution   if $b^2 - 4ac < 0$.

Some other types of equations can be solved by reducing them to quadratic equations.

EXAMPLE 6   Solve the equation $\dfrac{x^2 - 4}{x + 1} = 0$.

Solution

$$\frac{x^2 - 4}{x + 1} = 0$$

Multiply both sides by $x + 1$ $x^2 - 4 = 0$

Factor $(x - 2)(x + 2) = 0$

$x - 2 = 0$ and $x + 2 = 0$

$x = 2$ $x = -2.$

The solutions are $x = 2$ and $x = -2.$

EXAMPLE 7   Solve the equation $\dfrac{x}{x + 1} = \dfrac{4}{x - 1}.$

Solution

$$\frac{x}{x + 1} = \frac{4}{x - 1}$$

Multiply both sides by $x + 1$ $x = \dfrac{4(x + 1)}{x - 1}$

Multiply both sides by $x - 1$ $x(x - 1) = 4(x + 1)$

$x^2 - x = 4x + 4$

$x^2 - 5x - 4 = 0.$

Attempts to factor the polynomial are fruitless. The quadratic formula yields

$$b^2 - 4ac = (-5)^2 - 4(1)(-4) = 25 + 16 = 41$$

$$x = \frac{-(-5) \pm \sqrt{41}}{2(1)} = \frac{5 \pm \sqrt{41}}{2}.$$

Hence the equation has two solutions, $x = \dfrac{5 + \sqrt{41}}{2}$ and $x = \dfrac{5 - \sqrt{41}}{2}.$

Derivation of the Quadratic Formula

$ax^2 + bx + c = 0$

$ax^2 + bx = -c$

$4a^2x^2 + 4abx = -4ac$ (both sides were multiplied by $4a$),

$4a^2x^2 + 4abx + b^2 = b^2 - 4ac$ (b^2 added to both sides).

Now note that $4a^2x^2 + 4abx + b^2 = (2ax + b)^2$. To check this, simply multiply out the right-hand side! Therefore,

$$(2ax + b)^2 = b^2 - 4ac$$

$$2ax + b = \pm\sqrt{b^2 - 4ac}$$

$$2ax = -b \pm \sqrt{b^2 - 4ac}$$

$$x = \frac{-b \pm \sqrt{b^2 - 4ac}}{2a}.$$

PRACTICE PROBLEMS 6

Solve.

1. $x^2 + 6x + 3 = 0$

2. $\dfrac{150}{11 - x} = 5x$

EXERCISES 6

Solve the quadratic equations in Exercises 1–18 by factoring.

1. $x^2 - 8x + 15 = 0$

2. $x^2 - 4x - 12 = 0$

3. $x^2 - 8x + 16 = 0$

4. $x^2 + 10x + 16 = 0$

5. $x^2 - 17x = 0$

6. $8x - x^2 = 0$

7. $3x^2 + 3x - 6 = 0$

8. $5x^2 - 20x - 25 = 0$

9. $32 - 2x^2 = 0$

10. $-4x^2 + 20x - 24 = 0$

11. $\frac{1}{2}x^2 - \frac{1}{2}x - 10 = 0$

12. $3 - x - \frac{1}{4}x^2 = 0$

13. $6 - x - \frac{1}{3}x^2 = 0$

14. $\frac{1}{2}x^2 + x - 12 = 0$

15. $.2x^2 - .8x - 1 = 0$

16. $.1x^2 - .9x - 1 = 0$

17. $2 - .5x - .25x^2 = 0$

18. $5 - .2x^2 = 0$

Use the quadratic formula to solve the quadratic equations in Exercises 19–30.

19. $2x^2 - 7x + 6 = 0$

20. $3x^2 + 2x - 1 = 0$

21. $4x^2 - 12x + 9 = 0$

22. $\frac{1}{4}x^2 + x + 1 = 0$

23. $-2x^2 + 3x - 4 = 0$

24. $11x^2 - 2x + 1 = 0$

25. $5x^2 - 4x - 1 = 0$

26. $x^2 - 4x + 5 = 0$

27. $15x^2 - 135x + 300 = 0$

28. $x^2 - \sqrt{2}x - \frac{5}{4} = 0$

29. $\frac{3}{2}x^2 - 6x + 5 = 0$

30. $9x^2 - 12x + 4 = 0$

Solve the equations in Exercises 31–48.

31. $x^2 - 6 = x$

32. $\dfrac{2}{x} - 1 = x$

33. $\dfrac{x^2 + x - 12}{x + 1} = 0$

34. $x^2 + x = 12$

35. $x + \dfrac{1}{x} = 3$

36. $\dfrac{x^2 + x + 1}{1 + x^2} = 0$

37. $\dfrac{x - 1}{x - 2} = \dfrac{4}{x - 3}$

38. $x = \dfrac{1}{8 - 4x}$

39. $x^3 - 36x = 0$

40. $\dfrac{x}{x^2 - 1} = \dfrac{3}{x + 1}$

41. $(x^2 - 9x + 20)^2 = 0$

42. $x^3 + x^2 + 2x = 0$

43. $3 - x = \dfrac{1}{x}$

44. $5x^3 - 5(x^3 - x^2 - 2) = 0$

45. $x^3 - x(x^2 + 2x + 1) = 0$

46. $-x^2 = 0$

47. $x + \dfrac{5}{x} = -1$

48. $-16x^2 - 32x + 128 = 0$

SOLUTIONS TO PRACTICE PROBLEMS 6

1. Since the polynomial does not factor easily, we use the quadratic formula.

$$b^2 - 4ac = 36 - 4 \cdot 1 \cdot 3 = 36 - 12 = 24.$$

Therefore, the solutions are $x = \dfrac{-6 \pm \sqrt{24}}{2}$. Now, since $\sqrt{24} = \sqrt{4 \cdot 6} = \sqrt{4} \cdot \sqrt{6} = 2\sqrt{6}$, the solution can be simplified.

$$x = \frac{-6 \pm 2\sqrt{6}}{2} = -3 \pm \sqrt{6}.$$

2. When solving an equation involving a fraction it is usually a good idea to get rid of the fraction by multiplying both sides of the equation by the denominator of the fraction.

$$\frac{150}{11 - x} = 5x$$

$$150 = 5x(11 - x)$$

$$150 = 55x - 5x^2$$

$$5x^2 - 55x + 150 = 0.$$

Since 5, -55, and 150 are all divisible by 5, we can simplify the equation by multiplying both sides by $\frac{1}{5}$.

$$x^2 - 11x + 30 = 0$$

$$(x - 6)(x - 5) = 0.$$

Hence the solutions are $x = 6$ and $x = 5$.

1.7. Logarithms

Logarithms are often useful when working with equations involving exponents. In fact, as we shall see, the logarithm of a number *is* an exponent. In this section we introduce the basic concept of a logarithm, and in the next section we study algebraic properties of logarithms.

EXAMPLE 1   Find the exponent to which 10 must be raised to get:

(a) 100 (b) 1000 (c) 10

Solution   (a) Since $100 = 10^2$, the exponent is 2.
(b) Since $1000 = 10^3$, the exponent is 3.
(c) Since $10 = 10^1$, the exponent is 1.

In each of the computations above, we were asked to determine *the exponent to which 10 must be raised in order to get* a given number. Mathematicians use a shorthand for this italicized phrase. Suppose that A is a given positive number. Then

> log A is the exponent to which 10 must be raised in order to get A.

The number log A is called the logarithm of A.

EXAMPLE 2   Find the value of:

(a) log 100 (b) log 1000 (c) log 10

Solution   By Example 1 we have

(a) $100 = 10^2$. So log 100 = 2.

(b) $1000 = 10^3$. So log 1000 = 3.

(c) $10 = 10^1$. So log 10 = 1.

EXAMPLE 3   Determine.

(a) log 1 (b) log .1 (c) log 1,000,000

Solution   In each case we must write the given number (1, .1, or 1,000,000) as 10 raised to some exponent.

(a) Since $1 = 10^0$, we have log 1 = 0. That is, the exponent to which 10 must be raised in order to get 1 is 0.

(b) log .1 = -1 because $.1 = 10^{-1}$.

(c) log 1,000,000 = 6 because $1,000,000 = 10^6$.

The numbers used in Examples 2 and 3 were specially chosen for ease of computation. Usually, logarithms of numbers must be determined from a calculator or a table of logarithms. Table 1 gives the values of log A for a few choices of A. These values have been rounded off to four decimal places.

TABLE 1

A	$\log A$	A	$\log A$	A	$\log A$
.5	$-.3110$	4	.6021	7.5	.8751
1	.0000	4.5	.6532	8	.9031
1.5	.1761	5	.6990	8.5	.9294
2	.3010	5.5	.7404	9	.9542
2.5	.3979	6	.7782	9.5	.9777
3	.4771	6.5	.8129	10	1.0000
3.5	.5441	7	.8451	10.5	1.0212

In general, $\log A = c$ means that 10 raised to the exponent c gives A. That is, $10^c = A$.

$$\log A = c \quad \text{means that} \quad A = 10^c$$

The intimate connection between exponents and logarithms is most directly expressed by the following relationships.

$$\log 10^c = c \tag{1}$$
$$10^{\log A} = A \tag{2}$$

These two relationships are actually just reiterations of the definition of log A. Relationship (1) says that the exponent to which 10 must be raised in order to get 10^c is c. Relationship (2) says that if you raise 10 to the exponent to which 10 must be raised in order to get A, then indeed you do get A.

Some illustrations of these relationships are:

$$\log 10^5 = 5 \qquad 10^{\log 3} = 3$$
$$\log 10^{4x} = 4x \qquad 10^{\log(2x + 1)} = 2x + 1$$

These relationships are extremely useful in solving equations involving exponents.

EXAMPLE 4   Use Table 1 to solve.

(a) $10^{3x} = 4$ (b) $\log(3x + 1) = 2$ (c) $6 \cdot 10^{1-4x} = 27$

Solution (a) Take the logarithm of each side.

$$10^{3x} = 4$$

$$\log 10^{3x} = \log 4$$

$$3x = .6021, \qquad \text{by (1) and Table 1}$$

$$x = .2007.$$

(b)
$$\log(3x + 1) = 2$$

$$10^{\log(3x+1)} = 10^2$$

$$3x + 1 = 100, \qquad \text{by (2)}$$

$$3x = 99$$

$$x = 33.$$

(c) First divide both sides of the equation by 6 and then take the logarithm of each side.

$$6 \cdot 10^{1-4x} = 27$$

$$10^{1-4x} = 4.5$$

$$\log 10^{1-4x} = \log 4.5$$

$$1 - 4x = .6532, \qquad \text{by (1) and Table 1}$$

$$4x = .3468$$

$$x = .0867.$$

The logarithms discussed so far are based on the number 10 and are referred to as *common logarithms*. The number 10 is often highlighted by writing $\log_{10} A$ instead of $\log A$, and calling it *log to the base 10*. Actually, any positive number b ($\neq 1$) can be used as the base of a logarithm.

> For any positive number A, $\log_b A$ is the exponent to which b must be raised in order to get A.

EXAMPLE 5 Determine.

(a) $\log_2 16$ (b) $\log_5(.04)$ (c) $\log_7 1$

Solution (a) Since $16 = 2^4$, $\log_2 16 = 4$.

(b) Since $.04 = 5^{-2}$, $\log_5(.04) = -2$.

(c) Since $1 = 7^0$, $\log_7 1 = 0$.

The numbers used in Example 5 were specially chosen. For more complicated numbers we can compute logarithms to the base b from common logarithms by using the following formula.

$$\log_b A = \frac{\log_{10} A}{\log_{10} b} \qquad (3)$$

Formula (3) will be derived in the exercises of the next section.

EXAMPLE 6   Use Table 1 to determine $\log_2 7$.

Solution   By (3),

$$\log_2 7 = \frac{\log_{10} 7}{\log_{10} 2}$$

$$= \frac{.8451}{.3010}$$

$$\approx 2.808.$$

The results obtained for common logarithms carry over to logarithms with other bases.

Let A and b be positive numbers. Then $\log_b A = c$ means that $A = b^c$.

$$\log_b b^c = c,$$

$$b^{\log_b A} = A.$$

These results are helpful when solving equations. The next example uses logarithms to solve an applied problem.

EXAMPLE 7   Suppose that a rapidly growing bacteria culture contains $1000 \times 2^{.05t}$ bacteria after t minutes.

(a) How many bacteria will be present after 20 minutes?

(b) When will the size of the culture reach 7000?

Solution   (a) Set $t = 20$. Then

$$1000 \times 2^{.05t} = 1000 \times 2^{.05(20)}$$

$$= 1000 \times 2^1$$

$$= 2000.$$

There will be 2000 bacteria after 20 minutes.

(b) We want to know the value of t for which $1000 \times 2^{.05t} = 7000$. To solve this equation, we divide both sides by 1000 and then take $\log_2$ of each side.

$$2^{.05t} = 7$$

$$\log_2 2^{.05t} = \log_2 7$$

$$.05t = 2.808, \qquad \text{by Example 6}$$

$$t = \frac{2.808}{.05} = 56.16.$$

Therefore, the size will reach 7000 bacteria after 56.16 minutes.

PRACTICE PROBLEMS 7

1. Rewrite the expression $10^{.9031} = 8$ using logarithms.

2. Simplify $\log \sqrt[3]{10}$.

3. Solve the equation $3 \cdot \log_2(x^2 - 1) = 9$.

EXERCISES 7

In Exercises 1–12, rewrite each statement using logarithms.

1. $10^4 = 10,000$	**2.** $10^{.6021} = 4$	**3.** $10^{-.311} = .5$
4. $10^{-3} = .001$	**5.** $2^3 = 8$	**6.** $(.2)^3 = .008$
7. $8^{1/3} = 2$	**8.** $25^{.5} = 5$	**9.** $2^{-3} = .125$
10. $5^0 = 1$	**11.** $10^{3/2} = 31.62$	**12.** $10^{-.6} = 3.981$

In Exercises 13–24, rewrite each statement using exponents.

13. $\log_{10} \frac{1}{10} = -1$	**14.** $\log_{10} 5 = .699$	**15.** $\log_{10} 2.5 = .3979$
16. $\log_{10} 1 = 0$	**17.** $\log_2 16 = 4$	**18.** $\log_5 125 = 3$
19. $\log_{36} 6 = \frac{1}{2}$	**20.** $\log_4 \frac{1}{16} = -2$	**21.** $\log_8 1 = 0$
22. $\log_2 3.249 = 1.7$	**23.** $\log_3 1.316 = \frac{1}{4}$	**24.** $\log_5 .5848 = -\frac{1}{3}$

Simplify the expressions in Exercises 25–42.

25. $\log_{10} .01$	**26.** $\log_{10} 10,000$	**27.** $\log_2 32$
28. $\log_2 \frac{1}{4}$	**29.** $\log_{10} 10^{1.5}$	**30.** $\log_{10} 10^{-.2}$
31. $\log_2 \sqrt{2}$	**32.** $\log_2 \frac{1}{8}$	**33.** $10^{\log_{10} 8}$
34. $10^{\log_{10} .4}$	**35.** $2^{\log_2 5}$	**36.** $5^{\log_5 2}$
37. $\log_{10} 10^{(1-x)}$	**38.** $\log_{10} 10^{1/x}$	**39.** $\log_2 2^{(x^2+1)}$
40. $\log_2 2^{(t^2+2t)}$	**41.** $10^{\log_{10}(1+t)}$	**42.** $10^{\log_{10}(10t)}$

Solve the equations in Exercises 43–54. (Use Table 1 if appropriate.)

43. $10^{x+1} = 7$ **44.** $10^{2x} = 8.5$ **45.** $5 \cdot 10^{5x} = 45$

46. $\frac{1}{3} \cdot 10^{1-2x} = .233$ **47.** $\log_{10}(1 + 3x) = 3$ **48.** $\log_2(x^2 - 1) = 0$

49. $\log_2(3x + 4) = 5$ **50.** $5 \log_{10}(x^2 - 21x) = 10$ **51.** $3 \cdot 10^{x^2 - 2x} = 30$

52. $2^{x^2 + x} = 4$ **53.** $\log_{10} x = .7782$ **54.** $\log_2 x = -4$

Use Table 1 and formula (3) to determine the logarithms in Exercises 55–60.

55. $\log_5 8$ **56.** $\log_6 4$ **57.** $\log_{2.5} 3$

58. $\log_9 2$ **59.** $\log_7 6$ **60.** $\log_{.5} 9$

61. One thousand dollars deposited into a certain savings account grows to $1000 \times 2^{.2t}$ dollars after t years.

(a) What will the balance be after 5 years?

(b) When will the balance reach 7000 dollars?

62. The world population is currently 4 billion and will be $4 \times 10^{.0087t}$ billion after t years. When will the population reach 6 billion?

SOLUTIONS TO PRACTICE PROBLEMS 7

1. When we write log without specifying a base, we assume that the base is 10. Recall that

$$10^c = A \quad \text{is equivalent to} \quad \log A = c.$$

   With $c = .9031$ and $A = 8$,

$$10^{.9031} = 8 \quad \text{is equivalent to} \quad \log 8 = .9031.$$

   Answer: $\log 8 = .9031$.

2. Since $\sqrt[3]{10} = 10^{1/3}$,

$$\log \sqrt[3]{10} = \log 10^{1/3}$$
$$= \tfrac{1}{3}.$$

   In the last step we applied the relationship $\log_{10} 10^c = c$, with $c = \frac{1}{3}$.

3. Multiply both sides of the equation by $\frac{1}{3}$ so that just $\log_2(x^2 - 1)$ appears on the left side. Then, to strip away "$\log_2$," raise 2 to the $\log_2(x^2 - 1)$ power. Of course, this operation must be done to both sides of the equation.

$$3 \log_2(x^2 - 1) = 9$$
$$\log_2(x^2 - 1) = 3$$
$$2^{\log_2(x^2 - 1)} = 2^3$$
$$x^2 - 1 = 8 \quad \text{(on the left we used the property } 2^{\log_2 A} = A\text{)}$$
$$x^2 - 9 = 0$$
$$(x - 3)(x + 3) = 0.$$

   Therefore, $x = 3$ and $x = -3$ are the solutions.

1.8. Properties of Logarithms

Logarithms exhibit some special properties which traditionally made them an indispensable tool for computation. Nowadays, these properties are used mainly to simplify algebraic expressions and to better understand certain applications involving exponents and logarithms.

Properties of Logarithms

I. $\log_b A \cdot C = \log_b A + \log_b C$

II. $\log_b (A^r) = r \cdot \log_b A$

III. $\log_b \left(\dfrac{A}{C} \right) = \log_b A - \log_b C$

We shall illustrate each property for common logarithms (that is, $b = 10$) and give a general proof.

□ **Property I**   This property states that the logarithm of the product of two numbers is the sum of the logarithms of the two numbers. It is useful in both directions. That is, Property I may also be read from right to left, to obtain

$$\log_b A + \log_b C = \log_b AC.$$

For instance, to compute $\log 2 + \log 5$, proceed as follows:

$$\log 2 + \log 5 = \log 2 \cdot 5 = \log 10 = 1.$$

Without Property I it would have been necessary to look up the values of $\log 2$ and $\log 5$ in a table, and then add them.

To verify Property I, let $u = \log_b A$ and $v = \log_b C$. By the definition of u and v, we have $b^u = A$ and $b^v = C$. Then, by a property of exponents,

$$b^{u+v} = b^u \cdot b^v = AC.$$

That is, $u + v$ is the exponent to which b must be raised in order to get AC. By the definition of the logarithm of a number, $u + v = \log_b AC$. That is, $\log_b A + \log_b C = \log_b AC$.

□ **Property II**   This property states that the logarithm of a power of a number is the power times the logarithm of the number. As an application we compute $\log \sqrt{1000}$.

$$\log \sqrt{1000} = \log 1000^{1/2} = \tfrac{1}{2} \log 1000 = \tfrac{1}{2} \cdot 3 = \tfrac{3}{2}.$$

To verify Property II, let $u = \log_b A$. This means that $b^u = A$. By a property of exponents, $b^{ru} = (b^u)^r = A^r$. Thus ru is the exponent to which b must be raised in order to get A^r, that is, $ru = \log_b A^r$. Therefore, $r \log_b A = \log_b A^r$.

☐ **Property III** This property states that the logarithm of the quotient of two numbers is the difference of the logarithms of the two numbers. It is useful in both directions. For instance, to compute $\log 30 - \log 3$, proceed as follows:

$$\log 30 - \log 3 = \log \frac{30}{3} = \log 10 = 1.$$

Property III is a consequence of Properties I and II.

$$\log_b \frac{A}{C} = \log_b(AC^{-1})$$

$$= \log_b A + \log_b C^{-1} \qquad \text{(by Property I)}$$

$$= \log_b A + (-1)\log_b C \qquad \text{(by Property II)}$$

$$= \log_b A - \log_b C.$$

EXAMPLE 1 Express as a single logarithm.

(a) $\log_3 5 + \log_3 6$ (b) $\log 45 - 2 \log 3$ (c) $\log_2 x + \log_2 x^3 y$.

Solution   (a) $\log_3 5 + \log_3 6 = \log_3 5 \cdot 6$,   by Property I

$$= \log_3 30.$$

(b) $\log 45 - 2 \log 3 = \log 45 - \log 3^2$,   by Property II

$$= \log 45 - \log 9$$

$$= \log \tfrac{45}{9}, \qquad \text{by Property III}$$

$$= \log 5.$$

(c) $\log_2 x + \log_2 x^3 y = \log_2(x \cdot x^3 y)$,   by Property I

$$= \log_2 x^4 y.$$

EXAMPLE 2 Express in terms of $\log x$ and $\log y$.

(a) $\log \dfrac{x}{y^3}$ (b) $\log y\sqrt{x}$.

Solution   (a) $\log \dfrac{x}{y^3} = \log x - \log y^3$,   by Property III

$$= \log x - 3 \log y \qquad \text{by Property II}$$

(b) $\log y\sqrt{x} = \log y + \log \sqrt{x}$,   by Property I

$$= \log y + \log x^{1/2}$$

$$= \log y + \tfrac{1}{2} \log x, \qquad \text{by Property II}$$

EXAMPLE 3 Solve for x: $\log(x^2 - 9) = 2 + \log(x + 3)$.

Solution

$$\log(x^2 - 9) - \log(x + 3) = 2$$

$$\log \frac{x^2 - 9}{x + 3} = 2, \qquad \text{by Property III}$$

$$\log(x - 3) = 2, \qquad \text{since } x^2 - 9 = (x - 3)(x + 3)$$

$$10^{\log(x - 3)} = 10^2$$

$$x - 3 = 100$$

$$x = 103.$$

EXAMPLE 4 Solve for y in terms of x: $\log y = 1 + \log(2 - xy)$.

Solution

$$\log y - \log(2 - xy) = 1$$

$$\log \frac{y}{2 - xy} = 1, \qquad \text{by Property III}$$

$$10^{\log y/(2 - xy)} = 10^1$$

$$\frac{y}{2 - xy} = 10$$

$$y = 20 - 10xy$$

$$y + 10xy = 20$$

$$y(1 + 10x) = 20$$

$$y = \frac{20}{1 + 10x}.$$

PRACTICE PROBLEMS 8

1. Write $2 \log_2 3 - \log_2 3x$ as a single logarithm.

2. Use Table 1 of Section 1.7 to determine $\log 15$.

EXERCISES 8

In Exercises 1–16, write each expression as a single logarithm.

1. $\log_2 3 + \log_2 4$

2. $\log 9 + \log 3$

3. $\log 5 - \log 2$

4. $\log_2 \sqrt{5} + \log_2 \sqrt{3}$

5. $\log_2 10 + 3 \log_2 5$

6. $2 \log 3 - 3 \log 2$

7. $\log_5 8 - 2 \log_5 4$

8. $4 \log_2 3 + 3 \log_2 9$

9. $\log xy + \log y$

10. $\log_2 x^2 y^3 - \log_2 xy^4$

11. $3 \log_2 x + 4 \log_2 y$

12. $\log(x - 2) + \log(x + 2)$

13. $\frac{1}{2} \log x - \log x^2$

14. $5 \log_2 x - 2 \log_2 xy$

15. $3 \log_2 \sqrt{x} + \frac{1}{2} \log_2 x$

16. $3 \log y - 2 \log xy$

In Exercises 17–22, write each expression in terms of $\log x$ and $\log y$.

17. $\log x^2 y$

18. $\log \dfrac{x^2}{y}$

19. $\log \sqrt{xy}$

20. $\log(xy)^3$

21. $\log \dfrac{\sqrt{x}}{y}$

22. $\log \sqrt[3]{x}\, y^2$

In Exercises 23–28, solve for x.

23. $\log x = 1 - \log(x + 3)$

24. $\log(x + 1) = 1 + \log x$

25. $\log(x^2 + 2x + 1) = 3 + \log(x + 1)$

26. $\log(x^2 - 6) = 1$

27. $3 \log x - \log x^2 = 4$

28. $\log\left(\dfrac{1}{x}\right) = -1$

In Exercises 29–34, solve for y in terms of x.

29. $\log(xy + 1) - \log y = 3$

30. $\log \dfrac{x^3}{y^2} - \log \dfrac{x^5}{y^3} = 2$

31. $\log(x^3 y - x^2 y^2) = 1 + \log(x^2 y)$

32. $\log \dfrac{1}{x} + \log(y + 1) = 1$

33. $\log(x + 1) - \log y = 2$

34. $\log(x + y) - \log x = \log y$

In Exercises 35–40, use Table 1 of Section 1.7 to determine the given logarithms.

35. $\log 30$

36. $\log 36$

37. $\log \sqrt{6}$

38. $\log 4.25$

39. $\log \dfrac{8}{3}$

40. $\log \sqrt{70}$

41. Show that
$$\log_b A = \frac{\log A}{\log b}.$$

[*Hint:* Start with the equation $A = b^{\log_b A}$ and take the common log of both sides.]

SOLUTIONS TO PRACTICE PROBLEMS 8

1. $2 \log_2 3 - \log_2 3x = \log_2 3^2 - \log_2 3x, \quad$ by Property II

$\qquad = \log_2 9 - \log_2 3x$

$\qquad = \log_2 \dfrac{9}{3x}, \qquad$ by Property III

$\qquad = \log_2 \dfrac{3}{x}.$

2. Since log 15 is not in Table 1, we must use a property of logarithms to express it in terms of logarithms of numbers that are in the table.

$$\log 15 = \log 3 \cdot 5$$

$$= \log 3 + \log 5, \qquad \text{by Property I}$$

$$= .4771 + .6990, \qquad \text{by Table 1}$$

$$= 1.1761.$$

Chapter 1: CHECKLIST

- ☐ Positive, negative numbers
- ☐ Integers, rational, irrational numbers
- ☐ Absolute value
- ☐ Laws of signs
- ☐ Linear equation (in one variable)
- ☐ Linear inequality (in one variable)
- ☐ Laws of exponents
- ☐ Monomial
- ☐ Polynomial
- ☐ Arithmetic of fractions
- ☐ Quadratic formula
- ☐ Properties of logarithms

Chapter 1: SUPPLEMENTARY EXERCISES

1. Calculate $\left| -2^{-3} \right|$.

2. Express "a is negative" using an inequality sign.

Solve the inequalities in Exercises 3–6.

3. $-2x + 15 \leq 9$

4. $.2x - .3 \geq .7$

5. $3(1 - x) \geq 2x + 8$

6. $\dfrac{1}{2}x + 2 \geq \dfrac{3 + x}{4}$

7. Calculate $(81)^{3/4}$.

8. Calculate 5^{-2}.

In Exercises 9–12, use the laws of exponents to simplify the algebraic expressions.

9. $(\sqrt{x + 1})^4$

10. $\dfrac{xy^3}{x^{-5}y^6}$

11. $\dfrac{x^{3/2}}{\sqrt{x}}$

12. $\sqrt[3]{x}(8x^{2/3})$

In Exercises 13–18, perform the indicated algebraic operations.

13. $x(4 - x) + x^2$

14. $x(x + 1)(x + 2)$

15. $\dfrac{x + 1}{x} \cdot \dfrac{x^2}{x^2 - 1}$

16. $\dfrac{1}{x} + \dfrac{1}{x+1}$　　　　**17.** $3 - \dfrac{x}{x+5}$　　　　**18.** $\dfrac{x + \dfrac{1}{x}}{2x}$

19. Factor $24 + 6x - 3x^2$.

20. Factor $x^3 + x^2 - 30x$.

Solve the equations in Exercises 21–36.

21. $6x - 1 = 2(x + 2)$　　　　　　　**22.** $\frac{2}{3}x - 3 = 1$

23. $\dfrac{10}{2x+1} + 2 = 7$　　　　　　　**24.** $\dfrac{12}{4-x} + 5 = 11$

25. $x^2 + 5x - 14 = 0$　　　　　　**26.** $4x - x^2 = 4$

27. $x^2 - 3x + 1 = 0$　　　　　　　**28.** $x^2 - 2x + 2 = 0$

29. $4x^2 - 8x + 1 = 0$　　　　　　**30.** $\sqrt{2}x^2 - 4x + 2\sqrt{2} = 0$

31. $x + \dfrac{7}{x} = 8$　　　　　　　　**32.** $\dfrac{x^2 + 4}{x - 1} = 7$

33. $\log(3x + 1) = 1$　　　　　　**34.** $\log_2 x = 16$

35. $\log(x^2 - 5x + 6) = 1 + \log(x - 3)$　　**36.** $5\log x - \log x^3 = 2$

37. Rewrite $\log 7.5 = .8751$ using exponents.

38. Rewrite $8^{1/3} = 2$ using logarithms.

39. Simplify $\log 10^{3-x}$.　　　　　**40.** Simplify $10^{\log x^2}$.

41. Express $3\log x\sqrt{y} - \log y^{3/2}$ as a single logarithm.

42. Express $\log 2 + \log 500$ as a single logarithm.

Linear Equations and Inequalities

Chapter 2

Many applications considered later will involve linear equations and their geometric counterparts—straight lines. So we begin by studying the basic facts about these two important concepts.

2.1. Graphs and Linear Equations

A *Cartesian* (rectangular) *coordinate system* in a plane consists of two perpendicular lines, one horizontal and one vertical, with a unit of measurement marked on each line. These lines are called the *coordinate axes,* and the point where they meet is called the *origin* of the coordinate system. Often we call these axes the *x-axis* and the *y-axis.* The positive *x*-axis is to the right of the origin, the negative *x*-axis is to the left. The positive *y*-axis is above the origin, the negative *y*-axis is below. (See Fig. 1.)

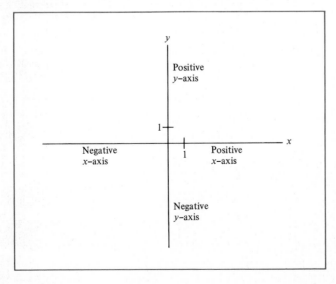

FIGURE 1

Each point in the plane is identified by a unique pair of numbers, (a, b), called the *coordinates* of the point. The coordinates describe the position of the point with respect to the origin. For example, in Fig. 2 the point $(4, 2)$ is located 4 units to the right and 2 units above the origin. We may think of $(4, 2)$ as the point obtained by moving (from the origin) 4 units in the x-direction and 2 units in the y-direction. Accordingly, we call 4 the x-*coordinate* of the point and 2 the y-*coordinate* of the point. The point $(-4, 2)$ is 4 units to the *left* and 2 units *above* the origin. The point $(-4, -2)$ is 4 units to the *left* and 2 units *below* the origin. Where should the point $(4, -2)$ be located?

Points on the x-axis all have y-coordinate 0, while points on the y-axis all have x-coordinate 0. (See Fig. 3.)

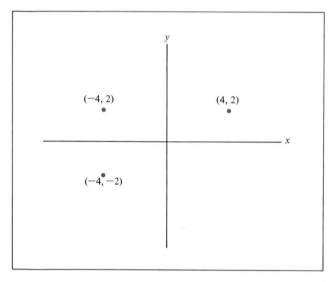

FIGURE 2

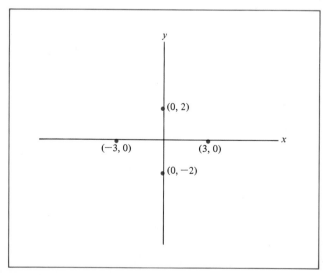

FIGURE 3

In many applications one encounters equations expressing relationships between variables x and y. Some typical equations are

$$y = 2x - 1$$
$$5x^2 + 3y^3 = 7$$
$$y = 2x^2 - 5x + 8.$$

To any equation in x and y one can associate a certain collection of points in the plane. Namely, the point (a, b) belongs to the collection provided that the equation is satisfied when we substitute a for each occurrence of x and substitute b for each occurrence of y. This collection of points is usually a curve of some sort and is called the *graph of the equation*.

EXAMPLE 1 Are the following points on the graph of the equation $y = 2x - 1$?

(a) $(3, 5)$ (b) $(5, 17)$

Solution (a) When we substitute 3 for each occurrence of x and 5 for each occurrence of y in the equation, we have

$$5 = 2(3) - 1.$$

This is certainly a true statement, so $(3, 5)$ is on the graph of the equation.

(b) If we replace x by 5 and y by 17 in the equation, we obtain

$$17 = 2(5) - 1.$$

This is not a true statement, so the equation is not satisfied by $x = 5$, $y = 17$. Hence $(5, 17)$ is *not* on the graph of the equation.

EXAMPLE 2 Sketch the graph of the equation $y = 2x - 1$.

Solution First find some points on the graph by choosing various values for x and determining the corresponding values for y:

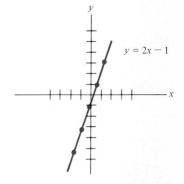

x	y
-2	$2 \cdot (-2) - 1 = -5$
-1	$2 \cdot (-1) - 1 = -3$
0	$2 \cdot 0 - 1 = -1$
1	$2 \cdot 1 - 1 = 1$
2	$2 \cdot 2 - 1 = 3$

Thus, the points $(-2, -5), (-1, -3), (0, -1), (1, 1)$, and $(2, 3)$ are all on the graph. Plot these points. It appears that the points lie on a straight line. By taking more values for x and plotting the corresponding points, it is easy to become convinced that the graph of $y = 2x - 1$ is, indeed, a straight line.

EXAMPLE 3 Sketch the graph of the equation $x = 3$.

Solution It is clear that the x-coordinate of any point on the graph must be 3. The y-coordinate can be anything. So some points on the graph are $(3, 0)$, $(3, 5)$, $(3, -4)$, $(3, -2)$. Again the graph is a straight line, as in the accompanying sketch.

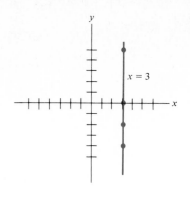

EXAMPLE 4 Sketch the graph of the equation $x = a$, where a is any given number.

Solution The x-coordinate of any point on the graph must be a. Reasoning as in Example 3, the graph is a vertical line a units away from the y-axis. (Of course, if a is negative, then the line lies to the left of the y-axis.)

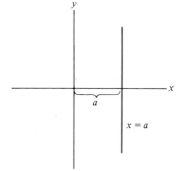

The equations in Examples 2 through 4 are all special cases of the general equation

$$cx + dy = e,$$

corresponding to particular choices of the constants c, d, e. Any such equation is called a *linear equation* (in the two variables x and y).

The *standard form* of a linear equation is obtained by solving for y if y appears, and for x if y does not appear. In the former case the standard form looks like

$$y = mx + b \qquad (m, b \text{ constants}),$$

whereas in the latter it looks like

$$x = a \qquad (a \text{ constant}).$$

EXAMPLE 5 Find the standard form of the following equations.

(a) $8x - 4y = 4$ (b) $2x + 3y = 3$ (c) $2x = 6$

Solution (a) Since y appears, we obtain the standard form by solving for y in terms of x:

$$8x - 4y = 4$$
$$-4y = -8x + 4$$
$$y = 2x - 1.$$

Thus the standard form of $8x - 4y = 4$ is $y = 2x - 1$—that is, $y = mx + b$ with $m = 2, b = -1$.

(b) Again y occurs, so we solve for y:

$$2x + 3y = 3$$
$$3y = -2x + 3$$
$$y = -\tfrac{2}{3}x + 1.$$

So the standard form is $y = -\tfrac{2}{3}x + 1$. Here $m = -\tfrac{2}{3}$ and $b = 1$.

(c) Here y does not occur, so solve for x:

$$2x = 6$$
$$x = 3.$$

Thus the standard form of $2x = 6$ is $x = 3$—that is, $x = a$ where a is 3.

We have seen that any linear equation has one of the two standard forms $y = mx + b, x = a$. From Example 4, the graph of $x = a$ is a vertical line, a units from the y-axis. What can be said about the graph of $y = mx + b$? In Example 2, we saw that the graph is a straight line in the special case $y = 2x - 1$. Actually, the graph of $y = mx + b$ is always a straight line. To sketch the graph, we need only locate two points. Two convenient points to locate are the *intercepts*, the points where the line crosses the x- and y-axes. When x is 0, $y = m \cdot 0 + b = b$. Thus, $(0, b)$ is on the graph of $y = mx + b$ and is the y-intercept of the line. The x-intercept is found as follows: A point on the x-axis has y-coordinate 0. So the x-coordinate of the x-intercept can be found by setting $y = 0$—that is, $mx + b = 0$—and solving this equation for x.

EXAMPLE 6   Sketch the graph of the equation $y = 2x - 1$.

Solution   Here $m = 2, b = -1$. The y-intercept is $(0, b) = (0, -1)$. To find the x-intercept, we must solve $2x - 1 = 0$. But then $2x = 1$ and $x = \tfrac{1}{2}$. So the x-intercept is $(\tfrac{1}{2}, 0)$. Plot the two points $(0, -1)$ and $(\tfrac{1}{2}, 0)$ and draw the straight line through them.

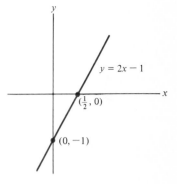

Note   The line in Example 6 had different x- and y-intercepts. Two other circumstances can occur, however. First, the two intercepts may be the same, as in $y = 3x$. Second, there may be no x-intercept, as in $y = 1$. In both of these circumstances, we must plot some point other than an intercept in order to graph the line.

To summarize, then:

> To graph the equation $y = mx + b$:
>
> 1. Plot the y-intercept $(0, b)$.
> 2. Plot some other point. (The most convenient choice is often the x-intercept $(x, 0)$, where x is determined by solving $mx + b = 0$.)
> 3. Draw a line through the two points.

The next example gives an application of linear equations.

EXAMPLE 7 (*Linear Depreciation*)   For tax purposes, businesses must keep track of the current values of each of their assets. A common mathematical model is to assume that the current value y is related to the age x of the asset by a linear equation. A moving company buys a 40-foot van with a useful lifetime of 5 years. After x months of use, the value y of the van is estimated by the linear equation

$$y = 25{,}000 - 400x.$$

(a) Sketch the graph of this linear equation.

(b) What is the value of the van after 5 years?

(c) What economic interpretation can be given to the y-intercept of the graph?

Solution (a) The y-intercept is $(0, 25{,}000)$. The x-intercept is found from the equation

$$25{,}000 - 400x = 0$$

$$x = \frac{125}{2},$$

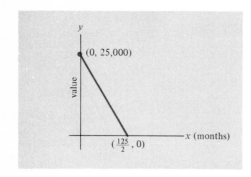

FIGURE 4

so the x-intercept is $\left(\dfrac{125}{2}, 0\right)$. The graph of the linear equation is sketched in Fig. 4. Note how the value decreases as the age of the truck increases. The value of the truck reaches 0 after 125/2 months. Note also that we have only sketched the portion of the graph which has physical meaning, namely the portion for x between 0 and 125/2.

(b) After 5 years (or 60 months), the value of the van is

$$y = 25{,}000 - 400(60)$$

$$= 25{,}000 - 24{,}000$$

$$= 1000.$$

Since the useful life of the van is 5 years, this value represents the *salvage value* of the van.

(c) The *y*-intercept corresponds to the value of the truck at $x = 0$ months, that is, the initial value of the truck, $25,000.

PRACTICE PROBLEMS 1

1. Plot the point (500, 200).

2. Is the point $(4, -7)$ on the graph of the linear equation $2x - 3y = 1$? The point $(5, 3)$?

3. Graph the linear equation $5x + y = 10$.

4. Graph the straight line $y = 3x$.

EXERCISES 1

In Exercises 1–8, plot the given point.

1. $(2, 3)$ 2. $(-1, 4)$ 3. $(0, -2)$ 4. $(2, 0)$

5. $(-2, 1)$ 6. $(-1, -\frac{5}{2})$ 7. $(-20, 40)$ 8. $(25, 30)$

In Exercises 9–12, each linear equation is in the standard form $y = mx + b$. Identify m and b.

9. $y = 5x + 8$ 10. $y = -2x - 6$ 11. $y = 3$ 12. $y = \frac{2}{3}x$

In Exercises 13–18, put the linear equations into standard form.

13. $14x + 7y = 21$ 14. $x - y = 3$ 15. $3x = 5$

16. $-\frac{1}{2}x + \frac{2}{3}y = 10$ 17. $\dfrac{x}{5} - \dfrac{y}{6} = 1$ 18. $x = 4 - 3y$

In Exercises 19–22, find the *x*-intercept and the *y*-intercept of each line.

19. $y = -4x + 8$ 20. $y = 5$ 21. $x = 7$ 22. $y = -8x$

In Exercises 23–28, graph the given linear equation.

23. $y = \frac{1}{3}x - 1$ 24. $y = 2x$ 25. $y = \frac{5}{2}$

26. $x = 0$ 27. $3x + 4y = 24$ 28. $x + y = 3$

29. What is the equation of the *x*-axis?

SOLUTIONS TO PRACTICE PROBLEMS 1

1. Since the numbers are large, make each hatchmark correspond to 100. Then the point (500, 200) is found by starting at the origin, moving 500 units to the right and then 200 units up.

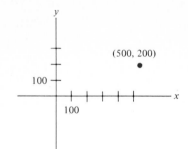

2. Substitute the point $(4, -7)$ into the equation $2x - 3y = 1$.

$$2(4) - 3(-7) = 1$$

$$8 + 21 = 1$$

$$29 = 1.$$

The equation does not hold, so $(4, -7)$ is not on the graph. Similarly, if we substitute (5.3) into the equation, we find that

$$2 \cdot 5 - 3 \cdot 3 = 1$$

$$10 - 9 = 1$$

$$1 = 1.$$

So the equation holds and (5, 3) is on the graph.

3. The standard form is obtained by solving for y:

$$y = -5x + 10.$$

Thus $m = -5$ and $b = 10$. The y-intercept is $(0, 10)$. To get the x-intercept, set $y = 0$:

$$0 = -5x + 10$$

$$5x = 10$$

$$x = 2.$$

So the x-intercept is $(2, 0)$.

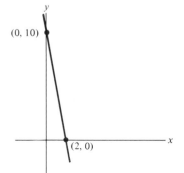

4. To find the y-intercept set $x = 0$; then $y = 3 \cdot 0 = 0$. To find the x-intercept set $y = 0$, then $3x = 0$ or $x = 0$. The two intercepts are the same point $(0, 0)$. We must therefore plot some other point. Setting $x = 1$, $y = 3 \cdot 1 = 3$, so another point on the line is $(1, 3)$.

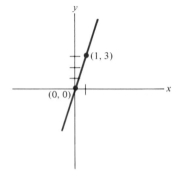

2.2. Linear Inequalities

When the equality sign in a linear equation $cx + dy = e$ is replaced by an inequality sign ($<$, $\leq$, $>$, or $\geq$), the resulting expression is called a *linear inequality* (in two variables). Our goal in this section is to represent a linear inequality by a graph.

Operations with inequalities were discussed in Section 1.2. For reference, we list the following properties of inequalities:

1.   Suppose that $a \leq b$ and c is any number; then $a + c \leq b + c$ and $a - c \leq b - c$.
2A. If $a \leq b$ and c is positive, then $ac \leq bc$.
2B. If $a \leq b$ and c is negative, then $ac \geq bc$.

In other words, any number can be added to or subtracted from both sides of an inequality, and an inequality can be multiplied by a positive number, just as in the case of an equation. But multiplying an inequality by a negative number (as in 2B) reverses the inequality sign.

We shall concentrate here on linear inequalities of the form $cx + dy \leq e$ and $cx + dy \geq e$, where c and d are not both zero. When $d \neq 0$ (that is, when y actually appears), the inequality can be put into one of the *standard forms* $y \leq mx + b$ or $y \geq mx + b$. When $d = 0$, the inequality can be put into one of the *standard forms* $x \leq a$ or $x \geq a$. The procedure for putting a linear inequality into standard form is analogous to that for putting a linear equation into standard form.

EXAMPLE 1   Put the linear inequality $2x - 3y \geq -9$ into standard form.

Solution   Since we want the y term on the left and all other terms on the right, begin by subtracting $2x$ from both sides:

$$-3y \geq -2x - 9.$$

Next, multiply by $-\frac{1}{3}$, remembering to change the inequality sign, since $-\frac{1}{3}$ is negative:

$$y \leq -\tfrac{1}{3}(-2x - 9)$$

$$y \leq \tfrac{2}{3}x + 3.$$

The last inequality is in standard form.

EXAMPLE 2   Find the standard form of the inequality $\frac{1}{2}x \geq 4$.

Solution   Note that y does not appear in the inequality. Just as was the case in finding the standard form of a linear equation when y does not appear, solve for x. To do this, multiply by 2 to get

$$x \geq 8,$$

the standard form.

Graphing Linear Inequalities Associated with every linear inequality, there is a set of points of the plane, the set of all those points which satisfy the inequality. This set of points is called the *graph* of the inequality.

EXAMPLE 3 Determine whether or not the given point satisfies the inequality $y \geq -\frac{2}{3}x + 4$.

(a) $(3, 4)$ (b) $(0, 0)$

Solution Substitute the x-coordinate of the point for x and the y-coordinate for y and determine if the resulting inequality is correct or not.

(a) $4 \geq -\frac{2}{3}(3) + 4$ (b) $0 \geq -\frac{2}{3}(0) + 4$
    $4 \geq -2 + 4$ $0 \geq 0 + 4$
    $4 \geq 2$ (correct) $0 \geq 4$ (not correct)

Therefore, the point $(3, 4)$ satisfies the inequality and the point $(0, 0)$ does not.

It is easiest to determine the graph of a given inequality after it has been written in standard form. Therefore, let us describe the graphs of each of the standard forms. The easiest to handle are the forms $x \geq a$ and $x \leq a$.

A point satisfies the inequality $x \geq a$ if and only if its x-coordinate is greater than or equal to a. The y-coordinate can be anything. Therefore, the graph of $x \geq a$ consists of all points to the right of and on the vertical line $x = a$. We will display the graph by crossing out the portion of the plane to the left of the line. (See Fig. 1.) Similarly, the graph of $x \leq a$ consists of the points to the left of and on the line $x = a$. This graph is shown in Fig. 2.

FIGURE 1   FIGURE 2

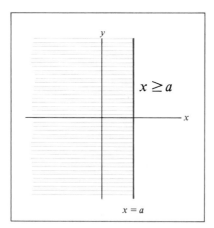

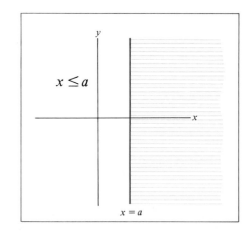

Here is a simple procedure for graphing the other two standard forms.

To graph $y \geq mx + b$ or $y \leq mx + b$:

1. Draw the graph of $y = mx + b$.
2. Throw away, that is, cross out the portion of the plane not satisfying the inequality. The graph of $y \geq mx + b$ consists of all points above or on the line. The graph of $y \leq mx + b$ consists of all points below or on the line.

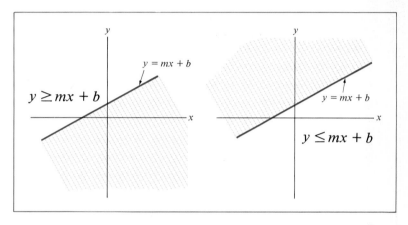

FIGURE 3

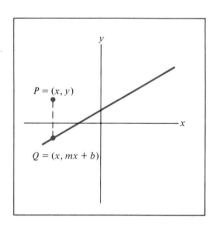

FIGURE 4

The graphs of the inequalities $y \geq mx + b$ and $y \leq mx + b$ are shown in Fig. 3. Some simple reasoning suffices to show why these graphs are correct. Draw the line $y = mx + b$, as in Fig. 4, and pick any point P above the line. Suppose P has coordinates (x, y). Let Q be the point on the line that lies directly below P. Then Q has the same first coordinate as P, that is, x. Since Q lies on the line, the second coordinate of Q is $mx + b$. Since the second coordinate of P is clearly larger than the second coordinate of Q, we must have $y \geq mx + b$. Similarly, any point below the line satisfies $y \leq mx + b$. Thus the two graphs are as given in Fig. 3.

EXAMPLE 4 Graph the inequality $2x + 3y \geq 15$.

Solution   In order to apply the procedure above, the inequality must first be put into standard form:

$$2x + 3y \geq 15$$

$$3y \geq -2x + 15$$

$$y \geq -\tfrac{2}{3}x + 5.$$

The last inequality is in standard form. Next, we graph the line $y = -\tfrac{2}{3}x + 5$. Its intercepts are $(0, 5)$ and $(7.5, 0)$. Since the inequality is "$y \geq$" we cross out the region below the line and label the region above with the inequality. The graph consists of all points above or on the line (Fig. 5).

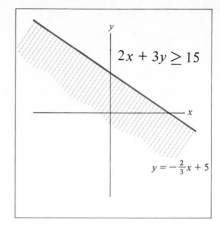

FIGURE 5

EXAMPLE 5   Graph the inequality $4x - 2y \geq 12$.

Solution   First put the inequality in standard form:

$$4x - 2y \geq 12$$

$$-2y \geq -4x + 12$$

$$y \leq 2x - 6$$

(note the change in the inequality sign!). Next, graph $y = 2x - 6$. The intercepts are $(0, -6)$ and $(3, 0)$. Since the inequality is "$y \leq$" the graph consists of all points below or on the line (Fig. 6).

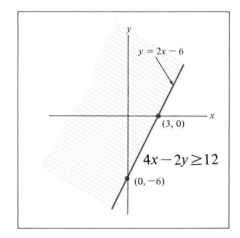

FIGURE 6

So far, we have been concerned only with graphing single inequalities. The next example concerns graphing a system of inequalities. That is, it asks us to determine all points of the plane which *simultaneously* satisfy all inequalities of a system.

EXAMPLE 6   Graph the system of inequalities

$$\begin{cases} 2x + 3y \geq 15 \\ 4x - 2y \geq 12 \\ y \geq 0. \end{cases}$$

Solution   The first two inequalities have already been graphed in Examples 4 and 5. The graph of $y \geq 0$ consists of all points above the x-axis. In Fig. 7, any point that is

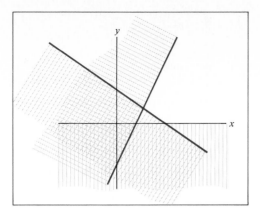

FIGURE 7

crossed out is *not* on the graph of at least one inequality. So the points which simultaneously satisfy all three inequalities are those in the remaining clear region.

Remark At first, our convention of crossing out those points *not* on the graph of an inequality (instead of shading the points *on* the graph) may have seemed odd. However, the real advantage of this convention becomes apparent when graphing a system of inequalities. Imagine trying to find the graph of the system of Example 6 if the points *on* the graph of each inequality had been shaded. It would have been necessary to locate the points that had been shaded three times. This is hard to do.

The graph of a system of inequalities is called a *feasible set*. The feasible set associated to the system of Example 6 is a three-sided, unbounded region.

PRACTICE PROBLEMS 2

1. Graph the inequality $3x - y \geq 3$.

2. Graph the feasible set for the system of inequalities

$$\begin{cases} x \geq 0, \quad\ \ y \geq 0 \\ x + 2y \leq 4 \\ 4x - 4y \geq -4. \end{cases}$$

EXERCISES 2

In Exercises 1–4, tell whether the inequality is true or false.

1. $2 \leq -3$ 2. $-2 \leq 0$ 3. $7 \leq 7$ 4. $0 \geq \frac{1}{2}$

In Exercises 5–8, solve for x.

5. $2x - 5 \geq 3$ 6. $3x - 7 \leq 2$

7. $-5x + 13 \leq -2$ 8. $-x + 1 \leq 3$

In Exercises 9–14, put the linear inequality into standard form.

9. $2x + y \leq 5$ 10. $-3x + y \geq 1$ 11. $5x - \frac{1}{3}y \leq 6$

12. $\frac{1}{2}x - y \leq -1$ 13. $4x \geq -3$ 14. $-2x \leq 4$

In Exercises 15–22, determine whether or not the given point satisfies the given inequality.

15. $3x + 5y \leq 12, (2, 1)$ 16. $-2x + y \geq 9, (3, 15)$

17. $y \geq -2x + 7, (3, 0)$ 18. $y \leq \frac{1}{2}x + 3, (4, 6)$

19. $y \le 3x - 4, (3, 5)$

20. $y \ge x, (-3, -2)$

21. $x \ge 5, (7, -2)$

22. $x \le 7, (0, 0)$

In Exercises 23–26, graph the given inequality by crossing out (i.e., discarding) the points not satisfying the inequality.

23. $y \le \frac{1}{3}x + 1$

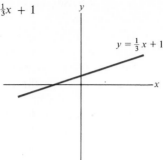

24. $y \ge -x + 1$

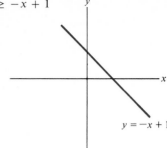

25. $x \ge 4$

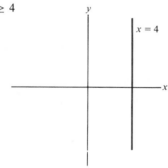

26. $y \le 2$

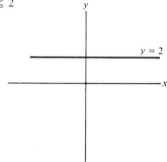

In Exercises 27–32, graph the given inequality.

27. $y \le 2x + 1$

28. $y \ge -3x + 6$

29. $x \ge 2$

30. $x \ge 0$

31. $x + 4y \ge 12$

32. $4x - 4y \ge 8$

In Exercises 33–38, graph the feasible set for the system of inequalities.

33. $\begin{cases} y \le 2x - 4 \\ y \ge 0 \end{cases}$

34. $\begin{cases} y \ge -\frac{1}{3}x + 1 \\ x \ge 0 \end{cases}$

35. $\begin{cases} x + 2y \ge 2 \\ 3x - y \ge 3 \end{cases}$

36. $\begin{cases} 3x + 6y \ge 24 \\ 3x + y \ge 6 \end{cases}$

37. $\begin{cases} x + 5y \le 10 \\ x + y \le 3 \\ x \ge 0, y \ge 0 \end{cases}$

38. $\begin{cases} x + 2y \ge 6 \\ x + y \ge 5 \\ x \ge 1 \end{cases}$

In Exercises 39–42, determine whether the given point is in the feasible set of the system of inequalities:

$$\begin{cases} 6x + 3y \le 96 \\ x + y \le 18 \\ 2x + 6y \le 72 \\ x \ge 0, \quad y \ge 0. \end{cases}$$

39. $(8, 7)$

40. $(14, 3)$

41. $(9, 10)$

42. $(16, 0)$

1. Linear inequalities are easiest to graph if they are first put into standard form. Subtract $3x$ from both sides and multiply by -1:

$$3x - y \geq 3$$

$$-y \geq -3x + 3$$

$$y \geq 3x - 3.$$

Now graph the line $y = 3x - 3$. The graph of the inequality is the portion of the plane below and on the line ("$\leq$" corresponds to below), so throw away (that is, cross out) the portion above the line.

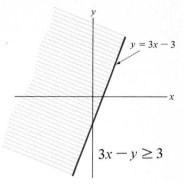

2. Begin by putting the linear inequalities into standard form and then graphing them all on the same coordinate system.

$$\begin{cases} x \geq 0, \quad y \geq 0 \\ x + 2y \leq 4 \\ 4x - 4y \geq -4 \end{cases} \quad \text{has standard form} \quad \begin{cases} x \geq 0, \quad y \geq 0 \\ y \leq -\frac{1}{2}x + 2 \\ y \leq x + 1. \end{cases}$$

A good procedure to follow is to graph all of the linear equations and then cross out the regions to be thrown away one at a time.

The inequalities $x \geq 0$ and $y \geq 0$ arise frequently in applications. The first has the form $x \geq a$, where $a = 0$, and the second has the form $y \geq mx + b$, where $m = 0$ and $b = 0$. To graph them, just cross out all points to the left of the y-axis and all points below the x-axis, respectively.

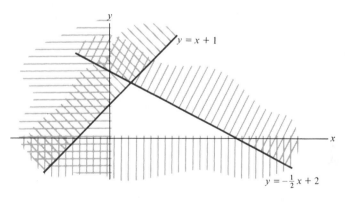

2.3. The Point of Intersection of a Pair of Lines

Suppose that we are given a pair of intersecting straight lines L and M. Let us consider the problem of determining the coordinates of the point of intersection $S = (x, y)$ (see Fig. 1). We may as well assume that the equations of L and M are

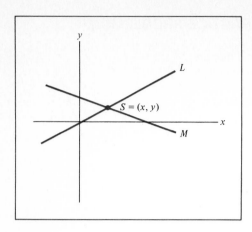

FIGURE 1

given in standard form. First, let us assume that both lines are in the first standard form—that is, that the equations are

$$L: y = mx + b, \qquad M: y = nx + c.$$

Since the point S is on both lines, its coordinates satisfy both equations. In particular, we have two expressions for its y-coordinate:

$$y = mx + b = nx + c.$$

The last equality gives an equation from which x can easily be determined. Then the value of y can be determined as $mx + b$. Let us see how this works in a particular example.

EXAMPLE 1 Find the point of intersection of the two lines $y = 2x - 3$, $y = x + 1$.

Solution To find the x-coordinate of the point of intersection, equate the two expressions for y and solve for x:

$$2x - 3 = x + 1$$
$$2x - x = 1 + 3$$
$$x = 4.$$

To find the value of y, set $x = 4$ in either equation, say the first. Then

$$y = 2 \cdot 4 - 3 = 5.$$

So the point of intersection is $(4, 5)$.

EXAMPLE 2 Find the point of intersection of the two lines $x + 2y = 6$, $5x + 2y = 18$.

Solution To use the method described, the equations must be in standard form. Solving both equations for y, we get the standard forms

$$y = -\tfrac{1}{2}x + 3$$
$$y = -\tfrac{5}{2}x + 9.$$

Equating the expressions for y gives

$$-\tfrac{1}{2}x + 3 = -\tfrac{5}{2}x + 9$$
$$\tfrac{5}{2}x - \tfrac{1}{2}x = 9 - 3$$
$$2x = 6$$
$$x = 3.$$

Setting $x = 3$ in the first equation gives

$$y = -\tfrac{1}{2}(3) + 3 = \tfrac{3}{2}.$$

So the intersection point is $(3, \tfrac{3}{2})$.

The method above works when both equations have the first standard form. In case one equation has the standard form $x = a$, things are much simpler. The value of x is then given directly without any work, namely $x = a$. The value of y can be found by substituting a for x in the other equation.

EXAMPLE 3   Find the point of intersection of the lines $y = 2x - 1$, $x = 2$.

Solution   The x-coordinate of the intersection point is 2, and the y-coordinate is $y = 2 \cdot 2 - 1 = 3$. Therefore, the intersection point is $(2, 3)$.

The method introduced above may be used to solve systems of two equations in two variables.

EXAMPLE 4   Solve the following system of linear equations:

$$\begin{cases} 2x + 3y = 7 \\ 4x - 2y = 9. \end{cases}$$

Solution   First convert the equations to standard form:

$$2x + 3y = 7$$
$$3y = -2x + 7$$
$$y = -\tfrac{2}{3}x + \tfrac{7}{3};$$
$$4x - 2y = 9$$
$$-2y = -4x + 9$$
$$y = 2x - \tfrac{9}{2}.$$

Now equate the two expressions for y:

$$2x - \tfrac{9}{2} = -\tfrac{2}{3}x + \tfrac{7}{3}$$
$$\tfrac{8}{3}x = \tfrac{7}{3} + \tfrac{9}{2} = \tfrac{14}{6} + \tfrac{27}{6} = \tfrac{41}{6}$$
$$x = \tfrac{3}{8} \cdot \tfrac{41}{6} = \tfrac{41}{16}$$
$$y = 2x - \tfrac{9}{2} = 2(\tfrac{41}{16}) - \tfrac{9}{2} = \tfrac{5}{8}.$$

So the solution of the given system is $x = \tfrac{41}{16}$, $y = \tfrac{5}{8}$.

EXAMPLE 5   (*Break-Even Analysis*) Suppose that a manufacturer sells a product for $3 per unit. If x units are sold, the manufacturer receives $3x$ dollars. The equation

$$y = 3x \qquad \text{(revenue equation)}$$

describes this relation between the revenue y and the number x of units sold. Suppose that during a certain time period the manufacturer has fixed overhead costs of $750, no matter how many units are produced and sold. In addition, for each unit sold, the manufacturer has "variable costs" of $1.50 per unit for labor and materials. Then the total cost y of producing and selling x units is given by the equation

$$y = 750 + 1.5x \qquad \text{(cost equation)}.$$

The graphs of the revenue equation and the cost equation are shown in Fig. 2. The point at which the graphs intersect is called the *break-even point* because at this value of x the manufacturer's revenue and cost are equal, and the manufacturer "breaks even." At a lower sales level x, the manufacturer will experience a loss, and at a higher sales level the manufacturer will make a profit. Find the break-even point in Fig. 2.

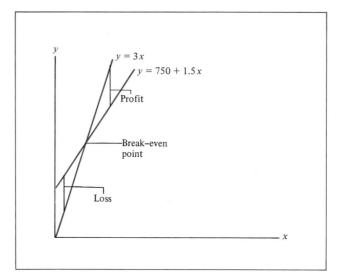

FIGURE 2

Solution We must find the value of x for which the revenue and cost associated with x units are equal:

$$[\text{revenue}] = [\text{cost}]$$
$$3x = 750 + 1.5x$$
$$1.5x = 750$$
$$x = \frac{750}{1.5} = 500.$$

When $x = 500$, $y = 3x = 3(500) = 1500$. Thus the break-even point is $(500, 1500)$. At a sales level of 500 units, the revenue and cost both equal 1500.

Supply and Demand Curves Let p denote the price of a commodity and q the quantity. Economists study two sorts of graphs which express relationships between p and q. To describe these graphs, let us plot price along the horizontal axis and quantity along the vertical axis. The first graph relating p and q is called a *supply curve* and expresses the relationship between p and q from a manufacturer's point of view. For every price p, the supply curve specifies the quantity q which the manufacturer is willing to produce at the market price p. The higher the price, the more the manufacturer is willing to supply. So supply curves rise when viewed from left to right.

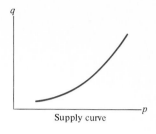

Supply curve

The second curve relating p and q is called a *demand curve* and expresses the relationship between p and q from the consumers' viewpoint. For each price p, the demand curve expresses the quantity q which consumers will buy if the market price is p. The higher the price, the less consumers will buy. So demand curves fall when viewed from left to right.

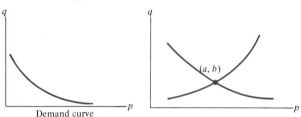

Demand curve

Suppose that the supply and demand curves for a commodity are drawn on a single coordinate system. The intersection point (a, b) of the two curves has an economic significance: prices will stabilize at an "equilibrium price" of a dollars per unit and the quantity produced will be b units.

EXAMPLE 6   Suppose that the supply curve for a certain commodity is the straight line whose equation is $q = 5000p - 10,000$ (p in dollars). Suppose that the demand curve for the same commodity is the straight line whose equation is $q = -2000p + 11,000$. Determine the equilibrium price at which the commodity will sell and determine the quantity produced.

Solution   We must solve the system of linear equations

$$\begin{cases} q = 5000p - 10,000 \\ q = -2000p + 11,000. \end{cases}$$

$$5000p - 10,000 = -2000p + 11,000$$
$$7000p = 21,000$$
$$p = 3$$
$$q = 5000(3) - 10,000$$
$$= 5000.$$

Thus the commodity will sell for \$3 and 5000 units will be produced.

PRACTICE PROBLEMS 3

Figure 3 shows the feasible set of a system of linear inequalities; its four vertices are labeled A, B, C, and D.

1. Use the method of this section to find the co-ordinates of the point C.

2. Determine the coordinates of the points A and B by inspection.

3. Find the coordinates of the point D.

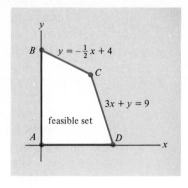

FIGURE 3

EXERCISES 3

In Exercises 1–6, find the point of intersection of the given pair of straight lines.

1. $\begin{cases} y = 4x - 5 \\ y = -2x + 7 \end{cases}$

2. $\begin{cases} y = 3x - 15 \\ y = -2x + 10 \end{cases}$

3. $\begin{cases} x - 4y = -2 \\ x + 2y = 4 \end{cases}$

4. $\begin{cases} x - 3y = -6 \\ 3x - 2y = 10 \end{cases}$

5. $\begin{cases} y = \frac{1}{3}x - 1 \\ x = 12 \end{cases}$

6. $\begin{cases} 2x - 3y = 3 \\ x = 6 \end{cases}$

Solve the following systems of linear equations.

7. $\begin{cases} 2x + y = 7 \\ x - y = 3 \end{cases}$

8. $\begin{cases} x + 2y = 4 \\ \frac{1}{2}x + \frac{1}{2}y = 3 \end{cases}$

9. $\begin{cases} 5x - 2y = 1 \\ 2x + y = -4 \end{cases}$

10. $\begin{cases} x + 2y = 6 \\ x - \frac{1}{3}y = 4 \end{cases}$

Find the coordinates of the vertices of the following feasible sets.

11.

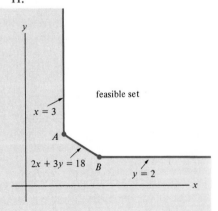

12.

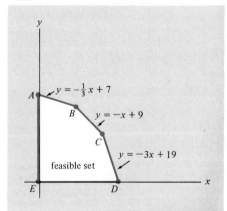

13.

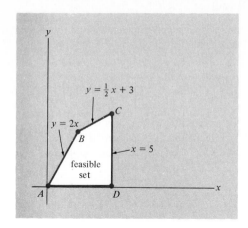

14.

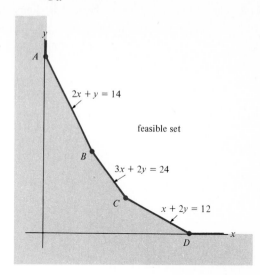

In Exercises 15–20, graph the feasible set for the system of inequalities and find the coordinates of the vertices.

15.
$$\begin{cases} 2y - x \le 6 \\ x + 2y \ge 10 \\ x \le 6 \end{cases}$$

16.
$$\begin{cases} 2x + y \ge 10 \\ x \ge 2 \\ y \ge 2 \end{cases}$$

17.
$$\begin{cases} x + 3y \le 18 \\ 2x + y \le 16 \\ x \ge 0, \quad y \ge 0 \end{cases}$$

18.
$$\begin{cases} 5x + 2y \ge 14 \\ x + 3y \ge 8 \\ x \ge 0, \quad y \ge 0 \end{cases}$$

19.
$$\begin{cases} 4x + y \ge 8 \\ x + y \ge 5 \\ x + 3y \ge 9 \\ x \ge 0, \quad y \ge 0 \end{cases}$$

20.
$$\begin{cases} x + 4y \le 28 \\ x + y \le 10 \\ 3x + y \le 24 \\ x \ge 0, \quad y \ge 0 \end{cases}$$

In Exercises 21 and 22, find the break-even point for the given revenue and cost equations.

21. (revenue) $y = 17.50x$
(cost) $y = 12x + 660$

22. (revenue) $y = 4.60x$
(cost) $y = 3.20x + 630$

23. The supply curve for a certain commodity is given by the equation $q = 10,000p - 500$.

(a) Suppose that the market price is $2 per unit. How many units of the commodity will be offered for sale?

(b) What is the highest price at which a quantity of 0 will be supplied?

24. The demand curve for a certain commodity is given by the equation $q = -1000p + 32,500$.

(a) Suppose that the market price is $1 per unit. How many units of the commodity will be demanded?

(b) What is the lowest price at which a quantity of 0 will be demanded?

25. The demand curve for a certain commodity is approximately the line whose equation is $q = 16,000 - 3500p$. The supply curve for the same commodity is given by the equation

$q = 1500p - 1400$. Find the equilibrium price at which the quantity demanded and the quantity supplied are equal.

26. Find the equilibrium price at which supply and demand are equal, if the supply equation is $q = 126p - 209$ and the demand equation is $q = 531 - 74p$.

SOLUTIONS TO PRACTICE PROBLEMS 3

1. The point C is the point of intersection of the lines with equations $y = -\frac{1}{2}x + 4$ and $3x + y = 9$. In order to use the method of this section, the second equation must first be put into its standard form $y = -3x + 9$. Now equate the two expressions for y and solve:

$$-\tfrac{1}{2}x + 4 = -3x + 9$$

$$\tfrac{5}{2}x = 5$$

$$x = \tfrac{2}{5} \cdot 5 = 2$$

$$y = -\tfrac{1}{2}(2) + 4 = 3.$$

Therefore, $C = (2, 3)$.

2. $A = (0, 0)$, since the point A is the origin. $B = (0, 4)$, since it is the y-intercept of the line with equation $y = -\frac{1}{2}x + 4$.

3. D is the x-intercept of the line $3x + y = 9$. Its first coordinate is found by setting $y = 0$ and solving for x.

$$3x + (0) = 9$$

$$x = 3.$$

Therefore, $D = (3, 0)$.

2.4. The Slope of a Straight Line

As we have seen, any linear equation can be put into one of the two standard forms $y = mx + b$ or $x = a$. In this section, let us exclude linear equations whose standard form is of the latter type. *Geometrically, this means that we will consider only nonvertical lines.*

Suppose that we are given a nonvertical line L whose equation is $y = mx + b$. The number m is called the *slope of L*. That is, the slope is the coefficient of x in the standard form of the equation of the line.

EXAMPLE 1 Find the slopes of the lines having the following equations:

(a) $y = 2x + 1$ (b) $y = -\frac{3}{4}x + 2$ (c) $y = 3$ (d) $-8x + 2y = 4$

Solution (a) $m = 2$.

(b) $m = -\frac{3}{4}$.

(c) When we write the equation in the form $y = 0 \cdot x + 3$, we see that $m = 0$.

(d) First, we put the equation in standard form:

$$-8x + 2y = 4$$
$$2y = 8x + 4$$
$$y = 4x + 2.$$

Thus $m = 4$.

The definition of the slope given is in terms of the standard form of the equation of the line. Let us give an alternative definition.

□ **Geometric Definition of Slope** Let L be a line passing through the points (x_1, y_1) and (x_2, y_2). Then the slope of L is given by the formula

$$m = \frac{y_2 - y_1}{x_2 - x_1}. \tag{1}$$

That is, the slope is the difference in the y-coordinates divided by the difference in the x-coordinates, with both differences formed in the same order.

Before proving this definition equivalent to the first one given, let us show how it can be used.

EXAMPLE 2 Find the slope of the line passing through the points $(1, 3)$ and $(4, 6)$.

Solution We have

$$m = \frac{[\text{difference in } y\text{-coordinates}]}{[\text{difference in } x\text{-coordinates}]} = \frac{6 - 3}{4 - 1} = \frac{3}{3} = 1.$$

Thus $m = 1$. [Note that if we reverse the order of the points and use the formula (1) to compute the slope, then we get

$$m = \frac{3 - 6}{1 - 4} = \frac{-3}{-3} = 1,$$

which is the same answer. The order of the points is immaterial. The important concern is to make sure that the differences in the x- and y-coordinates are formed in the same order.]

Justification of Formula 1 Since (x_1, y_1) and (x_2, y_2) are both on the line, both points satisfy the equation of the line, which has the form $y = mx + b$. Thus

$$y_2 = mx_2 + b$$
$$y_1 = mx_1 + b.$$

Subtracting these two equations gives

$$y_2 - y_1 = mx_2 - mx_1 = m(x_2 - x_1).$$

Dividing by $x_2 - x_1$, we have

$$m = \frac{y_2 - y_1}{x_2 - x_1},$$

which is formula (1). So the two definitions of slope lead to the same number.

Let us now study four of the most important properties of the slope of a straight line. We begin with the so-called steepness property, since it provides us with a geometric interpretation for the number m.

> **Steepness Property** Let the line L have slope m. If we start at any point on the line and move 1 unit to the right, then we must move m units vertically in order to return to the line (Fig. 1). (Of course, if m is positive, then we move up; and if m is negative, we move down.)

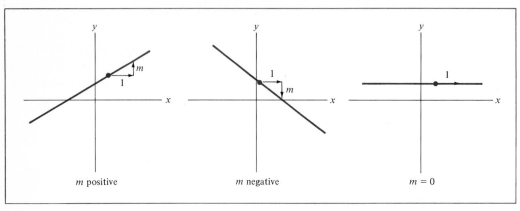

m positive m negative $m = 0$

FIGURE 1

EXAMPLE 3 Illustrate the steepness property for each of the lines.

(a) $y = 2x + 1$ (b) $y = -\frac{3}{4}x + 2$

(c) $y = 3$

FIGURE 2

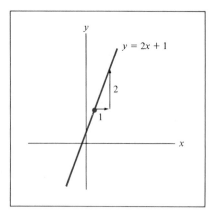

Solution (a) Here $m = 2$. So starting from any point on the line, proceeding 1 unit to the right, we must go 2 units up to return to the line (Fig. 2).

(b) Here $m = -\frac{3}{4}$. So starting from any point on the line, proceeding 1 unit to the right, we must go $\frac{3}{4}$ unit down to return to the line (Fig. 3).

(c) Here $m = 0$. So going 1 unit to the right requires going 0 units vertically to return to the line (Fig. 4).

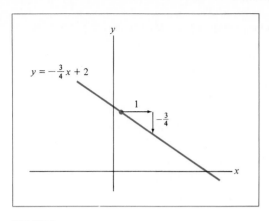

FIGURE 3

FIGURE 4

In the next example we introduce a new method for graphing a linear equation. This method relies on the steepness property and is much more efficient than finding two points on the line (e.g., the two intercepts).

EXAMPLE 4   Use the steepness property to draw the graph of $y = \frac{1}{2}x + \frac{3}{2}$.

Solution   The y-intercept is $(0, \frac{3}{2})$, as we read off from the equation. We can find another point on the line using the steepness property. Start at $(0, \frac{3}{2})$. Go 1 unit to the right. Since the slope is $\frac{1}{2}$, we must move vertically $\frac{1}{2}$ unit to return to the line. But this locates a second point on the line. So we draw the line through the two points. The entire procedure is illustrated in Fig. 5.

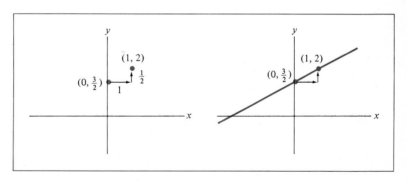

FIGURE 5

Actually, in order to use the steepness property to graph an equation, all that is needed is the slope plus *any* point (not necessarily the y-intercept).

EXAMPLE 5   Graph the line of slope -1 and passing through the point $(2, 2)$.

Solution   Start at (2, 2), move 1 unit to the right and then −1 unit vertically, that is, 1 unit down. The line through (2, 2) and the resulting point is the desired line (see Fig. 6).

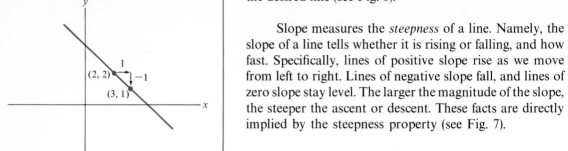

Slope measures the *steepness* of a line. Namely, the slope of a line tells whether it is rising or falling, and how fast. Specifically, lines of positive slope rise as we move from left to right. Lines of negative slope fall, and lines of zero slope stay level. The larger the magnitude of the slope, the steeper the ascent or descent. These facts are directly implied by the steepness property (see Fig. 7).

FIGURE 6

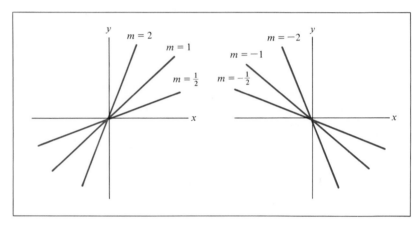

FIGURE 7

Justification of the Steepness Property

FIGURE 8

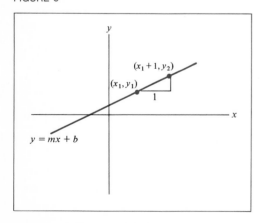

Consider a line with equation $y = mx + b$, and let (x_1, y_1) be any point on the line. If we start from this point and move 1 unit to the right, the first coordinate of the new point is $x_1 + 1$, since the x-coordinate is increased by 1. Now go far enough vertically to return to the line. Denote the y-coordinate of this new point by y_2 (see Fig. 8). We must show that to get y_2, we add m to y_1. That is, $y_2 = y_1 + m$. By equation (1), we can compute m as

$$m = \frac{[\text{difference in } y\text{-coordinates}]}{[\text{difference in } x\text{-coordinates}]}$$

$$= \frac{y_2 - y_1}{1} = y_2 - y_1.$$

In other words, $y_2 = y_1 + m$, which is what we desired to show.

Often the slopes of the straight lines which occur in applications have interesting and significant interpretations. An application in the field of economics is illustrated in the next example.

EXAMPLE 6 Suppose a manufacturer finds that the cost y of producing x units of a certain commodity is given by the formula $y = 2x + 5000$. What interpretation can be given the slope of the graph of this equation?

Solution Suppose that the firm is producing at a certain level and increases production by 1 unit. That is, x is increased by 1 unit. By the steepness property, the value of y then increases by 2, which is the slope of the line whose equation is $y = 2x + 5000$. Thus, *each additional unit of production costs $2*. (The graph of $y = 2x + 5000$ is called a *cost curve*. It relates the size of production to total cost. The graph is a straight line, and economists call its slope the *marginal cost of production*. The y-coordinate of the y-intercept is called the *fixed cost*. In this case the fixed cost is $5000, and it includes costs such as rent and insurance which are incurred even if no units are produced.)

In applied problems having time as a variable, the letter t is often used in place of the letter x. If so, straight lines have equations of the form $y = mt + b$ and are graphed on a ty-coordinate system.

EXAMPLE 7 *(Depreciation)* The federal government allows an income tax deduction for the decrease in value (or *depreciation*) of capital assets (such as buildings and equipment). One method of calculating the depreciation is to take equal amounts over the expected lifetime of the asset. This method is called *straight-line depreciation*. For tax purposes the value V of an asset t years after purchase is figured according to the equation $V = -100{,}000t + 700{,}000$. The expected life of the asset is 5 years.

(a) How much did the asset originally cost?

(b) What is the annual deduction for depreciation?

(c) What is the salvage value of the asset? (That is, what is the value of the asset after 5 years?)

Solution (a) The original cost is the value of V at $t = 0$, namely

$$V = -100{,}000(0) + 700{,}000 = 700{,}000.$$

That is, the asset originally cost $700,000.

(b) By the steepness property, each increase of 1 in t causes a decrease in V of 100,000. That is, the value is decreasing by $100,000 per year. So the depreciation deduction is $100,000 each year.

(c) After 5 years, the value V is given by

$$V = -100{,}000(5) + 700{,}000 = 200{,}000.$$

The salvage value is $200,000.

We have seen in Example 5 how to sketch a straight line when given its slope and one point on it. Let us now see how to find the equation of the line from this same data.

Point-Slope Formula The equation of the straight line passing through (x_1, y_1) and having slope m is given by $y - y_1 = m(x - x_1)$.

EXAMPLE 8 Find the equation of the line that passes through $(2, 3)$ and has slope $\frac{1}{2}$.

Solution Here $x_1 = 2$, $y_1 = 3$, $m = \frac{1}{2}$. So the equation is

$$y - 3 = \tfrac{1}{2}(x - 2)$$
$$y - 3 = \tfrac{1}{2}x - 1$$
$$y = \tfrac{1}{2}x + 2.$$

EXAMPLE 9 Find the equation of the line through the points $(3, 1)$ and $(6, 0)$.

Solution We can compute the slope from equation (1):

$$[\text{slope}] = \frac{[\text{difference in } y\text{-coordinates}]}{[\text{difference in } x\text{-coordinates}]} = \frac{1 - 0}{3 - 6} = -\tfrac{1}{3}.$$

Now we can determine the equation from the point-slope formula with $(x_1, y_1) = (3, 1)$ and $m = -\frac{1}{3}$:

$$y - 1 = -\tfrac{1}{3}(x - 3)$$
$$y = -\tfrac{1}{3}x + 2.$$

[Question: What would the equation be if we had chosen $(x_1, y_1) = (6, 0)$?]

Verification of the Point-Slope Formula Let (x, y) be any point on the line passing through the point (x_1, y_1) and having slope m. Then, by equation (1), we have

$$m = \frac{y - y_1}{x - x_1}.$$

Multiplying through by $x - x_1$ gives

$$y - y_1 = m(x - x_1). \tag{2}$$

Thus, every point (x, y) on the line satisfies equation (2). So (2) gives the equation of the line through (x_1, y_1) and having slope m.

The next property of slope relates the slopes of two perpendicular lines.

> **Perpendicular Property** When two lines are perpendicular, their slopes are negative reciprocals of one another. That is, if two lines with slopes m and n are perpendicular to one another, then*
>
> $$m = -\frac{1}{n}.$$
>
> Conversely, if two lines have slopes that are negative reciprocals of one another, then they are perpendicular.

A proof of the perpendicular property is outlined in Exercise 58. Let us show how it can be used to help find equations of lines.

EXAMPLE 10 Find the equation of the line perpendicular to the graph of $y = 2x - 5$ and passing through $(1, 2)$.

Solution The slope of the graph of $y = 2x - 5$ is 2. By the perpendicular property, the slope of a line perpendicular to it is $-\frac{1}{2}$. If a line has slope $-\frac{1}{2}$ and passes through $(1, 2)$, then it has equation

$$y - 2 = -\tfrac{1}{2}(x - 1)$$

or

$$y = -\tfrac{1}{2}x + \tfrac{5}{2}$$

(by the point-slope formula).

The final property of slopes gives the relationship between slopes of parallel lines. A proof is outlined in Exercise 57.

> **Parallel Property** Parallel lines have the same slope. Conversely, if two lines have the same slope, then they are parallel.

EXAMPLE 11 Find the equation of the line through $(2, 0)$ and parallel to the line whose equation is $y = \tfrac{1}{3}x - 11$.

Solution The slope of the line having equation $y = \tfrac{1}{3}x - 11$ is $\tfrac{1}{3}$. Therefore, any line parallel to it also has slope $\tfrac{1}{3}$. Therefore, the desired line passes through $(2, 0)$ and has slope $\tfrac{1}{3}$, so its equation is

$$y - 0 = \tfrac{1}{3}(x - 2)$$

or

$$y = \tfrac{1}{3}x - \tfrac{2}{3}.$$

---

* If $n = 0$, this formula does not say anything, since $1/0$ is undefined. However, in this case, one line is horizontal and one vertical, the vertical one having an undefined slope.

PRACTICE PROBLEMS 4

Suppose that the revenue y from selling x units of a certain commodity is given by the formula $y = 4x$. (Revenue is the amount of money received from the sale of the commodity.)

1. What interpretation can be given to the slope of the graph of this equation?

2. (See Example 6.) Find the coordinates of the point of intersection of $y = 4x$ and $y = 2x + 5000$.

3. What interpretation can be given to the value of the x-coordinate of the point found in Problem 2?

EXERCISES 4

In Exercises 1–4, find the slope of the line having the given equation.

1. $y = \frac{2}{3}x + 7$

2. $y = -4$

3. $y - 3 = 5(x + 4)$

4. $7x + 5y = 10$

In Exercises 5–8, find the slope of the line passing through the given points.

5. $(3, 4), (7, 9)$

6. $(-2, 1), (3, -3)$

7. $(0, 0), (4, 5)$

8. $(4, 17), (-2, 17)$

In Exercises 9–12, graph the given linear equation by beginning at the y-intercept, moving 1 unit to the right and m units in the y-direction.

9. $y = -2x + 1$ 10. $y = 4x - 2$ 11. $y = 3x$ 12. $y = -2$

In Exercises 13–20, find the equation of line L.

13.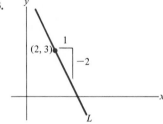

14.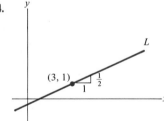

15.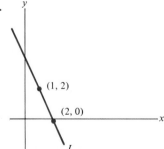

16.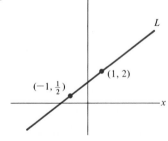

17.

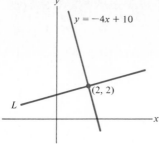

L perpendicular to $y = -4x + 10$

18.

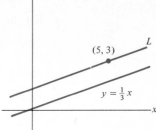

L parallel to $y = \frac{1}{3}x$

19.

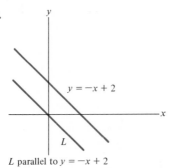

L parallel to $y = -x + 2$

20.

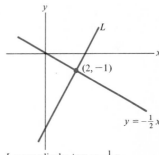

L perpendicular to $y = -\frac{1}{2}x$

21. Find the equation of the line passing through the point (2, 3) and parallel to the x-axis.

22. Find the equation of the line passing through (0, 0) and having slope 1.5.

23. Find the y-intercept of the line passing through the point (5, 6) and having slope $\frac{3}{5}$.

24. Find the slope of the line passing through the point (1, 4) and having y-intercept (0, 4).

25. A salesperson's weekly pay depends on the volume of sales. If she sells x units of goods, then her pay is $y = 5x + 60$ dollars. Give an interpretation to the slope and the y-intercept of this straight line.

26. A manufacturer has fixed costs (such as rent and insurance) of $2000 per month. The cost of producing each unit of goods is $4. Give the linear equation for the cost of producing x units per month.

27. The demand equation for a monopolist is $y = -.02x + 7$, where x is the number of units produced and y is the price. That is, in order to sell x units of goods, the price must be $y = -.02x + 7$ dollars. Interpret the slope and y-intercept of this line.

An apartment complex has a storage tank to hold its heating oil. The tank was filled on January 1, but no more deliveries of oil will be made until sometime in March. Let t denote the number of days after January 1 and let y denote the number of gallons of fuel oil in the tank. Current records show that y and t will be related by the equation $y = 30,000 - 400t$.

28. Graph the equation $y = 30,000 - 400t$.

29. How much oil will be in the tank on February 1?

30. How much oil will be in the tank on February 15?

31. Determine the y-intercept of the graph. Explain its significance.

32. Determine the t-intercept of the graph. Explain its significance.

A corporation receives payment for a large contract on July 1, bringing its cash reserves to $2.3 million. Let y denote its cash reserves (in millions) t days after July 1. The corporation's accountants estimate that y and t will be related by the equation $y = 2.3 - .15t$.

33. Graph the equation $y = 2.3 - .15t$.

34. How much cash does the corporation have on the morning of July 16?

35. Determine the y-intercept of the graph. Explain its significance.

36. Determine the t-intercept of the graph. Explain its significance.

37. Determine the cash reserves on July 4.

38. When will the cash reserves be $.8 million?

Find the equations of the following lines.

39. Slope is 3; y-intercept is $(0, -1)$

40. Slope is $-\frac{1}{2}$; y-intercept is $(0, 0)$

41. Slope is 1; $(1, 2)$ on line

42. Slope is $-\frac{1}{3}$; $(6, -2)$ on line

43. Slope is -7; $(5, 0)$ on line

44. Slope is $\frac{1}{2}$; $(2, -3)$ on line

45. Slope is 0; $(7, 4)$ on line

46. Slope is $-\frac{2}{5}$; $(0, 5)$ on line

47. $(2, 1)$ and $(4, 2)$ on line

48. $(5, -3)$ and $(-1, 3)$ on line

49. $(0, 0)$ and $(1, -2)$ on line

50. $(2, -1), (3, -1)$ on line

In each of Exercises 51–54 we specify a line by giving the slope and one point on the line. We give the first coordinate of some points on the line. Without deriving the equation of the line, find the second coordinate of each of the points.

51. Slope is 2, $(1, 3)$ on line; $(2, \quad)$; $(0, \quad)$; $(-1, \quad)$

52. Slope is -3, $(2, 2)$ on line; $(3, \quad)$; $(4, \quad)$; $(1, \quad)$

53. Slope is $-\frac{1}{4}$, $(-1, -1)$ on line; $(0, \quad)$; $(1, \quad)$; $(-2, \quad)$

54. Slope is $\frac{1}{3}$; $(-5, 2)$ on line; $(-4, \quad)$; $(-3, \quad)$; $(-2, \quad)$

For each pair of lines in the figures below, determine the one with the greater slope.

55.

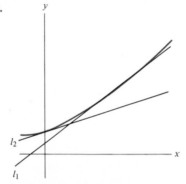

56.

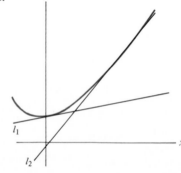

57. Prove the parallel property. [*Hint:* If $y = mx + b$ and $y = m'x + b'$ are the equations of two lines, then the two lines have a point in common if and only if the equation $mx + b = m'x + b'$ has a solution x.]

58. Prove the perpendicular property. [*Hint:* Without loss of generality, assume that both lines pass through the origin. Use the point-slope formula, the Pythagorean theorem, and the accompanying figure.]

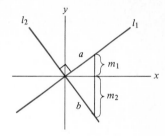

SOLUTIONS TO PRACTICE PROBLEMS 4

1. By the steepness property, whenever x is increased by 1 unit, the value of y is increased by 4 units. Therefore, *each additional unit of production brings in $4 of revenue.* (The graph of $y = 4x$ is called a *revenue curve* and its slope is called the *marginal revenue of production.*)

2. $\begin{cases} y = 4x \\ y = 2x + 5000. \end{cases}$ 　　Equate expressions for y: $4x = 2x + 5000$

$$2x = 5000$$

$$x = 2500$$

$$y = 4(2500) = 10{,}000.$$

3. When producing 2500 units, the revenue equals the cost. This value of x is called the *break-even point.* Since (profit) = (revenue) − (cost), the company will make a profit only if its level of production is greater than the break-even point.

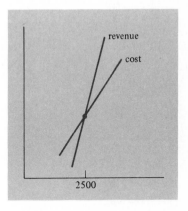

Chapter 2: CHECKLIST

☐ Cartesian coordinate system for the plane
☐ Origin, *x*-axis, *y*-axis
☐ Graph of an equation
☐ Linear equation in two variables

- Standard form of a linear equation in two variables
- y-intercept, x-intercept of a line
- Linear inequalities in two variables
- Graph of a linear inequality
- Feasible set of a system of linear inequalities
- Method to find point of intersection of pair of straight lines
- Slope of a line

- $m = \dfrac{y_2 - y_1}{x_2 - x_1}$

- Steepness property of slope
- Point-slope formula: $y - y_1 = m(x - x_1)$
- Perpendicular property
- Parallel property

Chapter 2: SUPPLEMENTARY EXERCISES

1. What is the equation of the y-axis?

2. Graph the linear equation $y = -\frac{1}{2}x$.

3. Find the point of intersection of the pair of straight lines $x - 5y = 6$ and $3x = 6$.

4. Find the slope of the line having the equation $3x - 4y = 8$.

5. Find the equation of the line having y-intercept $(0, 5)$ and x-intercept $(10, 0)$.

6. Graph the linear inequality $x - 3y \geq 12$.

7. Does the point $(1, 2)$ satisfy the linear inequality $3x + 4y \geq 11$?

8. Find the point of intersection of the pair of straight lines $2x - y = 1$ and $x + 2y = 13$.

9. Find the equation of the straight line passing through the point $(15, 16)$ and parallel to the line $2x - 10y = 7$.

10. Find the y-coordinate of the point having x-coordinate 1 and lying on the line $y = 3x + 7$.

11. Find the x-intercept of the straight line with equation $x = 5$.

12. Graph the linear inequality $y \leq 6$.

13. Solve the following system of linear equations:

$$\begin{cases} 3x - 2y = 1 \\ 2x + y = 24. \end{cases}$$

14. Graph the feasible set for the following system of inequalities:

$$\begin{cases} 2y + 7x \geq 30 \\ 4y - x \geq 0 \\ y \leq 8. \end{cases}$$

15. Find the y-intercept of the line passing through the point $(4, 9)$ and having slope $\frac{1}{2}$.

16. The fee charged by a local moving company depends on the amount of time required for the move. If x hours is required, then the fee is $y = 35x + 20$ dollars. Give an interpretation of the slope and y-intercept of this line.

17. How many units must a manufacturer produce and sell per day in order to just break even, if each unit sells for $5.00 and the total cost of producing x units per day is $360 + 4.20x$ dollars?

18. A company plans to market a new product for $18.50 per unit. The estimated total cost of producing and selling x units per week is $2205 + 14.00x$ dollars. How many units must the company produce and sell each week in order to break even?

19. If the price of a certain commodity is p dollars, the market demand will be $2750 - 5p$ units, while suppliers will be willing to sell $10p - 400$ units. Determine the price at which the supply and demand will be equal, and give the corresponding number of units demanded.

20. Suppose that the supply and demand equations for a certain industrial machine are $q = .1p - 40$ and $q = 710 - .5p$, respectively. How many machines will be sold if the price is set so that the supply and demand are equal?

Functions

Chapter 3

3.1. Functions and Their Graphs

A *function* of a variable x is a *rule* f that assigns to each value of x a unique number $f(x)$, called *the value of the function at* x. [We read "$f(x)$" as "f of x."]

The functions we shall meet in this book will usually be defined by algebraic formulas. For example, the function

$$f(x) = 3x - 1$$

is the rule that takes a number, multiplies it by 3, and then subtracts 1. If we specify a value of x, say $x = 2$, then we find the value of the function at 2 by substituting 2 for x in the formula:

$$f(2) = 3(2) - 1 = 5.$$

EXAMPLE 1 Let f be the function defined by

$$f(x) = 3x^3 - 4x^2 - 3x + 7.$$

Find $f(2)$ and $f(-2)$.

Solution To find $f(2)$ we substitute 2 for every occurrence of x in the formula for $f(x)$:

$$f(2) = 3(2)^3 - 4(2)^2 - 3(2) + 7$$
$$= 3(8) - 4(4) - 3(2) + 7$$
$$= 24 - 16 - 6 + 7$$
$$= 9.$$

The calculation of $f(-2)$ is similar.

$$f(-2) = 3(-2)^3 - 4(-2)^2 - 3(-2) + 7$$
$$= 3(-8) - 4(4) - 3(-2) + 7$$
$$= -24 - 16 + 6 + 7$$
$$= -27.$$

EXAMPLE 2 If x represents the temperature of an object in degrees Celsius, then the temperature in degrees Fahrenheit is a function of x, given by $f(x) = \frac{9}{5}x + 32$.

(a) Water freezes at $0°C$ (C = Celsius) and boils at $100°C$. What are the corresponding temperatures in degrees Fahrenheit?

(b) Aluminum melts at $660°C$. What is its melting point in degrees Fahrenheit?

Solution (a) $f(0) = \frac{9}{5}(0) + 32 = 32$. Water freezes at $32°F$.
$f(100) = \frac{9}{5}(100) + 32 = 180 + 32 = 212$. Water boils at $212°F$.

(b) $f(660) = \frac{9}{5}(660) + 32 = 1188 + 32 = 1220$. Aluminum melts at $1220°F$.

EXAMPLE 3 (*A Voting Model*) Let x be the proportion of the total popular vote which a Democratic candidate for president receives in a U.S. national election (so x is a number between 0 and 1). Political scientists have observed that a good estimate of the proportion of seats in the House of Representatives going to Democratic candidates is given by

$$f(x) = \frac{x^3}{x^3 + (1 - x)^3}, \qquad 0 \le x \le 1.$$

This formula is called the *cube law*. Compute $f(.6)$ and interpret the result.

Solution We must substitute .6 for every occurrence of x in $f(x)$:

$$f(.6) = \frac{(.6)^3}{(.6)^3 + (1 - .6)^3} = \frac{(.6)^3}{(.6)^3 + (.4)^3}$$

$$= \frac{.216}{.216 + .064} = \frac{.216}{.280} \approx .77.$$

This calculation shows that the cube law function predicts that if .6 (or 60%) of the total popular vote is for the Democratic candidate for president, then approximately .77 (or 77%) of the seats in the House of Representatives will be won by Democratic candidates; that is, about 335 of the 435 seats will be won by Democrats.

In the preceding examples, the functions involved only a single algebraic formula. However, to define some functions it is necessary to use two or more formulas. Here is an illustration of this phenomenon.

EXAMPLE 4   A leading brokerage firm charges a 6% commission on gold purchases in amounts from $50 to $300. For purchases exceeding $300, the firm charges 2% of the amount purchased plus $12.00. Let x denote the amount of gold purchased (in dollars) and let $f(x)$ be the commission charge as a function of x.

(a) Describe $f(x)$.

(b) Find $f(100)$ and $f(500)$.

Solution   (a) The formula for $f(x)$ depends on whether $50 \leq x \leq 300$ or $300 < x$. When $50 \leq x \leq 300$, the charge is $.06x$ dollars. When $300 < x$, the charge is $.02x + 12$. We write

$$f(x) = \begin{cases} .06x & \text{for } 50 \leq x \leq 300 \\ .02x + 12 & \text{for } 300 < x. \end{cases}$$

(b) Since $x = 100$ satisfies $50 \leq x \leq 300$, we use the first formula for $f(x)$: $f(100) = .06(100) = 6$. Since $x = 500$ satisfies $300 < x$, we use the second formula for $f(x)$: $f(500) = .02(500) + 12 = 22$.

In calculus, it is often necessary to substitute an algebraic expression for x and simplify the result.

EXAMPLE 5   If $f(x) = (4 - x)/(x^2 + 3)$, what is $f(a)$? $f(a + 1)$?

Solution   Here a represents some number. To find $f(a)$, we substitute a for x wherever x appears in the formula defining $f(x)$:

$$f(a) = \frac{4 - a}{a^2 + 3}$$

To evaluate $f(a + 1)$, we substitute $a + 1$ for each occurrence of x in the formula for $f(x)$:

$$f(a + 1) = \frac{4 - (a + 1)}{(a + 1)^2 + 3}.$$

The expression for $f(a + 1)$ may be simplified, using the fact that $(a + 1)^2 = (a + 1)(a + 1) = a^2 + 2a + 1$:

$$f(a + 1) = \frac{4 - (a + 1)}{(a + 1)^2 + 3} = \frac{4 - a - 1}{a^2 + 2a + 1 + 3} = \frac{3 - a}{a^2 + 2a + 4}.$$

The Domain of a Function   When defining a function, it is necessary to specify the set of acceptable values of the variable. This set is called the *domain* of the function. For instance, the cube law function of Example 3 was given by

$$f(x) = \frac{x^3}{x^3 + (1 - x)^3}, \qquad 0 \leq x \leq 1.$$

The domain of this function consists of those x for which $0 \leq x \leq 1$.

If no domain is specified, we will understand that the intended domain consists of all numbers for which the formula defining the function makes sense. For example, consider the function

$$f(x) = x^2 - x + 1.$$

The expression on the right may be evaluated for any value of x. So in the absence of any explicit restrictions on x, the domain is understood to consist of all numbers. As a second example consider the function

$$f(x) = \frac{1}{x}.$$

Here x may be any number except zero. (Division by zero is not permissible.) So the domain intended is the set of nonzero numbers. Similarly, when we write

$$f(x) = \sqrt{x}$$

we understand the domain of $f(x)$ to be the set of all nonnegative numbers, since the square root of a number x is defined if and only if $x \geq 0$.

Graphs of Functions   Often it is helpful to describe a function f geometrically, using a rectangular xy-coordinate system. Given any x in the domain of f, we can plot the point $(x, f(x))$. This is the point in the xy-plane whose y-coordinate is the value of the function at x. The set of *all* such points $(x, f(x))$ usually forms a curve in the xy-plane and is called the *graph of the function $f(x)$*.

It is possible to approximate the graph of $f(x)$ by plotting the points $(x, f(x))$ for a representative set of values of x and joining them by a smooth curve. (See Fig. 1.) The more closely spaced the values of x, the closer the approximation.

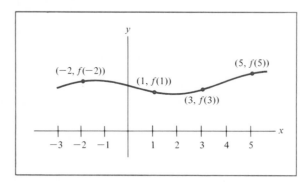

FIGURE 1

EXAMPLE 6   Sketch the graph of the function $f(x) = x^3$.

Solution   The domain consists of all numbers x. We choose some representative values of x and tabulate the corresponding values of $f(x)$. We then plot the points $(x, f(x))$ and sketch the graph indicated. (See Fig. 2.)

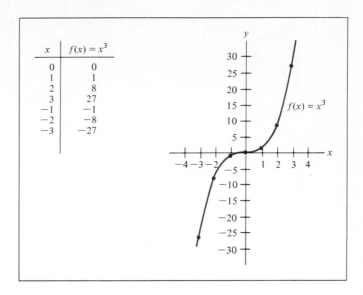

x	$f(x) = x^3$
0	0
1	1
2	8
3	27
−1	−1
−2	−8
−3	−27

FIGURE 2

EXAMPLE 7   Sketch the graph of the function $f(x) = 1/x$.

Solution   The domain of the function consists of all numbers except zero. The table in Fig. 3 lists some representative values of x and the corresponding values of $f(x)$. A function often has interesting behavior for x near a number not in the domain. So when we chose representative values of x from the domain, we included some values close to zero. The points $(x, f(x))$ are plotted and the graph sketched in Fig. 3.

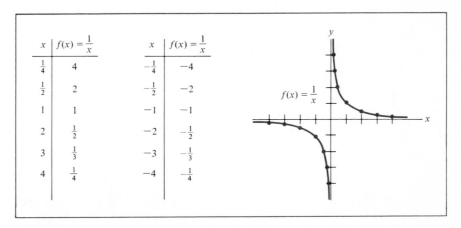

x	$f(x) = \dfrac{1}{x}$		x	$f(x) = \dfrac{1}{x}$
$\frac{1}{4}$	4		$-\frac{1}{4}$	−4
$\frac{1}{2}$	2		$-\frac{1}{2}$	−2
1	1		−1	−1
2	$\frac{1}{2}$		−2	$-\frac{1}{2}$
3	$\frac{1}{3}$		−3	$-\frac{1}{3}$
4	$\frac{1}{4}$		−4	$-\frac{1}{4}$

FIGURE 3

Graphing functions by plotting points is a tedious procedure. Moreover, we have as yet no way of knowing how many points or which points are sufficient to represent accurately all of the essential features of a graph. In calculus, we learn procedures for sketching graphs of functions that greatly ease the burden of plotting and yet guarantee that the sketch of the graph has the correct shape.

EXAMPLE 8 Suppose that f is the function whose graph is given in Fig. 4. Notice that the point $(x, y) = (3, 2)$ is on the graph of f.

(a) What is the value of the function when $x = 3$?

(b) Find $f(-2)$.

(c) What is the domain of f?

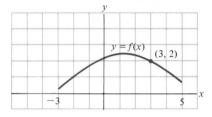

FIGURE 4

Solution (a) Since $(3, 2)$ is on the graph of f, the y-coordinate 2 must be the value of f at the x-coordinate 3. That is, $f(3) = 2$.

(b) To find $f(-2)$ we look at the y-coordinate of the point on the graph where $x = -2$. From Fig. 4 we see that $(-2, 1)$ is on the graph of f. Thus $f(-2) = 1$.

(c) The points on the graph of $f(x)$ all have x-coordinates between -3 and 5 inclusive; and for each value of x between -3 and 5 there is a point $(x, f(x))$ on the graph. So the domain consists of those x for which $-3 \leq x \leq 5$.

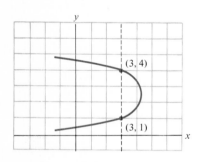

FIGURE 5 A curve that is *not* the graph of a function.

To every x in the domain, a function assigns one and only one value of y, namely, the function value $f(x)$. This implies, among other things, that not every curve is the graph of a function. To see this, refer first to the curve in Fig. 4, which *is* the graph of a function. It has the following important property: For each x between -3 and 5 inclusive there is a *unique* y such that (x, y) is on the curve. Now refer to the curve in Fig. 5. It cannot be the graph of a function, because a function f must assign to each x in its domain a *unique* value $f(x)$. However, for the curve of Fig. 5 there corresponds to $x = 3$ (for example) more than one y-value, namely, $y = 1$ and $y = 4$.

The essential difference between the curves in Figs. 4 and 5 leads us to the following test.

> *The Vertical Line Test*   A curve in the xy-plane is the graph of a function if and only if each vertical line cuts or touches the curve at no more than one point.

EXAMPLE 9   Which of the following curves are graphs of functions?

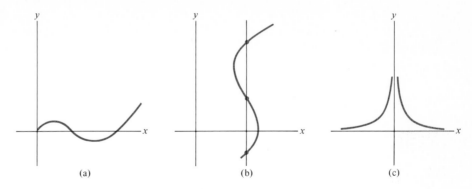

(a)   (b)   (c)

Solution   The curve in (a) is the graph of a function. It appears that vertical lines to the left of the y-axis do not touch the curve at all. This simply means that the function represented in (a) is defined only for $x \geq 0$. The curve in (b) is *not* the graph of a function because some vertical lines cut the curve in three places. The curve in (c) is the graph of a function whose domain is all nonzero x. [There is no point on the curve in (c) whose x-coordinate is 0.]

There is another notation for functions which we will find useful. Suppose that $f(x)$ is a function. When $f(x)$ is graphed on an xy-coordinate system, the values of $f(x)$ give the y-coordinates of points of the graph. For this reason, the function is often abbreviated by the letter y and we find it convenient to speak of "the function $y = f(x)$." For example, the function $y = 2x^2 + 1$ refers to the function $f(x)$ for which $f(x) = 2x^2 + 1$. The graph of a function $f(x)$ is often called *the graph of the equation* $y = f(x)$.

The equations arising in connection with functions are all of the form

$$y = [\text{an expression in } x].$$

However, not all equations connecting the variables x and y are of this sort. For example, consider these equations:

$$2x + 3y = 5$$

$$x = 3$$

$$x^2 + y^2 = 1.$$

It is possible to graph an equation by plotting points just as for functions. The only difference is that the resulting graph may not satisfy the vertical line test. For example, the graphs of the three equations above are shown in Fig. 6. Only the first graph is the graph of a function.

Letters other than f can be used to denote functions and letters other than x and y can be used to denote variables. This is especially common in applied problems where letters are chosen to suggest the quantities they depict. For instance, the revenue of a company as a function of time might be written $R(t)$.

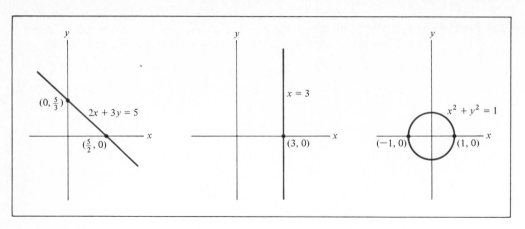

FIGURE 6

PRACTICE PROBLEMS 1

1. Is the point (3, 12) on the graph of the function $g(x) = x^2 + 5x - 10$?

2. Sketch the graph of the function $h(t) = t^2 - 2$.

EXERCISES 1

1. If $f(x) = x^2 - 3x$, find $f(0)$, $f(5)$, $f(3)$, and $f(-7)$.

2. If $f(x) = 9 - 6x + x^2$, find $f(0)$, $f(2)$, $f(3)$, and $f(-13)$.

3. If $f(x) = x^3 + x^2 - x - 1$, find $f(1)$, $f(-1)$, $f(\frac{1}{2})$, and $f(a)$.

4. If $g(t) = t^3 - 3t^2 + t$, find $g(2)$, $g(-\frac{1}{2})$, $g(\frac{2}{3})$, and $g(a)$.

5. If $h(s) = s/(1 + s)$, find $h(\frac{1}{2})$, $h(-\frac{3}{2})$, and $h(a + 1)$.

6. If $f(x) = x^2/(x^2 - 1)$, find $f(\frac{1}{2})$, $f(-\frac{1}{2})$, and $f(a + 1)$.

7. If $f(x) = x^2 - 2x$, find $f(a + 1)$ and $f(a + 2)$.

8. If $f(x) = x^2 + 4x + 3$, find $f(a - 1)$ and $f(a - 2)$.

9. An office supply firm finds that the number of Remington typewriters sold in year x is approximately given by the function $f(x) = 50 + 4x + \frac{1}{2}x^2$, where $x = 0$ corresponds to 1980.

    (a) What does $f(0)$ represent?

    (b) Find the number of Remington typewriters sold in 1982.

10. When a solution of acetylcholine is introduced into the heart muscle of a frog, it diminishes the force with which the muscle contracts. The data from experiments of A. J. Clark are closely approximated by a function of the form

$$R(x) = \frac{100x}{b + x}, \qquad x \geq 0,$$

where x is the concentration of acetylcholine (in appropriate units), b is a positive constant that depends on the particular frog, and $R(x)$ is the response of the muscle to the acetylcholine, expressed as a percentage of the maximum possible effect of the drug.

(a) Suppose that $b = 20$. Find the response of the muscle when $x = 60$.

(b) Determine the value of b if $R(50) = 60$—that is, if a concentration of $x = 50$ units produces a 60% response.

In Exercises 11–14, describe the domain of the function.

11. $f(x) = \dfrac{8x}{(x - 1)(x - 2)}$

12. $f(t) = \dfrac{1}{\sqrt{t}}$

13. $g(x) = \dfrac{1}{\sqrt{3 - x}}$

14. $g(x) = \dfrac{4}{x(x + 2)}$

In Exercises 15–20, decide which curves are graphs of functions.

15.

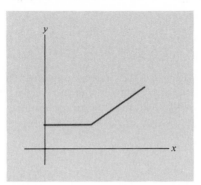

16.

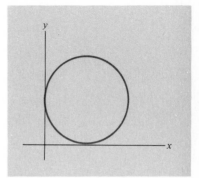

17.

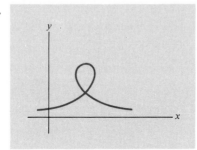

18.

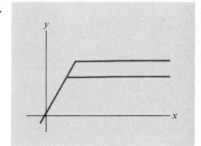

19.

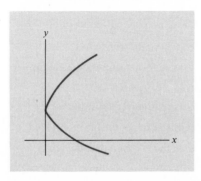

20.

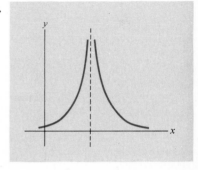

Exercises 21–30 relate to the function whose graph is sketched in Fig. 7.

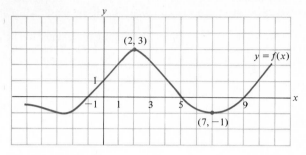

FIGURE 7

21. Find $f(0)$.

22. Find $f(7)$.

23. Find $f(2)$.

24. Find $f(-1)$.

25. Is $f(4)$ positive or negative?

26. Is $f(6)$ positive or negative?

27. Is $f(-\frac{1}{2})$ positive or negative?

28. Is $f(1)$ greater than $f(6)$?

29. For what values of x is $f(x) = 0$?

30. For what values of x is $f(x) \geq 0$?

Exercises 31–34 relate to Fig. 8. When a drug is injected into a person's muscle tissue, the concentration y of the drug in the blood is a function of the time elapsed since the injection. The graph of a typical time-concentration function f is given in Fig. 8, where $t = 0$ corresponds to the time of the injection.

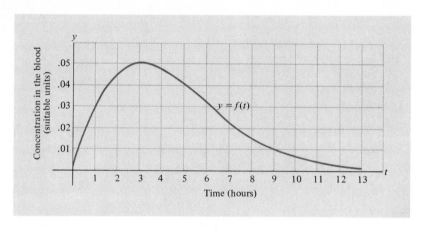

FIGURE 8

31. What is the concentration of the drug when $t = 1$?

32. What is the value of the time-concentration function f when $t = 6$?

33. Find $f(5)$.

34. At what time does $f(t)$ attain its largest value?

35. Is the point $(3, 12)$ on the graph of the function $f(x) = (x - \frac{1}{2})(x + 2)$?

36. Is the point $(-2, 12)$ on the graph of the function $f(x) = x(5 + x)(4 - x)$?

37. Is the point $(\frac{1}{2}, \frac{2}{5})$ on the graph of the function $g(x) = (3x - 1)/(x^2 + 1)$?

38. Is the point $(\frac{2}{3}, \frac{5}{3})$ on the graph of the function $g(x) = (x^2 + 4)/(x + 2)$?

39. Find the y-coordinate of the point $(a + 1, \ )$ if this point lies on the graph of the function $f(x) = x^3$.

40. Find the y-coordinate of the point $(2 + h, \ )$ if this point lies on the graph of the function $f(x) = (5/x) - x$.

In Exercises 41–44, compute $f(1)$, $f(2)$, and $f(3)$.

41. $f(x) = \begin{cases} \sqrt{x} & \text{for } 0 \le x < 2 \\ 1 + x & \text{for } 2 \le x \le 5 \end{cases}$

42. $f(x) = \begin{cases} \dfrac{1}{x} & \text{for } 1 \le x \le 2 \\ x^2 & \text{for } 2 < x \end{cases}$

43. $f(x) = \begin{cases} \pi x^2 & \text{for } x < 2 \\ 1 + x & \text{for } 2 \le x \le 2.5 \\ 4x & \text{for } 2.5 < x \end{cases}$

44. $f(x) = \begin{cases} \dfrac{3}{4 - x} & \text{for } x < 2 \\ 2x & \text{for } 2 \le x < 3 \\ \sqrt{x^2 - 5} & \text{for } 3 \le x \end{cases}$

45. Suppose that the brokerage firm in Example 4 decides to keep the commission charges unchanged for purchases up to and including $600, but to charge only 1.5% plus $15 for gold purchases exceeding $600. Express the brokerage commission as a function of the amount x of gold purchased.

SOLUTIONS TO PRACTICE PROBLEMS 1

1. If $(3, 12)$ is on the graph of $g(x) = x^2 + 5x - 10$, then we must have $g(3) = 12$. This is not the case, however, because

$$g(3) = 3^2 + 5(3) - 10$$
$$= 9 + 15 - 10 = 14.$$

Thus $(3, 12)$ is *not* on the graph of $g(x)$.

2. Choose some representative values for t, say $t = 0, \pm1, \pm2, \pm3$. For each value of t, calculate $h(t)$ and plot the point $(t, h(t))$.

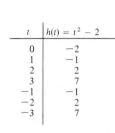

t	$h(t) = t^2 - 2$
0	-2
1	-1
2	2
3	7
-1	-1
-2	2
-3	7

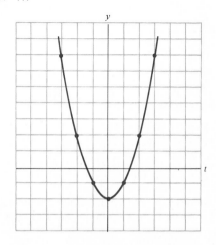

3.2. Some Important Functions

In this section we discuss some of the functions that play a prominent role in finite mathematics and calculus.

Linear Functions   As we shall see in Chapter 6, a knowledge of the algebraic and geometric properties of straight lines is essential for linear programming.

The straight line of Fig. 1(a) is the graph of the function $f(x) = mx + b$. Such a function, which is defined for all x, is called a *linear function*. Note that the straight line of Fig. 1(b) is not the graph of a function since the vertical line test is violated.

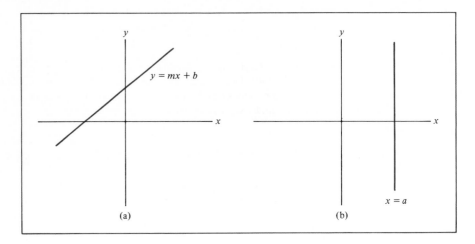

FIGURE 1

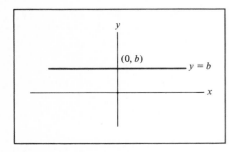

FIGURE 2   Graph of the constant function $f(x) = b$.

An important special case of a linear function occurs if the value of m is zero, that is, if $f(x) = b$ for some number b. In this case, $f(x)$ is called a *constant function* since it assigns the same number, b, to every value of x. Its graph is the horizontal line whose equation is $y = b$. (See Fig. 2.)

Linear functions often arise in real-life situations, as the next example shows.

EXAMPLE 1   When the U.S. Environmental Protection Agency found a certain company dumping sulfuric acid into the Mississippi River, it fined the company $125,000, plus

$1000 per day until the company complied with federal water pollution regulations. Express the total fine as a function of the number x of days the company continued to violate the federal regulations.

Solution The variable fine for x days of pollution, at $1000 per day, is $1000x$ dollars. The total fine is therefore given by the function

$$f(x) = 125,000 + 1000x.$$

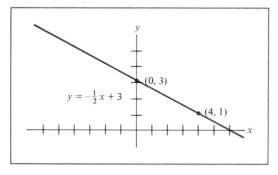

$y = -\frac{1}{2}x + 3$

(0, 3)

(4, 1)

FIGURE 3

Since the graph of a linear function is a line, we may sketch it by locating any two points on the graph and drawing the line through them. For example, to sketch the graph of the function $f(x) = -\frac{1}{2}x + 3$, we may select two convenient values of x, say 0 and 4, and compute $f(0) = -\frac{1}{2}(0) + 3 = 3$ and $f(4) = -\frac{1}{2}(4) + 3 = 1$. The line through the points $(0, 3)$ and $(4, 1)$ is the graph of the function. (See Fig. 3.)

Quadratic Functions Economists utilize average-cost curves which relate the average unit cost of manufacturing a commodity to the number of units to be produced. (See Fig. 4.) Ecologists use curves that relate the net primary production of nutrients in a plant to the surface area of the foliage. (See Fig. 5.) Each of the curves is bowl-shaped, opening either up or down. The simplest functions whose graphs resemble these curves are the quadratic functions.

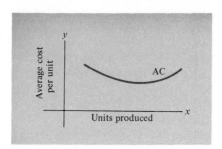

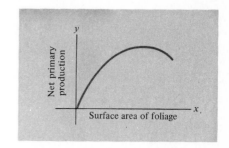

FIGURE 4   Average-cost curve.   FIGURE 5   Production of nutrients.

A *quadratic function* is a function of the form

$$f(x) = ax^2 + bx + c,$$

where a, b, and c are constants and $a \neq 0$. The domain of such a function consists of all numbers. The graph of a quadratic function is called a *parabola*. Two typical parabolas are drawn in Figs. 6 and 7. Graphs of quadratic functions are discussed

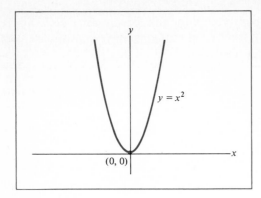

FIGURE 6 Graph of $f(x) = x^2$.

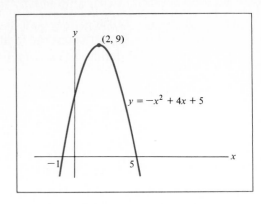

FIGURE 7 Graph of $f(x) = -x^2 + 4x + 5$.

in Section 4 and again later in the text when calculus techniques are applied to curve-sketching problems.

Polynomial and Rational Functions A *polynomial function* $f(x)$ is one of the form

$$f(x) = a_n x^n + a_{n-1} x^{n-1} + \cdots + a_0,$$

where n is a nonnegative integer and $a_0, a_1, \ldots, a_n$ are given numbers. Some examples of polynomial functions are

$$f(x) = 5x^3 - 3x^2 - 2x + 4$$

$$g(x) = x^4 - x + 1.$$

Of course, linear and quadratic functions are special cases of polynomial functions. The domain of a polynomial function consists of all numbers.

A function expressed as the quotient of two polynomials is called a *rational function*. Some examples are

$$h(x) = \frac{x^2 + 1}{x}$$

$$k(x) = \frac{x + 3}{x^2 - 4}.$$

The domain of a rational function excludes all values of x for which the denominator is zero. Thus, for example, the domain of $h(x)$ excludes $x = 0$, whereas the domain of $k(x)$ excludes $x = 2$ and $x = -2$. As we shall see, both polynomial and rational functions arise in applications of calculus.

Power Functions Functions of the form $f(x) = x^r$ are called *power functions*. Table 1 gives the definition and domain for certain special cases. Many familiar functions are power functions. For instance, $\sqrt{x}$ is $x^{1/2}$, and $1/x$ is x^{-1}.

TABLE 1

r	x^r	Domain
n (n is a positive integer)	x^n	all x
$\dfrac{1}{n}$ (n is a positive integer)	$\sqrt[n]{x}$	$x \geq 0$ if n is even, all x if n is odd
$\dfrac{m}{n}$ (m, n positive integers)	$(\sqrt[n]{x})^m$	
$-s$ (s a positive number)	$\dfrac{1}{x^s}$	same as for x^s except for $x = 0$, which is excluded

EXAMPLE 2   Use Table 1 to evaluate the following power functions at $x = 125$.

(a) $f(x) = x^{1/3}$ 　　　　(b) $g(x) = x^{2/3}$ 　　　　(c) $h(x) = x^{-(2/3)}$

Solution   (a) Here $r = \frac{1}{3}$. By the second line of the table, $x^{1/3}$ is $\sqrt[3]{x}$. Therefore,

$$f(125) = 125^{1/3} = \sqrt[3]{125} = 5.$$

(b) By the third line of the table, $x^{2/3}$ is $(\sqrt[3]{x})^2$. Therefore,

$$g(125) = 125^{2/3} = (\sqrt[3]{125})^2 = 5^2 = 25.$$

(c) By the last line of the table, $x^{-(2/3)}$ is $\dfrac{1}{x^{2/3}}$. Therefore,

$$h(125) = 125^{-(2/3)} = \frac{1}{125^{2/3}} = \frac{1}{25}.$$

The Absolute Value Function   The absolute value of a number x is denoted by $|x|$ and is defined by

FIGURE 8

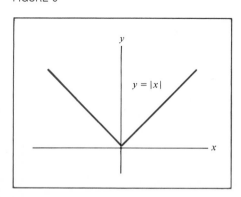

$$|x| = \begin{cases} x & \text{if } x \text{ is positive or zero,} \\ -x & \text{if } x \text{ is negative.} \end{cases}$$

Thus, for example, $|5| = 5$, $|0| = 0$, and $|-3| = -(-3) = 3$. The function defined for all numbers x by

$$f(x) = |x|$$

is called the *absolute value function*. Its graph coincides with the graph of the equation $y = x$ for $x \geq 0$ and with the graph of the equation $y = -x$ for $x < 0$. (See Fig. 8.)

PRACTICE PROBLEMS 2

1. A photocopy service has a fixed cost of $2000 per month (for rent, depreciation of equipment, etc.), and variable costs of $.04 for each page it reproduces for customers. Express its total cost as a (linear) function of the number of pages copied per month.

2. Consider the power function $f(x) = x^{2/5}$. Evaluate (a) $f(32)$; (b) $f(-1)$.

EXERCISES 2

Graph the following functions.

1. $f(x) = 2x - 1$ 2. $f(x) = 3$ 3. $f(x) = 3x + 1$

4. $f(x) = -\frac{1}{2}x - 4$ 5. $f(x) = -2x + 3$ 6. $f(x) = \frac{1}{4}$

Determine the x- and y-intercepts of the graphs of the following functions.

7. $f(x) = 9x + 3$ 8. $f(x) = -\frac{1}{2}x - 1$ 9. $f(x) = 5$

10. $f(x) = 14$ 11. $f(x) = -\frac{1}{4}x + 3$ 12. $f(x) = 6x - 4$

13. In biochemistry, such as in the study of enzyme kinetics, one encounters a linear function of the form $f(x) = (K/V)x + 1/V$, where K and V are constants.

    (a) If $f(x) = .2x + 50$, find K and V so that $f(x)$ may be written in the form $f(x) = (K/V)x + 1/V$.

    (b) Find the x- and y-intercepts of the line $y = (K/V)x + 1/V$ (in terms of K and V).

14. The constants K and V in Exercise 13 are often determined from experimental data. Suppose a line is drawn through data points and has x-intercept $(-500, 0)$ and y-intercept $(0, 60)$. Determine K and V so that the line is the graph of the function $f(x)=(K/V)x+1/V$. [*Hint:* Use Exercise 13(b).]

15. In some cities you can still rent a car for $12 per day and $.15 per mile.

    (a) Find the cost of renting the car for 1 day and driving 200 miles.

    (b) Suppose that the car is to be rented for 1 day, and express the total rental expense as a function of the number x of miles driven. (Assume that for each fraction of a mile driven, the same fraction of $.15 is charged.)

16. A gas company will pay a property owner $5000 for the right to drill on the land for natural gas, and $.10 for each thousand cubic feet of gas extracted from the land. Express the amount of money the landowner will receive as a function of the amount of gas extracted from the land.

17. In 1978, Blue Cross paid $135 per day for a semiprivate hospital room and $150 for an appendectomy operation. Express the total amount Blue Cross paid for an appendectomy as a function of the number of days of hospital confinement.

18. The R-rating of fiberglass home insulation is directly proportional to the thickness of the insulation. If the R-rating of a 6-inch batt of insulation is 19, express the R-rating of the insulation as a function of the thickness (in inches).

Each of the quadratic functions in Exercises 19–24 has the form $y = ax^2 + bx + c$. Identify a, b, and c.

19. $y = 3x^2 - 4x$

20. $y = \dfrac{x^2 - 6x + 2}{3}$

21. $y = 3x - 2x^2 + 1$

22. $y = 3 - 2x + 4x^2$

23. $y = 1 - x^2$

24. $y = \frac{1}{2}x^2 + \sqrt{3}x - \pi$

Evaluate each of the functions in Exercises 25–36 at the given value

25. $f(x) = x^{-3}$, $x = 2$

26. $f(x) = x^3$, $x = -\frac{1}{5}$

27. $f(x) = x^{1/4}$, $x = 16$

28. $f(x) = x^{3/2}$, $x = 100$

29. $f(x) = x^{2/3}$, $x = 8$

30. $f(x) = x^{-4}$, $x = -1$

31. $f(x) = x^{-1/2}$, $x = 25$

32. $f(x) = x^{1/3}$, $x = \frac{1}{27}$

33. $f(x) = |x|$, $x = -2.5$

34. $f(x) = |x|$, $x = \pi$

35. $f(x) = |x|$, $x = 10^{-2}$

36. $f(x) = |x|$, $x = -\frac{2}{3}$

Sketch the graphs of the following functions.

37. $f(x) = \begin{cases} 3x & \text{for } 0 \le x \le 1 \\ 5 - 2x & \text{for } x > 1 \end{cases}$

38. $f(x) = \begin{cases} 1 + x & \text{for } x \le 3 \\ 4 & \text{for } x > 3 \end{cases}$

39. $f(x) = \begin{cases} 3 & \text{for } x < 2 \\ 2x + 1 & \text{for } x \ge 2 \end{cases}$

40. $f(x) = \begin{cases} \frac{1}{2}x & \text{for } 0 \le x < 4 \\ 2x - 3 & \text{for } 4 \le x \le 5 \end{cases}$

41. $f(x) = \begin{cases} 4 - x & \text{for } 0 \le x < 2 \\ 2x - 2 & \text{for } 2 \le x < 3 \\ x + 1 & \text{for } x \ge 3 \end{cases}$

42. $f(x) = \begin{cases} 4x & \text{for } 0 \le x < 1 \\ 8 - 4x & \text{for } 1 \le x < 2 \\ 2x - 4 & \text{for } x \ge 2 \end{cases}$

SOLUTIONS TO PRACTICE PROBLEMS 2

1. If x represents the number of pages copied per month, then the variable cost is $.04x$ dollars. Now [total cost] = [fixed cost] + [variable cost]. If we define

$$f(x) = 2000 + .04x,$$

then $f(x)$ gives the total cost per month.

2. By Table 1, $x^{2/5}$ is $(\sqrt[5]{x})^2$.

(a) $f(32) = (\sqrt[5]{32})^2 = (2)^2 = 4$

(b) $f(-1) = (\sqrt[5]{-1})^2 = (-1)^2 = 1$

3.3. The Algebra of Functions

Many functions we shall encounter later in the text may be viewed as combinations of other functions. For example, let $P(x)$ represent the profit a company makes on the sale of x units of some commodity. If $R(x)$ denotes the revenue received

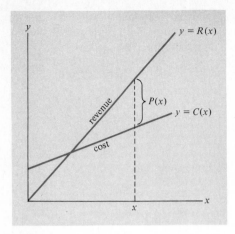

FIGURE 1

from the sale of x units, and if $C(x)$ is the cost of producing x units, then

$$P(x) = R(x) - C(x).$$

$$[\text{profit}] = [\text{revenue}] - [\text{cost}]$$

Writing the profit function in this way makes it possible to predict the behavior of $P(x)$ from properties of $R(x)$ and $C(x)$. (See Fig. 1.)

The first two examples below review the algebraic techniques needed to combine functions by addition, subtraction, multiplication, and division.

EXAMPLE 1 Let $f(x) = 3x + 4$ and $g(x) = 2x - 6$. Find $f(x) + g(x)$, $f(x) - g(x)$, $f(x)g(x)$, and $\dfrac{f(x)}{g(x)}$.

Solution

$$f(x) + g(x) = (3x + 4) + (2x - 6) = 5x - 2.$$

$$f(x) - g(x) = (3x + 4) - (2x - 6) = x + 10.$$

$$f(x)g(x) = (3x + 4)(2x - 6) = 6x^2 - 10x - 24.$$

$$\frac{f(x)}{g(x)} = \frac{3x - 4}{2x - 6}, \qquad x \neq 3.$$

When two functions are added, subtracted, or multiplied the domain of the resulting function consists of all numbers that are in the domains of both of the original functions. The same is true for the domain of $\dfrac{f(x)}{g(x)}$ except that numbers x for which $g(x) = 0$ must also be excluded.

EXAMPLE 2 Let $f(x) = x^2 - 4$ and $g(x) = x - 2$. Find $\dfrac{f(x)}{g(x)}$.

FIGURE 2

Solution

$$\frac{f(x)}{g(x)} = \frac{x^2 - 4}{x - 2} = \frac{(x + 2)(x - 2)}{x - 2} = x + 2, \qquad x \neq 2.$$

The graph of $\dfrac{f(x)}{g(x)}$ is shown in Fig. 2.

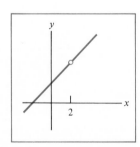

The point where $x = 2$ is omitted from the graph because $\dfrac{x^2 - 4}{x - 2}$ is not defined for $x = 2$. The function $h(x) = x + 2$ has essentially the same graph as $\dfrac{x^2 - 4}{x - 2}$, except that $h(x)$ *is* defined for $x = 2$.

Composition of Functions   Another important way of combining two functions $f(x)$ and $g(x)$ is to substitute the function $g(x)$ for every occurrence of the variable x in $f(x)$. The resulting function is called the *composition* (or *composite*) of $f(x)$ and $g(x)$, and is denoted by $f(g(x))$.

EXAMPLE 3   Let $f(x) = x^2 + 3x + 1$ and $g(x) = x^2 - 5$. Find $f(g(x))$.

Solution   We substitute $g(x)$ in place of each x in $f(x)$.

$$f(g(x)) = [g(x)]^2 + 3g(x) + 1$$
$$= (x^2 - 5)^2 + 3(x^2 - 5) + 1$$
$$= (x^4 - 10x^2 + 25) + (3x^2 - 15) + 1$$
$$= x^4 - 7x^2 + 11.$$

Later in the text we shall need to study expressions of the form $f(x + h)$, where $f(x)$ is a given function and h represents some number. The meaning of $f(x + h)$ is that $x + h$ is to be substituted for each occurrence of x in the formula for $f(x)$. In fact, $f(x + h)$ is just a special case of $f(g(x))$, where $g(x) = x + h$.

EXAMPLE 4   Let $f(x) = x^2$. Find $f(x + h) - f(x)$.

Solution

$$f(x + h) - f(x) = (x + h)^2 - x^2$$
$$= (x^2 + 2xh + h^2) - x^2$$
$$= 2xh + h^2.$$

EXAMPLE 5   In a certain lake, the bass feed primarily on minnows and the minnows feed on plankton. Suppose that the size of the bass population is a function $f(n)$ of the number n of minnows in the lake, and the number of minnows is a function $g(x)$ of the amount x of plankton in the lake. Express the size of the bass population as a function of the amount of plankton, if $f(n) = 50 + \sqrt{n/150}$ and $g(x) = 4x + 3$.

Solution   We have $n = g(x)$. Substituting $g(x)$ for n in $f(n)$, we find that the size of the bass population is given by

$$f(g(x)) = 50 + \sqrt{\dfrac{g(x)}{150}} = 50 + \sqrt{\dfrac{4x + 3}{150}}.$$

Translation of Graphs   Special cases of function addition and composition result in the translation of the graph of a function.

If $g(x) = C$, a constant function, the graph of $f(x) + g(x)$ is the graph of $f(x)$ moved C units vertically. Figure 3 shows the graphs of $x^2 + C$ for various choices of C.

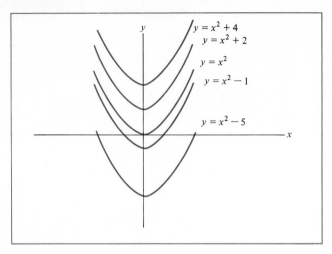

$y = x^2 + 4$
$y = x^2 + 2$
$y = x^2$
$y = x^2 - 1$
$y = x^2 - 5$

FIGURE 3

If $g(x) = x - a$, where a is a specific number, the graph of $f(g(x))$ is the graph of $f(x)$ moved a units horizontally. Figure 4 shows the graph of $(x - a)^2$ for various values of a. Note that for $a = -4$ we have written $[x - (-4)]^2$ in the form $(x + 4)^2$.

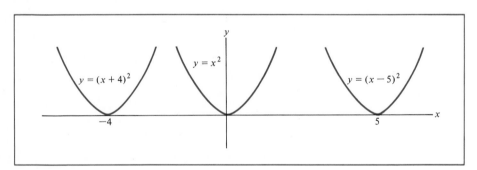

$y = (x + 4)^2$
$y = x^2$
$y = (x - 5)^2$
-4
5

FIGURE 4

The two operations above can be combined to translate a graph to any location in the plane.

EXAMPLE 6 Find the equation of the graph that is obtained from the graph of $f(x) = x^3$ by a translation of 2 units to the right and 1 unit up.

Solution The graph of $y = x^3$ was sketched in Section 3.1. This graph is reproduced in Fig. 5(a), and the translated graph is shown in Fig. 5(b). A translation of 2 units to the right corresponds to replacing x by $x - 2$ and a translation of 1 unit up corresponds to adding 1 to a function. Therefore, the equation of the resulting

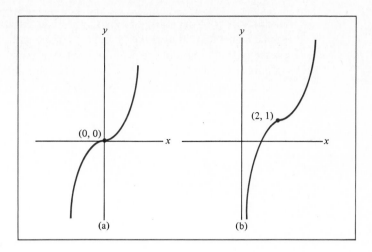

FIGURE 5

graph in Fig. 5(b) is

$$y = f(x - 2) + 1$$
$$= (x - 2)^3 + 1$$
$$= (x^3 - 2x^2 + 4x - 8) + 1$$
$$= x^3 - 2x^2 + 4x - 7.$$

EXAMPLE 7 Sketch the graph of $y = |x + 4| - 2$.

Solution The function $g(x) = |x + 4| - 2$ can be obtained by starting with the function $f(x) = |x|$, replacing x by $x+4$ and subtracting 2. Here the graph of $y = |x+4| - 2$ is obtained by starting with the graph of $y = |x|$, translating 4 units to the left [since $x + 4 = x - (-4)$] and translating 2 units down. (See Fig. 6.)

FIGURE 6

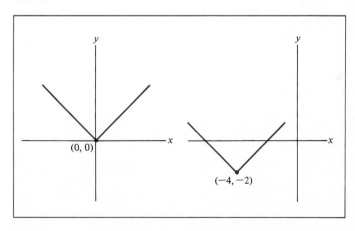

Reduction of Expressions to Functions Later in the text we shall encounter applied problems that involve some variable, call it A, which is given as an expression involving x and y. That is, A is a function of x and y. When there is also an equation relating x and y, we can often use algebraic techniques to express A as a function of the variable x alone. The final example in this section illustrates the procedure.

EXAMPLE 8   Suppose that $A = 3x^2 + 2xy$ and $x^2y = 5$. Express A as a function of x.

Solution   Since the variables x and y are related by the equation $x^2y = 5$, we may solve this equation for y as a function of x.

$$x^2y = 5$$

$$y = \frac{5}{x^2}.$$

Now substitute $\dfrac{5}{x^2}$ for y in the expression for A. (This substitution is actually a composition of functions.) Then A becomes a function of x alone.

$$A = 3x^2 + 2xy$$

$$= 3x^2 + 2x\left(\frac{5}{x^2}\right)$$

$$= 3x^2 + \frac{10}{x}.$$

PRACTICE PROBLEMS 3

1. Let $f(x) = x^5$ and $g(x) = x^3 - 4x^2 + x - 8$.

    (a) Find $f(g(x))$.   (b) Find $g(f(x))$.

2. Let $f(x) = x^3 + x^2$ and $h(x) = f(x + 2) - 5 = (x + 2)^3 + (x + 2)^2 - 5$. Describe how the graph of $h(x)$ may be obtained from the graph of $f(x)$ by translation.

EXERCISES 3

Let $f(x) = x^2 + 4x$, $g(x) = 9x$ and $h(x) = 5 - 2x^2$. Calculate the following functions.

1. $f(x) + g(x)$ 2. $f(x) - h(x)$ 3. $f(x)g(x)$

4. $g(x)h(x)$ 5. $f(t)/g(t)$ 6. $g(t)/h(t)$

Let $f(x) = x^6$, $g(x) = \dfrac{x}{1 - x}$ and $h(x) = x^3 - 5x^2 + 1$. Calculate the following functions.

7. $f(g(x))$ 8. $h(f(t))$ 9. $h(g(x))$

10. $g(f(x))$ 11. $g(h(t))$ 12. $f(h(x))$

13. If $f(x) = x^3$, find $f(x + h) - f(x)$ and simplify.

14. If $f(x) = 1/x$, find $f(x + h) - f(x)$ and simplify.

15. If $g(t) = 4t - t^2$, find $\dfrac{g(t + h) - g(t)}{h}$ and simplify.

16. If $g(x) = x^3 + 5$, find $\dfrac{g(x + h) - g(x)}{h}$ and simplify.

17. After t hours of operation, an assembly line has assembled $A(t) = 20t - \frac{1}{2}t^2$ power lawn mowers, $0 \le t \le 10$. Suppose that the factory's cost of manufacturing x units is $C(x)$ dollars, where $C(x) = 3000 + 80x$.

   (a) Express the factory's cost as a (composite) function of the number of hours of operation of the assembly line.

   (b) What is the cost of the first 2 hours of operation?

18. During the first half-hour, the employees of a machine shop prepare the work area for the day's work. After that, they turn out 10 precision machine parts per hour, so that the output after t hours is $f(t)$ machine parts, where $f(t) = 10(t - \frac{1}{2}) = 10t - 5$, $\frac{1}{2} \le t \le 8$. The total cost of producing x machine parts is $C(x)$ dollars, where $C(x) = .1x^2 + 25x + 200$.

   (a) Express the total cost as a (composite) function of t.

   (b) What is the cost of the first four hours of operation?

In Exercises 19–22, describe how the graph of $f(x)$ can be translated to obtain the graph of $g(x)$.

19. $f(x) = 3x^2 + 2$, $g(x) = 3x^2 + 7$

20. $f(x) = x^2 - 3x + 5$, $g(x) = (x - 4)^2 - 3(x - 4) + 5$

21. $f(x) = (x - 3)^2$, $g(x) = (x - 5)^2$. [*Hint:* Compare with $h(x) = x^2$.]

22. $f(x) = 5 - x^2$, $g(x) = 2 - x^2$

23. Find the equation of the parabola that is obtained from the graph of $f(x) = 4x^2 + 1$ by a translation of 3 units up and 5 units to the right.

24. Find the equation of the parabola that is obtained from the graph of $f(x) = \frac{1}{2}x^2 + 3x$ by a translation of 4 units to the left.

25. Find the equation of the curve that is obtained from the graph of $f(x) = x^3 - 3x$ by a translation of 1 unit to the left and 2 units up.

26. Find the equation of the parabola that is obtained from the graph of $f(x) = x^2 + 6x + 1$ by a translation of 3 units to the left and 1 unit down.

In Exercises 27–36, express A as a function of the variable x.

27. $A = 3xy$; $2x + y = 5$

28. $A = 3x + 4y$; $xy = 2$

29. $A = \frac{1}{5}x + \frac{1}{3}y$; $4xy = 1$

30. $A = \frac{1}{2}xy$; $3x - 5y = 5$

31. $A = 3x^2y$; $6y = 5$

32. $A = x - \frac{1}{2}y$; $x^2y = 5$

33. $A = 2y^2 + 3x^2$; $y - x = 1$

34. $A = x^2y$; $2x^2 + xy = 4$

35. $A = 3x^2y$; $\frac{1}{2}x^2 + 3xy = 1$

36. $A = \pi x^2 + 3xy$; $\frac{1}{2}x^2y = 5$

SOLUTIONS TO PRACTICE PROBLEMS 3

1. (a) $f(g(x)) = [g(x)]^5 = (x^3 - 4x^2 + x - 8)^5$

    (b) $g(f(x)) = [f(x)]^3 - 4[f(x)]^2 + f(x) - 8$

    $= (x^5)^3 - 4(x^5)^2 + x^5 - 8$

    $= x^{15} - 4x^{10} + x^5 - 8$

2. Since $x + 2 = x - (-2)$, the graph of $f(x + 2) = (x + 2)^3 + (x + 2)^2$ is obtained from the graph of $f(x)$ by a translation of -2 units horizontally, that is, 2 units to the left. When 5 is subtracted from the function, it translates the graph -5 units vertically, that is, 5 units down. Hence the graph of $h(x) = f(x + 2) - 5$ is produced by translating the graph of $f(x)$ 2 units to the left and 5 units down.

3.4. Graphs of Quadratic Functions

A quadratic function is a function that can be written in the form $f(x) = ax^2 + bx + c$ where a, b, and c are given numbers and $a \neq 0$. Some examples are:

$$f(x) = 3x^2 + 4x + 5$$

$$f(x) = 5 - x^2 \qquad [= (-1)x^2 + 0 \cdot x + 5]$$

$$f(x) = \tfrac{1}{2}x^2 \qquad [= \tfrac{1}{2} \cdot x^2 + 0 \cdot x + 0].$$

As we mentioned in Section 3.2, the graph of a quadratic function is called a *parabola*. The graph is symmetric about a vertical line, called the *axis of symmetry*, which passes through the lowest or highest point on the parabola. This lowest or highest point is called the *vertex* of the parabola. (See Fig. 1.)

The simplest quadratic functions have the form $f(x) = ax^2$. Several examples are shown in Fig. 2. The graphs all have the y-axis as the axis of symmetry and the point $(0, 0)$ as the vertex. Note that the graph of $f(x) = ax^2$ opens up if the

FIGURE 1

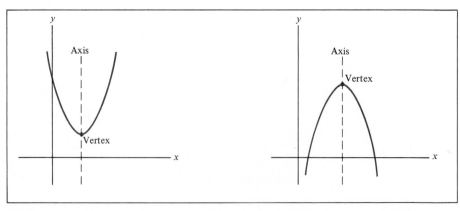

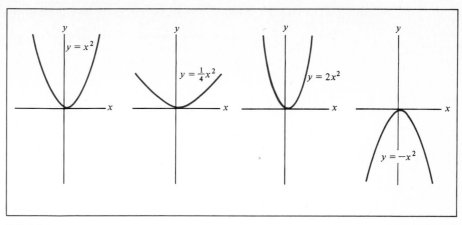

FIGURE 2

coefficient a is positive and down if a is negative. When $|a| < 1$, the graph of $f(x) = ax^2$ is wider than the graph of $f(x) = x^2$, and when $|a| > 1$, the graph of $f(x) = ax^2$ is narrower.

EXAMPLE 1 Sketch the graph of $f(x) = -\frac{1}{2}x^2$.

Solution The graph of $y = \frac{1}{2}x^2$ is similar to the graph of $y = x^2$, just a little wider. Some points on this graph are $(0, 0)$, $(1, \frac{1}{2})$, and $(3, \frac{9}{2})$. [See Fig. 3(a).] We flip this graph upside down to obtain the graph of $f(x) = -\frac{1}{2}x^2$. [See Fig. 3(b).]

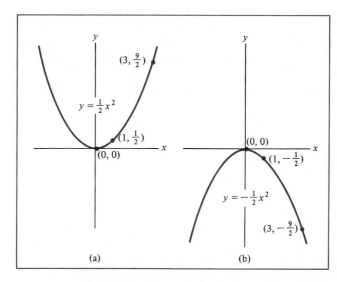

FIGURE 3

It is easy to see that any function of the form $f(x) = a(x - h)^2 + k$, where $a \neq 0$, is a quadratic function. We can sketch the graph of such a function using the translation technique of Section 3.3 if we have the graph of $y = ax^2$ at hand.

EXAMPLE 2 Sketch the graph of $f(x) = -\frac{1}{2}(x - 2)^2 + 3$.

Solution Translate the graph of $y = -\frac{1}{2}x^2$ to the right 2 units and up 3 units. (See Fig. 4.)

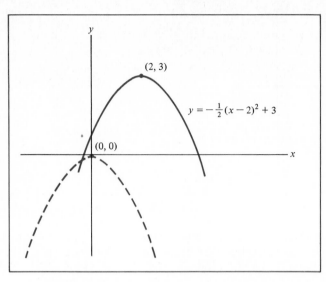

$$y = -\frac{1}{2}(x - 2)^2 + 3$$

(2, 3)

(0, 0)

FIGURE 4

Notice in Fig. 4 how the vertex (0, 0) of the graph of $y = -\frac{1}{2}x^2$ moved to the vertex (2, 3) of the graph of $y = -\frac{1}{2}(x - 2)^2 + 3$. This illustrates the following useful fact.

> The graph of a function of the form
> $$f(x) = a(x - h)^2 + k$$
> is a parabola with vertex at (h, k). The parabola opens up if a is positive and down if a is negative.

EXAMPLE 3 Determine the lowest point on the graph of the function $f(x) = \frac{1}{4}(x - 3)^2 + 1$.

Solution

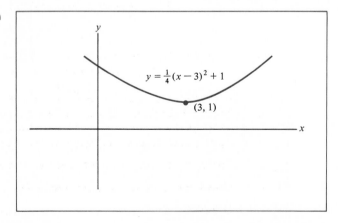

$$y = \frac{1}{4}(x - 3)^2 + 1$$

(3, 1)

FIGURE 5

Here $a = \frac{1}{4}$, $h = 3$, and $k = 1$. Since a is positive, the parabola opens up. Its lowest point, of course, is the vertex, which is at $(h, k) = (3, 1)$. (See Fig. 5.)

Thus far we have concentrated on quadratic functions of the form

$$f(x) = a(x - h)^2 + k. \tag{1}$$

What about the general case? It turns out that *every* quadratic function can be written in form (1). This is often done by a technique called *completing the square*. Here are two simple examples:

$$x^2 - 10x = x^2 - 10x + 25 - 25$$
$$= (x - 5)^2 - 25$$
$$x^2 + 3x = x^2 + 3x + (\tfrac{3}{2})^2 - (\tfrac{3}{2})^2$$
$$= (x + \tfrac{3}{2})^2 - \tfrac{9}{4}$$

In each case, we divided the coefficient of x by 2, squared the result, and then added and subtracted that quantity in order to get an expression that could be written in the form $(x - h)^2$. In the next example there is a constant term which we ignore at first. We complete the square inside the brackets:

$$x^2 + 3x + 5 = [x^2 + 3x \qquad] + 5$$
$$= [x^2 + 3x + (\tfrac{3}{2})^2] + 5 - (\tfrac{3}{2})^2$$
$$= (x + \tfrac{3}{2})^2 + 5 - (\tfrac{3}{2})^2$$
$$= (x + \tfrac{3}{2})^2 + \tfrac{11}{4}.$$

The case in which the coefficient of x^2 is not 1 is more complicated, and we present it in Example 4.

EXAMPLE 4   Complete the square for the quadratic function $f(x) = 4x^2 + 12x + 20$.

Solution   Think of the polynomial as consisting of two parts:

$$f(x) = [4x^2 + 12x \qquad] + 20$$
$$= \underbrace{4[x^2 + 3x \qquad]}_{\text{1st part}} \underbrace{+ 20}_{\text{2nd part}}$$

The first part contains the terms involving x with the coefficient of x^2 as a factor outside. The second part contains just the constant term. We need to add $(\tfrac{3}{2})^2$ inside the brackets in order to complete the square. However, since there is a 4 multiplying everything inside the brackets, we will have to compensate for the $(\tfrac{3}{2})^2$ by subtracting $4 \cdot (\tfrac{3}{2})^2$ from the constant term.

$$f(x) = 4[x^2 + 3x + (\tfrac{3}{2})^2] + 20 - 4 \cdot (\tfrac{3}{2})^2$$
$$= 4(x + \tfrac{3}{2})^2 + 20 - 4 \cdot \tfrac{9}{4}$$
$$= 4(x + \tfrac{3}{2})^2 + 11.$$

EXAMPLE 5 Find the vertex of the graph of $f(x) = 3x^2 + 5x + 4$.

Solution
$$f(x) = 3[x^2 + \tfrac{5}{3}x \qquad] + 4$$
$$= 3[x^2 + \tfrac{5}{3}x + (\tfrac{5}{6})^2] + 4 - 3 \cdot (\tfrac{5}{6})^2$$
$$= 3(x + \tfrac{5}{6})^2 + 4 - 3 \cdot \tfrac{25}{36}$$
$$= 3(x + \tfrac{5}{6})^2 + \tfrac{23}{12}.$$

Think of $x + \tfrac{5}{6}$ as $x - (-\tfrac{5}{6})$. The vertex is $(-\tfrac{5}{6}, \tfrac{23}{12})$.

A common way to sketch the graph of a quadratic function is to locate the vertex and two or three other convenient points. Usually, one such point is the y-intercept, that is, the point where the parabola crosses the y-axis.

EXAMPLE 6 Sketch the graph of $f(x) = \tfrac{1}{2}x^2 - 4x + 9$.

Solution
$$f(x) = \tfrac{1}{2}[x^2 - 8x \qquad] + 9$$
$$= \tfrac{1}{2}[x^2 - 8x + 16] + 9 - \tfrac{1}{2} \cdot 16$$
$$= \tfrac{1}{2}(x - 4)^2 + 1.$$

The vertex is (4, 1). Since the coefficient $\tfrac{1}{2}$ is positive and less than 1, we know that the parabola opens up and is wider than the graph of $y = x^2$. To get a more precise picture of the curve we plot more points. When $x = 0$,

$$f(0) = \tfrac{1}{2}(0)^2 - 4(0) + 9,$$

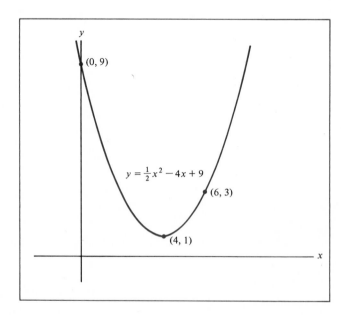

$$y = \tfrac{1}{2}x^2 - 4x + 9$$

FIGURE 6

so the y-intercept is $(0, 9)$. The form $f(x) = \frac{1}{2}(x - 4)^2 + 1$ is useful when computing $f(x)$ for values of x near 4. For instance,

$$f(6) = \tfrac{1}{2}(6 - 4)^2 + 1 = \tfrac{1}{2}(2)^2 + 1 = 3.$$

Hence $(6, 3)$ is on the graph. (See Fig. 6.)

When the graph of $f(x) = ax^2 + bx + c$ crosses the x-axis, it may be desirable to find the x-intercepts, that is, the points $(x, 0)$ where x satisfies $ax^2 + bx + c = 0$. Methods for solving such quadratic equations were discussed in Section 1.6.

PRACTICE PROBLEMS 4

1. Find the vertex of the graph of $f(x) = 2(x + 6)^2$.

2. Complete the square for the function $f(x) = x^2 - 7x - 3$.

3. Does the graph of $f(x) = 3x^2 + 8x - 1$ open up or down?

EXERCISES 4

In Exercises 1–6, state whether the given parabola is wider or narrower than the parabola $y = x^2$.

1. $y = 5x^2$ 2. $y = -\frac{2}{3}x^2$ 3. $y = \frac{1}{3}x^2 + 4$

4. $y = \frac{1}{3} - \frac{5}{2}x^2$ 5. $y = -2(x + 1)^2$ 6. $y = \frac{3}{4}(x - 6)$

In Exercises 7–12, state whether the given parabola opens up or down.

7. $y = -4x^2 + 3$ 8. $y = \frac{1}{3}x^2 - 5$ 9. $y = \frac{1}{2}(x - 3)^2 + 1$

10. $y = -5(x + 2)^2 + \frac{1}{2}$ 11. $y = 5 - 3(x + 1)^2$ 12. $y = 1 + (x - 2)^2$

In Exercises 13–24, find the vertex of the parabola that is the graph of the given function.

13. $f(x) = -x^2$ 14. $f(x) = 3x^2 + 2$

15. $f(x) = (x - 3)^2 + 4$ 16. $f(x) = \frac{1}{3}(x + 1)^2 - 2$

17. $f(x) = -5(x + \frac{1}{2})^2 + 2$ 18. $f(x) = 3(x - 4)^2 - \frac{1}{3}$

19. $f(x) = x^2 - 2x + 3$ 20. $f(x) = x^2 + 4x + 1$

21. $f(x) = 4x^2 - 12x + 7$ 22. $f(x) = 9x^2 + 6x + 5$

23. $f(x) = \frac{1}{2}x^2 + 3x$ 24. $f(x) = -\frac{1}{5}x^2 + x + 1$

In Exercises 25–36, sketch the graph of the given function.

25. $f(x) = -\frac{1}{3}x^2$ 26. $f(x) = \frac{5}{2}x^2$

27. $f(x) = 2(x + 3)^2 - 4$ 28. $f(x) = -5(x - 2)^2 - 3$

29. $f(x) = -\frac{1}{5}(x - \frac{2}{3})^2 + 1$

30. $f(x) = \frac{1}{3}(x + \frac{5}{4})^2 - \frac{4}{5}$

31. $f(x) = x^2 - 6x + 2$

32. $f(x) = x^2 + 5x - 1$

33. $f(x) = 2x^2 + 8x - 1$

34. $f(x) = -\frac{1}{4}x^2 + x + 2$

35. $f(x) = -\frac{2}{3}x^2 + 4x + 1$

36. $f(x) = 4x^2 - 10x + 2$

SOLUTIONS TO PRACTICE PROBLEMS 4

1. $f(x) = 2(x + 6)^2$
$$= 2(x - (-6))^2 + 0.$$
The vertex is $(h, k) = (-6, 0)$.

2. $f(x) = x^2 - 7x - 3$
$$= x^2 - 7x + (-\tfrac{7}{2})^2 - 3 - (-\tfrac{7}{2})^2$$
$$= (x - \tfrac{7}{2})^2 - 3 - \tfrac{49}{4}$$
$$= (x - \tfrac{7}{2})^2 - \tfrac{61}{4}.$$

3. If we start to complete the square,

$$f(x) = 3x^2 + 8x - 1$$
$$= 3[x^2 + \tfrac{8}{3}x \qquad] - 1,$$

we see that $f(x)$ will have the form

$$f(x) = 3(x - h)^2 + k.$$

This function is obtained from $f(x) = 3x^2$ by a suitable translation. Because $a = 3$ (a positive number), both parabolas open up. In general, if

$$f(x) = ax^2 + bx + c,$$

the graph of $f(x)$ opens up if a is positive and down if a is negative.

Chapter 3: CHECKLIST

☐ Function
☐ Value of a function at x
☐ Domain of a function
☐ Graph of a function
☐ Vertical line test
☐ Graph of an equation
☐ Linear function
☐ Constant function
☐ x- and y-intercepts
☐ Quadratic function
☐ Polynomial and rational functions
☐ Power function
☐ Absolute value function
☐ Addition, subtraction, multiplication, and division of functions

☐ Composition of functions
☐ Translation of graphs
☐ Vertex of a parabola
☐ Completing the square

Chapter 3: SUPPLEMENTARY EXERCISES

1. Let $f(x) = x^3 + 1/x$. Evaluate $f(1)$, $f(3)$, $f(-1)$, $f(-\frac{1}{2})$, $f(\sqrt{2})$.

2. Let $f(x) = 2x + 3x^2$. Evaluate $f(0)$, $f(-\frac{1}{4})$, $f(1/\sqrt{2})$.

3. Let $f(x) = x^2 - 2$. Evaluate $f(a - 2)$.

4. Let $f(x) = [1/(x + 1)] - x^2$. Evaluate $f(a + 1)$.

Determine the domains of the following functions.

5. $f(x) = \dfrac{1}{x(x + 3)}$

6. $f(x) = \sqrt{x - 1}$

7. $f(x) = \sqrt{x^2 + 1}$

8. $f(x) = 1/\sqrt{3x}$

9. Is the point $(\frac{1}{2}, -\frac{3}{5})$ on the graph of the function $h(x) = (x^2 - 1)/(x^2 + 1)$?

10. Is the point $(1, -2)$ on the graph of the function $k(x) = x^2 + (2/x)$?

Let $f(x) = x^2 - 2x$, $g(x) = 3x - 1$, $h(x) = \sqrt{x}$. Find the following functions.

11. $f(x) + g(x)$

12. $f(x) - g(x)$

13. $f(x)h(x)$

14. $f(x)g(x)$

15. $f(x)/h(x)$

16. $g(x)h(x)$

Let $f(x) = x/(x^2 - 1)$, $g(x) = (1 - x)/(1 + x)$, $h(x) = 2/(3x + 1)$. Express the following as rational functions.

17. $f(x) - g(x)$

18. $f(x) - g(x + 1)$

19. $g(x) - h(x)$

20. $f(x) + h(x)$

21. $g(x) - h(x - 3)$

22. $f(x) + g(x)$

Let $f(x) = x^2 - 2x + 4$, $g(x) = 1/x^2$, $h(x) = 1/(\sqrt{x} - 1)$. Determine the following functions.

23. $f(g(x))$

24. $g(f(x))$

25. $g(h(x))$

26. $h(g(x))$

27. $f(h(x))$

28. $h(f(x))$

29. Simplify $(81)^{3/4}$, $8^{5/3}$, $(.25)^{-1}$.

30. Simplify 5^{-2}, $(100)^{3/2}$, $(.001)^{1/3}$.

31. The population of a city is estimated to be $750 + 25t + .1t^2$ thousand people t years from the present. Ecologists estimate that the average level of carbon monoxide in the air above the city will be $1 + .4x$ ppm (parts per million) when the population is x thousand people. Express the carbon monoxide level as a function of the time t.

32. The revenue $R(x)$ (in thousands of dollars) a company receives from the sale of x thousand units is given by $R(x) = 5x - x^2$. The sales level x is in turn a function $f(d)$ of the number d of dollars spent on advertising, where

$$f(d) = 6[1 - 200/(d + 200)].$$

Express the revenue as a function of the amount spent on advertising.

33. $A = x^2 - 3xy; \frac{1}{2}x^2y = 4$. Express A as a function of x.

34. $A = \frac{1}{2}y^2 + \frac{5}{2}x^2; 6x + 3y = 3$. Express A as a function of x.

35. $A = \pi r^2 l; l + 2\pi r = 84$. Express A as a function of r.

36. $A = t^2 + 3st; 4s + 2t = 2$. Express A as a function of t.

37. Find the equation of the curve that is obtained from the graph of $f(x) = \dfrac{1}{x} + x - 5$ by a translation of 2 units to the left and 3 units up.

38. Find the equation of the parabola that is obtained from the graph of $f(x) = x^2 - 2x - 3$ by a translation of 4 units to the right and 1 unit down.

In Exercises 39–44, sketch the graph of the given function.

39. $f(x) = 3(x - 2)^2 - 4$

40. $f(x) = \frac{1}{2}(x + 1)^2 + 2$

41. $f(x) = -\frac{1}{4}(x + 3)^2 + 2$

42. $f(x) = 2(x - 2)^2 - 3$

43. $f(x) = x^2 - 8x + 21$

44. $f(x) = 3x^2 - 2x + \frac{1}{3}$

The Mathematics of Finance

Chapter 4

4.1. Interest

When you deposit money into a savings account, the bank pays you a fee for the use of your money. This fee is called *interest* and is determined by the amount deposited, the duration of the deposit, and the quoted interest rate. The amount deposited is called the *principal* and the amount to which the principal grows (after the addition of interest) is called the *compound amount*.

Here are the entries in a hypothetical bank passbook:

Date	Deposits	Withdrawals	Interest	Balance
1/1/80	$100.00			$100.00
4/1/80			$2.00	102.00
7/1/80			2.04	104.04
10/1/80			2.08	106.12
1/1/81			2.12	108.24

Note the following facts about this passbook:

1. The principal is $100.00. The compound amount after 1 year is $108.24.

2. Interest is being paid four times per year (or in financial language, *quarterly*).

3. Each quarter, the interest is 2% of the previous balance. That is, $2.00 is 2% of $100.00, $2.04 is 2% of $102.00, and so on. Since $4 \times 2\%$ is 8%, we say that the passbook is earning 8% *annual interest compounded quarterly*.

As in the passbook above, interest rates are usually stated as annual percentage rates, with the interest to be *compounded* (i.e., computed) a certain number of times per year. Some common frequencies for compounding are:

Number of interest periods per year	Length of each interest period	Interest called:
1	One year	Compounded annually
2	Six months	Compounded semiannually
4	Three months	Compounded quarterly
12	One month	Compounded monthly
52	One week	Compounded weekly
365	One day	Compounded daily

Of special importance is the *interest per period*, denoted i, which is calculated by dividing the annual percentage by the number of interest periods per year. For example, in our passbook above, the annual percentage rate was 8%, the interest was compounded quarterly, and the interest per period was $8\%/4 = 2\%$.

EXAMPLE 1   Determine the interest per period for each of the following interest rates.

(a) 10% interest compounded semiannually

(b) 6% interest compounded monthly

Solution   (a) The annual percentage rate is 10% and the number of interest periods is 2. Therefore,

$$i = \frac{10\%}{2} = 5\%.$$

Note In decimal form, $i = .05$.

(b) The annual percentage rate is 6% and the number of interest periods is 12. Therefore,

$$i = \frac{6\%}{12} = \frac{1}{2}\%.$$

Note In decimal form, $i = .005$.

Consider a savings account in which the interest per period is i. Then the interest earned for the next period is i times the current balance. The next balance, B_{next}, is then computed by adding this interest to the current balance, B_{current}. That is,

$$B_{\text{next}} = B_{\text{current}} + i \cdot B_{\text{current}}. \tag{1}$$

EXAMPLE 2   Compute the interest and the balance for the first two interest periods for a deposit of $1000 at 10% interest compounded semiannually.

Solution   Here $i = 5\%$ or $.05$. The interest for the first period is 5% of 1000, or 50. By formula (1), the balance after the first interest period is

$$B_{\text{first}} = 1000 + .05(1000)$$

$$= 1000 + 50$$

$$= \$1050.$$

Similarly, since 5% of 1050 is 52.50, the interest for the second period is $\$52.50$ and the balance is

$$B_{\text{second}} = B_{\text{first}} + .05B_{\text{first}}$$

$$= 1050 + .05(1050)$$

$$= \$1050 + 52.50$$

$$= \$1102.50.$$

Let us compute B_{second} using another method.

$$B_{\text{second}} = B_{\text{first}} + .05B_{\text{first}}$$

$$= 1 \cdot B_{\text{first}} + .05B_{\text{first}}$$

$$= (1 + .05)B_{\text{first}}$$

$$= (1.05)B_{\text{first}}$$

$$= (1.05)1050$$

$$= \$1102.50.$$

The alternative method for computing B_{second} just presented can be generalized. Namely, we always have

$$B_{\text{next}} = B_{\text{current}} + i \cdot B_{\text{current}}$$

$$= 1 \cdot B_{\text{current}} + i \cdot B_{\text{current}}.$$

Therefore,

$$B_{\text{next}} = (1 + i)B_{\text{current}}.$$

This last result says that balances for successive time periods are computed by multiplying by $1 + i$.

The formula for the balance after any number of interest periods is now easily derived:

Principal	P
Balance after 1 interest period	$(1 + i)P$
Balance after 2 interest periods	$(1 + i) \cdot (1 + i)P$ or $(1 + i)^2 P$
Balance after 3 interest periods	$(1 + i) \cdot (1 + i)^2 P$ or $(1 + i)^3 P$
Balance after 4 interest periods	$(1 + i)^4 P$
$\vdots$	
Balance after n interest periods	$(1 + i)^n P$

Denote the compound amount by the letter F (suggestive of "future value"). Then, the compound amount after n interest periods is given by the formula

$$F = (1 + i)^n P, \qquad (2)$$

where i is the interest rate per period and P is the principal.

Values of $(1 + i)^n$ for specific values of i and n are easily determined using either a calculator or tables. A brief table of useful values has been included as Table 2 of the Appendix.

EXAMPLE 3 Apply formula (2) to the savings account passbook discussed at the beginning of this section and calculate the compound amount after 1 year and after 5 years.

Solution The principal P is $100. Since the interest rate is 8% compounded quarterly, we have $i = 2\%$ or .02. One year consists of four interest periods, so $n = 4$. Therefore, the compound amount after 1 year is

$$F = (1 + .02)^4 \cdot 100$$
$$= (1.02)^4 \cdot 100$$
$$= (1.08243216) \cdot 100 \qquad \text{(from Table 2)}$$
$$= 108.243216$$
$$= \$108.24 \qquad \text{(after rounding off to the nearest cent)}.$$

Five years consists of $n = 5 \times 4 = 20$ interest periods. Therefore, the compound amount after 5 years is

$$F = (1.02)^{20} \cdot 100$$
$$= (1.48594740) \cdot 100$$
$$= \$148.59.$$

The next example is a variation on the previous one and introduces a new concept, present value.

EXAMPLE 4 How much money must be deposited now in order to have $1000 after 5 years if interest is paid at an 8% annual rate compounded quarterly?

Solution As in Example 3, we have $i = .02$ and $n = 20$. However, now we are given F and are asked to solve for P.

$$F = (1 + i)^n P$$
$$1000 = (1.02)^{20} P$$
$$P = \frac{1000}{(1.02)^{20}}.$$

From Table 2, $(1.02)^{20} = 1.48594740$. However, the cumbersome arithmetic can be avoided by using Table 3 of the Appendix which tabulates values of $1/(1 + i)^n$ for various values of i and n.

$$P = \frac{1}{(1.02)^{20}} \cdot 1000$$

$$= (.67297133) \cdot 1000$$

$$= \$672.97.$$

We say that \$672.97 is the present value of \$1000 5 years from now at 8% interest compounded quarterly.

In general, the present value of F dollars at a given interest rate and given length of time is the amount of money P that must be deposited now in order for the compound amount to grow to F dollars in the given length of time. From formula (2), we see that the present value P may be computed from the following formula:

$$P = \frac{1}{(1 + i)^n} \cdot F. \tag{3}$$

EXAMPLE 5 Determine the present value of a \$10,000 payment to be received on January 1, 1990, if it is now May 1, 1981, and money can be invested at 6% interest compounded monthly.

Solution Here $F = 10,000$, $n = 104$ (the number of months between the two given dates), and $i = \frac{1}{2}\%$. By formula (3),

$$P = \frac{1}{(1 + i)^n} \cdot F$$

$$= (.59529136) \cdot 10,000$$

$$= \$5952.91$$

Therefore, \$5952.91 invested on May 1, 1981, will grow to \$10,000 by January 1, 1990.

The interest that we have been discussing so far is the most prevalent type of interest and is known as *compound interest*. There is another type of interest, called *simple interest*, which is used in some financial circumstances. Let us now discuss this type of interest.

Interest rates for simple interest are given as an annual percentage rate r. Interest is earned *only* on the principal P and the interest is rP for each year. Therefore, the interest earned in n years is nrP. So the amount A after n years is the original amount plus the interest earned. That is,

$$A = P + nrP$$

$$= 1 \cdot P + nrP$$

$$= (1 + nr)P. \tag{4}$$

EXAMPLE 6 Calculate the amount after 4 years if $1000 is invested at 7% simple interest.

Solution Apply formula (4) with $P = \$1000$, $n = 4$, and $r = 7\%$ or .07.

$$A = (1 + nr)P$$
$$= [1 + 4(.07)]1000$$
$$= (1.28)1000$$
$$= \$1280.$$

In Example 6, had the money been invested at 7% compound interest, with annual compounding, then the compound amount would have been $1310.80. Money invested at simple interest is earning interest only on the principal amount. However, with compound interest, after the first interest period, the interest is also earning interest. Thus, if we compare two savings accounts, each earning interest at the same stated annual rate, but with one earning simple interest and one earning compound interest, the latter will grow at a faster rate.

Let us close this section with a summary of the key formulas we have developed so far.

Compound Interest

Compound amount: $F = (1 + i)^n P$

Present value: $P = \left[\dfrac{1}{(1 + i)^n} \right] F$

where i is the interest rate per period and n is the number of interest periods.

Simple Interest

Amount: $A = (1 + nr)P$

PRACTICE PROBLEMS 1

1. (a) In Table 3 of the Appendix, look up the value of $1/(1 + i)^n$ for $i = 6\%$ and $n = 10$.

    (b) Calculate the present value of $1000 to be received 10 years in the future at 6% interest compounded annually.

2. Calculate the compound amount after 2 years of $1 at 26% interest compounded weekly.

3. Calculate the amount of $2000 after 6 months if invested at 10% simple interest.

EXERCISES 1

1. Determine i and n for the following situations.

    (a) 12% interest compounded monthly for 2 years.

    (b) 8% interest compounded quarterly for 5 years.

    (c) 10% interest compounded semiannually for 20 years.

2. Determine i and n for the following situations.

    (a) 6% interest compounded annually for 3 years.

    (b) 6% interest compounded monthly from January 1, 1980, to July 1, 1981.

    (c) 9% interest compounded quarterly from January 1, 1980, to October 1, 1980.

3. Determine i, n, P, and F for the following situations.

    (a) $500 invested at 6% interest compounded annually grows to $631.24 in 4 years.

    (b) $800 invested on January 1, 1981, at 24% interest compounded monthly will grow to $8612.13 by January 1, 1991.

    (c) $2974.62 is deposited on January 1, 1981. The balance on July 1, 1990, is $9000 and the interest is 12% interest compounded semiannually.

4. Determine i, n, P, and F for each of the following situations.

    (a) The amount of money that must be deposited now at 26% interest compounded weekly in order to have $7500 in 1 year is $5786.63.

    (b) $3000 deposited at 12% interest compounded monthly will grow to $107,848.92 in 30 years.

    (c) In 1626, Peter Minuit, the first director-general of New Netherlands province, purchased Manhattan Island for trinkets and cloth valued at about $24. Had this money been invested at 8% interest compounded quarterly, it would have amounted to $39,125,072,000,000 by 1981.

5. Calculate the compound amount of $1000 after 2 years if deposited at 6% interest compounded monthly.

6. Calculate the present value of $10,000 payable in 5 years at 12% interest compounded semiannually.

7. Calculate the present value of $100,000 payable in 25 years at 12% interest compounded monthly.

8. Calculate the compound amount of $1000 after 1 year if deposited at 7.3% interest compounded daily.

9. If you had invested $10,000 on January 1, 1980, at 24% interest compounded quarterly, how much would you have on April 1, 1986?

10. In order to have $10,000 on his twenty-fifth birthday, how much would a person just turned 21 have to invest if the money will earn 24% interest compounded monthly?

11. Mr. Smith wishes to purchase a $10,000 sailboat upon his retirement in 3 years. He has just won the state lottery and would like to set aside enough cash in a savings account paying 4% interest compounded quarterly to buy the boat upon retirement. How much should he deposit?

12. Ms. Jones has just invested $100,000 at 6% interest compounded annually. How much money will she have in 20 years?

13. Consider the following savings account passbook:

Date	Deposits	Withdrawals	Interest	Balance
1/1/84	$1000.00			$1000.00
2/1/84			$10.00	1010.00
3/1/84			10.10	1020.10

(a) What annual interest rate is this bank paying?

(b) Give the interest and balance on 4/1/84.

(c) Give the interest and balance on 1/1/86.

14. Consider the following saving account passbook:

Date	Deposits	Withdrawals	Interest	Balance
1/1/84	$10,000.00			$10,000.00
7/1/84			$600.00	10,600.00
1/1/85			636.00	11,236.00

(a) Give the interest and balance on 7/1/85.

(b) Give the interest and balance on 1/1/93.

15. Is it more profitable to receive $1000 now or $1700 in 9 years? Assume that money can earn 6% interest compounded annually.

16. Is it more profitable to receive $7000 now or $10,000 in 9 years? Assume that money can earn 4% interest compounded quarterly.

17. Would you rather earn 30% interest compounded annually or 26% interest compounded weekly?

18. Would you rather earn 8% interest compounded annually or 7.3% interest compounded daily?

19. On January 1, 1981, a deposit was made into a savings account paying interest compounded quarterly. The balance on January 1, 1984, was $10,000 and the balance on April 1, 1984, was $10,200. How large was the deposit?

20. During the 1970s, a deposit was made into a savings account paying 4% interest compounded quarterly. On January 1, 1984, the balance was $2020. What was the balance on October 1, 1983?

Exercises 21–32 concern simple interest.

21. Determine r, n, P, and A for each of the following situations.

   (a) $500 invested at 7% simple interest grows to $517.50 in 6 months.

   (b) In order to have $580 after 2 years at 8% simple interest, $500 must be deposited.

22. Determine r, n, P, and A for each of the following situations.

   (a) At 15% simple interest, $1000 deposited on January 1, 1984, will be worth $1100 on September 1, 1984.

   (b) At 20% simple interest, in order to have $6000 in 5 years, $3000 must be deposited now.

23. Calculate the amount after 3 years if $1000 is deposited at 9% simple interest.

24. Calculate the amount after 18 months if $2000 is deposited at 8% simple interest.

25. Find the present value of $3000 in 2 years at 25% simple interest.

26. Find the present value of $2000 in 4 years at 15% simple interest.

27. Determine the (simple) interest rate at which $800 grows to $1000 in 6 months.

28. How many years are required for $1000 to grow to $1210 at 14% simple interest?

29. Determine the amount of time required for money to double at 10% simple interest.

30. Derive the formula for the (simple) interest rate at which P dollars grows to F dollars in n years. That is, express r in terms of P, A, and n.

31. Derive the formula for the present value of A dollars in n years at simple interest rate r.

32. Derive the formula for the number of years required for P dollars to grow to A dollars at simple interest rate r.

33. Compute the compound amount after 1 year for $100 invested at 24% interest compounded quarterly. What simple interest rate will yield the same amount in 1 year?

34. Compute the compound amount after 1 year for $100 invested at 12% interest compounded monthly. What simple interest rate will yield the same amount in 1 year?

Suppose that interest is compounded. Then the annual percentage rate is called the *nominal rate*. The simple interest rate that yields the same amount after 1 year is called the *effective rate*. For example, refer to Exercise 33. The nominal rate there is 24% and corresponds to an effective rate of 26.25%.

35. Calculate the effective rate for 12% interest compounded semiannually.

36. Calculate the effective rate for 26% interest compounded weekly.

37. Suppose that an investment earns a nominal rate of a (expressed as a decimal) compounded twice a year. What is the effective rate?

38. Suppose that an investment of P dollars earns a nominal rate of a compounded n times per year. What is the compound amount after 1 year?

39. Suppose that an investment of P dollars earns a nominal rate of a compounded n times per year. What is the effective rate?

SOLUTIONS TO PRACTICE PROBLEMS 1

1. (a) Look up the entry in Table 3 of the Appendix in the row labeled "10" and the column labeled 6%. That entry is .55839478.

   (b) Here we are given the value in the future $F =$ \$1000 and are asked to find the present value P. Interest compounded annually has just one interest period per year, so $n = 10/1 = 10$ and $i = 6\%/1 = 6\%$. By formula (3),

$$P = \frac{1}{(1 + i)^n} F$$

$$= \frac{1}{(1 + i)^n} \cdot 1000 \qquad \text{(with } i = 6\% \text{ and } n = 10)$$

$$= (.55839478) \cdot 1000 \qquad [\text{from part (a)}]$$

$$= 558.39478$$

$$= \$558.39 \qquad \text{(rounding off to the nearest cent).}$$

2. Here we are given the present value, $P = 1$, and are asked to find the value F at a future time. Interest compounded weekly has 52 interest periods each year, so $n = 2 \times 52 = 104$ and $i = 26\%/52 = \frac{1}{2}\%$. By formula (2), we have

$$F = (1 + i)^n P$$

$$= (1 + i)^n \cdot 1 \qquad \text{(with } i = \frac{1}{2}\%, \quad n = 104)$$

$$= (1.67984969) \cdot 1, \qquad \text{(using Table 2)}$$

$$= \$1.68.$$

[*Note:* In general, whenever $P =$ \$1, then $F = (1 + i)^n \cdot 1 = (1 + i)^n$. This explains why $(1 + i)^n$ is often referred to as the *compound amount of* \$1.]

3. In simple interest problems, time should be expressed in terms of years. Since 6 months is $\frac{1}{2}$ year, formula (4) gives

$$A = (1 + nr)P$$

$$= [1 + \tfrac{1}{2}(.10)]2000$$

$$= (1.05)2000$$

$$= \$2100.$$

4.2. Annuities

An annuity is a sequence of equal payments made at regular intervals of time. Here are two illustrations.

1. As the proud parent of a newborn daughter, you decide to save for her college education by depositing \$30 at the end of each month into a savings account paying 6% interest compounded monthly. Eighteen years from now after you make the last of 216 payments, the account will contain \$11,620.60.

2. Having just won the state lottery, you decide not to work for the next 5 years. You want to deposit enough money into the bank so that you can withdraw $900 at the end of each month for 60 months. If the bank pays 6% interest compounded monthly, you must deposit $46,553.

The payments in the foregoing financial transactions are called *rent*. The amount of a typical rent payment is denoted by the letter R. Thus, in the examples above, we have $R = \$30$ and $R = \$900$, respectively.

In illustration 1, you make equal payments to a bank in order to generate a large sum of money in the future. This sum, namely $11,620.60, is called the *future value of the annuity*.

In illustration 2, the bank will make equal payments to you in order to pay back the sum of money that you currently deposit. The value of the current deposit, namely $46,553, is called the *present value of the annuity*.

In this section we will derive formulas for future values and present values of annuities. However, as a mathematical preliminary, we must first discuss sums of geometric progressions.

Let a be any number. Then the *geometric progression* determined by a is the sequence of nonnegative powers of a:

$$1, a, a^2, a^3, a^4, \ldots$$

Let us assume that $a \neq 1$. Then in this case, the sum of the first n terms of the geometric progression determined by a is given by the formula

$$1 + a + a^2 + \cdots + a^{n-1} = \frac{a^n - 1}{a - 1} \tag{1}$$

For instance, with $a = 2$ and $n = 3$, formula (1) states that

$$1 + 2 + 2^2 = \frac{2^3 - 1}{2 - 1}.$$

Doing the arithmetic shows that this result is correct since both sides equal 7.

Formula (1) is derived by multiplying $a^{n-1} + \cdots + a^2 + a + 1$ by $a - 1$.

$$
\begin{array}{r}
a^{n-1} + \cdots + a^2 + a + 1 \\
a - 1 \\
\hline
-a^{n-1} - \cdots - a^2 - a - 1 \\
a^n + a^{n-1} + \cdots + a^2 + a \\
\hline
a^n \qquad\qquad\qquad\qquad\quad - 1
\end{array}
$$

That is,

$$(a - 1)(1 + a + a^2 + \cdots + a^{n-1}) = a^n - 1.$$

Divide both sides of this last equation by $a - 1$ (which is possible since a is not equal to 1) and formula (1) results.

Let us now apply formula (1) to annuities. Suppose that an annuity consists of a sequence of n equal payments, each of R dollars. Suppose that the annuity payments are deposited into an account paying compound interest at the rate of i

per interest period. We will further suppose that there is a single annuity payment per interest period and that the payment is made at the end of the interest period. Let us derive a formula for the future value of the annuity: that is, a formula for the balance of the account immediately after the last payment.

Each payment accumulates interest for a different number of interest periods, so let us calculate the balance in the account as the sum of n compound amounts, one corresponding to each payment.

Payment	Amount	Number of Interest Periods on Deposit	Compound Amount
1	R	$n - 1$	$(1 + i)^{n-1}R$
2	$R \cdot$	$n - 2$	$(1 + i)^{n-2}R$
$\vdots$			
$n - 2$	R	2	$(1 + i)^2 R$
$n - 1$	R	1	$(1 + i)R$
n	R	0	R

Denote the future value of the annuity by F. Then F is the sum of the numbers in the right-hand column:

$$F = R + (1 + i)R + (1 + i)^2 R + \cdots + (1 + i)^{n-1}R$$
$$= [1 + (1 + i) + (1 + i)^2 + \cdots + (1 + i)^{n-1}]R$$
$$= \frac{(1 + i)^n - 1}{(1 + i) - 1} R,$$

where we have applied formula (1) with $a = 1 + i$. Therefore, we have the result

$$F = \frac{(1 + i)^n - 1}{i} \cdot R.$$

The expression $[(1 + i)^n - 1]/i$ occurs often in financial analysis and is denoted by the special symbol $s_{\overline{n}|i}$ (read "s sub n angle i"). Values of $s_{\overline{n}|i}$ may be computed using a financial calculator or by consulting an appropriate table. We have included a brief table as Table 4 of the Appendix. We may summarize our calculation above as follows.

> Suppose that an annuity consists of n payments of R dollars each, deposited at the ends of consecutive interest periods into an account with interest compounded at a rate i per period. Then the future value F of the annuity is given by the formula
>
> $$F = s_{\overline{n}|i}R.$$ (2)

EXAMPLE 1 Calculate the future value of an annuity of \$100 per month for 5 years at 6% interest compounded monthly.

Solution Here $R = 100$ and $i = \frac{1}{2}\%$. Since a payment is made at the end of each month for 5 years, there will be $5 \times 12 = 60$ payments. So $n = 60$. Therefore,

$$F = s_{\overline{n}|i}R$$

$$= s_{\overline{60}|1/2\%} \cdot 100$$

$$= (69.77003051) \cdot 100 \qquad \text{(by Table 4)}$$

$$= \$6977.00$$

Formula (2) can also be used to determine the rent necessary to achieve a certain future value.

$$F = s_{\overline{n}|i}R$$

$$R = \frac{F}{s_{\overline{n}|i}} = \frac{1}{s_{\overline{n}|i}}F.$$

Thus we have established the following result.

> Suppose that an annuity of n payments has future value F and has interest compounded at the rate i per period. Then the rent R is given by
>
> $$R = \frac{1}{s_{\overline{n}|i}} \cdot F. \qquad (3)$$

Table 5 of the Appendix gives values of $1/s_{\overline{n}|i}$ for various n and i.

EXAMPLE 2 Ms. Adams would like to buy a \$30,000 airplane when she retires in 8 years. How much should she deposit at the end of each half-year into an account paying 12% interest compounded semiannually so that she will have enough money to purchase the airplane?

Solution Here $n = 16$, $i = 6\%$ and $F = 30,000$. Therefore,

$$R = \frac{1}{s_{\overline{n}|i}} \cdot F$$

$$= \frac{1}{s_{\overline{16}|6\%}} \cdot 30,000$$

$$= (.03895214) \cdot 30,000 \qquad \text{(by Table 5)}$$

$$= \$1168.56.$$

She should deposit \$1168.56 at the end of each half-year period.

The *present value* of an annuity is the amount of money it would take to finance the sequence of annuity payments. More specifically, the present value of an annuity is the amount you would need to deposit in order to provide the desired sequence of annuity payments and leave a balance of zero at the end of the term. Let us now find a formula for the present value of an annuity. There are two ways to proceed. On the one hand, the desired present value could be computed as the sum of the present values of the various annuity payments. This computation would make use of the formula for the sum of the first n terms of a geometric progression. However, there is a much "cleaner" derivation which proceeds indirectly to obtain a formula relating the present value P and the rent R.

As before, let us assume that our annuity consists of n payments made at the ends of interest periods, with interest compounded at a rate i per interest period.

Situation 1: Suppose that the P dollars were just left in the account and that the annuity payments were not withdrawn. At the end of the n interest periods there would be $(1 + i)^n P$ dollars in the account.

Situation 2: Suppose that the payments are withdrawn but immediately are redeposited into another account having the same rate of interest. At the end of the n interest periods, there would be $s_{\overline{n}|i} R$ dollars in the new account.

In both of these situations, P dollars is deposited and it, together with all the interest generated, is earning income at the same interest rate for the same amount of time. Therefore, the final amounts of money in the accounts should be the same. That is,

$$(1 + i)^n P = s_{\overline{n}|i} R$$

$$P = \frac{s_{\overline{n}|i}}{(1 + i)^n} \cdot R.$$

Let us now substitute the value of $s_{\overline{n}|i}$, namely $[(1 + i)^n - 1]/i$, into the formula above. If we denote by $a_{\overline{n}|i}$ the expression

$$a_{\overline{n}|i} = \frac{(1 + i)^n - 1}{i(1 + i)^n},$$

then the formula above may be written in the simple form

$$P = a_{\overline{n}|i} R; \qquad R = \frac{1}{a_{\overline{n}|i}} P$$

Tables 6 and 7 of the Appendix give, respectively, the values of $a_{\overline{n}|i}$ and $1/a_{\overline{n}|i}$ for various values of n and i. Let us record the main result of our discussion above.

> The present value P and the rent R of an annuity of n payments with interest compounded at a rate i per interest period are related by the formulas
>
> $$P = a_{\overline{n}|i} R; \qquad R = \frac{1}{a_{\overline{n}|i}} P$$

EXAMPLE 3 How much money must you deposit now at 6% interest compounded quarterly in order to be able to withdraw $3000 at the end of each quarter year for 2 years?

Solution Here $R = 3000$, $i = 1.5\%$, and $n = 8$. We are asked to calculate the present value of the sequence of payments.

$$P = a_{\overline{n}|i} R$$

$$= a_{\overline{8}|1.5\%} \cdot 3000$$

$$= (7.48592508) \cdot 3000 \qquad \text{(by Table 6)}$$

$$= \$22{,}457.78.$$

EXAMPLE 4 If you deposit $1000 into a fund paying 26% interest compounded weekly, how much can you withdraw at the end of each week for 1 year?

Solution Here $P = 1000$, $i = \frac{1}{2}\%$, and $n = 52$. We are asked to calculate the periodic payment, that is, the rent.

$$R = \frac{1}{a_{\overline{n}|i}} P$$

$$= \frac{1}{a_{\overline{52}|1/2\%}} \cdot 1000$$

$$= (.02188675) \cdot 1000$$

$$= \$21.89.$$

Remark In this section we have considered only annuities with payments made at the end of each interest period. Such annuities are called *ordinary annuities*. Annuities that have payments at the beginning of the interest period are called *annuities due*. Annuities whose payment period is different from the interest period are called *general annuities*.

PRACTICE PROBLEMS 2

Decide whether or not each of the following annuities are ordinary annuities, that is, the type of annuities considered in this section. If so, identify n, i, and R and calculate the present value or the future value, whichever is appropriate.

1. You make a deposit at 24% interest compounded monthly into a fund that pays you $1 at the end of each month for 5 years.

2. At the end of each week for 2 years you deposit $10 into a savings account earning 18% interest compounded monthly.

3. At the end of each month for 2 years, you deposit $10 into a savings account earning 18% interest compounded monthly.

EXERCISES 2

1. Determine the following sums by using formula (1).

    (a) $1 + 2 + 4 + 8 + 16$

    (b) $1 - 2 + 4 - 8 + 16$

    (c) $1 + \frac{1}{2} + \frac{1}{4} + \frac{1}{8} + \frac{1}{16}$

    (d) $3 + \frac{3}{2} + \frac{3}{4} + \frac{3}{8} + \frac{3}{16}$

2. Determine the following sums by using formula (1).

    (a) $1 + 11 + 121 + 1331 + 14{,}641 \; [Note: 11^5 = 161{,}051.]$

    (b) $1 - 9 + 81 - 729 + 6561 \; [Note: 9^5 = 59{,}049.]$

    (c) $1 + .5 + .25 + .125 + .0625 \; [Note: (.5)^5 = .03125.]$

    (d) $4 + 2 + 1 + \frac{1}{2} + \frac{1}{4}$

3. For each of the following annuities, specify i, n, R, and F.

    (a) If at the end of each month, $50 is deposited into a savings account paying 6% interest compounded monthly, the balance after 10 years will be $8193.97.

    (b) Mr. Smith is saving to buy a $65,000 yacht in 1990. Since 1980, he has been depositing $1767 at the end of each half-year into a fund paying 12% interest compounded semiannually.

4. For each of the following annuities, specify i, n, R, and P.

    (a) A retiree deposits $72,582 into a bank paying 6% interest compounded monthly and withdraws $520 at the end of each month for 20 years.

    (b) In order to receive $700 at the end of each quarter-year from 1981 until 1986, Ms. Jones deposited $11,446 into a savings account paying 8% interest compounded quarterly.

5. Calculate the future value of an annuity of $100 per month for 5 years at 12% interest compounded monthly.

6. Calculate the rent of an annuity at 12% interest compounded semiannually if payments are made every half-year and the future value after 7 years is $10,000.

7. Calculate the rent of an annuity at 8% interest compounded quarterly with payments made every quarter-year for 7 years and present value $100,000.

8. Calculate the present value of an annuity of $1000 per year for 10 years at 6% interest compounded annually.

9. A city has a debt of $1,000,000 falling due in 15 years. How much money must it deposit at the end of each half-year into a savings fund at 12% interest compounded semiannually in order to raise this amount?

10. On January 1, 1980, Tom decided to save for exactly 1 year for a 10-speed bike by depositing $10 at the end of each month into a savings account paying 6% interest compounded monthly. How much did he accumulate?

11. During Jack's freshman year at college, his father had been sending him $100 per month for incidental expenses. For the sophomore year, his father decided instead to make a deposit into a savings account on August 1 and have his son withdraw $100 on the first

of each month until May 1. If the bank pays 6% interest compounded monthly, how much should Jack's father deposit?

12. Suppose that a magazine subscription costs $9 per year and that you receive a magazine at the end of each month. At an interest rate of 12% compounded monthly, how much are you actually paying for each issue?

13. Is it more profitable to receive $1000 at the end of each month for 10 years or to receive a lump sum of $230,000 at the end of 10 years? Assume that money can earn 12% interest compounded monthly.

14. Is it more profitable to receive a lump sum of $10,000 at the end of 3 years or to receive $750 at the end of each quarter-year for 3 years? Assume that money can earn 8% interest compounded quarterly.

15. Suppose that you deposited $1000 into a savings account on January 1, 1980, and deposited an additional $100 into the account at the end of each quarter-year. If the bank pays 8% interest compounded quarterly, how much will be in the account on January 1, 1989?

16. Suppose that you opened a savings account on January 1, 1980, and made a deposit of $100. In 1981, you began depositing $10 into the account at the end of each month. If the bank pays 6% interest compounded monthly, how much money will be in the account on January 1, 1985?

17. Ms. Jones deposited $100 at the end of each month for 10 years into a savings account earning 6% interest compounded monthly. However, she deposited an additional $1000 at the end of the seventh year. How much money was in the account at the end of the tenth year?

18. Redo Exercise 17 for the situation where Ms. Jones withdrew $1000 at the end of the seventh year instead of depositing it.

A *perpetuity* is an annuity whose payments are to continue forever. Exercises 19 and 20 concern such annuities.

19. A grateful alumnus decides to donate a permanent scholarship of $1200 per year. How much money should be deposited in the bank at 12% interest compounded annually in order to be able to supply the money for the scholarship at the end of each year?

20. Show that to establish a perpetuity paying R dollars at the end of each interest period, it requires a deposit of R/i dollars, where i is the interest rate per interest period.

A *deferred annuity* is an annuity whose term is to start at some future date. Exercises 21–24 concern such annuities.

21. On his tenth birthday, a boy inherits $10,000, which is to be used for his college education. The money is deposited into a trust fund which will pay him R dollars on his 18th, 19th, 20th, and 21st birthdays. Find R if the money earns 6% interest compounded annually.

22. Refer to Exercise 21. Find the size of inheritance that would result in $10,000 per year during the college years (ages 18–21, inclusive).

23. On December 1, 1980, a philanthropist set up a permanent trust fund to buy Christmas presents for needy children. The fund will provide $6000 each year beginning on De-

cember 1, 1990. How much must have been set aside in 1980 if the money earns 6% interest compounded annually?

24. Show that the rent paid by a deferred annuity of n payments which are deferred by m interest periods is given by the formula

$$R = \frac{i(1 + i)^{n+m}}{(1 + i)^n - 1} \cdot P.$$

25. One dollar is deposited in a savings account with an interest rate of i per interest period. At the end of each interest period, the earned interest is withdrawn from the account and deposited into a second account earning the same rate of interest.

   (a) How much money will be in the first account after n interest periods?

   (b) How much money will be in the second account after n interest periods?

   (c) Since both the original deposit and the interest are all on deposit and earning interest throughout the entire n interest periods, the amounts in parts (a) and (b) must add up to $(1 + i)^n$. Use this fact to derive $s_{\overline{n}|i} = [(1 + i)^n - 1]/i$.

26. Show that $\dfrac{1}{a_{\overline{n}|i}} - \dfrac{1}{s_{\overline{n}|i}} = i$.

27. Show that $s_{\overline{n+1}|i} = (1 + i)s_{\overline{n}|i} + 1$.

28. One hundred dollars is deposited in the bank at the end of each year for 20 years. During the first 5 years, the bank paid 6% interest compounded annually and after that paid 7% interest compounded annually. Show that the balance after 20 years is $(1.07)^{15} \cdot 100s_{\overline{5}|6\%} + 100s_{\overline{15}|7\%}$.

29. A municipal bond pays 8% interest compounded semiannually on its face value of $5000. The interest is paid at the end of every half-year period. Fifteen years from now the face value of $5000 will be returned. The current market interest rate is 12% compounded semiannually. How much should you pay for the bond?

30. A businessman wishes to lend money for a second mortgage on some local real estate. Suppose that the mortgage pays $500 per month for a 5-year period. Suppose that you can invest your money in certificates of deposit paying 12% interest compounded monthly. How much should you offer for the mortgage?

31. Suppose that a business note for $50,000 carries an interest rate of 18% compounded monthly. Suppose that the business pays only interest for the first 5 years and then repays the loan amount plus interest in equal monthly installments for the next 5 years.

   (a) Calculate the monthly payments during the second 5-year period.

   (b) Assume that the current market interest rate for such loans is 12%. How much should you be willing to pay for such a note?

SOLUTIONS TO PRACTICE PROBLEMS 2

1. An ordinary annuity with $n = 60$, $i = 2\%$, and $R = 1$. You will make a deposit now, in the present, and then withdraw money each month. The amount of this deposit is the present value of the annuity.

$$P = a_{\overline{n}|i}R$$

$$= a_{\overline{60}|2\%} \cdot 1$$

$$= (34.76088668) \cdot 1 \qquad \text{(by Table 6)}$$

$$= \$34.76.$$

[*Notes:* (1) When $R = 1$, we have $P = a_{\overline{n}|i}$. This explains why $a_{\overline{n}|i}$ is often called the *present value of $1*. (2) For this transaction, the future value of the annuity has no significance. At the end of 5 years, the fund will have a balance of 0.]

2. Not an ordinary annuity since the payment period (1 week) is different from the interest period (1 month).

3. An ordinary annuity with $n = 24$, $i = 1.5\%$, and $R = \$10$. There is no money in the account now, in the present. However, in 2 years, in the future, money will have accumulated. So for this annuity, only the future value has significance.

$$F = s_{\overline{n}|i}R$$

$$= s_{\overline{24}|1.5\%} \cdot 10$$

$$= (28.63352080) \cdot 10$$

$$= \$286.34.$$

4.3. Amortization of Loans

In this section we analyze the mathematics of paying off loans. The loans we shall consider will all be repaid in a sequence of equal payments at regular time intervals, with the payment intervals coinciding with the interest periods. The process of paying off such a loan is called *amortization*. In order to obtain a feeling for the amortization process, let us consider a particular case, the amortization of a $563 loan to buy a color television set. Suppose that this loan charges interest at a 12% rate with interest compounded monthly and that the monthly payments are $116 for 5 months. The repayment process is summarized in the following chart.

Payment number	Amount	Interest	Applied to Principal	Unpaid balance
1	$116	$5.63	$110.37	$452.63
2	116	4.53	111.47	341.16
3	116	3.41	112.59	228.57
4	116	2.29	113.71	114.85
5	116	1.15	114.85	0.00

Note the following facts about the financial transactions above.

1. Payments are made at the end of each month. The payments have been carefully calculated to pay off the debt, with interest in the specified time interval.

2. Since $i = 1\%$, the interest to be paid each month is 1% of the unpaid balance at the end of the previous month. That is, 5.63 is 1% of 563, 4.53 is 1% of 452.63, and so on.

3. Although we write just one check each month for \$116, we regard part of the check as being applied to payment of that month's interest. The remainder, namely $116 -$ [interest], is regarded as being applied to repayment of part of the principal amount.

4. The unpaid balance at the end of each month is the previous unpaid balance minus the portion of the payment applied to the principal. A loan can be paid off early by just paying the current unpaid balance.

The four factors that describe the amortization process above are:

the principal	\$563
the interest rate	12% compounded monthly
the term	5 months
the monthly payment	\$116

The important fact to recognize is that the sequence of payments in the amortization above constitute an annuity, with the person taking out the loan paying the interest. Therefore, the mathematical tools developed in Section 2 suffice to analyze the amortization. In particular, we could determine the monthly payment or the principal once the other three factors have been specified.

EXAMPLE 1 Suppose that a loan has an interest rate of 12% compounded monthly and a term of 5 months.

(a) Given that the principal is \$563, calculate the monthly payment.

(b) Given that the monthly payment is \$116, calculate the principal.

(c) Given that the monthly payment is \$116, calculate the unpaid balance after 3 months.

Solution The sequence of payments constitute an annuity with the monthly payments as rent and the principal as the present value. Also, $n = 5$ and $i = 1\%$.

(a) Since $R = (1/a_{\overline{n}|i})P$, we see that

$$R = \frac{1}{a_{\overline{5}|1\%}} \cdot 563$$

$$= (.20603980) \cdot 563 \qquad \text{(by Table 7)}$$

$$= \$116.$$

(b) Since $P = a_{\overline{n}|i}R$, we see that

$$P = a_{\overline{5}|1\%} \cdot 116$$

$$= (4.85343124) \cdot 116$$

$$= \$563.00.$$

(c) The unpaid balance is most easily calculated by regarding it as the amount necessary to retire the debt. Therefore, it must be sufficient to generate, with interest, the sequence of two remaining payments. That is, the unpaid balance is the present value of an annuity of two payments of $116. So we see that

$$[\text{unpaid balance after 3 months}] = a_{\overline{2}|1\%} \cdot 116$$

$$= (1.97039506) \cdot 116$$

$$= \$228.57.$$

A mortgage is a long-term loan used to purchase real estate. The real estate is used as collateral to guarantee the loan.

EXAMPLE 2 On December 31, 1980, a house is purchased with the buyer taking out a 30-year $60,050 mortgage at 18% interest, compounded monthly. The mortgage payments are made at the end of each month.

(a) Calculate the amount of the monthly payment.

(b) Calculate the unpaid balance of the loan on December 31, 2006, just after the 312th payment.

(c) How much interest will be paid during the month of January 2007?

(d) How much of the principal will be paid off during the year 2006?

(e) How much interest will be paid during the year 2006?

Solution (a) Denote the monthly payment by R. Then $60,050 is the present value of an annuity of $n = 360$ payments, with $i = 1.5\%$. Therefore,

$$R = \frac{1}{a_{\overline{n}|i}} P$$

$$= \frac{1}{a_{\overline{360}|1.5\%}} \cdot 60{,}050$$

$$= (.01507085) \cdot 60{,}050 \qquad \text{(by Table 7)}$$

$$= \$905.00.$$

(b) The remaining payments constitute an annuity of 48 payments. Therefore, the unpaid balance is the present value of that annuity.

$$[\text{unpaid balance}] = a_{\overline{n}|i} R$$

$$= a_{\overline{48}|1.5\%} \cdot 905$$

$$= (34.04255365) \cdot 905 \qquad \text{(by Table 6)}$$

$$= \$30{,}808.51$$

(c) The interest paid during 1 month is i, the interest rate per month, times the unpaid balance at the end of the preceding month. Therefore,

$$[\text{interest for January 2007}] = 1.5\% \text{ of } \$30,808.51$$

$$= (.015) \cdot 30,808.51$$

$$= \$462.13.$$

(d) Since the portions of the monthly payments applied to repay the principal have the effect of reducing the unpaid balance, this question may be answered by calculating how much the unpaid balance will be reduced during 2006. Reasoning as in part (b), we determine that the unpaid balance on December 31, 2005 (just after the 300th payment), is equal to \$35,639.14. Therefore,

$$[\text{amount of principal repaid in 2006}]$$

$$= [\text{unpaid balance Dec. 31, 2005}] - [\text{unpaid balance Dec. 31, 2006}]$$

$$= \$35,639.14 - \$30,808.51 = \$4830.63.$$

(e) During the year 2006, the total amount paid is $12 \times 905 = 10,860$ dollars. But by part (d), \$4830.63 is applied to repayment of principal, the remainder being applied to interest.

$$[\text{interest in 2006}] = [\text{total amount paid}] - [\text{principal repaid in 2006}]$$

$$= 10,860 - 4830.63$$

$$= \$6029.37.$$

Note that in the early years of a mortgage, most of each monthly payment is applied to interest. For the mortgage above, the interest portion will exceed the principal portion until the twenty-sixth year.

We can easily derive a formula which illuminates exactly how the unpaid balance of a mortgage changes from period to period. Denote the current unpaid balance by B_{cur}. Then

$$B_{\text{next}} = B_{\text{cur}} - (R - [\text{interest}]_{\text{next}})$$

$$= B_{\text{cur}} - (R - iB_{\text{cur}})$$

$$= B_{\text{cur}} + iB_{\text{cur}} - R$$

$$= (1 + i)B_{\text{cur}} - R. \tag{1}$$

Successive unpaid balances are computed by multiplying by $1 + i$ and subtracting R.

EXAMPLE 3   Refer to Example 2. Compute the unpaid balance of the loan on January 31, 2007, just after the 313th payment.

Solution By formula (1),

$$B_{\text{Jan}} = (1 + i)B_{\text{Dec}} - R$$

$$= (1.015) \cdot 6029.37 - 905$$

$$= \$5214.81.$$

Sometimes amortized loans stipulate a *balloon payment* at the end of the term. For instance, you might pay $200 at the end of each quarter year for 3 years and an additional $1000 at the end of the third year. The $1000 is a balloon payment.

EXAMPLE 4 How much money can you borrow at 24% interest compounded quarterly if you agree to pay $200 at the end of each quarter-year for 3 years and in addition a balloon payment of $1000 at the end of the third year?

Solution Here you are borrowing in the present and repaying in the future. The amount of the loan will be the present value of *all* the future payments. The future payments consist of an annuity and a lump-sum payment. Let us calculate the present values of each of these separately. Now, $i = 6\%$ and $n = 12$.

$$[\text{present value of annuity}] = a_{\overline{n}|i}R$$

$$= a_{\overline{12}|6\%} \cdot 200$$

$$= (8.38384394) \cdot 200 \qquad \text{(by Table 6)}$$

$$= \$1676.77.$$

$$[\text{present value of balloon payment}] = \frac{1}{(1 + i)^n} \cdot F$$

$$= \frac{1}{(1 + i)^n} \cdot 1000 \qquad \text{(for } i = 6\%, n = 12)$$

$$= .49696936 \cdot (1000)$$

$$= \$496.97.$$

Therefore, the amount you can borrow is $1676.77 + $496.97 = $2173.74.

PRACTICE PROBLEMS 3

1. The word "amortization" comes from the French "a mort" meaning "at the point of death." Justify the word.

2. Explain why only present values and not future values arise in amortization problems.

EXERCISES 3

1. A loan of $10,000 is to be repaid with monthly payments for 5 years at 24% interest compounded monthly. Calculate the monthly payment.

2. Find the monthly payment on a $100,000 25-year mortgage at 12% interest compounded monthly.

3. How much money can you borrow at 12% interest compounded semiannually if the loan is to be repaid at half-year intervals for 10 years and you can afford to pay $1000 per half-year?

4. You buy a car with a down payment of $500 and $100 per month for 3 years. If the interest rate is 18% compounded monthly, how much did the car cost?

5. Consider a $58,331 30-year mortgage at interest rate 12% compounded monthly with a $600 monthly payment.

   (a) How much interest is paid the first month?

   (b) How much of the first month's payment is applied to paying off the principal?

   (c) What is the unpaid balance after 1 month?

   (d) What is the unpaid balance at the end of 25 years?

   (e) How much of the principal is repaid during the twenty-sixth year?

   (f) How much interest is paid during the 301st month?

6. Consider a $13,406.16 loan for 7 years at 24% interest compounded quarterly and a payment of $1000 per quarter-year.

   (a) Compute the unpaid balance after 5 years.

   (b) How much interest is paid during the fifth year?

   (c) How much principal is repaid in the first payment?

   (d) What is the total amount of interest paid on the loan?

7. A mortgage at 15% interest compounded monthly with a monthly payment of $1125 has an unpaid balance of $10,000 after 350 months. Find the unpaid balance after 351 months.

8. A loan with a quarterly payment of $1500 has an unpaid balance of $10,000 after 30 quarters and an unpaid balance of $9000 after 31 quarters. If interest is compounded quarterly, find the interest rate.

9. A loan is to be amortized over an 8-year term at 12% interest compounded semiannually, payments of $1000 every 6 months, and a balloon payment of $10,000 at the end of the term. Calculate the amount of the loan.

10. A loan of $105,504.50 is to be amortized over a 5-year term at 12% interest compounded monthly with monthly payments and a $10,000 balloon payment at the end of the term. Calculate the monthly payment.

11. Write out a complete amortization schedule (as on page 130) for the amortization of a $1000 loan with monthly payments at 12% interest compounded monthly for 4 months.

12. Write out a complete schedule (as on page 130) for the amortization of a $10,000 loan with payments every 6 months at 12% interest compounded semiannually for 1 year.

13. You purchase a $120,000 house, pay $20,000 down, and take out a 30-year mortgage with monthly payments, at an interest rate of 24% compounded monthly. How much money will you be paying each month?

14. In 1970, you purchased a house and took out a 25-year $50,000 mortgage at 6% interest compounded monthly. In 1980, you sold the house for $150,000. How much money did you have left after you paid the bank the unpaid balance on the mortgage?

15. You are considering the purchase of a condominium to use as a rental property. You estimate that you can rent the condominium for $1200 per month and that taxes, insurance, and maintenance costs will run about $200 per month. If interest rates are 18% compounded monthly, how large a 25-year mortgage can you assume and still have the rental income cover the monthly expenses?

16. Consider formula (1). Derive an analogous formula for the balance of an annuity for which regular payments are deposited into a savings account and accrue with interest.

17. A car is purchased for $5548.88 with $2000 down and a loan to be repaid at $100 a month for 3 years followed by a balloon payment. If the interest rate is 24% compounded monthly, how large will the balloon payment be?

18. A real estate speculator purchases a tract of land for $1 million and assumes a 25-year mortgage at 18% interest compounded monthly.

    (a) What is his monthly payment?

    (b) Suppose that at the end of 5 years the mortgage is changed to a 10-year term for the remaining balance. What is the new monthly payment?

    (c) Suppose that after 5 more years, the mortgage is required to be repaid in full. How much will then be due?

19. Suppose that you make annual payments of $5000 for 20 years into an annuity paying 6% interest compounded annually.

    (a) What is the value of the annuity at the end of the twentieth year?

    (b) Suppose that you elect to have your annuity repaid to you over a 10-year period in annual installments. What is the annual payment you will receive?

    (c) Suppose that after you receive payments for 5 years you elect to have the remainder of your annuity paid to you in a lump sum. How much will you receive?

A *sinking fund* is a pool of money accumulated by a corporation or government to repay a specific debt at some future date.

20. A corporation wishes to deposit money into a sinking fund at the end of each half year in order to repay $50 million in bonds in 10 years. It can expect to receive a 12% (compounded semiannually) return on its deposits to the sinking fund. How much should the deposits be?

21. The Federal National Mortgage Association ("Fannie Mae") puts $30 million dollars at the end of each year into a sinking fund paying 12% interest compounded monthly. The sinking fund is to be used to repay debentures that mature 15 years from the creation

of the fund. How large is the face amount of the debentures assuming the sinking fund will exactly pay them off?

22. A corporation borrows $5 million to erect a new headquarters. The financing is arranged using industrial development bonds, to be repaid in 20 years. How much should it deposit into a sinking fund at the end of each quarter if the sinking fund earns 8% interest compounded quarterly?

23. A corporation sets up a sinking fund to replace some aging machinery. It deposits $100,000 into the fund at the end of each month for 10 years. The sinking fund earns 12% interest compounded monthly. The equipment originally cost $6 million. However, the cost of the equipment is rising 6% each year. Will the sinking fund be adequate to replace the equipment? If not, how much additional money is needed?

24. A corporation sets up a sinking fund to replace an aging warehouse. The cost of the warehouse today would be $8 million. However, the corporation plans to replace the warehouse in 5 years. It estimates that the cost of the warehouse will increase 10% annually. The sinking fund will earn 12% interest compounded monthly. What should be the monthly payments to the sinking fund?

SOLUTIONS TO PRACTICE PROBLEMS 3

1. A portion of each payment is applied to reducing the debt and by the end of the term, the debt is totally annihilated.

2. The debt is formed when the creditor gives you a lump sum of money now, in the present. The lump sum of money is gradually repaid by you, with interest, thereby generating the annuity. At the end of the term, in the future, the loan is totally paid off, so there is no more debt. That is, the future value is always zero!

Remark You are actually functioning like a savings bank, since you are paying the interest. Think of the creditor as depositing the lump sum with you and then making regular withdrawals until the balance is 0.

Chapter 4: CHECKLIST

☐ Interest
☐ Principal
☐ Compound amount
☐ Compound interest
☐ Interest per period
☐ Present value
☐ Simple interest
☐ Annuity
☐ $s_{\overline{n}|i}$
☐ $a_{\overline{n}|i}$
☐ Future value of an annuity
☐ Present value of an annuity
☐ Geometric progression

□ Rent of an annuity
□ Amortization
□ Mortgage
□ Balloon payment

Chapter 4: SUPPLEMENTARY EXERCISES

1. Mr. West wishes to purchase a condominium for $80,000 cash upon his retirement 10 years from now. How much should he deposit at the end of each month into an annuity paying 12% interest compounded monthly in order to accumulate the required savings?

2. What is the monthly payment on a $150,000 30-year mortgage at 18% interest?

3. The income of a typical family in a certain city is currently $19,200 per year. Family finance experts recommend that mortgage payments not exceed 25% of a family's income. Assuming a current mortgage interest rate of 18% for a 30-year mortgage with monthly payments, how large a mortgage can the typical family in that city afford?

4. Calculate the compound amount of $50 after a year if deposited at 7.3% compounded daily.

5. Which is a better investment: 10% compounded annually or $9\frac{1}{8}$% compounded daily? [*Useful fact:* $(1.00025)^{365} = 1.09553$.]

6. Ms. Smith deposits $200 per month into a bond fund yielding 12% interest compounded semiannually. How much are her holdings worth after 5 years?

7. A real estate investor takes out a $200,000 mortgage subject to the following terms: For the first 5 years, the payments will be the same as the monthly payments on a 15-year mortgage at 12% interest compounded monthly. The balance will then be payable in full.

   (a) What are the monthly payments for the first 5 years?

   (b) What balance will be owed after 5 years?

8. College expenses at a private college currently average $12,000 per year. It is estimated that these expenses are increasing at the rate of 1% per month. What is the estimated cost of a year of college 10 years from now?

9. What is the present value of $50,000 in 10 years at 12% interest compounded monthly?

10. An investment will pay $10,000 in 2 years and $5,000 in 3 years. If the current market interest rate is 12% compounded monthly, what should a rational person be willing to pay for the investment?

11. A woman purchases a car for $12,000. She pays $3,000 as a down payment and finances the remaining amount at 24% interest compounded monthly for 4 years. What is her monthly car payment?

12. A businessman buys a $100,000 piece of manufacturing equipment on the following terms: Interest will be charged at a rate of 12% compounded semiannually, but no payments will be made until 2 years after purchase. Beginning at that time equal semiannual payments will be made for 5 years. Determine the semiannual payment.

13. A retired person has set aside a fund of $105,003.50 for his retirement. This fund is in a bank account paying 12% interest compounded monthly. How much can he draw out of the account at the end of each month so that there is a balance of $30,000 at the end of 15 years? [*Hint:* First compute the present value of the $30,000.]

14. A business loan of $509,289.22 is to be paid off in monthly payments for 10 years with a $100,000 balloon payment at the end of the tenth year. The interest rate on the loan is 24% compounded monthly. Calculate the monthly payment.

15. Ms. Jones saves $100 per month for 30 years at 12% interest compounded monthly. How much are her accumulated savings worth?

16. An apartment building is currently generating an income of $2000 per month. Its owners are considering a 10-year second mortgage at 18% interest compounded monthly in order to pay for repairs. How large a second mortgage can the income of the apartment house support?

17. Investment A generates $1000 at the end of each year for 10 years. Investment B generates $5000 at the end of the fifth year and $5000 at the end of the tenth year. Assume a market rate of interest of 6% compounded annually. Which is the better investment?

18. A 5-year bond has a face value of $1000 and is currently selling for $800. The bond pays $10 interest at the end of each month and, in addition, will repay the $1000 face value at the end of the fifth year. The market rate of interest is currently 18% compounded monthly. Is the bond a bargain? Why or why not? [*Note:* $1.015^{-60} = .40930.$]

19. A person makes an initial deposit of $10,000 into a savings account and then deposits $1000 at the end of each quarter-year for 15 years. If the interest rate is 8% compounded quarterly, how much money will be in the account after 15 years?

20. A $7000 car loan at 24% interest compounded monthly is to be repaid with 36 equal monthly payments. Write out an amortization schedule for the first 6 months of the loan.

21. A person pays $200 at the end of each month for 10 years into an account paying 1% interest per month compounded monthly. At the end of the tenth year, the payments cease, but the balance continues to earn interest. What is the value of the balance at the end of the twentieth year?

22. A savings fund currently contains $300,000. It is decided to pay out this amount with 12% interest compounded monthly over a 5-year period. What are the monthly payments?

23. Calculate the following sum:

$$1 + 3 + 3^2 + 3^3 + 3^4 + \cdots + 3^{11}.$$

[*Hint:* $3^{12} = 531,441.$]

24. A small loan company makes a $1000 loan at 2.5% simple interest per month. The monthly payments are $75. What is the balance immediately after the tenth payment?

Matrices

Chapter 5

We begin this chapter by developing a method for solving systems of linear equations in any number of variables. Our discussion of this method will lead naturally into the study of mathematical objects called *matrices*. The arithmetic and applications of matrices are the main topics of the chapter. We will discuss in detail the application of matrix arithmetic to input-output analysis, which can be (and is) used to make production decisions for large businesses and entire economies.

5.1. Solving Systems of Linear Equations, I

In an earlier chapter we presented a method for solving systems of linear equations in two variables. The method of Chapter 2 is very efficient for determining the solutions. Unfortunately, it works only for systems of linear equations having *two* variables. In many applications we meet systems having more than two variables, as the following example illustrates.

EXAMPLE 1 A bacteria culture contains three species of bacteria. Each species requires certain amounts of three different nutrients: a nitrogen source, a carbon source, and a phosphate source. The daily requirements (expressed in appropriate units) of each nutrient are summarized in the accompanying chart.

	Nutrient		
Species	Nitrogen source	Carbon source	Phosphate source
A	1 unit/day	3 units/day	4 units/day
B	2 units/day	5 units/day	7 units/day
C	2 units/day	5 units/day	8 units/day

Every day the culture is supplied with 15,000 units of a nitrogen source, 40,000 units of a carbon source, and 59,000 units of a phosphate source. Assume that all the nutrients are utilized. How many of each species of bacteria can the culture simultaneously support?

Solution Let us express the given data by a system of equations. Let x be the number of bacteria of species A, y the number of bacteria of species B, and z the number of bacteria of species C. The first piece of data is that there are 15,000 units per day of the nitrogen source. Therefore, we must have

[amount of nitrogen source for species A]

$$+ \text{[amount of nitrogen source for species B]}$$

$$+ \text{[amount of nitrogen source for species C]} = 15,000.$$

Thus, from the first column of the chart, we have

$$1 \cdot \text{[number of species A bacteria]} + 2 \cdot \text{[number of species B bacteria]}$$

$$+ 2 \cdot \text{[number of species C bacteria]} = 15,000.$$

That is,

$$x + 2y + 2z = 15,000.$$

In a similar way, we derive equations for carbon and phosphate:

$$3x + 5y + 5z = 40,000$$

$$4x + 7y + 8z = 59,000.$$

So we see that the numbers x, y, z of the three species in the culture must simultaneously satisfy these linear equations in three variables:

$$\begin{cases} x + 2y + 2z = 15,000 \\ 3x + 5y + 5z = 40,000 \\ 4x + 7y + 8z = 59,000. \end{cases} \tag{1}$$

Later we shall give a method for determining the solution of this system. It will be easy to show that the only solution is $x = 5000$, $y = 1000$, $z = 4000$. Once we are given the solution, it is easy to check that these values of x, y, z make all three equations true:

$$5000 + 2 \cdot 1000 + 2 \cdot 4000 = 15,000$$

$$3 \cdot 5000 + 5 \cdot 1000 + 5 \cdot 4000 = 40,000$$

$$4 \cdot 5000 + 7 \cdot 1000 + 8 \cdot 4000 = 59,000.$$

Thus the culture can support 5000 bacteria of species A, 1000 bacteria of species B, and 4000 bacteria of species C.

In this section we develop a step-by-step procedure for solving systems of linear equations such as (1). The procedure, called the *Gaussian elimination method*, consists of repeatedly simplifying the system, using so-called elementary row operations, until the solution stares us in the face!

In the system of linear equations (1) the equations have been written in such a way that the *x*-terms, the *y*-terms, and the *z*-terms lie in different columns. We shall always be careful to display systems of equations with separate columns for each variable. One of the key ideas of the Gaussian elimination method is to think of the solution as a system of linear equations in its own right. For example, we can write the solution of the system (1) as:

$$\begin{cases} x & & = 5000 \\ & y & = 1000 \\ & & z = 4000. \end{cases} \tag{2}$$

This is just a system of linear equations in which the coefficients of most terms are zero! Since the only terms with nonzero coefficients are arranged on a diagonal, such a system is said to be in *diagonal form*.

Our method for solving a system of linear equations consists of repeatedly using three operations which alter the system but do not change the solutions. The operations are used to transform the system into a system in diagonal form. Since the operations involve only elementary arithmetic and are applied to entire equations (i.e., rows of the system), they are called *elementary row operations*. Let us begin our study of the Gaussian elimination method by introducing these operations.

> **Elementary Row Operation 1**   Rearrange the equations in any order.

This operation is harmless enough. It certainly does not change the solutions of the system.

> **Elementary Row Operation 2**   Multiply an equation by a nonzero number.

For example, if we are given the system of linear equations

$$\begin{cases} 2x - 3y + 4z = 11 \\ 4x - 19y + z = 31 \\ 5x + 7y - z = 12, \end{cases}$$

then we may replace it by a new system obtained by leaving the last two equations unchanged and multiplying the first equation by 3. To accomplish this, multiply each term of the first equation by 3. The transformed system is

$$\begin{cases} 6x - 9y + 12z = 33 \\ 4x - 19y + z = 31 \\ 5x + 7y - z = 12. \end{cases}$$

The operation of multiplying an equation by a nonzero number does not change the solutions of the system. For if a particular set of values of the variables satisfy the original equation, they satisfy the resulting equation, and vice versa.

Elementary row operation 2 may be used to make the coefficient of a particular variable 1.

EXAMPLE 2 Replace the system

$$\begin{cases} -5x + 10y + 20z = 4 \\ x \qquad - 12z = 1 \\ x + y + z = 0 \end{cases}$$

by an equivalent system in which the coefficient of x in the first equation is 1.

Solution The coefficient of x in the first equation is -5, so we use elementary row operation 2 to multiply the first equation by $-\frac{1}{5}$. Multiplying each term of the first equation by $-\frac{1}{5}$ gives

$$\begin{cases} x - 2y - 4z = -\frac{4}{5} \\ x \qquad - 12z = 1 \\ x + y + z = 0. \end{cases}$$

Another operation that can be performed on a system without changing its solutions is to replace one equation by its sum with some other equation. For example, consider this system of equations:

$$A: \begin{cases} x + y - 2z = 3 \\ x + 2y - 5z = 4 \\ 5x + 8y - 18z = 14. \end{cases}$$

We can replace the second equation by the sum of the first and the second. Since

$$\begin{array}{r} x + y - 2z = 3 \\ + \quad x + 2y - 5z = 4 \\ \hline 2x + 3y - 7z = 7, \end{array}$$

the resulting system is

$$B: \begin{cases} x + y - 2z = 3 \\ 2x + 3y - 7z = 7 \\ 5x + 8y - 18z = 14. \end{cases}$$

If a particular choice of x, y, z satisfies system A, it also satisfies system B. This is because system B results from adding equations. Similarly, system A can be derived from system B by subtracting equations. So any particular solution of system A is a solution of system B, and vice versa.

The operation of adding equations is usually used in conjunction with elementary row operation 2. That is, an equation is changed by adding to it a nonzero multiple of another equation. For example, consider the system

$$\begin{cases} x + y - 2z = 3 \\ x + 2y - 5z = 4 \\ 5x + 8y - 18z = 14. \end{cases}$$

Let us change the second equation by adding to it twice the first. Since

$$\begin{array}{ll} 2(\text{first}) & 2x + 2y - 4z = 6 \\ + \ (\text{second}) & x + 2y - 5z = 4 \\ \hline & 3x + 4y - 9z = 10, \end{array}$$

the new second equation is

$$3x + 4y - 9z = 10$$

and the transformed system is

$$\begin{cases} x + y - 2z = 3 \\ 3x + 4y - 9z = 10 \\ 5x + 8y - 18z = 14. \end{cases}$$

Since addition of equations and elementary row operation 2 are often used together, let us define a third elementary row operation to be the combination:

Elementary Row Operation 3   Change an equation by adding to it a multiple of another equation.

For reference, let us summarize the elementary row operations we have just defined.

Elementary Row Operations

1. Rearrange the equations in any order.
2. Multiply an equation by a nonzero number.
3. Change an equation by adding to it a multiple of another equation.

The idea of the Gaussian elimination method is to transform an arbitrary system of linear equations into diagonal form by repeated application of the three elementary row operations. To see how the method works, consider the following example.

EXAMPLE 3 Solve the following system by the Gaussian elimination method.

$$\begin{cases} x - 3y = 7 \\ -3x + 4y = -1. \end{cases}$$

Solution Let us transform this system into diagonal form by examining one column at a time, starting from the left. Examine the first column:

The coefficient of the top x is 1, which is exactly what it should be for the system to be in diagonal form. So we do nothing to this term. Now examine the next term in the column, $-3x$. In diagonal form this term must be absent. In order to accomplish this we add a multiple of the first equation to the second. Since the coefficient of x in the second is -3, we add three times the first equation to the second equation in order to cancel the x-term. (Abbreviation: $[2] + 3[1]$. The $[2]$ means that we are changing equation 2. The expression $[2] + 3[1]$ means that we are replacing equation 2 by the original equation plus three times equation 1.)

$$\begin{cases} x - 3y = 7 \\ -3x + 4y = -1 \end{cases} \xrightarrow{\ [2] + 3[1]\ } \begin{cases} x - 3y = 7 \\ \quad\ - 5y = 20. \end{cases}$$

The first column now has the proper form, so we proceed to the second column. In diagonal form that column will have one nonzero term, namely the second, and the coefficient of y in that term must be 1. To bring this about, multiply the second equation by $-\frac{1}{5}$ (abbreviation: $-\frac{1}{5}[2]$):

$$\begin{cases} x - 3y = 7 \\ \quad\ - 5y = 20 \end{cases} \xrightarrow{\ -\frac{1}{5}[2]\ } \begin{cases} x - 3y = 7 \\ \qquad y = -4. \end{cases}$$

The second column still does not have the correct form. We must get rid of the $-3y$ term in the first equation. We do this by adding a multiple of the second equation to the first. Since the coefficient of the term to be canceled is -3, we add three times the second equation to the first:

$$\begin{cases} x - 3y = 7 \\ \qquad y = -4 \end{cases} \xrightarrow{\ [1] + 3[2]\ } \begin{cases} x \qquad = -5 \\ \qquad y = -4. \end{cases}$$

The system is now in diagonal form and the solution can be read off: $x = -5$, $y = -4$.

EXAMPLE 4 Use the Gaussian elimination method to solve the system

$$\begin{cases} 2x - 6y = -8 \\ -5x + 13y = 1. \end{cases}$$

Solution We can perform the calculations in a mechanical way, proceeding column by column from the left:

$$\begin{cases} 2x - 6y = -8 \\ -5x + 13y = 1 \end{cases} \xrightarrow{\frac{1}{2}[1]} \begin{cases} x - 3y = -4 \\ -5x + 13y = 1 \end{cases}$$

$$\xrightarrow{[2] + 5(1)} \begin{cases} x - 3y = -4 \\ -2y = -19 \end{cases}$$

$$\xrightarrow{-\frac{1}{2}[2]} \begin{cases} x - 3y = -4 \\ y = \frac{19}{2} \end{cases}$$

$$\xrightarrow{[1] + 3[2]} \begin{cases} x = \frac{49}{2} \\ y = \frac{19}{2}. \end{cases}$$

So the solution of the system is $x = \frac{49}{2}$, $y = \frac{19}{2}$.

The calculation becomes easier to follow if we omit writing down the variables at each stage and work only with the coefficients. At each stage of the computation the system is represented by a rectangular array of numbers. For instance, the original system is written*

$$\left[\begin{array}{cc|c} 2 & -6 & -8 \\ -5 & 13 & 1 \end{array}\right].$$

The elementary row operations are performed on the rows of this rectangular array just as if the variables were there. So, for example, the first step above is to multiply the first equation by $\frac{1}{2}$. This corresponds to multiplying the first row of the array by $\frac{1}{2}$ to get

$$\left[\begin{array}{cc|c} 1 & -3 & -4 \\ -5 & 13 & 1 \end{array}\right].$$

The diagonal form just corresponds to the array

$$\left[\begin{array}{cc|c} 1 & 0 & \frac{49}{2} \\ 0 & 1 & \frac{19}{2} \end{array}\right].$$

Note that this array has ones down the diagonal and zeros everywhere else on the left. The solution of the system appears on the right.

A rectangular array of numbers is called a *matrix* (plural: *matrices*). In the next example, we use matrices to carry out the Gaussian elimination method.

---

* The vertical line between the second and third columns is a placemarker which separates the data obtained from the right- and left-hand sides of the equations. It is inserted for visual convenience.

EXAMPLE 5 Use the Gaussian elimination method to solve the system

$$\begin{cases} 3x - 6y + 9z = 0 \\ 4x - 6y + 8z = -4 \\ -2x - y + z = 7. \end{cases}$$

Solution The initial array corresponding to the system is

$$\left[\begin{array}{ccc|c} 3 & -6 & 9 & 0 \\ 4 & -6 & 8 & -4 \\ -2 & -1 & 1 & 7 \end{array} \right].$$

We must use elementary row operations to transform this array into diagonal form—that is, with ones down the diagonal and zeros everywhere else on the left:

$$\left[\begin{array}{ccc|c} 1 & 0 & 0 & * \\ 0 & 1 & 0 & * \\ 0 & 0 & 1 & * \end{array} \right].$$

We proceed one column at a time.

$$\left[\begin{array}{ccc|c} 3 & -6 & 9 & 0 \\ 4 & -6 & 8 & -4 \\ -2 & -1 & 1 & 7 \end{array} \right] \xrightarrow{\frac{1}{3}[1]} \left[\begin{array}{ccc|c} 1 & -2 & 3 & 0 \\ 4 & -6 & 8 & -4 \\ -2 & -1 & 1 & 7 \end{array} \right] \xrightarrow{[2]+(-4)[1]}$$

$$\left[\begin{array}{ccc|c} 1 & -2 & 3 & 0 \\ 0 & 2 & -4 & -4 \\ -2 & -1 & 1 & 7 \end{array} \right] \xrightarrow{[3]+2[1]} \left[\begin{array}{ccc|c} 1 & -2 & 3 & 0 \\ 0 & 2 & -4 & -4 \\ 0 & -5 & 7 & 7 \end{array} \right] \xrightarrow{\frac{1}{2}[2]}$$

$$\left[\begin{array}{ccc|c} 1 & -2 & 3 & 0 \\ 0 & 1 & -2 & -2 \\ 0 & -5 & 7 & 7 \end{array} \right] \xrightarrow{[1]+2[2]} \left[\begin{array}{ccc|c} 1 & 0 & -1 & -4 \\ 0 & 1 & -2 & -2 \\ 0 & -5 & 7 & 7 \end{array} \right] \xrightarrow{[3]+5[2]}$$

$$\left[\begin{array}{ccc|c} 1 & 0 & -1 & -4 \\ 0 & 1 & -2 & -2 \\ 0 & 0 & -3 & -3 \end{array} \right] \xrightarrow{(-\frac{1}{3})[3]} \left[\begin{array}{ccc|c} 1 & 0 & -1 & -4 \\ 0 & 1 & -2 & -2 \\ 0 & 0 & 1 & 1 \end{array} \right] \xrightarrow{[1]+1[3]}$$

$$\left[\begin{array}{ccc|c} 1 & 0 & 0 & -3 \\ 0 & 1 & -2 & -2 \\ 0 & 0 & 1 & 1 \end{array} \right] \xrightarrow{[2]+2[3]} \left[\begin{array}{ccc|c} 1 & 0 & 0 & -3 \\ 0 & 1 & 0 & 0 \\ 0 & 0 & 1 & 1 \end{array} \right].$$

The last array is in diagonal form, so we just put back the variables and read off the solution:

$$x = -3, \qquad y = 0, \qquad z = 1.$$

Because so much arithmetic has been performed, it is a good idea to check the solution by substituting the values for x, y, z into each of the equations of the original system. This will uncover any arithmetic errors that may have occurred.

$$\begin{cases} 3x - 6y + 9z = 0 \\ 4x - 6y + 8z = -4 \\ -2x - y + z = 7 \end{cases} \qquad \begin{cases} 3(-3) - 6(0) + 9(1) = 0 \\ 4(-3) - 6(0) + 8(1) = -4 \\ -2(-3) - (0) + (1) = 7 \end{cases}$$

$$\begin{cases} -9 - 0 + 9 = 0 \\ -12 - 0 + 8 = -4 \\ 6 - 0 + 1 = 7 \end{cases}$$

$$\begin{cases} 0 = 0 \\ -4 = -4 \\ 7 = 7. \end{cases}$$

So we have indeed found a solution of the system.

Remark Note that so far we have not had to use elementary row operation 1, which allows interchange of equations. But in some examples it is definitely needed. Consider this system:

$$\begin{cases} y + z = 0 \\ 3x - y + z = 6 \\ 6x \qquad - z = 3. \end{cases}$$

The first step of the Gaussian elimination method consists of making the x-coefficient 1 in the first equation. But we cannot do this, since the first equation does not involve x. To remedy this difficulty, just interchange the first two equations to guarantee that the first equation involves x. Now proceed as before. Of course, in terms of the matrix of coefficients, interchanging equations corresponds to interchanging rows of the matrix.

PRACTICE PROBLEMS 1

1. Determine whether the following systems of linear equations are in diagonal form.

   (a) $\begin{cases} x \qquad + z = 3 \\ y \qquad = 2 \\ z = 7 \end{cases}$ (b) $\begin{cases} x \qquad = 3 \\ y = 5 \\ z \qquad = 7 \end{cases}$ (c) $\begin{cases} x \qquad = -1 \\ y \qquad = 0 \\ 3z = 4 \end{cases}$

2. Perform the indicated elementary row operation.

   (a) $\begin{cases} x - 3y = 2 \\ 2x + 3y = 5 \end{cases} \xrightarrow{[2] + (-2)[1]}$ (b) $\begin{cases} x + y = 3 \\ -x + 2y = 5 \end{cases} \xrightarrow{[2] + (1)[1]}$

3. State the next elementary row operation which should be performed when applying the Gaussian elimination method.

   (a) $\begin{bmatrix} 0 & 2 & 4 & | & 1 \\ 0 & 3 & -7 & | & 0 \\ 3 & 6 & -3 & | & 3 \end{bmatrix}$ (b) $\begin{bmatrix} 1 & -3 & 4 & | & 5 \\ 0 & 2 & 3 & | & 4 \\ -6 & 5 & -7 & | & 0 \end{bmatrix}$

EXERCISES 1

In Exercises 1–8, perform the indicated elementary row operations and give their abbreviations.

1. Operation 2: multiply the first equation by 2.
$$\begin{cases} \frac{1}{2}x - 3y = 2 \\ 5x + 4y = 1. \end{cases}$$

2. Operation 2: multiply the second equation by -1.
$$\begin{cases} x + 4y = 6 \\ \quad -y = 2. \end{cases}$$

3. Operation 3: change the second equation by adding to it 5 times the first equation.
$$\begin{cases} x + 2y = 3 \\ -5x + 4y = 1. \end{cases}$$

4. Operation 3: change the second equation by adding to it $(-\frac{1}{2})$ times the first equation.
$$\begin{cases} x - 6y = 4 \\ \frac{1}{2}x + 2y = 1. \end{cases}$$

5. Operation 3: change the third equation by adding to it (-4) times the first equation.
$$\begin{cases} x - 2y + \quad z = 0 \\ \qquad y - 2z = 4 \\ 4x + \quad y + 3z = 5. \end{cases}$$

6. Operation 3: change the third equation by adding to it 3 times the second equation.
$$\begin{cases} x + 6y - 4z = 1 \\ \qquad y + 3z = 1 \\ \quad -3y + 7z = 2. \end{cases}$$

7. Operation 3: change the first row by adding to it $\frac{1}{2}$ times the second row.
$$\left[\begin{array}{cc|c} 1 & -\frac{1}{2} & 3 \\ 0 & 1 & 4 \end{array}\right].$$

8. Operation 3: change the third row by adding to it (-4) times the second row.
$$\left[\begin{array}{ccc|c} 1 & 0 & 7 & 9 \\ 0 & 1 & -2 & 3 \\ 0 & 4 & 8 & 5 \end{array}\right].$$

In Exercises 9–16, state the next elementary row operation which must be performed in order to put the matrix into diagonal form. Do not perform the operation.

9. $\left[\begin{array}{cc|c} 1 & -5 & 1 \\ -2 & 4 & 6 \end{array}\right]$

10. $\left[\begin{array}{cc|c} 1 & 3 & 4 \\ 0 & 2 & 6 \end{array}\right]$

11. $\left[\begin{array}{cc|c} 1 & 2 & 3 \\ 0 & 1 & 4 \end{array}\right]$

12. $\begin{bmatrix} 1 & -2 & 5 & | & 7 \\ 0 & -3 & 6 & | & 9 \\ 4 & 5 & -6 & | & 7 \end{bmatrix}$

13. $\begin{bmatrix} 0 & 5 & -3 & | & 6 \\ 2 & -3 & 4 & | & 5 \\ 4 & 1 & -7 & | & 8 \end{bmatrix}$

14. $\begin{bmatrix} 1 & 4 & -2 & | & 5 \\ 0 & -3 & 6 & | & 9 \\ 0 & 4 & 3 & | & 1 \end{bmatrix}$

15. $\begin{bmatrix} 1 & 0 & 3 & | & 4 \\ 0 & 1 & 2 & | & 5 \\ 0 & 0 & 1 & | & 6 \end{bmatrix}$

16. $\begin{bmatrix} 1 & 2 & 4 & | & 5 \\ 0 & 0 & 3 & | & 6 \\ 0 & 1 & 1 & | & 7 \end{bmatrix}$

Solve the following linear systems by using the Gaussian elimination method.

17. $\begin{cases} 3x + 9y = 6 \\ 2x + 8y = 6 \end{cases}$

18. $\begin{cases} \frac{1}{3}x + 2y = 1 \\ -2x - 4y = 6 \end{cases}$

19. $\begin{cases} x - 3y + 4z = 1 \\ 4x - 10y + 10z = 4 \\ -3x + 9y - 5z = -6 \end{cases}$

20. $\begin{cases} \frac{1}{2}x + y = 4 \\ -4x - 7y + 3z = -31 \\ 6x + 14y + 7z = 50 \end{cases}$

21. $\begin{cases} 2x - 2y = -4 \\ 3x + 4y = 1 \end{cases}$

22. $\begin{cases} 2x + 3y = 4 \\ -x + 2y = -2 \end{cases}$

23. $\begin{cases} 4x - 4y + 4z = -8 \\ x - 2y - 2z = -1 \\ 2x + y + 3z = 1 \end{cases}$

24. $\begin{cases} x + 2y + 2z = 11 \\ x - y - z = -4 \\ 2x + 5y + 9z = 39 \end{cases}$

25. A bank wishes to invest a $100,000 trust fund in three sources—bonds paying 8%, certificates of deposit paying 7%, and first mortgages paying 10%. The bank wishes to realize an $8000 annual income from the investment. A condition of the trust is that the total amount invested in bonds and certificates of deposit must be triple the amount invested in mortgages. How much should the bank invest in each possible category? Let x, y, and z, respectively, be the amounts invested in bonds, certificates of deposit, and first mortgages. Solve the system of equations by the Gaussian elimination method.

26. A dietician wishes to plan a meal around three foods. Each ounce of food I contains 10% of the daily requirements for carbohydrates, 10% for protein, and 15% for vitamin C. Each ounce of food II contains 10% of the daily requirements for carbohydrates, 5% for protein, and 0% for vitamin C. Each ounce of food III contains 10% of the daily requirements for carbohydrates, 25% for protein, and 10% for vitamin C. How many ounces of each food should be served in order to supply exactly the daily requirements for each of carbohydrates, protein and vitamin C? Let x, y, and z, respectively, be the number of ounces of foods I, II, and III.

SOLUTIONS TO PRACTICE PROBLEMS 1

1. (a) Not in diagonal form, since the first equation contains both x and z.

(b) Not in diagonal form, since the variables are not arranged in diagonal fashion.

(c) Not in diagonal form, since the coefficient of z is not 1.

2. (a) Change the system into another system in which the second equation is altered by having (-2) (first equation) added to it. The new system is

$$\begin{cases} x - 3y = 2 \\ 9y = 1. \end{cases}$$

The equation $9y = 1$ was obtained as follows:

$$\begin{array}{ll} (-2)(\text{first equation}) & -2x + 6y = -4 \\ +\ (\text{second equation}) & \underline{2x + 3y = 5} \\ & 9y = 1. \end{array}$$

(b) Replace the second equation by the first equation multiplied by 1 and added to the second. This is the same as adding the first equation to the second. The result is

$$\begin{cases} x + y = 3 \\ 3y = 8. \end{cases}$$

3. (a) The first row should contain a nonzero number as its first entry. This can be accomplished by interchanging the first and third rows.

(b) The first column can be put into proper form by eliminating the -6. To accomplish this, multiply the first row by 6 and add this product to the third row. The notation for this operation is

$$\xrightarrow{[3] + 6[1]}$$

5.2. Solving Systems of Linear Equations, II

In this section we introduce the operation of pivoting and consider systems of linear equations which do not have unique solutions.

Roughly speaking, the Gaussian elimination method applied to a matrix proceeds as follows: Consider the columns one at a time, from left to right. For each column use the elementary row operations to transform the appropriate entry to a one and the remaining entries in the column to zeros. (The "appropriate" entry is the first entry in the first column, the second entry in the second column, and so forth.) This sequence of elementary row operations performed for each column is called *pivoting*. More precisely:

> **Method**   *To pivot a matrix about a given nonzero entry:*
>
> 1. Transform the given entry into a one.
> 2. Transform all other entries in the same column into zeros.

Pivoting is used in solving problems other than systems of linear equations. As we shall see in Chapter 4, it is the basis for the simplex method of solving linear programming problems.

EXAMPLE 1 Pivot the matrix about the circled element.

$$\begin{bmatrix} 18 & \boxed{-6} & | & 15 \\ 5 & -2 & | & 4 \end{bmatrix}$$

Solution The first step is to transform the -6 to a 1. We do this by multiplying the first row by $-\frac{1}{6}$:

$$\begin{bmatrix} 18 & -6 & | & 15 \\ 5 & -2 & | & 4 \end{bmatrix} \xrightarrow{-\frac{1}{6}[1]} \begin{bmatrix} -3 & 1 & | & -\frac{5}{2} \\ 5 & -2 & | & 4 \end{bmatrix}.$$

Next, we transform the -2 (the only remaining entry in column 2) into a 0:

$$\begin{bmatrix} -3 & 1 & | & -\frac{5}{2} \\ 5 & -2 & | & 4 \end{bmatrix} \xrightarrow{[2] + 2[1]} \begin{bmatrix} -3 & 1 & | & -\frac{5}{2} \\ -1 & 0 & | & -1 \end{bmatrix}.$$

The last matrix is the result of pivoting the original matrix about the circled entry.

In terms of pivoting, we can give the following summary of the Gaussian elimination method.

> **Gaussian Elimination Method to Transform a System of Linear Equations into Diagonal Form**
>
> 1. Write down the matrix corresponding to the linear system.
> 2. Make sure that the first entry in the first column is nonzero. Do this by interchanging the first row with one of the rows below it, if necessary.
> 3. Pivot the matrix about the first entry in the first column.
> 4. Make sure that the second entry in the second column is nonzero. Do this by interchanging the second row with one of the rows below it, if necessary.
> 5. Pivot the matrix about the second entry in the second column.
> 6. Continue in this manner.

All the systems considered in the preceding section had only a single solution. In this case we say that the solution is *unique*. Let us now use the Gaussian elimination method to study the various possibilities other than a unique solution. We first experiment with an example.

EXAMPLE 2 Determine all solutions of the system

$$\begin{cases} 2x + 2y + 4z = 8 \\ x - y + 2z = 2 \\ -x + 5y - 2z = 2. \end{cases}$$

We set up the matrix corresponding to the system and perform the appropriate pivoting operations. (The elements pivoted about are circled.)

$$\begin{bmatrix} ② & 2 & 4 & | & 8 \\ 1 & -1 & 2 & | & 2 \\ -1 & 5 & -2 & | & 2 \end{bmatrix} \begin{array}{c} \frac{1}{2}[1] \\ \underrightarrow{[2] + (-1)[1]} \\ [3] + (1)[1] \end{array} \begin{bmatrix} 1 & 1 & 2 & | & 4 \\ 0 & ⊖2 & 0 & | & -2 \\ 0 & 6 & 0 & | & 6 \end{bmatrix}$$

$$\begin{array}{c} (-\frac{1}{2})[2] \\ \underrightarrow{[1] + (-1)[2]} \\ [3] + (-6)[2] \end{array} \begin{bmatrix} 1 & 0 & 2 & | & 3 \\ 0 & 1 & 0 & | & 1 \\ 0 & 0 & 0 & | & 0 \end{bmatrix}.$$

Note that our method must terminate here, since there is no way to transform the third entry in the third column into a 1 without disturbing the columns already in appropriate form. The equations corresponding to the last matrix read

$$\begin{cases} x & + 2z = 3 \\ & y & = 1 \\ & 0 & = 0. \end{cases}$$

The last equation does not involve any of the variables and so may be omitted. This leaves the two equations

$$\begin{cases} x & + 2z = 3 \\ & y & = 1. \end{cases}$$

Now, taking the $2z$ term in the first equation to the right side, we can write the equations

$$\begin{cases} x = 3 - 2z \\ y = 1. \end{cases}$$

The value of y is given: $y = 1$. The value of x is given in terms of z. To find a solution to this system, assign any value to z. Then the first equation gives a value for x and thereby a specific solution to the system. For example, if we take $z = 1$, then the corresponding specific solution is

$$z = 1$$
$$x = 3 - 2(1) = 1$$
$$y = 1.$$

If we take $z = -3$, the corresponding specific solution is

$$z = -3$$
$$x = 3 - 2(-3) = 9$$
$$y = 1.$$

Thus, we see that the original system has infinitely many specific solutions, corresponding to the infinitely many possible different choices for z.

We say that the *general solution* of the system is

$$z = \text{any value}$$

$$x = 3 - 2z$$

$$y = 1.$$

When a linear system cannot be *completely* diagonalized:

1. Apply the Gaussian elimination method to as many columns as possible. Proceed from left to right, but do not disturb columns that have already been put into proper form.
2. Variables corresponding to columns not in proper form can assume any value.
3. The other variables can be expressed in terms of the variables of step 2.

EXAMPLE 3 Find all solutions of the linear system

$$\begin{cases} x + 2y - z + 3w = 5 \\ y + 2z + w = 7. \end{cases}$$

Solution The Gaussian elimination method proceeds as follows:

$$\begin{bmatrix} 1 & 2 & -1 & 3 & | & 5 \\ 0 & ① & 2 & 1 & | & 7 \end{bmatrix} \qquad \text{(The first column is already in proper form.)}$$

$$\xrightarrow{\quad [1] + (-2)[2] \quad} \begin{bmatrix} 1 & 0 & -5 & 1 & | & -9 \\ 0 & 1 & 2 & 1 & | & 7 \end{bmatrix}.$$

We cannot do anything further with the last two columns (without disturbing the first two columns), so the corresponding variables, z and w, can assume any values. Writing down the equations corresponding to the last matrix yields

$$\begin{cases} x \qquad - 5z + w = -9 \\ \quad\; y + 2z + w = \;\;\; 7 \end{cases}$$

or

$$z = \text{any value}$$

$$w = \text{any value}$$

$$x = -9 + 5z - w$$

$$y = \quad 7 - 2z - w.$$

To determine an example of a specific solution, let $z = 1$, $w = 2$. Then a specific solution of the original system is

$$z = 1$$

$$w = 2$$

$$x = -9 + 5(1) - (2) = -6$$

$$y = \quad 7 - 2(1) - (2) = \quad 3.$$

EXAMPLE 4 Find all solutions of the system of equations

$$\begin{cases} x - 7y + z = 3 \\ 2x - 14y + 3z = 4. \end{cases}$$

Solution The first pivot operation is routine:

$$\begin{bmatrix} \textcircled{1} & -7 & 1 & \bigm| & 3 \\ 2 & -14 & 3 & \bigm| & 4 \end{bmatrix} \xrightarrow{[2] + (-2)[1]} \begin{bmatrix} 1 & -7 & 1 & \bigm| & 3 \\ 0 & 0 & 1 & \bigm| & -2 \end{bmatrix}.$$

However, it is impossible to pivot about the zero in the second column. So skip the second column and pivot about the second entry in the third column to get

$$\begin{bmatrix} 1 & -7 & 0 & \bigm| & 5 \\ 0 & 0 & 1 & \bigm| & -2 \end{bmatrix}.$$

This is as far as we can go. The variable corresponding to the second column, namely y, can assume any value, and the general solution of the system is obtained from the equations

$$\begin{cases} x - 7y \quad = \quad 5 \\ \qquad\quad z = -2. \end{cases}$$

Therefore, the general solution of the system is

$$y = \text{any value}$$

$$x = 5 + 7y$$

$$z = -2.$$

We have seen that a linear system may have a unique solution or it may have infinitely many solutions. But another phenomenon can occur: A system may have no solutions at all, as the next example shows.

EXAMPLE 5 Find all solutions of the system

$$\begin{cases} x - y + z = 3 \\ x + y - z = 5 \\ -2x + 4y - 4z = 1. \end{cases}$$

Solution We apply the Gaussian elimination method to the matrix of the system.

$$
\begin{bmatrix}
① & -1 & 1 & \bigm| & 3 \\
1 & 1 & -1 & \bigm| & 5 \\
-2 & 4 & -4 & \bigm| & 1
\end{bmatrix}
\xrightarrow[\;[3]+2[1]\;]{[2]+(-1)[1]}
\begin{bmatrix}
1 & -1 & 1 & \bigm| & 3 \\
0 & ② & -2 & \bigm| & 2 \\
0 & 2 & -2 & \bigm| & 7
\end{bmatrix}
$$

$$
\xrightarrow[\;[3]+(-2)[2]\;]{\substack{\frac{1}{2}[2] \\ [1]+(1)[2]}}
\begin{bmatrix}
1 & 0 & 0 & \bigm| & 4 \\
0 & 1 & -1 & \bigm| & 1 \\
0 & 0 & 0 & \bigm| & 5
\end{bmatrix}.
$$

We cannot pivot about the last zero in the third column, so we have carried the method as far as we can. Let us write out the equations corresponding to the last matrix:

$$
\begin{cases}
x & = 4 \\
y - z & = 1 \\
0 & = 5.
\end{cases}
$$

Note that the last equation is a built-in contradiction. In mathematical terms, the last equation is *inconsistent*. So the last equation can never be satisfied, no matter what the values of x, y, z. Thus, the original system has no solutions. Systems with no solutions can always be detected by the presence of inconsistent equations in the last matrix resulting from the Gaussian elimination method.

At first it might seem strange that some systems have no solutions, some have one, and yet others have infinitely many. The reason for the difference can be explained geometrically. For simplicity, consider the case of systems of two equations in two variables. Each equation in this case has a graph in the xy-plane, and the graph is a straight line. As we have seen, solving the system corresponds to finding the points lying on both lines. There are three possibilities. First, the two lines may intersect. In this case the solution is unique. Second, the two lines may be parallel. Then the two lines do not intersect and the system has no solutions. Finally, the two equations may represent the same line, as, for example, do the equations $2x + 3y = 1$, $4x + 6y = 2$. In this case every point on the line is a solution of the system; that is, there are infinitely many solutions (Fig. 1).

FIGURE 1

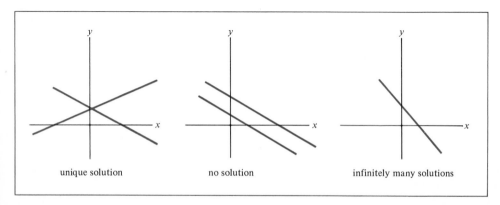

unique solution   no solution   infinitely many solutions

PRACTICE PROBLEMS 2

1. Find a specific solution to a system of linear equations whose general solution is

$$w = \text{any value}$$

$$y = \text{any value}$$

$$z = 7 + 6w$$

$$x = 26 - 2y + 14w.$$

2. Find all solutions of this system of linear equations.

$$\begin{cases} 2x + 4y - 4z - 4w = 24 \\ -3x - 6y + 10z - 18w = -8 \\ -x - 2y + 4z - 10w = 2. \end{cases}$$

EXERCISES 2

Pivot each of the following matrices about the circled element.

1. $\begin{bmatrix} ② & -4 & 6 \\ 3 & 7 & 1 \end{bmatrix}$

2. $\begin{bmatrix} 1 & 2 & 3 \\ 4 & ⑧ & -12 \end{bmatrix}$

3. $\begin{bmatrix} 7 & 1 & 4 & 5 \\ -1 & 1 & ② & 6 \\ 4 & 0 & 2 & 3 \end{bmatrix}$

4. $\begin{bmatrix} 5 & 10 & -10 & 12 \\ 4 & 3 & 6 & 12 \\ 4 & ④ & 4 & -16 \end{bmatrix}$

5. $\begin{bmatrix} ② & 3 \\ 6 & 0 \\ 1 & 5 \end{bmatrix}$

6. $\begin{bmatrix} 2 & 1 \\ ① & 0 \end{bmatrix}$

7. $\begin{bmatrix} 4 & 3 & 0 \\ \frac{2}{3} & 0 & -2 \\ 1 & 3 & ⑥ \end{bmatrix}$

8. $\begin{bmatrix} 1 & 0 & 2 \\ -1 & 1 & ② \\ 1 & 2 & 6 \end{bmatrix}$

Use the Gaussian elimination method to find all solutions of the following systems of linear equations.

9. $\begin{cases} 2x - 4y = 6 \\ -x + 2y = -3 \end{cases}$

10. $\begin{cases} -\frac{1}{2}x + y = \frac{3}{2} \\ -3x + 6y = 10 \end{cases}$

11. $\begin{cases} x + 2y = 5 \\ 3x - y = 1 \\ -x + 3y = 5 \end{cases}$

12. $\begin{cases} x - 6y = 12 \\ -\frac{1}{2}x + 3y = -6 \\ \frac{1}{3}x - 2y = 4 \end{cases}$

13. $\begin{cases} x - y + 3z = 3 \\ -2x + 3y - 11z = -4 \\ x - 2y + 8z = 6 \end{cases}$

14. $\begin{cases} x - 3y + z = 5 \\ -2x + 7y - 6z = -9 \\ x - 2y - 3z = 6 \end{cases}$

15. $\begin{cases} x + y + z = -1 \\ 2x + 3y + 2z = 3 \\ 2x + y + 2z = -7 \end{cases}$

16. $\begin{cases} x - 3y + 2z = 10 \\ -x + 3y - z = -6 \\ -x + 3y + 2z = 6 \end{cases}$

17. $\begin{cases} x + 2y + 3z = 4 \\ 5x + 6y + 7z = 8 \\ x + 2y + 3z = 5 \end{cases}$

18. $\begin{cases} x + 3y = 7 \\ x + 2y = 5 \\ -x + y = 2 \end{cases}$

19. $\begin{cases} x + y - 2z + 2w = 5 \\ 2x + y - 4z + w = 5 \\ 3x + 4y - 6z + 9w = 20 \\ 4x + 4y - 8z + 8w = 20 \end{cases}$

20. $\begin{cases} 2y + z - w = 1 \\ x - y + z + w = 14 \\ -x - 9y - z + 4w = 11 \\ x + y + z = 9 \end{cases}$

21. In a laboratory experiment, a researcher wants to provide a rabbit with exactly 1000 units of vitamin A, exactly 1600 units of vitamin C, and exactly 2400 units of vitamin E. The rabbit is fed a mixture of three foods. Each gram of food 1 contains 2 units of vitamin A, 3 units of vitamin C, and 5 units of vitamin E. Each gram of food 2 contains 4 units of vitamin A, 7 units of vitamin C, and 9 units of vitamin E. Each gram of food 3 contains 6 units of vitamin A, 10 units of vitamin C, and 14 units of vitamin E. How many grams of each food should the rabbit be fed?

22. Rework Exercise 21 with the requirement for vitamin E changed to 2000 units.

SOLUTIONS TO PRACTICE PROBLEMS 2

1. Since w and y can each assume any value, select any numbers, say $w = 1$ and $y = 2$. Then $z = 7 + 6(1) = 13$ and $x = 26 - 2(2) + 14(1) = 36$. So $x = 36$, $y = 2$, $z = 13$, $w = 1$ is a specific solution. There are infinitely many different specific solutions, since there are infinitely many different choices for w and y.

2. We apply the Gaussian elimination method to the matrix of the system.

$$\begin{bmatrix} ② & 4 & -4 & -4 & | & 24 \\ -3 & -6 & 10 & -18 & | & -8 \\ -1 & -2 & 4 & -10 & | & 2 \end{bmatrix} \xrightarrow[\substack{\frac{1}{2}[1] \\ [2] + 3[1] \\ [3] + 1[1]}]{} \begin{bmatrix} 1 & 2 & -2 & -2 & | & 12 \\ 0 & 0 & ④ & -24 & | & 28 \\ 0 & 0 & 2 & -12 & | & 14 \end{bmatrix}$$

$$\xrightarrow[\substack{\frac{1}{4}[2] \\ [1] + 2[2] \\ [3] + (-2)[2]}]{} \begin{bmatrix} 1 & 2 & 0 & -14 & | & 26 \\ 0 & 0 & 1 & -6 & | & 7 \\ 0 & 0 & 0 & 0 & | & 0 \end{bmatrix}.$$

The corresponding system of equation is

$$\begin{cases} x + 2y \quad\quad - 14w = 26 \\ \quad\quad z - 6w = 7. \end{cases}$$

The general solution is

$$w = \text{any value}$$

$$y = \text{any value}$$

$$z = 7 + 6w$$

$$x = 26 - 2y + 14w.$$

5.3. Arithmetic Operations on Matrices

We introduced matrices in the preceding sections in order to display the coefficients of a system of linear equations. For example, the linear system

$$\begin{cases} 5x - 3y = \frac{1}{2} \\ 4x + 2y = -1 \end{cases}$$

is represented by the matrix

$$\left[\begin{array}{cc|c} 5 & -3 & \frac{1}{2} \\ 4 & 2 & -1 \end{array}\right].$$

After we have become accustomed to using such matrices in solving linear systems, we may omit the vertical line which separates the left and the right sides of the equations. We need only remember that the right side of the equations is recorded in the right column. So, for example, we would write the matrix above in the form

$$\begin{bmatrix} 5 & -3 & \frac{1}{2} \\ 4 & 2 & -1 \end{bmatrix}.$$

A matrix is *any* rectangular array of numbers and may be of any size. Here are some examples of matrices of various sizes:

$$\begin{bmatrix} 3 & 7 \\ 0 & -1 \end{bmatrix}, \quad \begin{bmatrix} 1 \\ 2 \end{bmatrix}, \quad [2 \ 1], \quad [6], \quad \begin{bmatrix} 5 & 7 & -1 \\ 0 & 3 & 5 \\ 6 & 0 & 5 \end{bmatrix}.$$

Examples of matrices abound in everyday life. For example, the newspaper stock-market report is a large matrix with several thousand rows, one for each listed stock. The columns of the matrix give various data about each stock, such as opening and closing price, number of shares traded, and so forth. Another example of a matrix is a mileage charge on a road map. The rows and columns are labeled with the names of cities. The number at a given row and column gives the distance between the corresponding cities.

In these everyday examples, matrices are used only to display data. However, the most important applications involve arithmetic operations on matrices—namely, addition, subtraction, and multiplication of matrices. The major goal of this section is to discuss these operations. Before we can do so, however, we need some vocabulary with which to describe matrices.

A matrix is described by the number of rows and columns it contains. For example, the matrix

$$\begin{bmatrix} 7 & 5 \\ \frac{1}{2} & -2 \\ 2 & -11 \end{bmatrix}$$

has three rows and two columns and is referred to as a 3 × 2 (read: "three-by-two") *matrix*. The matrix $[4 \ \ 5 \ \ 0]$ has one row and three columns and is a 1 × 3 *matrix.* A matrix with only one row is often called a *row matrix* (sometimes also called a *row vector*). A matrix, such as $\begin{bmatrix} 2 \\ 7 \end{bmatrix}$, which has only one column is called a *column matrix* or *column vector*. If a matrix has the same number of rows and columns, it is called a *square matrix.* Here are some square matrices of various sizes:

$$[5], \quad \begin{bmatrix} 1 & 2 \\ 3 & 4 \end{bmatrix}, \quad \begin{bmatrix} 2 & -1 & 0 \\ 3 & 5 & 4 \\ 0 & 3 & -7 \end{bmatrix}.$$

The rows of a matrix are numbered from the top down, and the columns are numbered from left to right. For example, the first row of the matrix

$$\begin{bmatrix} 1 & -1 & 0 \\ 2 & 1 & 7 \\ -3 & 2 & 4 \end{bmatrix}$$

is $\begin{bmatrix} 1 & -1 & 0 \end{bmatrix}$, and its third column is

$$\begin{bmatrix} 0 \\ 7 \\ 4 \end{bmatrix}.$$

The numbers in a matrix, called *entries,* may be identified in terms of the row and column containing the entry in question. For example, the entry in the first row, third column of the following matrix is 0:

$$\begin{bmatrix} 1 & -1 & \boxed{0} \\ 2 & 1 & 7 \\ -3 & 2 & 4 \end{bmatrix};$$

the entry in the second row, first column is 2:

$$\begin{bmatrix} 1 & -1 & 0 \\ \boxed{2} & 1 & 7 \\ -3 & 2 & 4 \end{bmatrix};$$

and the entry in the third row, third column is 4:

$$\begin{bmatrix} 1 & -1 & 0 \\ 2 & 1 & 7 \\ -3 & 2 & \boxed{4} \end{bmatrix}.$$

We use the double-subscripted letters to indicate the locations of the entries of a matrix. We denote the entry in the ith row, jth column by a_{ij}. For instance, in the above matrix we have $a_{13} = 0$, $a_{21} = 2$, and $a_{33} = 4$.

We say that two matrices A and B are *equal,* denoted $A = B$, provided that they have the same size and that all their corresponding entries are equal.

Addition and Subtraction of Matrices We define the sum $A + B$ of two matrices A and B only if A and B are two matrices of the same size—that is, if A and B

have the same number of rows and the same number of columns. In this case $A + B$ is the matrix formed by adding the corresponding entries of A and B. Thus, for example,

$$\begin{bmatrix} 2 & 0 \\ 1 & 1 \\ 5 & 3 \end{bmatrix} + \begin{bmatrix} 5 & 4 \\ 0 & 2 \\ 2 & 6 \end{bmatrix} = \begin{bmatrix} 2+5 & 0+4 \\ 1+0 & 1+2 \\ 5+2 & 3+6 \end{bmatrix} = \begin{bmatrix} 7 & 4 \\ 1 & 3 \\ 7 & 9 \end{bmatrix}.$$

We subtract matrices of the same size by subtracting corresponding entries. Thus, we have

$$\begin{bmatrix} 7 \\ 1 \end{bmatrix} - \begin{bmatrix} 3 \\ 2 \end{bmatrix} = \begin{bmatrix} 7-3 \\ 1-2 \end{bmatrix} = \begin{bmatrix} 4 \\ -1 \end{bmatrix}.$$

Multiplication of Matrices It might seem that to define the product of two matrices, one would start with two matrices of like size and multiply the corresponding entries. But this definition is not useful, since the calculations that arise in applications require a somewhat more complex multiplication. In the interests of simplicity, we start by defining the product of a row matrix times a column matrix.

If A is a row matrix and B is a column matrix, then we can form the product $A \cdot B$ provided that the two matrices have the same length. The product $A \cdot B$ is the 1×1 matrix obtained by multiplying corresponding entries of A and B and then forming the sum.

We may put this definition into algebraic terms as follows. Suppose that A is the row matrix

$$A = \begin{bmatrix} a_1 & a_2 & \cdots & a_n \end{bmatrix},$$

and B is the column matrix

$$B = \begin{bmatrix} b_1 \\ b_2 \\ \vdots \\ b_n \end{bmatrix}.$$

Note that A and B are both of the same length, namely n. Then

$$A \cdot B = \begin{bmatrix} a_1 & a_2 & \cdots & a_n \end{bmatrix} \cdot \begin{bmatrix} b_1 \\ b_2 \\ \vdots \\ b_n \end{bmatrix}$$

is calculated by multiplying corresponding entries of A and B and forming the sum; that is,

$$A \cdot B = \begin{bmatrix} a_1 b_1 + a_2 b_2 + \cdots + a_n b_n \end{bmatrix}.$$

Notice that the product is a 1×1 matrix, namely a single number in brackets.

Here are some examples of the product of a row matrix times a column matrix:

$$[3 \quad \tfrac{1}{2}] \cdot \begin{bmatrix} 1 \\ 4 \end{bmatrix} = [3 \cdot 1 + \tfrac{1}{2} \cdot 4] = [5],$$

$$[2 \quad 0 \quad -1] \cdot \begin{bmatrix} 6 \\ 5 \\ 3 \end{bmatrix} = [2 \cdot 6 + 0 \cdot 5 + (-1) \cdot 3] = [9].$$

In multiplying a row matrix times a column matrix, it helps to use both hands. Use your left index finger to point to an element of the row matrix and your right to point to the corresponding element of the column. Multiply the elements you are pointing to and keep a running total of the products in your head. After each multiplication move your fingers to the next elements of each matrix. With a little practice you should be able to multiply a row times a column quickly and accurately.

The definition of multiplication above may seem strange. But products of this sort occur in many down-to-earth problems. Consider, for instance, the next example.

EXAMPLE 1 A dairy farm produces three items—milk, eggs, and cheese. The prices of these items are $1.50 per gallon, $.80 per dozen, and $2 per pound, respectively. In a certain week the dairy farm sells 30,000 gallons of milk, 2000 dozen eggs, and 5000 pounds of cheese. Represent its total revenue as a matrix product.

Solution The total revenue equals

$$(1.50)(30,000) + (.80)(2000) + (2)(5000).$$

This suggests that we define two matrices: the first displays the prices of the various produce:

$$[1.50 \quad .80 \quad 2].$$

The second represents the production:

$$\begin{bmatrix} 30,000 \\ 2000 \\ 5000 \end{bmatrix}.$$

Then the revenue for the week, when placed in a 1×1 matrix, equals

$$[1.50 \quad .80 \quad 2] \begin{bmatrix} 30,000 \\ 2000 \\ 5000 \end{bmatrix} = [56,600].$$

The principle behind Example 1 is this: any sum of products of the form $a_1b_1 + a_2b_2 + \cdots + a_nb_n$, when placed in a 1×1 matrix, can be written as the matrix product

$$[a_1b_1 + a_2b_2 + \cdots + a_nb_n] = [a_1 \quad a_2 \quad \cdots \quad a_n] \begin{bmatrix} b_1 \\ b_2 \\ \vdots \\ b_n \end{bmatrix}.$$

Let us illustrate the procedure for multiplying more general matrices by working out a typical product:

$$\begin{bmatrix} 2 & 1 \\ 0 & 1 \\ 1 & 0 \end{bmatrix} \cdot \begin{bmatrix} 1 & 1 \\ 4 & 2 \end{bmatrix}.$$

To obtain the entries of the product, we multiply the rows of the left matrix by the columns of the right matrix, taking care to arrange the products in a specific way to yield a matrix, as follows. Start with the first row on the left, $[2 \quad 1]$, and the first column on the right, $\begin{bmatrix} 1 \\ 4 \end{bmatrix}$. Their product is $[6]$, so we enter 6 as the element in the first row, first column of the product:

$$\begin{bmatrix} 2 & 1 \\ 0 & 1 \\ 1 & 0 \end{bmatrix} \cdot \begin{bmatrix} 1 & 1 \\ 4 & 2 \end{bmatrix} = \begin{bmatrix} 6 & \\ & \end{bmatrix}.$$

The product of the first row of the left matrix and the second column of the right matrix is $[4]$, so we put a 4 in the first row, second column of the product:

$$\begin{bmatrix} 2 & 1 \\ 0 & 1 \\ 1 & 0 \end{bmatrix} \cdot \begin{bmatrix} 1 & 1 \\ 4 & 2 \end{bmatrix} = \begin{bmatrix} 6 & 4 \\ & \end{bmatrix}.$$

There are no more columns which can be multiplied by the first row, so let us move to the second row and shift back to the first column. Correspondingly, we move one row down in the product:

$$\begin{bmatrix} 2 & 1 \\ 0 & 1 \\ 1 & 0 \end{bmatrix} \cdot \begin{bmatrix} 1 & 1 \\ 4 & 2 \end{bmatrix} = \begin{bmatrix} 6 & 4 \\ 4 & \end{bmatrix}$$

$$\begin{bmatrix} 2 & 1 \\ 0 & 1 \\ 1 & 0 \end{bmatrix} \cdot \begin{bmatrix} 1 & 1 \\ 4 & 2 \end{bmatrix} = \begin{bmatrix} 6 & 4 \\ 4 & 2 \end{bmatrix}.$$

We have now exhausted the second row of the left matrix, so we shift to the third row and correspondingly move down one row in the product.

$$\begin{bmatrix} 2 & 1 \\ 0 & 1 \\ 1 & 0 \end{bmatrix} \cdot \begin{bmatrix} 1 & 1 \\ 4 & 2 \end{bmatrix} = \begin{bmatrix} 6 & 4 \\ 4 & 2 \\ 1 & \end{bmatrix}$$

$$\begin{bmatrix} 2 & 1 \\ 0 & 1 \\ 1 & 0 \end{bmatrix} \cdot \begin{bmatrix} 1 & 1 \\ 4 & 2 \end{bmatrix} = \begin{bmatrix} 6 & 4 \\ 4 & 2 \\ 1 & 1 \end{bmatrix}.$$

Note that we have now multiplied every row of the left matrix by every column of the right matrix. This completes computation of the product:

$$\begin{bmatrix} 2 & 1 \\ 0 & 1 \\ 1 & 0 \end{bmatrix} \cdot \begin{bmatrix} 1 & 1 \\ 4 & 2 \end{bmatrix} = \begin{bmatrix} 6 & 4 \\ 4 & 2 \\ 1 & 1 \end{bmatrix}.$$

EXAMPLE 2   Calculate the following product:

$$\begin{bmatrix} 1 & 5 \\ 3 & 2 \end{bmatrix} \cdot \begin{bmatrix} 1 & 2 \\ 1 & 0 \end{bmatrix}.$$

Solution

$$\begin{bmatrix} 1 & 5 \\ 3 & 2 \end{bmatrix} \cdot \begin{bmatrix} 1 & 2 \\ 1 & 0 \end{bmatrix} = \begin{bmatrix} 6 & \\ & \end{bmatrix}$$

$$\begin{bmatrix} 1 & 5 \\ 3 & 2 \end{bmatrix} \cdot \begin{bmatrix} 1 & 2 \\ 1 & 0 \end{bmatrix} = \begin{bmatrix} 6 & 2 \\ & \end{bmatrix}$$

$$\begin{bmatrix} 1 & 5 \\ 3 & 2 \end{bmatrix} \cdot \begin{bmatrix} 1 & 2 \\ 1 & 0 \end{bmatrix} = \begin{bmatrix} 6 & 2 \\ 5 & \end{bmatrix}$$

$$\begin{bmatrix} 1 & 5 \\ 3 & 2 \end{bmatrix} \cdot \begin{bmatrix} 1 & 2 \\ 1 & 0 \end{bmatrix} = \begin{bmatrix} 6 & 2 \\ 5 & 6 \end{bmatrix}$$

Thus

$$\begin{bmatrix} 1 & 5 \\ 3 & 2 \end{bmatrix} \cdot \begin{bmatrix} 1 & 2 \\ 1 & 0 \end{bmatrix} = \begin{bmatrix} 6 & 2 \\ 5 & 6 \end{bmatrix}.$$

Notice that we cannot use the method above to compute the product $A \cdot B$ of *any* matrices A and B. For the procedure to work, it is crucial that the number of entries of each row of A be the same as the number of entries of each column of B. (Or, to put it another way, the number of columns of the left matrix must equal the number of rows of the right matrix.) Therefore, in order for us to form the product $A \cdot B$, the sizes of A and B must match up in a special way. If A is $m \times n$ and B is $p \times q$, then the product $A \cdot B$ is defined only in case the "inner" dimensions n and p are equal. In that case, the product is determined by the "outer" dimensions m and q. It is an $m \times q$ matrix.

$$\begin{array}{ccccc} A & \cdot & B & = & C. \\ m \times n & & p \times q & & m \times q \end{array}$$

equal

So, for example,

$$\begin{bmatrix} & & & \\ & & & \\ & & & \end{bmatrix} \begin{bmatrix} & \\ & \\ & \\ & \end{bmatrix} = \begin{bmatrix} & \\ & \\ & \end{bmatrix}$$

$$\underset{3 \times 4}{} \qquad \underset{4 \times 2}{} \qquad \underset{3 \times 2}{}$$

$$\begin{bmatrix} & \\ & \end{bmatrix} \begin{bmatrix} \\ \end{bmatrix} = \begin{bmatrix} \\ \end{bmatrix}.$$

$$\underset{2 \times 2}{} \quad \underset{2 \times 1}{} \quad \underset{2 \times 1}{}$$

If the sizes of A and B do not match up in the way just described, the product $A \cdot B$ is not defined.

EXAMPLE 3 Calculate the following products, if defined.

(a) $\begin{bmatrix} 3 & -1 \\ 2 & 0 \\ 1 & 5 \end{bmatrix} \begin{bmatrix} 1 & 0 \\ 5 & -4 \\ 2 & -1 \end{bmatrix}$

(b) $\begin{bmatrix} 3 & -1 \\ 2 & 0 \\ 1 & 5 \end{bmatrix} \begin{bmatrix} 5 & 4 \\ -2 & 3 \end{bmatrix}$

Solution (a) The matrices to be multiplied are 3×2 and 3×2. The inner dimensions do not match, so the product is undefined.

(b) We are asked to multiply a 3×2 matrix times a 2×2 matrix. The inner dimensions match, so the product is defined and has size determined by the outer dimensions, that is, 3×2.

$$\begin{bmatrix} 3 & -1 \\ 2 & 0 \\ 1 & 5 \end{bmatrix} \cdot \begin{bmatrix} 5 & 4 \\ -2 & 3 \end{bmatrix} = \begin{bmatrix} 3 \cdot 5 + (-1) \cdot (-2) & 3 \cdot 4 + (-1) \cdot 3 \\ 2 \cdot 5 + 0 \cdot (-2) & 2 \cdot 4 + 0 \cdot 3 \\ 1 \cdot 5 + 5 \cdot (-2) & 1 \cdot 4 + 5 \cdot 3 \end{bmatrix}$$

$$= \begin{bmatrix} 17 & 9 \\ 10 & 9 \\ -5 & 19 \end{bmatrix}.$$

Multiplication of matrices has many properties in common with multiplication of ordinary numbers. However, there is at least one important difference. With matrix multiplication, the order of the factors is usually important. For example, the product of a 2×3 matrix times a 3×2 matrix is defined and the product is a 2×2 matrix. If the order is reversed to a 3×2 matrix times a 2×3 matrix, the product is a 3×3 matrix. So reversing the order may change the size of the product. Even when it does not, reversing the order may still change the entries in the product, as the following two products demonstrate.

$$\begin{bmatrix} 1 & 5 \\ 3 & 2 \end{bmatrix} \begin{bmatrix} 1 & 2 \\ 1 & 0 \end{bmatrix} = \begin{bmatrix} 6 & 2 \\ 5 & 6 \end{bmatrix}$$

$$\begin{bmatrix} 1 & 2 \\ 1 & 0 \end{bmatrix} \begin{bmatrix} 1 & 5 \\ 3 & 2 \end{bmatrix} = \begin{bmatrix} 7 & 9 \\ 1 & 5 \end{bmatrix}.$$

EXAMPLE 4 An investment trust has investments in three states. Its deposits in each state are divided among bonds, mortgages, and consumer loans. On January 1 the amount (in millions of dollars) of money invested in each category by state is given by the matrix

$$
\begin{array}{c}
 \\
\text{State A} \\
\text{State B} \\
\text{State C}
\end{array}
\begin{array}{ccc}
\text{Bonds} & \text{Mortgages} & \begin{array}{c}\text{Consumer}\\\text{loans}\end{array} \\
\left[\begin{array}{ccc}
10 & 5 & 20 \\
30 & 12 & 10 \\
15 & 6 & 25
\end{array}\right]
\end{array}.
$$

The current average yields are 7% for bonds, 9% for mortgages, and 15% for consumer loans. Determine the earnings of the trust from its investments in each state.

Solution Define the matrix of investment yields by

$$
\begin{bmatrix}
.07 \\
.09 \\
.15
\end{bmatrix}
\begin{array}{l}
\text{Bonds} \\
\text{Mortgages} \\
\text{Consumer loans.}
\end{array}
$$

The amount earned in state A, for instance, is

[amount of bonds][yield of bonds]

$\quad\quad$ + [amount of mortgages][yield of mortgages]

$\quad\quad$ + [amount of consumer loans][yield of consumer loans]

$\quad = (10)(.07) + (5)(.09) + (20)(.15).$

And this is just the first entry of the product

$$
\begin{bmatrix}
10 & 5 & 20 \\
30 & 12 & 10 \\
15 & 6 & 25
\end{bmatrix}
\begin{bmatrix}
.07 \\
.09 \\
.15
\end{bmatrix}.
$$

Similarly, the earnings for the other states are the second and third entries of the product. Carrying out the arithmetic, we find that

$$
\begin{bmatrix}
10 & 5 & 20 \\
30 & 12 & 10 \\
15 & 6 & 25
\end{bmatrix}
\begin{bmatrix}
.07 \\
.09 \\
.15
\end{bmatrix}
=
\begin{bmatrix}
4.15 \\
4.68 \\
5.34
\end{bmatrix}.
$$

Therefore, the trust earns $4.15 million in state A, $4.68 million in state B, and $5.34 million in state C.

EXAMPLE 5 A clothing manufacturer has factories in Los Angeles, San Antonio, and Newark. Sales (in thousands) during the first quarter of last year are summarized in the production matrix

	Los Angeles	San Antonio	Newark
Coats	12	13	38
Shirts	25	5	26
Sweaters	11	8	8
Ties	5	0	12

During this period the selling price of a coat was $100, of a shirt $10, of a sweater $25, and of a tie $5.

(a) Use a matrix calculation to determine the total revenue produced by each of the factories.

(b) Suppose that the prices had been $110, $8, $20, and $10, respectively. How would this have affected the revenue of each factory?

Solution (a) For each factory, we wish to multiply the price of each item by the number produced to arrive at revenue. Since the production figures for the various items of clothing are arranged down the columns, we arrange the prices in a row matrix, ready for multiplication; the price matrix is

$$[100 \quad 10 \quad 25 \quad 5].$$

The revenues of the various factories are then the entries of the product

$$[100 \quad 10 \quad 25 \quad 5] \begin{bmatrix} 12 & 13 & 38 \\ 25 & 5 & 26 \\ 11 & 8 & 8 \\ 5 & 0 & 12 \end{bmatrix} = \begin{matrix} \text{Los Angeles} & \text{San Antonio} & \text{Newark} \\ [1750 & 1550 & 4320]. \end{matrix}$$

Since the production figures are in thousands, the revenue figures are in thousands of dollars. That is, the Los Angeles factory has revenues of $1,750,000, the San Antonio factory $1,550,000, and the Newark factory $4,320,000.

(b) In a similar way, we determine the revenue of each factory if the price matrix had been $[110 \quad 8 \quad 20 \quad 10]$.

$$[110 \quad 8 \quad 20 \quad 10] \begin{bmatrix} 12 & 13 & 38 \\ 25 & 5 & 26 \\ 11 & 8 & 8 \\ 5 & 0 & 12 \end{bmatrix} = \begin{matrix} \text{Los Angeles} & \text{San Antonio} & \text{Newark} \\ [1790 & 1630 & 4668]. \end{matrix}$$

The change in revenue at each factory can be read from the difference of the revenue matrices:

$$[1790 \quad 1630 \quad 4668] - [1750 \quad 1550 \quad 4320] = [40 \quad 80 \quad 348].$$

That is, if prices had been as given in part (b), then revenues of the Los Angeles factory would have increased by 40, revenues at San Antonio would have increased by 80, and revenues at Newark would have increased by 348.

There are special matrices analogous to the number 1. Such matrices are called *identity matrices*. The identity matrix I_n of size n is the $n \times n$ square matrix with all zeros, except for ones down the upper-left-to-lower-right diagonal. Here are the identity matrices of sizes 2, 3, and 4:

$$I_2 = \begin{bmatrix} 1 & 0 \\ 0 & 1 \end{bmatrix}, \qquad I_3 = \begin{bmatrix} 1 & 0 & 0 \\ 0 & 1 & 0 \\ 0 & 0 & 1 \end{bmatrix}, \qquad I_4 = \begin{bmatrix} 1 & 0 & 0 & 0 \\ 0 & 1 & 0 & 0 \\ 0 & 0 & 1 & 0 \\ 0 & 0 & 0 & 1 \end{bmatrix}.$$

The characteristic property of an identity matrix is that it plays the role of the number 1; that is,

$$I_n \cdot A = A \cdot I_n = A$$

for all $n \times n$ matrices A.

One of the principal uses of matrices is in dealing with systems of linear equations. Matrices provide a compact way of writing systems, as the next example shows.

EXAMPLE 6 Write the system of linear equations

$$\begin{cases} -2x + 4y = 2 \\ -3x + 7y = 7 \end{cases}$$

as a matrix equation.

Solution The system of equations can be written in the form

$$\begin{bmatrix} -2x + 4y \\ -3x + 7y \end{bmatrix} = \begin{bmatrix} 2 \\ 7 \end{bmatrix}.$$

So consider the matrices

$$A = \begin{bmatrix} -2 & 4 \\ -3 & 7 \end{bmatrix}, \qquad X = \begin{bmatrix} x \\ y \end{bmatrix}, \qquad B = \begin{bmatrix} 2 \\ 7 \end{bmatrix}.$$

Notice that

$$AX = \begin{bmatrix} -2 & 4 \\ -3 & 7 \end{bmatrix} \begin{bmatrix} x \\ y \end{bmatrix} = \begin{bmatrix} -2x + 4y \\ -3x + 7y \end{bmatrix}.$$

Thus AX is a 2×1 column matrix whose entries correspond to the left side of the given system of linear equations. Since the entries of B correspond to the right side of the system of equations, we can rewrite the given system in the form

$$AX = B$$

—that is,

$$\begin{bmatrix} -2 & 4 \\ -3 & 7 \end{bmatrix} \begin{bmatrix} x \\ y \end{bmatrix} = \begin{bmatrix} 2 \\ 7 \end{bmatrix}.$$

The matrix A of the example above displays the coefficients of the variables x and y, so it is called the *coefficient matrix* of the system.

PRACTICE PROBLEMS 3

1. Compute

$$\begin{bmatrix} 3 & 1 & 2 \\ -1 & 0 & \frac{1}{2} \\ 0 & 4 & 1 \end{bmatrix} \begin{bmatrix} 7 & -1 & 0 \\ 5 & 4 & 2 \\ -6 & 0 & 4 \end{bmatrix}.$$

2. Give the system of linear equations which is equivalent to the matrix equation

$$\begin{bmatrix} 3 & -6 \\ 2 & 1 \end{bmatrix} \begin{bmatrix} x \\ y \end{bmatrix} = \begin{bmatrix} 5 \\ 0 \end{bmatrix}.$$

3. Give a matrix equation equivalent to this system of equations:

$$\begin{cases} 8x + 3y = 7 \\ 9x - 2y = -5. \end{cases}$$

EXERCISES 3

In Exercises 1–6, give the size and special characteristics of the given matrix (such as square, column, row, identity).

1. $\begin{bmatrix} 3 & 2 & .4 \\ \frac{1}{2} & 0 & 6 \end{bmatrix}$
2. $\begin{bmatrix} 3 \\ -1 \end{bmatrix}$
3. $[2 \quad \frac{1}{3} \quad 0]$

4. $\begin{bmatrix} 1 & 0 \\ 0 & 1 \end{bmatrix}$
5. $[-2]$
6. $\begin{bmatrix} 0 & 0 & 0 & 0 \\ 0 & 0 & 0 & 0 \end{bmatrix}$

In Exercises 7–14, perform the indicated matrix calculations.

7. $\begin{bmatrix} 4 & -2 \\ 3 & 0 \end{bmatrix} + \begin{bmatrix} 5 & 5 \\ 4 & -1 \end{bmatrix}$
8. $\begin{bmatrix} 8 \\ -3 \end{bmatrix} + \begin{bmatrix} 5 \\ 6 \end{bmatrix}$

9. $\begin{bmatrix} 2 & 8 \\ \frac{4}{3} & 4 \\ 1 & -2 \end{bmatrix} - \begin{bmatrix} 1 & 5 \\ \frac{1}{3} & 2 \\ -3 & 0 \end{bmatrix}$
10. $\begin{bmatrix} 1 & 0 \\ 0 & 1 \end{bmatrix} - \begin{bmatrix} .8 & .5 \\ .2 & .5 \end{bmatrix}$

11. $[5 \quad 3] \begin{bmatrix} 1 \\ 2 \end{bmatrix}$
12. $[1 \quad 0 \quad 0] \begin{bmatrix} \frac{1}{2} \\ 6 \\ 2 \end{bmatrix}$

13. $[6 \quad 1 \quad 5] \begin{bmatrix} \frac{1}{2} \\ -3 \\ 2 \end{bmatrix}$
14. $[0 \quad 0] \begin{bmatrix} 5 \\ -3 \end{bmatrix}$

In Exercises 15–20, the sizes of two matrices are given. Tell whether or not the product AB is defined. If so, give its size.

15. $A, 3 \times 4$; $B, 4 \times 5$ 16. $A, 3 \times 3$; $B, 3 \times 4$ 17. $A, 3 \times 2$; $B, 3 \times 2$

18. $A, 1 \times 1$; $B, 1 \times 1$ 19. $A, 3 \times 3$; $B, 3 \times 1$ 20. $A, 4 \times 2$; $B, 3 \times 4$

In Exercises 21–30, perform the multiplication.

21. $\begin{bmatrix} 3 & 1 \\ 0 & 2 \end{bmatrix} \begin{bmatrix} 1 & 4 \\ 3 & 5 \end{bmatrix}$

22. $\begin{bmatrix} 4 & -1 \\ 2 & \frac{1}{2} \end{bmatrix} \begin{bmatrix} 3 \\ 2 \end{bmatrix}$

23. $\begin{bmatrix} 4 & 1 & 0 \\ -2 & 0 & 3 \\ 1 & 5 & -1 \end{bmatrix} \begin{bmatrix} 5 \\ 1 \\ 2 \end{bmatrix}$

24. $\begin{bmatrix} 0 & 0 \\ 0 & 0 \\ 0 & 0 \end{bmatrix} \begin{bmatrix} 1 & 2 \\ 3 & 4 \end{bmatrix}$

25. $\begin{bmatrix} 1 & 0 \\ 0 & 1 \end{bmatrix} \begin{bmatrix} 5 & 6 \\ 7 & 8 \end{bmatrix}$

26. $\begin{bmatrix} 1 & 2 \\ 1 & 3 \end{bmatrix} \begin{bmatrix} 3 & -2 \\ -1 & 1 \end{bmatrix}$

27. $\begin{bmatrix} .6 & .3 \\ .4 & .7 \end{bmatrix} \begin{bmatrix} .6 & .3 \\ .4 & .7 \end{bmatrix}$

28. $\begin{bmatrix} 0 & 1 & 2 \\ -1 & 4 & \frac{1}{2} \\ 1 & 3 & 0 \end{bmatrix} \begin{bmatrix} 3 & -1 & 5 \\ 0 & 2 & 2 \\ 4 & -6 & 0 \end{bmatrix}$

29. $\begin{bmatrix} 2 & -1 & 4 \\ 0 & 1 & 0 \\ \frac{1}{2} & 3 & -2 \end{bmatrix} \begin{bmatrix} 4 & 8 & 0 \\ 3 & -1 & 2 \\ 5 & 0 & 1 \end{bmatrix}$

30. $\begin{bmatrix} 1 & 0 & 0 \\ 0 & 1 & 0 \\ 0 & 0 & 1 \end{bmatrix} \begin{bmatrix} 1 \\ 2 \\ 3 \end{bmatrix}$

In Exercises 31–34, give the system of linear equations that is equivalent to the matrix equation. Do not solve.

31. $\begin{bmatrix} 2 & 3 \\ 4 & 5 \end{bmatrix} \begin{bmatrix} x \\ y \end{bmatrix} = \begin{bmatrix} 6 \\ 7 \end{bmatrix}$

32. $\begin{bmatrix} -3 & 4 \\ 0 & 1 \end{bmatrix} \begin{bmatrix} x \\ y \end{bmatrix} = \begin{bmatrix} 1 \\ 1 \end{bmatrix}$

33. $\begin{bmatrix} 1 & 2 & 3 \\ 4 & 5 & 6 \\ 7 & 8 & 9 \end{bmatrix} \begin{bmatrix} x \\ y \\ z \end{bmatrix} = \begin{bmatrix} 10 \\ 11 \\ 12 \end{bmatrix}$

34. $\begin{bmatrix} 1 & 0 & 0 \\ 0 & 1 & 0 \\ 0 & 0 & 1 \end{bmatrix} \begin{bmatrix} x \\ y \\ z \end{bmatrix} = \begin{bmatrix} 1 \\ 2 \\ 3 \end{bmatrix}$

In Exercises 35–38, write the given system of linear equations in matrix form.

35. $\begin{cases} 3x + 2y = -1 \\ 7x - y = 2 \end{cases}$

36. $\begin{cases} 5x - 2y = 6 \\ -3x + 4y = 0 \end{cases}$

37. $\begin{cases} x - 2y + 3z = 5 \\ y + z = 6 \\ z = 2 \end{cases}$

38. $\begin{cases} -2x + 4y - z = 5 \\ x + 6y + 3z = -1 \\ 7x + 4z = 8 \end{cases}$

The distributive law says that $(A + B)C = AC + BC$. That is, adding A and B and then multiplying on the right by C gives the same result as first multiplying each of A and B on the right by C and then adding. In Exercises 39 and 40, verify the distributive law for the given matrices.

39. $A = \begin{bmatrix} 1 & 2 \\ 0 & 3 \end{bmatrix}$, $B = \begin{bmatrix} 3 & -2 \\ 4 & 5 \end{bmatrix}$, $C = \begin{bmatrix} 1 & 6 \\ 2 & 0 \end{bmatrix}$

40. $A = \begin{bmatrix} 1 & 0 & 0 \\ 0 & 1 & 0 \\ 0 & 0 & 1 \end{bmatrix}, B = \begin{bmatrix} 2 & 1 & 3 \\ 0 & 5 & -1 \\ 3 & 6 & 0 \end{bmatrix}, C = \begin{bmatrix} 0 \\ 3 \\ -4 \end{bmatrix}$

Two $n \times n$ matrices A and B are called *inverses* (of one another) if both products AB and BA equal I_n. Check that the pairs of matrices in Exercises 41 and 42 are inverses.

41. $\begin{bmatrix} 3 & -1 \\ -1 & \frac{1}{2} \end{bmatrix}, \begin{bmatrix} 1 & 2 \\ 2 & 6 \end{bmatrix}$

42. $\begin{bmatrix} 2 & 8 & -11 \\ -1 & -5 & 7 \\ 1 & 2 & -3 \end{bmatrix}, \begin{bmatrix} 1 & 2 & 1 \\ 4 & 5 & -3 \\ 3 & 4 & -2 \end{bmatrix}$

43. In a certain town the percentages of voters voting Democratic and Republican by various age groups is summarized by this matrix:

$$
\begin{array}{c}
\\
\text{Under 30} \\
\text{30–50} \\
\text{Over 50}
\end{array}
\begin{array}{cc}
\text{Dem.} & \text{Rep.} \\
\begin{bmatrix} .65 & .35 \\ .55 & .45 \\ .45 & .55 \end{bmatrix} & = A.
\end{array}
$$

The population of voters in the town by age group is given by the matrix

$$B = \underbrace{[6000}_{\substack{\text{Under} \\ 30}} \quad \underbrace{8000}_{\text{30–50}} \quad \underbrace{4000]}_{\substack{\text{Over} \\ 50}}.$$

Interpret the entries of the matrix product BA.

44. Refer to Exercise 43.

(a) Using the given data, which party would win and what would be the percentage of the winning vote?

(b) Suppose that the population of the town shifted toward older residents as reflected in the population matrix $B = [2000 \quad 4000 \quad 12{,}000]$. What would be the result of the election now?

45. Suppose that a contractor employs carpenters, bricklayers, and plumbers, working three shifts per day. The number of man-hours employed in each of the shifts is summarized in the matrix

$$
\begin{array}{c}
\\
\\
\text{Carpenters} \\
\text{Bricklayers} \\
\text{Plumbers}
\end{array}
\begin{array}{c}
\text{Shift} \\
\hline
\begin{array}{ccc} 1 & 2 & 3 \end{array} \\
\begin{bmatrix} 50 & 20 & 10 \\ 30 & 30 & 15 \\ 20 & 20 & 5 \end{bmatrix}.
\end{array}
$$

Labor in shift 1 costs $10 per hour, in shift 2 $15 per hour, and in shift 3 $20 per hour.

(a) Without using matrix multiplication, compute the amount spent on labor in each of the shifts.

(b) Use matrix multiplication to compute the amount spent on each type of labor.

46. A flu epidemic hits a large city. Each resident of the city is either sick, well, or a carrier. The proportion of people in each of the categories is expressed by the matrix

	Age		
	0–10	10–30	Over 30
Well	.70	.70	.60
Sick	.10	.20	.30
Carrier	.20	.10	.10

$$= A.$$

The population of the city is distributed by age and sex as follows:

		Male	Female
	0–10	60,000	65,000
Age	10–30	100,000	110,000
	Over 30	200,000	230,000

$$= B.$$

(a) Compute AB.

(b) How many sick males are there?

(c) How many female carriers are there?

SOLUTIONS TO PRACTICE PROBLEMS 3

1. Answer:

$$\begin{bmatrix} 3 & 1 & 2 \\ -1 & 0 & \frac{1}{2} \\ 0 & 4 & 1 \end{bmatrix} \begin{bmatrix} 7 & -1 & 0 \\ 5 & 4 & 2 \\ -6 & 0 & 4 \end{bmatrix} = \begin{bmatrix} 14 & 1 & 10 \\ -10 & 1 & 2 \\ 14 & 16 & 12 \end{bmatrix}.$$

The systematic steps to be taken are:

(a) Determine the size of the product matrix.
Since we have a ③ × 3 times a 3 × ③, the size of the product is given by the outer
— outer dimensions —
dimensions or 3 × 3. Begin by drawing a 3 × 3 rectangular array.

(b) Find the entries one at a time.
To find the entry in the first row, first column of the product, look at the first row of the left given matrix and the first column of the right given matrix and form their product.

$$\begin{bmatrix} 3 & 1 & 2 \\ -1 & 0 & \frac{1}{2} \\ 0 & 4 & 1 \end{bmatrix} \begin{bmatrix} 7 & -1 & 0 \\ 5 & 4 & 2 \\ -6 & 0 & 4 \end{bmatrix} = \begin{bmatrix} 14 & & \\ & & \\ & & \end{bmatrix},$$

since $3 \cdot 7 + 1 \cdot 5 + 2(-6) = 14$. In general, to find the entry in the ith row, jth column of the product, put one finger on the ith row of the left given matrix and another finger on the jth column of the right given matrix. Then multiply the row matrix times the column matrix to get the desired entry.

2. Denote the three matrices by A, X, and B, respectively. Since b_{11} (the entry of the first row, first column of B) is 5, this means that

$$[\text{first row of } A] \begin{bmatrix} \text{first} \\ \text{column} \\ \text{of } X \end{bmatrix} = [b_{11}].$$

That is,

$$[3 \quad -6]\begin{bmatrix} x \\ y \end{bmatrix} = [5]$$

or

$$3x - 6y = 5.$$

Similarly, $b_{21} = 0$ says that $2x + y = 0$. Therefore, the corresponding system of linear equations is

$$\begin{cases} 3x - 6y = 5 \\ 2x + y = 0. \end{cases}$$

3. The coefficient matrix is

$$\begin{bmatrix} 8 & 3 \\ 9 & -2 \end{bmatrix}.$$

So the system of equations is equivalent to the matrix equation

$$\begin{bmatrix} 8 & 3 \\ 9 & -2 \end{bmatrix}\begin{bmatrix} x \\ y \end{bmatrix} = \begin{bmatrix} 7 \\ -5 \end{bmatrix}.$$

5.4. The Inverse of a Matrix

In the preceding section we introduced the operations of addition, subtraction, and multiplication of matrices. In this section let us pursue the algebra of matrices a bit further and consider equations involving matrices. Specifically, we shall consider equations of the form

$$AX = B, \tag{1}$$

where A and B are given matrices and X is an unknown matrix whose entries are to be determined. Such equations among matrices are intimately bound up with the theory of systems of linear equations. Indeed, we described the connection in a special case in Example 6 of the preceding section. In that example we wrote the system of linear equations

$$\begin{cases} -2x + 4y = 2 \\ -3x + 7y = 7 \end{cases}$$

as a matrix equation of the form (1), where

$$A = \begin{bmatrix} -2 & 4 \\ -3 & 7 \end{bmatrix}, \qquad B = \begin{bmatrix} 2 \\ 7 \end{bmatrix}, \qquad X = \begin{bmatrix} x \\ y \end{bmatrix}.$$

Note that by determining the entries (x and y) of the unknown matrix X, we solve the system of linear equations. We will return to this example after we have made a complete study of the matrix equation (1).

As motivation for our solution of equation (1), let us consider the analogous equation among numbers:

$$ax = b,$$

where a and b are given numbers* and x is to be determined. Let us examine its solution in great detail. Multiply both sides by $1/a$. (Note that $1/a$ makes sense, since $a \neq 0$.)

$$\left(\frac{1}{a}\right) \cdot (ax) = \frac{1}{a} \cdot b$$

$$\left(\frac{1}{a} \cdot a\right) \cdot x = \frac{1}{a} \cdot b$$

$$1 \cdot x = \frac{1}{a} \cdot b$$

$$x = \frac{1}{a} \cdot b.$$

Let us model our solution of equation (1) on the calculation above. To do so, we wish to multiply both sides of the equation by a matrix that plays the same role in matrix arithmetic as $1/a$ plays in ordinary arithmetic. Our first task then will be to introduce this matrix and study its properties.

The number $1/a$ has the following relationship to the number a:

$$\frac{1}{a} \cdot a = a \cdot \frac{1}{a} = 1. \tag{2}$$

The matrix analog of the number 1 is an identity matrix I. This prompts us to generalize equation (2) to matrices as follows. Suppose that we are given a square matrix A. Then the *inverse* of A, denoted A^{-1}, is a square matrix with the property

$$A^{-1}A = I \quad \text{and} \quad AA^{-1} = I, \tag{3}$$

where I is an identity matrix of the same size as A. The matrix A^{-1} is the matrix analogue of the number $1/a$. It can be shown that a matrix A has at most one inverse. (However, A may not have an inverse at all; see below.)

If we are given a matrix A, then it is easy to determine whether or not a given matrix is its inverse. Merely check equation (3) with the given matrix substituted for A^{-1}. For example, if

$$A = \begin{bmatrix} -2 & 4 \\ -3 & 7 \end{bmatrix},$$

---

* We may as well assume that $a \neq 0$. Otherwise, x does not occur.

then

$$A^{-1} = \begin{bmatrix} -\frac{7}{2} & 2 \\ -\frac{3}{2} & 1 \end{bmatrix}.$$

Indeed, we have

$$\underset{A^{-1}}{\begin{bmatrix} -\frac{7}{2} & 2 \\ -\frac{3}{2} & 1 \end{bmatrix}} \underset{A}{\begin{bmatrix} -2 & 4 \\ -3 & 7 \end{bmatrix}} = \begin{bmatrix} 7-6 & -14+14 \\ 3-3 & -6+7 \end{bmatrix} = \underset{I_2}{\begin{bmatrix} 1 & 0 \\ 0 & 1 \end{bmatrix}}$$

and

$$\underset{A}{\begin{bmatrix} -2 & 4 \\ -3 & 7 \end{bmatrix}} \underset{A^{-1}}{\begin{bmatrix} -\frac{7}{2} & 2 \\ -\frac{3}{2} & 1 \end{bmatrix}} = \begin{bmatrix} 7-6 & -4+4 \\ \frac{21}{2}-\frac{21}{2} & -6+7 \end{bmatrix} = \underset{I_2}{\begin{bmatrix} 1 & 0 \\ 0 & 1 \end{bmatrix}}.$$

The inverse of a matrix can be calculated using Gaussian elimination, as the next example illustrates.

EXAMPLE 1   Let $A = \begin{bmatrix} 3 & 1 \\ 5 & 2 \end{bmatrix}$. Determine A^{-1}.

Solution   Since A is a 2×2 matrix, A^{-1} is also a 2×2 matrix and satisfies

$$AA^{-1} = I_2 \quad \text{and} \quad A^{-1}A = I_2, \tag{4}$$

where $I_2 = \begin{bmatrix} 1 & 0 \\ 0 & 1 \end{bmatrix}$ is a 2×2 identity matrix. Suppose that

$$A^{-1} = \begin{bmatrix} x & y \\ z & w \end{bmatrix}.$$

Then the first equation of (4) reads

$$\begin{bmatrix} 3 & 1 \\ 5 & 2 \end{bmatrix} \begin{bmatrix} x & y \\ z & w \end{bmatrix} = \begin{bmatrix} 1 & 0 \\ 0 & 1 \end{bmatrix}.$$

Multiplying out the matrices on the left gives

$$\begin{bmatrix} 3x + z & 3y + w \\ 5x + 2z & 5y + 2w \end{bmatrix} = \begin{bmatrix} 1 & 0 \\ 0 & 1 \end{bmatrix}.$$

Now equate corresponding elements in the two matrices to obtain the equations

$$\begin{cases} 3x + z = 1 \\ 5x + 2z = 0, \end{cases} \qquad \begin{cases} 3y + w = 0 \\ 5y + 2w = 1. \end{cases}$$

Notice that the equations break up into two pairs of linear equations, each pair involving only two variables. Solving these two systems of linear equations yields $x = 2, z = -5, y = -1, w = 3$. Therefore,

$$A^{-1} = \begin{bmatrix} 2 & -1 \\ -5 & 3 \end{bmatrix}.$$

Indeed, we may readily verify that

$$\begin{bmatrix} 3 & 1 \\ 5 & 2 \end{bmatrix}\begin{bmatrix} 2 & -1 \\ -5 & 3 \end{bmatrix} = \begin{bmatrix} 1 & 0 \\ 0 & 1 \end{bmatrix}$$

$$\begin{bmatrix} 2 & -1 \\ -5 & 3 \end{bmatrix}\begin{bmatrix} 3 & 1 \\ 5 & 2 \end{bmatrix} = \begin{bmatrix} 1 & 0 \\ 0 & 1 \end{bmatrix}.$$

The method above can be used to calculate the inverse of matrices of any size, although it involves considerable calculation. We shall provide a rather efficient computational method for calculating A^{-1} in the next section. For now, however, let us be content with the above method. Using it, we can derive a general formula for A^{-1} in case A is a 2×2 matrix.

To determine the inverse of a 2×2 matrix Let

$$A = \begin{bmatrix} a & b \\ c & d \end{bmatrix}.$$

Let $\Delta = ad - bc$ and assume that $\Delta \neq 0$. Then A^{-1} is given by the formula

$$A^{-1} = \begin{bmatrix} \dfrac{d}{\Delta} & -\dfrac{b}{\Delta} \\[2ex] -\dfrac{c}{\Delta} & \dfrac{a}{\Delta} \end{bmatrix}. \tag{5}$$

We will omit the derivation of this formula. It proceeds along lines similar to those of Example 1. Notice that formula (5) involves division by Δ. Since division by 0 is not permissible, it is necessary that $\Delta \neq 0$ for formula (5) to be applied. We will discuss the case $\Delta = 0$ below.

Equation (5) can be reduced to a simple step-by-step procedure.

To determine the inverse of $\begin{bmatrix} a & b \\ c & d \end{bmatrix}$ *if* $\Delta = ad - bc \neq 0$:

1. Interchange a and d to get $\begin{bmatrix} d & b \\ c & a \end{bmatrix}$.

2. Change the signs of b and c to get $\begin{bmatrix} d & -b \\ -c & a \end{bmatrix}$.

3. Divide all entries by Δ to get $\begin{bmatrix} \dfrac{d}{\Delta} & -\dfrac{b}{\Delta} \\[2ex] -\dfrac{c}{\Delta} & \dfrac{a}{\Delta} \end{bmatrix}.$

EXAMPLE 2   Calculate the inverse of $\begin{bmatrix} -2 & 4 \\ -3 & 7 \end{bmatrix}$.

Solution $\Delta = (-2) \cdot 7 - 4 \cdot (-3) = -2$, so $\Delta \neq 0$, and we may use the computation above.

  1. Interchange a and d:

$$\begin{bmatrix} 7 & 4 \\ -3 & -2 \end{bmatrix}$$

  2. Change signs of b and c:

$$\begin{bmatrix} 7 & -4 \\ 3 & -2 \end{bmatrix}$$

  3. Divide all entries by $\Delta = -2$:

$$\begin{bmatrix} -\frac{7}{2} & 2 \\ -\frac{3}{2} & 1 \end{bmatrix}$$

Thus

$$\begin{bmatrix} -2 & 4 \\ -3 & 7 \end{bmatrix}^{-1} = \begin{bmatrix} -\frac{7}{2} & 2 \\ -\frac{3}{2} & 1 \end{bmatrix}.$$

Not every square matrix has an inverse. Indeed, it may be impossible to satisfy equations (3) for any choice of A^{-1}. This phenomenon can even occur in the case of 2×2 matrices. Here, one can show that *if $\Delta = 0$, then the matrix does not have an inverse.* The next example illustrates this phenomenon in a special case.

EXAMPLE 3   Show that $\begin{bmatrix} 1 & 1 \\ 1 & 1 \end{bmatrix}$ does not have an inverse.

Solution   Note first that $\Delta = 1 \cdot 1 - 1 \cdot 1 = 0$, so the inverse cannot be computed via equation (5). Suppose that the given matrix did have an inverse, say

$$\begin{bmatrix} s & t \\ u & v \end{bmatrix}.$$

Then the following equation would hold:

$$\begin{bmatrix} s & t \\ u & v \end{bmatrix} \begin{bmatrix} 1 & 1 \\ 1 & 1 \end{bmatrix} = \begin{bmatrix} 1 & 0 \\ 0 & 1 \end{bmatrix}.$$

On multiplying out the two matrices on the left, we get the equation

$$\begin{bmatrix} s + t & s + t \\ u + v & u + v \end{bmatrix} = \begin{bmatrix} 1 & 0 \\ 0 & 1 \end{bmatrix},$$

or, on equating entries in the first row:

$$s + t = 1, \qquad s + t = 0.$$

But $s + t$ cannot equal both 1 and 0. So we reach a contradiction, and therefore the original matrix cannot have an inverse.

We were led to introduce the inverse of a matrix from a discussion of the matrix equation $AX = B$. Let us now return to that discussion. Suppose that A and B are given matrices and that we wish to solve the matrix equation

$$AX = B$$

for the unknown matrix X. Suppose further that A has an inverse A^{-1}. Multiply both sides of the equation on the left by A^{-1} to obtain

$$A^{-1} \cdot AX = A^{-1}B.$$

Because $A^{-1} \cdot A = I$, we have

$$IX = A^{-1}B$$

$$X = A^{-1}B.$$

Thus the matrix X is found by simply multiplying B on the left by A^{-1}, and we can summarize our findings as follows:

> *Solving a Matrix Equation* If the matrix A has an inverse, then the solution of the matrix equation
>
> $$AX = B$$
>
> is given by
>
> $$X = A^{-1}B.$$

Matrix equations can be used to solve systems of linear equations, as illustrated in the next example.

EXAMPLE 4   Use a matrix equation to solve the system of linear equations

$$\begin{cases} -2x + 4y = 2 \\ -3x + 7y = 7. \end{cases}$$

Solution   In Example 6 of the preceding section we saw that the system could be written as a matrix equation:

$$\underset{A}{\begin{bmatrix} -2 & 4 \\ -3 & 7 \end{bmatrix}} \underset{X}{\begin{bmatrix} x \\ y \end{bmatrix}} = \underset{B}{\begin{bmatrix} 2 \\ 7 \end{bmatrix}}.$$

We happen to know A^{-1} from Example 2, namely

$$A^{-1} = \begin{bmatrix} -\frac{7}{2} & 2 \\ -\frac{3}{2} & 1 \end{bmatrix}.$$

So we may compute the matrix $X = A^{-1}B$:

$$X = \begin{bmatrix} x \\ y \end{bmatrix} = \begin{bmatrix} -\frac{7}{2} & 2 \\ -\frac{3}{2} & 1 \end{bmatrix} \begin{bmatrix} 2 \\ 7 \end{bmatrix} = \begin{bmatrix} 7 \\ 4 \end{bmatrix}.$$

Thus the solution of the system is $x = 7$, $y = 4$.

Here is an application of matrix questions which is a preview of the discussion of stochastic matrices in Chapter 8.

EXAMPLE 5 Let x and y denote the number of married and single adults in a certain town as of January 1. Let m and s denote the corresponding numbers for the following year. A statistical survey shows that x, y, m, and s are related by the equations

$$.9x + .2y = m$$

$$.1x + .8y = s.$$

In a given year there were found to be 490,000 married adults and 147,000 single adults.

(a) How many married adults were there in the preceding year?

(b) How many married adults were there 2 years ago?

Solution (a) The given equations can be written in the matrix form

$$AX = B,$$

where

$$A = \begin{bmatrix} .9 & .2 \\ .1 & .8 \end{bmatrix}, \qquad X = \begin{bmatrix} x \\ y \end{bmatrix}, \qquad B = \begin{bmatrix} m \\ s \end{bmatrix}.$$

We are given that $B = \begin{bmatrix} 490{,}000 \\ 147{,}000 \end{bmatrix}$. So, since

$$X = A^{-1}B$$

and

$$A^{-1} = \begin{bmatrix} \frac{8}{7} & -\frac{2}{7} \\ -\frac{1}{7} & \frac{9}{7} \end{bmatrix},$$

we have

$$X = \begin{bmatrix} \frac{8}{7} & -\frac{2}{7} \\ -\frac{1}{7} & \frac{9}{7} \end{bmatrix} \begin{bmatrix} 490{,}000 \\ 147{,}000 \end{bmatrix} = \begin{bmatrix} 518{,}000 \\ 119{,}000 \end{bmatrix}.$$

Thus last year there were 518,000 married adults and 119,000 single adults.

(b) We deduce x and y for two years ago from the values of m and s for last year, namely $m = 518{,}000$, $s = 119{,}000$.

$$X = A^{-1}B = \begin{bmatrix} \frac{8}{7} & -\frac{2}{7} \\ -\frac{1}{7} & \frac{9}{7} \end{bmatrix} \begin{bmatrix} 518{,}000 \\ 119{,}000 \end{bmatrix} = \begin{bmatrix} 558{,}000 \\ 79{,}000 \end{bmatrix}.$$

That is, 2 years ago there were 558,000 married adults and 79,000 single adults.

EXAMPLE 6   In the next section we will show that if

$$A = \begin{bmatrix} 4 & -2 & 3 \\ 8 & -3 & 5 \\ 7 & -2 & 4 \end{bmatrix}, \quad \text{then} \quad A^{-1} = \begin{bmatrix} -2 & 2 & -1 \\ 3 & -5 & 4 \\ 5 & -6 & 4 \end{bmatrix}.$$

(a) Use this fact to solve the system of linear equations

$$\begin{cases} 4x - 2y + 3z = 1 \\ 8x - 3y + 5z = 4 \\ 7x - 2y + 4z = 5. \end{cases}$$

(b) Solve the system of equations

$$\begin{cases} 4x - 2y + 3z = 4 \\ 8x - 3y + 5z = 7 \\ 7x - 2y + 4z = 6. \end{cases}$$

Solution   (a) The system can be written in the matrix form

$$\underset{A}{\begin{bmatrix} 4 & -2 & 3 \\ 8 & -3 & 5 \\ 7 & -2 & 4 \end{bmatrix}} \underset{X}{\begin{bmatrix} x \\ y \\ z \end{bmatrix}} = \underset{B}{\begin{bmatrix} 1 \\ 4 \\ 5 \end{bmatrix}}.$$

The solution of this matrix equation is $X = A^{-1}B$ or

$$\begin{bmatrix} x \\ y \\ z \end{bmatrix} = \begin{bmatrix} -2 & 2 & -1 \\ 3 & -5 & 4 \\ 5 & -6 & 4 \end{bmatrix} \begin{bmatrix} 1 \\ 4 \\ 5 \end{bmatrix} = \begin{bmatrix} 1 \\ 3 \\ 1 \end{bmatrix}.$$

Thus the solution of the system is $x = 1$, $y = 3$, $z = 1$.

(b) This system has the same left-hand side as the preceding system, so its solution is

$$\begin{bmatrix} x \\ y \\ z \end{bmatrix} = \begin{bmatrix} -2 & 2 & -1 \\ 3 & -5 & 4 \\ 5 & -6 & 4 \end{bmatrix} \begin{bmatrix} 4 \\ 7 \\ 6 \end{bmatrix} = \begin{bmatrix} 0 \\ 1 \\ 2 \end{bmatrix}.$$

That is, the solution of the system is $x = 0$, $y = 1$, $z = 2$.

Using the method of matrix equations to solve a system of linear equations is especially efficient if one wishes to solve a number of systems all having the same left-hand sides but different right-hand sides. For then A^{-1} must be computed only once for all the systems under consideration. (This point is useful in Exercises 17–20.)

PRACTICE PROBLEMS 4

1. Show that the inverse of

$$\begin{bmatrix} -4 & 1 & 2 \\ 7 & -1 & -4 \\ -\frac{1}{2} & 0 & \frac{1}{2} \end{bmatrix} \text{ is } \begin{bmatrix} 1 & 1 & 4 \\ 3 & 2 & 4 \\ 1 & 1 & 6 \end{bmatrix}.$$

2. Use the method of this section to solve the system of linear equations

$$\begin{cases} .8x + .6y = 5 \\ .2x + .4y = 2. \end{cases}$$

EXERCISES 4

In Exercises 1 and 2, use the fact that

$$\begin{bmatrix} 2 & 2 \\ \frac{1}{2} & 1 \end{bmatrix}^{-1} = \begin{bmatrix} 1 & -2 \\ -\frac{1}{2} & 2 \end{bmatrix}.$$

1. Solve $\begin{cases} 2x + 2y = 4 \\ \frac{1}{2}x + y = 1. \end{cases}$

2. Solve $\begin{cases} 2x + 2y = 14 \\ \frac{1}{2}x + y = 4. \end{cases}$

In Exercises 3–10, find the inverse of the given matrix.

3. $\begin{bmatrix} 7 & 2 \\ 3 & 1 \end{bmatrix}$

4. $\begin{bmatrix} 2 & 3 \\ 5 & 7 \end{bmatrix}$

5. $\begin{bmatrix} 6 & 2 \\ 5 & 2 \end{bmatrix}$

6. $\begin{bmatrix} 1 & .5 \\ 0 & .5 \end{bmatrix}$

7. $\begin{bmatrix} .7 & .2 \\ .3 & .8 \end{bmatrix}$

8. $\begin{bmatrix} 0 & 1 \\ 1 & 0 \end{bmatrix}$

9. $[3]$

10. $[.2]$

In Exercises 11–14, use the method of this section to solve the system of linear equations.

11. $\begin{cases} x + 2y = 3 \\ 2x + 6y = 5 \end{cases}$

12. $\begin{cases} 5x + 3y = 1 \\ 7x + 4y = 2 \end{cases}$

13. $\begin{cases} \frac{1}{2}x + 2y = 4 \\ 3x + 16y = 0 \end{cases}$

14. $\begin{cases} .8x + .6y = 2 \\ .2x + .4y = 1 \end{cases}$

15. It is found that the number of married and single adults in a certain town are subject to the following statistics. Suppose that x and y denote the number of married and single adults, respectively, in a given year (say as of January 1) and let m, s denote the corre-

sponding numbers for the following year. Then

$$.8x + .3y = m$$

$$.2x + .7y = s.$$

(a) Write this system of equations in matrix form.

(b) Solve the resulting matrix equation for $X = \begin{bmatrix} x \\ y \end{bmatrix}$.

(c) Suppose that in a given year there were found to be 100,000 married adults and 50,000 single adults. How many married (resp. single) adults were there the preceding year?

(d) How many married (resp. single) adults were there 2 years ago?

16. A flu epidemic is spreading through a town of 48,000 people. It is found that if x and y denote the numbers of people sick and well in a given week, respectively, and if s and w denote the corresponding numbers for the following week, then

$$\tfrac{1}{3}x + \tfrac{1}{4}y = s$$

$$\tfrac{2}{3}x + \tfrac{3}{4}y = w.$$

(a) Write this system of equations in matrix form.

(b) Solve the resulting matrix equation for $X = \begin{bmatrix} x \\ y \end{bmatrix}$.

(c) Suppose that 13,000 people are sick in a given week. How many were sick the preceding week?

(d) Same question as part (c), except assume that 14,000 are sick.

In Exercises 17 and 18, use the fact that

$$\begin{bmatrix} 1 & 2 & 2 \\ 1 & 3 & 2 \\ 1 & 2 & 3 \end{bmatrix}^{-1} = \begin{bmatrix} 5 & -2 & -2 \\ -1 & 1 & 0 \\ -1 & 0 & 1 \end{bmatrix}.$$

17. Solve $\begin{cases} x + 2y + 2z = 1 \\ x + 3y + 2z = -1 \\ x + 2y + 3z = -1. \end{cases}$

18. Solve $\begin{cases} x + 2y + 2z = 1 \\ x + 3y + 2z = 0 \\ x + 2y + 3z = 0. \end{cases}$

In Exercises 19 and 20, use the fact that

$$\begin{bmatrix} 9 & 0 & 2 & 0 \\ -20 & -9 & -5 & 5 \\ 4 & 0 & 1 & 0 \\ -4 & -2 & -1 & 1 \end{bmatrix}^{-1} = \begin{bmatrix} 1 & 0 & -2 & 0 \\ 0 & 1 & 0 & -5 \\ -4 & 0 & 9 & 0 \\ 0 & 2 & 1 & -9 \end{bmatrix}.$$

19. Solve $\begin{cases} 9x \quad\quad + 2z \quad\quad = 1 \\ -20x - 9y - 5z + 5w = 0 \\ 4x \quad\quad + z \quad\quad = 0 \\ -4x - 2y - z + w = -1. \end{cases}$

20. Solve $\begin{cases} 9x \quad\quad + 2z \quad\quad = 2 \\ -20x - 9y - 5z + 5w = 1 \\ 4x \quad\quad + z \quad\quad = 3 \\ -4x - 2y - z + w = 0. \end{cases}$

1. To see if this matrix is indeed the inverse, multiply it by the original matrix and find out if the products are identity matrices.

$$\begin{bmatrix} 1 & 1 & 4 \\ 3 & 2 & 4 \\ 1 & 1 & 6 \end{bmatrix} \begin{bmatrix} -4 & 1 & 2 \\ 7 & -1 & -4 \\ -\frac{1}{2} & 0 & \frac{1}{2} \end{bmatrix} = \begin{bmatrix} 1 & 0 & 0 \\ 0 & 1 & 0 \\ 0 & 0 & 1 \end{bmatrix}, \qquad \text{an identity matrix.}$$

$$\begin{bmatrix} -4 & 1 & 2 \\ 7 & -1 & -4 \\ -\frac{1}{2} & 0 & \frac{1}{2} \end{bmatrix} \begin{bmatrix} 1 & 1 & 4 \\ 3 & 2 & 4 \\ 1 & 1 & 6 \end{bmatrix} = \begin{bmatrix} 1 & 0 & 0 \\ 0 & 1 & 0 \\ 0 & 0 & 1 \end{bmatrix}.$$

2. The matrix form of this system is

$$\begin{bmatrix} .8 & .6 \\ .2 & .4 \end{bmatrix} \begin{bmatrix} x \\ y \end{bmatrix} = \begin{bmatrix} 5 \\ 2 \end{bmatrix}.$$

Therefore, the solution is

$$\begin{bmatrix} x \\ y \end{bmatrix} = \begin{bmatrix} .8 & .6 \\ .2 & .4 \end{bmatrix}^{-1} \begin{bmatrix} 5 \\ 2 \end{bmatrix}.$$

To compute the inverse of the 2×2 matrix, first compute Δ.

$$\Delta = ad - bc = (.8)(.4) - (.6)(.2) = .32 - .12 = .2.$$

Thus

$$\begin{bmatrix} .8 & .6 \\ .2 & .4 \end{bmatrix}^{-1} = \begin{bmatrix} .4/.2 & -.6/.2 \\ -.2/.2 & .8/.2 \end{bmatrix} = \begin{bmatrix} 2 & -3 \\ -1 & 4 \end{bmatrix}.$$

Therefore,

$$\begin{bmatrix} x \\ y \end{bmatrix} = \begin{bmatrix} 2 & -3 \\ -1 & 4 \end{bmatrix} \begin{bmatrix} 5 \\ 2 \end{bmatrix} = \begin{bmatrix} 4 \\ 3 \end{bmatrix}.$$

So the solution is $x = 4$, $y = 3$.

5.5. The Gauss-Jordan Method for Calculating Inverses

Of the several popular methods for finding the inverse of a matrix, the Gauss-Jordan method is probably the easiest to describe. It can be used on square matrices of any size. Also, the mechanical nature of the computations allows this method to be programmed for a computer with relative ease. We shall illustrate the procedure with a 2×2 matrix, whose inverse can also be calculated using the method of the previous section. Let

$$A = \begin{bmatrix} \frac{1}{2} & 1 \\ 1 & 3 \end{bmatrix}.$$

It is simple to check that

$$A^{-1} = \begin{bmatrix} 6 & -2 \\ -2 & 1 \end{bmatrix}.$$

Let us now derive this result using the Gauss-Jordan method.

> *Step 1* Write down the matrix A, and on its right an identity matrix of the same size.

This is most conveniently done by placing I_2 beside A in a single matrix.

$$\left[\begin{array}{cc|cc} \frac{1}{2} & 1 & 1 & 0 \\ 1 & 3 & 0 & 1 \end{array}\right].$$
$$\underbrace{}_{A} \quad \underbrace{}_{I_2}$$

> *Step 2* Perform elementary row operations on the left-hand matrix so as to transform it into an identity matrix. Each operation performed on the left-hand matrix is also performed on the right-hand matrix.

This step proceeds exactly like the Gaussian elimination method and may be most conveniently expressed in terms of pivoting.

$$\left[\begin{array}{cc|cc} \boxed{\tfrac{1}{2}} & 1 & 1 & 0 \\ 1 & 3 & 0 & 1 \end{array}\right], \quad \left[\begin{array}{cc|cc} 1 & 2 & 2 & 0 \\ 0 & \boxed{1} & -2 & 1 \end{array}\right], \quad \left[\begin{array}{cc|cc} 1 & 0 & 6 & -2 \\ 0 & 1 & -2 & 1 \end{array}\right].$$

> *Step 3* When the matrix on the left becomes an identity matrix, the matrix on the right is the desired inverse.

So, from the last matrix of our calculation above, we have

$$A^{-1} = \begin{bmatrix} 6 & -2 \\ -2 & 1 \end{bmatrix}.$$

This is the same result obtained earlier.

We will demonstrate why the method above works after some further examples.

EXAMPLE 1 Find the inverse of the matrix

$$A = \begin{bmatrix} 4 & -2 & 3 \\ 8 & -3 & 5 \\ 7 & -2 & 4 \end{bmatrix}.$$

Solution

$$\begin{bmatrix} ④ & -2 & 3 & | & 1 & 0 & 0 \\ 8 & -3 & 5 & | & 0 & 1 & 0 \\ 7 & -2 & 4 & | & 0 & 0 & 1 \end{bmatrix}$$

$$\begin{bmatrix} 1 & -\frac{1}{2} & \frac{3}{4} & | & \frac{1}{4} & 0 & 0 \\ 0 & ① & -1 & | & -2 & 1 & 0 \\ 0 & \frac{3}{2} & -\frac{5}{4} & | & -\frac{7}{4} & 0 & 1 \end{bmatrix}$$

$$\begin{bmatrix} 1 & 0 & \frac{1}{4} & | & -\frac{3}{4} & \frac{1}{2} & 0 \\ 0 & 1 & -1 & | & -2 & 1 & 0 \\ 0 & 0 & ①{\scriptstyle\frac{1}{4}} & | & \frac{5}{4} & -\frac{3}{2} & 1 \end{bmatrix}$$

$$\begin{bmatrix} 1 & 0 & 0 & | & -2 & 2 & -1 \\ 0 & 1 & 0 & | & 3 & -5 & 4 \\ 0 & 0 & 1 & | & 5 & -6 & 4 \end{bmatrix}.$$

Therefore,

$$A^{-1} = \begin{bmatrix} -2 & 2 & -1 \\ 3 & -5 & 4 \\ 5 & -6 & 4 \end{bmatrix}.$$

Not all square matrices have inverses. If a matrix does not have an inverse, this will become apparent when applying the Gauss-Jordan method. At some point there will be no way to continue transforming the left-hand matrix into an identity matrix. This is illustrated in the next example.

EXAMPLE 2 Find the inverse of the matrix

$$A = \begin{bmatrix} 1 & 3 & 2 \\ 0 & 1 & 4 \\ 1 & 5 & 10 \end{bmatrix}.$$

Solution

$$\begin{bmatrix} 1 & 3 & 2 & | & 1 & 0 & 0 \\ ⓪ & 1 & 4 & | & 0 & 1 & 0 \\ 1 & 5 & 10 & | & 0 & 0 & 1 \end{bmatrix}$$

$$\begin{bmatrix} 1 & 3 & 2 & | & 1 & 0 & 0 \\ 0 & ① & 4 & | & 0 & 1 & 0 \\ 0 & 2 & 8 & | & -1 & 0 & 1 \end{bmatrix}$$

$$\begin{bmatrix} 1 & 0 & -10 & | & 1 & -3 & 0 \\ 0 & 1 & 4 & | & 0 & 1 & 0 \\ 0 & 0 & 0 & | & -1 & -2 & 1 \end{bmatrix}.$$

Since the third row of the left-hand matrix has only zero entries, it is impossible to complete the Gauss-Jordan method. Therefore, the matrix A has no inverse matrix.

In the preceding section we showed how to calculate the inverse by solving several systems of linear equations. Actually, the Gauss-Jordan method is just an organized way of going about the calculation. To see why, let us consider a concrete example:

$$A = \begin{bmatrix} 4 & -2 & 3 \\ 8 & -3 & 5 \\ 7 & -2 & 4 \end{bmatrix}.$$

We wish to determine A^{-1}, so regard it as a matrix of unknowns:

$$A^{-1} = \begin{bmatrix} x_1 & x_2 & x_3 \\ y_1 & y_2 & y_3 \\ z_1 & z_2 & z_3 \end{bmatrix}.$$

The statement $AA^{-1} = I_3$ is

$$\begin{bmatrix} 4 & -2 & 3 \\ 8 & -3 & 5 \\ 7 & -2 & 4 \end{bmatrix} \begin{bmatrix} x_1 & x_2 & x_3 \\ y_1 & y_2 & y_3 \\ z_1 & z_2 & z_3 \end{bmatrix} = \begin{bmatrix} 1 & 0 & 0 \\ 0 & 1 & 0 \\ 0 & 0 & 1 \end{bmatrix}.$$

Multiplying out the matrices on the left and comparing the result with the matrix on the right gives us nine equations, namely

$$\begin{cases} 4x_1 - 2y_1 + 3z_1 = 1 \\ 8x_1 - 3y_1 + 5z_1 = 0 \\ 7x_1 - 2y_1 + 4z_1 = 0 \end{cases}$$

$$\begin{cases} 4x_2 - 2y_2 + 3z_2 = 0 \\ 8x_2 - 3y_2 + 5z_2 = 1 \\ 7x_2 - 2y_2 + 4z_2 = 0 \end{cases}$$

$$\begin{cases} 4x_3 - 2y_3 + 3z_3 = 0 \\ 8x_3 - 3y_3 + 5z_3 = 0 \\ 7x_3 - 2y_3 + 4z_3 = 1. \end{cases}$$

Notice that each system of equations corresponds to one column of unknowns in A^{-1}. More precisely, if we set

$$X_1 = \begin{bmatrix} x_1 \\ y_1 \\ z_1 \end{bmatrix}, \qquad X_2 = \begin{bmatrix} x_2 \\ y_2 \\ z_2 \end{bmatrix}, \qquad X_3 = \begin{bmatrix} x_3 \\ y_3 \\ z_3 \end{bmatrix},$$

then the three systems above have the respective matrix forms

$$AX_1 = \begin{bmatrix} 1 \\ 0 \\ 0 \end{bmatrix}, \qquad AX_2 = \begin{bmatrix} 0 \\ 1 \\ 0 \end{bmatrix}, \qquad AX_3 = \begin{bmatrix} 0 \\ 0 \\ 1 \end{bmatrix}.$$

Now imagine the process of applying Gaussian elimination to solve these three systems. We apply elementary row operations to the matrices

$$\left[\begin{array}{c|c} A & \begin{array}{c} 1 \\ 0 \\ 0 \end{array} \end{array}\right], \quad \left[\begin{array}{c|c} A & \begin{array}{c} 0 \\ 1 \\ 0 \end{array} \end{array}\right], \quad \left[\begin{array}{c|c} A & \begin{array}{c} 0 \\ 0 \\ 1 \end{array} \end{array}\right].$$

The process ends when we convert A into the identity matrix, at which point the solutions may be read off the right column. So the procedure ends with the matrices

$$[I_3 \mid X_1], \quad [I_3 \mid X_2], \quad [I_3 \mid X_3].$$

Realize, however, that at each step of the three Gaussian eliminations we are performing the same operations, since all three start with the matrix A on the left. So, in order to save calculation, perform the three Gaussian eliminations simultaneously by performing the row operations on the composite matrix

$$\left[\begin{array}{c|ccc} A & 1 & 0 & 0 \\ & 0 & 1 & 0 \\ & 0 & 0 & 1 \end{array}\right] = [A \mid I_3].$$

The procedure ends when this matrix is converted into

$$[I_3 \mid X_1 \; X_2 \; X_3].$$

That is, since $A^{-1} = [X_1 \; X_2 \; X_3]$, the procedure ends with A^{-1} on the right. This is the reasoning behind the Gauss-Jordan method of calculating inverses.

PRACTICE PROBLEMS 5

1. Use the Gauss-Jordan method to calculate the inverse of the matrix

$$\begin{bmatrix} 1 & 0 & 2 \\ 0 & 1 & -4 \\ 0 & 0 & 2 \end{bmatrix}.$$

2. Solve the system of linear equations

$$\begin{cases} x & + 2z = 4 \\ y - 4z = 6 \\ 2z = 9. \end{cases}$$

EXERCISES 5

Use the Gauss-Jordan method to compute the inverses of the following matrices.

1. $\begin{bmatrix} 7 & 3 \\ 5 & 2 \end{bmatrix}$

2. $\begin{bmatrix} 5 & -2 \\ 6 & 2 \end{bmatrix}$

3. $\begin{bmatrix} 10 & 12 \\ 3 & -4 \end{bmatrix}$

4. $\begin{bmatrix} 1 & -3 \\ 0 & 1 \end{bmatrix}$

5. $\begin{bmatrix} 2 & -4 \\ -1 & 2 \end{bmatrix}$

6. $\begin{bmatrix} 1 & 3 & 1 \\ -1 & 2 & 0 \\ 2 & 11 & 3 \end{bmatrix}$

7. $\begin{bmatrix} 1 & 2 & -2 \\ 1 & 1 & 1 \\ 0 & 0 & 1 \end{bmatrix}$

8. $\begin{bmatrix} 2 & 2 & 0 \\ 0 & -2 & 0 \\ 3 & 0 & 1 \end{bmatrix}$

9. $\begin{bmatrix} -2 & 5 & 2 \\ 1 & -3 & -1 \\ -1 & 2 & 1 \end{bmatrix}$

10. $\begin{bmatrix} 1 & 0 & 0 \\ 2 & 1 & -2 \\ -1 & 2 & 1 \end{bmatrix}$

11. $\begin{bmatrix} 1 & 6 & 0 & 0 \\ 1 & 5 & 0 & 0 \\ 0 & 0 & 4 & 2 \\ 0 & 0 & 50 & 2 \end{bmatrix}$

12. $\begin{bmatrix} 6 & 0 & 2 & 0 \\ -6 & 1 & 0 & 1 \\ 1 & 0 & 1 & 0 \\ -9 & 0 & -1 & 1 \end{bmatrix}$

In Exercises 13-16, use matrix inversion to solve the system of linear equations.

13. $\begin{cases} x + y + 2z = 3 \\ 3x + 2y + 2z = 4 \\ x + y + 3z = 5 \end{cases}$

14. $\begin{cases} x + 2y + 3z = 4 \\ 3x + 5y + 5z = 3 \\ 2x + 4y + 2z = 4 \end{cases}$

15. $\begin{cases} x \quad\;\; - 2z - 2w = 0 \\ y \quad\;\; - 5w = 1 \\ -4x \quad\;\; + 9z + 9w = 2 \\ 2y + z - 8w = 3 \end{cases}$

16. $\begin{cases} y + 2z = 1 \\ 2x + y + 3z = 2 \\ x + y + 2z = 3 \end{cases}$

SOLUTIONS TO PRACTICE PROBLEMS 5

1. First write the given matrix beside an identity matrix of the same size

$$\left[\begin{array}{ccc|ccc} 1 & 0 & 2 & 1 & 0 & 0 \\ 0 & 1 & -4 & 0 & 1 & 0 \\ 0 & 0 & 2 & 0 & 0 & 1 \end{array}\right].$$

The object is to use elementary row operations to transform the 3×3 matrix on the left into the identity matrix. The first two columns are already in the correct form.

$$\left[\begin{array}{ccc|ccc} 1 & 0 & 2 & 1 & 0 & 0 \\ 0 & 1 & -4 & 0 & 1 & 0 \\ 0 & 0 & 2 & 0 & 0 & 1 \end{array}\right]$$

$\xrightarrow{\;\frac{1}{2}[3]\;}$ $\left[\begin{array}{ccc|ccc} 1 & 0 & 2 & 1 & 0 & 0 \\ 0 & 1 & -4 & 0 & 1 & 0 \\ 0 & 0 & 1 & 0 & 0 & \frac{1}{2} \end{array}\right]$

$\xrightarrow{\;[1] + (-2)[3]\;}$ $\left[\begin{array}{ccc|ccc} 1 & 0 & 0 & 1 & 0 & -1 \\ 0 & 1 & -4 & 0 & 1 & 0 \\ 0 & 0 & 1 & 0 & 0 & \frac{1}{2} \end{array}\right]$

$\xrightarrow{\;[2] + (4)[3]\;}$ $\left[\begin{array}{ccc|ccc} 1 & 0 & 0 & 1 & 0 & -1 \\ 0 & 1 & 0 & 0 & 1 & 2 \\ 0 & 0 & 1 & 0 & 0 & \frac{1}{2} \end{array}\right].$

Thus the inverse of the given matrix is

$$\begin{bmatrix} 1 & 0 & -1 \\ 0 & 1 & 2 \\ 0 & 0 & \frac{1}{2} \end{bmatrix}.$$

2. The matrix form of this system of equations is $AX = B$, where A is the matrix whose inverse was found in Problem 1, and

$$B = \begin{bmatrix} 4 \\ 6 \\ 9 \end{bmatrix}.$$

Therefore, $X = A^{-1}B$, so that

$$\begin{bmatrix} x \\ y \\ z \end{bmatrix} = \begin{bmatrix} 1 & 0 & -1 \\ 0 & 1 & 2 \\ 0 & 0 & \frac{1}{2} \end{bmatrix}\begin{bmatrix} 4 \\ 6 \\ 9 \end{bmatrix} = \begin{bmatrix} -5 \\ 24 \\ \frac{9}{2} \end{bmatrix}.$$

So the solution of the system is $x = -5$, $y = 24$, $z = \frac{9}{2}$.

5.6. Input-Output Analysis

In recent years matrix arithmetic has played an ever-increasing role in economics, especially in that branch of economics called *input-output analysis*. Pioneered by the Harvard economist Vassily Leontieff, input-output analysis is used to analyze an economy in order to determine how much output must be produced by each segment of the economy in order to meet given consumption and export demands. As we shall see, such analysis leads into matrix calculations and in particular to inverses of matrices. Input-output analysis has been of such great significance that Leontieff was awarded the 1973 Nobel prize in economics for his fundamental work in the subject.

Suppose that we divide an economy up into a number of industries—transportation, agriculture, steel, and so on. Each industry produces a certain output using certain raw materials (or input). The input of each industry is made up in part by the outputs of other industries. For example, in order to produce food, agriculture uses as input the output of many industries, such as transportation (tractors and trucks) and oil (gasoline and fertilizers). This interdependence among the industries of the economy is summarized in a matrix—a so-called *input-output matrix*. There is one column for each industry's input requirements. The entries in the column reflect the amount of input required from each of the industries. A typical input-output matrix looks like this:

Input requirements of:

		Industry 1	Industry 2	Industry 3	...
	Industry 1				
From	Industry 2				
	Industry 3				
	$\vdots$				

It is most convenient to express the entries of this matrix in monetary terms. That is, each column gives the dollar values of the various inputs needed by an industry in order to produce $1 worth of output.

There are consumers (other than the industries themselves) who want to purchase some of the output of these industries. The quantity of goods that these consumers want (or demand) is called the *final demand* on the economy. The final demand can be represented by a column matrix, with one entry for each industry, indicating the amount of consumable output demanded from the industry:

$$[\text{final demand}] = \begin{bmatrix} \text{amount from industry 1} \\ \text{amount from industry 2} \\ \vdots \end{bmatrix}.$$

We shall consider the situation in which the final-demand matrix is given and it is necessary to determine how much output should be produced by each industry in order to provide the needed inputs of the various industries and also to satisfy the final demand. The proper level of output can be computed using matrix calculations as illustrated in the next example.

EXAMPLE 1 Suppose that an economy is composed of only three industries—coal, steel, and electricity. Each of these industries depends on the others for some of its raw materials. Suppose that to make $1 of coal, it takes no coal, but $.02 of steel and $.01 of electricity; to make $1 of steel, it takes $.15 of coal, $.03 of steel, and $.08 of electricity; and to make $1 of electricity, it takes $.43 of coal, $.20 of steel, and $.05 of electricity. How much should each industry produce to allow for consumption (not used for production) at these levels: $2 billion coal, $1 billion steel, $3 billion electricity?

Solution Put all the data indicating the interdependence of the industries in a matrix. In each industry's column put the amount of input from each of the industries needed to produce $1 of output in that particular industry:

$$\begin{array}{c} \\ \text{Coal} \\ \text{Steel} \\ \text{Electricity} \end{array} \begin{array}{ccc} \text{Coal} & \text{Steel} & \text{Electricity} \\ \begin{bmatrix} 0 & .15 & .43 \\ .02 & .03 & .20 \\ .01 & .08 & .05 \end{bmatrix} \end{array} = A.$$

This matrix is the *input-output matrix* corresponding to the economy. Let D denote the final-demand matrix. Then, letting the numbers in D stand for billions of dollars, we have

$$D = \begin{bmatrix} 2 \\ 1 \\ 3 \end{bmatrix}.$$

Suppose that the coal industry produces x billion dollars of output, the steel industry y billion dollars, and the electrical industry z billion dollars. Our problem is to determine the x, y, and z that yield the desired amounts left over from the production process. As an example, consider coal. The amount of coal that can be consumed or exported is just

$$x - [\text{amount of coal used in production}].$$

To determine the amount of coal used in production, refer to the input-output matrix. Production of x billion dollars of coal takes $0 \cdot x$ billion dollars of coal, production of y billion dollars of steel takes $.15y$ billion dollars of coal, and production of z billion dollars of electricty takes $.43z$ billion dollars of coal. Thus,

$$[\text{amount of coal used in production}] = 0 \cdot x + .15y + .43z.$$

This quantity should be recognized as the first entry of a matrix product. Namely, if we let

$$X = \begin{bmatrix} x \\ y \\ z \end{bmatrix},$$

then

$$\begin{bmatrix} \text{coal} \\ \text{steel} \\ \text{electricity} \end{bmatrix}_{\text{used in production}} = \begin{bmatrix} 0 & .15 & .43 \\ .02 & .03 & .20 \\ .01 & .08 & .05 \end{bmatrix} \begin{bmatrix} x \\ y \\ z \end{bmatrix}.$$

$$= AX.$$

But then the amount of each output available for purposes other than production is $X - AX$. That is, we have the matrix equation

$$X - AX = D.$$

To solve this equation for X, proceed as follows. Since $IX = X$, write the equation in the form

$$IX - AX = D$$

$$(I - A)X = D$$

$$\boxed{X = (I - A)^{-1}D.} \tag{1}$$

So, in other words, X may be found by multiplying D on the left by $(I - A)^{-1}$. Let us now do the arithmetic.

$$I - A = \begin{bmatrix} 1 & 0 & 0 \\ 0 & 1 & 0 \\ 0 & 0 & 1 \end{bmatrix} - \begin{bmatrix} 0 & .15 & .43 \\ .02 & .03 & .20 \\ .01 & .08 & .05 \end{bmatrix} = \begin{bmatrix} 1 & -.15 & -.43 \\ -.02 & .97 & -.20 \\ -.01 & -.08 & .95 \end{bmatrix}.$$

Applying the Gauss-Jordan method, we find

$$(I - A)^{-1} = \begin{bmatrix} 1.01 & .20 & .50 \\ .02 & 1.05 & .23 \\ .01 & .09 & 1.08 \end{bmatrix},$$

where all figures are carried to two decimal places (Exercise 5). Therefore,

$$X = (I - A)^{-1}D = \begin{bmatrix} 1.01 & .20 & .50 \\ .02 & 1.05 & .23 \\ .01 & .09 & 1.08 \end{bmatrix} \begin{bmatrix} 2 \\ 1 \\ 3 \end{bmatrix} = \begin{bmatrix} 3.72 \\ 1.78 \\ 3.35 \end{bmatrix}.$$

In other words, coal should produce $3.72 billion worth of output, steel $1.78 billion, and electricity $3.35 billion. This output will meet the required final demands from each industry.

The analysis above is useful in studying not only entire economies but also segments of economies and even individual companies.

EXAMPLE 2 A conglomerate has three divisions, which produce computers, semiconductors, and business forms. For each $1 of output, the computer division needs $.02 worth of computers, $.20 worth of semiconductors, and $.10 worth of business forms. For each $1 of output, the semiconductor division needs $.02 worth of computers, $.01 worth of semiconductors, and $.02 worth of business forms. For each $1 of output, the business forms division requires $.10 worth of computers and $.01 worth of business forms. The conglomerate estimates the sales demand to be $300,000,000 for the computer division, $100,000,000 for the semiconductor division, and $200,000,000 for the business forms division. At what level should each division produce in order to satisfy this demand?

Solution The conglomerate can be viewed as a miniature economy and its sales as the final demand. The input-output matrix for this "economy" is

	Computers	Semiconductors	Business forms
Computers	.02	.02	.10
Semiconductors	.20	.01	0
Business forms	.10	.02	.01

$$= A.$$

The final-demand matrix is

$$D = \begin{bmatrix} 3 \\ 1 \\ 2 \end{bmatrix},$$

where the demand is expressed in hundreds of millions of dollars. By equation (1) the matrix X, giving the desired levels of production for the various divisions, is given by

$$X = (I - A)^{-1}D.$$

But

$$I - A = \begin{bmatrix} .98 & -.02 & -.10 \\ -.20 & .99 & 0 \\ -.10 & -.02 & .99 \end{bmatrix},$$

so that (Exercise 6)

$$(I - A)^{-1} = \begin{bmatrix} 1.04 & .02 & .10 \\ .21 & 1.01 & .02 \\ .11 & .02 & 1.02 \end{bmatrix},$$

and

$$(I - A)^{-1}D = \begin{bmatrix} 3.34 \\ 1.68 \\ 2.39 \end{bmatrix}.$$

Therefore,

$$X = \begin{bmatrix} 3.34 \\ 1.68 \\ 2.39 \end{bmatrix}.$$

That is, the computer division should produce $334,000,000, the semiconductor division $168,000,000, and the business forms division $239,000,000.

Input-output analysis is usually applied to the entire economy of a country having hundreds of industries. The resulting matrix equation $(I - A)X = D$ could be solved by the Gaussian elimination method. However, it is best to find the inverse of $I - A$ and solve for X as we have done in the examples of this section. Over a short period, D might change but A is unlikely to change. Therefore, the proper outputs to satisfy the new demand can easily be determined by using the already computed inverse of $I - A$.

The Closed Leontieff Model   The foregoing description of an economy is usually called the *Leontieff open model*, since it views exports as an activity that takes place external to the economy. However, it is possible to consider export as but another industry in the economy. Instead of describing exports by a demand column D, we describe it by a column in the input-output matrix. That is, the export column describes how each dollar of exports is divided among the various industries. Since exports are now regarded as another industry, each of the original columns has an additional entry, namely the amount of output from the export industry (that is, imports) used to produce $1 of goods (of the industry corresponding to the column). If A denotes the expanded input-output matrix and X the production matrix (as before), then AX is the matrix describing the total demand experienced by each of the industries. In order for the economy to function efficiently, the total amount demanded by the various industries should equal the amount produced. That is, the production matrix must satisfy the equation

$$AX = X.$$

By studying the solutions to this equation, it is possible to determine the equilibrium states of the economy—that is, the production matrices X for which the amounts produced exactly equal the amounts needed by the various industries. The model just described is called *Leontieff's closed model*.

We may expand the Leontieff closed model to include the effects of labor and monetary phenomena by considering labor and banking as yet further industries to be incorporated in the input-output matrix.

PRACTICE PROBLEMS 6

1. Let

$$I = \begin{bmatrix} 1 & 0 & 0 \\ 0 & 1 & 0 \\ 0 & 0 & 1 \end{bmatrix}, \quad A = \begin{bmatrix} .1 & 0 & .1 \\ .2 & .1 & .1 \\ .1 & .2 & 0 \end{bmatrix}, \quad X = \begin{bmatrix} x \\ y \\ z \end{bmatrix}, \quad D = \begin{bmatrix} 100 \\ 200 \\ 50 \end{bmatrix}.$$

Solve the matrix equation

$$(I - A)X = D.$$

2. Let I, A, X be as in Problem 1, but let

$$D = \begin{bmatrix} 300 \\ 100 \\ 100 \end{bmatrix}.$$

Solve the matrix equation $(I - A)X = D$.

EXERCISES 6

1. Suppose that in the economy of Example 1 the demand for electricity triples and the demand for coal doubles, whereas the demand for steel increases only by 50%. At what levels should the various industries produce in order to satisfy the new demand?

2. Suppose that the conglomerate of Example 2 is faced with an increase of 50% in demand for computers, a doubling in demand for semiconductors, and a decrease of 50% in demand for business forms. At what levels should the various divisions produce in order to satisfy the new demand?

3. Suppose that the conglomerate of Example 2 experiences a doubling in the demand for business forms. At what levels should the computer and semiconductor divisions produce?

4. A multinational corporation does business in the United States, Canada, and England. Its branches in one country purchase goods from the branches in other countries according to the matrix

Branch in:

		United States	Canada	England
Purchase from	United States	.02	0	.02
	Canada	.01	.03	.01
	England	.03	0	.01

where the entries in the matrix represent percentages of total sales by the respective branch. The external sales by each of the offices are $800,000,000 for the U.S. branch,

$300,000,000 for the Canadian branch, and $1,400,000,000 for the English branch. At what level should each of the branches produce in order to satisfy the total demand?

5. Show that to two decimal places

$$\begin{bmatrix} 1 & -.15 & -.43 \\ -.02 & .97 & -.20 \\ -.01 & -.08 & .95 \end{bmatrix}^{-1} = \begin{bmatrix} 1.01 & .20 & .50 \\ .02 & 1.05 & .23 \\ .01 & .09 & 1.08 \end{bmatrix}.$$

(A hand calculator would help a great deal here.)

6. Show that to two decimal places

$$\begin{bmatrix} .98 & -.02 & -.10 \\ -.2 & .99 & 0 \\ -.10 & -.02 & .99 \end{bmatrix}^{-1} = \begin{bmatrix} 1.04 & .02 & .10 \\ .21 & 1.01 & .02 \\ .11 & .02 & 1.02 \end{bmatrix}.$$

7. A corporation has a plastics division and an industrial equipment division. For each $1 worth of output, the plastics division needs $.02 worth of plastics and $.10 worth of equipment. For each $1 worth of output, the industrial equipment division needs $.01 worth of plastics and $.05 worth of equipment. At what level should the divisions produce to meet a demand for $930,000 worth of plastics and $465,000 worth of industrial equipment?

8. Rework Exercise 7 under the condition that the demand for plastics is $1,860,000 and the demand for industrial equipment is $2,790,000.

SOLUTIONS TO PRACTICE PROBLEMS 6

1. The equation $(I - A)X = D$ has the form $CX = D$, where C is the matrix $I - A$. From Section 4 we know that $X = C^{-1}D$. That is,

$$X = (I - A)^{-1}D.$$

Now

$$I - A = \begin{bmatrix} 1 & 0 & 0 \\ 0 & 1 & 0 \\ 0 & 0 & 1 \end{bmatrix} - \begin{bmatrix} .1 & 0 & .1 \\ .2 & .1 & .1 \\ .1 & .2 & 0 \end{bmatrix} = \begin{bmatrix} .9 & 0 & -.1 \\ -.2 & .9 & -.1 \\ -.1 & -.2 & 1 \end{bmatrix}.$$

Using the Gauss-Jordan method to find the inverse of this matrix, we have (to two decimal places)

$$(I - A)^{-1} = \begin{bmatrix} 1.13 & .03 & .12 \\ .27 & 1.14 & .14 \\ .17 & .23 & 1.04 \end{bmatrix}.$$

Therefore, rounding to the nearest integer, we have

$$X = (I - A)^{-1}D = \begin{bmatrix} 1.13 & .03 & .12 \\ .27 & 1.14 & .14 \\ .17 & .23 & 1.04 \end{bmatrix}\begin{bmatrix} 100 \\ 200 \\ 50 \end{bmatrix} = \begin{bmatrix} 125 \\ 262 \\ 115 \end{bmatrix}.$$

2. We have $X = (I - A)^{-1}D$, where $(I - A)^{-1}$ is as computed in Problem 1. So

$$X = \begin{bmatrix} 1.13 & .03 & .12 \\ .27 & 1.14 & .14 \\ .17 & .23 & 1.04 \end{bmatrix} \begin{bmatrix} 300 \\ 100 \\ 100 \end{bmatrix} = \begin{bmatrix} 354 \\ 209 \\ 178 \end{bmatrix}.$$

Chapter 5: CHECKLIST

- ☐ System of linear equations
- ☐ Elementary row operations
- ☐ Diagonal form
- ☐ Gaussian elimination method
- ☐ Matrix
- ☐ Pivoting
- ☐ Row matrix
- ☐ Column matrix
- ☐ Square matrix
- ☐ The ijth entry, a_{ij}
- ☐ Addition and subtraction of matrices
- ☐ Multiplication of matrices
- ☐ Identity matrix, I_n
- ☐ Inverse of a matrix, A^{-1}
- ☐ Formula for inverse of a 2×2 matrix
- ☐ Solution of matrix equation $AX = B$
- ☐ Use of inverse matrix to solve system of linear equations
- ☐ Gauss-Jordan method for calculating inverse of a matrix
- ☐ Input-output analysis

Chapter 5: SUPPLEMENTARY EXERCISES

Pivot each of the following matrices around the circled element.

1. $\begin{bmatrix} ③ & -6 & 1 \\ 2 & 4 & 6 \end{bmatrix}$

2. $\begin{bmatrix} -5 & -3 & 1 \\ 4 & ② & 0 \\ 0 & 6 & 7 \end{bmatrix}$

Use the Gaussian elimination method to find all solutions of the following systems of linear equations.

3. $\begin{cases} \frac{1}{2}x - y = -3 \\ 4x - 5y = -9 \end{cases}$

4. $\begin{cases} 3x \quad\quad + 9z = 42 \\ 2x + y + 6z = 30 \\ -x + 3y - 2z = -20 \end{cases}$

5. $\begin{cases} 3x - 6y + 6z = -5 \\ -2x + 3y - 5z = \frac{7}{3} \\ x + y + 10z = 3 \end{cases}$

6. $\begin{cases} 3x + 6y - 9z = 1 \\ 2x + 4y - 6z = 1 \\ 3x + 4y + 5z = 0 \end{cases}$

7. $\begin{cases} x + 2y - 5z + 3w = 16 \\ -5x - 7y + 13z - 9w = -50 \\ -x + y - 7z + 2w = 9 \\ 3x + 4y - 7z + 6w = 33 \end{cases}$

8. $\begin{cases} 5x - 10y = 5 \\ 3x - 8y = -3 \\ -3x + 7y = 0 \end{cases}$

Perform the indicated matrix operations.

9. $\begin{bmatrix} 2 \\ -1 \\ 0 \end{bmatrix} + \begin{bmatrix} 3 \\ 4 \\ 7 \end{bmatrix}$

10. $\begin{bmatrix} 1 & 3 & -2 \\ 4 & 0 & -1 \end{bmatrix} \begin{bmatrix} 3 & 5 \\ 1 & 0 \\ 0 & -6 \end{bmatrix}$

11. Find the inverse of the appropriate matrix and use it to solve the system of equations

$$\begin{cases} 3x + 2y = 0 \\ 5x + 4y = 2. \end{cases}$$

12. The matrices

$$\begin{bmatrix} 4 & -2 & 3 \\ 8 & -3 & 5 \\ 7 & -2 & 4 \end{bmatrix} \quad \text{and} \quad \begin{bmatrix} -2 & 2 & -1 \\ 3 & -5 & 4 \\ 5 & -6 & 4 \end{bmatrix}$$

are inverses of each other. Use these matrices to solve the following systems of linear equations.

(a) $\begin{cases} -2x + 2y - z = 1 \\ 3x - 5y + 4z = 0 \\ 5x - 6y + 4z = 3 \end{cases}$

(b) $\begin{cases} 4x - 2y + 3z = 0 \\ 8x - 3y + 5z = -1 \\ 7x - 2y + 4z = 2 \end{cases}$

Use the Gauss-Jordan method to calculate the inverses of the following matrices.

13. $\begin{bmatrix} 2 & 6 \\ 1 & 2 \end{bmatrix}$

14. $\begin{bmatrix} 1 & 1 & 1 \\ 3 & 4 & 3 \\ 1 & 1 & 2 \end{bmatrix}$

15. The economy of a small country can be regarded as consisting of two industries, I and II, whose input-output matrix is

$$A = \begin{bmatrix} .4 & .2 \\ .1 & .3 \end{bmatrix}.$$

How many units should be produced by each industry in order to meet a demand for 8 units from industry I and 12 units from industry II?

Linear Programming,
A Geometric Approach

Chapter 6

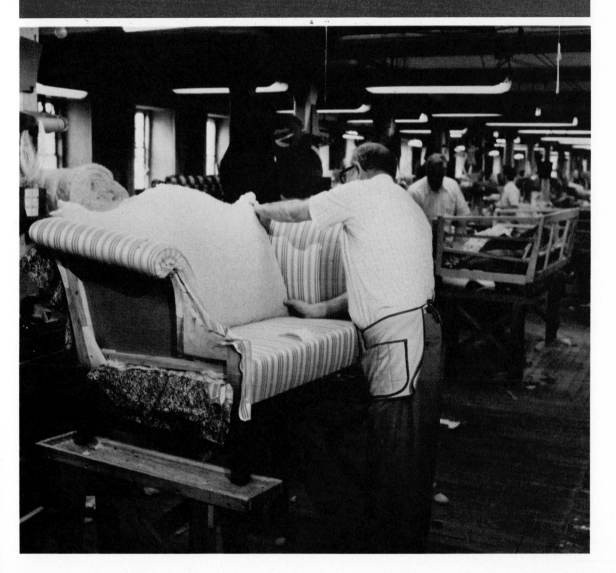

Linear programming is a method for solving problems in which a linear function (representing cost, profit, distance, weight, or the like) is to be maximized or minimized. Such problems are called *optimization problems*. As we shall see, these problems, when translated into mathematical language, involve systems of linear inequalities, systems of linear equations, and eventually (in Chapter 7) matrices.

6.1. A Linear Programming Problem

Let us begin with a detailed discussion of a typical problem that can be solved by linear programming.

Furniture Manufacturing Problem

A furniture manufacturer makes two types of furniture—chairs and sofas. For simplicity, divide the production process into three distinct operations—carpentry, finishing, and upholstery. The amount of labor required for each operation varies. Manufacture of a chair requires 6 hours of carpentry, 1 hour of finishing, and 2 hours of upholstery. Manufacture of a sofa requires 3 hours of carpentry, 1 hour of finishing, and 6 hours of upholstery. Owing to limited availability of skilled labor as well as of tools and equipment, the factory has available each day 96 man-hours for carpentry, 18 man-hours for finishing, and 72 man-hours for upholstery. The profit per chair is $80 and the profit per sofa $70. How many chairs and how many sofas should be produced each day in order to maximize the profit?

It is often helpful to tabulate data given in verbal problems. Our first step, then, is to construct a chart.

	Chair	Sofa	Available labor
Carpentry	6 hours	3 hours	96 man-hours
Finishing	1 hour	1 hour	18 man-hours
Upholstery	2 hours	6 hours	72 man-hours
Profit	$80	$70	

The next step is to translate the problem into mathematical language. As you know, this is done by identifying what is unknown and denoting the unknown quantities by letters. Since the problem asks for the optimum number of chairs and sofas to be produced each day, there are two unknowns—the number of chairs produced each day and the number of sofas produced each day. Let x denote the former and y the latter.

In order to achieve a large profit, one need only manufacture a large number of chairs and sofas. But, owing to restricted availability of tools and labor, the factory cannot manufacture an unlimited quantity of furniture. Let us translate the restrictions into mathematical language. Each row of the chart gives one restriction. The first row says that the amount of carpentry required is 6 hours for each chair and 3 hours for each sofa. Also, there are available only 96 man-hours of carpentry per day. We can compute the total number of man-hours of carpentry required per day to produce x chairs and y sofas as follows:

[number of man-hours per day of carpentry]

$$= \text{(number of hours carpentry per chair)} \cdot \text{(number of chairs per day)}$$

$$+ \text{(number of hours carpentry per sofa)} \cdot \text{(number of sofas per day)}$$

$$= 6 \cdot x + 3 \cdot y.$$

The requirement that at most 96 man-hours of carpentry be used per day means that x and y must satisfy the inequality

$$6x + 3y \leq 96. \tag{1}$$

The second row of the chart gives a restriction imposed by finishing. Since 1 hour of finishing is required for each chair and sofa, and since at most 18 man-hours of finishing are available per day, the same reasoning as used to derive inequality (1) yields

$$x + y \leq 18. \tag{2}$$

Similarly, the third row of the chart gives the restriction due to upholstery:

$$2x + 6y \leq 72. \tag{3}$$

A further restriction is given by the fact that the numbers of chairs and sofas must be nonnegative:

$$x \geq 0, \quad y \geq 0. \tag{4}$$

Now that we have written down the restrictions which constrain x and y, let us express the profit (which is to be maximized) in terms of x and y. The profit comes from two sources—chairs and sofas. Therefore,

$$[\text{profit}] = [\text{profit from chairs}] + [\text{profit from sofas}]$$

$$= [\text{profit per chair}] \cdot [\text{number of chairs}]$$

$$+ [\text{profit per sofa}] \cdot [\text{number of sofas}]$$

$$= 80x + 70y \tag{5}$$

Combining (1) to (5), we arrive at the following:

□ **Furniture Manufacturing Problem—Mathematical Formulation** Find numbers x and y for which $80x + 70y$ is as large as possible, and for which all the following inequalities hold simultaneously:

$$\begin{cases} 6x + 3y \le 96 \\ x + y \le 18 \\ 2x + 6y \le 72 \\ x \ge 0, \quad y \ge 0. \end{cases} \tag{6}$$

We may describe this mathematical problem in the following general way. We are required to maximize an expression in a certain number of variables, where the variables are subject to restrictions in the form of one or more inequalities. Problems of this sort are called *mathematical programming problems*. Actually, general mathematical programming problems can be quite involved, and their solutions may require very sophisticated mathematical ideas. However, this is not the case with the furniture manufacturing problem. What makes it a rather simple mathematical programming problem is that both the expression to be maximized and the inequalities are linear. For this reason the furniture manufacturing problem is called a *linear programming problem*. The theory of linear programming is a fairly recent advance in mathematics. It was developed over the last 40 years to deal with the increasingly more complicated problems of our technological society. The 1975 Nobel prize in economics was awarded to Kantorovich and Koopmans for their pioneering work in the field of linear programming.

We will solve the furniture manufacturing problem in Section 2, where we will develop a general technique for handling similar linear programming problems. At this point it is worthwhile to attempt to gain some insights into the problem and possible methods for attacking it.

It seems clear that a factory will operate most efficiently when its labor is fully utilized. Let us therefore take the operations one at a time and determine the conditions on x and y that fully utilize the three kinds of labor. The restriction on carpentry asserts that

$$6x + 3y \le 96.$$

If x and y were chosen so that $6x + 3y$ is actually *less* than 96, we would leave the carpenters idle some of the time, a waste of labor. Thus, it would seem reasonable to choose x and y to satisfy

$$6x + 3y = 96.$$

Similarly, to utilize all the finishers' time, x and y must satisfy

$$x + y = 18,$$

and to utilize all the upholsterers' time, we must have

$$2x + 6y = 72.$$

Thus, if no labor is to be wasted, then x and y must satisfy the system of equations

$$\begin{cases} 6x + 3y = 96 \\ x + y = 18 \\ 2x + 6y = 72. \end{cases} \tag{7}$$

Let us now graph the three equations of (7), which represent the conditions for full utilization of all forms of labor. (See chart below and Fig. 1.)

Equation	Standard form	x-intercept	y-intercept
$6x + 3y = 96$	$y = -2x + 32$	$(16, 0)$	$(0, 32)$
$x + y = 18$	$y = -x + 18$	$(18, 0)$	$(0, 18)$
$2x + 6y = 72$	$y = -\frac{1}{3}x + 12$	$(36, 0)$	$(0, 12)$

What does Fig. 1 say about the furniture manufacturing problem? Each particular pair of numbers (x, y) is called a *production schedule*. Each of the lines in Fig. 1 gives the production schedules which fully utilize one of the types of labor. Notice that the three lines do not have a common intersection point. This means that there is *no* production schedule which *simultaneously* makes full use of all three types of labor. In any production schedule at least some of the man-hours

FIGURE 1

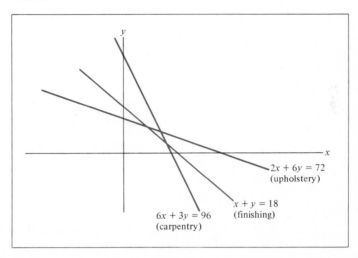

of labor must be wasted. This is not a solution to the furniture manufacturing problem, but is a valuable insight. It says that in the inequalities of (6) not all of the corresponding equations can hold. This suggests that we take a closer look at the system of inequalities.

The standard forms of the inequalities (6) are

$$\begin{cases} y \leq -2x + 32 \\ y \leq -x + 18 \\ y \leq -\frac{1}{3}x + 12 \\ x \geq 0, \quad y \geq 0. \end{cases}$$

By using the techniques of Section 2.2, we arrive at a feasible set for the system of inequalities above, as shown in Fig. 2.

The feasible set for the furniture manufacturing problem is a bounded, five-sided region. The points on and inside the boundary of this feasible set give the production schedules which satisfy all the restrictions. In the next section we will show how to pick out the particular point of the feasible set that corresponds to a maximum profit.

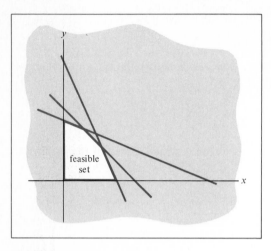

FIGURE 2

PRACTICE PROBLEMS 1

1. Determine whether the following points are in the feasible set of the furniture manufacturing problem: (a) (10, 9); (b) (14, 4).

2. A physical fitness enthusiast decides to devote her exercise time to a combination of jogging and cycling. She wants to earn aerobic points (a measure of the benefit of the exercise to strengthening the heart and lungs) and also to achieve relaxation and enjoyment. She jogs at 6 miles per hour and cycles at 18 miles per hour. An hour of jogging earns 12 aerobic points and an hour of cycling earns 9 aerobic points. Each week she would like to earn at least 36 aerobic points, cover at least 54 miles, and cycle at least as much as she jogs.

   (a) Fill in the chart below.

	One hour of jogging	One hour of cycling	Requirement
Miles covered			
Aerobic points			
Time required			

   (b) Let x be the number of hours of jogging and y the number of hours of cycling each week. Referring to the chart, give the inequalities that x and y must satisfy due to miles covered and aerobic points.

(c) Give the inequalities that x and y must satisfy due to her preference for cycling and also due to the fact that x and y cannot be negative.

(d) Express the time required as a linear function of x and y.

(e) Graph the feasible set for the system of linear inequalities.

EXERCISES 1

In Exercises 1–4, determine whether the given point is in the feasible set of the furniture manufacturing problem. (The inequalities are given below.)

$$\begin{cases} 6x + 3y \le 96 \\ x + y \le 18 \\ 2x + 6y \le 72 \\ x \ge 0, \quad y \ge 0. \end{cases}$$

1. $(8, 7)$ 2. $(14, 3)$ 3. $(9, 10)$ 4. $(16, 0)$

5. (*Shipping Problem*)   A truck traveling from New York to Baltimore is to be loaded with two types of cargo. Each crate of cargo A is 4 cubic feet in volume, weights 100 pounds, and earns \$13 for the driver. Each crate of cargo B is 3 cubic feet in volume, weighs 200 pounds, and earns \$9 for the driver. The truck can carry no more than 300 cubic feet of crates and no more than 10,000 pounds. Also, the number of crates of cargo B must be less than or equal to twice the number of crates of cargo A.

(a) Fill in the chart below.

	A	B	Truck capacity
Volume			
Weight			
Earnings			

(b) Let x be the number of crates of cargo A and y the number of crates of cargo B. Referring to the chart, give the two inequalities that x and y must satisfy because of the truck's capacity for volume and weight.

(c) Give the inequalities that x and y must satisfy because of the last sentence of the problem and also because x and y cannot be negative.

(d) Express the earnings from carrying x crates of cargo A and y crates of cargo B.

(e) Graph the feasible set for the shipping problem.

6. (*Mining Problem*)   A coal company owns mines in two different locations. Each day mine 1 produces 4 tons of anthracite (hard coal), 4 tons of ordinary coal, and 7 tons of bituminous (soft) coal. Each day mine 2 produces 10 tons of anthracite, 5 tons of ordinary coal, and 5 tons of bituminous coal. It costs the company \$150 per day to operate mine 1 and \$200 per day to operate mine 2. An order is received for 80 tons of anthracite, 60 tons of ordinary coal, and 75 tons of bituminous coal.

(a) Fill in the chart below.

	Mine 1	Mine 2	Ordered
Anthracite			
Ordinary			
Bituminous			
Daily cost			

(b) Let x be the number of days mine 1 should be operated and y the number of days mine 2 should be operated. Refer to the chart and give three inequalities that x and y must satisfy to fill the order.

(c) Give other requirements that x and y must satisfy.

(d) Find the cost of operating mine 1 for x days and mine 2 for y days.

(e) Graph the feasible set for the mining problem.

SOLUTIONS TO PRACTICE PROBLEMS 1

1. A point is in the feasible set of a system of inequalities if it satisfies every inequality. Either the original form or the standard form of the inequalities may be used. The original form of the inequalities of the furniture manufacturing problem is

$$\begin{cases} 6x + 3y \le 96 \\ x + y \le 18 \\ 2x + 6y \le 72 \\ x \ge 0, \quad y \ge 0. \end{cases}$$

(a) (10, 9)

$$\begin{cases} 6(10) + 3(9) \le 96 \\ 10 + 9 \le 18 \\ 2(10) + 6(9) \le 72 \\ 10 \ge 0, \quad 9 \ge 0; \end{cases} \qquad \begin{cases} 87 \le 96 & \text{true} \\ 19 \le 18 & \text{false} \\ 74 \le 72 & \text{false} \\ 10 \ge 0, \quad 9 \ge 0. & \text{true} \end{cases}$$

(b) (14, 4)

$$\begin{cases} 6(14) + 3(4) \le 96 \\ 14 + 4 \le 18 \\ 2(14) + 6(4) \le 72 \\ 14 \ge 0, \quad 4 \ge 0; \end{cases} \qquad \begin{cases} 96 \le 96 & \text{true} \\ 18 \le 18 & \text{true} \\ 52 \le 72 & \text{true} \\ 14 \ge 0, \quad 4 \ge 0. & \text{true} \end{cases}$$

Therefore, (14, 4) is in the feasible set and (10, 9) is not.

2. (a)

	One hour of jogging	One hour of cycling	Requirement
Miles covered	6	18	54
Aerobic points	12	9	36
Time required	1	1	

(b) Miles covered: $6x + 18y \geq 54$.
   Aerobic points: $12x + 9y \geq 36$.

(c) $y \geq x, x \geq 0$. It is not necessary to list $y \geq 0$ since this is automatically assured if the other two inequalities hold.

(d) $x + y$. (An objective of the exercise program might be to minimize $x + y$.)

(e)

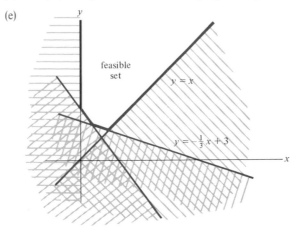

6.2. Linear Programming, I

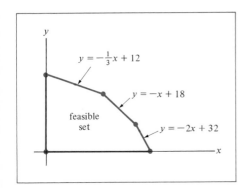

FIGURE 1

We have shown that the feasible set for the furniture manufacturing problem—that is, the set of points corresponding to production schedules satisfying all five restriction inequalities—consists of the points in and on the interior and boundary of the five-sided region drawn in Fig. 1. For reference, we have labeled each line segment with the equation of the line to which it belongs. The line segments intersect in five points, each of which is a corner of the feasible set. Such a corner is called a *vertex*. Somehow, we must pick out of the feasible set an *optimum point*—that is, a point corresponding to a production schedule which yields a maximum profit. To assist us in this task, we have the following result:*

□ **Fundamental Theorem of Linear Programming**   The maximum (or minimum) value of the objective function is achieved at one of the vertices of the feasible set.

This result does not completely solve the furniture manufacturing problem for us, but it comes close. It tells us that an optimum production schedule (a, b)

---

* For a proof, see Section 6.3.

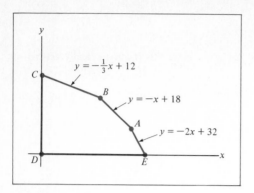

corresponds to one of the five points labeled A–E in Fig. 2. So to complete the solution of the furniture manufacturing problem, it suffices to find the coordinates of the five points, evaluate the profit at each, and then choose the point corresponding to the maximum profit.

FIGURE 2

Solution of the Furniture Manufacturing Problem

Let us begin by determining the coordinates of the points A–E in Fig. 2. Remembering that the x-axis has the equation $y = 0$ and the y-axis the equation $x = 0$, we see from Fig. 2 that the coordinates of A–E can be found as intersections of the following lines:

$$A: \begin{cases} y = -x + 18 \\ y = -2x + 32 \end{cases} \qquad B: \begin{cases} y = -x + 18 \\ y = -\frac{1}{3}x + 12 \end{cases}$$

$$C: \begin{cases} y = -\frac{1}{3}x + 12 \\ x = 0 \end{cases} \qquad D: \begin{cases} y = 0 \\ x = 0 \end{cases}$$

$$E: \begin{cases} y = 0 \\ y = -2x + 32. \end{cases}$$

The point D is clearly $(0, 0)$, and C is clearly the point $(0, 12)$. We obtain A from

$$-x + 18 = -2x + 32$$

$$x = 14$$

$$y = -14 + 18 = 4.$$

Hence $A = (14, 4)$. Similarly, we obtain B from

$$-x + 18 = -\frac{1}{3}x + 12$$

$$-\frac{2}{3}x = -6$$

$$x = 9$$

$$y = -9 + 18 = 9,$$

so $B = (9, 9)$. Finally, E is obtained from

$$0 = -2x + 32$$

$$x = 16$$

$$y = 0,$$

and thus $E = (16, 0)$. We have listed the vertices in Table 1. In the second column we have evaluated the profit, which is given by $80x + 70y$, at each of the vertices.

TABLE 1

Vertex	Profit $= 80x + 70y$
$(14, 4)$	$80(14) + 70(4) = 1400$
$(9, 9)$	$80(9) + 70(9) = 1350$
$(0, 12)$	$80(0) + 70(12) = 840$
$(0, 0)$	$80(0) + 70(0) = 0$
$(16, 0)$	$80(16) + 70(0) = 1280$

Note that the largest profit occurs at the vertex $(14, 4)$, so the solution of the linear programming problem is $x = 14$, $y = 4$. In other words, the factory should produce 14 chairs and 4 sofas each day in order to achieve maximum profit, and the maximum profit is $1400 per day.

The furniture manufacturing problem is one particular example of a linear programming problem. Generally, such problems involve finding the values of x and y which maximize (or minimize) a particular linear expression in x and y and where x and y are chosen so as to satisfy one or more restrictions in the form of linear inequalities. The expression that is to be maximized (or minimized) is called the *objective function*. On the basis of our experience with the furniture manufacturing problem, we can summarize the steps to be followed in approaching *any* linear programming problem.

Step 1 Translate the problem into mathematical language.

A. Organize the data.
B. Identify the unknown quantities and define corresponding variables.
C. Translate the restrictions into linear inequalities.
D. Form the objective function.

Step 2 Graph the feasible set.

A. Put the inequalities in standard form.
B. Graph the straight line corresponding to each inequality.
C. Determine the side of the line belonging to the graph of each inequality. Cross out the other side. The remaining region is the feasible set.

Step 3 Determine the vertices of the feasible set.
Step 4 Evaluate the objective function at each vertex. Determine the optimum point.

Linear programming can be applied to many problems. The Army Corps of Engineers has used linear programming to plan the location of a series of dams so as to maximize the resulting hydroelectric power production. The restrictions were to provide adequate flood control and irrigation. Public transit companies have used linear programming to plan routes and schedule buses in order to maximize services. The restrictions in this case arose from the limitations on manpower, equipment, and funding. The petroleum industry uses linear programming in the refining and blending of gasoline. Profit is maximized subject to restrictions on availability of raw materials, refining capacity, and product specifications. Some large advertising firms have used linear programming in media selection. The problem consists of determining how much to spend in each medium in order to maximize the number of consumers reached. The restrictions come from limitations on the budget and the relative costs of different media. Linear programming has been also used by psychologists to design an optimum battery of tests. The problem is to maximize the correlation between test scores and the characteristic that is to be predicted. The restrictions are imposed by the length and cost of the testing.

Linear programming is also used by dieticians in planning meals for large numbers of people. The object is to minimize the cost of the diet, and the restrictions reflect the minimum daily requirements of the various nutrients considered in the diet. The next example is representative of this type of problem. Whereas in actual practice many nutritional factors are considered, we shall simplify the problem by considering only three: protein, calories, and riboflavin.

EXAMPLE 1   (*Nutrition Problem*)   Suppose that in a developing nation the government wants to encourage everyone to make rice and soybeans part of his staple diet. The object is to design a lowest-cost diet which provides certain minimum levels of protein, calories, and vitamin B_2 (riboflavin). Suppose that one cup of uncooked rice costs 21 cents and contains 15 grams of protein, 810 calories, and $\frac{1}{9}$ milligram of riboflavin. On the other hand, one cup of uncooked soybeans costs 14 cents and contains 22.5 grams of protein, 270 calories, and $\frac{1}{3}$ millgram of riboflavin. Suppose that the minimum daily requirements are 90 grams of protein, 1620 calories, and 1 milligram of riboflavin. Design the lowest-cost diet meeting these specifications.

Solution   We solve the problem by following steps 1–4. The first step is to translate the problem into mathematical language, and the first part of this step is to organize the data, preferably into a chart (Table 2).

Now that we have organized the data, we ask for the unknowns. We wish to know how many cups each of rice and soybeans should comprise the diet, so we identify appropriate variables:

$$x = \text{number of cups of rice per day}$$

$$y = \text{number of cups of soybeans per day.}$$

Next, we obtain the restrictions on the variables. There is one restriction corresponding to each nutrient. That is, there is one restriction for each row of the

TABLE 2

	Rice	Soybeans	Required level per day
Protein (grams/cup)	15	22.5	90
Calories (per cup)	810	270	1620
Riboflavin (milligrams/cup)	$\frac{1}{9}$	$\frac{1}{3}$	1
Cost (cents/cup)	21	14	

chart. If x cups of rice and y cups of soybeans are consumed, then the amount of protein is $15x + 22.5y$ grams. Thus, from the first row of the chart, $15x + 22.5y \geq 90$, a restriction expressing the fact that there must be at least 90 grams of protein per day. Similarly, the restrictions for calories and riboflavin lead to the inequalities $810x + 270y \geq 1620$ and $\frac{1}{9}x + \frac{1}{3}y \geq 1$, respectively. As in the furniture manufacturing problem, x and y cannot be negative, so there are two further restrictions: $x \geq 0$, $y \geq 0$. In all there are five restrictions:

$$
\begin{cases}
15x + 22.5y \geq 90 \\
810x + 270y \geq 1620 \\
\frac{1}{9}x + \frac{1}{3}y \geq 1 \\
x \geq 0, \quad y \geq 0.
\end{cases}
\tag{1}
$$

Now that we have the restrictions, we form the objective function, which tells what we are out to maximize or minimize. Since we wish to minimize cost, we express cost in terms of x and y. Now x cups of rice costs $21x$ cents and y cups of soybeans costs $14y$ cents, so the objective function is given by

$$
[\text{cost}] = 21x + 14y.
\tag{2}
$$

And the problem can finally be stated in mathematical form: Minimize the objective function (2) subject to the restrictions (1). This completes the first step of the solution process.

The second step requires that we graph each of the inequalities (1). In Table 3 we have summarized all the steps necessary to obtain the information from which to draw the graphs. We have sketched the graphs in Fig. 3. From Fig. 3(b) we see

TABLE 3

			Intercepts		
Inequality	Standard form	Line	x	y	Graph
$15x + 22.5y \geq 90$	$y \geq -\frac{2}{3}x + 4$	$y = -\frac{2}{3}x + 4$	$(6, 0)$	$(0, 4)$	above
$810x + 270y \geq 1620$	$y \geq -3x + 6$	$y = -3x + 6$	$(2, 0)$	$(0, 6)$	above
$\frac{1}{9}x + \frac{1}{3}y \geq 1$	$y \geq -\frac{1}{3}x + 3$	$y = -\frac{1}{3}x + 3$	$(9, 0)$	$(0, 3)$	above
$x \geq 0$	$x \geq 0$	$x = 0$	$(0, 0)$	—	right
$y \geq 0$	$y \geq 0$	$y = 0$	—	$(0, 0)$	above

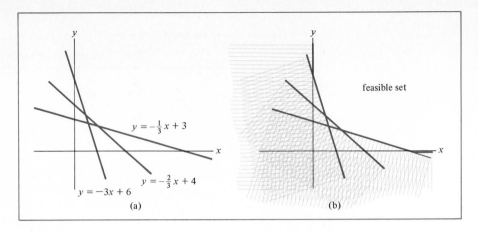

$y = -\frac{1}{3}x + 3$

$y = -\frac{2}{3}x + 4$

$y = -3x + 6$

(a)

feasible set

(b)

FIGURE 3

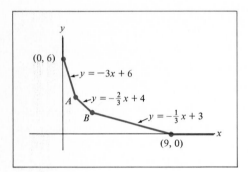

$(0, 6)$

$y = -3x + 6$

A $\quad y = -\frac{2}{3}x + 4$

B

$y = -\frac{1}{3}x + 3$

$(9, 0)$

FIGURE 4

that the feasible set is an unbounded, five-sided region. There are four vertices, two of which are known from Table 3, since they are intercepts of boundary lines. Label the remaining two vertices A and B (Fig. 4).

The third step of the solution process consists of determining the coordinates of A and B. From Fig. 4, these coordinates can be found by solving the following systems of equations:

$$A: \begin{cases} y = -3x + 6 \\ y = -\frac{2}{3}x + 4 \end{cases} \quad B: \begin{cases} y = -\frac{2}{3}x + .4 \\ y = -\frac{1}{3}x + 3. \end{cases}$$

To solve the first system, equate the two expressions for y:

$$-\tfrac{2}{3}x + 4 = -3x + 6$$

$$3x - \tfrac{2}{3}x = 6 - 4$$

$$\tfrac{7}{3}x = 2$$

$$x = \tfrac{6}{7}$$

$$y = -3x + 6 = -3\left(\tfrac{6}{7}\right) + 6 = \tfrac{24}{7}$$

$$A = \left(\tfrac{6}{7}, \tfrac{24}{7}\right).$$

Similarly, we find B:

$$-\tfrac{2}{3}x + 4 = -\tfrac{1}{3}x + 3$$

$$-\tfrac{1}{3}x = -1$$

$$x = 3$$

$$y = -\tfrac{2}{3}(3) + 4 = 2$$

$$B = (3, 2).$$

TABLE 4

Vertex	$Cost = 21x + 14y$
$(0, 6)$	$21 \cdot 0 + 14 \cdot 6 = 84$
$(\frac{6}{7}, \frac{24}{7})$	$21 \cdot \frac{6}{7} + 14 \cdot \frac{24}{7} = 66$
$(3, 2)$	$21 \cdot 3 + 14 \cdot 2 = 91$
$(9, 0)$	$21 \cdot 9 + 14 \cdot 0 = 189$

The fourth step consists of evaluating the objective function, in this case $21x + 14y$, at each vertex. From Table 4, we see that the minimum cost is achieved at the vertex $(\frac{6}{7}, \frac{24}{7})$. So the optimum diet—that is, the one which gives nutrients at the desired levels but at minimum cost—is the one which has $\frac{6}{7}$ cup of rice per day and $\frac{24}{7}$ cups of soybeans per day.

PRACTICE PROBLEMS 2

1. The feasible set for the nutrition problem is shown in the accompanying sketch. The cost is $21x + 14y$. *Without* using the fundamental theorem of linear programming, explain why the cost could not possibly be minimized at the point $(4, 4)$.

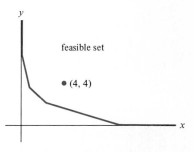

2. Rework the nutrition problem assuming that the cost of rice is changed to 7 cents per cup.

EXERCISES 2

For each of the feasible sets in Exercises 1–4, determine x and y so that the objective function $4x + 3y$ is maximized.

1.

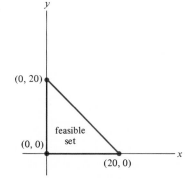

2.

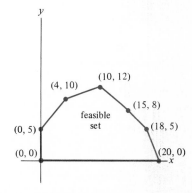

3.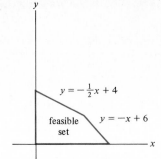

4.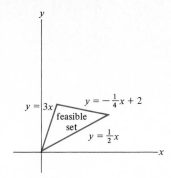

5. (*Shipping Problem*) Refer to Exercises 1, Problem 5. How many crates of each cargo should be shipped in order to satisfy the shipping requirements and yield the greatest earnings? (See the graph of the feasible set below.)

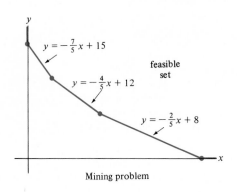

Mining problem   Shipping problem

6. (*Mining Problem*) Refer to Exercises 1, Problem 6. Find the number of days that each mine should be operated in order to fill the order at the least cost. (See the graph of the feasible set above.)

In Exercises 7 and 8, rework the furniture manufacturing problem, where everything is the same except that the profit per chair is changed to the given value. (See Table 1 for vertices.)

7. $150   8. $60

9. Minimize the objective function $7x + 4y$ subject to the restrictions

$$\begin{cases} y \geq -2x + 11 \\ y \leq -x + 10 \\ y \leq -\frac{1}{3}x + 6 \\ y \geq -\frac{1}{4}x + 4. \end{cases}$$

10. Maximize the objective function $x + 2y$ subject to the restrictions

$$\begin{cases} y \leq -x + 100 \\ y \geq \frac{1}{3}x + 20 \\ y \leq x. \end{cases}$$

11. Maximize the objective function $100x + 150y$ subject to the constraints

$$\begin{cases} x + 3y \leq 120 \\ 35x + 10y \leq 780 \\ x \leq 20 \\ x \geq 0, \quad y \geq 0. \end{cases}$$

12. Minimize the objective function $\frac{1}{2}x + \frac{3}{4}y$ subject to the constraints

$$\begin{cases} 2x + 2y \geq 8 \\ 3x + 5y \geq 16 \\ x \geq 0, \quad y \geq 0. \end{cases}$$

13. A contractor builds two types of homes. The first type requires one lot, $12,000 capital, and 150 man-days of labor to build and is sold for a profit of $2400. The second type of home requires one lot, $32,000 capital, and 200 man-days of labor to build and is sold for a profit of $3400. The contractor owns 150 lots and has available for the job $2,880,000 capital and 24,000 man-days of labor. How many homes of each type should she build in order to realize the greatest profit?

14. A nutritionist, working for NASA, must meet certain nutritional requirements and yet keep the weight of the food at a minimum. He is considering a combination of two foods which are packaged in tubes. Each tube of food A contains 4 units of protein, 2 units of carbohydrate, 2 units of fat, and weighs 3 pounds. Each tube of food B contains 3 units of protein, 6 units of carbohydrate, 1 unit of fat, and weighs 2 pounds. The requirement calls for 42 units of protein, 30 units of carbohydrate, and 18 units of fat. How many tubes of each food should be supplied to the astronauts?

15. The Beautiful Day Fruit Juice Company makes two varieties of fruit drink. Each can of Fruit Delight contains 10 ounces of pineapple juice, 3 ounces of orange juice, and 1 ounce of apricot juice, and makes a profit of 20 cents. Each can of Heavenly Punch contains 10 ounces of pineapple juice, 2 ounces of orange juice, and 2 ounces of apricot juice, and makes a profit of 30 cents. Each week, the company has available 9000 ounces of pineapple juice, 2400 ounces of orange juice, and 1400 ounces of apricot juice. How many cans of Fruit Delight and of Heavenly Punch should be produced each week in order to maximize profits?

16. The Bluejay Lacrosse Stick Company makes two kinds of lacrosse sticks. Type A sticks require 2 man-hours for cutting, 1 man-hour for stringing, and 2 man-hours for finishing, and are sold for a profit of $8. Type B sticks require 1 man-hour for cutting, 3 man-hours for stringing, and 2 man-hours for finishing, and are sold for a profit of $10. Each day the company has available 120 man-hours for cutting, 150 man-hours for stringing, and 140 man-hours for finishing. How many lacrosse sticks of each kind should be manufactured each day in order to maximize profits?

SOLUTIONS TO PRACTICE PROBLEMS 2

1. The point P has a smaller value of x and a smaller value of y than $(4, 4)$ and is still in the feasible set. It therefore corresponds to a lower cost than $(4, 4)$ and still meets the requirements. We conclude that no interior point of the feasible set could possibly be an optimum point. This geometric argument indicates that an optimum point might be one that juts out far—that is, a vertex.

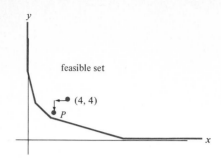

feasible set

(4, 4)

P

2. The system of linear inequalities, feasible set, and vertices will all be the same as before. Only the objective function changes. The new objective function is $7x + 14y$. The minimum cost occurs when using 3 cups of rice and 2 cups of soybeans.

Vertex	Cost = $7x + 14y$
(0, 6)	84
$(\frac{6}{7}, \frac{24}{7})$	54
(3, 2)	49
(9, 0)	63

6.3. Linear Programming, II

In this section we apply the technique of linear programming to the design of a portfolio for a retirement fund and to the transportation of goods from warehouses to retail outlets. The significant new feature of each of these problems is that, on the surface, they appear to involve more than two variables. However, they can be translated into mathematical language so that only two variables are required.

EXAMPLE 1 (*Investment Analysis*)   A pension fund has $30 million to invest. The money is to be divided among Treasury notes, bonds, and stocks. The rules for administration of the fund require that at least $3 million be invested in each type of investment, at least half the money be invested in Treasury notes and bonds, and the amount invested in bonds not exceed twice the amount invested in Treasury notes. The annual yields for the various investments are 7% for Treasury notes, 8% for bonds, and 9% for stocks. How should the money be allocated among the various investments to produce the largest return?

Solution First, let us agree that all numbers stand for millions. That is, we write 30 to stand for 30 million. This will save us from writing too many zeros. In examining the problem, we find that very little organization needs to be done. The rules for administration of the fund are written in a form from which inequalities can be read right off. Let us just summarize the remaining data in the first row of a chart (Table 1).

TABLE 1

	Treasury notes	*Bonds*	*Stocks*
Yield	.07	.08	.09
Variables	x	y	$30 - (x + y)$

There appear to be three variables—the amounts to be invested in each of the three categories. However, since the three investments must total 30, we need only two variables. Let x = the amount to be invested in Treasury notes and y = the amount to be invested in bonds. Then the amount invested in stocks is $30 - (x + y)$. We have displayed the variables in Table 1.

Now for the restrictions. Since at least 3 (million dollars) must be invested in each category, we have the three inequalities

$$x \geq 3$$

$$y \geq 3$$

$$30 - (x + y) \geq 3.$$

Moreover, since at least half the money, or 15, must be invested in Treasury notes and bonds, we must have

$$x + y \geq 15.$$

Finally, since the amount invested in bonds must not exceed twice the amount invested in Treasury notes, we must have

$$y \leq 2x.$$

(In this example we do not need to state that $x \geq 0$, $y \geq 0$, since we have already required that they be greater than or equal to 3.) Thus there are five restriction inequalities:

$$\begin{cases} x \geq 3, \quad y \geq 3 \\ 30 - (x + y) \geq 3 \\ \qquad x + y \geq 15 \\ \qquad\qquad y \leq 2x. \end{cases} \tag{1}$$

Next, we form the objective function, which in this case equals the total return on the investment. Since x dollars is invested at 7%, y dollars at 8%, and $30 - (x + y)$ dollars at 9%, the total return is

$$[\text{return}] = .07x + .08y + .09[30 - (x + y)]$$

$$= .07x + .08y + 2.7 - .09x - .09y$$

$$= 2.7 - .02x - .01y. \tag{2}$$

So the mathematical statement of the problem is: Maximize the objective function (2) subject to the restrictions (1).

TABLE 2

Inequality	Standard form	Equation	Intercepts x	Intercepts y	Graph
$x \geq 3$	$x \geq 3$	$x = 3$	$(3, 0)$	—	Right of line
$y \geq 3$	$y \geq 3$	$y = 3$	—	$(0, 3)$	Above line
$30 - (x + y) \geq 3$	$y \leq -x + 27$	$y = -x + 27$	$(27, 0)$	$(0, 27)$	Below line
$x + y \geq 15$	$y \geq -x + 15$	$y = -x + 15$	$(15, 0)$	$(0, 15)$	Above line
$y \leq 2x$	$y \leq 2x$	$y = 2x$	$(0, 0)$	$(0, 0)$	Below line

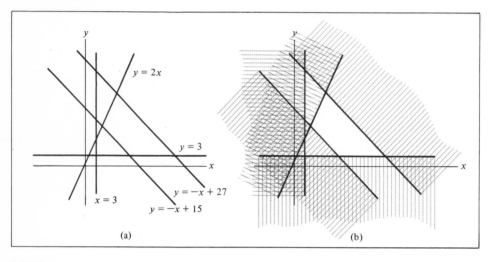

(a)   (b)

FIGURE 1

The next step of the solution is to graph the inequalities (1). The necessary information is tabulated in Table 2.

One point about the chart is worth noting: It contains enough data to graph each of the lines, with the exception of $y = 2x$. The reason is that the x- and y-intercepts of this line are the same, $(0, 0)$. So to graph $y = 2x$, we must find an additional point on the line. For example, if we set $x = 2$, then $y = 4$, so $(2, 4)$ is on the line. In Fig. 1(a) we have drawn the various lines, and in Fig. 1(b) we have crossed out the appropriate regions to produce the graph of the system. The feasible set, as well as the equations of the various lines that make up its boundary, are shown in Fig. 2. From Fig. 2 we find the pairs of equations that determine each of the vertices A–D. This is the third step of the solution procedure.

$$A: \begin{cases} y = 3 \\ y = -x + 15 \end{cases} \qquad B: \begin{cases} y = 3 \\ y = -x + 27 \end{cases}$$

$$C: \begin{cases} y = -x + 27 \\ y = 2x \end{cases} \qquad D: \begin{cases} y = 2x \\ y = -x + 15. \end{cases}$$

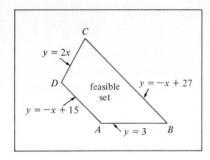

FIGURE 2

A and B are the easiest to determine. To find A, we must solve

$$3 = -x + 15$$

$$x = 12$$

$$y = 3$$

$$A = (12, 3).$$

Similarly, $B = (24, 3)$. To find C, we must solve

$$2x = -x + 27$$

$$3x = 27$$

$$x = 9$$

$$y = 2(9) = 18$$

$$C = (9, 18).$$

Similarly, $D = (5, 10)$.

Finally, we list the four vertices A, B, C, D and evaluate the objective function (2) at each one. The results are summarized in Table 3.

It is clear that the largest return occurs when $x = 5$, $y = 10$. In other words, \$5 million should be invested in Treasury notes, \$10 million in bonds, and $30 - (x + y) = 30 - (5 + 10) = \$15$ million in stocks.

Linear programming is of use not only in analyzing investments but in the fields of transportation and shipping. It is often used to plan routes, determine locations of warehouses, and develop efficient procedures for getting goods to people. Many linear programming problems of this variety can be formulated as *transportation problems*. A typical transportation problem involves determining the least-cost scheme for delivering a commodity stocked in a number of different warehouses to a number of different locations, say retail stores. Of course, in practical applications, it is necessary to consider problems involving perhaps dozens or even hundreds of warehouses, and possibly just as many delivery locations. For problems on such a grand scale, the methods developed so far are inadequate. For one thing, the number of variables required is usually more than two. We must wait until Chapter 7 for methods that apply to such problems. However, the next example gives an instance of a transportation problem which

TABLE 3

Vertex	Return $= 2.7 - .02x - .01y$
(5, 10)	$2.7 - .02(5) - .01(10) = \$2.5$ million
(9, 18)	$2.7 - .02(9) - .01(18) = \$2.34$ million
(24, 3)	$2.7 - .02(24) - .01(3) = \$2.19$ million
(12, 3)	$2.7 - .02(12) - .01(3) = \$2.43$ million

does not involve too many warehouses or too many delivery points. It gives the flavor of general transportation problems.

EXAMPLE 2 Suppose that a Maryland TV dealer has stores in Annapolis and Rockville and warehouses in College Park and Baltimore. The cost of shipping a set from College Park to Annapolis is $6; from College Park to Rockville, $3; from Baltimore to Annapolis, $9; and from Baltimore to Rockville, $5. Suppose that the Annapolis store orders 25 TV sets and the Rockville store 30. Suppose further that the College Park warehouse has a stock of 45 sets and the Baltimore warehouse 40. What is the most economical way to supply the requested TV sets to the two stores?

Solution The first step in solving a linear programming problem is to translate it into mathematical language. And the first part of this step is to organize the information given, preferably in the form of a chart. In this case, since the problem is geographic, we draw a schematic diagram, as in Fig. 3, which shows the flow of goods between warehouses and retail stores. By each route, we have written the cost. Below each warehouse we have written down its stock and below each retail store the number of TV sets it ordered.

Next, let us determine the variables. It appears initially that four variables are required, namely the number of TV sets to be shipped over each route. However, a closer look shows that only two variables are required. For if x denotes the number of TV sets to be shipped from College Park to Rockville, then since Rockville ordered 30 sets, the number shipped from Baltimore to Rockville is $30 - x$. Similarly, if y denotes the number of sets shipped from College Park to Annapolis, then the number shipped from Baltimore to Annapolis is $25 - y$. We have written the appropriate shipment sizes beside the various routes in Fig. 3.

As the third part of the translation process let us write down the restrictions on the variables. Basically, there are two kinds of restrictions: none of x, y, $30 - x$, $25 - y$ can be negative, and a warehouse cannot ship more TV sets than it has in stock. Referring to Fig. 3, we see that College Park ships $x + y$ sets, so that $x + y \leq 45$. Similarly, Baltimore ships $(30 - x) + (25 - y)$ sets, so that $(30 - x) + (25 - y) \leq 40$. Simplifying this inequality, we get

$$55 - x - y \leq 40$$

$$-x - y \leq -15$$

$$x + y \geq 15.$$

The inequality $30 - x \geq 0$ can be simplified to $x \leq 30$, and the inequality $25 -$

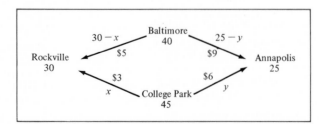

FIGURE 3

$y \geq 0$ can be written $y \leq 25$. So our restriction inequalities are these:

$$\begin{cases} x \geq 0, \quad y \geq 0 \\ x \leq 30, \quad y \leq 25 \\ x + y \geq 15 \\ x + y \leq 45. \end{cases} \tag{3}$$

The final step in the translation process is to form the objective function. In this problem we are attempting to minimize cost, so the objective function must express the cost in terms of x and y. Refer again to Fig. 3. There are x sets going from College Park to Rockville, and each costs \$3 to transport, so the cost of delivering these x sets is $3x$. Similarly, the costs of making the other deliveries are $6y$, $5(30 - x)$, and $9(25 - y)$, respectively. Thus the objective function is

$$\begin{aligned} [\text{cost}] &= 3x + 6y + 5(30 - x) + 9(25 - y) \\ &= 3x + 6y + 150 - 5x + 225 - 9y \\ &= 375 - 2x - 3y. \end{aligned} \tag{4}$$

FIGURE 4

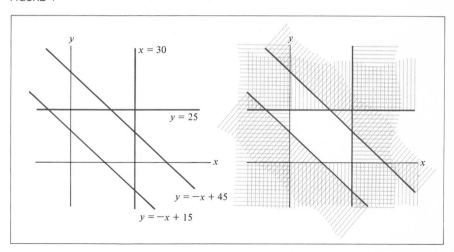

FIGURE 5

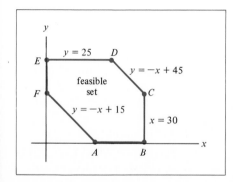

So the mathematical problem we must solve is: Find x and y that minimize the objective function (4) and satisfy the restrictions (3).

To solve the mathematical problem, we must graph the system of inequalities in (3). Four of the inequalities have graphs determined by horizontal and vertical lines. The only inequalities involving any work are $x + y \geq 15$ and $x + y \leq 45$. And even these are very easy to graph. The result is the graph in Fig. 4.

In Fig. 5 we have drawn the feasible set and have labeled each boundary line with its equation. The vertices A–F are now simple to determine. First, A and F are the intercepts of the line $y = -x + 15$. Therefore, $A = (15, 0)$ and $F = (0, 15)$. Since B is the x-intercept of the line $x = 30$, we have $B = (30, 0)$. Similarly, $E = (0, 25)$. Since C is on the line $x = 30$, its x-coordinate is 30. Its y-coordinate is $y = -30 + 45 = 15$, so $C = (30, 15)$. Similarly, since D has y-coordinate 25, its x-coordinate is given by $25 = -x + 45$ or $x = 20$. Thus $D = (20, 25)$.

We have listed in Table 4 the vertices A–F, as well as the cost corresponding to each one. The minimum cost occurs at the vertex $(20, 25)$. So $x = 20$, $y = 25$ yields the minimum of the objective function. In other words, 20 TV sets should be shipped from College Park to Rockville and 25 from College Park to Annapolis, $30 - x = 10$ from Baltimore to Rockville, and $25 - y = 0$ from Baltimore to Annapolis. This solves our problem.

TABLE 4

Vertex	Cost $= 375 - 2x - 3y$
(0, 25)	300
(0, 15)	330
(15, 0)	345
(30, 0)	315
(30, 15)	270
(20, 25)	260

Remarks Concerning the Transportation Problem   Note that the highest-cost route is the one from Baltimore to Annapolis. The solution we have obtained eliminates any shipments over this route. One might infer from this that one should always avoid the most expensive route. But this is not correct reasoning. To see why, reconsider Example 2, except change the cost of transporting a TV set from Baltimore to Annapolis from \$9 to \$7. The Baltimore-Annapolis route is still the most expensive. However, in this case the minimum cost is not obtained by eliminating the Baltimore-Annapolis route. For the revised problem, the linear inequalities stay the same. So the feasible set and the vertices remain the same. The only change is in the objective function, which now is given by

$$[\text{cost}] = 3x + 6y + 5(30 - x) + 7(25 - y)$$
$$= 325 - 2x - y.$$

Therefore, the costs at the various vertices are as given in Table 5. So the minimum cost of \$250 is achieved when $x = 30$, $y = 15$, $30 - x = 0$, and $25 - y = 10$. Note that 10 sets are being shipped from Baltimore to Annapolis, even though this is the most expensive route.

It is even possible for the cost function to be optimized simultaneously at two different vertices. For example, if the cost from Baltimore to Annapolis is \$8 and all other data are the same as in Example 2, then the optimum cost is \$260 and is

TABLE 5

Vertex	Cost $= 325 - 2x - y$
(0, 25)	300
(0, 15)	310
(15, 0)	295
(30, 0)	265
(30, 15)	250
(20, 25)	260

achieved at both vertices (30, 15) and (20, 25). We leave the calculations to the reader.

Verification of the Fundamental Theorem

The fundamental theorem of linear programming asserts that the objective function assumes its optimum value at a vertex of the feasible set. Let us verify this fact. For simplicity, we give the argument only in a special case, namely for the furniture manufacturing problem. However, this is for convenience of exposition only. The same argument as given below may be used to prove the fundamental theorem in general. Our argument relies on the parallel property for straight lines, which asserts that parallel lines have the same slope.

EXAMPLE 3 Prove the fundamental theorem of linear programming in the special case of the furniture manufacturing problem.

Solution The profit derived from producing x chairs and y sofas is $80x + 70y$ dollars. Let us examine all those production schedules having a given profit. As an example, consider a profit of \$2800. Then x and y must satisfy $80x + 70y = 2800$. That is, (x, y) must lie on the line whose equation is $80x + 70y = 2800$, or in standard form, $y = -\frac{8}{7}x + 40$. The slope of this line is $-\frac{8}{7}$ and its y-intercept is (0, 40). We have drawn this line in Fig. 6(a), in which we have also drawn the feasible set for

FIGURE 6

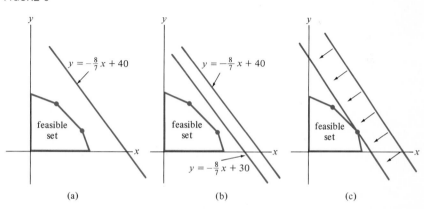

(a) (b) (c)

the furniture manufacturing problem. Note two fundamental facts: (1) Every production schedule on the line corresponds to a profit of $2800. (2) The line lies above the feasible set. In particular, no production schedule on the line satisfies all the restrictions of the problem. The difficulty is that $2800 is too high a profit for which to ask.

So now lower the profit, say, to $2100. In this case the production schedule (x, y) lies on the line $80x + 70y = 2100$, or in standard form, $y = -\frac{8}{7}x + 30$. This line is drawn in Fig. 6(b). Note that since both lines have slope $-\frac{8}{7}$, they are parallel by the parallel property. Actually, if we look at the production schedules yielding any fixed profit p, then they will lie along a line of slope $-\frac{8}{7}$, which is then parallel to the two lines already drawn. For if the production schedule (x, y) yields a profit p, then $80x + 70y = p$ or $y = -\frac{8}{7}x + p/70$. In other words, (x, y) lies on a line of slope $-\frac{8}{7}$ and y-intercept $(0, p/70)$. In particular, all the "lines of constant profit" are parallel to one another. So let us go back to the line of $2800 profit. It does not touch the feasible set. So now lower the profit and therefore translate the line downward parallel to itself. Next lower the profit until we first touch the feasible set. This line now touches the feasible set at a vertex [Fig. 6(c)]. And this vertex corresponds to the optimum production schedule, since any other point of the feasible set lies on a "line of constant profit" corresponding to an even lower profit. This shows why the fundamental theorem of linear programming is true.

PRACTICE PROBLEMS 3

Problems 1 and 2 refer to Example 1. Translate the statement into an inequality.

1. The amount to be invested in bonds is at most $5 million more than the amount to be invested in Treasury notes.

2. No more than $25 million should be invested in stocks and bonds.

3. Rework Example 1, assuming that the yield for Treasury notes goes up to 8%.

4. A linear programming problem has objective function: $[\text{cost}] = 5x + 10y$ which is to be minimized. Figure 7 shows the feasible set and the straight line of all combinations of x and y for which $[\text{cost}] = \$20$.

FIGURE 7

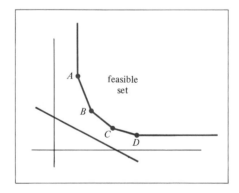

(a) Give the linear equation (in standard form) of the line of constant cost c.

(b) As c increases, does the line of constant cost c move up or down?

(c) By inspection, find the vertex of the feasible set that gives the optimum solution.

EXERCISES 3

1. Mr. Smith decides to feed his pet Doberman pinscher a combination of two dog foods. Each can of brand A contains 3 units of protein, 1 unit of carbohydrates, and 2 units of fat and costs 80 cents. Each can of brand B contains 1 unit of protein, 1 unit of carbohydrates, and 6 units of fat and costs 50 cents. Mr. Smith feels that each day his dog should have at least 6 units of protein, 4 units of carbohydrates, and 12 units of fat. How many cans of each dog food should he give to his dog each day in order to provide the minimum requirements at the least cost?

2. An oil company owns two refineries. Refinery I produces each day 100 barrels of high-grade oil, 200 barrels of medium-grade oil, and 300 barrels of low-grade oil and costs $10,000 to operate. Refinery II produces each day 200 barrels of high-grade, 100 barrels of medium-grade, and 200 barrels of low-grade oil and costs $9000 to operate. An order is received for 1000 barrels of high-grade oil, 1000 barrels of medium-grade oil, and 1800 barrels of low-grade oil. How many days should each refinery be operated in order to fill the order at the least cost?

3. A produce dealer in Florida ships oranges, grapefruits, and avocados to New York by truck. Each truckload consists of 100 crates, of which at least 20 crates must be oranges, at least 10 crates must be grapefruits, at least 30 crates must be avocados, and there must be at least as many crates of oranges as grapefruits. The profit per crate is $5 for oranges, $6 for grapefruits, and $4 for avocados. How many crates of each type should be shipped in order to maximize the profit? [Hint: Let $x =$ number of crates of oranges, $y =$ number of crates of grapefruit. Then $100 - x - y =$ number of crates of avocados.]

4. Mr. Jones has $9000 to invest in three types of stocks: low-risk, medium-risk, and high-risk. He invests according to three principles. The amount invested in low-risk stocks will be at most $1000 more than the amount invested in medium-risk stocks. At least $5000 will be invested in low- and medium-risk stocks. No more than $7000 will be invested in medium- and high-risk stocks. The expected yields are 6% for low-risk stock, 7% for medium-risk stocks, and 8% for high-risk stocks. How much money should Mr. Jones invest in each type of stock in order to maximize his total expected yield?

5. An automobile manufacturer has assembly plants in Detroit and Cleveland, each of which can assemble cars and trucks. The Detroit plant can assemble at most 800 vehicles in one day at a cost of $1200 per car and $2100 per truck. The Cleveland plant can assemble at most 500 vehicles in one day at a cost of $1000 per car and $2000 per truck. A rush order is received for 600 cars and 300 trucks. How many vehicles of each type should each plant produce in order to fill the order at the least cost? [Hint: Let $x =$ number of cars to be produced in Detroit, $y =$ number of trucks to be produced in Detroit, $600 - x =$ number of cars to be produced in Cleveland, $300 - y =$ number of trucks to be produced in Cleveland.]

6. A foreign car dealer with warehouses in New York and Baltimore receives orders from dealers in Philadelphia and Trenton. The dealer in Philadelphia needs 4 cars and the dealer in Trenton needs 7. The New York warehouse has 6 cars and the Baltimore warehouse has 8. The cost of shipping cars from Baltimore to Philadelphia is $120 per car,

from Baltimore to Trenton $90 per car, from New York to Philadelphia $100 per car, from New York to Trenton $70 per car. Find the number of cars to be shipped from each warehouse to each dealer in order to minimize the shipping cost.

7. Figure 8(a) shows the feasible set of the nutrition problem of Section 2 and the straight line of all combinations of rice and soybeans for which the cost is 42 cents.

(a) The objective function is $21x + 14y$. Give the linear equation (in standard form) of the line of constant cost c.

(b) As c increases, does the line of constant cost move up or down?

(c) By inspection, find the vertex of the feasible set that gives the optimum solution.

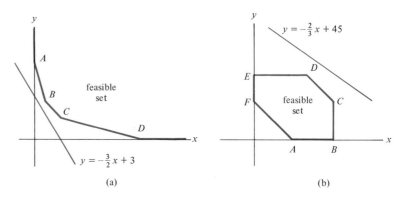

(a)   (b)

FIGURE 8

8. Figure 8(b) shows the feasible set of the transportation problem of Example 2 and the straight line of all combinations of shipments for which the transportation cost is $240.

(a) The objective function is $[\text{cost}] = 375 - 2x - 3y$. Give the linear equation (in standard form) of the line of constant cost c.

(b) As c increases, does the line of constant cost move up or down?

(c) By inspection, find the vertex of the feasible set that gives the optimum solution.

9. An oil refinery produces gasoline, jet fuel, and diesel fuel. The profits per gallon from the sale of these fuels are $.15, $.12, and $.10, respectively. The refinery has a contract with an airline to deliver a minimum of 20,000 gallons per day of jet fuel and/or gasoline (or some of each). It has a contract with a trucking firm to deliver a minimum of 50,000 gallons per day of diesel fuel and/or gasoline (or some of each). The refinery can produce 100,000 gallons of fuel per day, distributed among the fuels in any fashion. It wishes to produce at least 5000 gallons per day of each fuel. How many gallons of each should be produced in order to maximize the profit?

10. Suppose that a price war reduces the profits of gasoline in Problem 9 to $.05 per gallon and that the profits on jet fuel and diesel fuel are unchanged. How many gallons of each fuel should now be produced to maximize the profit?

SOLUTIONS TO PRACTICE PROBLEMS 3

1. Amount invested in bonds $= y$. Five million dollars more than the amount invested in Treasury notes is $x + 5$. Therefore, $y \le x + 5$.

2. Amount invested in stocks $= 30 - (x + y)$. Amount invested in bonds $= y$. Therefore,

$$30 - (x + y) + y \le 25$$

$$30 - x \le 25$$

$$x \ge 5.$$

3. The feasible set stays the same but the return becomes

$$[\text{return}] = .08x + .08y + .09[30 - (x + y)]$$

$$= .08x + .08y + 2.7 - .09x - 0.9y$$

$$= 2.7 - .01x - .01y.$$

When the return is evaluated at each of the vertices of the feasible set, the greatest return is achieved at two vertices. Either of these vertices yields an optimum solution.

	$2.7 - .01x - .01y$
(5, 10)	2.55
(12, 3)	2.55
(24, 3)	2.43
(9, 18)	2.43

4. (a) The values of x and y for which the cost is c dollars satisfy $5x + 10y = c$. The standard form of this linear equation is $y = -\frac{1}{2}x + c/10$.

   (b) The line $y = -\frac{1}{2}x + c/10$ has slope $-\frac{1}{2}$ and y-intercept $(0, c/10)$. As c increases, the slope stays the same, but the y-intercept moves up. Therefore, the line moves up.

   (c) The line of constant cost $20 does not contain any points of the feasible set, so such a low cost cannot be achieved. Increase the cost until the line of constant cost just touches the feasible set. As c increases, the line moves up (keeping the same slope) and first touches the feasible set at vertex C. Therefore, taking x and y to be the coordinates of C yields the minimum cost.

Chapter 6: CHECKLIST

☐ Objective function
☐ Fundamental theorem of linear programming
☐ Four-step procedure for solving linear programming problems

Chapter 6: SUPPLEMENTARY EXERCISES

1. Terrapin Airlines wants to fly 1400 members of a ski club to Colorado. The airline owns two types of planes. Type A can carry only 50 passengers, requires three stewards, and costs $14,000 for the trip. Type B can carry 300 passengers, requires four stewards, and costs $90,000 for the trip. If the airline must use at least as many type A planes as type B and has available only 42 stewards, how many planes of each type should be used to minimize the cost for the trip?

2. A nutritionist is designing a new breakfast cereal using wheat germ and enriched oat flour as the basic ingredients. Each ounce of wheat germ contains 2 milligrams of niacin, 3 milligrams of iron, .5 milligram of thiamin, and costs 3 cents. Each ounce of enriched oat flour contains 3 milligrams of niacin, 3 milligrams of iron, .25 milligram of thiamin, and costs 4 cents. The nutritionist wants the cereal to have at least 7 milligrams of niacin, 8 milligrams of iron, and 1 milligram of thiamin. How many ounces of wheat germ and how many ounces of enriched oat flour should be used in each serving in order to meet the nutritional requirements at the least cost?

3. An automobile manufacturer makes hardtops and sports cars. Each hardtop requires 8 man-hours to assemble, 2 man-hours to paint, 2 man-hours to upholster, and is sold for a profit of $90. Each sports car requires 18 man-hours to assemble, 2 man-hours to paint, 1 man-hour to upholster, and is sold for a profit of $100. During each day 360 man-hours are available to assemble, 50 man-hours to paint, and 40 man-hours to upholster auto-mobiles. How many hardtops and sports cars should be produced each day in order to maximize the profit?

4. A confectioner makes two raisin-nut mixtures. A box of mixture A contains 6 ounces of peanuts, 1 ounce of raisins, 4 ounces of cashews, and sells for 50 cents. A box of mixture B contains 12 ounces of peanuts, 3 ounces of raisins, 2 ounces of cashews, and sells for 90 cents. He has available 5400 ounces of peanuts, 1200 ounces of raisins, and 2400 ounces of cashews. How many boxes of each mixture should he make in order to maximize revenue?

5. A textbook publisher puts out 72 new books each year, which are classified as elementary, intermediate, and advanced. The company's policy for new books is to publish at least four advanced books, at least three times as many elementary books as intermediate books, and at least twice as many intermediate books as advanced books. On the average, the annual profits are $8000 for each elementary book, $7000 for each intermediate book, and $1000 for each advanced book. How many new books of each type should be published in order to maximize the annual profit while conforming to company policy?

The Simplex Method

Chapter 7

In Chapter 6 we introduced a graphical method for solving linear programming problems. This method, although very simple, is of limited usefulness, since it applies only to problems which involve (or can be reduced to) two variables. On the other hand, linear programming applications in business and economics can involve dozens or even hundreds of variables. In this chapter we describe a method for handling such applications. This method, called the *simplex method* (or *simplex algorithm*), was developed by the mathematician George B. Dantzig in the late 1940s and today is the principal method used in solving complex linear programming problems. The simplex method can be used for problems in any number of variables and is easily adapted to computer calculations.

7.1. Slack Variables and the Simplex Tableau

In this section and the next, we explain how the simplex method can be used to solve linear programming problems such as the following:

PROBLEM A Maximize the objective function $3x + 4y$ subject to the constraints

$$\begin{cases} x + y \le 20 \\ x + 2y \le 25 \\ x \ge 0 \\ y \ge 0. \end{cases}$$

PROBLEM B Maximize the objective function $x + 2y + z$ subject to the constraints

$$\begin{cases} x - y + 2z \le 10 \\ 2x + y + 3z \le 12 \\ x \ge 0 \\ y \ge 0 \\ z \ge 0. \end{cases}$$

Each of these problems exhibits certain features that make it particularly convenient to work with.

1. The objective function is to be maximized.
2. Each variable is constrained to be ≥ 0.
3. All other constraints are of the form

$$[\text{linear polynomial}]^* \leq [\text{nonnegative constant}].$$

A linear programming problem satisfying these conditions is said to be in *standard form*. Our initial discussion of the simplex method will involve only such problems. Then, in Section 3, we will consider problems in nonstandard form.

The first step of the simplex method is to convert the given linear programming problem into a system of linear *equations*. To see how this is done, consider Problem A above. It specifies that the variables x and y are subject to the constraint

$$x + y \leq 20.$$

Let us introduce another variable, u, which turns the inequality into an equation:

$$x + y + u = 20.$$

The variable u "takes up the slack" between $x + y$ and 20 and is therefore called a *slack variable*. Moreover, since $x + y$ is at most 20, the variable u must be ≥ 0. In a similar way the constraint

$$x + 2y \leq 25$$

can be turned into the equation

$$x + 2y + v = 25,$$

where v is a slack variable and $v \geq 0$. Let us even turn our objective function $3x + 4y$ into an equation by introducing the new variable M defined by $M = 3x + 4y$. Then M is the variable we want to maximize. Moreover, it satisfies the equation

$$-3x - 4y + M = 0.$$

Thus, Problem A can be restated in terms of a system of linear equations as follows:

PROBLEM A′ Among all solutions of the system of linear equations

$$\begin{cases} x + y + u \phantom{{}+v+M} = 20 \\ x + 2y \phantom{{}+u} + v \phantom{{}+M} = 25 \\ -3x - 4y \phantom{{}+u+v} + M = 0, \end{cases}$$

find one for which $x \geq 0$, $y \geq 0$, $u \geq 0$, $v \geq 0$, and for which M is as large as possible.

---

* A linear polynomial is an expression of the form $ax + by + cz + \cdots + dw$, where $a, b, c, \ldots,$ d are specific numbers and $x, y, z, \ldots, w$ are variables. Some examples are $2x - 3y + z$, $x + 2y + 3z - 4w$, and $-x + 3z - 2w$.

In a similar way, any linear programming problem in standard form can be reduced to that of determining a certain type of solution of a system of linear equations.

EXAMPLE 1   Formulate Problem B in terms of a system of linear equations.

Solution   The two constraints $x - y + 2z \leq 10$, $2x + y + 3z \leq 12$ yield the equations

$$x - y + 2z + u \qquad = 10$$
$$2x + y + 3z \qquad + v = 12.$$

The objective function yields the equation $M = x + 2y + z$—that is,

$$-x - 2y - z + M = 0.$$

So Problem B can be reformulated: Among all solutions of the system of linear equations

$$\begin{cases} x - \ y + 2z + u \qquad\qquad = 10 \\ 2x + \ y + 3z \qquad + v \qquad = 12 \\ -x - 2y - \ z \qquad\qquad + M = 0, \end{cases}$$

find one for which $x \geq 0, y \geq 0, z \geq 0, u \geq 0, v \geq 0$, and M is as large as possible.

We shall now discuss a scheme for solving systems of equations like those just encountered. For the moment, we will not worry about maximizing M or keeping the variables ≥ 0. Rather, let us concentrate on a particular method for determining solutions. In order to be concrete, consider the system of linear equations from Problem A':

$$\begin{cases} x + \ y + u \qquad\qquad = 20 \\ x + 2y \qquad + v \qquad = 25 \\ -3x - 4y \qquad\qquad + M = 0, \end{cases} \qquad (1)$$

This system has an infinite number of solutions. In fact, we can solve the equations for u, v, M in terms of x and y:

$$u = 20 - \ x - \ y$$
$$v = 25 - \ x - 2y$$
$$M = \qquad 3x + 4y.$$

Given any values of x and y, we can determine corresponding values for u, v, M. For example, if $x = 0$ and $y = 0$, then

$$u = 20 - \quad 0 - \quad 0 = 20$$
$$v = 25 - \quad 0 - 2 \cdot 0 = 25$$
$$M = \qquad 3 \cdot 0 + 4 \cdot 0 = 0.$$

Therefore, $x = 0$, $y = 0$, $u = 20$, $v = 25$, $M = 0$ is a specific solution of the system. Note that these values for u, v, M are precisely the numbers that appear to the right of the equality signs in our system of linear equations. Therefore, this particular solution could have been read off directly from the system (1) without any computation. This method of generating solutions is used in the simplex method, so let us explore further the special properties of the system which allowed us to read off a specific solution so easily.

Note that the system of linear equations has five variables: x, y, u, v, M. These variables may be divided into two groups. Group I consists of those which were set equal to 0, namely x and y. Group II consists of those whose particular values were read off from the right-hand sides of the equations, namely u, v, and M. Note also that the system has a special form which allows the particular values of the group II variables to be read off: each of the equations involves exactly one of the group II variables, and these variables always appear with coefficient 1. Thus, for example, the first equation involves the group II variable u:

$$x + y + u = 20.$$

Therefore, when all group I variables (x and y) are set equal to 0, only the term u remains on the left, and the particular value of u can then be read off from the right-hand side.

The special form of the system can best be described in matrix form. Write the system in the usual way as a matrix, but add column headings corresponding to the variables:

$$
\begin{array}{ccccc}
x & y & u & v & M \\
\end{array}
$$
$$
\left[
\begin{array}{ccccc|c}
1 & 1 & 1 & 0 & 0 & 20 \\
1 & 2 & 0 & 1 & 0 & 25 \\
-3 & -4 & 0 & 0 & 1 & 0
\end{array}
\right].
$$

Note closely the columns corresponding to the group II variables u, v, M.

$$
\begin{array}{ccccc}
x & y & u & v & M \\
\end{array}
$$
$$
\left[
\begin{array}{ccccc|c}
1 & 1 & 1 & 0 & 0 & 20 \\
1 & 2 & 0 & 1 & 0 & 25 \\
-3 & -4 & 0 & 0 & 1 & 0
\end{array}
\right].
$$

The presence of these columns gives the system the special form discussed above. Indeed, the u column asserts that u appears only in the first equation and is coefficient there is 1, and similarly for the v and M columns.

The property of allowing us to read off a particular solution from the right-hand column is shared by all linear systems whose matrices contain the columns

$$
\begin{array}{cccc}
1 & 0 & 0 & \cdots & 0 \\
0 & 1 & 0 & \cdots & 0 \\
0 & 0 & 1 & \cdots & 0 \\
\vdots & \vdots & \vdots & & \vdots \\
0 & 0 & 0 & & 1.
\end{array}
$$

(These columns need not appear in exactly the order shown.) The variables corresponding to these columns are called the group II variables. The group I variables consist of all the others. To get one particular solution to the system, set all the group I variables equal to zero and read off the values of the group II variables from the right-hand side of the system. This procedure is illustrated in the following example.

EXAMPLE 2 Determine by inspection one set of solutions to each of these systems of linear equations:

(a) $\begin{cases} x - 5y + u & = 3 \\ -2x + 8y & + v & = 11 \\ -\frac{1}{2}x & + M = 0 \end{cases}$ (b) $\begin{cases} -y + 2u + v & = 12 \\ x + \frac{1}{2}y - 6u & = -1 \\ 3y + 8u & + M = 4 \end{cases}$

Solution (a) The matrix of the system is

$$
\begin{array}{ccccc}
x & y & u & v & M \\
\end{array}
$$
$$
\left[\begin{array}{ccccc|c}
1 & -5 & 1 & 0 & 0 & 3 \\
-2 & 8 & 0 & 1 & 0 & 11 \\
-\frac{1}{2} & 0 & 0 & 0 & 1 & 0
\end{array}\right].
$$

We look for each variable whose column contains one entry of 1 and all the other entries 0.

$$
\begin{array}{ccccc}
x & y & u & v & M \\
\end{array}
$$
$$
\left[\begin{array}{ccccc|c}
1 & -5 & 1 & 0 & 0 & 3 \\
-2 & 8 & 0 & 1 & 0 & 11 \\
-\frac{1}{2} & 0 & 0 & 0 & 1 & 0
\end{array}\right].
$$

The group II variables should be u, v, M, with x, y as the group I variables. Set all group I variables equal to 0. The corresponding values of the group II variables may then be read off the last column: $u = 3$, $v = 11$, $M = 0$. So one solution of the system is

$$x = 0, \qquad y = 0, \qquad u = 3, \qquad v = 11, \qquad M = 0.$$

(b) The matrix of the system is

$$
\begin{array}{ccccc}
x & y & u & v & M \\
\end{array}
$$
$$
\left[\begin{array}{ccccc|c}
0 & -1 & 2 & 1 & 0 & 12 \\
1 & \frac{1}{2} & -6 & 0 & 0 & -1 \\
0 & 3 & 8 & 0 & 1 & 4
\end{array}\right].
$$

The shaded columns show that the group II variables should be v, x, M, with y, u, as the group I variables. So the corresponding solution is

$$x = -1, \qquad y = 0, \qquad u = 0, \qquad v = 12, \qquad M = 4.$$

A *simplex tableau* is a matrix (corresponding to a linear system) in which each of the columns

$$\begin{matrix} 1 & 0 & \cdots & 0 \\ 0 & 1 & \cdots & 0 \\ \vdots & \vdots & & \vdots \\ 0 & 0 & & 1 \end{matrix}$$

is present exactly once (in some order) to the left of the vertical line. We have seen how to construct a simplex tableau corresponding to a linear programming problem in standard form. From this initial simplex tableau we can read off one particular solution of the linear system by using the method described above. This particular solution may or may not correspond to the solution of the original optimization problem. If it does not, we replace the initial tableau with another one whose corresponding solution is "closer" to the optimum. How do we replace the initial simplex tableau with another? Just pivot it about a nonzero entry! Indeed, one of the key reasons the simplex method works is that pivoting transforms one simplex tableau into another. Note also that since pivoting consists of elementary row operations, the solution corresponding to a transformed tableau is a solution of the original linear system. The next example illustrates how pivoting transforms a tableau into another one.

EXAMPLE 3 Consider the simplex tableau obtained from Problem A:

$$\begin{array}{ccccc} x & y & u & v & M \\ \left[\begin{array}{ccccc|c} 1 & 1 & 1 & 0 & 0 & 20 \\ 1 & ② & 0 & 1 & 0 & 25 \\ -3 & -4 & 0 & 0 & 1 & 0 \end{array}\right]. \end{array}$$

(a) Pivot this tableau around the entry 2.

(b) Calculate the particular solution corresponding to the transformed tableau which results from setting the new group I variables equal to 0.

Solution (a) The first step in pivoting is to replace the pivot element 2 by a 1. To do this, multiply the second row of the tableau by $\frac{1}{2}$ to get

$$\left[\begin{array}{ccccc|c} 1 & 1 & 1 & 0 & 0 & 20 \\ \frac{1}{2} & 1 & 0 & \frac{1}{2} & 0 & \frac{25}{2} \\ -3 & -4 & 0 & 0 & 1 & 0 \end{array}\right].$$

Next, we must replace all nonpivot elements in the second column by zeros. Do this by adding to the first row (-1) times the second row:

$$\left[\begin{array}{ccccc|c} \frac{1}{2} & 0 & 1 & -\frac{1}{2} & 0 & \frac{15}{2} \\ \frac{1}{2} & 1 & 0 & \frac{1}{2} & 0 & \frac{25}{2} \\ -3 & -4 & 0 & 0 & 1 & 0 \end{array}\right],$$

and by adding to the third row 4 times the second row:

$$
\begin{array}{ccccc}
x & y & u & v & M \\
\end{array}
$$

$$
\begin{bmatrix}
\frac{1}{2} & 0 & 1 & -\frac{1}{2} & 0 & \bigm| & \frac{15}{2} \\
\frac{1}{2} & 1 & 0 & \frac{1}{2} & 0 & \bigm| & \frac{25}{2} \\
-1 & 0 & 0 & 2 & 1 & \bigm| & 50
\end{bmatrix}.
$$

Note that we indeed get a new simplex tableau. The new group II variables are now u, y, M. The new group I variables are x, v.

$$
\begin{array}{ccccc}
x & y & u & v & M \\
\end{array}
$$

$$
\begin{bmatrix}
\frac{1}{2} & 0 & 1 & -\frac{1}{2} & 0 & \bigm| & \frac{15}{2} \\
\frac{1}{2} & 1 & 0 & \frac{1}{2} & 0 & \bigm| & \frac{25}{2} \\
-1 & 0 & 0 & 2 & 1 & \bigm| & 50
\end{bmatrix}.
$$

(b) Set the group I variables equal to 0:

$$x = 0, \qquad v = 0.$$

Read off the particular values of the group II variables from the right-hand column:

$$u = \tfrac{15}{2}, \qquad y = \tfrac{25}{2}, \qquad M = 50.$$

So the particular solution corresponding to the transformed tableau is

$$x = 0, \qquad y = \tfrac{25}{2}, \qquad u = \tfrac{15}{2}, \qquad v = 0, \qquad M = 50.$$

PRACTICE PROBLEMS 1

1. Determine by inspection a particular solution of the following system of linear equations:

$$
\begin{cases}
x + 2y + 3u & = 6 \\
y \quad\;\; + v & = 4 \\
5y + 2u \quad\;\; + M & = 0.
\end{cases}
$$

2. Pivot the simplex tableau about the circled element:

$$
\begin{bmatrix}
2 & 4 & 1 & 0 & 0 & \bigm| & 6 \\
3 & \boxed{1} & 0 & 1 & 0 & \bigm| & 0 \\
1 & 1 & 0 & 0 & 1 & \bigm| & 1
\end{bmatrix}.
$$

EXERCISES 1

For each of the following linear programming problems, determine the corresponding linear system and restate the linear programming problem in terms of the linear system.

1. Maximize $8x + 13y$ subject to the constraints

$$
\begin{cases}
20x + 30y \le 3500 \\
50x + 10y \le 5000 \\
x \ge 0 \\
y \ge 0.
\end{cases}
$$

2. Maximize $x + 15y$ subject to the constraints

$$\begin{cases} 3x + 2y \le 10 \\ x \qquad\quad \le 15 \\ \qquad y \le 3 \\ x + \; y \le 5 \\ x \ge 0 \\ y \ge 0. \end{cases}$$

3. Maximize $x + 2y - 3z$ subject to the constraints

$$\begin{cases} x + \quad y + z \le 100 \\ 3x \qquad\quad + z \le 200 \\ 5x + 10y \qquad \le 100 \\ x \ge 0 \\ y \ge 0 \\ z \ge 0. \end{cases}$$

4. Maximize $2x + y + 50$ subject to the constraints

$$\begin{cases} x + 3y \le 24 \\ \quad y \le 5 \\ x + 7y \le 10 \\ x \ge 0 \\ y \ge 0. \end{cases}$$

5-8. For each of the linear programming problems in Exercises 1–4:

(a) Set up the simplex tableau.

(b) Determine the particular solution corresponding to the tableau.

Find the particular solutions corresponding to these tableaux.

9.

x	y	u	v	M	
0	2	1	0	0	10
1	3	0	12	0	15
0	−1	0	17	1	20

10.

x	y	u	v	M	
1	0	3	11	0	6
0	1	10	17	0	16
0	0	5	−1	1	3

11.

x	y	z	u	v	w	M	
0	3	1	0	1	15	0	15
1	−1	0	0	2	−5	0	10
0	2	0	1	−5	4	0	23
0	11	0	0	11	6	1	−11

12.

$$\begin{array}{ccccccc} x & y & z & u & v & w & M \end{array}$$

$$\left[\begin{array}{ccccccc|c} 6 & 0 & 1 & 0 & 5 & -1 & 0 & \frac{1}{4} \\ 5 & 1 & 0 & 0 & 3 & \frac{1}{3} & 0 & 100 \\ 4 & 0 & 0 & 1 & 8 & \frac{1}{2} & 0 & 11 \\ 2 & 0 & 0 & 0 & 6 & \frac{1}{7} & 1 & -\frac{1}{2} \end{array}\right]$$

13. Pivot the simplex tableau

$$\begin{array}{ccccc} x & y & u & v & M \end{array}$$

$$\left[\begin{array}{ccccc|c} 2 & 3 & 1 & 0 & 0 & 12 \\ 1 & 1 & 0 & 1 & 0 & 10 \\ -10 & -20 & 0 & 0 & 1 & 0 \end{array}\right]$$

about the indicated element and compute the particular solution corresponding to the new tableau.

(a) 2

(b) 3

(c) 1 (second row, first column)

(d) 1 (second row, second column).

14. Pivot the simplex tableau

$$\begin{array}{ccccc} x & y & u & v & M \end{array}$$

$$\left[\begin{array}{ccccc|c} 5 & 4 & 1 & 0 & 0 & 100 \\ 10 & 6 & 0 & 1 & 0 & 1200 \\ -1 & 2 & 0 & 0 & 1 & 0 \end{array}\right]$$

about the indicated element and compute the solution corresponding to the new tableau.

(a) 5   (b) 4   (c) 10   (d) 6.

15. Determine which of the pivot operations in Exercise 13 increases M the most.

16. Determine which of the pivot operations in Exercise 14 increases M the most.

SOLUTIONS TO PRACTICE PROBLEMS 1

1. The matrix of the system is

$$\begin{array}{ccccc} x & y & u & v & M \end{array}$$

$$\left[\begin{array}{ccccc|c} 1 & 2 & 3 & 0 & 0 & 6 \\ 0 & 1 & 0 & 1 & 0 & 4 \\ 0 & 5 & 2 & 0 & 1 & 0 \end{array}\right],$$

from which we see that the group II variables are x, v, M, and the group I variables y, u. To obtain a solution, we set the group I variables equal to 0. We obtain from the first equation that $x = 6$, from the second that $v = 4$, and from the third that $M = 0$. Thus a solution of the system is $x = 6, y = 0, u = 0, v = 4, M = 0$.

2. We must use elementary row operations to transform the second column into

$$\begin{bmatrix} 0 \\ 1 \\ 0 \end{bmatrix}.$$

$$\begin{bmatrix} 2 & 4 & 1 & 0 & 0 & | & 6 \\ 3 & 1 & 0 & 1 & 0 & | & 0 \\ 1 & 1 & 0 & 0 & 1 & | & 1 \end{bmatrix} \xrightarrow{[1] + (-4)[2]} \begin{bmatrix} -10 & 0 & 1 & -4 & 0 & | & 6 \\ 3 & 1 & 0 & 1 & 0 & | & 0 \\ 1 & 1 & 0 & 0 & 1 & | & 1 \end{bmatrix}$$

$$\xrightarrow{[3] + (-1)[2]} \begin{bmatrix} -10 & 0 & 1 & -4 & 0 & | & 6 \\ 3 & 1 & 0 & 1 & 0 & | & 0 \\ -2 & 0 & 0 & -1 & 1 & | & 1 \end{bmatrix}.$$

7.2. The Simplex Method, I—Maximum Problems

We can now describe the simplex method for solving linear programming problems. The procedure will be illustrated as we solve Problem A of Section 1. Recall that we must maximize the objective function $3x + 4y$ subject to the constraints

$$\begin{cases} x + y \le 20 \\ x + 2y \le 25 \\ x \ge 0, \quad y \ge 0. \end{cases}$$

> *Step 1* Introduce slack variables and state the problem in terms of a system of linear equations.

We carried out this step in the preceding section. The result was the following restatement of the problem.

PROBLEM A' Among the solutions of the system of linear equations

$$\begin{cases} x + y + u & = 20 \\ x + 2y & + v & = 25 \\ -3x - 4y & + M = 0, \end{cases}$$

find one for which $x \ge 0$, $y \ge 0$, $u \ge 0$, $v \ge 0$, and for which M is as large as possible.

> *Step 2* Construct the simplex tableau corresponding to the linear system.

This step was alo carried out in Section 1. The tableau is

$$\begin{array}{c} \\ u \\ v \\ M \end{array} \begin{array}{cccccc} x & y & u & v & M & \\ \begin{bmatrix} 1 & 1 & 1 & 0 & 0 & | & 20 \\ 1 & 2 & 0 & 1 & 0 & | & 25 \\ -3 & -4 & 0 & 0 & 1 & | & 0 \end{bmatrix} \end{array}. \qquad (1)$$

Note that we have made two additions to the previously found tableau. First, we have separated the last row from the others by means of a horizontal line. This is because the last row, which corresponds to the objective function in the original problem, will play a special role in what follows. The second addition is that we have labeled each row with one of the group II variables—namely, the variable whose value is determined by the row. Thus, for example, the first row gives the particular value of u, which is 20, so the row is labeled with a u. We will find these labels convenient.

Corresponding to this tableau, there is a particular solution to the linear system, namely the one obtained by setting all group I variables equal to 0. Reading the values of the group II variables from the last column, we obtain

$$x = 0, \qquad y = 0, \qquad u = 20, \qquad v = 25, \qquad M = 0.$$

Our object is to make M as large as possible. How can the value of M be increased? Look at the equation corresponding to the last row of the tableau. It reads

$$-3x - 4y + M = 0.$$

Note that two of the coefficients, -3 and -4, are negative. Or, what amounts to the same thing, if we solve for M and get

$$M = 3x + 4y,$$

then the coefficients on the right-hand side are *positive*. This fact is significant. It says that M can be increased by increasing either the value of x or the value of y. A unit change in x will increase M by 3 units, whereas a unit change in y will increase M by 4 units. And since we wish to increase M by as much as possible, it is reasonable to attempt to increase the value of y. Let us indicate this by drawing an arrow pointing to the y column of the tableau:

$$
\begin{array}{c}
 \\ u \\ v \\ M
\end{array}
\begin{array}{c}
\begin{array}{ccccc}
x & y & u & v & M
\end{array} \\
\left[
\begin{array}{ccccc|c}
1 & 1 & 1 & 0 & 0 & 20 \\
1 & 2 & 0 & 1 & 0 & 25 \\
\hline
-3 & -4 & 0 & 0 & 1 & 0
\end{array}
\right].
\end{array}
$$

In order to increase y (from its present value, zero), we will pivot about one of the entries (above the horizontal line) in the y column. In this way, y will become a group II variable and hence will not necessarily be zero in our next particular solution. But which entry should we pivot around? To find out, let us experiment. The results from pivoting about the 1 and the 2 in the y column are, respectively,

$$
\begin{array}{c}
 \\ y \\ v \\ M
\end{array}
\begin{array}{c}
\begin{array}{ccccc}
x & y & u & v & M
\end{array} \\
\left[
\begin{array}{ccccc|c}
1 & 1 & 1 & 0 & 0 & 20 \\
-1 & 0 & -2 & 1 & 0 & -15 \\
\hline
1 & 0 & 4 & 0 & 1 & 80
\end{array}
\right]
\end{array}
$$

Pivot about 1

$$\begin{array}{c} & \begin{array}{ccccc} x & y & u & v & M \end{array} \\ \begin{array}{c} u \\ y \\ M \end{array} & \left[\begin{array}{ccccc|c} \frac{1}{2} & 0 & 1 & -\frac{1}{2} & 0 & \frac{15}{2} \\ \frac{1}{2} & 1 & 0 & \frac{1}{2} & 0 & \frac{25}{2} \\ \hline -1 & 0 & 0 & 2 & 1 & 50 \end{array}\right]. \end{array}$$

Pivot about 2

Note that the labels on the rows have *changed* because the group II variables are now *different*. The solutions corresponding to these tableaux are, respectively,

$$x = 0, \quad y = 20, \quad u = 0, \quad v = -15, \quad M = 80$$

and

$$x = 0, \quad y = \tfrac{25}{2}, \quad u = \tfrac{15}{2}, \quad v = 0, \quad M = 50.$$

The first solution violates the requirement that all variables be ≥ 0. Thus we use the second solution, which arose from pivoting about 2. Using this solution, we have increased the value of M to 50 and have replaced our original tableau by

$$\begin{array}{c} & \begin{array}{ccccc} x & y & u & v & M \end{array} \\ \begin{array}{c} u \\ y \\ M \end{array} & \left[\begin{array}{ccccc|c} \frac{1}{2} & 0 & 1 & -\frac{1}{2} & 0 & \frac{15}{2} \\ \frac{1}{2} & 1 & 0 & \frac{1}{2} & 0 & \frac{25}{2} \\ \hline -1 & 0 & 0 & 2 & 1 & 50 \end{array}\right]. \end{array}$$

Can M be increased further? To answer this question, look at the last row of the tableau, corresponding to the equation

$$-x + 2v + M = 50.$$

There is a negative coefficient for the variable x in this equation. Correspondingly, when the equation is solved for M, there is a positive coefficient for x:

$$M = 50 + x - 2v.$$

Now it is clear that we should try to increase x. So we pivot about one of the entries in the x column. A quick calculation shows that pivoting about the second entry leads to a solution having some negative values. Therefore, we pivot about the top entry. The result is

$$\begin{array}{c} & \begin{array}{ccccc} x & y & u & v & M \end{array} \\ \begin{array}{c} x \\ y \\ M \end{array} & \left[\begin{array}{ccccc|c} 1 & 0 & 2 & -1 & 0 & 15 \\ 0 & 1 & -1 & 1 & 0 & 5 \\ \hline 0 & 0 & 2 & 1 & 1 & 65 \end{array}\right]. \end{array}$$

The corresponding solution is

$$x = 15, \quad y = 5, \quad u = 0, \quad v = 0, \quad M = 65.$$

Note that with this second pivot operation we have increased the value of M from 50 to 65.

Can we increase M still further? Let us reason as before. Use the last row of the current tableau to write M in terms of the other variables:

$$2u + v + M = 65, \qquad M = 65 - 2u - v.$$

Note, however, that in contrast to the previous expressions for M, this one has *no positive coefficients*. And since u and v are ≥ 0, this means that M can be *at most* 65. But M is already 65. So M cannot be increased further. Thus we have shown that the maximum value of M is 65, and this occurs when $x = 15$, $y = 5$. This solves the original linear programming problem.

Based on the discussion above, we can state several general principles. First of all, the following criterion determines when a simplex tableau yields a maximum.

Condition for a Maximum   The particular solution derived from a simplex tableau is a maximum if and only if the bottom row contains no negative entries except perhaps the entry in the last column.*

We saw this condition illustrated in the example above. Each of the first two tableaux had negative entries in the last row, and, as we showed, their corresponding solutions were not maxima. However, the third tableau, with no negative entries in the last row, did yield a maximum.

The crucial point of the simplex method is the correct choice of a pivot element. In the example above we decided to choose a pivot element from the column corresponding to the most negative entry in the last row. It can be proved that this is the proper choice in general; that is, we have the following rule:

Choosing the Pivot Column   The pivot element should be chosen from that column to the left of the vertical line which has the most negative entry in the last row.†

Choosing the correct pivot element from the designated column is somewhat more complicated. Our approach above was to calculate the tableau associated with each element and observe that only one corresponded to a solution with nonnegative elements. However, there is a simpler way to make the choice. As an illustration, let us reconsider tableau (1). We have already decided to pivot around some entry in the second column. For each *positive* entry in the pivot column we compute a ratio: the corresponding entry in the right-hand column divided by the entry in the pivot column. So for example, for the first entry the ratio is $\frac{20}{1}$ and

---

*In Section 3 we shall encounter maximum problems whose final tableaux have a negative number in the lower right-hand corner.

† In case two or more columns are tied for the honor of being pivot column, an arbitrary choice among them may be made.

for the second $\frac{25}{2}$. We write these ratios to the right of the matrix as follows:

$$
\begin{array}{c}
 \\
u \\
v \\
M
\end{array}
\begin{array}{ccccc}
x & y & u & v & M \\
\end{array}
\left[
\begin{array}{ccccc|c}
1 & 1 & 1 & 0 & 0 & 20 \\
1 & 2 & 0 & 1 & 0 & 25 \\
\hline
-3 & -4 & 0 & 0 & 1 & 0
\end{array}
\right]
\begin{array}{l}
20/1 \\
25/2 \\

\end{array}
$$

It is possible to prove the following rule, which allows us to determine the pivot element from the above display:

Choosing the Pivot Element For each positive entry of the pivot column, compute the appropriate ratio. Choose as pivot element the one corresponding to the least (nonnegative) ratio.

For instance, consider the choice of pivot element in the example above. The least among the ratios is $\frac{25}{2}$. So we choose 2 as the pivot element.

At first, this method for choosing the pivot element might seem very odd. However, it is just a way of guaranteeing that the last column of the new tableau will have entries ≥ 0. And that is just the basis on which we chose the pivot element earlier. To obtain further insight, let us analyze the example above yet further.

Suppose that we pivot our tableau about the 1 in column 2. The first step in pivoting is to divide the pivot row by the pivot element (in this case 1). This gives the array

$$
\left[
\begin{array}{ccccc|c}
1 & 1 & 1 & 0 & 0 & \frac{20}{1} \\
1 & 2 & 0 & 1 & 0 & 25 \\
\hline
-3 & -4 & 0 & 0 & 1 & 0
\end{array}
\right],
$$

where we have written $\frac{20}{1}$ rather than 20 to emphasize that we have divided by the pivot element. The next step in the pivot procedure is to replace the second row by the second row plus (-2) times the first row. The result is

$$
\left[
\begin{array}{ccccc|c}
1 & 1 & 1 & 0 & 0 & \frac{20}{1} \\
-1 & 0 & -2 & 1 & 0 & 25 + (-2)\frac{20}{1} \\
\hline
-3 & -4 & 0 & 0 & 1 & 0
\end{array}
\right].
$$

The final step of the pivot process is to replace the third row by $[3] + 4[1]$, to obtain

$$
\left[
\begin{array}{ccccc|c}
1 & 1 & 1 & 0 & 0 & \frac{20}{1} \\
-1 & 0 & -2 & 1 & 0 & 25 + (-2)\frac{20}{1} \\
\hline
1 & 0 & 4 & 0 & 1 & 80
\end{array}
\right].
$$

Notice that the two upper entries in the last column may be written

$$
\frac{20}{1}, \qquad 2 \cdot \left(\frac{25}{2} - \frac{20}{1} \right).
$$

Notice the difference of the ratios appearing in the second entry! If we similarly pivot about the 2 in the second column of the original tableau, we obtain

$$\begin{bmatrix} \frac{1}{2} & 0 & 1 & -\frac{1}{2} & 0 & 20 + (-1)\frac{25}{2} \\ \frac{1}{2} & 1 & 0 & \frac{1}{2} & 0 & \frac{25}{2} \\ -1 & 0 & 0 & 2 & 1 & 50 \end{bmatrix}.$$

The upper entries in the last column are

$$\frac{20}{1} - \frac{25}{2}, \quad \frac{25}{2}.$$

Again notice the difference of the ratios, this time the reverse of the previous difference. Note that when we pivot about 1, we subtract the ratio $20/1$, whereas if we pivot about 2, we subtract the ratio $25/2$. In order to arrive at nonnegative entries in the last column, we should subtract off as little as possible. That is, we should pivot about the entry corresponding to the smallest ratio. This is the rationale governing our choice of pivot element!

Now that we have assembled all the components of the simplex method, we can summarized it as follows:

The Simplex Method for Problems in Standard Form

1. Introduce slack variables and state the problem in terms of a system of linear equations.
2. Construct the simplex tableau corresponding to the system.
3. Determine if the left part of the bottom row contains negative entries. If none are present, the solution corresponding to the tableau yields a maximum and the problem is solved.
4. If the left part of the bottom row contains negative entries, construct a new simplex tableau.
   (a) Choose the pivot column by inspecting the entries of the last row of the current tableau, excluding the right-hand entry. The pivot column is the one containing the most negative of these entries.
   (b) Choose the pivot element by computing ratios associated with the positive entries of the pivot column. The pivot element is the one corresponding to the smallest (nonnegative) ratio.
   (c) Construct the new simplex tableau by pivoting around the selected element.
5. Return to step 3. Steps 3 and 4 are repeated as many times as necessary to find a maximum.

Let us now work some problems to see how this method is applied.

EXAMPLE 1 Maximize the objective function $10x + y$ subject to the constraints

$$\begin{cases} x + 2y \leq 10 \\ 3x + 4y \leq 6 \\ x \geq 0, \quad y \geq 0. \end{cases}$$

Solution The corresponding system of linear equations with slack variables is

$$\begin{cases} x + 2y + u & = 10 \\ 3x + 4y \quad + v & = 6 \\ -10x - y \qquad\quad + M & = 0, \end{cases}$$

and we must find that solution of the system for which $x \geq 0, y \geq 0, u \geq 0, v \geq 0$, and M is as large as possible. Here is the initial simplex tableau:

$$\begin{array}{c} \\ u \\ v \\ M \end{array} \begin{array}{ccccc} x & y & u & v & M \\ \left[\begin{array}{ccccc|c} 1 & 2 & 1 & 0 & 0 & 10 \\ 3 & 4 & 0 & 1 & 0 & 6 \\ \hline -10 & -1 & 0 & 0 & 1 & 0 \end{array}\right] \end{array}$$

$\uparrow$

Note that this tableau does not correspond to a maximum, since the left part of the bottom row has negative entries. So we pivot to create a new tableau. Since -10 is the most negative entry in the last row, we choose the first column as the pivot column. To determine the pivot element, we compute ratios:

$$\begin{array}{c} \\ u \\ v \\ M \end{array} \begin{array}{ccccc} x & y & u & v & M \\ \left[\begin{array}{ccccc|c} 1 & 2 & 1 & 0 & 0 & 10 \\ ③ & 4 & 0 & 1 & 0 & 6 \\ \hline -10 & -1 & 0 & 0 & 1 & 0 \end{array}\right] \end{array} \begin{array}{l} \text{Ratios} \\ 10/1 = 10 \\ 6/3 = 2\,^{\textbf{\'}} \\ \end{array}$$

$\uparrow$

The smallest ratio is 2, so we pivot about 3, which we have circled. The new tableau is therefore

$$\begin{array}{c} \\ u \\ x \\ M \end{array} \begin{array}{ccccc} x & y & u & v & M \\ \left[\begin{array}{ccccc|c} 0 & \frac{1}{3} & 1 & -\frac{1}{3} & 0 & 8 \\ 1 & \frac{4}{3} & 0 & \frac{1}{3} & 0 & 2 \\ 0 & \frac{37}{3} & 0 & \frac{10}{3} & 1 & 20 \end{array}\right] \end{array}$$

Note that this tableau corresponds to a maximum, since there are no negative entries in the left part of the last row. The solution corresponding to the tableau is

$$x = 2, \qquad y = 0, \qquad u = 8, \qquad v = 0, \qquad M = 20.$$

Therefore, the objective function assumes its maximum value of 20 when $x = 2$ and $y = 0$.

Let us now check that the simplex method yields the same result as the graphical method of Chapter 6 by reworking a problem from there.

EXAMPLE 2 Use the simplex method to solve the furniture manufacturing problem of Section 1 of Chapter 6.

Solution   Here x represents the number of chairs and y the number of sofas to be produced each day. The daily profit is $80x + 70y$ dollars, and the limitations imposed by available labor are expressed by the constraints

$$\begin{cases} 6x + 3y \le 96 \\ x + y \le 18 \\ 2x + 6y \le 72 \\ x \ge 0, \quad y \ge 0. \end{cases}$$

The linear system is then

$$\begin{cases} 6x + 3y + u && = 96 \\ x + y && + v && = 18 \\ 2x + 6y &&& + w & = 72 \\ -80x - 70y &&&& + M = 0. \end{cases}$$

The simplex method then proceeds as follows:

	x	y	u	v	w	M		
u	⑥	3	1	0	0	0	96	$96/6 = 16$
v	1	1	0	1	0	0	18	$18/1 = 18$
w	2	6	0	0	1	0	72	$72/2 = 36$
M	-80	-70	0	0	0	1	0	
	↑							

	x	y	u	v	w	M		
x	1	$\frac{1}{2}$	$\frac{1}{6}$	0	0	0	16	$16/\frac{1}{2} = 32$
v	0	$\left(\frac{1}{2}\right)$	$-\frac{1}{6}$	1	0	0	2	$2/\frac{1}{2} = 4$
w	0	5	$-\frac{1}{3}$	0	1	0	40	$40/5 = 8$
M	0	-30	$\frac{40}{3}$	0	0	1	1280	
		↑						

	x	y	u	v	w	M	
x	1	0	$\frac{1}{3}$	-1	0	0	14
y	0	1	$-\frac{1}{3}$	2	0	0	4
w	0	0	$\frac{4}{3}$	-10	1	0	20
M	0	0	$\frac{10}{3}$	60	0	1	1400

This last tableau corresponds to maximum profit, and the solution is

$$x = 14, \quad y = 4, \quad u = 0, \quad v = 0, \quad w = 20, \quad M = 1400.$$

Thus the maximum profit of $1400 occurs when $x = 14$ and $y = 4$.

The simplex method can be used to solve problems in any number of variables. Let us illustrate the method for three variables.

EXAMPLE 3 Maximize the objective function $x + 2y + z$ subject to the constraints

$$\begin{cases} x - y + 2z \leq 10 \\ 2x + y + 3z \leq 12 \\ x \geq 0, \quad y \geq 0, \quad z \geq 0. \end{cases}$$

Solution We determined the corresponding linear system in Example 1 of Section 1:

$$\begin{cases} x - y + 2z + u = 10 \\ 2x + y + 3z + v = 12 \\ -x - 2y - z + M = 0. \end{cases}$$

So the simplex method works as follows:

	x	y	z	u	v	M	
u	1	-1	2	1	0	0	10
v	2	①	3	0	1	0	12
M	-1	-2	-1	0	0	1	0

$12/1 = 12$ (smallest ratio)

	x	y	z	u	v	M	
u	3	0	5	1	1	0	22
y	2	1	3	0	1	0	12
M	3	0	5	0	2	1	24

Thus the solution of the original problem is: $x = 0$, $y = 12$, $z = 0$ yields the maximum value of the objective function $x + 2y + z$. The maximum value is 24.

PRACTICE PROBLEMS 2

1. Which of these simplex tableaux has a solution which corresponds to a maximum for the associated linear programming problem?

(a)

	x	y	u	v	M	
	3	1	0	1	0	5
	2	0	0	0	1	0
	-1	-2	1	0	0	3

(b)

	x	y	u	v	M	
	2	1	0	11	0	10
	1	0	1	7	0	1
	1	0	0	4	1	-2

2. Suppose that in the solution of a linear programming problem by the simplex method we encounter the following simplex tableau. What is the next step in the solution?

$$
\begin{array}{ccccc}
x & y & u & v & M \\
\begin{bmatrix} 0 & 4 & 1 & 2 & 0 \\ 1 & 5 & 0 & 1 & 0 \\ 0 & 2 & 0 & -3 & 1 \end{bmatrix} & & & & \begin{matrix} 4 \\ 9 \\ 6 \end{matrix}
\end{array}
$$

EXERCISES 2

For each of the following simplex tableaux:

(a) Compute the next pivot element.

(b) Determine the next tableau.

(c) Determine the particular solution corresponding to the tableau of part (b).

1.
$$
\begin{array}{ccccc}
x & y & u & v & M \\
\begin{bmatrix} 6 & 2 & 1 & 0 & 0 \\ 1 & 3 & 0 & 1 & 0 \\ -4 & -12 & 0 & 0 & 1 \end{bmatrix} & & & & \begin{matrix} 10 \\ 6 \\ 0 \end{matrix}
\end{array}
$$

2.
$$
\begin{array}{ccccc}
x & y & u & v & M \\
\begin{bmatrix} 1 & 0 & 3 & 1 & 0 \\ 0 & 1 & 2 & 0 & 0 \\ -6 & 0 & 5 & 0 & 1 \end{bmatrix} & & & & \begin{matrix} 5 \\ 12 \\ 10 \end{matrix}
\end{array}
$$

3.
$$
\begin{array}{ccccc}
x & y & u & v & M \\
\begin{bmatrix} 5 & 12 & 1 & 0 & 0 \\ 15 & 10 & 0 & 1 & 0 \\ 4 & -2 & 0 & 0 & 1 \end{bmatrix} & & & & \begin{matrix} 12 \\ 5 \\ 0 \end{matrix}
\end{array}
$$

4.
$$
\begin{array}{ccccc}
x & y & u & v & M \\
\begin{bmatrix} 0 & 6 & 3 & 1 & 0 \\ 1 & -5 & 2 & 0 & 0 \\ 0 & 20 & -10 & 0 & 1 \end{bmatrix} & & & & \begin{matrix} 5 \\ 8 \\ 22 \end{matrix}
\end{array}
$$

Solve the following linear programming problems using the simplex method.

5. Maximize $x + 3y$ subject to the constraints

$$
\begin{cases}
x + y \le 7 \\
x + 2y \le 10 \\
x \ge 0, \quad y \ge 0.
\end{cases}
$$

6. Maximize $x + 2y$ subject to the constraints

$$
\begin{aligned}
-x + y &\le 100 \\
6x + 6y &\le 1200 \\
x \ge 0, \quad y &\ge 0.
\end{aligned}
$$

7. Maximize $4x + 2y$ subject to the constraints

$$\begin{cases} 5x + y \le 80 \\ 3x + 2y \le 76 \\ x \ge 0, \quad y \ge 0. \end{cases}$$

8. Maximize $2x + 6y$ subject to the constraints

$$\begin{cases} -x + 8y \le 160 \\ 3x - y \le 3 \\ x \ge 0, \quad y \ge 0. \end{cases}$$

9. Maximize $x + 3y + 5z$ subject to the constraints

$$\begin{cases} x + 2z \le 10 \\ 3y + z \le 24 \\ x \ge 0, \quad y \ge 0, \quad z \ge 0. \end{cases}$$

10. Maximize $-x + 8y + z$ subject to the constraints

$$\begin{cases} x - 2y + 9z \le 10 \\ y + 4z \le 12 \\ x \ge 0, \quad y \ge 0, \quad z \ge 0. \end{cases}$$

11. Maximize $2x + 3y$ subject to the constraints

$$\begin{cases} 5x + y \le 30 \\ 3x + 2y \le 60 \\ x + y \le 50 \\ x \ge 0, \quad y \ge 0. \end{cases}$$

12. Maximize $10x + 12y + 10z$ subject to the constraints

$$\begin{cases} x - 2y \le 6 \\ 3x + z \le 9 \\ y + 3z \le 12 \\ x \ge 0, \quad y \ge 0, \quad z \ge 0. \end{cases}$$

13. Maximize $6x + 7y + 300$ subject to the constraints

$$\begin{cases} 2x + 3y \le 400 \\ x + y \le 150 \\ x \ge 0, \quad y \ge 0. \end{cases}$$

14. Maximize $10x + 20y + 50$ subject to the constraints

$$\begin{cases} x + y \le 10 \\ 5x + 2y \le 20 \\ x \ge 0, \quad y \ge 0. \end{cases}$$

15. Suppose that a furniture manufacturer makes chairs, sofas, and tables. The amounts of labor of various types as well as the relative availability of each type is summarized by the following chart:

	Chair	Sofa	Table	Daily labor available (man-hours)
Carpentry	6	3	8	768
Finishing	1	1	2	144
Upholstery	2	5	0	216

The profit per chair is $80, per sofa $70, and per table $120. How many pieces of each type of furniture should be manufactured each day in order to maximize the profit?

16. A stereo store sells three brands of stereo system, brands A, B, and C. It can sell a total of 100 stereo systems per month. Brands A, B, and C take up, respectively, 5, 4, and 4 cubic feet of warehouse space and a maximum of 480 cubic feet of warehouse space is available. Brands A, B, and C generate sales commissions of $40, $20, and $30, respectively, and $3200 is available to pay the sales commissions. The profit generated from the sale of each brand is $70, $210, and $140, respectively. How many of each brand of stereo system should be sold to maximize the profit?

17. A furniture manufacturer produces small sofas, large sofas, and chairs. The profits per item are, respectively, $60, $60, and $50. The pieces of furniture require the following numbers of man-hours for their manufacture:

	Carpentry	Upholstery	Finishing
Small sofas	10	30	20
Large sofas	10	30	0
Chairs	10	10	10

The following amounts of labor are available per month: carpentry at most 1200 hours, upholstery at most 3000 hours, and finishing at most 1800 hours. How many each of small sofas, large sofas, and chairs should be manufactured to maximize the profit?

18. Maximize $60x + 90y + 300z$ subject to the constraints

$$\begin{cases} x + y + z \le 600 \\ x + 3y \le 600 \\ 2x + z \le 900 \\ x \ge 0, \quad y \ge 0, \quad z \ge 0. \end{cases}$$

19. Maximize $200x + 500y$ subject to the constraints

$$\begin{cases} x + 4y \le 300 \\ x + 2y \le 200. \end{cases}$$

SOLUTIONS TO PRACTICE PROBLEMS 2

1. (a) Does not correspond to a maximum, since among the entries $-1, -2, 1, 0, 0$ on the last row, at least one is negative.

   (b) Corresponds to a maximum since none of the entries $1, 0, 0, 4, 1$ of the last row is negative. Note that it does not matter that the entry -2 in the right-hand corner of the matrix is negative. This number gives the value of M. In this example -2 is as large as M can become.

2. First choose the column corresponding to the most negative entry of the final row, that is, the fourth column. For each entry in the fourth column which is above the horizontal line, compute the ratio with the sixth column. The smallest ratio is 2 and appears in the first row, so the next operation is to pivot around the 2 in the first row of the fourth column.

$$
\begin{bmatrix}
0 & 4 & 1 & ② & 0 & 4 \\
1 & 5 & 0 & 1 & 0 & 9 \\
\hline
0 & 2 & 0 & -3 & 1 & 6
\end{bmatrix}
\begin{matrix}
4/2 = 2 \\
9/1 = 9 \\
\\
\end{matrix}
$$

$\uparrow$

7.3. The Simplex Method, II—Minimum Problems

In the preceding section we developed the simplex method and applied it to a number of problems. However, throughout we restricted ourselves to linear programming problems in standard form. Recall that such problems satisfied three properties: (1) the objective function is to be maximized; (2) each variable must be ≥ 0; and (3) all constraints other than those implied by (2) must be of the form

[linear polynomial] $\leq$ [nonnegative constant].

In this section we shall do what we can to relax these restrictions.

Let us begin with restriction (3). This could be violated in two ways. First, the constant on the right-hand side of one or more constraints could be negative. Thus, for example, one constraint might be

$$x - y \leq -2.$$

A second way in which restriction (3) can be violated is for some constraints to involve $\geq$ rather than $\leq$. An example of such a constraint is

$$2x + 3y \geq 5.$$

However, we can convert such a constraint into one involving $\leq$ by multiplying both sides of the inequality by -1:

$$-2x - 3y \leq -5.$$

Of course, the right-hand constant is no longer nonnegative. Thus, if we allow negative constants on the right, we can write all constraints in the form

[linear polynomial] $\leq$ [constant].

Henceforth, the first step in solving a linear programming problem will be to write the constraints in this form. Let us now see how to deal with the phenomenon of negative constants.

EXAMPLE 1　Maximize the objective function $5x + 10y$ subject to the constraints

$$\begin{cases} x + y \leq 20 \\ 2x - y \geq 10 \\ x \geq 0, \quad y \geq 0. \end{cases}$$

Solution　The first step is to put the second constraint into $\leq$ form. Multiply the second inequality by -1 to obtain

$$\begin{cases} x + y \leq 20 \\ -2x + y \leq -10 \\ x \geq 0, \quad y \geq 0. \end{cases}$$

Just as before, write this as a linear system:

$$\begin{cases} x + y + u = 20 \\ -2x + y + v = -10 \\ -5x - 10y + M = 0. \end{cases}$$

From the linear system construct the simplex tableau:

$$
\begin{array}{c}
 \\ u \\ v \\ M
\end{array}
\begin{array}{cccccc}
x & y & u & v & M & \\
\left[\begin{array}{ccccc|c} 1 & 1 & 1 & 0 & 0 & 20 \\ -2 & 1 & 0 & 1 & 0 & -10 \\ -5 & -10 & 0 & 0 & 1 & 0 \end{array}\right].
\end{array}
$$

Everything would proceed exactly as before, except that the right-hand column has a -10 in it. This means that the initial value for v is -10, which violates the condition that all variables be ≥ 0. Before we can apply the simplex method of Section 2, we must first put the tableau into standard form. This can be done by pivoting so as to remove the negative entry in the right column.

We choose the pivot element as follows. Look along the left side of the -10 row of the tableau and locate any negative entry. There is only one: -2. Use the column containing the -2—column 1—as the pivot column. Now compute ratios as before:*

$$
\begin{array}{c}
 \\ u \\ v \\ M
\end{array}
\begin{array}{ccccccc}
x & y & u & v & M & \\
\left[\begin{array}{ccccc|c} 1 & 1 & 1 & 0 & 0 & 20 \\ \boxed{-2} & 1 & 0 & 1 & 0 & -10 \\ -5 & -10 & 0 & 0 & 1 & 0 \end{array}\right] & \begin{array}{l} 20/1 = 20 \\ -10/-2 = 5 \end{array}
\end{array}
$$

————

* Note, however, that in this circumstance we compute ratios corresponding to both positive *and* negative entries (except the last) in the pivot column, considering further only those ratios which are positive.

The smallest positive ratio is 5, so we choose -2 as the pivot element. The new tableau is

$$
\begin{array}{c}
 \\
u \\
x \\
M
\end{array}
\begin{array}{c}
\begin{array}{ccccc}
x & y & u & v & M
\end{array} \\
\left[
\begin{array}{ccccc|c}
0 & \frac{3}{2} & 1 & \frac{1}{2} & 0 & 15 \\
1 & -\frac{1}{2} & 0 & -\frac{1}{2} & 0 & 5 \\
0 & -\frac{25}{2} & 0 & -\frac{5}{2} & 1 & 25
\end{array}
\right].
\end{array}
$$

Note that all entries in the right-hand column are now nonnegative;* that is, the corresponding solution has all variables ≥ 0. From here on we follow the simplex method for tableaux in standard form:

$$
\begin{array}{c}
 \\
u \\
x \\
M
\end{array}
\begin{array}{c}
\begin{array}{ccccc}
x & y & u & v & M
\end{array} \\
\left[
\begin{array}{ccccc|c}
0 & \left(\frac{3}{2}\right) & 1 & \frac{1}{2} & 0 & 15 \\
1 & -\frac{1}{2} & 0 & -\frac{1}{2} & 0 & 5 \\
0 & -\frac{25}{2} & 0 & -\frac{5}{2} & 1 & 25
\end{array}
\right]
\end{array}
\quad 15/\tfrac{3}{2} = 10
$$

$$
\begin{array}{c}
 \\
y \\
x \\
M
\end{array}
\begin{array}{c}
\begin{array}{ccccc}
x & y & u & v & M
\end{array} \\
\left[
\begin{array}{ccccc|c}
0 & 1 & \frac{2}{3} & \frac{1}{3} & 0 & 10 \\
1 & 0 & \frac{1}{3} & -\frac{1}{3} & 0 & 10 \\
0 & 0 & \frac{25}{3} & \frac{5}{3} & 1 & 150
\end{array}
\right].
\end{array}
$$

So the maximum value of M is 150, which is attained for $x = 10$, $y = 10$.

In summary:

The Simplex Method for Problems in Nonstandard Form

1. If necessary, convert all inequalities (except $x \geq 0$, $y \geq 0$), into the form

    [linear polynomial] $\leq$ [constant].

2. If a negative number appears in the upper part of the last column of the simplex tableau, remove it by pivoting.
    (a) Select one of the negative entries in its row. The column containing the entry will be the pivot column.
    (b) Select the pivot element by determining the least of the positive ratios associated to entries in the pivot column (except the bottom entry).
    (c) Pivot.

3. Repeat step 2 until there are no negative entries in the upper part of the right-hand column of the simplex tableau.

4. Proceed to apply the simplex method for tableaux in standard form.

---

* In general, it may be necessary to pivot several times before all elements in the last column are ≥ 0.

The method we have just developed can be used to solve *minimum* problems as well as maximum problems. Minimizing the objective function f is the same as maximizing $(-1) \cdot f$. This is so since multiplying an inequality by -1 reverses the direction of the inequality sign. Thus, in order to apply our method to a minimum problem, we merely multiply the objective function by -1 and turn the problem into a maximum problem.

EXAMPLE 2   Minimize the objective function $3x + 2y$ subject to the constraints

$$\begin{cases} x + y \geq 10 \\ x - y \leq 15 \\ x \geq 0, \quad y \geq 0. \end{cases}$$

Solution   First transform the problem so that the first two constraints are in $\leq$ form:

$$\begin{cases} -x - y \leq -10 \\ x - y \leq 15 \\ x \geq 0, \quad y \geq 0. \end{cases}$$

Instead of minimizing $3x + 2y$, let us maximize $-3x - 2y$. Let $M = -3x - 2y$. Then our initial simplex tableau reads

$$
\begin{array}{c}
 \\ u \\ v \\ M
\end{array}
\begin{array}{ccccc}
x & y & u & v & M \\
\end{array}
\left[\begin{array}{ccccc|c}
-1 & \boxed{-1} & 1 & 0 & 0 & -10 \\
1 & -1 & 0 & 1 & 0 & 15 \\
3 & 2 & 0 & 0 & 1 & 0
\end{array}\right]
\quad -10/-1 = 10
$$

We first eliminate the -10 in the right-hand column. We have a choice of two negative entries in the -10 row. Let us choose the one in the y column. The ratios are then tabulated as above, and we pivot around the circled element. The new tableau is*

$$
\begin{array}{c}
 \\ y \\ v \\ M
\end{array}
\begin{array}{ccccc}
x & y & u & v & M \\
\end{array}
\left[\begin{array}{ccccc|c}
1 & 1 & -1 & 0 & 0 & 10 \\
2 & 0 & -1 & 1 & 0 & 25 \\
1 & 0 & 2 & 0 & 1 & -20
\end{array}\right].
$$

Since all entries in the bottom row, except the last, are positive, this tableau corresponds to a maximum. Thus the maximum value of $-3x - 2y$ (subject to the constraints) is -20 and this value occurs for $x = 0$, $y = 10$. Thus the *minimum* value of $3x + 2y$ subject to the constraints is 20.

Let us now rework an applied problem previously treated (see Example 2 in Section 3 of Chapter 6), this time using the simplex method. For easy reference we restate the problem.

---

* Note that we do not need the last entry in the last column positive. We require *only* that x, y, u, and v be ≥ 0.

EXAMPLE 3 (*Transportation Problem*) Suppose that a TV dealer has stores in Annapolis and Rockville and warehouses in College Park and Baltimore. The cost of shipping sets from College Park to Annapolis is $6 per set; from College Park to Rockville, $3; from Baltimore to Annapolis, $9; and from Baltimore to Rockville, $5. Suppose that the Annapolis store orders 25 TV sets and the Rockville store 30. Further suppose that the College Park warehouse has a stock of 45 sets, and the Baltimore warehouse 40. What is the most economical way to supply the requested TV sets to the two stores?

Solution via the Simplex Method   As in the previous solution, let x be the number of sets shipped from College Park to Rockville, and y the number shipped from College Park to Annapolis. The flow of sets is depicted in Fig. 1.

Exactly as in our previous solution, we reduce the problem to the following algebraic form: Minimize $375 - 2x - 3y$ subject to the constraints

$$\begin{cases} x \le 30, \quad y \le 25 \\ x + y \ge 15 \\ x + y \le 45 \\ x \ge 0, \quad y \ge 0. \end{cases}$$

Two changes are needed. First, instead of minimizing $375 - 2x - 2y$ we maximize $-(375 - 2x - 3y) = 2x + 3y - 375$. Second, we write the constraint $x + y \ge 15$ in the form

$$-x - y \le -15.$$

With these changes made, we can write down the linear system:

$$\begin{cases} x \qquad\quad + t \qquad\qquad\qquad\qquad = 30 \\ \quad\; y \qquad + u \qquad\qquad\qquad = 25 \\ -x - \;y \qquad\quad + v \qquad\qquad = -15 \\ \;\; x + \;y \qquad\qquad\quad + w \qquad = 45 \\ -2x - 3y \qquad\qquad\qquad\quad + M = -375. \end{cases}$$

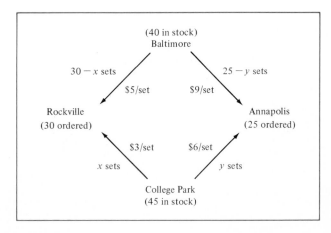

From here on we follow our routine procedure in a mechanical way:

$$
\begin{array}{c}
\begin{array}{ccccccc} x & y & t & u & v & w & M \end{array} \\
\begin{array}{c} t \\ u \\ v \\ w \\ M \end{array}
\left[
\begin{array}{ccccccc|c}
1 & 0 & 1 & 0 & 0 & 0 & 0 & 30 \\
0 & 1 & 0 & 1 & 0 & 0 & 0 & 25 \\
\boxed{-1} & -1 & 0 & 0 & 1 & 0 & 0 & -15 \\
1 & 1 & 0 & 0 & 0 & 1 & 0 & 45 \\
\hline
-2 & -3 & 0 & 0 & 0 & 0 & 1 & -375
\end{array}
\right]
\begin{array}{l}
30/1 = 30 \\
\\
-15/-1 = 15 \\
45/1 = 45
\end{array}
\end{array}
$$

$$
\begin{array}{c}
\begin{array}{ccccccc} x & y & t & u & v & w & M \end{array} \\
\begin{array}{c} t \\ u \\ x \\ w \\ M \end{array}
\left[
\begin{array}{ccccccc|c}
0 & -1 & 1 & 0 & \boxed{1} & 0 & 0 & 15 \\
0 & 1 & 0 & 1 & 0 & 0 & 0 & 25 \\
1 & 1 & 0 & 0 & -1 & 0 & 0 & 15 \\
0 & 0 & 0 & 0 & 1 & 1 & 0 & 30 \\
\hline
0 & -1 & 0 & 0 & -2 & 0 & 1 & -345
\end{array}
\right]
\begin{array}{l}
15/1 = 15 \\
\\
\\
30/1 = 30
\end{array}
\end{array}
$$

$$
\begin{array}{c}
\begin{array}{ccccccc} x & y & t & u & v & w & M \end{array} \\
\begin{array}{c} v \\ u \\ x \\ w \\ M \end{array}
\left[
\begin{array}{ccccccc|c}
0 & -1 & 1 & 0 & 1 & 0 & 0 & 15 \\
0 & 1 & 0 & 1 & 0 & 0 & 0 & 25 \\
1 & 0 & 1 & 0 & 0 & 0 & 0 & 30 \\
0 & \boxed{1} & -1 & 0 & 0 & 1 & 0 & 15 \\
\hline
0 & -3 & 2 & 0 & 0 & 0 & 1 & -315
\end{array}
\right]
\begin{array}{l}
\\
25/1 = 25 \\
\\
15/1 = 15
\end{array}
\end{array}
$$

$$
\begin{array}{c}
\begin{array}{ccccccc} x & y & t & u & v & w & M \end{array} \\
\begin{array}{c} v \\ u \\ x \\ y \\ M \end{array}
\left[
\begin{array}{ccccccc|c}
0 & 0 & 0 & 0 & 1 & 1 & 0 & 30 \\
0 & 0 & \boxed{1} & 1 & 0 & -1 & 0 & 10 \\
1 & 0 & 1 & 0 & 0 & 0 & 0 & 30 \\
0 & 1 & -1 & 0 & 0 & 1 & 0 & 15 \\
\hline
0 & 0 & -1 & 0 & 0 & 3 & 1 & -270
\end{array}
\right]
\begin{array}{l}
\\
10/1 = 10 \\
30/1 = 30
\end{array}
\end{array}
$$

$$
\begin{array}{c}
\begin{array}{ccccccc} x & y & t & u & v & w & M \end{array} \\
\begin{array}{c} v \\ t \\ x \\ y \\ M \end{array}
\left[
\begin{array}{ccccccc|c}
0 & 0 & 0 & 0 & 1 & 1 & 0 & 30 \\
0 & 0 & 1 & 1 & 0 & -1 & 0 & 10 \\
1 & 0 & 0 & -1 & 0 & 1 & 0 & 20 \\
0 & 1 & 0 & 1 & 0 & 0 & 0 & 25 \\
\hline
0 & 0 & 0 & 1 & 0 & 2 & 1 & -260
\end{array}
\right].
\end{array}
$$

This last tableau corresponds to a maximum. So $2x + 3y - 375$ has a maximum value -260, and therefore $375 - 2x - 3y$ has a minimum value 260. This value occurs when $x = 20$ and $y = 25$. This is in agreement with our previous graphical solution of the problem.

The calculations used in the preceding example are not that much simpler than those in the original solution. Why, then, should we concern ourselves with the simplex method? For one thing, the simplex method is so mechanical in its execution that it is much easier to program for a computer. For another, our previous method was restricted to problems in two variables. However, suppose that the two warehouses were to deliver their TV sets to three or four or perhaps even 100 stores? Our previous method could not be applied. However, the simplex method, although yielding very large matrices and very tedious calculations, is applicable. Indeed, this is the method used by many industrial concerns to optimize distribution of their products.

Some Further Comments on the Simplex Method Our discussion has omitted some of the technical complications arising in the simplex method. A complete discussion of these is beyond the scope of this book. However, let us mention two. First of all, it is possible that a given linear programming problem has more than one solution. This can occur, for example, if there are ties for the choice of pivot column. For instance, if the bottom row of the simplex tableau is

$$\begin{bmatrix} -3 & -7 & 4 & -7 & 1 & 3 \end{bmatrix},$$
$$\qquad\quad\uparrow\qquad\qquad\uparrow$$

then -7 is the most negative entry and we may choose as pivot column either the second or fourth. In such a circumstance the pivot column may be chosen arbitrarily. Different choices, however, may lead to different solutions of the problem.

A second difficulty is that a given linear programming problem may have no solution at all. In this case the method will break down at some point. For example, among the ratios at a given stage there may be no nonnegative ones to consider. Then we cannot choose a pivot element. Such a breakdown of the method indicates that the associated linear programming problem has no solution.

PRACTICE PROBLEMS 3

1. Convert the following minimum problem into a maximum problem in standard form: Minimize $3x + 4y$ subject to the constraints
$$\begin{cases} x - y \geq 0 \\ 3x - 4y \geq 0 \\ x \geq 0, \quad y \geq 0. \end{cases}$$

2. Suppose that the solution of a minimum problem yields the final simplex tableau

x	y	u	v	M	
1	6	-1	0	0	11
0	5	3	1	0	16
0	2	4	0	1	-40

What is the minimum value sought in the original problem?

EXERCISES 3

Solve the following linear programming problems by the simplex method.

1. Maximize $40x + 30y$ subject to the constraints
$$\begin{cases} x + y \leq 5 \\ -2x + 3y \geq 12 \\ x \geq 0, \quad y \geq 0. \end{cases}$$

2. Maximize $3x - y$ subject to the constraints
$$\begin{cases} 2x + 5y \leq 100 \\ x \qquad \geq 10 \\ \qquad y \geq 0. \end{cases}$$

3. Minimize $3x + y$ subject to the constraints
$$\begin{cases} x + y \geq 3 \\ 2x \qquad \geq 5 \\ x \geq 0, \quad y \geq 0. \end{cases}$$

4. Minimize $3x + 5y + z$ subject to the constraints
$$\begin{cases} x + y + z \geq 20 \\ y + 2z \geq 10 \\ x \geq 0, \quad y \geq 0, \quad z \geq 0. \end{cases}$$

5. Minimize $13x + 4y$ subject to the constraints
$$\begin{cases} y \geq -2x + 11 \\ y \leq -x + 10 \\ y \leq -\frac{1}{3}x + 6 \\ y \geq -\frac{1}{4}x + 4 \\ x \geq 0, \quad y \geq 0. \end{cases}$$

6. Minimize $500 - 10x - 3y$ subject to the constraints
$$\begin{cases} x + y \leq 20 \\ 3x + 2y \geq 50 \\ x \geq 0, \quad y \geq 0. \end{cases}$$

7. A dietician is designing a daily diet that is to contain at least 60 units of protein, 40 units of carbohydrate, and 120 units of fat. The diet is to consist of two types of foods. One serving of food A contains 30 units of protein, 10 units of carbohydrate, and 20 units of fat and costs $3. One serving of food B contains 10 units of protein, 10 units of carbohydrate, and 60 units of fat and costs $1.50. Design the diet that provides the daily requirements at the least cost.

8. A manufacturing company has two plants, each capable of producing radios, television sets, and stereo systems. The daily production capacities of each plant are as follows:

	Plant I	Plant II
Radios	10	20
Television sets	30	20
Stereo systems	20	10

Plant I costs $1500 per day to operate, whereas plant II costs $1200. How many days should each plant be operated to fill an order for 1000 radios, 1800 television sets, and 1000 stereo systems at the minimum cost?

9. An appliance store sells three brands of color television sets, brands A, B, and C. The profit per set is $30 for brand A, $50 for brand B, and $60 for brand C. The total warehouse space allocated to all brands is sufficient for 600 sets and the inventory is delivered only once per month. At least 100 customers per month will demand brand A, at least 50 will demand brand B, and at least 200 will demand either brand B or brand C. How can the appliance store satisfy all these demands and earn maximum profit?

10. A citizen decides to campaign for the election of a candidate for city council. Her goal is to generate at least 210 votes by a combination of door-to-door canvassing, letter writing, and phone calls. She figures that each hour of door-to-door canvassing will generate four votes, each hour of letter writing will generate two votes, and each hour on the phone will generate three votes. She would like to devote at least seven hours to phone calls and spend at most half of her time at door-to-door canvassing. How much time should she allocate to each task in order to achieve her goal in the least amount of time?

SOLUTIONS TO PRACTICE PROBLEMS 3

1. To minimize $3x + 4y$, we maximize $-(3x + 4y) = -3x - 4y$. So the associated maximum problem is: Maximize $-3x - 4y$ subject to the constraints

$$\begin{cases} -x + y \le 0 \\ -3x + 4y \le 0 \\ x \ge 0, \quad y \ge 0. \end{cases}$$

2. The value -40 in the lower right corner gives the solution of the associated *maximum* problem. The minimum value originally sought is the negative of the maximum value—that is, $-(-40) = 40$.

7.4. Duality

Each linear programming problem may be converted into a related linear programming problem called its *dual*. The dual problem is sometimes easier to solve than the original problem and moreover has the same optimum value. Furthermore, the solution of the dual problem often can provide valuable insights into the original problem. In order to understand the relationship between a linear programming problem and its dual, it is best to begin with a concrete example.

PROBLEM A Maximize the objective function $6x + 5y$ subject to the constraints

$$\begin{cases} 4x + 8y \le 32 \\ 3x + 2y \le 12 \\ x \ge 0, \quad y \ge 0. \end{cases}$$

The dual of Problem A is the following problem.

PROBLEM B   Minimize the objective function $32u + 12v$ subject to the constraints

$$\begin{cases} 4u + 3v \geq 6 \\ 8u + 2v \geq 5 \\ u \geq 0, \quad v \geq 0. \end{cases}$$

The relationship between the two problems is easiest to see if we first display the data from Problem A in a matrix, allowing each nontrivial constraint (i.e., other than $x \geq 0$, $y \geq 0$) to occupy one row. The objective function is written in the last row.

$$\begin{array}{cc} x & y \\ \begin{bmatrix} 4 & 8 & 32 \\ 3 & 2 & 12 \\ 6 & 5 & 0 \end{bmatrix} \end{array}$$

The dual problem is obtained from the columns of the matrix. The final column corresponds to the new objective function and the remaining columns give the constraints. Note that the $\leq$ signs of Problem A become $\geq$ signs in the dual problem.

In a similar fashion, we may form the dual of any linear programming problem using the following procedure.

The Dual Problem

1. If the given problem is a maximum problem, write all nontrivial constraints using only $\leq$. If the given problem is a minimum problem, write all constraints using only $\geq$. (If an inequality points in the wrong direction, we need only multiply it by -1.)

2. Display the data of the given problem in matrix form, with each nontrivial constraint occupying one row and the objective function occupying the final row.

3. The objective function of the dual problem is formed from the final column of the matrix. If the original problem involved a maximum, then the dual problem involves a minimum; if the original problem involved a minimum, then the dual problem involves a maximum.

4. The nontrivial constraints of the dual problem are formed from the remaining columns of the matrix. The inequality signs are the reverse of those in the original problem. In addition, the variables are constrained to be nonnegative.

EXAMPLE 1   Determine the dual of the following linear programming problem. Minimize $18x + 20y + 2z$ subject to the constraints

$$\begin{cases} 3x - 5y - 2z \leq 4 \\ 6x \quad\quad - 8z \geq 9 \\ x \geq 0, \quad y \geq 0, \quad z \geq 0. \end{cases}$$

Solution Since the given problem is a minimization, all inequalities must be written using the inequality sign $\geq$. To put the first inequality in this form, we must multiply by -1 to obtain

$$-3x + 5y + 2z \geq -4.$$

We now display the data in a matrix.

$$\begin{array}{ccc} x & y & z \end{array}$$
$$\begin{bmatrix} -3 & 5 & 2 & -4 \\ 6 & 0 & -8 & 9 \\ 18 & 20 & 2 & 0 \end{bmatrix}.$$

We form the dual problem from the columns of this matrix. The objective function is obtained from the last column: $-4u + 9v$. And since the given problem is a minimization, the objective function of the dual problem is to be maximized. The constraints are obtained from the remaining columns:

$$\begin{cases} -3u + 6v \leq 18 \\ 5u \quad\;\;\; \leq 20 \\ 2u - 8v \leq 2 \\ u \geq 0, \quad v \geq 0. \end{cases}$$

Let us now return to Problems A and B in order to examine the connection between the solutions of a linear programming problem and its dual problem. Problems A and B both involve two variables and hence can be solved by the geometric method of Chapter 6. Figure 1 shows their respective feasible sets and the vertices that yield the optimum values of the objective functions. The feasible sets do not look alike and the optimal vertices are different. However, both problems have the same optimum value, 27. The relationship between the two problems is brought into even sharper focus by looking at the final tableaux that

FIGURE 1

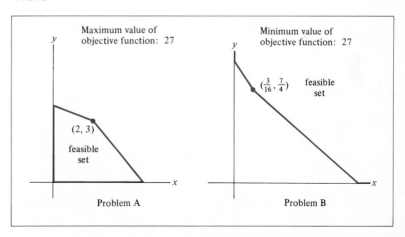

arise when the two problems are solved by the simplex method. (*Note:* In Problem B, the original variables are u and v and the slack variables have been named x and y.)

Final Tableaux

$$
\begin{array}{c}
\begin{array}{cccccc}
& x & y & u & v & M \\
\end{array} \\
\begin{array}{c}
y \\
x \\
\\
\end{array}
\left[
\begin{array}{ccccc|c}
0 & 1 & \frac{3}{16} & -\frac{1}{4} & 0 & 3 \\
1 & 0 & -\frac{1}{8} & \frac{1}{2} & 0 & 2 \\
0 & 0 & \frac{3}{16} & \frac{7}{4} & 1 & 27 \\
\end{array}
\right]
\end{array}
$$

Problem A

$$
\begin{array}{c}
\begin{array}{cccccc}
& u & v & x & y & M \\
\end{array} \\
\begin{array}{c}
v \\
u \\
\\
\end{array}
\left[
\begin{array}{ccccc|c}
0 & 1 & -\frac{1}{2} & \frac{1}{4} & 0 & \frac{7}{4} \\
1 & 0 & \frac{1}{8} & -\frac{3}{16} & 0 & \frac{3}{16} \\
0 & 0 & 2 & 3 & 1 & -27 \\
\end{array}
\right]
\end{array}
$$

Problem B

The final tableau for Problem A contains the solution to Problem B ($u = \frac{3}{16}$, $v = \frac{7}{4}$) in the final entries of the u and v columns. Similarly, the final tableau for Problem B gives the solution to Problem A ($x = 2$, $y = 3$) in the final entries of its x and y columns. This situation always occurs. The solutions to a linear programming problem and its dual problem may be obtained simultaneously by solving just one of the problems using the simplex method and applying the following theorem

> *Fundamental Theorem of Duality* At their respective optimal points, the objective functions of a linear programming problem and its dual problem have the same values. Furthermore, the solution of one of these problems by the simplex method yields the solution of the other as the final entries in the columns for the slack variables.

EXAMPLE 2 Solve the linear programming problem of Example 1 by applying the simplex method to its dual problem.

Solution In the solution to Example 1, the dual problem is: Maximize $-4u + 9v$ subject to the constraints

$$
\begin{cases}
-3u + 6v \le 18 \\
5u \qquad \le 20 \\
2u - 8v \le 2 \\
u \ge 0, \quad v \ge 0.
\end{cases}
$$

Since there are three nontrivial inequalities, the simplex method calls for three slack variables. Denote the slack variables by x, y, and z. Let $M = -4u + 9v$ and apply the simplex method.

$$\begin{array}{c c}& \begin{array}{c c c c c c} u & v & x & y & z & M \end{array} \\ \begin{array}{c} x \\ y \\ z \\ {} \end{array} & \left[\begin{array}{c c c c c c|c} -3 & ⑥ & 1 & 0 & 0 & 0 & 18 \\ 5 & 0 & 0 & 1 & 0 & 0 & 20 \\ 2 & -8 & 0 & 0 & 1 & 0 & 2 \\ \hline 4 & -9 & 0 & 0 & 0 & 1 & 0 \end{array}\right] \end{array}$$

$$\begin{array}{c c}& \begin{array}{c c c c c c} u & v & x & y & z & M \end{array} \\ \begin{array}{c} v \\ y \\ z \\ {} \end{array} & \left[\begin{array}{c c c c c c|c} -\frac{1}{2} & 1 & \frac{1}{6} & 0 & 0 & 0 & 3 \\ ⑤ & 0 & 0 & 1 & 0 & 0 & 20 \\ -2 & 0 & \frac{4}{3} & 0 & 1 & 0 & 26 \\ \hline -\frac{1}{2} & 0 & \frac{3}{2} & 0 & 0 & 1 & 27 \end{array}\right] \end{array}$$

$$\begin{array}{c c}& \begin{array}{c c c c c c} u & v & x & y & z & M \end{array} \\ \begin{array}{c} v \\ u \\ z \\ {} \end{array} & \left[\begin{array}{c c c c c c|c} 0 & 1 & \frac{1}{6} & \frac{1}{10} & 0 & 0 & 5 \\ 1 & 0 & 0 & \frac{1}{5} & 0 & 0 & 4 \\ 0 & 0 & \frac{4}{3} & \frac{2}{5} & 1 & 0 & 34 \\ \hline 0 & 0 & \frac{3}{2} & \frac{1}{10} & 0 & 1 & 29 \end{array}\right] \end{array}$$

Since the maximum value of the dual problem is 29, we know that the minimum value of the original problem is also 29. Looking at the last row of the final tableau, we conclude that this minimum value is assumed when $x = \frac{3}{2}$, $y = \frac{1}{10}$, and $z = 0$.

In Example 2, the dual problem was easier to solve than the original problem given in Example 1. Thus, we see how consideration of the dual problem may simplify the solution of linear programming problems in some cases.

Economic Significance of the Solution of the Dual It may seem that the entire discussion of duality is a mathematical curiosity which is useful for calculation, but has no further "real-world" meaning. But it would be very irresponsible if we left you with this mistaken impression. To illustrate the meaning of the solution of the dual problem, let's reconsider the furniture manufacturing problem of the preceding chapter. Recall that this problem asked us to maximize the profit from the sale of x chairs and y sofas subject to limitations on the amount of labor available for carpentry, upholstery, and finishing. In mathematical terms, the problem required us to maximize $80x + 70y$ subject to the constraints

$$\begin{cases} 6x + 3y \le 96 \\ x + y \le 18 \\ 2x + 6y \le 72 \\ x \ge 0, \quad y \ge 0. \end{cases}$$

The furniture manufacturing problem was solved by the simplex method in Section 2. From the last row of the final tableau, we conclude that the optimum value of the dual problem occurs when $u = \frac{10}{3}$, $v = 60$, and $w = 0$. (Here u was the slack variable for the first inequality, v for the second, and w for the third.) The numbers $\frac{10}{3}$, 60, and 0 have the following significance: In the event that one additional man-hour becomes available for carpentry, the profit could be increased by $\frac{10}{3}$ dollars. Similarly, an additional man-hour for upholstery could result in an increase in profit of 60 dollars. However, there can be no increase in profit due to the application of an additional man-hour of finishing.

More precisely, suppose that the number 96 were changed by h man-hours, where h is either positive or negative. If h is not too large,* then the maximum profit will change by $\frac{10}{3}$ times h. For instance, a decrease of $\frac{1}{2}$ man-hour available for carpentry ($h = -\frac{1}{2}$) results in a change in maximum profit of $(-\frac{1}{2})(\frac{10}{3}) = -\frac{5}{3}$ dollars.

An Economic Interpretation of the Dual Problem

Here, also, we shall consider the furniture manufacturing problem. Its dual problem is to minimize $96u + 18v + 72w$ subject to the constraints

$$\begin{cases} 6u + v + 2w \geq 80 \\ 3u + v + 6w \geq 70 \\ u \geq 0, \quad v \geq 0, \quad w \geq 0. \end{cases}$$

The variables u, v, and w can be assigned a meaning so that the dual problem has a significant interpretation in terms of the original problem.

First, recall the following table of data (man-hours except as noted):

	Chair	Sofa	Available labor
Carpentry	6	3	96
Finishing	1	1	18
Upholstery	2	6	72
Profit	\$80	\$70	

Suppose that we have an opportunity to hire out all our workers. Suppose that hiring out the carpenters will yield a profit of u dollars per hour, the finishers v dollars per hour, and the upholsterers w dollars per hour. Of course, u, v, and w must all be ≥ 0. However, there are other constraints which we should reasonably impose. Any scheme for hiring out the workers should generate at least as much profit as is currently being generated in the construction of chairs and sofas. In terms of the potential profits from hiring the workers out, the labor involved in constructing a chair will generate

$$6u + v + 2w$$

dollars of profit. And this amount should be at least equal to the \$80 profit that could be earned by using the labor to construct a chair. That is, we have the

---

* The number h must be such that the new problem will have the same group II variables in its final tableau as the original problem. Sophisticated computer programs for solving linear programming problems often provide the acceptable range of values for h.

constraint

$$6u + v + 2w \geq 80.$$

Similarly, considering the labor involved in building a sofa, we derive the constraint

$$3u + v + 6w \geq 70.$$

Since there are available 96 hours of carpentry, 18 hours of finishing, and 72 hours of upholstery, the total profit from hiring out the workers would be

$$96u + 18v + 72w.$$

Thus the problem of determining the least acceptable profit from hiring out the workers is equivalent to the following: Minimize $96u + 18v + 72w$ subject to the constraints

$$\begin{cases} 6u + v + 2w \geq 80 \\ 3u + v + 6w \geq 70 \\ u \geq 0, \quad v \geq 0, \quad w \geq 0. \end{cases}$$

This is just the dual of the furniture manufacturing problem. The values u, v, and w are measures of the value of an hour's labor by each type of worker. Economists often refer to them as *shadow profits*. The fundamental theorem of duality asserts that the minimum acceptable profit that can be achieved by hiring the workers out is equal to the maximum profit that can be generated if they make furniture.

Matrices, Linear Programming, and Duality   Linear programming problems can be neatly formulated in terms of matrices. Such a formulation is easily transformed into a matrix statement of the dual problem. To introduce the matrix formulation of a linear programming problem, we need the concept of inequality for matrices.

Let A and B be two matrices of the same size. We say that A less than or equal to B (denoted $A \leq B$) if each entry of A is less than or equal to the corresponding entry of B. For instance, we have the following matrix inequalities:

$$\begin{bmatrix} 2 & -3 \\ \frac{1}{2} & 0 \end{bmatrix} \leq \begin{bmatrix} 5 & -1 \\ 1 & 0 \end{bmatrix} \quad \text{and} \quad \begin{bmatrix} 5 \\ 6 \end{bmatrix} \leq \begin{bmatrix} 8 \\ 9 \end{bmatrix}.$$

The symbol $\geq$ has an analogous meaning for matrices.

EXAMPLE 3   Let

$$A = \begin{bmatrix} 6 & 3 \\ 1 & 1 \\ 2 & 6 \end{bmatrix}, \quad B = \begin{bmatrix} 96 \\ 18 \\ 72 \end{bmatrix}, \quad C = [80 \quad 70], \quad X = \begin{bmatrix} x \\ y \end{bmatrix}.$$

Carry out the indicated matrix multiplications in the following statement: Maximize CX subject to the constraints $AX \leq B$, $X \geq \mathbf{0}$. (Here $\mathbf{0}$ is the zero matrix.)

Solution $CX = \begin{bmatrix} 80 & 70 \end{bmatrix} \begin{bmatrix} x \\ y \end{bmatrix} = \begin{bmatrix} 80x + 70y \end{bmatrix}.$

$$AX = \begin{bmatrix} 6 & 3 \\ 1 & 1 \\ 2 & 6 \end{bmatrix} \begin{bmatrix} x \\ y \end{bmatrix} = \begin{bmatrix} 6x + 3y \\ x + y \\ 2x + 6y \end{bmatrix}.$$

$$AX \le B \text{ means } \begin{bmatrix} 6x + 3y \\ x + y \\ 2x + 6y \end{bmatrix} \le \begin{bmatrix} 96 \\ 18 \\ 72 \end{bmatrix} \quad \text{or} \quad \begin{cases} 6x + 3y \le 96 \\ x + y \le 18 \\ 2x + 6y \le 72. \end{cases}$$

$$X \ge 0 \text{ means } \begin{bmatrix} x \\ y \end{bmatrix} \ge \begin{bmatrix} 0 \\ 0 \end{bmatrix} \quad \text{or} \quad \begin{cases} x \ge 0 \\ y \ge 0. \end{cases}$$

Hence the statement "Maximize CX subject to the constraints $AX \le B$, $X \ge \mathbf{0}$" is a matrix formulation of the furniture manufacturing problem.

EXAMPLE 4   Express the dual of the furniture manufacturing problem in matrix form using the matrices of Example 3.

Solution   The dual problem is: Minimize $96u + 18v + 72w$ subject to the constraints

$$\begin{cases} 6u + v + 2w \ge 80 \\ 3u + v + 6w \ge 70 \\ u \ge 0, \quad v \ge 0, \quad w \ge 0. \end{cases}$$

Let U be the matrix $\begin{bmatrix} u & v & w \end{bmatrix}$ and A, B, C the matrices of Example 3. Then

$$UB = \begin{bmatrix} u & v & w \end{bmatrix} \begin{bmatrix} 96 \\ 18 \\ 72 \end{bmatrix} = \begin{bmatrix} 96u + 18v + 72w \end{bmatrix}$$

and

$$UA = \begin{bmatrix} u & v & w \end{bmatrix} \begin{bmatrix} 6 & 3 \\ 1 & 1 \\ 2 & 6 \end{bmatrix} = \begin{bmatrix} 6u + v + 2w & 3u + v + 6w \end{bmatrix}.$$

Hence the dual problem may be stated: Minimize UB subject to the constraints $UA \ge C$, $U \ge \mathbf{0}$.

Example 4 is a particular instance of the following result.

Matrix formulation of the dual   Let

$$X = \begin{bmatrix} x \\ y \\ z \\ \vdots \end{bmatrix}$$

and A, B, C be matrices of appropriate sizes. The linear programming problem "Maximize CX subject to the constraints $AX \leq B$, $X \geq 0$" has as its dual "Minimize UB subject to the constraints $UA \geq C$, $U \geq 0$, where $U = \begin{bmatrix} u & v & w & \cdots \end{bmatrix}$. Similarly, the linear programming problem "Minimize CX subject to the constraints $AX \geq B$, $X \geq 0$" has as its dual "Maximize UB subject to the constraints $UA \leq C$, $U \geq 0$."

PRACTICE PROBLEMS 4

A linear programming problem involving three variables and four nontrivial inequalities has the number 52 as the maximum value of its objective function.

1. How many variables and nontrivial inequalities will the dual problem have?

2. What is the optimum value for the objective function of the dual problem?

EXERCISES 4

In Exercises 1–6, determine the dual problem of the given linear programming problem.

1. Maximize $4x + 2y$ subject to the constraints

$$\begin{cases} 5x + y \leq 80 \\ 3x + 2y \leq 76 \\ x \geq 0, \quad y \geq 0. \end{cases}$$

2. Minimize $30x + 60y + 50z$ subject to the constraints

$$\begin{cases} 5x + 3y + z \geq 2 \\ x + 2y + z \geq 3 \\ x \geq 0, \quad y \geq 0, \quad z \geq 0. \end{cases}$$

3. Minimize $10x + 12y$ subject to the constraints

$$\begin{cases} x + 2y \geq 1 \\ -x + y \geq 2 \\ 2x + 3y \geq 1 \\ x \geq 0, \quad y \geq 0. \end{cases}$$

4. Maximize $80x + 70y + 120z$ subject to the constraints

$$\begin{cases} 6x + 3y + 8z \leq 768 \\ x + y + 2z \leq 144 \\ 2x + 5y \quad\quad \leq 216 \\ x \geq 0, \quad y \geq 0, \quad z \geq 0. \end{cases}$$

5. Minimize $3x + 5y + z$ subject to the constraints

$$\begin{cases} 2x - 4y - 6z \leq 7 \\ y \geq 10 - 8x - 9z \\ x \geq 0, \quad y \geq 0, \quad z \geq 0. \end{cases}$$

6. Maximize $2x - 3y + 4z - 5w$ subject to the constraints

$$\begin{cases} x + y + z + w - 6 \le 10 \\ 7x + 9y - 4z - 3w \ge 5 \\ x \ge 0, \quad y \ge 0, \quad z \ge 0, \quad w \ge 0. \end{cases}$$

7. The final simplex tableau for the linear programming problem of Exercise 1 appears below. Give the solution to the problem and to its dual.

$$\begin{array}{c} \begin{array}{ccccc} x & y & u & v & M \end{array} \\ \begin{array}{c} x \\ y \\ M \end{array} \left[\begin{array}{ccccc|c} 1 & 0 & \frac{2}{7} & -\frac{1}{7} & 0 & 12 \\ 0 & 1 & -\frac{3}{7} & \frac{5}{7} & 0 & 20 \\ 0 & 0 & \frac{2}{7} & \frac{6}{7} & 1 & 88 \end{array} \right] \end{array}$$

8. The final simplex tableau for the *dual* of the linear programming problem of Exercise 2 appears below. Give the solution to the problem and to its dual.

$$\begin{array}{c} \begin{array}{cccccc} u & v & x & y & z & M \end{array} \\ \begin{array}{c} v \\ y \\ z \\ M \end{array} \left[\begin{array}{cccccc|c} 5 & 1 & 1 & 0 & 0 & 0 & 30 \\ -7 & 0 & -2 & 1 & 0 & 0 & 0 \\ -4 & 0 & -1 & 0 & 1 & 0 & 20 \\ 13 & 0 & 0 & 0 & 0 & 1 & 90 \end{array} \right] \end{array}$$

9. The final simplex tableau for the *dual* of the linear programming problem of Exercise 3 appears below. Give the solution to the problem and to its dual.

$$\begin{array}{c} \begin{array}{cccccc} u & v & w & x & y & M \end{array} \\ \begin{array}{c} x \\ v \\ M \end{array} \left[\begin{array}{cccccc|c} 3 & 0 & 5 & 1 & 1 & 0 & 22 \\ 2 & 1 & 3 & 0 & 1 & 0 & 12 \\ 3 & 0 & 5 & 0 & 2 & 1 & 24 \end{array} \right] \end{array}$$

10. The final simplex tableau for the linear programming problem of Exercise 4 appears below. Give the solution to the problem and to its dual.

$$\begin{array}{c} \begin{array}{ccccccc} x & y & z & u & v & w & M \end{array} \\ \begin{array}{c} x \\ z \\ y \\ M \end{array} \left[\begin{array}{ccccccc|c} 1 & 0 & 0 & \frac{5}{12} & -\frac{5}{3} & \frac{1}{12} & 0 & 98 \\ 0 & 0 & 1 & -\frac{1}{8} & 1 & -\frac{1}{8} & 0 & 21 \\ 0 & 1 & 0 & -\frac{1}{6} & \frac{2}{3} & \frac{1}{6} & 0 & 4 \\ 0 & 0 & 0 & \frac{20}{3} & \frac{100}{3} & \frac{10}{3} & 1 & 10{,}640 \end{array} \right] \end{array}$$

In Exercises 11–14, determine the dual problem. Solve either the original problem or its dual by the simplex method and then give the solutions to both.

11. Minimize $3x + y$ subject to the constraints

$$\begin{cases} x + y \ge 3 \\ 2x \ge 5 \\ x \ge 0, \quad y \ge 0. \end{cases}$$

12. Minimize $3x + 5y + z$ subject to the constraints

$$\begin{cases} x + y + z \ge 20 \\ y + 2z \ge 0 \\ x \ge 0, \quad y \ge 0, \quad z \ge 0. \end{cases}$$

13. Maximize $10x + 12y + 10z$ subject to the constraints

$$\begin{cases} x - 2y & \le 6 \\ 3x & + z \le 9 \\ y + 3z \le 12 \\ x \ge 0, \quad y \ge 0, \quad z \ge 0. \end{cases}$$

14. Maximize $x + 3y$ subject to the constraints

$$\begin{cases} x + y \le 7 \\ x + 2y \le 10 \\ x \ge 0, \quad y \ge 0. \end{cases}$$

Exercises 15–18 refer to the furniture manufacturing problem. Some of the data from the problem are:

[carpentry] $6x + 3y \le 96$ Solution: Maximum profit $= \$1400$

[finishing] $x + y \le 18$ Solution to dual problem: $u = \frac{10}{3}, v = 60, w = 0$

[upholstery] $2x + 6y \le 72$

15. What is the maximum profit possible if 99 man-hours of labor are available for carpentry?

16. What is the maximum profit possible if the number of man-hours of labor available for finishing is decreased by 1?

17. What is the maximum profit possible if 18.5 man-hours of labor are available for finishing?

18. What is the maximum profit possible if 74 man-hours of labor are available for upholstery?

Exercises 19–22 refer to the nutrition problem (Example 2) from Section 2 of the previous chapter. Some of the data from this problem are:

[protein] $15x + 22.5y \ge 90$ Solution: Minimum cost is 66 cents

[calories] $810x + 270y \ge 1620$ Solution to dual problem: $u = \frac{2}{5}, v = \frac{1}{54}, w = 0$

[riboflavin] $\frac{1}{9}x + \frac{1}{3}y \ge 1$

19. What would be the minimum cost possible if only 80 grams of protein were required?

20. How much could be saved by decreasing the calorie requirement to 1512?

21. How much could be saved by decreasing the riboflavin requirement by $\frac{1}{4}$ milligram?

22. What would be the minimum cost possible if 95 grams of protein were required?

23. Give an economic interpretation to the dual of the shipping problem of Exercise 5 of Section 3 of the preceding chapter.

24. Give an economic interpretation to the dual of the mining problem in Exercise 6 of Section 3 of the preceding chapter.

25–30. For each of the linear programming problems in Exercises 1–6, identify the matrices A, B, C, X and state the problem in terms of matrices. Then identify the matrix U and express the dual problem in terms of matrices.

SOLUTIONS TO PRACTICE PROBLEMS 4

1. Four variables and three nontrivial inequalities. The number of variables in the dual problem is always the same as the number of nontrivial inequalities in the original problem. The number of nontrivial inequalities in the dual problem is the same as the number of variables in the original problem.

2. Minimum value of 52. The original problem and the dual problem always have the same optimum values. However, if this value is a maximum for one of the problems, it will be a minimum for the other.

Chapter 7: CHECKLIST

☐ Standard form of linear programming problem
☐ Slack variable
☐ Group I, group II variables
☐ Simplex tableau
☐ Simplex method for problems in standard form
☐ Converting minimization problems to maximization problems
☐ Reduction of linear programming problems to standard form
☐ Dual problem
☐ Fundamental theorem of duality
☐ Matrix formulation of linear programming problem

Chapter 7: SUPPLEMENTARY EXERCISES

Use the simplex method to solve the following linear programming problems.

1. Maximize $3x + 4y$ subject to the constraints

$$\begin{cases} 2x + y \le 7 \\ -x + y \le 1 \\ x \ge 0, \quad y \ge 0. \end{cases}$$

2. Maximize $2x + 5y$ subject to the constraints

$$\begin{cases} x + y \le 7 \\ 4x + 3y \le 24 \\ x \ge 0, \quad y \ge 0. \end{cases}$$

3. Maximize $2x + 3y$ subject to the constraints

$$\begin{cases} x + 2y \le 14 \\ x + y \le 9 \\ 3x + 2y \le 24 \\ x \ge 0, \quad y \ge 0. \end{cases}$$

4. Maximize $3x + 7y$ subject to the constraints

$$\begin{cases} x + 2y \le 10 \\ 4x + 3y \le 30 \\ -2x + y \le 0 \\ x \ge 0, \quad y \ge 0. \end{cases}$$

5. Minimize $x + y$ subject to the constraints

$$\begin{cases} 7x + 5y \ge 40 \\ x + 4y \ge 9 \\ x \ge 0, \quad y \ge 0. \end{cases}$$

6. Minimize $3x + 2y$ subject to the constraints

$$\begin{cases} x + y \ge 6 \\ x + 2y \ge 0 \\ x \ge 0, \quad y \ge 0. \end{cases}$$

7. Minimize $20x + 30y$ subject to the constraints

$$\begin{cases} x + 4y \ge 8 \\ x + y \ge 5 \\ 2x + y \ge 7 \\ x \ge 0, \quad y \ge 0. \end{cases}$$

8. Minimize $5x + 7y$ subject to the constraints

$$\begin{cases} 2x + y \ge 10 \\ 3x + 2y \ge 18 \\ x + 2y \ge 10 \\ x \ge 0, \quad y \ge 0. \end{cases}$$

9. Maximize $36x + 48y + 70z$ subject to the constraints

$$\begin{cases} x \le 4 \\ y \le 6 \\ z \le 8 \\ 4x + 3y + 2z \le 38 \\ x \ge 0, \quad y \ge 0, \quad z \ge 0. \end{cases}$$

10. Maximize $3x + 4y + 5z + 4w$ subject to the constraints

$$\begin{cases} 6x + 9y + 12z + 15w \le 672 \\ x - y + 2z + 2w \le 92 \\ 5x + 10y - 5z + 4w \le 280 \\ x \ge 0, \quad y \ge 0, \quad z \ge 0, \quad w \ge 0. \end{cases}$$

11. Determine the dual problem of the linear programming problem in Exercise 3.

12. Determine the dual problem of the linear programming problem in Exercise 7.

13. The final simplex tableau for the linear programming problem of Exercise 3 appears below. Give the solution to the problem and to its dual.

	x	y	u	v	w	M	
y	0	1	1	-1	0	0	5
x	1	0	-1	2	0	0	4
w	0	0	1	-4	1	0	2
M	0	0	1	1	0	1	23

14. The final simplex tableau for the *dual* of the linear programming problem of Exercise 7 appears below. Give the solution to the problem and to its dual.

	u	v	w	x	y	M	
v	0	1	$\frac{7}{3}$	$\frac{4}{3}$	$-\frac{1}{3}$	0	$\frac{50}{3}$
u	1	0	$-\frac{1}{3}$	$-\frac{1}{3}$	$\frac{1}{3}$	0	$\frac{10}{3}$
	0	0	2	4	1	1	110

15, 16. For each of the linear programming problems in Exercises 3 and 7, identify the matrices A, B, C, X and state the problem in terms of matrices. Then identify the matrix U and express the dual problem in terms of matrices.

Sets and Counting

Chapter 8

In this chapter we introduce some ideas useful in the study of probability (Chapter 9). Our first topic, the theory of sets, will provide a convenient language and notation in which to discuss probability. Using set theory, we will develop a number of counting principles which can also be applied to computing probabilities.

8.1. Sets

In many applied problems one must consider collections of various sorts of objects. For example, a survey of unemployment might consider the collection of all U.S. cities with current unemployment greater than 7%. A study of birthrates might consider the collection of countries with a current birthrate less than 20 per 1000 population. Such collections are examples of sets. A *set* is any collection of objects. The objects, which may be countries, cities, years, numbers, letters, or anything else, are called the *elements* of the set. A set is often specified by a listing of its elements inside a pair of braces. For example, the set whose elements are the first six letters of the alphabet is written

$$\{a, b, c, d, e, f\}.$$

Similarly, the set whose elements are the even numbers between 1 and 11 is written

$$\{2, 4, 6, 8, 10\}.$$

We can also specify a set by giving a description of its elements (without actually listing the elements). For example, the set $\{a, b, c, d, e, f\}$ can also be written

$$\{\text{the first six letters of the alphabet}\},$$

and the set $\{2, 4, 6, 8, 10\}$ can be written

$$\{\text{all even numbers between 1 and 11}\}.$$

For convenience, we usually denote sets by capital letters A, B, C, and so on.

The great diversity of sets is illustrated by the following examples:

1. In a linear programming problem, the feasible set is the set of all points satisfying a system of linear inequalities. The feasible set of the furniture manufacturing problem is the set of all points on or inside the five-sided region in Fig. 1.

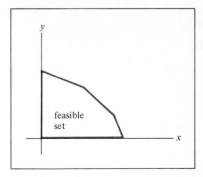

FIGURE 1

2. Let B = {license plate numbers consisting of three letters followed by three digits}. Some typical elements of B are

$$\text{SBG 602,} \qquad \text{GXZ 179,} \qquad \text{YHJ 006.}$$

The number of elements in B is sufficiently large so that listing all of them is impractical. However, in this chapter we will develop a technique that allows us to calculate the number of elements of B.

3. Let C = {possible sequences of outcomes of tossing a coin three times}. If we let H denote "heads" and T denote "tails," the various sequences can be easily described:

$$C = \{\text{HHH, THH, HTH, HHT, TTH, THT, HTT, TTT}\},$$

where, for instance, HTH means "first toss heads, second toss tails, third toss heads."

Sets arise in many practical contexts, as the next example shows.

EXAMPLE 1 The following table gives the rate of inflation, as measured by the percentage change in the consumer price index, for the years 1961–1975.

Year	Inflation (%)	Year	Inflation (%)
1961	1.0	1969	5.4
1962	1.1	1970	5.9
1963	1.2	1971	4.3
1964	1.3	1972	3.3
1965	1.7	1973	6.2
1966	2.9	1974	11.0
1967	2.9	1975	9.1
1968	4.2		

Let

$$A = \{\text{years from 1961 to 1975 in which inflation was above 5\%}\},$$

$$B = \{\text{years from 1961 to 1975 in which inflation was below 2\%}\}.$$

Determine the elements of A and B.

Solution   By reading the table, we see that

$$A = \{1969, 1970, 1973, 1974, 1975\},$$

$$B = \{1961, 1962, 1963, 1964, 1965\}.$$

Suppose we are given two sets, A and B. Then it is possible to form new sets from A and B as follows: The *union* of A and B, denoted $A \cup B$, is the set consisting of all those elements which belong to either A *or* B (or both). The *intersection* of A and B, denoted $A \cap B$, is the set consisting of those elements which belong to both A *and* B. For example, let

$$A = \{1, 2, 3, 4\}, \qquad B = \{1, 3, 5, 7, 11\}.$$

Then, since $A \cup B$ consists of those elements belonging to either A or B (or both), we have

$$A \cup B = \{1, 2, 3, 4, 5, 7, 11\}.$$

Moreover, since $A \cap B$ consists of those elements that belong to both A and B, we have

$$A \cap B = \{1, 3\}.$$

EXAMPLE 2   Here is a table giving the rates of unemployment and inflation for the years 1965–1975.

Year	Unemployment (%)	Inflation (%)
1965	4.5	1.7
1966	3.8	2.9
1967	3.9	2.9
1968	3.6	4.2
1969	3.5	5.4
1970	4.9	5.9
1971	5.9	4.3
1972	5.6	3.3
1973	4.9	6.2
1974	5.6	11.0
1975	8.5	9.1

Let

$A = \{$years from 1965 to 1975 in which unemployment is at least 5%$\}$,

$B = \{$years from 1965 to 1975 in which the inflation rate is at least 5%$\}$.

(a) Describe the sets $A \cap B$ and $A \cup B$.

(b) Determine the elements of A, B, $A \cap B$, and $A \cup B$.

Solution   (a) From the descriptions of A and B, we have

$A \cap B = \{$years from 1965 to 1975 in which unemployment is at least 5% and inflation is at least 5%$\}$

$A \cup B = \{$years from 1965 to 1975 in which either unemployment or inflation (or both) is at least 5%$\}$.

(b) From the table, we see that

$$A = \{1971, 1972, 1974, 1975\}$$

$$B = \{1969, 1970, 1973, 1974, 1975\}$$

$$A \cap B = \{1974, 1975\}$$

$$A \cup B = \{1969, 1970, 1971, 1972, 1973, 1974, 1975\}.$$

We have defined the union and the intersection of two sets. In a similar manner, we can define the union and intersection of any number of sets. For example, if A, B, C are three sets, then their union, denoted $A \cup B \cup C$, is the set whose elements are precisely those which belong to at least one of the sets A, B, C. Similarly, the intersection of A, B, C, denoted $A \cap B \cap C$, is the set consisting of those elements which belong to all the sets A, B, C. In a similar way, we may define the union and intersection of more than three sets.

Suppose that we are given a set A. We may form new sets by selecting elements from A. Sets formed in this way are called *subsets* of A. More precisely, a set B is called a *subset* of A provided that every element of B is also an element of A. For example, let $A = \{1, 2, 3\}$, $B = \{1, 3\}$. Since the elements of B (1 and 3) are also elements of A, B is a subset of A.

One set which is very often considered is the set which contains no elements at all. This set is called the *empty set* (or *null set*) and is written $\varnothing$. The empty set is a subset of every set.*

EXAMPLE 3 Let $A = \{a, b, c\}$. Find all subsets of A.

Solution Since A contains three elements, every subset of A has at most three elements. So We look for subsets according to the number of elements:

Number of elements in subset	Possible subsets
0	$\varnothing$
1	$\{a\}, \{b\}, \{c\}$
2	$\{a, b\}, \{a, c\}, \{b, c\}$
3	$\{a, b, c\}$

Thus, we see that A has eight subsets, namely those listed on the right. (Note that we count A as a subset of itself.)

It is usually convenient to regard all sets involved in a particular discussion as subsets of a single larger set. Thus, for example, if a problem involves the sets $\{a, b, c\}$, $\{e, f\}$, $\{g\}$, $\{b, x, y\}$, then we can regard all of these as subsets of the set

$$U = \{\text{all letters of the alphabet}\}.$$

---

* Here is why: Let A be any set. Every element of $\varnothing$ also belongs to A. If you do not agree, then you must produce an element of $\varnothing$ which does not belong to A. But you cannot, since $\varnothing$ has no elements. So $\varnothing$ is a subset of A.

Since U contains all sets being discussed, it is called a *universal set* (for the particular problem). In this book we shall always specify the particular universal set we have in mind.

Suppose that U is a universal set and A is a subset of U. The set of elements of U which are not in A is called the *complement* of A, denoted A'. For example, if

$$U = \{1, 2, 3, 4, 5, 6, 7, 8, 9\}$$

and

$$A = \{2, 4, 6, 8\},$$

then

$$A' = \{1, 3, 5, 7, 9\}.$$

EXAMPLE 4 Let $U = \{a, b, c, d, e, f, g\}$, $S = \{a, b, c\}$, $T = \{a, c, d\}$. List the elements of the sets

(a) S' (b) T' (c) $(S \cap T)'$ (d) $S' \cap T'$

Solution (a) S' consists of those elements of U which are not in S, so $S' = \{d, e, f, g\}$.

(b) Similarly, $T' = \{b, e, f, g\}$.

(c) To determine $(S \cap T)'$ we must first determine $S \cap T$:

$$S \cap T = \{a, c\}.$$

Then we determine the complement of this set:

$$(S \cap T)' = \{b, d, e, f, g\}.$$

(d) We determined S' and T' in parts (a) and (b). The set $S' \cap T'$ consists of the elements that belong to both S' and T'. Therefore, referring to (a) and (b), we have

$$S' \cap T' = \{e, f, g\}.$$

PRACTICE PROBLEMS 1

1. Let $U = \{a, b, c, d, e, f, g\}$, $R = \{a, b, c, d\}$, $S = \{c, d, e\}$, $T = \{c, e, g\}$. List the elements of the sets

   (a) R' (b) $R \cap S$

   (c) $(R \cap S) \cap T$ (d) $R \cap (S \cap T)$

2. Let $U = \{$all Nobel prize winners$\}$, $W = \{$women who have won Nobel prizes$\}$, $A = \{$Americans who have won Nobel prizes$\}$, $L = \{$winners of the Nobel prize in literature$\}$. Describe the following sets.

   (a) W' (b) $A \cap L'$ (c) $W \cap A \cap L'$

3. Refer to Problem 2. Use set-theoretic notation to describe $\{$Nobel prize winners who are American men or recipients of the prize in literature$\}$.

EXERCISES 1

1. Let $U = \{1, 2, 3, 4, 5, 6, 7\}$, $S = \{1, 2, 3, 4\}$, $T = \{1, 3, 5, 7\}$. List the elements of the sets

   (a) S' (b) $S \cup T$ (c) $S \cap T$ (d) $S' \cap T$

2. Let $U = \{1, 2, 3, 4, 5\}$, $S = \{1, 2, 3\}$, $T = \{5\}$. List the elements of the sets

   (a) S' (b) $S \cup T$ (c) $S \cap T$ (d) $S' \cap T$

3. Let $U = \{$all letters of the alphabet$\}$, $R = \{a, b, c\}$, $S = \{c, d, e, f\}$, $T = \{x, y, z\}$. List the elements of the sets

   (a) $R \cup S$ (b) $R \cap S$ (c) $S \cap T$

4. Let $U = \{a, b, c, d, e, f, g\}$, $R = \{a\}$, $S = \{a, b\}$, $T = \{b, d, e, f, g\}$. List the elements of the sets

   (a) $R \cup S$ (b) $R \cap S$ (c) T' (d) $T' \cup S$

5. List all subsets of the set $\{1, 2\}$.

6. List all subsets of the set $\{1\}$.

7. Let $U = \{$all college students$\}$, $S = \{$all male college students$\}$, $T = \{$all college students who like football$\}$. Describe the elements of the sets

   (a) $S \cap T$ (b) S' (c) $S' \cap T'$ (d) $S \cup T$

8. Let $U = \{$all corporations$\}$, $S = \{$all corporations with headquarters in New York City$\}$, $T = \{$all privately owned corporations$\}$. Describe the elements of the sets

   (a) S' (b) T' (c) $S \cap T$ (d) $S \cap T'$

9. The Standard and Poor's Index measures the price of a certain collection of 500 stocks. Table 1 compares the percentage change in the index during the first 5 days of certain

TABLE 1   Percentage Change in the Standard and Poor's Index

	Percent change for first 5 days	*Percent change for year*		*Percent change for first 5 days*	*Percent change for year*
1977	−2.3	−12.36	1963	2.6	18.9
1976	4.9	19.1	1962	−3.4	−11.8
1975	2.2	31.5	1961	1.2	23.1
1974	−1.5	−29.7	1960	−0.7	−3.0
1973	1.5	−17.4	1959	0.3	8.5
1972	1.4	15.6	1958	2.5	38.1
1971	0.0	10.8	1957	−0.9	−14.3
1970	0.7	−0.1	1956	−2.1	2.6
1969	−2.9	−11.4	1955	−1.8	26.4
1968	0.2	−7.7	1954	0.5	45.0
1967	3.1	20.1	1953	−0.9	−6.6
1966	0.8	−13.1	1952	0.6	11.8
1965	0.7	9.1	1951	2.3	16.5
1964	1.3	13.0	1950	2.0	21.8

years with the percentage change for the entire year. Let $U = \{$all years from 1950 to 1977$\}$, $S = \{$all years during which the Index increased by 2% or more during the first 5 days$\}$, $T = \{$all years for which the index increased by 16% or more during the entire year$\}$. List the elements of the sets

(a) S (b) T (c) $S \cap T$ (d) $S' \cap T$ (e) $S \cap T'$

10. Refer to Table 1. Let $U = \{$all years from 1950 to 1977$\}$, $A = \{$all years during which the Index declined during the first 5 days$\}$, $B = \{$all years during which the Index declined for the entire year$\}$. List the elements of the sets

(a) A (b) B (c) $A \cap B$ (d) $A' \cap B$ (e) $A \cap B'$

11. Refer to Exercise 9. Describe verbally the fact that $S \cap T' = \emptyset$.

12. Refer to Exercise 10. Describe verbally the fact that $A \cap B'$ has two elements.

13. Let $U = \{a, b, c, d, e, f\}$, $R = \{a, b, c\}$, $S = \{a, c, e\}$, $T = \{e, f\}$. List the elements of the following sets.

(a) $(R \cup S)'$ (b) $R \cup S \cup T$ (c) $R \cap S \cap T$

(d) $R \cap S \cap T'$ (e) $R' \cap S \cap T$ (f) $S \cup T$

14. Let $U = \{1, 2, 3, 4, 5\}$, $R = \{1, 3, 5\}$, $S = \{3, 4, 5\}$, $T = \{2, 4\}$. List the elements of the following sets.

(a) $R \cap S \cap T$ (b) $R \cap S \cap T'$ (c) $R \cap S' \cap T$

(d) $R' \cap T$ (e) $R \cup S$ (f) $R' \cup R$

In Exercises 15–20, simplify the given expression.

15. $(S')'$ **16.** $S \cap S'$ **17.** $S \cup S'$

18. $S \cap \emptyset$ **19.** $T \cap S \cap T'$ **20.** $S \cup \emptyset$

A large corporation classifies its many divisions by their performance in the preceding year. Let $P = \{$divisions that made a profit$\}$, $L = \{$divisions that had an increase in labor costs$\}$, and $T = \{$divisions whose total revenue increased$\}$. Describe the following sets using set-theoretic notation.

21. $\{$divisions that had increases in labor costs or total revenue$\}$

22. $\{$divisions that did not make a profit$\}$

23. $\{$divisions that made a profit despite an increase in labor costs$\}$

24. $\{$divisions that had an increase in labor costs and were either unprofitable or did not increase their total revenue$\}$

25. $\{$profitable divisions with increases in labor costs and total revenue$\}$

26. $\{$divisions that were unprofitable or did not have increases in either labor costs or total revenue$\}$

An automobile insurance company classifies applicants by their driving records for the previous three years. Let $S = \{$applicants who have received speeding tickets$\}$, $A = \{$applicants who have caused accidents$\}$, and $D = \{$applicants who have been arrested for driving while intoxicated$\}$. Describe the following sets using set-theoretic notation.

27. {applicants who have not received speeding tickets}

28. {applicants who have caused accidents and been arrested for drunk driving}

29. {applicants who received speeding tickets, caused accidents, or were arrested for drunk driving}

30. {applicants who have not been arrested for drunk driving but have received speeding tickets or have caused accidents}

31. {applicants who have not both caused accidents and received speeding tickets but who have been arrested for drunk driving}

32. {applicants who have not caused accidents or have not been arrested for drunk driving}

SOLUTIONS TO PRACTICE PROBLEMS 1

1. (a) $\{e, f, g\}$.

(b) $\{c, d\}$.

(c) $\{c\}$. This problem asks for the intersection of two sets. The first set is $R \cap S = \{c, d\}$ and the second set is $T = \{c, e, g\}$. The intersection of these sets is $\{c\}$.

(d) $\{c\}$. Here again the problem asks for the intersection of two sets. However, now the first set is $R = \{a, b, c, d\}$ and the second set is $S \cap T = \{c, e\}$. The intersection of these sets is $\{c\}$.

[*Note:* It should be expected that the set $(R \cap S) \cap T$ is the same as the set $R \cap (S \cap T)$, for each set consists of those elements that are in all three sets. Therefore, each of these sets equals the set $R \cap S \cap T$.]

2. (a) $W' = \{$men who have won Nobel prizes$\}$. This is so since W' consists of those elements of U that are not in W—that is, those Nobel prize winners who are not women.

(b) $A \cap L' = \{$Americans who have received Nobel prizes in fields other than literature$\}$.

(c) $W \cap A \cap L' = \{$American women who have received Nobel prizes in fields other than literature$\}$. This is so since to qualify for $W \cap A \cap L'$, a Nobel prize winner must simultaneously be in W, in A, and in L'—that is, a woman, an American, and not a winner of the Nobel prize in literature.

3. $(A \cap W') \cup L$.

8.2. A Fundamental Principle of Counting

A counting problem is one that requires us to determine the number of elements in a set S. Counting problems arise in many applications of mathematics and comprise the mathematical field of *combinatorics*. We shall study a number of different sorts of counting problems in the remainder of this chapter.

If S is any set, we will denote the number of elements in S by $n(S)$. Thus, for example, if $S = \{1, 7, 11\}$, then $n(S) = 3$, and if $S = \{a, b, c, d, e, f, g, h, i\}$, then $n(S) = 9$. Of course, if $S = \varnothing$, the empty set, then $n(S) = 0$. (The empty set contains no elements.)

Let us begin by stating one of the fundamental principles of counting, the so-called *inclusion-exclusion principle.*

Inclusion-Exclusion Principle   Let S and T be sets. Then

$$n(S \cup T) = n(S) + n(T) - n(S \cap T). \tag{1}$$

Note that formula (1) connects the four quantities $n(S \cup T)$, $n(S)$, $n(T)$, $n(S \cap T)$. Given any three, the remaining quantity can be determined by using this formula.

To test the plausibility of the inclusion-exclusion principle, consider this example. Let $S = \{a, b, c, d, e\}$, $T = \{a, c, g, h\}$. Then

$$S \cup T = \{a, b, c, d, e, g, h\} \qquad n(S \cup T) = 7$$

$$S \cap T = \{a, c\} \qquad\qquad\quad n(S \cap T) = 2.$$

In this case the inclusion-exclusion principle reads

$$n(S \cup T) = n(S) + n(T) - n(S \cap T),$$

$$7 \quad = \quad 5 \; + \; 4 \; - \quad 2$$

which is correct.

Here is the reason for the validity of the inclusion-exclusion principle: The left side of formula (1) is $n(S \cup T)$, the number of elements in either S or T (or both). As a first approximation to this number, add the number of elements in S to the number of elements in T, obtaining $n(S) + n(T)$. However, if an element lies in both S and T, it is counted twice—once in $n(S)$ and again in $n(T)$. To make up for this double counting we must subtract off the number of elements counted twice, namely $n(S \cap T)$. This gives us $n(S) + n(T) - n(S \cap T)$ as the number of elements in $S \cup T$.

The next example illustrates a typical use of the inclusion-exclusion principle in an applied problem.

EXAMPLE 1   In 1977, *Executive* magazine surveyed the presidents of the 500 largest corporations in the United States. Of these 500 people, 310 had degrees (of any sort) in business, 238 had undergraduate degrees in business, and 184 had graduate degrees in business. How many presidents had both undergraduate and graduate degrees in business?

Solution   Let

$$S = \{\text{presidents with an undergraduate degree in business}\},$$

$$T = \{\text{presidents with a graduate degree in business}\}.$$

Then

$S \cup T = \{\text{presidents with at least one degree in business}\}$,

$S \cap T = \{\text{presidents with both undergraduate and graduate degrees in business}\}$.

From the data given we have

$$n(S) = 238, \qquad n(T) = 184, \qquad n(S \cup T) = 310.$$

The problem asks for $n(S \cap T)$. By the inclusion-exclusion principle we have

$$n(S \cup T) = n(S) + n(T) - n(S \cap T)$$

$$310 = 238 + 184 - n(S \cap T)$$

$$n(S \cap T) = 112.$$

That is, exactly 112 of the presidents had both undergraduate and graduate degrees in business.

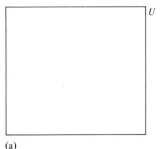 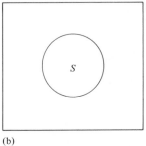 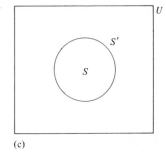

(a)　　　　　　　(b)　　　　　　　(c)

FIGURE 1

It is possible to visualize sets geometrically by means of drawings known as *Venn diagrams*. Such graphical representations of sets are very useful tools in solving counting problems. In order to describe Venn diagrams, let us begin with a single set S contained in a universal set U. Draw a rectangle and view its points as the elements of U [Fig. 1(a)]. To show that S is a subset of U we draw a circle inside the rectangle and view S as the set of points in the circle [Fig. 1(b)]. The resulting diagram is called a *Venn diagram* of S. It illustrates the proper relationship between S and U. Since S' consists of those elements of U which are not in S, we may view the portion of the rectangle which is outside of the circle as representing S' [Fig. 1(c)].

FIGURE 2

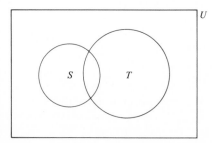

Venn diagrams are particularly useful for visualizing the relationship between two or more sets. Suppose that we are given two sets S and T in a universal set U. As before, we represent each of the sets by means of a circle inside the rectangle (Fig. 2).

We can now illustrate a number of sets by shading in appropriate regions of the rectangle. For instance, in Fig. 3 we have shaded the regions corresponding to T, $S \cup T$, and $S \cap T$.

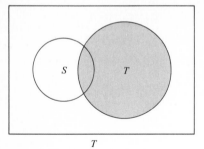

T

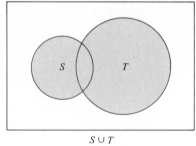

$S \cup T$

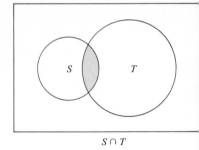

$S \cap T$

FIGURE 3

EXAMPLE 2 Shade the portions of the rectangle corresponding to the sets

(a) $S \cap T'$ (b) $(S \cap T')'$

Solution (a) $S \cap T'$ consists of the points in S and in T', that is, the points in S and not in T. So we shade the points that are in the circle S but are not in the circle T [Fig. 4(a)].

(b) $(S \cap T')'$ is the complement of the set $(S \cap T')$. Therefore, it consists of exactly those points not shaded in Fig. 4(a). [See Fig. 4(b).]

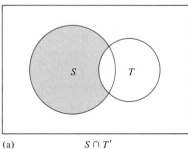

(a) $S \cap T'$

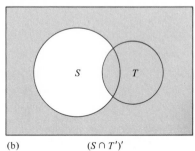

(b) $(S \cap T')'$

FIGURE 4

In a similar manner, Venn diagrams can illustrate intersections and unions of three sets. Some representative regions are shaded in Fig. 5.

There are many formulas expressing relationships between intersections and unions of sets. Possibly the most fundamental are the two formulas known as De Morgan's laws.

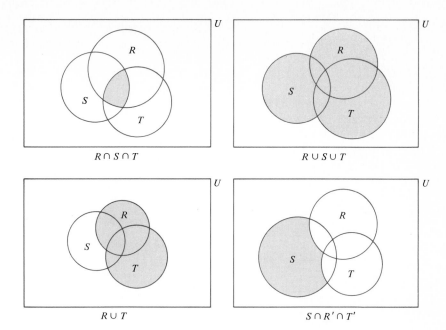

$R \cap S \cap T$

$R \cup S \cup T$

$R \cup T$

$S \cap R' \cap T'$

FIGURE 5

> *De Morgan's Laws*   Let S and T be sets. Then
> $$(S \cup T)' = S' \cap T'$$
> and
> $$(S \cap T)' = S' \cup T'.$$

In other words, De Morgan's laws state that to form the complement of a union (or intersection) form the complements of the individual sets and change unions to intersections (or intersections to unions).

Verification of De Morgan's Laws

Let us utilize Venn diagrams to describe $(S \cup T)'$. In Fig. 6(a) we have shaded the area corresponding to $S \cup T$. In Fig. 6(b) we have shaded the area corresponding to $(S \cup T)'$. On the other hand, in Fig. 6(c) we have shaded the area corresponding to S' and in Fig. 6(d) the area corresponding to T'. By considering the common shaded areas of Fig. 6(c) and (d) we arrive at the shaded area corresponding to $S' \cap T'$ [Fig. 6(e)]. Note that this is the same region as shaded in Fig. 6(b). Therefore,

$$(S \cup T)' = S' \cap T'.$$

This verifies the first of De Morgan's laws. The proof of the second law is similar.

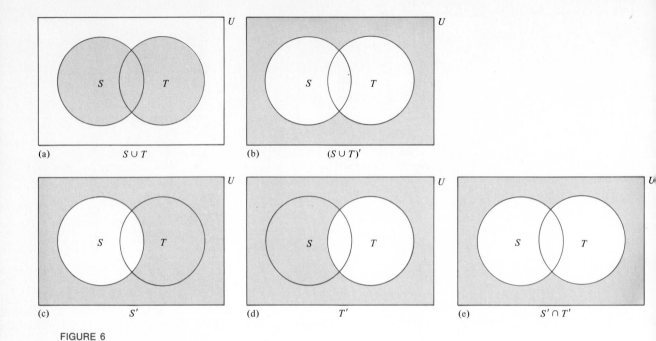

FIGURE 6

PRACTICE PROBLEMS 2

1. Draw a two-circle Venn diagram and shade the portion corresponding to the set $(S \cap T') \cup (S \cap T)$.

2. Suppose that $n(S) + n(T) = n(S \cup T)$. What can you conclude about S and T?

EXERCISES 2

1. Find $n(S \cup T)$, given that $n(S) = 5$, $n(T) = 4$, and $n(S \cap T) = 2$.

2. Find $n(S \cup T)$, given that $n(S) = 17$, $n(T) = 13$, and $n(S \cap T) = 9$.

3. Find $n(S \cap T)$, given that $n(S) = 7$, $n(T) = 8$, and $n(S \cup T) = 15$.

4. Find $n(S \cap T)$, given that $n(S) = 4$, $n(T) = 12$, and $n(S \cup T) = 15$.

5. Find $n(S)$, given that $n(T) = 7$, $n(S \cap T) = 5$, and $n(S \cup T) = 13$.

6. Find $n(T)$, given that $n(S) = 14$, $n(S \cap T) = 6$, and $n(S \cup T) = 14$.

7. If $n(S) = n(S \cap T)$, what can you conclude about S and T?

8. If $n(S) = n(S \cup T)$, what can you conclude about S and T?

9. Suppose that each of the 180 million adults in South America is fluent in Portuguese or Spanish. If 99 million are fluent in Portuguese and 95 million are fluent in Spanish, how many are fluent in both languages?

10. Suppose that all of the 1000 freshmen at a certain college are enrolled in a math or an English course. Suppose that 400 are taking both math and English and 600 are taking English. How many are taking a math course?

11. The combined membership of the MAA (Mathematical Association of America) and the AMS (American Mathematical Society) is approximately 29,000. Seventeen thousand people belong to the AMS and 7000 of them also belong to the MAA. How many people belong to the MAA?

12. A survey of employees in a certain company revealed that 300 people subscribe to *Newsweek*, 200 subscribe to *Time*, and 50 subscribe to both. How many people subscribe to at least one of these magazines?

In Exercises 13–24, draw a two-circle Venn diagram and shade the portion corresponding to the set.

13. $S' \cap T'$

14. $S' \cap T$

15. $S \cup T'$

16. $S' \cup T'$

17. $(S \cap T)'$

18. $(S' \cap T)'$

19. $(S \cap T') \cup (S' \cap T)$

20. $(S \cap T) \cup (S' \cap T')$

21. $S \cup (S \cap T)$

22. $S \cup (T' \cup S)$

23. $S \cup S'$

24. $S \cap S'$

In Exercises 25–36, draw a three-circle Venn diagram and shade the portion corresponding to the set.

25. $R \cap S \cap T'$

26. $R' \cap S' \cap T$

27. $R \cup (S \cap T)$

28. $R \cap (S \cup T)$

29. $R \cap (S' \cup T)$

30. $R' \cup (S \cap T')$

31. $R \cap T$

32. $S \cap T'$

33. $R' \cap S' \cap T'$

34. $(R \cup S \cup T)'$

35. $(R \cap T) \cup (S \cap T')$

36. $(R \cup S') \cap (R \cup T')$

In Exercises 37–42, use De Morgan's laws to simplify the given expression.

37. $S' \cup (S \cap T)'$

38. $T \cap (S \cup T)'$

39. $(S' \cup T)'$

40. $(S' \cap T')'$

41. $T \cup (S \cap T)'$

42. $(S' \cap T)' \cup S$

In Exercises 43–48, give a set-theoretic expression that describes the shaded portion of the Venn diagram.

43.
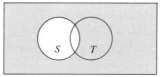

44.
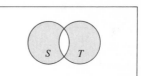

45.
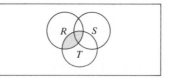

46.

47.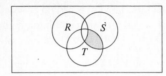

48.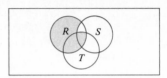

SOLUTIONS TO PRACTICE PROBLEMS 2

1. $(S \cap T') \cup (S \cap T)$ is given as a union of two sets, $S \cap T'$ and $S \cap T$. The Venn diagrams for these two sets are given in Fig. 7(a) and (b). The desired set consists of the elements which are in one or the other (or both) of the two sets. Therefore, its Venn diagram is obtained by shading everything that is shaded in either Fig. 7(a) or (b). [See Fig. 7(c).] [*Note:* Looking at Fig. 7(c) reveals that $(S \cap T') \cup (S \cap T)$ and S are the same set. Often Venn diagrams can be used to simplify complicated set-theoretic expressions.]

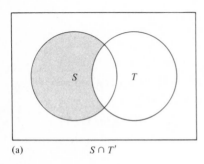

(a) $S \cap T'$

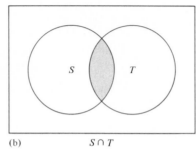

(b) $S \cap T$

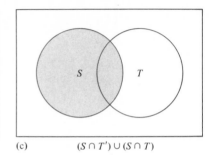

(c) $(S \cap T') \cup (S \cap T)$

FIGURE 7

2. From the inclusion-exclusion principle, we obtain $n(S \cap T) = 0$—that is, $S \cap T = \emptyset$. We conclude that S and T have no elements in common.

8.3. Venn Diagrams and Counting

In this section we discuss the use of Venn diagrams in solving counting problems. The techniques developed are especially useful in analyzing survey data.

Each Venn diagram divides the universal set U into a certain number of regions. For example, the Venn diagram for a single set divides U into two regions—the inside and outside of the circle [Fig. 1(a)]. The Venn diagram for two sets divides U into four regions [Fig. 1(b)]. And the Venn diagram for three sets divides U into eight regions [Fig. 1(c)]. Each of the regions is called a *basic region* for the Venn diagram. Knowing the number of elements in each basic region is of great use in many applied problems. As an illustration consider the following example.

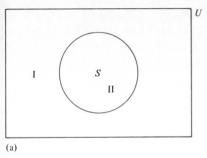

(a)

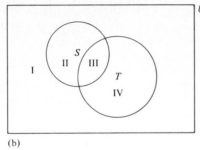

(b)

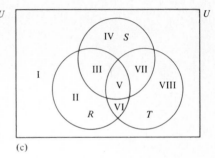

(c)

FIGURE 1

EXAMPLE 1 Let

U = {winners of the Nobel prize during the period 1901–1976},

A = {American winners of the Nobel prize during the period 1901–1976},

C = {winners of the Nobel prize in chemistry during the period 1901–1976},

P = {winners of the Nobel Peace Prize during the period 1901–1976}.

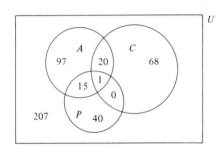

FIGURE 2

These sets are illustrated in the Venn diagram of Fig. 2 in which each basic region has been labeled with the number of elements in it.

(a) How many Americans received a Nobel prize during the period 1901–1976?

(b) How many Americans received Nobel prizes in fields other than chemistry and peace during this period?

(c) How many Americans received the Nobel Peace Prize during this period?

(d) How many people received Nobel prizes during this period?

Solution (a) The number of Americans who received a Nobel prize is the total number contained in the circle A, which is $97 + 15 + 1 + 20 = 133$ [Fig. 3(a)].

FIGURE 3

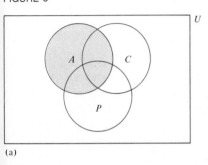

(a)

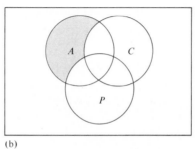

(b)

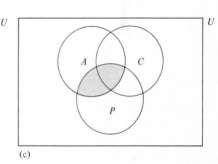

(c)

(b) The question asks for the number of Nobel laureates in A but not in C and not in P. So start with the A circle and eliminate those basic regions belonging to C or P [Fig. 3(b)]. There remains a single basic region with 97 Nobel laureates. Note that this region corresponds to $A \cap C' \cap P'$.

(c) The question asks for the number of elements in both A and P—that is, $n(A \cap P)$. But $A \cap P$ comprises two basic regions [Fig. 3(c)]. Thus, to compute $n(A \cap P)$ we add the numbers in these basic regions to obtain $15 + 1 = 16$ Americans who have won the Nobel Peace Prize.

(d) The number of recipients is just $n(U)$, and we obtain it by adding together the numbers corresponding to the basic regions. We obtain $207 + 97 + 20 + 1 + 15 + 68 + 0 + 40 = 448$.

One need not always be given the number of elements in each of the basic regions of a Venn diagram. Very often these data can be deduced from given information.

EXAMPLE 2 Consider the set of 500 corporate presidents of Example 1, Section 2.

(a) Draw a Venn diagram displaying the given data and determine the number of elements in each basic region.

(b) Determine the number of presidents having exactly one degree (graduate or undergraduate) in business.

Solution (a) Recall that we defined the following sets:

$$S = \{\text{presidents with an undergraduate degree in business}\}$$

$$T = \{\text{presidents with a graduate degree in business}\}.$$

We were given the following data:

$$n(S) = 238, \qquad n(T) = 184, \qquad n(S \cup T) = 310.$$

FIGURE 4

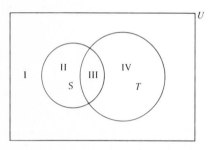

We draw a Venn diagram corresponding to S and T (Fig. 4). Notice that none of the given information corresponds to a basic region of the Venn diagram. So we must use our wits to determine the number of presidents in each of the regions I–IV. Region I is the complement of $S \cup T$, so it contains

$$n(U) - n(S \cup T) = 500 - 310 = 190$$

presidents. Region III is just $S \cap T$. By using the inclusion-exclusion principle, in Example 1, Section 2, we determined that $n(S \cap T) = 112$. Now the total number of presidents in II and III combined equals $n(S)$, or 238. Therefore, the number of presidents in II is

$$238 - 112 = 126.$$

Similarly, the number of presidents in IV is

$$184 - 112 = 72.$$

Thus, we may fill in the data determined to obtain a completed Venn diagram (Fig. 5).

(b) The number of people with exactly one business degree corresponds to the shaded region in Fig. 6. Adding together the number of presidents in each of these regions gives $126 + 72 = 198$ presidents with exactly one business degree.

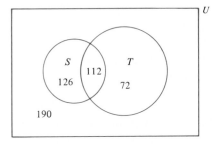

FIGURE 5

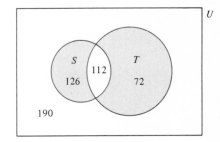

FIGURE 6

Here is another example illustrating the procedure for determining the number of elements in each of the basic regions of a Venn diagram.

EXAMPLE 3   An advertising agency finds that of its 170 clients, 115 use television (T), 100 use radio (R), 130 use magazines (M), 75 use television and radio, 95 use radio and magazines, 85 use television and magazines, and 70 use all three. Use these data to complete a Venn diagram displaying the use of mass media (Fig. 7).

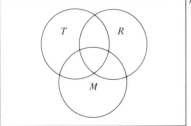

$n(U) = 170$

FIGURE 7

Solution   Of the various data given, only the last item corresponds to one of the eight basic regions of the Venn diagram, namely the "70" corresponding to the use of all three media. So we begin by entering this number in the diagram [Fig. 8(a)]. We can fill in the rest of the Venn diagram by working with the remaining information one piece at a time in the reverse order that it is given. Since 85 clients advertise in television and magazines, $85 - 70 = 15$ advertise in television and magazines but not on radio. The appropriate region is labeled in Fig. 8(b). In Fig. 8(c) the next two pieces of information have been used in the same way to fill in two more

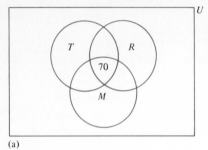

(a)

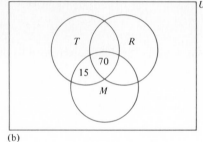

(b)

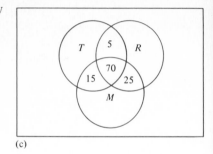

(c)

FIGURE 8

basic regions. In Fig. 8(c) we observe that three of the four basic regions comprising M have been filled in. Since $n(M) = 130$, we deduce that the number of clients advertising only in magazines is $130 - (15 + 70 + 25) = 130 - 110 = 20$ [Fig. 9(a)]. By similar reasoning the number of clients using only radio advertising and the number using only television advertising can be determined [Fig. 9(b)]. Adding together the numbers in the three circles gives the number of clients utilizing television, radio, or magazines as $25 + 5 + 0 + 15 + 70 + 25 + 20 = 160$. Since there were 170 clients in total, the remainder—or $170 - 160 = 10$ clients—use none of these media. Figure 9(c) gives a complete display of the data.

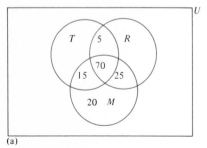

(a)

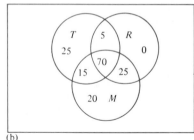

(b)

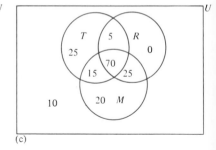

(c)

FIGURE 9

PRACTICE PROBLEMS 3

1. Of the 1000 freshmen at a certain college, 700 take mathematics courses, 300 take mathematics and economics courses, and 200 do not take any mathematics or economics courses. Represent these data in a Venn diagram.

2. Refer to the Venn diagram from Problem 1.

   (a) How many of the freshmen take an economics course?

   (b) How many take an economics course but not a mathematics course?

EXERCISES 3

In Exercises 1–4, draw an appropriate Venn diagram and use the given data to determine the number of elements in each basic region.

1. $n(U) = 14, n(S) = 5, n(T) = 6, n(S \cap T) = 2$.

2. $n(U) = 20, n(S) = 11, n(T) = 7, n(S \cap T) = 7$.

3. $n(U) = 44, n(R) = 17, n(S) = 17, n(T) = 17, n(R \cap S) = 7, n(R \cap T) = 6, n(S \cap T) = 5$, $n(R \cap S \cap T) = 2$.

4. $n(U) = 29, n(R) = 10, n(S) = 12, n(T) = 10, n(R \cap S) = 1, n(R \cap T) = 5, n(S \cap T) = 4$, $n(R \cap S \cap T) = 1$.

5. $n(U) = 75, n(S) = 15, n(T) = 25, n(S' \cap T') = 40$.

6. $n(S) = 9, n(T) = 11, n(S \cap T) = 5, n(S') = 13$.

7. $n(R') = 22, \quad n(R \cup S) = 21, \quad n(S) = 14, \quad n(T) = 22, \quad n(R \cap S) = 7, \quad n(S \cap T) = 9$, $n(R \cap T) = 11, n(R \cap S \cap T) = 5$.

8. $n(U) = 64, \quad n(R \cup S \cup T) = 45, \quad n(R) = 22, \quad n(T) = 26, \quad n(R \cap S) = 4, \quad n(S \cap T) = 6$, $n(R \cap T) = 8, n(R \cap S \cap T) = 1$.

9. A survey of 70 high school students revealed that 35 like folk music, 15 like classical music, and 5 like both. How many of the students surveyed do not like either folk or classical music?

10. A total of 450 Nobel prizes had been awarded by 1976. Fourteen of the 74 prizes in literature were awarded to Scandinavians. Scandinavians received a total of 40 awards. How many Nobel prizes outside of literature have been awarded to non-Scandinavians?

11. Out of 35 students in a finite math class, 22 are male, 19 are business majors, 27 are freshmen, 14 are male business students, 17 are male freshmen, 15 are freshmen business majors, and 11 are male freshmen business majors. How many upperclass women nonbusiness majors are in the class? How many women business majors are in the class?

12. A survey of 100 college faculty who exercise regularly found that 45 jog, 30 swim, 20 cycle, 6 jog and swim, 1 jogs and cycles, 5 swim and cycle, and 1 does all three. How many of the faculty members do not do any of these three activities? How many just jog?

13. One hundred college students were surveyed after voting in an election involving a Democrat and a Republican. Fifty of the students were freshmen, 55 voted Democratic, and 25 were nonfreshmen who voted Republican. How many freshmen voted Democratic?

14. A group of 100 workers were asked if they were college graduates and if they belonged to a union. Sixty were not college graduates, 20 were nonunion college graduates, and 30 were union members. How many of the workers were neither college graduates nor union members?

15. A class of 30 students was given a diagnostic test on the first day of a mathematics course. At the end of the semester, only 2 of the 21 students who had passed the diagnostic test failed the course. A total of 23 students passed the course. How many students managed to pass the course even though they had failed the diagnostic test?

16. A group of applicants for training as air-traffic controllers consisted of 35 pilots, 20 veterans, 30 pilots who were not veterans, and 50 people who were neither veterans nor pilots. How large was the group?

17. One hundred and eighty business executives were surveyed to determine if they regularly read *Fortune*, *Time*, or *Money* magazines. Seventy-five read *Fortune*, 70 read *Time*, 55 read *Money*, 45 read exactly two of the three magazines, 25 read *Fortune* and *Time*, 25 read *Time* and *Money*, and 5 read all three magazines. How many read none of the three magazines?

18. A survey of the characteristics of 100 small businesses which had failed revealed that 95 of them either were undercapitalized, had inexperienced management, or had a poor location. Four of the businesses had all three of these characteristics. Forty businesses were undercapitalized but had experienced management and good location. Fifteen businesses had inexperienced management but sufficient capitalization and good location. Seven were undercapitalized and had inexperienced management. Nine were undercapitalized and had poor location. Ten had inexperienced management and poor location. How many of the business had poor location? Which of the three characteristics was most prevalent in the failed businesses?

SOLUTIONS TO PRACTICE PROBLEMS 3

1. Draw a Venn diagram with two circles, one for mathematics (M) and one for economics (E) [Fig. 10(a)]. This Venn diagram has four basic regions, and our goal is to label each basic region with the proper number of students. The numbers for two of the basic regions are given directly. Since "300 take mathematics and economics," $n(M \cap E) = 300$. Since

FIGURE 10

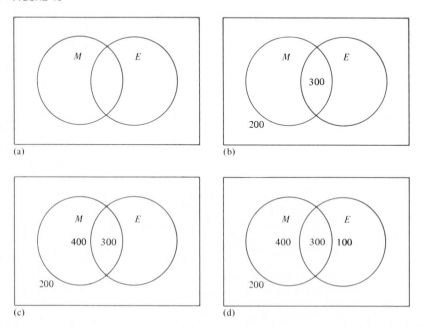

(a)   (b)   (c)   (d)

"200 do not take any mathematics or economics courses," $n((M \cup E)') = 200$ [Fig. 10(b)]. Now "700 take mathematics courses." Since M is made up of two basic regions and one region has 300 elements, the other basic region of M must contain 400 elements [Fig. 10(c)]. At this point all but one of the basic regions have been labeled and $400 + 300 + 200 = 900$ students have been accounted for. Since there is a total of 1000 students, the remaining basic region has 100 students [Fig. 10(d)].

2. (a) 400. "Economics" refers to the entire circle E which is made up of two basic regions, one having 300 elements and the other 100. (A common error is to interpret the question as asking for the number of freshmen who take economics exclusively and therefore give the answer 100. To say that a person takes an economics course does not imply anything about the person's enrollment in mathematics courses.)

(b) 100.

8.4. The Multiplication Principle

In this section we introduce a second fundamental principle of counting, the so-called *multiplication principle*. By way of motivation, consider the following example.

EXAMPLE 1   A medical researcher wishes to test the effect of a drug on a rat's perception by studying its ability to run a maze while under the influence of the drug. The maze is constructed so that, in order to arrive at the exit point C, the rat must pass through a central point B. There are five paths from the entry point A to B, and three paths from B to C. In how many different ways can the rat run the maze from A to C? (See Fig. 1.)

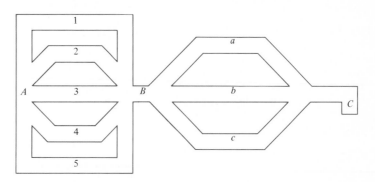

FIGURE 1   Paths through a maze

Solution   The paths from A to B have been labeled 1 through 5, and the paths from B to C have been labeled a through c. The various paths through the maze can be schematically represented as in Fig. 2. The diagram shows that there are five ways to go from A to B. For each of these there are three ways to go from B to C. So there are five groups of three paths each, and therefore $5 \cdot 3 = 15$ possible

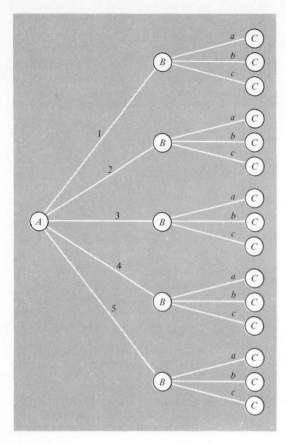

FIGURE 2

paths from A to C. (A diagram such as Fig. 2, called a *tree diagram*, is useful in enumerating the various possibilities in counting problems.)

In the problem above choosing a path is a task that can be broken up into two consecutive operations.

| Choose path from A to B | Choose path from B to C |

Operation 1   Operation 2

The first operation can be performed in five ways and, after the first operation has been carried out, the second can be performed in three ways. And we determined that the entire task can be performed in $5 \cdot 3 = 15$ ways. The same reasoning as used above yields the following useful counting principle:

Multiplication Principle Suppose that a task is composed of two consecutive operations. If operation 1 can be performed in m ways and, for each of these, operation 2 can be performed in n ways, then the complete task can be performed in $m \cdot n$ ways.

EXAMPLE 2 An airline passenger must fly from New York to Frankfurt via London. There are 8 flights leaving New York for London. All of these provide connections on any one of 19 flights from London to Frankfurt. In how many different ways can the passenger book reservations?

Solution The task "fly from New York to Frankfurt" is composed of two consecutive operations:

| Fly from New York to London | Fly from London to Frankfurt |

Operation 1   Operation 2

From the data given the multiplication principle implies that the task can be accomplished in $8 \cdot 19 = 152$ ways.

It is possible to generalize the multiplication principle to tasks consisting of more than two operations.

Generalized Multiplication Principle Suppose that a task consists of t operations performed consecutively. Suppose that operation 1 can be performed in m_1 ways; for each of these, operation 2 in m_2 ways; for each of these, operation 3 in m_3 ways; and so forth. Then the task can be performed in

$$m_1 \cdot m_2 \cdot m_3 \cdots m_t$$

ways.

EXAMPLE 3   A corporation has a board of directors consisting of 10 members. The board must select from among its members a chairman, vice-chairman, and secretary. In how many ways can this be done?

Solution   The task "select the three officers" can be divided into three consecutive operations:

Select chairman	Select vice-chairman	Select secretary

Since there are 10 directors, operation 1 can be performed in 10 ways. After the chairman has been selected, there are 9 directors left as possible candidates for vice-chairman, so that for each way of performing operation 1, operation 2 can be performed in 9 ways. After this has been done, there are 8 directors who are possible candidates for secretary, so operation 3 can be performed in 8 ways. By the generalized multiplication principle the number of possible ways to perform the sequence of three operations equals $10 \cdot 9 \cdot 8$ or 720. So the officers of the board can be selected in 720 ways.

In Example 3 we made important use of the phrase "for each of these" in the generalized multiplication principle. The operation "select a vice-chairman" can be performed in 10 ways, since any member of the board is eligible. However, when we view the selection process as a sequence of operations of which "select a vice-chairman" is the second operation, the situation has changed. *For each way that the first operation is performed one person will have been used up; hence there will be only 9 possibilities for choosing the vice-chairman.*

EXAMPLE 4   In how many ways can a baseball team of nine members arrange themselves in a line for a group picture?

Solution   Choose the players by their place in the picture, say from left to right. The first can be chosen in nine ways; for each of these choices the second can be chosen in eight ways; for each of these choices the third can be chosen in seven ways; and so forth. So the number of possible arrangements is

$$9 \cdot 8 \cdot 7 \cdot 6 \cdot 5 \cdot 4 \cdot 3 \cdot 2 \cdot 1 = 362,880.$$

EXAMPLE 5   A certain state uses automobile license plates which consist of three letters followed by three digits. How many such license plates are there?

Solution   The task in this case, "form a license plate," consists of a sequence of six operations: three for choosing letters and three for choosing digits. Each letter can be chosen in 26 ways and each digit in 10 ways. So the number of license plates is

$$26 \cdot 26 \cdot 26 \cdot 10 \cdot 10 \cdot 10 = 17,576,000.$$

PRACTICE PROBLEMS 4

1. There are six seats available in a sedan. In how many ways can six people be seated if only three can drive?

2. A multiple-choice exam contains 10 questions, each having 3 possible answers. How many different ways are there of completing the exam?

EXERCISES 4

1. If there are three routes from College Park to Baltimore and five routes from Baltimore to New York, how many routes are there from College Park to New York via Baltimore?

2. How many different outfits consisting of a coat and a hat can be chosen from two coats and three hats?

3. How many different two-letter words can be formed allowing repetition of letters?

4. How many different two-letter words can be formed such that the two letters are distinct?

5. A railway has 20 stations. If the names of the point of departure and the destination are printed on each ticket, how many different kinds of single tickets must be printed?

6. Refer to Exercise 5. How many different kinds of tickets are needed if each ticket may be used in either direction between two stations?

7. A man has five different pairs of gloves. In how many ways can he select a right-hand glove and a left-hand glove that do not match?

8. How many license plates consisting of two letters followed by four digits are possible?

9. How many ways can five people be arranged in a line for a group picture?

10. In how many different ways can four books be arranged on a bookshelf?

11. Toss a coin six times and observe the sequence of heads or tails that results. How many different sequences are possible?

12. Refer to Exercise 11. In how many of the sequences are the first and last tosses identical?

13. Twenty athletes enter an Olympic event. How many different possibilities are there for winning the Gold Medal, Silver Medal, and Bronze Medal?

14. A sportswriter is asked to rank eight teams. How many different orderings are possible?

15. In how many different ways can a 30-member football team select a captain and a cocaptain?

16. How many different outfits can be selected from two coats, three hats, and two scarves?

17. How many different words can be formed using the four letters of the word "MATH"?

18. If you can travel from Frederick, Maryland, to Baltimore, Maryland, by car, bus, or train and from Baltimore to London by airplane or ship, how many different ways are there to go from Frederick to London?

19. An exam contains five "true or false" questions. In how many different ways can the exam be completed?

20. A company has 700 employees. Explain why there must be two people with the same pair of initials.

21. A computer manufacturer assigns serial numbers to its computers. The first symbol of a serial number is either A, B, or C, indicating the manufacturing plant. The second and third symbols taken together are one of the numbers $01, 02, \ldots, 12$ indicating the month of manufacture. The final four symbols are digits. How many possible serial numbers are there?

22. How many four-letter words (including nonsense words) can be made from the letters of "statistics" assuming that each word may not have repeated letters?

SOLUTIONS TO PRACTICE PROBLEMS 4

1. 360. Pretend that you are given the task of seating the six people. This task consists of six operations performed consecutively, as shown in the accompanying chart.

Operation	Number of ways operation can be performed
1: select person to drive	3
2: select person for middle front seat	5
3: select person for right front seat	4
4: select person for left rear seat	3
5: select person for middle rear seat	2
6: select person for right rear seat	1

After you have performed operation 1, five people will remain, and any one of these five can be seated in the middle front seat. After operation 2, four people remain, and so on. By the generalized multiplication principle, the task can be performed in $3 \cdot 5 \cdot 4 \cdot 3 \cdot 2 \cdot 1 = 360$ ways.

2. 3^{10}. The task of answering the questions consists of 10 consecutive operations, each of which can be performed in three ways. Therefore, by the generalized multiplication principle, the task can be performed in $\underbrace{3 \cdot 3 \cdot 3 \cdot \ldots \cdot 3}_{10 \text{ terms}}$ ways.

[*Note:* The answer can be left as 3^{10} or can be multiplied out to 59,049.]

8.5. Permutations and Combinations

In preceding sections we have solved a variety of counting problems using Venn diagrams and the generalized multiplication principle. Let us now turn our attention to two types of counting problems which occur very frequently and which can be solved using formulas derived from the generalized multiplication principle. These problems involve what are called permutations and combinations, each of which are particular types of arrangements of elements of a set. The sort of arrangements we have in mind are illustrated in two problems:

> **PROBLEM A** How many words (by which we mean strings of letters) of two distinct letters can be formed from the letters $\{a, b, c\}$?
>
> **PROBLEM B** A construction crew has three members. A team of two must be chosen for a particular job. In how many ways can the team be chosen?

Each of the two problems can be solved by enumerating all possibilities.

Solution of Problem A   There are six possible words, namely

$$ab, ac, ba, bc, ca, cb.$$

Solution of Problem B   Designate the three crew members by a, b, and c. Then there are three possible two-man teams, namely

$$ab, ac, bc.$$

(Note that ba, the team consisting of b and a, is the same as the team ab.)

We deliberately set up both problems using the same letters in order to facilitate comparison. Both problems are concerned with counting the number of arrangements of the elements of the set $\{a, b, c\}$, taken two at a time, without allowing repetition (for example, aa was not allowed). However, in Problem A the order of the arrangement mattered, whereas in Problem B it did not. Arrangements of the sort considered in Problem A are called *permutations*, whereas those in Problem B are called *combinations*.

More precisely, suppose that we are given a set of n objects.* Then a *permutation of n objects taken r at a time* is an arrangement of r of the n objects in a specific

---

* All assumed to be different.

order. So, for example, Problem A was concerned with permutations of the three objects a, b, c ($n = 3$), taken two at a time ($r = 2$). A *combination of n objects taken r at a time* is a selection of r objects from among the n, with order disregarded. Thus, for example, in Problem B we considered combinations of the three objects a, b, c ($n = 3$), taken two at a time ($r = 2$).

It is convenient to introduce the following notation for counting permutations and combinations. Let

$P(n, r)$ = the number of permutations of n objects taken r at a time,

$C(n, r)$ = the number of combinations of n objects taken r at a time.

Thus, for example, from our solutions to Problems A and B we have

$$P(3, 2) = 6, \qquad C(3, 2) = 3.$$

There are very simple formulas for $P(n, r)$ and $C(n, r)$ which allow us to calculate these quantities for any n and r. Let us begin by stating the formula for $P(n, r)$. For $r = 1, 2, 3$ we have, respectively,

$$P(n, 1) = n$$

$$P(n, 2) = n(n - 1) \qquad \text{(two factors)}$$

$$P(n, 3) = n(n - 1)(n - 2) \qquad \text{(three factors)},$$

and, in general,

$$P(n, r) = n(n - 1)(n - 2) \cdots (n - r + 1) \qquad \text{(r factors).} \tag{1}$$

This formula is verified at the end of this section.

EXAMPLE 1 Compute the following numbers.

(a) $P(100, 2)$ (b) $P(6, 4)$ (c) $P(5, 5)$

Solution (a) Here $n = 100$, $r = 2$. So we take the product of two factors, beginning with 100:

$$P(100, 2) = 100 \cdot 99 = 9900.$$

(b) $P(6, 4) = 6 \cdot 5 \cdot 4 \cdot 3 = 360$

(c) $P(5, 5) = 5 \cdot 4 \cdot 3 \cdot 2 \cdot 1 = 120$

In order to state the formula for $C(n, r)$, we must introduce some further notation. Suppose that r is any positive integer. We denote by $r!$ (read "r factorial") the product of all positive integers from r down to 1:

$$r! = r \cdot (r - 1) \cdot \ldots \cdot 2 \cdot 1.$$

Thus, for instance,

$$1! = 1$$
$$2! = 2 \cdot 1 = 2$$
$$3! = 3 \cdot 2 \cdot 1 = 6$$
$$4! = 4 \cdot 3 \cdot 2 \cdot 1 = 24$$
$$5! = 5 \cdot 4 \cdot 3 \cdot 2 \cdot 1 = 120.$$

In terms of this notation we can state a very simple formula for $C(n, r)$, the number of combinations of n things taken r at a time.

$$C(n, r) = \frac{P(n, r)}{r!} = \frac{n(n - 1) \cdot \ldots \cdot (n - r + 1)}{r(r - 1) \cdot \ldots \cdot 1}. \qquad (2)$$

This formula is verified at the end of the section.

EXAMPLE 2   Compute the following numbers.

(a) $C(100, 2)$ (b) $C(6, 4)$ (c) $C(5, 5)$

Solution   (a) $C(100, 2) = \dfrac{P(100, 2)}{2!} = \dfrac{100 \cdot 99}{2 \cdot 1} = 4950$

(b) $C(6, 4) = \dfrac{P(6, 4)}{4!} = \dfrac{6 \cdot 5 \cdot 4 \cdot 3}{4 \cdot 3 \cdot 2 \cdot 1} = 15$

(c) $C(5, 5) = \dfrac{P(5, 5)}{5!} = \dfrac{5 \cdot 4 \cdot 3 \cdot 2 \cdot 1}{5 \cdot 4 \cdot 3 \cdot 2 \cdot 1} = 1$

EXAMPLE 3   Solve Problems A and B using formulas (1) and (2).

Solution   The number of two-letter words which can be formed from the three letters a, b, c is equal to $P(3, 2) = 3 \cdot 2 = 6$, in agreement with our previous solution.

The number of two-worker teams which can be formed from three individuals is equal to $C(3, 2)$, and

$$C(3, 2) = \frac{P(3, 2)}{2!} = \frac{3 \cdot 2}{2 \cdot 1} = 3,$$

in agreement with our previous result.

EXAMPLE 4   The board of directors of a corporation has 10 members. In how many ways can they choose a committee of 3 board members to negotiate a merger?

Solution   Since the committee of three involves no ordering of its members, we are concerned here with combinations. The number of combinations of 10 people taken 3 at a time is $C(10, 3)$, which is

$$C(10, 3) = \frac{10 \cdot 9 \cdot 8}{3 \cdot 2 \cdot 1} = 120.$$

Thus there are 120 choices for the committee.

EXAMPLE 5   Eight horses are entered in a race in which a first, second, and third prize will be awarded. Assuming no ties, how many different outcomes are possible?

Solution   In this example we are considering ordered arrangements of three horses, so we are dealing with permutations. The number of permutations of eight horses taken three at a time is

$$P(8, 3) = 8 \cdot 7 \cdot 6 = 336,$$

so the number of possible outcomes of the race is 336.

EXAMPLE 6   A political pollster wishes to survey 1500 individuals chosen from a sample of 5,000,000 adults. In how many ways can the 1500 individuals be chosen?

Solution   No ordering of the 1500 individuals is involved, so we are dealing with combinations. So the number in question is $C(5,000,000, 1500)$, a number too large to be written down in digit form. (It has several thousand digits!) But it could be calculated with the aid of a computer.

EXAMPLE 7   A club has 10 members. In how many ways can they choose a slate of four officers, consisting of a president, vice-president, secretary, and treasurer?

Solution   In this problem we are dealing with an ordering of four members. (The first is the president, the second the vice-president, and so on.) So we are dealing with permutations, and the number of ways of choosing the officers is

$$P(10, 4) = 10 \cdot 9 \cdot 8 \cdot 7 = 5040.$$

Verification of the Formulas for $P(n, r)$ and $C(n, r)$ Let us first derive the formula for $P(n, r)$, the number of permutations of n objects taken r at a time. The task of choosing r objects (in a given order) consists of r consecutive operations (Fig. 1). The first operation can be performed in n ways.

FIGURE 1

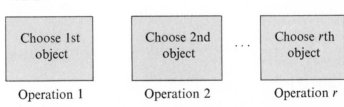

| Operation 1 | Operation 2 | Operation r |

For each way that the first operation is performed one object will have been used up and so we can perform the second operation in $n - 1$ ways, and so on. For each way of performing the sequence of operations $1, 2, 3, \ldots, r - 1$ the rth operation can be performed in $n - (r - 1) = n - r + 1$ ways. By the generalized multiplication principle, the task of choosing the r objects from among the n can be performed in

$$n(n - 1) \cdot \ldots \cdot (n - r + 1)$$

ways. That is,

$$P(n, r) = n(n - 1) \cdot \ldots \cdot (n - r + 1),$$

which is formula (1).

Let us now verify the formula for $C(n, r)$, the number of combinations of n objects taken r at a time. Each such combination is a set of r objects and therefore can be ordered in

$$P(r, r) = r(r - 1) \cdot \ldots \cdot 2 \cdot 1 = r!$$

ways by formula (1). In other words, each different combination of r objects gives rise to $r!$ permutations of the same r objects. On the other hand, each permutation of n objects taken r at a time gives rise to a combination of n objects taken r at a time, by simply ignoring the order of the permutation. Thus, if we start with the $P(n, r)$ permutations, we will have all the combinations of n objects taken r at a time, with each combination repeated $r!$ times. Thus

$$P(n, r) = r! \, C(n, r).$$

On dividing both sides of the equation by $r!$, we obtain formula (2).

PRACTICE PROBLEMS 5

1. Calculate:

   (a) 3! (b) 7! (c) $C(7, 3)$ (d) $P(7, 3)$

2. A newborn child is to be given a first name and a middle name from a selection of 10 names. How many different possibilities are there?

EXERCISES 5

Evaluate:

1. $P(4, 2)$ 2. $P(5, 1)$ 3. $P(6, 3)$ 4. $P(5, 4)$

5. $C(10, 3)$ 6. $C(12, 2)$ 7. $C(5, 4)$ 8. $C(6, 3)$

9. 6! 10. 10!/4!

11. How many different selections of two books can be made from a set of nine books?

12. A pizza parlor offers five toppings for the plain cheese base of the pizzas. How many different pizzas are possible which use three of the toppings?

13. In how many ways can four people line up in a row for a group picture?

14. How many different outcomes of "winner" and "runner-up" are possible if there are six contestants in a pie-eating contest?

15. If you are going on a trip and decide to take three of your seven sweaters, how many different possibilities are there?

16. A student must choose five courses out of seven that he would like to take. How many possibilities are there?

17. How many different three-letter words are there having no repetition of letters?

18. A sportswriter makes a preseason guess of the top 10 football teams (in order) from among 40 major teams. How many different possibilities are there?

19. In how many different ways can a committee of 15 senators be selected from the 100 members of the U.S. Senate?

20. Theoretically, how many possibilities are there for first, second, and third place in a marathon race with 1000 entries?

21. How many different poker hands are there? (A poker hand consists of five cards.)

22. How many different poker hands consist entirely of aces and kings?

SOLUTIONS TO PRACTICE PROBLEMS 5

1. (a) $3! = 3 \cdot 2 \cdot 1 = 6$ (b) $7! = 7 \cdot 6 \cdot 5 \cdot 4 \cdot 3 \cdot 2 \cdot 1 = 5040$

   (c) $C(7, 3) = \dfrac{7 \cdot 6 \cdot 5}{3 \cdot 2 \cdot 1} = \dfrac{7 \cdot \cancel{6} \cdot 5}{\cancel{3} \cdot \cancel{2} \cdot 1} = 35$

   [A convenient procedure to follow when calculating $C(n, r)$ is to first write the product expansion of $r!$ in the denominator and then to write in the numerator an integer from the descending expansion of $n!$ above each integer in the denominator.]

   (d) $P(7, 3) = \underbrace{7 \cdot 6 \cdot 5}_{3 \text{ factors}} = 210$

   [In general, $P(n, r)$ is the product of the first r factors in the descending expansion of $n!$.]

2. 90. The first question to be asked here is whether permutations or combinations are involved. Two names are to be selected, and the order of the names is important. (The name Amanda Beth is different from the name Beth Amanda.) Since the problem asks for arrangements of 10 names taken 2 at a time in a *specific order*, the number of arrangements if $P(10, 2) = 10 \cdot 9 = 90$. In general, order is important if a different outcome results when two items in the selection are interchanged.

8.6. Further Counting Problems

In the last section we introduced permutations and combinations and developed formulas for counting all permutations (or combinations) of a given type. Many counting problems can be formulated in terms of permutations or combinations. But in order to successfully use the formulas of Section 5, we must be able to recognize these problems when they occur and to translate them into a form in which the formulas may be applied. In this section we practice doing that. We consider three typical applications giving rise to permutations or combinations. The first two applications may seem, at first glance, to have little practical significance. However, they suggest a common way to "model" outcomes of real-life situations having two equally likely results.

As our first application, consider a coin-tossing experiment in which we toss a coin a fixed number of times. We can describe the outcome of the experiment as a sequence of "heads" and "tails." For instance, if a coin is tossed three times, then one possible outcome is "head on first toss, tail on second toss, and tail on third toss." This outcome can be abbreviated as HTT. We can use the methods of the preceding section to count the number of possible outcomes having various prescribed properties.

EXAMPLE 1   Suppose that an experiment consists of tossing a coin 10 times and observing the sequence of heads and tails.

(a) How many different outcomes are possible?

(b) How many different outcomes have exactly four heads?

Solution   (a) Visualize each outcome of the experiment as a sequence of 10 boxes, where each box contains one letter, H or T, with the first box recording the result of the first toss, the second box recording the result of the second toss, and so

H	T	H	T	T	T	H	T	H	T
1	2	3	4	5	6	7	8	9	10

forth. Each box can be filled in two ways. So by the generalized multiplication principle the sequence of 10 boxes can be filled in

$$\underbrace{2 \cdot 2 \cdot \ldots \cdot 2}_{10 \text{ factors}} = 2^{10}$$

ways. So there are $2^{10} = 1024$ different possible outcomes.

(b) An outcome with four heads corresponds to filling the boxes with 4 H's and 6 T's. A particular outcome is determined as soon as we decide where to place the H's. The 4 boxes to receive H's can be selected from the 10 boxes in $C(10, 4)$ ways. So the number of outcomes with 4 heads is

$$C(10, 4) = \frac{10 \cdot 9 \cdot 8 \cdot 7}{4 \cdot 3 \cdot 2 \cdot 1} = 210.$$

Ideas similar to those applied in Example 1 are useful in counting even more complicated sets of outcomes of coin-tossing experiments. The second part of our next example highlights a trick that can often save time and effort.

EXAMPLE 2 Consider the coin-tossing experiment of Example 1.

(a) How many different outcomes have at most two heads?

(b) How many different outcomes have at least three heads?

Solution (a) The outcomes with at most two heads are those having 0, 1, or 2 heads. Let us count the number of these outcomes separately:

0 heads: There is 1 outcome, namely TTTTTTTTTT.

1 head: To determine such an outcome we just select the box in which to put the single H. And this can be done in $C(10, 1) = 10$ ways.

2 heads: To determine such an outcome we just select the boxes in which to put the two H's. And this can be done in $C(10, 2) = (10 \cdot 9)/(2 \cdot 1) = 45$ ways.

Adding up all the possible outcomes, we see that the number of outcomes with at most two heads is equal to $1 + 10 + 45 = 56$.

(b) "At least three heads" refers to an outcome with either 3, 4, 5, 6, 7, 8, 9, or 10 heads. And the total number of such outcomes is

$$C(10, 3) + C(10, 4) + \cdots + C(10, 10).$$

This sum can, of course, be calculated, but there is a less tedious way to solve the problem. Just start with all outcomes [1024 of them by Example 1(a)] and subtract off those with at most two heads [56 of them by part (a)]. So the number of outcomes with at least three heads is $1024 - 56 = 968$.

Let us now turn to a different sort of counting problem, namely one that involves counting the number of paths between two points.

EXAMPLE 3 In Fig. 1 we have drawn a partial map of the streets in a certain city. A tourist wishes to walk from point A to point B. We have drawn two possible routes from A to B. What is the total number of routes (with no backtracking) from A to B?

Solution Any particular route can be described by giving the directions of each block walked in the appropriate order. For instance, the route on the left of Fig. 1 is described as "a block south, a block south, a block east, a block east, a block east, a block

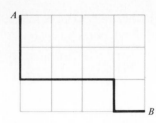 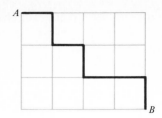

FIGURE 1

south, a block east." Using S for south and E for east, this route can be designated by the string of letters SSEEESE. Similarly, the route on the right is ESESEES. Note that each route is then described by a string of seven letters, of which three are S's (we must go three blocks south) and four E's (we must go four blocks east). Selecting a route is thus the same as placing three S's in a string of seven boxes:

	S		S			S

The three boxes to receive S's can be selected in $C(7, 3) = 35$ ways. So the number of paths from A to B is 35.

Let us now move on to a third type of counting problem. Suppose that we have an urn in which there are a certain number of red balls and a certain number of white balls. We perform an experiment which consists of selecting a number of balls from the urn and observing the color distribution of the sample selected. (This model may be used, for example, to describe the process of selecting people to be polled in a survey. The different colors correspond to different opinions.) By using familiar counting techniques we can calculate the number of possible samples having a given color distribution. The next example illustrates a typical computation.

EXAMPLE 4   An urn contains 25 numbered balls, of which 15 are red and 10 are white. A sample of 5 balls is to be selected.

(a) How many different samples are possible?

(b) How many samples contain all red balls?

(c) How many samples contain 3 red balls and 2 white balls?

(d) How many samples contain at least 4 red balls?

Solution   (a) A sample is just an unordered selection of 5 balls out of 25. There are $C(25, 5)$ such samples. Numerically, we have

$$C(25, 5) = \frac{25 \cdot 24 \cdot 23 \cdot 22 \cdot 21}{5 \cdot 4 \cdot 3 \cdot 2 \cdot 1} = 53{,}130$$

samples.

(b) To form a sample of all red balls we must select 5 balls from the 15 red ones. This can be done in $C(15, 5)$ ways—that is, in

$$C(15, 5) = \frac{15 \cdot 14 \cdot 13 \cdot 12 \cdot 11}{5 \cdot 4 \cdot 3 \cdot 2 \cdot 1} = 3003$$

ways.

(c) To answer this question we use both the multiplication principle and the formula for $C(n, r)$. We form a sample of 3 red balls and 2 white balls using a sequence of two operations:

Select 3 red balls	Select 2 white balls

| Operation 1 | Operation 2 |

The first operation can be performed in $C(15, 3)$ ways and the second in $C(10, 2)$ ways. Thus, the total number of samples having 3 red and 2 white balls is $C(15, 3) \cdot C(10, 2)$. However,

$$C(15, 3) = \frac{15 \cdot 14 \cdot 13}{3 \cdot 2 \cdot 1} = 455$$

$$C(10, 2) = \frac{10 \cdot 9}{2 \cdot 1} = 45$$

$$C(15, 3) \cdot C(10, 2) = 455 \cdot 45 = 20{,}475.$$

So the number of possible samples is 20,475.

(d) A sample with at least 4 red balls has either 4 or 5 red balls. By part (b) the number of samples with 5 red balls is 3003. Using the same reasoning as in part (c), the number of samples with 4 red balls is $C(15, 4) \cdot C(10, 1) = 1365 \cdot 10 = 13{,}650$. Thus the total number of samples having at least 4 red balls is $13{,}650 + 3003 = 16{,}653$.

PRACTICE PROBLEMS 6

1. A newspaper reporter wants an indication of how the 15 members of the school board feel about a certain proposal. She decides to question a sample of 6 of the board members.

   (a) How many different samples are possible?

   (b) Suppose that 10 of the board members support the proposal and 5 oppose it. How many of the samples reflect the distribution of the board? That is, in how many of the samples do 4 people support the proposal and 2 oppose it?

2. A basketball player shoots eight free throws and lists the sequence of results of each trial in order. Let S represent "success" and F represent "failure." Then, for instance, FFSSSSSS represents the outcome of missing the first two shots and hitting the rest.

   (a) How many different outcomes are possible?

   (b) How many of the outcomes have six successes?

EXERCISES 6

1. An experiment consists of tossing a coin six times and observing the sequence of heads and tails.

   (a) How many different outcomes are possible?

   (b) How many different outcomes have exactly three heads?

   (c) How many different outcomes have more heads than tails?

   (d) How many different outcomes have at least two heads?

2. Refer to the map in Fig. 2. How many routes are there from A to B?

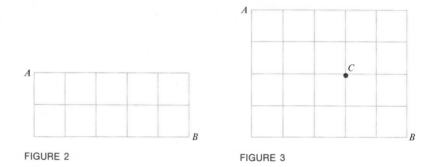

FIGURE 2   FIGURE 3

3. Refer to the map in Fig. 3. How many routes are there from A to B?

4. An urn contains 12 numbered balls, of which 8 are red and 4 are white. A sample of 4 balls is to be selected.

   (a) How many different samples are possible?

   (b) How many samples contain all red balls?

   (c) How many samples contain 2 red balls and 2 white balls?

   (d) How many samples contain at least 3 red balls?

   (e) How many samples contain a different number of red balls than white balls?

5. A bag of 10 apples contains 2 rotten apples and 8 good apples. A shopper selects a sample of 3 apples from the bag.

   (a) How many different samples are possible?

(b) How many samples contain all good apples?

(c) How many samples contain at least 1 rotten apple?

6. An experiment consists of tossing a coin 8 times and observing the sequence of heads and tails.

   (a) How many different outcomes are possible?

   (b) How many different outcomes have exactly 3 heads?

   (c) How many different outcomes have at least 2 heads?

   (d) How many different outcomes have 4 heads or 5 heads?

7. How many ways can a group of 100 students be assigned to dorms A, B, and C, with 25 assigned to dorm A, 40 to dorm B, and 35 to dorm C?

8. In the World Series the American League team ("A") and the National League team ("N") play until one team wins four games. If the sequence of winners is designated by letters (NAAAA means National League won the first game and lost the next four), how many different sequences are possible?

9. Refer to the map in Fig. 3. How many of the routes from A to B pass through the point C?

10. A package contains 100 fuses, of which 10 are defective. A sample of 5 fuses is selected at random.

    (a) How many different samples are there?

    (b) How many of the samples contain 2 defective fuses?

    (c) How many of the samples contain at least 1 defective fuse?

11. In how many ways can a committee of 15 senators be selected from the 100 members of the U.S. Senate so that no two committee members are from the same state?

12. An exam contains five "true or false" questions. How many of the 32 different ways of answering these questions contain 3 or more correct answers?

13. How many poker hands consist of 3 aces and 2 kings?

14. How many poker hands consist of 2 aces, 2 cards of another denomination, and 1 card of a third denomination?

15. How many poker hands consist of 3 cards of one denomination and 2 cards of another denomination? (Such a poker hand is called a "full house.")

16. How many poker hands consist of 2 cards of one denomination, 2 cards of another (different) denomination, and 1 card of a third denomination? (Such a poker hand is called "two pairs.")

SOLUTIONS TO PRACTICE PROBLEMS 6

1. (a) $C(15, 6)$. Each sample is an unordered selection of 15 objects taken 6 at a time.

   (b) $C(10, 4) \cdot C(5, 2)$. Asking for the number of samples of a certain type is the same as asking for the number of ways that the task of forming such a sample can be performed.

This task is composed of two consecutive operations. Operation 1, selecting 4 people from among the 10 that support the proposal, can be performed in $C(10, 4)$ ways. Operation 2, selecting 2 people from among the 5 people that oppose the proposal, can be performed in $C(5, 2)$ ways. Therefore, by the multiplication principle, the complete task can be performed in $C(10, 4) \cdot C(5, 2)$ ways.

[*Note:* $C(15, 6) = 5005$ and $C(10, 4) \cdot C(5, 2) = 2100$. Therefore, less than half of the possible samples reflect the true distribution of the school board.]

2. (a) 2^8. Apply the generalized multiplication principle.

   (b) $C(8, 6)$ or 28. Each outcome having 6 successes corresponds to a sequence of 8 letters of which 6 are S's and 2 are F's. Such an outcome is specified by selecting the 6 locations for the S's from among the 8 locations, and this has $C(8, 6)$ possibilities.

8.7. The Binomial Theorem

In the preceding two sections we have dealt with permutations and combinations and, in particular, have derived a formula for $C(n, r)$, the number of combinations of n objects taken r at a time. Namely, we have

$$C(n, r) = \frac{P(n, r)}{r!} = \frac{n(n - 1) \cdot \ldots \cdot (n - r + 1)}{r!}. \tag{1}$$

Actually, formula (1) was verified in case both n and r are positive integers. But it is useful to consider $C(n, r)$ also in case $r = 0$. In this case we are considering the number of combinations of n things taken 0 at a time. There is clearly only one such combination: the one containing no elements. Therefore,

$$C(n, 0) = 1. \tag{2}$$

Here is another convenient formula for $C(n, r)$:

$$C(n, r) = \frac{n!}{r! \, (n - r)!}. \tag{3}$$

For instance, according to formula (3),

$$C(8, 3) = \frac{8!}{3!(8 - 3)!} = \frac{8!}{3!5!} = \frac{8 \cdot 7 \cdot 6 \cdot \cancel{5} \cdot \cancel{4} \cdot \cancel{3} \cdot \cancel{2} \cdot \cancel{1}}{3 \cdot 2 \cdot 1 \cdot \cancel{5} \cdot \cancel{4} \cdot \cancel{3} \cdot \cancel{2} \cdot \cancel{1}} = \frac{8 \cdot 7 \cdot 6}{3 \cdot 2 \cdot 1}$$

which agrees with the result given by formula (1).

Verification of Formula (3)   Note that

$$n(n - 1) \cdot \ldots \cdot (n - r + 1)$$

$$= \frac{n(n - 1) \cdot \ldots \cdot (n - r + 1)\cancel{(n - r)}\cancel{(n - r - 1)} \cdot \ldots \cdot \cancel{2} \cdot \cancel{1}}{\cancel{(n - r)}\cancel{(n - r - 1)} \cdot \ldots \cdot \cancel{2} \cdot \cancel{1}} = \frac{n!}{(n - r)!}.$$

Then, by formula (1), we have

$$C(n, r) = \frac{n(n-1) \cdot \ldots \cdot (n-r+1)}{r!} = \frac{\dfrac{n!}{(n-r)!}}{r!} = \frac{n!}{r!(n-r)!},$$

which is formula (3).

Note that for $r = 0$, formula (3) reads

$$C(n, 0) = \frac{n!}{0!(n-0)!} = \frac{n!}{0!n!} = \frac{1}{0!}.$$

Let us agree that the value of $0!$ is 1. Then the right-hand side of the equation above is 1, so that formula (3) holds also for $r = 0$.

Formula (3) can be used to prove many facts about $C(n, r)$. For example, here is a formula which is useful in calculating $C(n, r)$ for large values of r:

$$C(n, r) = C(n, n-r). \tag{4}$$

For example, suppose that we wish to calculate $C(100, 98)$. If we apply formula (4), we have

$$C(100, 98) = C(100, 100 - 98) = C(100, 2) = \frac{100 \cdot 99}{2 \cdot 1} = 4950.$$

Verification of Formula (4) Apply formula (3) to evaluate $C(n, n-r)$:

$$C(n, n-r) = \frac{n!}{(n-r)!(n-(n-r))!} = \frac{n!}{(n-r)!r!}$$

$$= C(n, r) \quad \text{[by formula (3) again]}.$$

The formula is intuitively reasonable since each time we select a subset of r elements we are excluding a subset of $n - r$ elements. Thus there are as many subsets of $n - r$ elements as there are subsets of r elements.

An alternative notation for $C(n, r)$ often appears in books, namely $\dbinom{n}{r}$. Thus, for example,

$$\binom{5}{2} = C(5, 2) = \frac{5 \cdot 4}{2 \cdot 1} = 10.$$

The symbol $\dbinom{n}{r}$ is called a *binomial coefficient*. To discover why, let us tabulate the values of $\dbinom{n}{r}$ for some small values of n and r.

$$n = 2: \quad \binom{2}{0} = 1 \quad \binom{2}{1} = 2 \quad \binom{2}{2} = 1$$

$$n = 3: \quad \binom{3}{0} = 1 \quad \binom{3}{1} = 3 \quad \binom{3}{2} = 3 \quad \binom{3}{3} = 1$$

$$n = 4: \quad \binom{4}{0} = 1 \quad \binom{4}{1} = 4 \quad \binom{4}{2} = 6 \quad \binom{4}{3} = 4 \quad \binom{4}{4} = 1$$

$$n = 5: \quad \binom{5}{0} = 1 \quad \binom{5}{1} = 5 \quad \binom{5}{2} = 10 \quad \binom{5}{3} = 10 \quad \binom{5}{4} = 5 \quad \binom{5}{5} = 1.$$

Each row of the table above consists of the coefficients that arise in calculating $(x + y)^n$. To see this, inspect the results of calculating $(x + y)^n$ for $n = 2, 3, 4, 5$:

$$(x + y)^2 = x^2 + 2xy + y^2$$
$$(x + y)^3 = x^3 + 3x^2y + 3xy^2 + y^3$$
$$(x + y)^4 = x^4 + 4x^3y + 6x^2y^2 + 4xy^3 + y^4$$
$$(x + y)^5 = x^5 + 5x^4y + 10x^3y^2 + 10x^2y^3 + 5xy^4 + y^5.$$

Compare the coefficients in any row with the values in the corresponding row of binomial coefficients. Note that they are the same. Thus, we see that the binomial coefficients arise as coefficients in multiplying out powers of the binomial $x + y$; whence the name *binomial coefficient*.

What we observed above for the exponents $n = 2, 3, 4, 5$ holds true for any positive integer n. We have the following result, a proof of which is given at the end of this section.

Binomial Theorem

$$(x+y)^n = \binom{n}{0}x^n + \binom{n}{1}x^{n-1}y + \binom{n}{2}x^{n-2}y^2 + \cdots + \binom{n}{n-1}xy^{n-1} + \binom{n}{n}y^n.$$

EXAMPLE 1 Calculate $(x + y)^6$.

Solution By the binomial theorem

$$(x + y)^6 = \binom{6}{0}x^6 + \binom{6}{1}x^5y + \binom{6}{2}x^4y^2 + \binom{6}{3}x^3y^3$$
$$+ \binom{6}{4}x^2y^4 + \binom{6}{5}xy^5 + \binom{6}{6}y^6.$$

Furthermore,

$$\binom{6}{0} = 1 \qquad \binom{6}{1} = \frac{6}{1} = 6 \qquad \binom{6}{2} = \frac{6 \cdot 5}{2 \cdot 1} = 15$$

$$\binom{6}{3} = \frac{6 \cdot 5 \cdot 4}{3 \cdot 2 \cdot 1} = 20 \qquad \binom{6}{4} = \frac{6 \cdot 5 \cdot \cancel{4} \cdot \cancel{3}}{\cancel{4} \cdot \cancel{3} \cdot 2 \cdot 1} = 15$$

$$\binom{6}{5} = \frac{6 \cdot \cancel{5} \cdot \cancel{4} \cdot \cancel{3} \cdot \cancel{2}}{\cancel{5} \cdot \cancel{4} \cdot \cancel{3} \cdot 2 \cdot 1} = 6 \qquad \binom{6}{6} = \frac{\cancel{6} \cdot \cancel{5} \cdot \cancel{4} \cdot \cancel{3} \cdot \cancel{2} \cdot \cancel{1}}{\cancel{6} \cdot \cancel{5} \cdot \cancel{4} \cdot \cancel{3} \cdot \cancel{2} \cdot \cancel{1}} = 1.$$

Thus,

$$(x + y)^6 = x^6 + 6x^5y + 15x^4y^2 + 20x^3y^3 + 15x^2y^4 + 6xy^5 + y^6.$$

The binomial theorem can be used to count the number of subsets of a set, as shown in the next example.

EXAMPLE 2 Determine the number of subsets of a set with five elements.

Solution Let us count the number of subsets of each possible size. A subset of r elements can be chosen in $\binom{5}{r}$ ways, since $C(5, r) = \binom{5}{r}$. So the set has $\binom{5}{0}$ subsets with 0 elements, $\binom{5}{1}$ subsets with 1 element, $\binom{5}{2}$ subsets with 2 elements, and so on. Therefore, the total number of subsets is

$$\binom{5}{0} + \binom{5}{1} + \binom{5}{2} + \binom{5}{3} + \binom{5}{4} + \binom{5}{5}.$$

On the other hand, the binomial theorem for $n = 5$ gives

$$(x + y)^5 = \binom{5}{0}x^5 + \binom{5}{1}x^4y + \binom{5}{2}x^3y^2 + \binom{5}{3}x^2y^3 + \binom{5}{4}xy^4 + \binom{5}{5}y^5.$$

Set $x = 1$ and $y = 1$ in this formula, deriving

$$(1 + 1)^5 = \binom{5}{0}1^5 + \binom{5}{1}1^4 \cdot 1 + \binom{5}{2}1^3 \cdot 1^2 + \binom{5}{3}1^2 \cdot 1^3 + \binom{5}{4}1 \cdot 1^4 + \binom{5}{5}1^5$$

$$2^5 = \binom{5}{0} + \binom{5}{1} + \binom{5}{2} + \binom{5}{3} + \binom{5}{4} + \binom{5}{5}.$$

Thus the total number of subsets of a set with five elements (the right side) equals $2^5 = 32$.

There is nothing special about the number five in the preceding example. An analogous argument gives the following result.

> A set of n elements has 2^n subsets.

EXAMPLE 3   A pizza parlor offers a plain cheese pizza to which any number of six possible toppings can be added. How many different pizzas can be ordered?

Solution   Ordering a pizza requires selecting a subset of the six possible toppings. Since the set of six toppings has 2^6 different subsets, there are 2^6 or 64 different pizzas. (Note that the plain cheese pizza corresponds to selecting the empty subset of toppings.)

Proof of the Binomial Theorem   Note that

$$(x + y)^n = \underbrace{(x + y)(x + y) \cdot \ldots \cdot (x + y)}_{n \text{ factors}}.$$

Multiplying out these factors involves forming all products, where one term is selected from each factor, and then combining like products. For instance,

$$(x + y)(x + y)(x + y) = x \cdot x \cdot x + x \cdot x \cdot y + x \cdot y \cdot x + y \cdot x \cdot x + x \cdot y \cdot y$$
$$+ y \cdot x \cdot y + y \cdot y \cdot x + y \cdot y \cdot y.$$

The first product on the right, $x \cdot x \cdot x$, is obtained by selecting the x-term from each of the three factors. The next term, $x \cdot x \cdot y$, is obtained by selecting the x-terms from the first two factors and the y-term from the third. The next product, $x \cdot y \cdot x$, is obtained by selecting the x-terms from the first and third factors and the y-term from the second. And so on. There are as many products containing two x's and one y as there are ways of selecting the factor from which to pick the y-term—namely $\binom{3}{1}$.

In general, when multiplying the n factors $(x + y)(x + y) \cdots (x + y)$, the number of products having k y's (and therefore $n - k$ x's) is equal to the number of different ways of selecting the k factors from which to take the y-term—that is, $\binom{n}{k}$. Therefore, the coefficient of $x^{n-k}y^k$ is $\binom{n}{k}$. This proves the binomial theorem.

PRACTICE PROBLEMS 7

1. Calculate $\binom{12}{8}$.

2. An ice cream parlor offers 10 flavors of ice cream and 5 toppings. How many different servings are possible if each choice consists of one flavor of ice cream and as many toppings as desired?

EXERCISES 7

Calculate:

1. $\begin{pmatrix} 6 \\ 2 \end{pmatrix}$ 2. $\begin{pmatrix} 20 \\ 18 \end{pmatrix}$ 3. $0!$ 4. $\begin{pmatrix} 9 \\ 0 \end{pmatrix}$ 5. $\begin{pmatrix} 18 \\ 15 \end{pmatrix}$

6. $1!$ 7. $\begin{pmatrix} 500 \\ 0 \end{pmatrix}$ 8. $\begin{pmatrix} 25 \\ 24 \end{pmatrix}$ 9. $(x + y)^7$ 10. $(x + y)^8$

11. Calculate the first three terms in the binomial expansion of $(x + y)^{10}$.

12. Calculate the first three terms in the binomial expansion of $(x + y)^{20}$.

13. How many different subsets can be chosen from a set of six elements?

14. How many different subsets can be chosen from a set of 100 elements?

15. How many different tips could you leave in a restaurant if you had a nickel, a dime, a quarter, and a half-dollar?

16. A pizza parlor offers mushroom, green peppers, onions, and sausage as topping for the plain cheese base. How many different types of pizzas can be made?

17. An ice cream parlor offers four flavors of ice cream, three toppings, and two sizes of glasses. How many different servings consisting of a single flavor of ice cream are possible?

18. A salad bar offers a base of lettuce to which tomatoes, chick-peas, beets, pinto beans, olives, and green peppers can be added. How many different salads are possible?

19. How many batting orders are possible in a nine-member baseball team if the catcher must bat fourth and the pitcher last?

20. How many ways can a selection of at least one book be made from a set of eight books?

21. In how many ways can a committee of 5 people be chosen from 12 married couples if:

    (a) The committee must consist of two men and three women?

    (b) A man and his wife cannot both serve on the committee?

22. Suppose that you are voting in an election for state delegate. Two state delegates are to be elected from among seven candidates. In how many different ways can you cast your ballot? (*Note:* You may vote for two candidates. However, some people "single-shoot" and others don't pull any levers.)

23. Show that, for n a positive integer, we have

$$\begin{pmatrix} n \\ 0 \end{pmatrix} + \begin{pmatrix} n \\ 1 \end{pmatrix} + \cdots + \begin{pmatrix} n \\ n \end{pmatrix} = 2^n$$

[*Hint:* Write down the binomial theorem for $(1 + 1)^n$.]

24. Show that, for n a positive integer, we have

$$\begin{pmatrix} n \\ 0 \end{pmatrix} - \begin{pmatrix} n \\ 1 \end{pmatrix} + \cdots \pm \begin{pmatrix} n \\ n \end{pmatrix} = 0.$$

[*Hint:* Write down the binomial theorem for $(1 + (-1))^n$.]

25. Show that, for n, k positive integers, we have

$$\binom{n}{k-1} + \binom{n}{k} = \binom{n+1}{k}.$$

$$\left[Hint: \binom{n}{k-1} + \binom{n}{k} = \frac{n!}{(k-1)!(n-k)!}\left(\frac{1}{n-k+1} + \frac{1}{k}\right).\right]$$

26. In the following table (known as Pascal's triangle) the entries in the nth row are $\binom{n}{0}$, $\binom{n}{1}$, $\binom{n}{2}$, ..., $\binom{n}{n}$. Use Exercise 25 to explain how we can use each row to calculate the row below it by using only addition. Complete the first eight rows of Pascal's triangle.

$$
\begin{array}{ccccccc}
1 & 1 & & & & & \\
1 & 2 & 1 & & & & \\
1 & 3 & 3 & 1 & & & \\
1 & 4 & 6 & 4 & 1 & & \\
1 & 5 & 10 & 10 & 5 & 1 & \\
\vdots & & & & & &
\end{array}
$$

SOLUTIONS TO PRACTICE PROBLEMS 7

1. 495. $\binom{12}{8}$ is the same as $C(12, 8)$, which equals $C(12, 12 - 8)$ or $C(12, 4)$.

$$C(12, 4) = \frac{12 \cdot 11 \cdot 10 \cdot 9}{4 \cdot 3 \cdot 2 \cdot 1} = \frac{\cancel{12} \cdot 11 \cdot \overset{5}{\cancel{10}} \cdot 9}{\cancel{4} \cdot \cancel{3} \cdot \cancel{2} \cdot 1} = 495.$$

2. 320. The task of deciding what sort of serving to have consists of two operations. The first operation, selecting the flavor of ice cream, can be performed in 10 ways. The second operation, selecting the toppings, can be performed in 2^5 or 32 ways, since selecting the topping amounts to selecting a subset from the set of 5 toppings and a set of 5 elements has 2^5 subsets. (Notice that selecting the empty subset corresponds to ordering a plain dish of ice cream.) By the multiplication principle, the task can be performed in $10 \cdot 32 = 320$ ways.

Chapter 8: CHECKLIST

☐ Set
☐ Element
☐ Union and intersection of sets
☐ Subset
☐ Universal set
☐ Complement of a set
☐ Inclusion-exclusion principle
☐ Venn diagram
☐ Generalized multiplication principle

□ Permutations
□ Combinations
□ $P(n, r) = n(n - 1) \cdots (n - r + 1)$
□ $n!$
□ $C(n, r) = \dfrac{P(n, r)}{r!} = \dfrac{n!}{r! \, (n - r)!}$
□ $C(n, r) = C(n, n - r)$
□ $\dbinom{n}{r}$
□ Binomial theorem
□ Number of subsets of a set

Chapter 8: SUPPLEMENTARY EXERCISES

1. List all subsets of the set $\{a, b\}$.

2. Draw a two-circle Venn diagram and shade the portion corresponding to the set $(S \cup T')'$.

3. There are 16 contestants in a tennis tournament. How many different possibilities are there for the two people who will play in the final round?

4. In how many ways can a coach and five basketball players line up in a row for a picture if the coach insists on standing at one of the ends of the row?

5. Draw a three-circle Venn diagram and shade the portion corresponding to the set $R' \cap (S \cup T)$.

6. Calculate the first three terms in the binomial expansion of $(x + y)^{12}$.

7. An urn contains 14 numbered balls of which 8 are red and 6 are green. How many different possibilities are there for selecting a sample of 5 balls in which 3 are red and 2 are green?

8. Sixty people with a certain medical condition were given pills. Fifteen of these people received placebos. Forty people showed improvement, and 30 of these people received an actual drug. How many of the people who received the drug showed no improvement?

9. An appliance store carries seven different types of washing machines and five different types of dryers. How many different combinations are possible for a customer who wants to purchase a washing machine and a dryer?

10. There are 12 contestants in a contest. Two will receive trips around the world, 4 will receive cars, and 6 will receive color TV sets. In how many different ways can the prizes be awarded?

11. Out of a group of 115 applicants for jobs at the World Bank, 70 speak French, 65 speak Spanish, 65 speak German, 45 speak French and Spanish, 35 speak Spanish and German, 40 speak French and German, and 35 speak all three languages. How many of the people speak none of the three languages?

12. Calculate $\dbinom{17}{15}$.

13. How many different nine-letter words (i.e., sequences of letters) can be made using four S's and five T's?

14. Forty people take an exam. How many different possibilities are there for the set of people who pass the exam?

15. A survey at a small New England college showed that 400 students skied, 300 played ice hockey, and 150 did both. How many students participated in at least one of these sports?

16. A poker hand consists of five cards. How many different poker hands contain all cards of the same suit? (Such a hand is called a "flush.")

Probability

Chapter 9

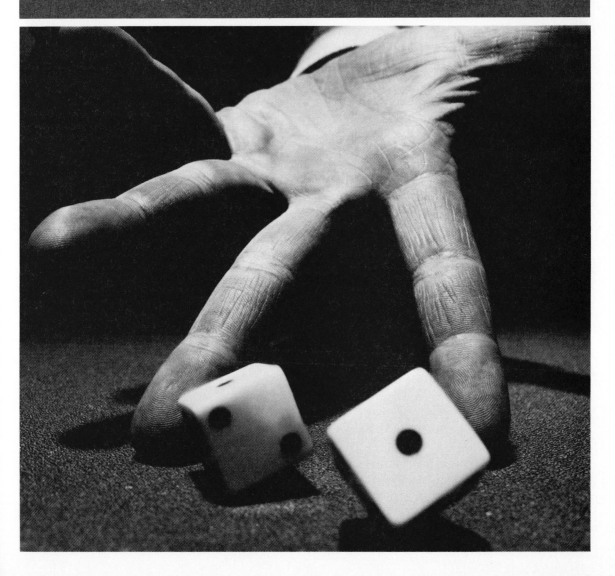

9.1. Introduction

Many events in the world around us exhibit a random character. Yet by repeated observations of such events we can often discern long-run patterns which persist despite random, short-term fluctuations. Probability is the branch of mathematics devoted to a study of such events. To obtain a clearer idea of the sort of events considered, let us discuss a concrete example from the field of medicine.

Suppose that we wish to analyze the reliability of a skin test for active pulmonary tuberculosis. Unfortunately, such a test is not completely reliable. On the one hand, the test may be negative even for a person with tuberculosis. On the other hand, the test may be positive for a person who does not have tuberculosis. For the moment, let us concentrate on errors of the first sort and consider only individuals actually having tuberculosis. Suppose that by observing the results of the test on increasingly large populations of tuberculosis patients we accumulate the following data.

Number of tuberculosis patients (N)	Number of positive test results (m)	Relative frequency of positive test results (m/N)
100	97	.97
500	494	.988
1,000	981	.981
10,000	9,806	.9806
50,000	49,008	.98016
100,000	98,005	.98005

Note that out of each group of tuberculosis patients the test fails to identify a certain number. However, the data does exhibit a pattern. It appears that out of a very large population of tuberculosis patients the skin test will successfully identify about 98% of them. In fact, it appears that as the size of the population is increased, the relative frequency m/N more and more closely approximates the number .98. In a situation like this, we say that the skin test detects tuberculosis with a 98% likelihood or that the *probability* that the test detects tuberculosis (when present) is .98.

More generally, the *probability of an event* is a number which expresses the long-run likelihood that the event will occur. Such numbers are always chosen to lie between 0 and 1. The smaller the probability, the less likely the event is to occur. So, for example, an event having probability .1 is rather unlikely to occur; an event with probability .9 is very likely to occur; and an event with probability .5 is just as likely to occur as not.

We assign probabilities to events on the following intuitive basis: The probability of an event should represent the long-run proportion of the time that the event can be expected to occur. Thus, for example, an event with probability .9 can be expected to occur 90% of the time and an event with probability .1 can be expected to occur 10% of the time.

As we shall see, many real-life problems require us to calculate probabilities from known data. Here is one example which arises in connection with the skin test for tuberculosis.

> **MEDICAL DIAGNOSIS**   A clinic tests for active pulmonary tuberculosis. If a person has tuberculosis, the probability of a positive test result is .98. If a person does not have tuberculosis, the probability of a negative test result is .99. The incidence of tuberculosis in a certain city is 2 cases per 10,000 population. Suppose that an individual is tested and a positive result is noted. What is the probability that that individual actually has active pulmonary tuberculosis?

Before we can solve this problem, it will be necessary to do considerable preliminary work. We begin this work in Section 2, where we introduce a convenient language for discussing events and the process of observing them. In Section 3 we introduce probabilities of events, and in Sections 4, 5, 6 and 7 we develop methods for calculating probabilities of various sorts of events. The solution of the medical diagnosis problem is presented in Section 6.

9.2. Experiments, Outcomes, and Events

The events whose probabilities we wish to compute all arise as outcomes of various experiments. So as our first step in developing probability theory let us describe, in mathematical terms, the notions of experiment, outcomes, and event.

For our purposes an *experiment* is an activity with an observable outcome. Here are some typical examples of experiments.

> **EXPERIMENT 1**   Flip a coin and observe the side that faces upward.
>
> **EXPERIMENT 2**   Allow a conditioned rat to run a maze and observe which one of the three possible paths it takes.

Choose a year and tabulate the amount of rainfall in New York City during that year.

We think of an experiment as being performed repeatedly. Each repetition of the experiment is called a *trial*. In each trial we observe the *outcome* of the experiment. For example, a possible outcome of Experiment 1 is "heads"; a possible outcome of Experiment 2 is "path 3"; and a possible outcome of Experiment 3 is "37.23 inches."

In order to describe an experiment in mathematical language, it is most convenient to form the set consisting of all possible outcomes of the experiment. This set is called the *sample space* of the experiment. For example, if S_1, S_2, S_3 are the sample spaces for Experiments 1, 2, 3, respectively, then we immediately see that

$$S_1 = \{\text{heads, tails}\}$$

$$S_2 = \{\text{path 1, path 2, path 3}\}.$$

Moreover, since any nonnegative number is a candidate for the amount of rainfall, we have

$$S_3 = \{\text{all numbers} \geq 0\}.$$

We can describe an experiment in terms of the sample space as follows:

> Suppose that an experiment has a sample space S. Then each trial has as its outcome one of the elements of S.

Thus, for example, each trial for Experiment 1 has as its outcome one of the elements of the set

$$S_1 = \{\text{heads, tails}\}.$$

Henceforth, we shall always describe experiments in terms of their respective sample spaces. So it is important to be able to recognize the appropriate sample space in each instance. The next few examples should help you obtain the necessary facility in doing this.

EXAMPLE 1 An experiment consists of tossing a die and observing the number on the uppermost face. Describe the sample space S for this experiment.

Solution There are six outcomes of the experiment, corresponding to the six possible numbers on the uppermost face. Therefore,

$$S = \{1, 2, 3, 4, 5, 6\}.$$

EXAMPLE 2 Once an hour a supermarket manager observes the number of people standing in checkout lines. The store has space for at most 30 customers to wait in line. What is the sample space S for this experiment?

Solution   The outcome of the experiment is the number of people standing in checkout lines. And this number may be 0, 1, 2, . . . , or 30. Therefore,

$$S = \{0, 1, 2, \ldots, 30\}.$$

EXAMPLE 3   An experiment consists of throwing two dice, one red and one green, and observing the uppermost face on each. What is the associated sample space S?

Solution   Each outcome of the experiment can be regarded as an ordered pair of numbers, the first representing the number on the red die and the second the number on the green die. Thus, for example, the pair of numbers $(3, 5)$ represents the outcome "3 on the red die, 5 on the green die." The sample space consists of all possible pairs of numbers (r, g), where r and g are each one of the numbers 1, 2, 3, 4, 5, 6. This sample space has 36 elements:

$$S = \{(1, 1), (1, 2), (1, 3), (1, 4), (1, 5), (1, 6),$$
$$(2, 1), (2, 2), (2, 3), (2, 4), (2, 5), (2, 6),$$
$$(3, 1), (3, 2), (3, 3), (3, 4), (3, 5), (3, 6),$$
$$(4, 1), (4, 2), (4, 3), (4, 4), (4, 5), (4, 6),$$
$$(5, 1), (5, 2), (5, 3), (5, 4), (5, 5), (5, 6),$$
$$(6, 1), (6, 2), (6, 3), (6, 4), (6, 5), (6, 6)\}.$$

EXAMPLE 4   The Environmental Protection Agency orders Middle States Edison Corporation to install "scrubbers" to remove pollutants from its smokestacks. To monitor the effectiveness of the scrubbers, the corporation installs monitoring devices to record the levels of sulfur dioxide, oxides of nitrogen, and particulate matter (in parts per million) in the smokestack emissions. Consider the monitoring operation as an experiment. Describe the associated sample space.

Solution   Each reading of the instruments consists of an ordered triple of numbers (x, y, z), where $x = $ the level of sulfur dioxide, $y = $ the level of oxides of nitrogen, and $z = $ the level of particulate matter. The sample space thus consists of all possible triples (x, y, z), where $x \geq 0, y \geq 0, z \geq 0$.

The sample spaces in Examples 1 to 3 are *finite*. That is, the associated experiments have only a finite number of possible outcomes. However, the sample space of Example 4 is *infinite*, since there are infinitely many triples (x, y, z), $x \geq 0, y \geq 0, z \geq 0$.

Now that we have discussed experiments and their outcomes, let us turn our attention to the notion of "event." In connection with our preceding discussion, we can define many events whose probabilities we might wish to know. For example, in connection with experiment 2, we can consider the event

"A conditioned rat chooses either path 2 or path 3."

Here are two events associated with experiment 3:

"The annual rainfall in New York City exceeds 50 inches."
"The annual rainfall in New York City is less than 35 inches."

It is easy to describe events in terms of the sample space. For example, let us consider the die-tossing experiment of Example 1 and the following events.

I. An even number occurs.
II. A number greater than 2 occurs.

We saw above that the sample space S for this experiment is

$$S = \{1, 2, 3, 4, 5, 6\}.$$

Assume that the experiment is performed. Then event I occurs precisely when the outcome of the experiment is 2, 4, or 6. That is, the event I occurs precisely when the outcome belongs to the set

$$E_I = \{2, 4, 6\}.$$

Note that this set is a subset of the sample space S. Similarly, we can describe event II by the set

$$E_{II} = \{3, 4, 5, 6\}.$$

Event II occurs precisely when the outcome of the experiment is an element of E_{II}.

The sets E_I and E_{II} contain all the information we need in order to completely describe the events I and II. This observation suggests the following definition of an event in terms of the sample space.

An *event E* is a subset of the sample space. We say that the event *occurs* when the outcome of the experiment is an element of E.

The next few examples provide some practice in describing events as subsets of the sample space.

EXAMPLE 5   Consider the supermarket of Example 2. Describe the following events as subsets of the sample space.

(a) Fewer than 5 people are waiting in line.

(b) More than 23 people are waiting in line.

(c) No people are waiting in line.

Solution   We saw that the sample space for this example is given by

$$S = \{0, 1, 2, \ldots, 30\}.$$

(a) If fewer than five people are waiting in line, then the number waiting is 0, 1, 2, 3, or 4. So the subset of S corresponding to event (a) is

$$\{0, 1, 2, 3, 4\}.$$

(b) In this case the number waiting must be 24, 25, 26, 27, 28, 29, or 30. So the event is just the subset

$$\{24, 25, 26, 27, 28, 29, 30\}.$$

(c) $\{0\}$.

EXAMPLE 6   Suppose that an experiment consists of tossing a coin three times and observing the sequence of heads and tails. (Order counts.)

(a) Determine the sample space S.

(b) Determine the event $E =$ "exactly two heads."

Solution   (a) Denote "heads" by H and "tails" by T. Then a typical outcome of the experiment is a sequence of H's and T's. So, for instance, the sequence HTT would stand for a head followed by two tails. We exhibit all such sequences and arrive at the sample space S:

$$S = \{\text{HHH, HHT, HTH, THH, HTT, THT, TTH, TTT}\}.$$

(b) Here are the outcomes in which exactly two heads occur: HHT, HTH, THH. Therefore, the event E is

$$E = \{\text{HHT, HTH, THH}\}.$$

EXAMPLE 7   A political poll surveys a group of people to determine their income levels and political affiliations. People are classified as either low-, middle-, or upper-level income, and as either Democrat, Republican, or Independent.

(a) Find the sample space corresponding to the poll.

(b) Determine the event $E_1 =$ "Independent."

(c) Determine the event $E_2 =$ "low income and not Independent."

(d) Determine the event $E_3 =$ "neither upper income nor Independent."

Solution   (a) Let us abbreviate low, middle, and upper income, respectively, by the letters L, M, and U, respectively. And let us abbreviate Democrat, Republican, and Independent by the letters D, R, and I, respectively. Then a response to the poll can be represented as a pair of letters. For example, the pair (L, D) refers to a lower-income-level Democrat. The sample space S is then given by

$$S = \{(\text{L, D}), (\text{L, R}), (\text{L, I}), (\text{M, D}), (\text{M, R}), (\text{M, I}), (\text{U, D}), (\text{U, R}), (\text{U, I})\}.$$

(b) For the event E_1 the income level may be anything, but the political affiliation is Independent. Thus,

$$E_1 = \{(\text{L, I}), (\text{M, I}), (\text{U, I})\}.$$

(c) For the event E_2 the income level is low and the political affiliation may be either Democrat or Republican, so that

$$E_2 = \{(L, D), (L, R)\}.$$

(d) For the event E_3 the income level may be either low or middle and the political affiliation may be Democrat or Republican. This,

$$E_3 = \{(L, D), (M, D), (L, R), (M, R)\}.$$

As we have seen, an event is a subset of the sample space. Two events are worthy of special mention. The first is the event corresponding to the empty set, $\varnothing$. This is called the *impossible event*, since it can never occur. The second special event is the set S itself. Every outcome is an element of S, so S always occurs. For this reason S is called the *certain event*.

One particular advantage of defining experiments and events in terms of sets is that it allows us to define new events from given ones by applying the operations of set theory. When so doing we always let the sample space S play the role of universal set. (All outcomes belong to the universal set.)

If E and F are events, then so are $E \cup F$, $E \cap F$, and E'. For example, consider the die-tossing experiment of Example 1. Then

$$S = \{1, 2, 3, 4, 5, 6\}.$$

Let E and F be the events given by

$$E = \{3, 4, 5, 6\} \qquad F = \{1, 4, 6\}.$$

Then we have

$$E \cup F = \{1, 3, 4, 5, 6\}$$

$$E \cap F = \{4, 6\}$$

$$E' = \{1, 2\}.$$

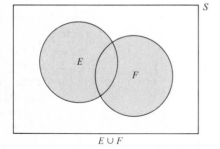

$E \cup F$

FIGURE 1

Let us interpret the events $E \cup F$, $E \cap F$, and E' using Venn diagrams. In Fig. 1 we have drawn a Venn diagram for $E \cup F$. Note that $E \cup F$ occurs precisely when the experimental outcome belongs to the shaded region—that is, to either E or F. Thus we have the following result.

The event $E \cup F$ occurs precisely when either E or F (or both) occurs.

Similarly, we can interpret the event $E \cap F$. This event occurs when the experimental outcome belongs to the shaded region of Fig. 2—that is, to both E and F. Thus, we have an interpretation for $E \cap F$:

The event $E \cap F$ occurs precisely when both E and F occur.

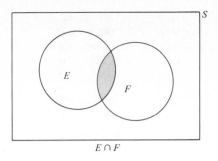

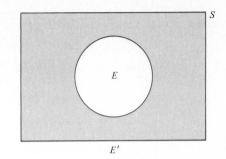

$$E \cap F$$

FIGURE 2

$$E'$$

FIGURE 3

Finally, the event E' consists of all those outcomes not in E (Fig. 3). Therefore, we have:

> The event E' occurs precisely when E does not occur.

EXAMPLE 8   Consider the pollution monitoring described in Example 4. Let E, F, and G be the events

$$E = \text{"level of sulfur dioxide} \geq 100,\text{"}$$

$$F = \text{"level of particulate matter} \leq 50,\text{"}$$

$$G = \text{"level of oxides of nitrogen} \leq 30.\text{"}$$

Describe the following events.

(a) $E \cap F$ (b) E' (c) $E \cup G$ (d) $E' \cap F \cap G$

Solution   (a) $E \cap F = \text{"level of sulfur dioxide} \geq 100$ *and* particulate matter $\leq 50.\text{"}$

(b) $E' = \text{"level of sulfur dioxide} < 100.\text{"}$

(c) $E \cup G = \text{"level of sulfur dioxide} \geq 100$ *or* oxides of nitrogen $\leq 30.\text{"}$

(d) $E' \cap F \cap G = \text{"level of sulfur dioxide} < 100$ *and* particulate matter ≤ 50 *and* oxides of nitrogen $\leq 30.\text{"}$

FIGURE 4

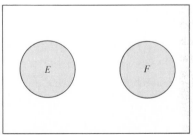

E and F are mutually exclusive

Suppose that E and F are events in a sample space S. We say that E and F are *mutually exclusive (or disjoint)* provided that $E \cap F = \emptyset$. In terms of Venn diagrams, we may represent a pair of mutually exclusive events as a pair of circles with no points in common (Figs. 4). If the events E and F are mutually exclusive, then E and F cannot simultaneously occur; if E occurs, then F does not; and if F occurs, then E does not.

EXAMPLE 9   Let $S = \{a, b, c, d, e, f, g\}$ be a sample space and let $E = \{a, b, c\}$, $F = \{e, f, g\}$, $G = \{c, d, f\}$.

(a) Are E and F mutually exclusive?

(b) Are F and G mutually exclusive?

Solution   (a) $E \cap F = \emptyset$, so E and F are mutually exclusive.

(b) $F \cap G = \{f\}$, so F and G are *not* mutually exclusive.

PRACTICE PROBLEMS 2

1. A machine produces light bulbs. As part of a quality control procedure, a sample of five light bulbs is collected each hour and the number of defective light bulbs among these is observed.

   (a) What is the sample space for this experiment?

   (b) Describe the event "there are at most two defective light bulbs" as a subset of the sample space.

2. Suppose that there are two crates of citrus fruit and each crate contains oranges, grapefruit, and tangelos. An experiment consists of selecting a crate and then selecting a piece of fruit from that crate. Both the crate and the type of fruit are noted. Refer to the crates as crate I and crate II.

   (a) What is the sample space for this experiment?

   (b) Describe the event "a tangelo is selected" as a subset of the sample space.

EXERCISES 2

1. A committee of two people is to be selected from five people, R, S, T, U, and V.

   (a) What is the sample space for this experiment?

   (b) Describe the event "R is on the committee" as a subset of the sample space.

   (c) Describe the event "neither R nor S is on the committee" as a subset of the sample space.

2. A letter is selected at random from the word "MISSISSIPPI."

   (a) What is the sample space for this experiment?

   (b) Describe the event "the letter chosen is a vowel" as a subset of the sample space.

3. An experiment consists of tossing a coin two times and observing the sequence of heads and tails.

   (a) What is the sample space of this experiment?

   (b) Describe the event "the first toss is a head" as a subset of the sample space.

4. A campus survey is taken to correlate the number of years that students have been on campus with their political leaning. Students are classified as freshman, sophomore, junior, or senior and as conservative or liberal.

    (a) Find the sample space corresponding to the poll.

    (b) Determine the event E_1 = "conservative."

    (c) Determine the event E_2 = "junior and liberal."

    (d) Determine the event E_3 = "neither freshman nor conservative."

5. Suppose that we have two urns—call them urn I and urn II—each containing red balls and white balls. An experiment consists of selecting an urn and then selecting a ball from that urn and noting its color.

    (a) What is a suitable sample space for this experiment?

    (b) Describe the event "Urn I is selected" as a subset of the sample space.

6. An experiment consists of tossing a coin four times and observing the sequence of heads and tails.

    (a) What is the sample space of this experiment?

    (b) Determine the event E_1 = "more heads than tails occur."

    (c) Determine the event E_2 = "the first toss is a head."

    (d) Determine the event $E_1 \cap E_2$.

7. A corporation efficiency expert records the time it takes an assembly line worker to perform a particular task. Let E be the event "more than 5 minutes," F the event "less than 8 minutes," and G the event "less than 4 minutes."

    (a) Describe the sample space for this experiment.

    (b) Describe the events $E \cap F, E \cap G, E', F', E' \cap F, E' \cap F \cap G, E \cup F$.

8. A manufacturer of kitchen appliances tests the reliability of its refrigerators by recording in a laboratory test the elapsed time between consecutive failures. Let E be the event "more than nine months," F the event "less than two years." Describe the events $E \cap F$, $E \cup F, E', F', (E \cup F)'$.

9. Let $S = \{1, 2, 3, 4, 5, 6\}$ be a sample space, $E = \{1, 2\}, F = \{2, 3\}, G = \{1, 5, 6\}$.

    (a) Are E and F mutually exclusive?

    (b) Are F and G mutually exclusive?

10. Show that if E is any event, E' its complement, then E and E' are mutually exclusive.

11. Let $S = \{a, b, c\}$ be a sample space. Determine all possible events associated with S.

12. Let S be a sample space with n outcomes. How many events are associated with S?

13. Let $S = \{1, 2, 3, 4\}$ be a sample space, $E = \{1\}, F = \{2, 3\}$. Are the events $E \cup F$ and $E' \cap F'$ mutually exclusive?

14. Let S be any sample space, E, F any events associated with S. Are the events $E \cup F$ and $E' \cap F'$ mutually exclusive? [*Hint:* Apply De Morgan's laws.]

15. Suppose that 10 coins are tossed and the number of heads observed.

   (a) Describe the sample space for this experiment.

   (b) Describe the event "more heads than tails" in terms of the sample space.

16. Suppose that five nickels and five dimes are tossed and the numbers of heads from each group recorded.

   (a) Describe the sample space for this experiment.

   (b) Describe the event "more heads on the nickels than on the dimes" in terms of the sample space.

17. An experiment consists of observing the eye color and sex of the students at a certain school. Let E be the event "blue eyes," F the event "male," G the event "brown eyes and female."

   (a) Are E and F mutually exclusive?

   (b) Are E and G mutually exclusive?

   (c) Are F and G mutually exclusive?

18. Consider the experiment and events of Exercise 17. Describe the following events.
   (a) $E \cup F$ (b) $E \cap G$ (c) E' (d) F' (e) $(G \cup F) \cap E$ (f) $G' \cap E$

19. Suppose that you observe the length of the line at a fast-food restaurant. Describe the sample space.

20. Suppose that you observe the time to be served at a fast-food restaurant. Describe the sample space.

21. Suppose that you observe the time between customer arrivals at a fast-food restaurant. Describe the sample space.

22. Suppose that you observe the length of the line when a customer arrives and the length of time it takes for him to be served in a fast-food restaurant. Describe the sample space.

SOLUTIONS TO PRACTICE PROBLEMS 2

1. (a) $\{0, 1, 2, 3, 4, 5\}$. The sample space is the set of all outcomes of the experiment. At first glance it might seem that each outcome is a set of five light bulbs. What is observed, however, is not the specific sample but rather the number of defective bulbs in the sample. Therefore, the outcome must be a number.

   (b) $\{0, 1, 2\}$. "At most 2" means "2 or less."

2. (a) {(crate I, orange), (crate I, grapefruit), (crate I, tangelo), (crate II, orange), (crate II, grapefruit), (crate II, tangelo)}. Two selections are being made and should both be recorded.

   (b) {(crate I, tangelo), (crate II, tangelo)}. This set consists of those outcomes in which a tangelo is selected.

9.3. Assignment of Probabilities

In the last section we introduced the sample space of an experiment and used it to describe events. We complete our description of experiments by introducing probabilities associated to events. For the remainder of this chapter* let us limit our discussion to experiments with only a finite number of outcomes.

Suppose that an experiment has a sample space S consisting of a finite number of outcomes $s_1, s_2, \ldots, s_N$. To each outcome we associate a number, called the *probability of the outcome*, which measures the relative likelihood that the outcome will occur. Suppose that to the outcome s_1 we associate the probability p_1, to the outcome s_2 the probability p_2, and so forth. We can summarize these data in a chart of the following sort:

Outcome	Probability
s_1	p_1
s_2	p_2
$\vdots$	$\vdots$
s_N	p_N

Such a chart is called the *probability distribution* for the experiment. The numbers $p_1, \ldots, p_N$ are chosen so that each probability represents the long-run proportion of trials in which the associated outcome can be expected to occur.

The next three examples illustrate some methods for determining probability distributions.

EXAMPLE 1   Toss an unbiased coin and observe the uppermost face. Determine the probability distribution for this experiment.

Solution   Since the coin is unbiased, we expect each of the outcomes "heads" and "tails" to be equally likely. So we assign the two outcomes equal probabilities, namely $\frac{1}{2}$. So the probability distribution is

Outcome	Probability
Heads	$\frac{1}{2}$
Tails	$\frac{1}{2}$

EXAMPLE 2   Toss a die and observe the uppermost face. Determine the probability distribution for this experiment.

---

* This restriction will remain in effect until our discussion of the normal distribution in Section 10.4.

Solution There are six possible outcomes, namely 1, 2, 3, 4, 5, 6. Assuming that the die is unbiased, these outcomes are equally likely. So we assign to each outcome the probability $\frac{1}{6}$. Here is the probability distribution for the experiment:

Outcome	Probability
1	$\frac{1}{6}$
2	$\frac{1}{6}$
3	$\frac{1}{6}$
4	$\frac{1}{6}$
5	$\frac{1}{6}$
6	$\frac{1}{6}$

EXAMPLE 3 Traffic engineers measure the traffic on a major highway during rush hour. By observing the number of cars for 300 consecutive rush hours, they arrive at the following data:

Number of cars observed	Frequency observed
≤ 1000	30
1001–3000	45
3001–5000	135
5001–7000	75
> 7000	15

(a) Describe the sample space associated to this experiment.

(b) Calculate the probability distribution for this experiment.

Solution (a) The experiment consists of counting the number of cars during rush hour and observing which one of five categories the number belongs to. The five outcomes are

$$s_1 = \text{``}\leq 1000 \text{ cars''}$$

$$s_2 = \text{``}1001\text{–}3000 \text{ cars''}$$

$$s_3 = \text{``}3001\text{–}5000 \text{ cars''}$$

$$s_4 = \text{``}5001\text{–}7000 \text{ cars''}$$

$$s_5 = \text{``}> 7000 \text{ cars.''}$$

And the sample space is

$$S = \{s_1, s_2, s_3, s_4, s_5\}.$$

(b) For each outcome we use the available statistics to compute its relative frequency. For example, the relative frequency of outcome s_1 is

$$\frac{[\text{number of times } s_1 \text{ occurs}]}{[\text{number of trials}]} = \frac{30}{300} = .1.$$

That is, s_1 occurred in 10% of the observations. If we assume that the 300 observations are representative of rush hours in general, then it would seem reasonable to assign the outcome s_1 the probability .1. Similarly, we can assign probabilities to the other outcomes and arrive at this probability distribution:

Outcome	Probability
s_1	$\frac{30}{300} = .1$
s_2	$\frac{45}{300} = .15$
s_3	$\frac{135}{300} = .45$
s_4	$\frac{75}{300} = .25$
s_5	$\frac{15}{300} = .05$

This method of assigning probabilities to outcomes is valid only insofar as the observed trials are "representative."

Let an experiment have outcomes $s_1, s_2, \ldots, s_N$ with respective probabilities $p_1, p_2, p_3, \ldots, p_N$. Then the numbers $p_1, p_2, \ldots, p_N$ must satisfy two basic properties.

Fundamental Property 1 Each of the numbers $p_1, p_2, \ldots, p_N$ is between 0 and 1.

Fundamental Property 2 $p_1 + p_2 + \cdots + p_N = 1$.

Roughly speaking, Fundamental Property 1 says that the likelihood of each outcome lies between 0% and 100%, whereas Fundamental Property 2 says that there is a 100% likelihood that one of the outcomes $s_1, s_2, \ldots, s_N$ will occur. The two fundamental properties may be easily verified for the probability distributions of Examples 1 to 3.

Suppose that we are given an experiment with a finite number of outcomes. Let us now assign to each event E a probability, which we denote by $\Pr(E)$. If E consists of a single outcome, say $E = \{s\}$, then E is called an *elementary event*. In this case we associate to E the probability of the outcome s. If E consists of more than one outcome, we may compute $\Pr(E)$ via the so-called addition principle.

Addition Principle Suppose that an event E consists of the finite number of outcomes $s, t, u, \ldots, z$. That is,

$$E = \{s, t, u, \ldots, z\}.$$

Then

$$\Pr(E) = \Pr(s) + \Pr(t) + \Pr(u) + \cdots + \Pr(z).$$

We supplement the addition principle with the convention that the probability of the impossible event ∅ is 0. This is certainly reasonable, since the impossible event never occurs.

EXAMPLE 4   Suppose that we toss a die and observe the uppermost face. What is the probability that an odd number will occur?

Solution   The event "odd number occurs" corresponds to the subset of the sample space given by

$$E = \{1, 3, 5\}.$$

That is, the event occurs if a 1, 3, or 5 appears on the uppermost face. By the addition principle

$$Pr(E) = Pr(1) + Pr(3) + Pr(5).$$

As we observed in Example 2, each of the elementary outcomes in the die-tossing experiment has probability $\frac{1}{6}$. Therefore,

$$Pr(E) = \tfrac{1}{6} + \tfrac{1}{6} + \tfrac{1}{6} = \tfrac{1}{2}.$$

So we expect an odd number to occur approximately half of the time.

EXAMPLE 5   Consider the traffic study of Example 3. What is the probability that at most 5000 cars will use the highway during rush hour?

Solution   The event

"at most 5000 cars"

is the same as

$$\{s_1, s_2, s_3\},$$

where we use the same notation for the outcomes as we used in Example 3. Thus the probability of the event is

$$Pr(s_1) + Pr(s_2) + Pr(s_3) = .1 + .15 + .45 = .7.$$

Therefore, we expect that traffic will involve ≤ 5000 cars in approximately 70% of the rush hours.

EXAMPLE 6   Suppose that we toss a red die and a green die and observe the numbers on the uppermost faces.

(a) Calculate the probabilities of the elementary events.

(b) Calculate the probability that the two dice show the same number.

Solution   (a) As shown in Example 3 of Section 2, the sample space consists of 36 pairs of numbers:

$$S = \{(1, 1), (1, 2), \ldots, (6, 5), (6, 6)\}.$$

Each of these pairs is equally likely to occur. (How could the dice show favoritism to a particular pair?) Therefore, each outcome is expected to occur about $\frac{1}{36}$ of the time and the probability of each elementary event is $\frac{1}{36}$.

(b) The event

$$E = \text{"both dice show the same number"}$$

consists of six outcomes:

$$E = \{(1, 1), (2, 2), (3, 3), (4, 4), (5, 5), (6, 6)\}.$$

Thus, by the addition principle,

$$\Pr(E) = \tfrac{1}{36} + \tfrac{1}{36} + \tfrac{1}{36} + \tfrac{1}{36} + \tfrac{1}{36} + \tfrac{1}{36} = \tfrac{6}{36} = \tfrac{1}{6}.$$

EXAMPLE 7 A person playing a certain lottery can win $100, $10, or $1, can break even, or can lose $10. These five outcomes with their corresponding probabilities are given by the probability distribution in Table 1.

(a) What outcome has the greatest probability?

(b) What outcome has the least probability?

(c) What is the probability that the person will win some money?

TABLE 1

Winnings	Probability
100	.02
10	.05
1	.40
0	.03
−10	.50

Solution (a) Table 1 reveals that the outcome −10 has the greatest probability, .50. (A person playing the lottery repeatedly can expect to lose $10 about 50% of the time.) This outcome is just as likely to occur as not.

(b) The outcome 100 has the least probability, .02. A person playing the lottery can expect to win $100 about 2% of the time. (This outcome is quite unlikely to occur.)

(c) We are asked for the probability that the event E occurs where $E = \{100, 10, 1\}$. By the addition principle

$$\begin{aligned}
\Pr(E) &= \Pr(100) + \Pr(10) + \Pr(1) \\
&= \quad .02 \quad + \quad .05 \quad + \quad .40 \\
&= \quad .47.
\end{aligned}$$

Here is a useful formula that relates $\Pr(E \cup F)$ to $\Pr(E \cap F)$:

> Inclusion-Exclusion Principle   Let E and F be any events. Then
>
> $$\Pr(E \cup F) = \Pr(E) + \Pr(F) - \Pr(E \cap F).$$
>
> In particular, if E and F are mutually exclusive, then
>
> $$\Pr(E \cup F) = \Pr(E) + \Pr(F).$$

The inclusion-exclusion principle will be verified after first giving an illustration of its use.

EXAMPLE 8   A factory needs two raw materials. The probability of not having an adequate supply of material A is .05, whereas the probability of not having an adequate supply of material B is .03. A study determines that the probability of a shortage of both A and B is .01. What proportion of the time can the factory operate?

Solution   Let E be the event "shortage of A," F the event "shortage of B." We are given that

$$\Pr(E) = .05, \qquad \Pr(F) = .03, \qquad \Pr(E \cap F) = .01.$$

Therefore, by the inclusion-exclusion principle

$$\Pr(E \cup F) = \Pr(E) + \Pr(F) - \Pr(E \cap F)$$
$$= \quad .05 \quad + \quad .03 \quad - \quad .01$$
$$= .07.$$

But $E \cup F$ is the event "shortage of A or B." Thus the factory is likely to be short of one raw material or the other 7% of the time. Therefore, the factory can expect to operate 93% of the time.

Verification of the Inclusion-Exclusion Principle   Let $E \cap F = \{t_1, \ldots, t_k\}$. Then

$$E = \{r_1, \ldots, r_n, t_1, \ldots, t_k\}$$
$$F = \{s_1, \ldots, s_m, t_1, \ldots, t_k\}$$
$$E \cup F = \{r_1, \ldots, r_n, t_1, \ldots, t_k, s_1, \ldots, s_m\}.$$

Therefore, by the addition principle we have

$$\Pr(E \cup F) = \Pr(r_1) + \cdots + \Pr(r_n) + \Pr(t_1) + \cdots + \Pr(t_k)$$
$$+ \Pr(s_1) + \cdots + \Pr(s_m)$$
$$= [\Pr(r_1) + \cdots + \Pr(r_n) + \Pr(t_1) + \cdots + \Pr(t_k)]$$
$$+ [\Pr(s_1) + \cdots + \Pr(s_m) + \Pr(t_1) + \cdots + \Pr(t_k)]$$
$$- [\Pr(t_1) + \cdots + \Pr(t_k)]$$
$$= \Pr(E) + \Pr(F) - \Pr(E \cap F).$$

If E and F are mutually exclusive, then $E \cap F = \varnothing$ and $\Pr(E \cap F) = 0$. Thus,

$$\Pr(E \cup F) = \Pr(E) + \Pr(F).$$

Odds   Frequently in applications we meet statements like these:

> The odds of a Republican victory are 3 to 2.
> The odds of a recession next year are 1 to 3.

Such statements may be readily translated into the language of probability. For example, consider the first statement above. It means that if the election were repeated often (a theoretical possibility), then for every 2 Democratic wins there would be 3 Republican wins. That is, the Republicans would win $\frac{3}{5}$ (or 60%) of the elections. In terms of probability, this means that the probability of a Republican win is .6. In a similar way, we can translate the second statement above into probabilistic terms. The statement says that if we consider a large number of years experiencing conditions identical to this year's, then for every one that is followed by a recession three years are not. That is, the probability of having a recession next year is $\frac{1}{4}$.

We may generalize our reasoning to obtain the following result:

> If the odds in favor of an event are a to b, then the probability of the event is $a/(a + b)$.

EXAMPLE 9   Suppose that the odds of rain tomorrow are 5 to 3. What is the probability that rain will occur?

Solution   The probability that rain will occur is

$$\frac{5}{5 + 3} = \frac{5}{8}.$$

FIGURE 1

cheese   shock

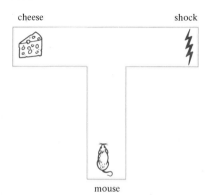

mouse

PRACTICE PROBLEMS 3

1. A mouse is put into a T-maze (a maze shapped like a "T") (Fig. 1). If he turns to the left he receives cheese, and if he turns to the right he receives a mild shock. This trial is done twice with the same mouse and the directions of the turns recorded.

    (a) What is the sample space for this experiment?

    (b) Why would it not be reasonable to assign each outcome the same probability?

2. (a) What are the odds in favor of an event that is just as likely to occur as not?

   (b) Is there any difference between the odds 6 to 4 and the odds 3 to 2?

3. Suppose that E and F are any events. Show that $\Pr(E) = \Pr(E \cap F) + \Pr(E \cap F')$.

EXERCISES 3

1. There are 774,746 words in the Bible. The word "and" occurs 46,277 times and the word "Lord" occurs 1855 times.* Suppose that a word is selected at random from the Bible.

   (a) What is the probability that the word is "and"?

   (b) What is the probability that the word is "and" or "Lord"?

2. An experiment consists of tossing a coin two times and observing the sequence of heads and tails. Each of the four outcomes has the same probability of occurring.

   (a) What is the probability that "HH" is the outcome?

   (b) What is the probability of the event "at least one head"?

3. Suppose that a red die and a green die are tossed and the numbers on the uppermost faces are observed. (See Example 6.)

   (a) What is the probability that the numbers add up to 8?

   (b) What is the probability that the sum of the numbers is less than 5?

4. A state is selected at random from the 50 states of the United States. What is the probability that it is one of the 6 New England states?

5. The modern American roulette wheel has 38 slots, which are labeled with 36 numbers evenly divided between red and black plus two green numbers 0 and 00. What is the probability that the ball lands on a green number?

6. An experiment consists of selecting a number at random from the set of numbers $\{1, 2, 3, 4, 5, 6, 7, 8, 9\}$. Find the probability that the number selected is:

   (a) Less than 4.   (b) Odd.   (c) Less than 4 or odd.

7. Three horses, call them A, B, and C, are going to race against each other. The probability that A will win is $\frac{1}{3}$ and the probability that B will win is $\frac{1}{2}$. What is the probability that C will win?

8. Which of the following probabilities are feasible for an experiment having sample space $\{s_1, s_2, s_3\}$?

   (a) $\Pr(s_1) = .4, \Pr(s_2) = .4, \Pr(s_3) = .4$

   (b) $\Pr(s_1) = .5, \Pr(s_2) = .7, \Pr(s_3) = -.2$

   (c) $\Pr(s_1) = 2, \Pr(s_2) = 1, \Pr(s_3) = \frac{1}{2}$

   (d) $\Pr(s_1) = \frac{1}{4}, \Pr(s_2) = \frac{1}{2}, \Pr(s_3) = \frac{1}{4}$

---

* According to *The People's Almanac* by Wallechinsky and Wallace (New York: Doubleday, 1975).

9. An experiment with outcomes s_1, s_2, s_3, s_4 is described by the probability table:

Outcome	Probability
s_1	.1
s_2	.6
s_3	.2
s_4	.1

(a) What is $Pr(\{s_1, s_2\})$?   (b) What is $Pr(\{s_2, s_4\})$?

10. An experiment with outcomes $s_1, s_2, s_3, s_4, s_5, s_6$ is described by the probability table:

Outcome	Probability
s_1	.05
s_2	.25
s_3	.05
s_4	.01
s_5	.63
s_6	.01

Let $E = \{s_1, s_2\}$, $F = \{s_3, s_5, s_6\}$.

(a) Determine $Pr(E)$, $Pr(F)$.   (b) Determine $Pr(E')$.

(c) Determine $Pr(E \cap F)$.   (d) Determine $Pr(E \cup F)$.

11. Convert the following odds to probabilities:

(a) 10 to 1   (b) 1 to 2   (c) 4 to 5

12. In poker, the probability of being dealt a hand containing a pair of jacks or better is about $\frac{1}{6}$. What are the corresponding odds?

13. Let E and F be events for which $Pr(E) = .6$, $Pr(F) = .5$ and $Pr(E \cap F) = .4$. Find

(a) $Pr(E \cup F)$

(b) $Pr(E \cap F')$ [Hint: See Practice Problem 3.]

14. Let E and F be events for which $Pr(E) = .4$, $Pr(F) = .5$ and $Pr(E \cap F') = .3$. Find

(a) $Pr(E \cap F)$ (b) $Pr(E \cup F)$

15. A statistical analysis of the wait (in minutes) at the checkout line of a certain supermarket yields the following probability distribution:

Wait	Probability
At most 3	.1
More than 3 and at most 5	.2
More than 5 and at most 10	.25
More than 10 and at most 15	.25
More than 15	.2

(a) What is the probability of waiting more than 3 minutes but at most 15?

(b) If you observed the waiting times of a representative sample of 10,000 supermarket customers, approximately how many would you expect to wait for more than 3 minutes but at most 15?

16. What is the probability of the certain event?

17. A computer firm analyzes the failures of circuit boards for their newest personal computer. It observes the following data.

Failures in	Probability
First month	.05
First 2 months	.10
First 3 months	.20
First 4 months	.25
First 5 months	.30
First 6 months	.32

Convert these data into a probability distribution.

18. Refer to the data of Exercise 17. What is the probability that a computer will not fail during its first 3 months of use? First 6 months of use?

SOLUTIONS TO PRACTICE PROBLEMS 3

1. (a) {LL, LR, RR, RL}. Here LL means that the mouse turned left both times, LR means that the mouse turned left the first time and right the second, and so on.

   (b) The mouse will learn something from the first trial. If he turned left the first time and got rewarded, then he is more likely to turn left again on the second trial. Hence LL should have a greater probability than LR. Similarly, RL should be more likely than RR.

2. (a) 1 to 1. An event that is just as likely to occur as not has probability $\frac{1}{2}$. So if the odds are a to b, then we may set $a = 1$, $a + b = 2$. Thus $a = 1$, $b = 1$. (The odds could also be given as 2 to 2, 3 to 3, etc.)

   (b) No. Odds of 6 to 4 correspond to a probability of $6/(6 + 4) = \frac{6}{10} = \frac{3}{5}$. Odds of 3 to 2 correspond to a probability of $3/(3 + 2) = \frac{3}{5}$. (There are always many different ways to express the same odds.)

3. The sets $(E \cap F)$ and $(E \cap F')$ have no elements in common and so are mutually exclusive. Since $E = (E \cap F) \cup (E \cap F')$, the result follows from the inclusion-exclusion principle.

9.4. Calculating Probabilities of Events

In this section we use the counting techniques of Chapter 8 to compute the probabilities of various events. In addition, we illustrate the exceptionally wide range of applications which make use of probability theory.

Experiments with Equally Likely Outcomes   In the experiments associated to many common applications, all outcomes are equally likely—that is, they all have the same probability. This is the case, for example, if we toss an unbiased coin or select a person at random from a population. If a sample space has N equally likely outcomes, then the probability of each outcome is $1/N$ (since the probabilities must add up to 1). Using this fact, the probability of any event is then easy to compute. Namely, suppose that E is an event consisting of M outcomes. Then, by the addition principle,

$$\Pr(E) = \underbrace{\frac{1}{N} + \frac{1}{N} + \cdots + \frac{1}{N}}_{M \text{ times}} = \frac{M}{N}.$$

We can restate this fundamental result as follows:

> Let S be a sample space consisting of N equally likely outcomes. Let E be any event. Then
>
> $$\Pr(E) = \frac{[\text{number of outcomes in } E]}{N}. \qquad (1)$$

In order to apply (1) in particular examples, it is necessary to compute N, the number of outcomes, and [number of outcomes in E]. Often these quantities can be determined using the counting techniques of Chapter 8. Some illustrative computations are provided in Examples 1 to 5 below.

We should mention that, although the urn and dice problems being considered in this section and the next might seem artificial and removed from applications, many applied problems can be described in mathematical terms as urn or dice-tossing experiments. We begin our discussion with two examples involving abstract urn problems. Then in two more examples we show the utility of urn models by applying them to quality control and medical screening problems.

EXAMPLE 1   An urn contains eight white balls and two green balls. A sample of three balls is selected at random. What is the probability of selecting only white balls?

Solution   The experiment consists of selecting three balls from among the 10. Since the order in which the three balls are selected is immaterial, the samples are combinations of 10 balls taken three at a time. The total number of samples is therefore $\binom{10}{3}$, and this is N, the number of elements in the sample space. Since the selection of the sample is random, all samples are equally likely, and thus we can use formula (1) to compute the probability of any event. The problem asks us to compute the probability of the event $E = $ "all three balls selected are white." Since there are

8 white balls, the number of different samples in which all are white is $\binom{8}{3}$. Thus,

$$\Pr(E) = \frac{[\text{number of outcomes in } E]}{N} = \frac{\binom{8}{3}}{\binom{10}{3}} = \frac{56}{120} = \frac{7}{15}.$$

EXAMPLE 2   An urn contains eight white balls and two green balls. A sample of three balls is selected at random. What is the probability that the sample contains at least one green ball?

Solution   Let F be the event "at least one green ball is selected." Let us determine the number of different outcomes in F. These outcomes contain either one or two green balls. There are $\binom{2}{1}$ ways to select one green ball from among two; and for each of these, there are $\binom{8}{2}$ ways to select two white balls from among eight. So the number of samples containing one green ball equals $\binom{2}{1}\binom{8}{2}$. Similarly, the number of samples containing two green balls equals $\binom{2}{2}\binom{8}{1}$. Therefore, the number of outcomes in F—namely, the number of samples having at least one green ball—equals

$$\binom{2}{1}\binom{8}{2} + \binom{2}{2}\binom{8}{1} = 2 \cdot 28 + 1 \cdot 8 = 64.$$

And so

$$\Pr(F) = \frac{[\text{number of outcomes in } F]}{N} = \frac{64}{\binom{10}{3}} = \frac{64}{120} = \frac{8}{15}.$$

EXAMPLE 3   (*Quality Control*)   A toy manufacturer inspects boxes of toys before shipment. Each box contains 10 toys. The inspection procedure consists of randomly selecting three toys from the box. If any are defective, the box is not shipped. Suppose that a given box has two defective toys. What is the probability that it will be shipped?

Solution   This problem is not really new! We solved it in disguise as Example 1. The urn can be regarded as a box of toys, and the balls as individual toys. The white balls are nondefective toys and the green balls defective toys. The random selection of three balls from the urn is just the inspection procedure. And the event "all three balls selected are white" corresponds to the box being shipped. As we calculated above, the probability of this event is $\frac{7}{15}$. (Since $\frac{7}{15} \approx .47$, there is approximately a 47% chance of shipping a box with two defective toys. This inspection procedure is not particularly effective!)

EXAMPLE 4 (*Medical Screening*)   Suppose that a cruise ship returns to the United States from the Far East. Unknown to anyone, four of its 600 passengers have contracted a rare disease. Suppose that the Public Health Service screens 20 passengers, selected at random, to see whether the disease is present aboard ship. What is the probability that the presence of the disease will escape detection?

Solution   The sample space consists of samples of 20 drawn from among the 600 passengers. There are $\binom{600}{20}$ such samples. The number of samples containing none of the sick passengers is $\binom{596}{20}$. Therefore, the probability of not detecting the disease is

$$\frac{\binom{596}{20}}{\binom{600}{20}} = \frac{\dfrac{596!}{20!\,576!}}{\dfrac{600!}{20!\,580!}} = \frac{596!}{600!} \cdot \frac{580!}{576!} = \frac{580 \cdot 579 \cdot 578 \cdot 577}{600 \cdot 599 \cdot 598 \cdot 597} \approx .87.$$

So there is approximately an 87% chance that the disease will escape detection.

The Complement Rule   The *complement rule* relates the probability of an event E to the probability of its complement E'. When applied together with counting techniques, it often simplifies computation of probabilities.

> **Complement Rule**   Let E be any event, E' its complement. Then
> $$\Pr(E) = 1 - \Pr(E').$$

For example, recall Example 1. We determined the probability of the event

$$E = \text{"all three balls selected are white"}$$

associated to the experiment of selecting three balls from an urn containing eight white balls and two green balls. We found that $\Pr(E) = \frac{7}{15}$. On the other hand, in Example 2 we determined the probability of the event

$$F = \text{"at least one green ball is selected."}$$

The event E is the complement of F:
$$E = F'.$$
So, by the complement rule,
$$\Pr(F) = 1 - \Pr(F') = 1 - \Pr(E)$$
$$= 1 - \tfrac{7}{15} = \tfrac{8}{15},$$
in agreement with the calculations of Example 2.

The complement rule is especially useful in situations where $\Pr(E')$ is easier to compute than $\Pr(E)$. One of these situations arises in the celebrated *birthday problem*.

EXAMPLE 5   A group of five people is to be selected at random. What is the probability that two or more of them have the same birthday?

Solution   For simplicity we ignore leap years. Furthermore, we assume that each of the 365 days in a year is an equally likely birthday. (Not an unreasonable assumption.) The experiment we have in mind is this. Pick out five people and observe their birthdays. The outcomes of this experiment are strings of five dates, corresponding to the birthdays. For example, one outcome of the experiment is

(June 2, April 6, Dec. 20, Feb. 12, Aug. 5).

Each date has 365 different possibilites. So, by the generalized multiplication principle, the total number N of possible outcomes of the experiment is

$$N = 365 \cdot 365 \cdot 365 \cdot 365 \cdot 365 = 365^5.$$

Let E be the event "at least two people have the same birthday." It is very difficult to calculate directly the number of outcomes in E. However, it is comparatively simple to compute the number of outcomes in E' and hence to compute $\Pr(E')$. This is because E' is the event "all five birthdays are different." An outcome in E' can be selected in a sequence of five steps:

Select a day	Select a different day	Select yet a different day	Select yet a different day	Select yet a different day

These five steps will result in a sequence of five different birthdays. The first step can be performed in 365 ways; for each of these, the next step in 364; for each of these, the next step in 363; for each of these, the next step in 362; and for each of these, the last step in 361. Therefore, E' contains $365 \cdot 364 \cdot 363 \cdot 362 \cdot 361$ outcomes, and

$$\Pr(E') = \frac{365 \cdot 364 \cdot 363 \cdot 362 \cdot 361}{365^5} \approx .973.$$

By the complement rule,

$$\Pr(E) = 1 - \Pr(E') = 1 - .973 = .027.$$

So the likelihood is about 2.7% that two or more of the five people will have the same birthday.

The experiment of Example 5 can be repeated using samples of 8, 10, 20, or any number of people. As before, let E be the event "at least two people have the same birthday," so that $E' =$ "all the birthdays are different." If a sample of r

people is used, then the same reasoning as used above yields

$$\Pr(E') = \frac{365 \cdot 364 \cdot \ldots \cdot (365 - r + 1)}{365^r}.$$

Table 1 gives the values of $\Pr(E') = 1 - \Pr(E')$ for various values of r. You may be surprised by the numbers in the table. Even with as few as 23 people it is more likely than not that at least two people have the same birthday. With a sample of 50 people we are almost certain to have two with the same birthday. (Try this experiment in your dormitory or class.)

TABLE 1 Probability That, in a Randomly Selected Group of r People, at Least Two Will Have the Same Birthday

r	5	10	15	20	22	23	25	30	40	50
$\Pr(E)$	.027	.117	.253	.411	.476	.507	.569	.706	.891	.970

PRACTICE PROBLEMS 4

1. A couple decides to have four children. What is the probability that among the children will be at least one boy and at least one girl?

2. (a) Find the probability that all the numbers are different in three spins of a roulette wheel. [*Note:* A roulette wheel has 38 numbers.]

   (b) Guess how many spins are required in order that the probability that all the numbers are different will be less than .5?

EXERCISES 4

1. An urn contains six white balls and five red balls. A sample of four balls is selected at random from the urn. What is the probability that the sample contains two white balls and two red balls?

2. A factory produces fuses, which are packaged in boxes of 10. Three fuses are selected at random from each box for inspection. The box is rejected if at least one of these three fuses is defective. What is the probability that a box containing five defective fuses will be rejected?

3. Of the nine members of the board of trustees of a college, five agree with the president on a certain issue. The president selects three trustees at random and asks for their opinions. What is the probability that at least two of them will agree with him?

4. An urn contains five red balls and four white balls. A sample of two balls is selected at random from the urn. What is the probability that at least one of the balls is red?

5. Without consultation, each of four organizations announces a one-day convention to be held during June. Find the probability that at least two organizations specify the same day for their convention.

6. Five letters are selected from the alphabet, one at a time with replacement, to form a five-letter word. What is the probability that the word has five different letters?

7. An exam contains five "true or false" questions. What is the probability that a student guessing at the answers will get three or more answers correct?

8. An airport limousine has four passengers and stops at six different hotels. What is the probability that two or more people will be staying at the same hotel? (Assume that each person is just as likely to stay in one hotel as another.)

9. In a certain agricultural region the probability of a drought during the growing season is .2, the probability of a severe cold spell is .15, and the probability of both is .1. Find the probability of

(a) Either a drought or a severe cold spell.

(b) Neither a drought nor a severe cold spell.

(c) Not having a drought.

10. A coin is to be tossed seven times. What is the probability of obtaining four heads and three tails?

11. Let E and F be events such that $Pr(E) = .3$, $Pr(F') = .6$, and $Pr(E \cup F) = .7$. What is $Pr(E \cap F)$?

12. In a certain manufacturing process the probability of a type I defect is .12, the probability of a type II defect is .22, and the probability of having both types of defects is .02. Find the probability of having neither type of defect.

13. A man has six pairs of socks, from which he selects two socks at random. What is the probability that the selected socks will match?

14. Of the 15 members on a Senate committee, 10 plan to vote "yes" and 5 plan to vote "no" on an important issue. A reporter attempts to predict the outcome of the vote by questioning 6 of the senators. Find the probability that this sample is precisely representative of the final vote. That is, find the probability that four of the six senators questioned planned to vote "yes."

15. A man, a woman, and their three children randomly stand in a row for a family picture. What is the probability that the parents will be standing next to each other?

16. In poker the probabilities of being dealt a flush, a straight, and a straight flush are .0019654, .0039246, and .0000153, respectively. What is the probability of being dealt a straight or a flush?

17. Figure 1 shows a partial map of the streets in a certain city. A tourist starts at point A and selects at random a path to point B. (We shall assume that he walks only south and east.) Find the probability that

FIGURE 1

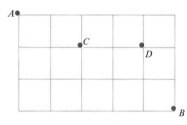

(a) He passes through point C.

(b) He passes through point D.

(c) He passes through point C and point D.

(d) He passes through point C or point D.

18. A couple decide to have four children. What is the probability that they will have more girls than boys?

19. A law firm has six senior and four junior partners. A committee of three partners is selected at random to represent the firm at a conference. What is the probability that at least one of the junior partners is on the committee?

20. A coin is to be tossed six times. What is the probability of obtaining exactly three heads?

21. A bag contains nine tomatoes, of which one is rotten. A sample of three tomatoes is selected at random. What is the probability that the sample contains the rotten tomato?

22. A vacationer has brought along four novels and four nonfiction books. One day the person selects two at random to take to the beach. What is the probability that both are novels?

SOLUTIONS TO PRACTICE PROBLEMS 4

1. Each possible outcome is a string of four letters composed of B's and G's. By the generalized multiplication principle, there are 2^4 or 16 possible outcomes. Let E be the event "children of both sexes." Then $E' = \{BBBB, GGGG\}$, and

$$\Pr(E') = \frac{[\text{number of outcomes in } E']}{[\text{total number of outcomes}]} = \frac{2}{16} = \frac{1}{8}.$$

Therefore,

$$\Pr(E) = 1 - \Pr(E') = 1 - \frac{1}{8} = \frac{7}{8}.$$

So the probability is 87.5% that they will have children of both sexes.

2. (a) Each sequence of three numbers is just as likely to occur as any other. Therefore,

$$\Pr(\text{numbers different}) = \frac{[\text{number of outcomes with numbers different}]}{[\text{number of possible outcomes}]}$$

$$= \frac{38 \cdot 37 \cdot 36}{38^3} \approx .92.$$

(b) 8

9.5. Conditional Probability and Independence

The probability of an event depends, often in a critical way, on the sample space in question. In this section we explore this dependence in some detail by introducing what are called conditional probabilities.

   To illustrate the dependence of probabilities on the sample space, consider the following example.

EXAMPLE 1 Suppose that a certain mathematics class contains 26 students. Of these, 15 are freshmen, 14 are business majors, and 7 are neither. Suppose that a person is selected at random from the class.

(a) What is the probability that the person is both a freshman and a business major?

(b) Suppose we are given the additional information that the person selected is a freshman. What is the probability that he is also a business major?

Solution Let B denote the set of business majors, F the set of freshmen. A complete Venn diagram of the class is given in Fig. 1.

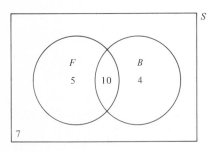

FIGURE 1

(a) In selecting a student from the class, the sample space consists of all 26 students. Since the choice is random, all students are equally likely to be selected. The event "freshman and business major" corresponds to the set $F \cap B$ of the Venn diagram. Therefore,

$$\Pr(F \cap B) = \frac{[\text{number of outcomes in } F \cap B]}{[\text{number of possible outcomes}]}$$
$$= \frac{10}{26} = \frac{5}{13}.$$

So the probability of selecting a freshman business major is $\frac{5}{13}$.

(b) If we know that the student selected is a freshman, then the possible outcomes of the experiment are restricted. They must belong to F. In other words, given the additional information we must alter the sample space from "all students" to "freshmen." Since each of the 15 freshmen is equally likely to be selected and since 10 of the 15 freshmen are business majors, the probability of choosing a business major under these circumstances is equal to $\frac{10}{15} = \frac{2}{3}$.

Example 1 may be generalized as follows. Suppose that we consider an experiment, call it experiment I, with associated events E and F. Suppose that we are given that the event F occurred. The probability that E also occurred is called the *conditional probability of E given F* and is denoted $\Pr(E|F)$.

All probabilities are defined with reference to a particular experiment. Let us be clear about the experiment associated with the conditional probability $\Pr(E|F)$. Call this one experiment II. A trial of experiment II consists of performing experiment I. If the event F occurs, then the outcome is observed. If the event F does not occur, then experiment I is repeated. And so on until F does occur. (Some trials of experiment II may last a long time!) In effect, this experiment is just experiment I, ignoring outcomes in which F does not occur. Now that we have specified the experiment, we must specify its associated probability distribution. That is, we must assign a value to the conditional probability $\Pr(E|F)$. This is done via the following formula:

$$
\boxed{\begin{array}{c}
\textit{Conditional Probability}\\[4pt]
\Pr(E|F) = \dfrac{\Pr(E \cap F)}{\Pr(F)}.
\end{array}}
\tag{1}
$$

We will provide an intuitive justification of this formula below. However, we will first give two applications.

EXAMPLE 2 Consider all families with two children (not twins). Assume that the different birth sequences

boy boy, boy girl, girl boy, girl girl

are equally likely. Let E be the event "boy boy" and F the event "at least one boy." Calculate $\Pr(E|F)$.

Solution Each of the four birth sequences is equally likely and has probability $\frac{1}{4}$. In particular, since $E \cap F = \{\text{boy boy}\}$, we have $\Pr(E \cap F) = \frac{1}{4}$. On the other hand, $F = \{\text{boy boy, boy girl, girl boy}\}$, so that $\Pr(F) = \frac{1}{4} + \frac{1}{4} + \frac{1}{4} = \frac{3}{4}$. The conditional probability $\Pr(E|F)$ equals the probability that a two-child family has two boys given that it has at least one. This conditional probability has the value

$$
\Pr(E|F) = \frac{\Pr(E \cap F)}{\Pr(F)} = \frac{\frac{1}{4}}{\frac{3}{4}} = \frac{1}{3}.
$$

[Note that this result is contrary to many people's intuitions. They reason that among families with at least one boy, the other child is just as likely to be a boy as a girl. Therefore, they conclude incorrectly that the probability $\Pr(E|F)$ is $\frac{1}{2}$.]

EXAMPLE 3 Ajax Steel Company employs 20% college graduates. Of all its employees, 25% earn more than \$30,000 per year, and 15% are college graduates earning more than \$30,000 per year. What is the probability that an employee selected at random earns more than \$30,000 per year, given that he is a college graduate?

Solution Let H and C be the events

$$
H = \text{"earns more than \$30,000 per year"}
$$

$$
C = \text{"college graduate."}
$$

We are asked to calculate $\Pr(H|C)$. The given data are

$$
\Pr(H) = .25, \qquad \Pr(C) = .20, \qquad \Pr(H \cap C) = .15.
$$

By formula (1), we have

$$Pr(H|C) = \frac{Pr(H \cap C)}{Pr(C)} = \frac{.15}{.20} = \tfrac{3}{4}.$$

Thus $\tfrac{3}{4}$ of all college graduates at Ajax Steel earn more than \$30,000 per year.

Suppose that an experiment has N equally likely outcomes. Then we may apply the following formula to calculate $Pr(E|F)$.

> *Conditional Probability in Case of Equally Likely Outcomes*
>
> $$Pr(E|F) = \frac{[\text{number of outcomes in } E \cap F]}{[\text{number of outcomes in } F]}.$$
>
> (2)

For instance, in Example 2, the different birth sequences were all equally likely. Moreover, the number of outcomes in $E \cap F$ (two boys) is 1, whereas the number of outcomes in F (at least one boy) is 3. Then formula (2) gives that $Pr(E|F) = \tfrac{1}{3}$, in agreement with the calculation of Example 2.

Let us now justify formulas (1) and (2).

Justification of Formula (1) Formula (1) is a definition of conditional probability and as such does not really need any justification. (We can make whatever definitions we choose!) However, let us proceed intuitively and show that the definition is a reasonable one, in the sense that formula (1) gives the expected long-run proportion of occurrences of E given that F occurs. Assume that our experiment is performed repeatedly, say for 10,000 trials. We would expect F to occur in approximately 10,000 $Pr(F)$ trials. Among these, the trials for which E also occurs are exactly those for which *both* E and F occur. In other words, the trials for which E also occurs are exactly those for which the event $E \cap F$ occurs; and this event has probability $Pr(E \cap F)$. Out of the original 10,000 trials, there should be approximately 10,000 $Pr(E \cap F)$ in which E and F both occur. Thus, considering only those trials in which F occurs, the proportion in which E also occurs is

$$\frac{10{,}000 \ Pr(E \cap F)}{10{,}000 \ Pr(F)} = \frac{Pr(E \cap F)}{Pr(F)}.$$

Thus, at least intuitively, it seems reasonable to define $Pr(E|F)$ by formula (1).

Justification of Formula (2) Suppose that the number of outcomes of the experiment is N. Then

$$Pr(F) = \frac{[\text{number of outcomes in } F]}{N},$$

$$Pr(E \cap F) = \frac{[\text{number of outcomes in } E \cap F]}{N}.$$

Therefore, using formula (1), we have

$$Pr(E|F) = \frac{Pr(E \cap F)}{Pr(F)}$$

$$= \frac{\dfrac{[\text{number of outcomes in } E \cap F]}{N}}{\dfrac{[\text{number of outcomes in } F]}{N}}$$

$$= \frac{[\text{number of outcomes in } E \cap F]}{[\text{number of outcomes in } F]}.$$

From formula (1), by multiplying both sides of the equation by $Pr(F)$, we can deduce the following useful fact.

Product Rule

$$Pr(E \cap F) = Pr(F) \cdot Pr(E|F). \tag{3}$$

The next example illustrates the use of this rule.

EXAMPLE 4   Assume that a certain school contains an equal number of female and male students and that 5% of the male population is color-blind. Find the probability that a randomly selected student is a color-blind male.

Solution   Let $M = $ "male" and $B = $ "color-blind." We wish to calculate $Pr(B \cap M)$. From the given data

$$Pr(M) = .5 \quad \text{and} \quad Pr(B|M) = .05.$$

Therefore, by the product rule

$$Pr(B \cap M) = Pr(M) \cdot Pr(B|M) = (.5)(.05) = .025.$$

Often an event G can be described as a sequence of two other events E and F. That is, G occurs if F occurs and then E occurs. The product rule allows us to compute the probability of G as the probability of F times the conditional probability $Pr(E|F)$. The next example illustrates this point.

EXAMPLE 5   A sequence of two playing cards is drawn at random (without replacement) from a standard deck of 52 cards. What is the probability that the first card is red and the second is black?

Solution   The event in question is a sequence of two events, namely

$$F = \text{"the first card is red"}$$

$$E = \text{"the second card is black."}$$

Since half the deck consists of red cards, $\Pr(F) = \frac{1}{2}$. If we are given that F occurs, then there are only 51 cards left in the deck, of which 26 are black, so

$$\Pr(E \mid F) = \tfrac{26}{51}.$$

By the product rule

$$\Pr(E \cap F) = \Pr(F) \cdot \Pr(E \mid F) = \tfrac{1}{2} \cdot \tfrac{26}{51} = \tfrac{13}{51}.$$

The product rule may be generalized to sequences of three events E_1, E_2, E_3:

$$\Pr(E_1 \cap E_2 \cap E_3) = \Pr(E_1) \cdot \Pr(E_2 \mid E_1) \cdot \Pr(E_3 \mid E_1 \cap E_2).$$

Similar formulas hold for sequences of four or more events.

One of the most important applications of conditional probability is in the discussion of independent events. Intuitively, two events are independent of each other if the occurrence of one has no effect on the likelihood that the other will occur. For example, suppose that we toss a die two times. Let the events E and F be

$$F = \text{"first throw is a 6"}$$

$$E = \text{"second throw is a 3."}$$

Then intuitively these events are independent of one another. Throwing a 6 on the first throw has no effect whatsoever on the outcome of the second throw. On the other hand, suppose that we draw a sequence of two cards at random from a deck. Then the events

$$F = \text{"first card is red"}$$

$$E = \text{"second card is black"}$$

are not independent of one another, at least intuitively. Indeed, whether or not we draw a red on the first card effects the likelihood of drawing a black on the second.

The notion of independence of events can be simply formulated in terms of conditional probability.

> Let E and F be events. We say that E and F are *independent* provided that
> $$\Pr(E \mid F) = \Pr(E) \qquad \text{and} \qquad \Pr(F \mid E) = \Pr(F).$$

That is, the additional information that F has occurred has no effect on the likelihood of E; and the additional information that E has occurred has no effect on the likelihood of F.

EXAMPLE 6   Suppose that an experiment consists of observing the outcome of two consecutive throws of a die. Let E and F be the events

$$E = \text{"the first throw is a 3,"}$$

$$F = \text{"the second throw is a 6."}$$

Show that these events are independent.

Solution   Clearly, $\Pr(E) = \Pr(F) = \frac{1}{6}$. To compute $\Pr(E|F)$, assume that F occurs. Then there are six possible outcomes:

$$F = \{(1, 6), (2, 6), (3, 6), (4, 6), (5, 6), (6, 6)\},$$

and all outcomes are equally likely. Moreover,

$$E \cap F = \{(3, 6)\},$$

so that

$$\Pr(E|F) = \frac{[\text{number of outcomes in } E \cap F]}{[\text{number of outcomes in } F]} = \frac{1}{6} = \Pr(E).$$

Similarly, $\Pr(F|E) = \Pr(F)$. So E and F are independent events, in agreement with our intuition.

EXAMPLE 7   Suppose that an experiment consists of observing the results of drawing two consecutive cards from a 52-card deck. Let E and F be the events

$$E = \text{"second card is black,"}$$

$$F = \text{"first card is red."}$$

Are these events independent?

Solution   There are the same number of outcomes with the second card red as with the second card black, so $\Pr(E) = \frac{1}{2}$. To compute $\Pr(E|F)$, note that if F occurs, then there are 51 equally likely choices for the second card, of which 26 are black, so that $\Pr(E|F) = \frac{26}{51}$. Note that $\Pr(E|F) \neq \Pr(E)$, so E and F are not independent, in agreement with our intuition.

By using the product rule we can arrive at an alternative formulation of independence. Indeed, if E and F are independent, then

$$\Pr(E|F) = \Pr(E),$$

so the product rule yields

$$\Pr(E \cap F) = \Pr(E|F) \cdot \Pr(F) = \Pr(E) \cdot \Pr(F). \qquad (4)$$

By following the same calculation in reverse we deduce that if formula (4) holds, then E and F are independent events. Thus we have arrived at the following useful test for independence.

> *Test for Independence*   The events E and F are independent if and only if
> $$\Pr(E \cap F) = \Pr(E) \cdot \Pr(F).$$

EXAMPLE 8   Suppose that we toss a coin three times and record the sequence of heads and tails. Let E be the event "at most one head occurs" and F the event "both heads and tails occur." Are E and F independent?

Solution   Using the abbreviations H for "heads" and T for "tails," we have

$$E = \{TTT, HTT, THT, TTH\}$$

$$F = \{HTT, HTH, HHT, THH, THT, TTH\}$$

$$E \cap F = \{HTT, THT, TTH\}.$$

The sample space contains eight equally likely outcomes, so that

$$\Pr(E) = \tfrac{1}{2}, \qquad \Pr(F) = \tfrac{3}{4}, \qquad \Pr(E \cap F) = \tfrac{3}{8}.$$

Moreover,

$$\Pr(E) \cdot \Pr(F) = \tfrac{1}{2} \cdot \tfrac{3}{4} = \tfrac{3}{8},$$

which equals $\Pr(E \cap F)$. So E and F are independent.

EXAMPLE 9   Suppose that a family has four children. Let E be the event "at most one boy" and F the event "at least one child of each sex." Are E and F independent?

Solution   Let B stand for "boy" and G for "girl." Then

$$E = \{GGGG, GGGB, GGBG, GBGG, BGGG\}$$

$$F = \{GGGB, GGBG, GBGG, BGGG, BBBG, BBGB, BGBB,$$

$$GBBB, BBGG, BGBG, BGGB, GBBG, GBGB, GGBB\},$$

and the sample space consists of 16 equally likely outcomes. Furthermore,

$$E \cap F = \{GGGB, GGBG, GBGG, BGGG\}.$$

Therefore,

$$\Pr(E) = \tfrac{5}{16}, \qquad \Pr(F) = \tfrac{7}{8}, \qquad \Pr(E \cap F) = \tfrac{1}{4}.$$

In this example

$$\Pr(E) \cdot \Pr(F) = \tfrac{5}{16} \cdot \tfrac{7}{8} = \tfrac{35}{128} \neq \Pr(E \cap F).$$

So E and F are *not* independent events.

EXAMPLE 10   A new hand calculator is designed to be ultrareliable by reason of its two independent calculating units. The probability that a given calculating unit fails within the first 1000 hours of operation is .001. What is the probability that at least one calculating unit will operate without failure for the first 1000 hours of operation?

Solution   Let

$$E = \text{``calculating unit 1 fails in first 1000 hours,''}$$

$$F = \text{``calculating unit 2 fails in first 1000 hours.''}$$

Then E and F are independent events, since the calculating units are independent of one another. Therefore,

$$\Pr(E \cap F) = \Pr(E) \cdot \Pr(F) = (.001)^2 = .000001$$

$$\Pr((E \cap F)') = 1 - .000001 = .999999.$$

Since $(E \cap F)' = $ "not both calculating units fail in first 1000 hours," the desired probability is .999999.

The concept of independent events can be extended to more than two events:

> A set of events is said to be *independent* if, for each collection of events chosen from among them, say $E_1, E_2, \ldots, E_n$, we have
>
> $$\Pr(E_1 \cap E_2 \cap \cdots \cap E_n) = \Pr(E_1) \cdot \Pr(E_2) \cdot \ldots \cdot \Pr(E_n).$$

EXAMPLE 11   Three events A, B, C are independent and $\Pr(A) = .5$, $\Pr(B) = .3$, $\Pr(C) = .2$.

(a) Calculate $\Pr(A \cap B \cap C)$.   (b) Calculate $\Pr(A \cap C)$.

Solution   (a) $\Pr(A \cap B \cap C) = \Pr(A) \cdot \Pr(B) \cdot \Pr(C) = (.5)(.3)(.2) = .03$.

(b) $\Pr(A \cap C) = \Pr(A) \cdot \Pr(C) = (.5)(.2) = .1$.

We shall leave as an exercise the intuitively reasonable result that if E and F are independent events, then so are E and F', E' and F, and E' and F'. This result also generalizes to any collection of independent events.

EXAMPLE 12   A company manufactures stereo components. Experience shows that defects in manufacture are independent of one another. Quality control studies reveal that

   2% of turntables are defective,
   3% of amplifiers are defective,
   7% of speakers are defective.

A system consists of a turntable, an amplifier, and two speakers. What is the probability that the system is not defective?

Solution   Let T, A, S_1, and S_2 be the events corresponding to defective turntable, amplifier, speaker 1, and speaker 2, respectively. Then

$$\Pr(T) = .02, \qquad \Pr(A) = .03, \qquad \Pr(S_1) = \Pr(S_2) = .07.$$

We wish to calculate $\Pr(T' \cap A' \cap S_1' \cap S_2')$. By the complement rule we have

$$\Pr(T') = .98, \qquad \Pr(A') = .97, \qquad \Pr(S_1') = \Pr(S_2') = .93.$$

Since we have assumed that T, A, S_1, S_2 are independent, so are T', A', S_1', S_2'. Therefore,

$$\Pr(T' \cap A' \cap S_1' \cap S_2') = \Pr(T') \cdot \Pr(A') \cdot \Pr(S_1') \cdot \Pr(S_2')$$

$$= (.98)(.97)(.93)^2 \approx .822.$$

Thus there is an 82.2% chance that the system is not defective.

PRACTICE PROBLEMS 5

1. Suppose there are three cards: one red on both sides, one white on both sides, and one having a side of each color. A card is selected at random and placed on a table. If the up side is red, what is the probability that the down side is red? (Try guessing at the answer before working it using the formula for conditional probability.)

2. Show that if events E and F are independent of each other, then so are E and F'. [*Hint*: Since $E \cap F$ and $E \cap F'$ are mutually exclusive, we have

$$\Pr(E) = \Pr(E \cap F) + \Pr(E \cap F').]$$

EXERCISES 5

In Exercises 1–4, let S be a sample space and E, F events associated with S. Suppose that $\Pr(E) = .5$, $\Pr(F) = .3$, $\Pr(E \cap F) = .1$.

1. Calculate $\Pr(E|F)$, $\Pr(F|E)$.

2. Are E and F independent events? Explain.

3. Calculate $\Pr(E|F')$.

4. Calculate $\Pr(E'|F')$.

5. A doctor studies the known cancer patients in a certain town. The probability that a randomly chosen resident has cancer is found to be .001. It is found that 30% of the town works for Ajax Chemical Company. The probability that an employee of Ajax has cancer is equal to .003. Are the events "has cancer" and "works for Ajax" independent of one another?

6. The proportion of individuals in a certain city earning more than $25,000 per year is .25. The proportion of individuals earning more than $25,000 and having a college degree is .10. Suppose that a person is randomly chosen and he turns out to be earning more than $25,000. What is the probability that he is a college graduate?

7. A medical screening program administers three independent fitness tests. Of the persons taking the tests, 80% pass test I, 75% pass test II, and 60% pass test III. A participant is chosen at random. What is the probability that he will pass all three tests?

8. A stereo system contains 50 transistors. The probability that a given transistor will fail in 100,000 hours of use is .0005. Assume that the failures of the various transistors are independent of one another. What is the probability that no transistor will fail during the first 100,000 hours of use?

9. A television set contains five circuit boards of type A, five of type B, and three of type C. The probability of failing in its first 5000 hours of use is .01 for a type A circuit board, .02 for a type B circuit board, and .025 for a type C circuit board. Assuming that the failures of the various circuit boards are independent of one another, compute the probability that no circuit board fails in the first 5000 hours of use.

10. A certain brand of long-life bulb has probability .01 of burning out in less than 1000 hours. Suppose that we wish to light a corridor with a number of independent bulbs in such a

way that at least one of the bulbs remains lit for 1000 consecutive hours. What is the minimum number of bulbs needed to assure that the probability of success is at least .99999?

11. Let E and F be events with $\Pr(E) = \frac{1}{2}$, $\Pr(F) = \frac{1}{3}$, $\Pr(E \cap F) = \frac{1}{4}$. Compute $\Pr(E|F)$ and $\Pr(F|E)$.

12. Let E and F be events with $\Pr(E) = .3$, $\Pr(F) = .6$, and $\Pr(E \cup F) = .7$. Find:

    (a) $\Pr(E \cap F)$ (b) $\Pr(E|F)$ (c) $\Pr(F|E)$

    (d) $\Pr(E' \cap F)$ (e) $\Pr(E'|F)$

13. Of the registered voters in a certain town, 50% are Democrats, 40% favor a school loan, and 30% are Democrats who favor a school loan. Suppose that a registered voter is selected at random from the town.

    (a) What is the probability that the person is not a Democrat and opposes the school loan?

    (b) What is the conditional probability that the person favors the school loan given that he is a Democrat?

    (c) What is the conditional probability that the person is a Democrat given that he favors the school loan?

14. Of the students at a certain college, 50% regularly attend the football games, 30% are freshmen, and 40% are upperclassmen who do not regularly attend football games. Suppose that a student is selected at random.

    (a) What is the probability that the person both is a freshman and regularly attends football games?

    (b) What is the conditional probability that the person regularly attends football games given that he is a freshman?

    (c) What is the conditional probability that the person is a freshman given that he regularly attends football games?

15. A coin is tossed three times. What is the conditional probability that the outcome is HHH given that at least two heads occur?

16. Two balls are selected at random from an urn containing two white balls and three red balls. What is the conditional probability that both balls are white given that at least one of them is white?

17. The probabilities that a person A and a person B will live an additional 15 years are .8 and .7, respectively. Assuming that their lifespans are independent, what is the probability that A or B will live an additional 15 years?

18. A sample of two balls is drawn from an urn containing two white balls and three red balls. Are the events "the sample contains at least one white ball" and "the sample contains balls of both colors" independent?

19. Show that if events E and F are independent of each other, then so are E' and F'.

20. Show that if E and F are independent events then

$$\Pr(E \cup F) = 1 - \Pr(E') \cdot \Pr(F').$$

A battery of tests is administered to a group of students with learning problems. The battery consists of two tests and all students take the entire battery. The probability that the first test is able to correctly classify the learning problem is .3, whereas the probability that the second is able to correctly classify the learning problem is .5. The probability that at least one test correctly classifies the learning problem is .6.

21. What is the probability that both tests will correctly classify the learning problem in a randomly chosen student?

22. What is the conditional probability that test I correctly classifies a learning problem given that test II correctly classifies it?

23. What is the conditional probability that test II correctly classifies a learning problem given that test I correctly classifies it?

24. Are the events "test I correctly classifies a learning problem" and "test II correctly classifies a learning problem" independent?

SOLUTIONS TO PRACTICE PROBLEMS 5

1. $\frac{2}{3}$. Let F be the event that the up side is red and E the event that the down side is red. $Pr(F) = \frac{1}{2}$, since half the faces are red. $F \cap E$ is the event that both sides of the card are red—that is, that the card which is red on both sides was selected, an event with probability $\frac{1}{3}$. By (2),

$$Pr(E|F) = \frac{Pr(E \cap F)}{Pr(F)} = \frac{\frac{1}{3}}{\frac{1}{2}} = \frac{2}{3}.$$

(This result may seem more intuitively evident when you realize that two-thirds of the time the card will have the same color on the bottom as on the top.)

2. By the hint,

$$Pr(E \cap F') = Pr(E) - Pr(E \cap F)$$

$$= Pr(E) - Pr(E) \cdot Pr(F) \qquad \text{(since } E \text{ and } F \text{ are independent)}$$

$$= Pr(E)[1 - Pr(F)]$$

$$= Pr(E) \cdot Pr(F'). \qquad \text{(by the complement rule).}$$

Therefore, E and F' are independent events.

9.6. Tree Diagrams

In solving many probability problems, it is helpful to represent the various events and their associated probabilities by a *tree diagram*. To explain this useful notion, suppose that we wish to compute the probability of an event which results from performing a sequence of experiments. The various outcomes of each experiment

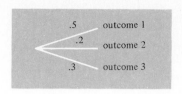

FIGURE 1

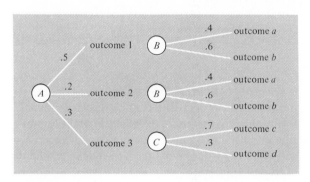

FIGURE 2

are represented as branches emanating from a point. For example, Fig. 1 represents an experiment with three outcomes. Notice that each branch has been labeled with the probability of the associated outcome. For example, the probability of outcome 1 is .5.

We represent experiments performed one after another by stringing together diagrams of the above sort, proceeding from left to right. For example, the diagram in Fig. 2 indicates that first we perform experiment A, having three outcomes, labeled 1–3. If the outcome is 1 or 2, we then perform experiment B. If the outcome is 3, we perform experiment C. The probabilities on the right are conditional probabilities. For example, the top probability is the probability of outcome a (of B) given outcome 1 (of A). The probability of a sequence of outcomes may then be computed by multiplying the probabilities along a path. For example, to calculate the probability of outcome 2 followed by outcome b, we must calculate $\Pr(2 \text{ and } b) = \Pr(2) \cdot \Pr(b|2)$. To carry out this calculation, trace out the sequence of outcomes. Multiplying the probabilities along the path gives $(.2)(.6) = .12$—the probability of outcome 2 followed by outcome b.

The next example illustrates the use of tree diagrams in calculating probabilities.

EXAMPLE 1 A pollster is hired by a presidential candidate to determine his support among the voters of Pennsylvania's two big cities: Philadelphia and Pittsburgh. The pollster designs the following sampling technique: Select one of the cities at random and then poll a voter selected at random from that city. Suppose that in Philadelphia two-fifths of the voters favor the Republican candidate and three-fifths favor the Democratic candidate. Suppose that in Pittsburgh two-thirds of the voters favor the Republican candidate and one-third favor the Democratic candidate.

(a) Draw a tree diagram describing the survey.

(b) Find the probability that the voter polled is from Philadelphia and favors the Republican candidate.

(c) Find the probability that the voter favors the Republican candidate.

(d) Find the probability that the voter is from Philadelphia, given that he favors the Republican candidate.

Solution (a) The survey proceeds in two steps: first, select a city, and second, poll a voter. Figure 3(a) shows the possible outcomes of the first step and the associated probabilities. For each outcome of the first step there are two possibilities for

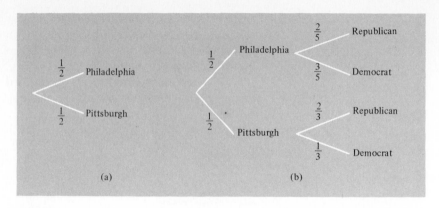

FIGURE 3

the second step: the person selected could favor the Republican or the Democrat. In Fig. 3(b) we have represented these possibilities by drawing branches emanating from each of the outcomes of the first step. The probabilities on the new branches are actually conditional probabilities. For instance, $\frac{2}{5} = $ Pr(Rep|Phila), the probability that the voter favors the Republican candidate, given that the voter is from Philadelphia.

(b) Pr(Phila $\cap$ Rep) = Pr(Phila) $\cdot$ Pr(Rep|Phila) = $\frac{1}{2} \cdot \frac{2}{5} = \frac{1}{5}$.

That is, the probability is $\frac{1}{5}$ that the combined outcome corresponds to the path highlighted in Fig. 4(a). We have written the probability $\frac{1}{5}$ at the end of the path to which it corresponds.

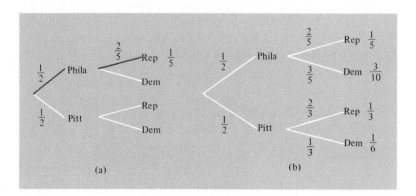

FIGURE 4

(c) In Fig. 4(b) we have computed the probabilities for each path of the tree as in part (b) above. Namely, the probability for a given path is the product of the probabilities for each of its segments. We are asked for Pr(Rep). There are two paths through the tree leading to Republican, namely Philadelphia-Republican and Pittsburgh-Republican. The probabilities of these two paths are $\frac{1}{5}$ and $\frac{1}{3}$, respectively. So the probability that the Republican is favored equals $\frac{1}{5} + \frac{1}{3} = \frac{8}{15}$.

(d) Here we are asked for $\Pr(\text{Phila}|\text{Rep})$. By the definition of conditional probability

$$\Pr(\text{Phila}|\text{Rep}) = \frac{\Pr(\text{Phila} \cap \text{Rep})}{\Pr(\text{Rep})} = \frac{\frac{1}{5}}{\frac{8}{15}} = \frac{3}{8}.$$

Note that from part (c) we might be led to conclude that the Republican candidate is leading, with $\frac{8}{15}$ of the vote. However, we must always be careful when interpreting surveys. The results depend heavily on the survey design. For example, the survey above drew half of its sample from each of the cities. However, Philadelphia is a much larger city and is leaning toward the Democratic candidate—so much so, in fact, that in terms of popular vote the Democratic candidate would win, contrary to our expectations drawn from (c). A pollster must be very careful in designing the procedure for selecting people.

We now finally solve the medical diagnosis problem mentioned in Section 1 and in the Introduction as Problem 2.

EXAMPLE 2 Suppose that the reliability of a skin test for active pulmonary tuberculosis (TB) is specified as follows: Of people with TB, 98% have a positive reaction and 2% have a negative reaction; of people free of TB, 99% have a negative reaction and 1% have a positive reaction. From a large population of which 2 per 10,000 persons have TB, a person is selected at random and given a skin test, which turns out to be positive. What is the probability that the person has active pulmonary tuberculosis?

Solution The given data are organized in Fig. 5. The procedure called for is as follows: First select a person at random from the population. There are two possible outcomes: The person has TB [$\Pr(\text{TB}) = 2/10{,}000 = .0002$] or the person does not have TB [$\Pr(\text{not TB}) = .9998$]. For each of these two possibilities the possible test results and conditional probabilities are given. Multiplying the probabilities along each of the paths through the tree gives the probabilities of the different outcomes. The resulting probabilities are written on the right in Fig. 5. The problem asks for the conditional probability that a person has TB, given that the test is positive. By definition

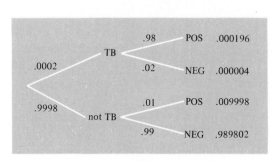

FIGURE 5

$$\Pr(\text{TB}|\text{POS}) = \frac{\Pr(\text{TB} \cap \text{POS})}{\Pr(\text{POS})} = \frac{.000196}{.000196 + .009998} = \frac{.000196}{.010194} \approx .02.$$

Therefore, the probability is .02 that a person with a positive skin test has TB. In other words, although the skin test is quite reliable, only about 2% of those with a positive test turn out to have active TB. This result must be taken into account when large-scale medical diagnostic tests are planned. Because the group of people without TB is so much larger than the group with TB, the small error in the former group is magnified to the point where it dominates the calculation.

Note The numerical data presented in Example 2 are only approximate. Variations in air quality for different localities within the United States cause variations in the incidence of TB and the reliability of skin tests.

Tree diagrams come in all shapes and sizes. Three or more branches might emanate from a single point, for example, and some trees may not have the symmetry of those in Examples 1 and 2. Tree diagrams arise whenever an activity can be thought of as a sequence of simpler activities.

EXAMPLE 3 A box contains five good light bulbs and two defective ones. Bulbs are selected one at a time (without replacement) until a good bulb is found. Find the probability that the number of bulbs selected is (i) one, (ii) two, (iii) three.

Solution The initial situation in the box is shown in Fig. 6(a). A bulb selected at random will be good (G) with probability $\frac{5}{7}$ and defective (D) with probability $\frac{2}{7}$. If a good bulb is selected, the activity stops. Otherwise, the situation is as shown in Fig. 6(b), and a bulb selected at random has probability $\frac{5}{6}$ of being good and probability $\frac{1}{6}$ of

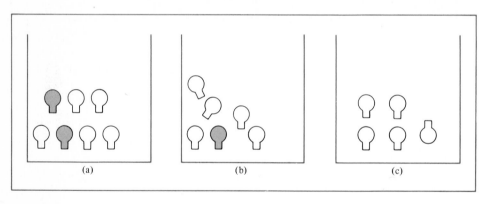

(a)   (b)   (c)

FIGURE 6

FIGURE 7

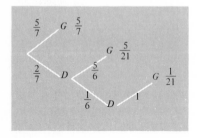

being defective. If the second bulb is good, the activity stops. If the second bulb is defective, then the situation is as shown in Fig. 6(c). At this point a bulb has probability 1 of being good.

The tree diagram corresponding to the sequence of activities is given in Fig. 7. Each of the three paths has a different length. The probability associated with each path has been computed by multiplying the probabilities for its branches. The first path corresponds to the situation where only one bulb is selected, the second path corresponds to two bulbs, and the third path to three bulbs. Therefore,

(i) $\Pr(1) = \frac{5}{7}$ (ii) $\Pr(2) = \frac{5}{21}$ (iii) $\Pr(3) = \frac{1}{21}$

PRACTICE PROBLEMS 6

Fifty percent of the students enrolled in a business statistics course had previously taken a finite math course. Thirty percent of these students received an A for the statistics course, whereas 20% of the other students received an A for the statistics course.

1. Draw a tree diagram and label it with the appropriate probabilities.

2. What is the probability that a student selected at random previously took a finite math course and did not receive an A in the statistics course?

3. What is the probability that a student selected at random from among the business statistics students received an A in the statistics course?

4. What is the conditional probability that a student previously took a finite math course, given that he received an A in the statistics course?

EXERCISES 6

In Exercises 1–4, draw trees representing the sequence of experiments.

1. Experiment I is performed. Outcome a occurs with probability .3 and outcome b occurs with probability .7. Then experiment II is performed. Its outcome c occurs with probability .6 and its outcome d occurs with probability .4.

2. Experiment I is performed twice. The three outcomes of experiment I are equally likely.

3. A stereo repair shop uses a two-step diagnostic procedure to repair amplifiers. Step I locates the problem in an amplifier with probability .8. Step II (which is executed only if step I fails to locate the problem) locates the problem with probability .6.

4. A training program is used by a corporation to direct hirees to appropriate jobs. The program consists of two steps. Step I identifies 30% as management trainees, 60% as nonmanagerial workers, and 10% as to be fired. In step II, 75% of the management trainees are assigned to managerial positions, 20% to nonmanagerial positions, and 5% are fired. In step II, 60% of the nonmanagerial workers are kept in the same category, 10% are assigned to management positions, and 30% are fired.

5. Refer to Exercise 3. What is the probability that the procedure will fail to locate the problem?

6. Refer to Exercise 4. What is the probability that a randomly chosen hiree will be assigned to a management position at the end of the training period?

7. Refer to Exercise 4. What is the probability that a randomly chosen hiree will be fired by the end of the training period?

8. Refer to Exercise 4. What is the probability that a randomly chosen hiree will be designated a management trainee but *not* be appointed to a management position?

9. Suppose that we have a white urn containing two white and one red balls and we have a red urn containing one white and three red balls. An experiment consists of selecting

at random a ball from the white urn and then (without replacing the first ball) selecting at random a ball from the urn having the color of the first ball. Find the probability that the second ball is red.

10. Color blindness is a sex-linked, inherited condition that is much more common among men than women. Suppose that 5% of all men and .4% of all women are color-blind. A person is chosen at random and found to be color-blind. What is the probability that the person is male? (You may assume that 50% of the population are men and 50% are women.)

11. A mouse is put into a T-maze (a maze shaped like a "T"). In this maze he has the choice of turning to the left and being rewarded with cheese or going to the right and receiving a mild shock. Before any conditioning takes place (i.e., on trial 1), the mouse is equally likely to go to the left or to the right. After the first trial his decision is influenced by what happened on the previous trial. If he receives cheese on any trial, the probabilities of his going to the left or right become .9 and .1, respectively, on the following trial. If he receives the electric shock on any trial, the probabilities of his going to the left or right on the next trial become .7 and .3, respectively. What is the probability that the mouse will turn left on the second trial?

12. Refer to Exercise 11. What is the probability that the mouse will turn left on the third trial?

13. A factory has two machines that produce bolts. Machine I produces 60% of the daily output of bolts, and 3% of its bolts are defective. Machine II produces 40% of the daily output, and 2% of its bolts are defective.

    (a) What is the probability that a bolt selected at random will be defective?

    (b) If a bolt is selected at random and found to be defective, what is the probability that it was produced by machine I?

14. Three ordinary quarters and a fake quarter with two heads are placed in a hat. One quarter is selected at random and tossed twice. If the outcome is "HH," what is the probability that the fake quarter was selected?

15. Suppose that the reliability of a test for hepatitis is specified as follows: of people with hepatitis, 95% have a positive reaction and 5% have a negative reaction; of people free of hepatitis, 90% have a negative reaction and 10% have a positive reaction. From a large population, of which .05% of the people have hepatitis, a person is selected at random and given the test. If the test is positive, what is the probability that the person actually has hepatitis?

16. A small citrus farmer has two groves. This season, grove I produced 3000 crates of oranges and 1000 crates of grapefruit. Grove II produced 5000 crates of oranges and 1000 crates of grapefruit. A crate is selected at random and is found to contain grapefruit. What is the probability that the crate is from grove I?

17. Suppose that during any year the probability of an accidental nuclear war is .0001 (provided, of course, that there hasn't been one in a previous year). Draw a tree diagram representing the possibilities for the next three years. What is the probability that there will be an accidental nuclear war during the next three years?

18. Refer to Exercise 9. What is the probability that there will be an accidental nuclear war during the next n years?

1.

			.3	A	.15
	.5	Finite math			
			.7	not A	.35
	.5	No finite math	.2	A	.10
			.8	not A	.40

2. The event "finite math and not A" corresponds to the second path of the tree diagram, which has probability .35.

3. This event is satisfied by the first and third paths and therefore has probability $.15 + .10 = .25$.

4. $\Pr(\text{finite math} \mid A) = \dfrac{\Pr(\text{finite math and A})}{\Pr(A)} = \dfrac{.15}{.25} = .6.$

9.7. Bayes' Theorem

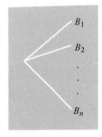

FIGURE 1

FIGURE 2

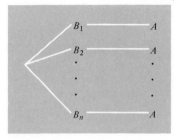

In the preceding section we analyzed multistage experiments by means of tree diagrams. As we have seen, a single event often results from any one of several different paths through the tree. For example, the event "Republican" in the tree of Fig. 3(b) of the previous section can be reached via two paths, one through Philadelphia and one through Pittsburgh. Often in applied problems we know that a given event (such a "Republican") has occurred and we must calculate the probability that a given path was taken. We solved a problem of this sort in Example 1(d) of the previous section, where we determined the probability that a voter is from Philadelphia, given that he favors the Republican candidate. The solution of this problem is a special case of a general principle called *Bayes' theorem*.

To state Bayes' theorem, let us consider a two-stage tree. Suppose that at the first stage there are events $B_1, B_2, \ldots, B_n$ which are mutually exclusive (Fig. 1). Let us examine only the paths of the tree which lead to one specific event, A (e.g., "Republican" in Example 1 of the preceding section). Suppose that we are given that A occurs. What is the probability that the path to A goes through B_1? In terms of conditional probabilities, we are asking for the value of $\Pr(B_1 \mid A)$. Similarly, we could ask for the probabilities that the path goes, respectively, through $B_2, B_3, \ldots,$ or B_n (Fig. 2)—namely

$$\Pr(B_2 \mid A), \Pr(B_3 \mid A), \ldots, \Pr(B_n \mid A).$$

These conditional probabilities may be computed using the following result:

Bayes' Theorem

$$\Pr(B_1|A) = \frac{\Pr(B_1)\Pr(A|B_1)}{\Pr(B_1)\Pr(A|B_1) + \Pr(B_2)\Pr(A|B_2) + \cdots + \Pr(B_n)\Pr(A|B_n)}$$

$$\Pr(B_2|A) = \frac{\Pr(B_2)\Pr(A|B_2)}{\Pr(B_1)\Pr(A|B_1) + \Pr(B_2)\Pr(A|B_2) + \cdots + \Pr(B_n)\Pr(A|B_n)}$$

and so forth.

We may easily verify Bayes' theorem using tree diagrams. Before doing so, however, let us illustrate the theorem.

EXAMPLE 1 Solve the tuberculosis skin test problem of Example 2 of the previous section by using Bayes' theorem.

Solution The observed event A is "positive skin test result." There are two possible events leading to A—namely

$$B_1 = \text{"person has tuberculosis"}$$

$$B_2 = \text{"person does not have tuberculosis."}$$

We wish to calculate $\Pr(B_1|A)$. From the data given we have

$$\Pr(B_1) = \frac{2}{10,000} = .0002$$

$$\Pr(B_2) = .9998$$

$$\Pr(A|B_1) = \Pr(\text{POS}|\text{TB}) = .98$$

$$\Pr(A|B_2) = \Pr(\text{POS}|\text{not TB}) = .01.$$

Therefore, by Bayes' theorem,

$$\Pr(B_1|A) = \frac{\Pr(B_1)\Pr(A|B_1)}{\Pr(B_1)\Pr(A|B_1) + \Pr(B_2)\Pr(A|B_2)}$$

$$= \frac{(.0002)(.98)}{(.0002)(.98) + (.9998)(.01)} \approx .02,$$

in agreement with our calculation of Example 2 of the previous section.

The advantages of Bayes' theorem over the use of tree diagrams are: (1) we do not need to draw the tree diagram to calculate the desired probability, and (2) we need not compute extraneous probabilities. These advantages become very significant in dealing with experiments with many outcomes.

EXAMPLE 2 A printer has seven book-binding machines. For each machine Table 1 gives the proportion of the total book production that it binds and the probability that the machine produces a defective binding.

TABLE 1

Machine	Proportion of books bound	Probability of defective binding
1	.10	.03
2	.05	.03
3	.20	.02
4	.15	.02
5	.25	.01
6	.15	.02
7	.10	.03

For instance, machine 1 binds 10% of the books and produces a defective binding with probability .03. Suppose that a book is selected at random and found to have a defective binding. What is the probability that it was bound by machine 1?

Solution Let B_i $(i = 1, 2, \ldots, 7)$ be the event that the book was bound by machine i and let A be the event that the book has a defective binding. Then, for example, $\Pr(B_1) = .10$ and $\Pr(A|B_1) = .03$. The problem asks for the reversed conditional probability, $\Pr(B_1|A)$. By Bayes' theorem

$$\Pr(B_1|A)$$

$$= \frac{\Pr(B_1) \Pr(A|B_1)}{\Pr(B_1) \Pr(A|B_1) + \Pr(B_2) \Pr(A|B_2) + \cdots + \Pr(B_7) \Pr(A|B_7)}$$

$$= \frac{(.10)(.03)}{(.10)(.03) + (.05)(.03) + (.20)(.02) + (.15)(.02) + (.25)(.01) + (.15)(.02) + (.10)(.03)}$$

$$= \frac{.003}{.02} = .15.$$

Verification of Bayes' Theorem The tree diagram in Fig. 3 shows all the paths that result in event A along with the associated probabilities. Looking at the first path, we see that $\Pr(B_1 \cap A) = \Pr(B_1) \cdot \Pr(A|B_1)$. Similarly, $\Pr(B_2 \cap A) = \Pr(B_2) \cdot \Pr(A|B_2)$, and so forth. By adding the probabilities for each path that leads to the event A, we obtain

$$\Pr(A) = \Pr(B_1) \Pr(A|B_1) + \Pr(B_2) \Pr(A|B_2) + \cdots + \Pr(B_n) \Pr(A|B_n).$$

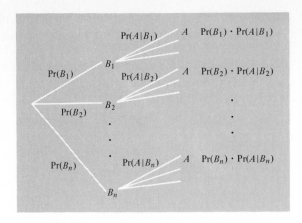

FIGURE 3

Now,

$$\Pr(B_1 | A) = \frac{\Pr(B_1 \cap A)}{\Pr(A)}$$

$$= \frac{\Pr(B_1) \Pr(A | B_1)}{\Pr(B_1) \Pr(A | B_1) + \Pr(B_2) \Pr(A | B_2) + \cdots + \Pr(B_n) \Pr(A | B_n)}.$$

An analogous derivation can be used for $\Pr(B_2 | A), \ldots, \Pr(B_n | A)$.

PRACTICE PROBLEMS 7

Refer to Example 2. Suppose that a book is selected at random and found to have a defective binding.

1. What is the probability that the book was bound by machine 2?

2. By what machine is the book most likely to have been bound?

EXERCISES 7

1. An automobile insurance company has determined the accident rate (probability of having at least one accident during a year) for various age groups. (See Table 2.) Suppose that a policyholder calls in to report an accident. What is the probability that he is over 60?

TABLE 2

Age group	Proportion of total insured	Accident rate
Under 21	.05	.06
21–30	.10	.04
31–40	.25	.02
41–50	.20	.015
51–60	.30	.025
Over 60	.10	.04

2. An electronic device has six different types of transistors. For each type of transistor, Table 3 gives the portion of the total number of transistors of that type and the failure rate (probability of failing within one year). If a transistor fails, what is the probability that it is type 1?

TABLE 3

Type	Proportion of total	Failure rate
1	.30	.0002
2	.25	.0004
3	.20	.0005
4	.10	.001
5	.05	.002
6	.10	.004

3. The enrollment in a certain course is 10% freshmen, 30% sophomores, 40% juniors, and 20% seniors. Past experience has shown that the likelihood of receiving an A in the course is .2 for freshmen, .4 for sophomores, .3 for juniors, and .1 for seniors. Find the probability that a student who receives an A is a sophomore.

4. A metropolitan police department maintains statistics of larcenies reported in the various precincts of the city. It records the proportion of the city population in each precinct and the precinct larceny rate ($=$ the proportion of the precinct population reporting a larceny within the past year). These statistics are summarized in Table 4. A larceny victim is randomly chosen from the city population. What is the probability he comes from Precinct 3?

TABLE 4

Precinct	Proportion of population	Larceny rate
1	.20	.01
2	.10	.02
3	.40	.05
4	.30	.04

5. Table 5 gives the distribution of incomes and that of two-car families by income level for a certain suburban county. Suppose that a randomly chosen family has two or more cars. What is the probability that its income is at least $25,000 per year?

6. Table 6 gives the distribution of voter registration and voter turnouts for a certain city. A randomly chosen person is questioned at the polls. What is the probability that the person is an Independent?

TABLE 5

Annual family income	Proportion of people	Proportion having two or more cars
< $10,000	.10	.2
$10,000–$14,999	.20	.5
$15,000–$19,999	.35	.6
$20,000–$24,999	.30	.75
≥ $25,000	.05	.9

TABLE 6

	Proportion registered	Proportion turnout
Democrat	.50	.4
Republican	.20	.5
Independent	.30	.7

SOLUTIONS TO PRACTICE PROBLEMS 7

1. The problem asks for $\Pr(B_2|A)$. Bayes' theorem gives this probability as a quotient with numerator $\Pr(B_2)\Pr(A|B_2)$ and the same denominator as in the solution to Example 2. Therefore,

$$\Pr(B_2|A) = \frac{\Pr(B_2)\,\Pr(A|B_2)}{.02} = \frac{(.05)(.03)}{.02} = .075.$$

2. To solve this problem we must compute the seven conditional probabilities $\Pr(B_1|A)$, $\Pr(B_2|A), \ldots, \Pr(B_7|A)$ and see which one is the largest. The first two have already been computed. Using the method of the preceding problem we find that $\Pr(B_3|A) = .20$, $\Pr(B_4|A) = .15$, $\Pr(B_5|A) = .125$, $\Pr(B_6|A) = .15$, and $\Pr(B_7|A) = .15$. Therefore, the book was most likely bound by machine 3.

Chapter 9: CHECKLIST

☐ Experiment
☐ Outcome
☐ Sample space
☐ Trial
☐ Event
☐ Impossible event
☐ Certain event
☐ Intersection, union, and complement of events

- ☐ Mutually exclusive events
- ☐ Elementary event
- ☐ Probability of an event
- ☐ Addition principle
- ☐ Inclusion-exclusion principle
- ☐ Odds
- ☐ Equally likely outcomes
- ☐ Complement rule
- ☐ $\Pr(E|F)$
- ☐ Independent events
- ☐ Tree diagram
- ☐ Bayes' theorem

Chapter 9: SUPPLEMENTARY EXERCISES

1. A coin is to be tossed five times. What is the probability of obtaining at least one head?

2. Suppose that we toss a coin three times and observe the sequence of heads and tails. Let E be the event that "the first toss lands heads" and F the event that "there are more heads than tails." Are E and F independent?

3. Each box of a certain brand of candy contains either a toy airplane or a toy gun. If one-third of the boxes contain an airplane and two-thirds contain a gun, what is the probability that a person who buys two boxes of candy will receive both an airplane and a gun?

4. A committee consists of five men and five women. If three people are selected at random from the committee, what is the probability that they will all be men?

5. Out of the 50 colleges in a certain state, 25 are private, 15 offer engineering majors, and 5 are private colleges offering engineering majors. If a college is selected at random, what is the conditional probability that it offers an engineering major given that it is a public college?

6. An auditing procedure for income tax returns has the following characteristics: If the return is incorrect, the probability is 90% that it will be rejected, and if the return is correct, the probability is 95% that it will be accepted. Suppose that 80% of all income tax returns are correct. If a return is audited and rejected, what is the probability that the return was actually correct?

7. Two archers shoot at a moving target. One can hit the target with probability $\frac{1}{4}$ and the other with probability $\frac{1}{3}$. Assuming that their efforts are independent events, what is the probability that

    (a) both will hit the target?   (b) at least one will hit the target?

8. If the odds in favor of an event are 7 to 5, what is the probability that the event will occur?

9. In an Olympic swimming event, two of the seven contestants are American. The contestants are randomly assigned to lanes 1 through 7. What is the probability that the Americans are assigned to the first two lanes?

10. An urn contains three balls numbered 1, 2, and 3. Balls are drawn one at a time without replacement until the sum of the numbers drawn is four or more. Find the probability of stopping after exactly two balls are drawn.

11. A red die and a green die are tossed as a pair. Let E be the event that "the red die shows a 2" and let F be the event that "the sum of the numbers is 8." Are the events E and F independent?

12. Let E and F be events with $\Pr(E) = .4$, $\Pr(F) = .3$, and $\Pr(E \cup F) = .5$. Find $\Pr(E|F)$.

13. A supermarket has three employees who package and weigh produce. Employee A records the correct weight 98% of the time. Employees B and C record the correct weight 97% and 95% of the time, respectively. Employees A, B, and C handle 40%, 40%, and 20% of the packaging, respectively. A customer complains about the incorrect weight recorded on a package he has purchased. What is the probability that the package was weighed by employee C?

14. Three people are chosen at random. What is the probability that at least two of them were born on the same day of the week?

Probability
and Statistics

Chapter 10

Statistics is the branch of mathematics which deals with data—their collection, description, analysis, and prediction. In this chapter we present some topics in statistics which can be used as a springboard to further study. Since we are presenting a series of topics, rather than a comprehensive survey, we shall bypass large areas of statistics without saying anything about them. However, the discussion should give you some feeling for the subject. In Section 1 we discuss the problem of describing data by means of a probability distribution and a histogram. We shall also introduce the language of random variables. In Sections 2 and 3 we introduce the two most frequently used descriptive statistics: the mean and standard deviation. In Section 3 we illustrate how Chebychev's inequality can be used to make estimations. In Sections 4 and 5 we discuss the probability distributions most frequently employed in statistical work: the binomial and the normal distributions.

10.1. Probability Distributions and Random Variables

Our goal in this section is to describe given data in terms that allow for interpretation and comparison. As we shall see, both graphical and tabular displays of data can be useful for this purpose.

Our modern technological society has a fetish about gathering statistical data. It is hardly possible to glance at a newspaper or a magazine and not be confronted with massive arrays of statistics gathered from the latest studies of schools, churches, the economy, and so forth. One of the chief tasks confronting us is to interpret in a meaningful way the data collected and to make decisions based on our interpretations. The mathematical tools for doing this belong to that part of mathematics called *statistics*. To get an idea of the problems considered in statistics, let us consider a concrete example.

Mr. Jones, a businessman, is interested in purchasing a car dealership. Two dealerships are for sale, and each dealer has provided him with data describing past sales. Dealership A provided 1 year's worth of data, dealership B 2 years' worth. The data are summarized in Table 1. The problem confronting Mr. Jones is that of analyzing the data to determine which car dealership to buy.

TABLE 1

Dealership A			Dealership B	
Weekly sales	Number of occurrences		Weekly sales	Number of occurrences
5	2		5	20
6	2		6	0
7	13		7	0
8	20		8	10
9	10		9	12
10	4		10	50
11	1		11	12

(a) (b)

These data are presented in a form often used in statistical surveys. For each possible value of a statistical variable (in this case the number of cars sold weekly) we have tabulated the number of occurrences. Such a tabulation is called a *frequency distribution.* Although a frequency distribution is a very useful way of displaying and summarizing survey data, it is by no means the most efficient form in which to analyze such data. For example, it is difficult to compare dealership A with dealership B using only Table 1. Comparisons are much more easily made if we use proportions rather than actual numbers of occurrences. For example, instead of recording that dealership A had weekly sales of 5 cars during two of the weeks of the year, let us record that the proportion of the observed weeks in which dealership A had weekly sales of 5 was $\frac{2}{52} \approx .04$. Similarly, by dividing each of the entries in the right column by 52, we obtain a new table describing the sales of dealership A (Table 2).*

We can similarly construct a new table for dealership B (Table 3).

We can interpret the proportions on the right sides of Tables 2 and 3 as probabilities relative to the following experiment. Pick a week at random and observe the number of cars sold that week. Then the possible outcomes are 5, 6, 7, 8, 9, 10, and 11. The probability of each outcome is listed in the column on the right. Tables such as 2 and 3 are called *probability distributions.* More precisely, suppose that we are given an experiment with numerical outcomes $x_1, x_2, x_3, \ldots,$ x_N. Suppose that the probability of x_1 is p_1, of x_2 is p_2, and so forth. Then the

---

* For simplification we shall round off the data of this example to two decimal places.

TABLE 2

Dealership A	
Weekly sales	Proportion of occurrences
5	$\frac{2}{52} \approx .04$
6	$\frac{2}{52} \approx .04$
7	$\frac{13}{52} = .25$
8	$\frac{20}{52} \approx .38$
9	$\frac{10}{52} \approx .19$
10	$\frac{4}{52} \approx .08$
11	$\frac{1}{52} \approx .02$

TABLE 3

Dealership B	
Weekly sales	Proportion of occurrences
5	$\frac{20}{104} \approx .19$
6	0
7	0
8	$\frac{10}{104} \approx .10$
9	$\frac{12}{104} \approx .12$
10	$\frac{50}{104} \approx .48$
11	$\frac{12}{104} \approx .12$

table is called the *probability distribution* for the experiment. The probability distribution lists all of the outcomes of the experiment and their probabilities. (For the sake of simplicity, we usually arrange $x_1, x_2, \ldots, x_N$ in increasing order.)

Outcome	Probability
x_1	p_1
x_2	p_2
$\vdots$	$\vdots$
x_N	p_N

EXAMPLE 1 A coin is tossed five times and the number of occurrences of heads is observed. Determine the probability distribution for this experiment.

Solution The number of heads is 0, 1, 2, 3, 4, or 5. The number of distinct sequences of 5 tosses is 2^5 or 32. The number of sequences having k heads (and $5 - k$ tails) is

$C(5, k) = \binom{5}{k}$. Thus

$$\Pr(k \text{ heads}) = \frac{\binom{5}{k}}{2^5} \qquad (k = 0, 1, 2, 3, 4, 5).$$

The probability distribution for the experiment is, therefore:

Number of heads	Probability
0	$\dfrac{\binom{5}{0}}{2^5} = \dfrac{1}{32}$
1	$\dfrac{\binom{5}{1}}{2^5} = \dfrac{5}{32}$
2	$\dfrac{\binom{5}{2}}{2^5} = \dfrac{10}{32}$
3	$\dfrac{\binom{5}{3}}{2^5} = \dfrac{10}{32}$
4	$\dfrac{\binom{5}{4}}{2^5} = \dfrac{5}{32}$
5	$\dfrac{\binom{5}{5}}{2^5} = \dfrac{1}{32}$

It is often possible to gain useful insight into an experiment by representing its probability distribution in graphical form. For instance, let us graph the probability distribution for car dealership A. Begin by drawing a number line (Fig. 1).

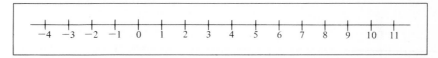

FIGURE 1

The numbers which represent possible outcomes of the experiment (weekly car sales) are 5, 6, 7, 8, 9, 10, 11. Locate each of these numbers on the number line. Above each number erect a rectangle whose height is the probability of that number. Above the number 5, for example, we draw a rectangle of height .04. The completed graph is shown in Fig. 2. For the sake of comparison we have also drawn a graph of the probability distribution for dealership B.

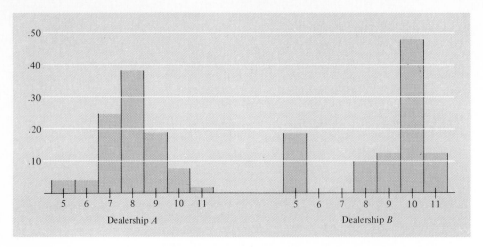

FIGURE 2

Graphs of the type just drawn are called *histograms*. They vividly describe the experiment being considered. For example, comparison of the histograms of Fig. 2 reveals significant differences between the two dealerships. On the one hand, dealership A is very consistent. Most weeks its sales are in the middle range of 7, 8, or 9. On the other hand, dealership B can often achieve very high sales (it had sales of 10 in 48% of the weeks) at the expense of a significant number of weeks of low sales (sales of 5 in 19% of the weeks).

It is possible to use histograms to represent the probabilities of events as areas. To illustrate the procedure, consider the histogram corresponding to dealership A. This histogram consists of a number of rectangles. Each rectangle is of width 1 and height equal to the probability of a particular outcome. For instance, the highest rectangle is centered over the number 8 and has height .38, the probability of the outcome 8. Note that the area of this rectangle is

$$\text{area} = (\text{height})(\text{width}) = (.38)(1) = .38.$$

In other words, the area of the rectangle also equals the probability of the corresponding outcome. We have verified this in the case of the outcome 8, but it is true generally. In a similar fashion we may represent the probabilities of more complicated events as areas. Consider, for example, the event E = "sales between 7 and 10 inclusive." This event consists of the set of outcomes $\{7, 8, 9, 10\}$, and so its probability is just the sum

$$\Pr(7) + \Pr(8) + \Pr(9) + \Pr(10).$$

And each of these probabilities is equal to the area of a rectangle in the histogram, namely those shaded in Fig. 3. Thus the area of the shaded region equals the probability of the event E. This result is a special instance of the following general fact.

FIGURE 3

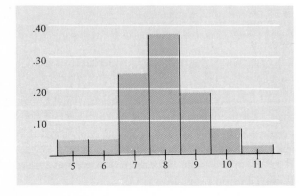

Suppose that the rectangles of a histogram of a probability distribution all have width 1. Then the probability of an event E is the sum of the areas of the rectangles corresponding to the outcomes in E.

EXAMPLE 2 Suppose that the histogram of a probability distribution is as given in Fig. 4. Shade in the portion of the histogram whose area is the probability of the event "more than 50."

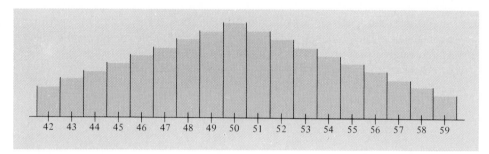

FIGURE 4

Solution The event "more than 50" consists of the outcomes $\{51, 52, 53, 54, 55, 56, 57, 58, 59\}$. So we shade in the portion of the histogram corresponding to these outcomes (Fig. 5). The area of the shaded region is the probability of the event "more than 50."

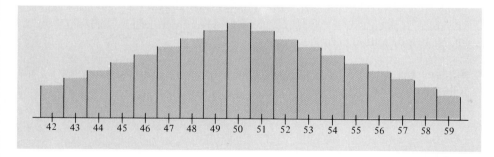

FIGURE 5

Random Variables   Consider an experiment with numerical outcomes. Denote the outcome of the experiment by the letter X. Thus, for example, if the experiment consists of observing the result of tossing a single die, then X assumes one of the six values 1, 2, 3, 4, 5, 6. Since the values of the variable X are determined by the unpredictable random outcomes of the experiment, X is called a *random variable* or, more specifically, the *random variable associated to the experiment*. The random-variable notation is often convenient and is in common use in probability and statistics texts.

If k is one of the possible outcomes of the experiment with associated random variable X, then we denote the probability of the outcome k by

$$\Pr(X = k).$$

For example, if we consider the die-tossing experiment described above, then

$$\Pr(X = 3)$$

denotes the probability of throwing a 3.

Rather than speak of the probability distribution associated to an experiment, we can speak of the *probability distribution associated to the corresponding random variable*. Such a probability distribution is a table listing the various values of X (i.e., outcomes of the experiment) and their associated probabilities:

k	$\Pr(X = k)$
x_1	p_1
x_2	p_2
$\vdots$	$\vdots$
x_N	p_N

The advantage of the random-variable notation is that the variable X can be treated algebraically and we can consider expressions such as X^2. This is just the random variable corresponding to the experiment whose outcomes are the squares of the outcomes of the original experiment. Similarly, we can consider random variables such as $X + 3$ and $(X - 2)^2$. In considering several different experiments, it is sometimes necessary to use letters other than X to stand for random variables. It is customary, however, to use only capital letters, such as X, Y, Z, W, U, V, for random variables.

EXAMPLE 3 Suppose that a random variable X has probability distribution given by the following table:

k	$\Pr(X = k)$
1	.2
2	.3
3	.1
4	.4

Determine the probability distribution of the random variable X^2.

Solution The outcomes of X^2 are the squares of the outcomes of X. In this case, the probabilities of the outcomes of X^2 are determined by the probabilities of the outcomes of X. The possible outcomes of X^2 are $k = 1, 4, 9$, and 16. So the probability distribution of X^2 is:

k	$\Pr(X^2 = k)$
1	.2
4	.3
9	.1
16	.4

EXAMPLE 4 Let X denote the random variable defined as the sum of the upper faces appearing when two dice are thrown. Determine the probability distribution of X.

Solution The experiment of throwing two dice leads to 36 possibilities, each having probability $\frac{1}{36}$.

(1, 1)	(1, 2)	(1, 3)	(1, 4)	(1, 5)	(1, 6)
(2, 1)	(2, 2)	(2, 3)	(2, 4)	(2, 5)	(2, 6)
(3, 1)	(3, 2)	(3, 3)	(3, 4)	(3, 5)	(3, 6)
(4, 1)	(4, 2)	(4, 3)	(4, 4)	(4, 5)	(4, 6)
(5, 1)	(5, 2)	(5, 3)	(5, 4)	(5, 5)	(5, 6)
(6, 1)	(6, 2)	(6, 3)	(6, 4)	(6, 5)	(6, 6)

The sum of the numbers in each pair gives the value of X. For example, the pair (3, 1) corresponds to $X = 4$. Note that the pairs corresponding to a given value of X lie on a diagonal, as shown above, where we have indicated all pairs corresponding to $X = 4$. It is now easy to calculate the number of pairs corresponding to a given value of X and from it the probability $\Pr(X = k)$. For example, there are three pairs adding to four, so

$$\Pr(X = 4) = \tfrac{3}{36} = \tfrac{1}{12}.$$

Performing this calculation for all k from 2 to 12 gives the following probability distribution:

k	$\Pr(X = k)$	k	$\Pr(X = k)$
2	$\frac{1}{36}$	8	$\frac{5}{36}$
3	$\frac{1}{18}$	9	$\frac{1}{9}$
4	$\frac{1}{12}$	10	$\frac{1}{12}$
5	$\frac{1}{9}$	11	$\frac{1}{18}$
6	$\frac{5}{36}$	12	$\frac{1}{36}$
7	$\frac{1}{6}$		

PRACTICE PROBLEMS 1

1. In a certain carnival game a wheel is divided into five equal parts, of which two are red and three are white. The player spins the wheel until the marker lands on "red" or until three spins have occurred. The number of spins is observed. Determine the probability distribution for this experiment.

2. Refer to the carnival game of Problem 1. Suppose that the player pays $1 to play this game and receives 50 cents for each spin. Determine the probability distribution for the experiment of playing the game and observing the player's earnings.

EXERCISES 1

1. Table 4(a) gives the frequency distribution for the final grades in a course. (Here $A = 4$, $B = 3$, $C = 2$, $D = 1$, $F = 0$.) Determine the probability distribution associated with these data.

TABLE 4

Grade	Number of occurrences		Number of cars waiting	Number of occurrences
0	2		0	0
1	3		1	9
2	10		2	21
3	6		3	15
4	4		4	12
			5	3

(a) (b)

2. The number of cars waiting to be served at a gas station was counted at the beginning of every minute during the morning rush hour. The frequency distribution is given in Table 4(b). Determine the probability distribution associated with these data and draw the histogram.

3. The telephone company counted the number of people dialing the weather each minute on a rainy morning from 5 A.M to 6 A.M. The frequency distribution is given in Table 5(a). Determine the probability distribution associated with these data.

4. A production manager counted the number of items produced each hour during a 40-hour workweek. The frequency distribution is given in Table 5(b). Determine the probability distribution associated with these data.

5. A fair coin is tossed three times and the number of heads is observed. Determine the probability distribution for this experiment and draw its histogram.

6. An urn contains three red balls and four white balls. A sample of three balls is selected at random and the number of red balls observed. Determine the probability distribution for this experiment and draw its histogram.

7. An archer can hit the bull's-eye of the target with probability $\frac{1}{3}$. She shoots until she hits the bull's-eye or until four shots have been taken. The number of shots is observed. Determine the probability distribution for this experiment.

8. A die is rolled and the number on the top face is observed. Determine the probability distribution for this experiment and draw its histogram.

TABLE 5

Number of calls during minute	Number of occurrences
20	3
21	3
22	0
23	6
24	18
25	12
26	0
27	9
28	6
29	3

(a)

Number produced during hour	Number of occurrences
50	2
51	0
52	4
53	6
54	14
55	8
56	4
57	0
58	0
59	2

(b)

9. In a certain carnival game the player selects two balls at random from an urn containing two red balls and four white balls. The player receives $5 if he draws two red balls and $1 if he draws one red ball. He loses $1 if no red balls are in the sample. Determine the probability distribution for the experiment of playing the game and observing the player's earnings.

10. In a certain carnival game a player pays a dollar and then tosses a fair coin until either a "head" occurs or he has tossed the coin four times. He receives 50 cents for each toss. Determine the probability distribution for the experiment of playing the game and observing the player's earnings.

11. Figure 6(a) is the histogram for a probability distribution. What is the probability that the outcome is between 5 and 7 inclusive?

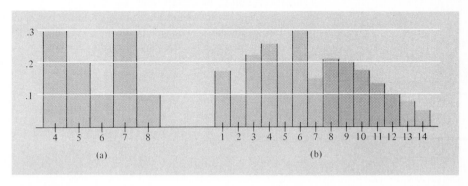

FIGURE 6

12. Figure 6(b) is the histogram for a probability distribution. To what event do the shaded rectangles correspond?

Let the random variables X and Y have the following probability distributions

k	$\Pr(X = k)$	k	$\Pr(Y = k)$
0	.1	5	.3
1	.2	10	.4
2	.3	15	.1
3	.2	20	.1
4	.2	25	.1

Determine the probability distributions of the following random variables.

13. X^2 **14.** Y^2 **15.** $X - 1$ **16.** $Y - 15$

17. $\frac{1}{5} Y$ **18.** $2X^2$ **19.** $(X + 1)^2$ **20.** $(\frac{1}{5} Y + 1)^2$

SOLUTIONS TO PRACTICE PROBLEMS 1

1. Since the outcomes are the number of spins, there are three possible outcomes: one, two, and three spins. The probabilities for each of these outcomes can be computed from a tree diagram. For instance, the outcome two (spins) occurs if the first spin lands on white and the second spin on red. The probability of this outcome is $\frac{3}{5} \cdot \frac{2}{5} = \frac{6}{25}$.

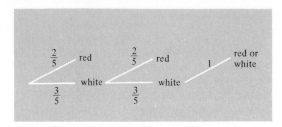

Outcome	Probability
1	$\frac{2}{5}$
2	$\frac{6}{25}$
3	$\frac{9}{25}$

2. The same game is being played as in Problem 1, except that now the outcome we are concentrating on is the player's financial situation at the end of the game. The player's earnings depend on the number of spins as follows: one spin results in $-\$.50$ earnings (i.e., a loss of 50 cents), two spins result in $\$0$ earnings (i.e., breaking even) and 3 spins result in $\$.50$ earnings (i.e., the player ends up ahead by 50 cents). The probabilities for these three situations are the same as before.

Earnings	Probability
$-\$.50$	$\frac{2}{5}$
0	$\frac{6}{25}$
$\$.50$	$\frac{9}{25}$

10.2. The Mean of a Probability Distribution

Let us now introduce the *average* (also called the *mean* or *expected value*) of a probability distribution. The average of a set of numbers is familiar from elementary mathematics. For example, to form the average of 11, 17, 18, and 10 we add these numbers and divide by 4:

$$[\text{average}] = \frac{11 + 17 + 18 + 10}{4} = 14.$$

More generally, we have the following result:

The *average* of the numbers $x_1, x_2, \ldots, x_N$ is

$$\frac{x_1 + x_2 + \cdots + x_N}{N}.$$

EXAMPLE 1   Compute the average weekly sales of dealership A of the preceding section.

Solution   Recall that the weekly sales of dealership A are given in the table:

Weekly sales	Number of occurrences
5	2
6	2
7	13
8	20
9	10
10	4
11	1

So we may form the average of the weekly sales figures as follows:

$$\tfrac{1}{52}\big[(5+5)+(6+6)+\underbrace{(7+\cdots+7)}_{13\text{ times}}+\underbrace{(8+\cdots+8)}_{20\text{ times}}+\underbrace{(9+\cdots+9)}_{10\text{ times}}+\underbrace{(10+\cdots+10)}_{4\text{ times}}+11\big]$$

$$= \frac{5 \cdot 2 + 6 \cdot 2 + 7 \cdot 13 + 8 \cdot 20 + 9 \cdot 10 + 10 \cdot 4 + 11}{52}$$

$$= 5 \cdot \tfrac{2}{52} + 6 \cdot \tfrac{2}{52} + 7 \cdot \tfrac{13}{52} + 8 \cdot \tfrac{20}{52} + 9 \cdot \tfrac{10}{52} + 10 \cdot \tfrac{4}{52} + 11 \cdot \tfrac{1}{52} \quad (*)$$

$$\approx 7.96.$$

Thus the average weekly sales is approximately 7.96 cars.

Let us reexamine the calculation above. The average is given by the expression $(*)$. Another way of writing this expression is:

$$[\text{average weekly sales}] = 5 \cdot \text{Pr(weekly sales} = 5) + 6 \cdot \text{Pr(weekly sales} = 6)$$

$$+ 7 \cdot \text{Pr(weekly sales} = 7) + 8 \cdot \text{Pr(weekly sales} = 8)$$

$$+ 9 \cdot \text{Pr(weekly sales} = 9) + 10 \cdot \text{Pr(weekly sales} = 10)$$

$$+ 11 \cdot \text{Pr(weekly sales} = 11).$$

That is, to compute the average weekly sales we need only add up the products [number of sales][probability of occurrence] over all possible sales figures. This suggests a method for calculating the average value of any probability distribution:

Average Value   Suppose that an experiment has as possible outcomes the numbers $x_1, x_2, \ldots, x_N$. Further suppose that the probability of x_1 is p_1, of x_2 is p_2, and so forth. Then the *average value* of the probability distribution for the experiment is

$$[\text{average value}] = x_1 p_1 + x_2 p_2 + \cdots + x_N p_N.$$

The average value of a probability distribution is also called its *mean* or its *expected value*. The three terms are used interchangeably. Most probability and statistics texts denote the average value of a probability distribution by the Greek letter μ (mu). The average value has the following meaning: If the experiment is performed a large number of times, then the average of all outcomes of the experiment should be approximately the number μ. This coincides with our intuitive notion of average.

EXAMPLE 2   An ecologist observes the life expectancy of a certain species of deer. Based on a population of 1000 deer, he observes the following data.

Age at death (years)	Number observed
1	0
2	60
3	180
4	250
5	200
6	120
7	50
8	120
9	20

What is the average life expectancy of this species of deer?

Solution   We convert the given data into a probability distribution by replacing observed frequencies by relative frequencies [= (observed frequency)/1000].

Age at death	Probability
1	0
2	.06
3	.18
4	.25
5	.20
6	.12
7	.05
8	.12
9	.02

The expected value of this probability distribution is

$$\mu = 1 \cdot 0 + 2 \cdot (.06) + 3 \cdot (.18) + 4 \cdot (.25) + 5 \cdot (.20) + 6 \cdot (.12)$$
$$+ 7 \cdot (.05) + 8 \cdot (.12) + 9 \cdot (.02) = 4.87.$$

So the average life expectancy of a deer of this species is 4.87 years.

EXAMPLE 3 Which car dealership should Mr. Jones buy if he wants the one which will, on the average, sell the most cars?

Solution We have seen above that the expected value of the probability distribution for dealership A is just $\mu_A = 7.96$. On the other hand, associated to the probability distribution for dealership B—namely

Weekly sales	Probability
5	$\frac{20}{104}$
6	0
7	0
8	$\frac{10}{104}$
9	$\frac{12}{104}$
10	$\frac{50}{104}$
11	$\frac{12}{104}$

—we find the average value:

$$\mu = 5 \cdot \left(\tfrac{20}{104}\right) + 6 \cdot 0 + 7 \cdot 0 + 8 \cdot \left(\tfrac{10}{104}\right) + 9 \cdot \left(\tfrac{12}{104}\right) + 10 \cdot \left(\tfrac{50}{104}\right) + 11 \cdot \left(\tfrac{12}{104}\right)$$
$$B = 8.85.$$

Thus, the average sales of dealership A are 7.96 cars per week, whereas those of dealership B are 8.85 cars per week. Mr. Jones should buy dealership B.

The notion of expected value may be expressed in the language of random variables. Suppose that X is a random variable with the following probability distribution:

k	$\Pr(X = k)$
x_1	p_1
x_2	p_2
$\vdots$	$\vdots$
x_N	p_N

Then the values of X (namely $x_1, x_2, \ldots, x_N$) are the outcomes of an experiment. The *expected value* of X, denoted $E(X)$, is defined as the expected value of that experiment. That is, we have the following formula for $E(X)$.

EXAMPLE 4   Let the random variable X denote the sum of the faces appearing after throwing two dice. Determine $E(X)$.

Solution   We determined the probability distribution of X in Example 4 of Section 1:

k	$\Pr(X = k)$	k	$\Pr(X = k)$
2	$\frac{1}{36}$	8	$\frac{5}{36}$
3	$\frac{1}{18}$	9	$\frac{1}{9}$
4	$\frac{1}{12}$	10	$\frac{1}{12}$
5	$\frac{1}{9}$	11	$\frac{1}{18}$
6	$\frac{5}{36}$	12	$\frac{1}{36}$
7	$\frac{1}{6}$		

Therefore,

$$E(X) = 2 \cdot \tfrac{1}{36} + 3 \cdot \tfrac{1}{18} + 4 \cdot \tfrac{1}{12} + 5 \cdot \tfrac{1}{9} + 6 \cdot \tfrac{5}{36} + 7 \cdot \tfrac{1}{6}$$
$$+ 8 \cdot \tfrac{5}{36} + 9 \cdot \tfrac{1}{9} + 10 \cdot \tfrac{1}{12} + 11 \cdot \tfrac{1}{18} + 12 \cdot \tfrac{1}{36} = 7.$$

The expected value of a random variable may be used to analyze games of chance, as the next two examples show.

EXAMPLE 5   Two people play a dice game. A single die is thrown. If the outcome is 1 or 2, then A pays B \$2. If the outcome is 3, 4, 5, or 6, then B pays A \$4. What is the long-run expectation of the game for A?

Solution   Let X be the random variable giving the payoff to A. Then X assumes the possible values -2 and 4. Moreover, since the probability of 1 or 2 on the die is $\frac{1}{3}$, we have

$$\Pr(X = -2) = \tfrac{1}{3}.$$

Similarly,

$$\Pr(X = 4) = \tfrac{2}{3}.$$

Therefore,

$$E(X) = (-2) \cdot \tfrac{1}{3} + 4 \cdot \tfrac{2}{3} = 2.$$

In other words, the expected payoff to A is \$2 per play. If the game is repeated a large number of times, then on the average A should profit \$2 per play.

In evaluating a game of chance we use the expected value of the winnings to determine how fair the game is. The expected value of a completely fair game is zero. Let us compute the expected value of the winnings for two variations of the game roulette. American and European roulette games differ in both the nature of the wheel and the rules for playing.

American roulette wheels have 38 numbers (1 through 36 plus 0 and 00), of which 18 are red, 18 are black, and 2 are green. Many different types of bets are possible. We shall consider the "red" bet. When you bet $1 on red, you win $1 if a red number appears and lose $1 otherwise.

European roulette wheels have 37 numbers (1 through 36 plus 0). The rules of European roulette differ from the American rules. One variation is as follows: When you bet $1 on "red," then you win $1 if the ball lands on a red number and lose $1 if the ball lands on a black number. However, if the ball lands on the green number (0), then your bet stays on the table (the bet is said to be "imprisoned") and the payoff is determined by the result of the next spin. If a red number appears, you recieve your $1 bet back, and if a black number appears, you lose your $1 bet. However, if the green number (0) appears, you get back half of your bet, $.50. The tree diagrams for American and European roulette are given in Fig. 1.

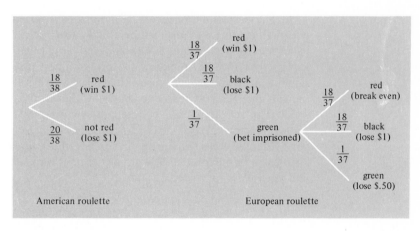

FIGURE 1

EXAMPLE 6 (a) Set up the probability distribution tables for the earnings in American roulette and European roulette for the $1 bet on red.

(b) Compute the expected values for the probability distributions in part (a).

Solution (a) For American roulette there are only two possibilities: earnings of $1 or of -1. These occur with probabilities $\frac{18}{38}$ and $\frac{20}{38}$, respectively.

For European roulette the possible earnings are $1, $0, $-.50$, or -1. There are two ways in which to lose $1, one with probability $\frac{18}{37}$ and the other with probability $\frac{1}{37} \cdot \frac{18}{37} = \frac{18}{1369}$. Therefore, $\Pr(\text{lose } \$1) = \frac{18}{37} + \frac{18}{1369} = \frac{684}{1369}$.

TABLE 1

American roulette		European roulette	
Earnings	Probability	Earnings	Probability
1	$\frac{18}{38}$	1	$\frac{18}{37}$
-1	$\frac{20}{38}$	0	$\frac{18}{1369}$
		$-\frac{1}{2}$	$\frac{1}{1369}$
		-1	$\frac{684}{1369}$

These probability distributions are tabulated in Table 1.

(b) American roulette: $\mu = 1 \cdot \frac{18}{38} + (-1) \cdot \frac{20}{38} = -\frac{2}{38} \approx -.0526.$

European roulette: $\mu = 1 \cdot \frac{18}{37} + 0 \cdot \frac{18}{1369} + (-\frac{1}{2})\frac{1}{1369} + (-1)\frac{684}{1369}$

$$= -\frac{1}{74} \approx -.0135.$$

The secrets of Nicholas Dandolas, one of the famous gamblers of the twentieth century, are revealed by Ted Thackrey, Jr., in *Gambling Secrets of Nick the Greek* (Rand McNally, 1968). The chapter entitled "Roulette" is subtitled "For Europeans Only." Looking at the probabilities of winning does not reveal any significant advantage of European roulette over American roulette: $\frac{18}{37}$ is not much bigger than $\frac{18}{38}$. Also, the chance in European roulette to break even is very small. The real difference between the two games is revealed by the expected values. Someone playing American roulette will lose, on the average, about $5\frac{1}{4}$ cents per \$1 bet, whereas for European roulette the average loss is about $1\frac{1}{3}$ cents. In both cases you expect to lose money in the long run, but in American roulette you lose nearly four times as much.

PRACTICE PROBLEMS 2

1. A 74-year-old man pays \$100 for a 1-year life insurance policy, which pays \$2000 in the event that he dies during the next year. According to life insurance tables, the probability of a 74-year-old man's living one additional year is .95. Write down the probability distribution for the possible financial outcomes and determine its expected value.

2. According to life insurance tables, the probability that a 74-year-old man will live an additional 5 years is .7. How much should a 74-year-old be willing to pay for a policy that pays \$2000 in the event of death any time within the next 5 years?

EXERCISES 2

1. Find the expected value for the probability distribution in Table 2(a).

2. Find the expected value for the probability distribution in Table 2(b).

TABLE 2

Values	Probability
0	.15
1	.2
2	.1
3	.25
4	.3

(a)

Values	Probability
-1	.1
$-\frac{1}{2}$	.4
0	.25
$\frac{1}{2}$	.2
1	.05

(b)

3. A college student received the following course grades for 10 (three-credit) courses during his freshman year: 4, 4, 4, 3, 3, 3, 3, 2, 2, 1.

   (a) Find his grade point average by adding the grades and dividing by 10.

   (b) Write down the probability distribution table.

   (c) Find the mean of the probability distribution in part (b).

4. An Olympic gymnast received the following scores from six judges: 9.8, 9.8, 9.4, 9.2, 9.2, 9.0.

   (a) Find the average score by adding the scores and dividing by 6.

   (b) Write down the probability distribution table.

   (c) Find the mean of the probability distribution in part (b).

5. Table 3 gives the probability distributions of the number of cavities for two groups of children trying different brands of toothpaste. Calculate the means of the probability distributions to determine which group had the fewer cavities.

   TABLE 3

Group A		Group B	
Number of cavities	Probability	Number of cavities	Probability
0	.3	0	.2
1	.3	1	.3
2	.2	2	.3
3	.1	3	.1
4	0	4	.1
5	.1	5	0

6. Table 4 gives the possible returns of two different investments and their probabilities. Calculate the means of the probability distribution to determine which investment has the greater expected return.

TABLE 4

Investment A	
Return	Probability
$1000	.2
$2000	.5
$3000	.3

Investment B	
Return	Probability
− $3000	.1
0	.3
$4000	.6

7. In American roulette, a bettor may place a $1 bet on any one of the 38 numbers on the roulette wheel. He wins $35 (plus the return of his bet) if the ball lands on his number; otherwise, he loses his bet. Write down the probability distribution for the earnings from this type of bet and find the expected value.

8. In American roulette, a dollar may be bet on a pair of numbers. The expected earnings for this type of bet is $-\$\frac{1}{19}$. How much money does the bettor receive if the ball lands on one of the two numbers?

9. In a carnival game, the player selects balls one at a time, without replacement, from an urn containing two red and four white balls. The game proceeds until a red ball is drawn. The player pays $1 to play the game and receives $\$\frac{1}{2}$ for each ball drawn. Write down the probability distribution for the player's earnings and find its expected value.

10. In a carnival game, the player selects two coins from a bag containing two silver dollars and six slugs. Write down the probability distribution for the winnings and determine how much the player would have to pay so that he would break even, on average, over many repetitions of the game.

11. Using life insurance tables, a retired man determines that the probability of living 5 more years is .9. He decides to take out a life insurance policy which will pay $10,000 in the event that he dies during the next 5 years. How much should he be willing to pay for this policy? (Do not take account of interest rates or inflation.)

12. Using life insurance tables, a retired couple determine that the probability of living 5 more years is .9 for the man and .95 for the woman. They decide to take out a life insurance policy which will pay $10,000 if either one dies during the next 5 years and $15,000 if both die during that time. How much should they be willing to pay for this policy? (Assume that their life spans are independent events.)

13. A pair of dice is tossed and the larger of the two numbers showing is recorded. Find the expected value of this experiment.

SOLUTIONS TO PRACTICE PROBLEMS 2

1. There are two possibilities. If the man lives until the end of the year, he loses $100. If he dies during the year, his estate gains by $1900 (the $2000 settlement minus the $100 premium).

Outcome	Probability
−$100	.95
$1900	.05

$$\mu = (-100)(.95) + (1900)(.05) = 0.$$

(Thus, if the insurance company insures a large number of people, it should break even. Its profits will result from the interest that it earns on the money being held.)

2. Let x denote the cost of the policy. The probability distribution is given below.

Outcome	Probability
−x	.7
$2000 - x$	.3

$$\mu = (-x)(.7) + (2000 - x)(.3)$$

$$= -.7x + 600 - .3x$$

$$= 600 - x.$$

The expected value will be zero if $x = 600$. Therefore, the man should be willing to pay $600 for his policy.

10.3. The Variance and Standard Deviation of a Probability Distribution

In the preceding section we introduced the mean (or expected value) of a probability distribution. The mean is probably the single most important number which can be used to describe a probability distribution. The next most important number is the variance. Roughly speaking, the variance measures the dispersal or spread of a probability distribution about its mean. The more closely concentrated the distribution about its mean, the smaller the variance; the more spread out, the larger the variance. Thus, for example, the probability distribution whose histogram is drawn in Fig. 1(a) has a smaller variance than that in Fig. 1(b).

FIGURE 1

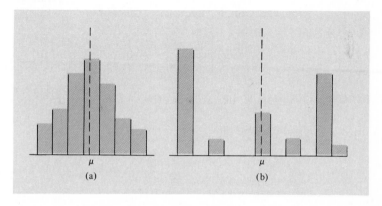

(a)　　　(b)

The importance of the notion of dispersal about the mean is illustrated by the following example. Suppose that a lumber mill is considering the purchase of one of two semiautomatic saws for cutting framing lumber. Owing to imprecisions in the machinery and irregularities in the logs, each saw cuts lumber only to within a certain tolerance. The lumber mill tests out the two saws by means of an experiment in which each saw is used to cut 10 pieces of lumber, each supposed to be precisely 10 feet long. The actual lengths of the cut lumber are then measured. Here are the results:

Saw I: 10.1, 9.8, 9.9, 10.1, 10.1, 10.0, 10.1, 9.9, 9.9, 10.1
Saw II: 10.0, 10.3, 9.9, 9.7, 10.2, 10.1, 9.9, 10.0, 9.6, 10.3.

The mean of the lengths for each saw is 10.0 feet. However, it is clear that saw II exhibits a great deal more irregularity than saw I. The lengths of lumber cut by saw II show much greater dispersion about the mean than those of saw I. The variance provides a numerical measure of this dispersion.

Let us now define the variance of a probability distribution. Suppose that we are considering an experiment with numerical outcomes $x_1, x_2, \ldots, x_N$ and respective probabilities $p_1, p_2, \ldots, p_N$. Suppose that the mean is μ. Then the deviations of the various outcomes from the mean are given by the N differences

$$x_1 - \mu, x_2 - \mu, \ldots, x_N - \mu.$$

Since we want to give weight to the various deviations according to their likelihood of occurrence, it is tempting to multiply each deviation by its probability of occurrence. However, this will not lead to a very satisfactory measure of deviation from the mean. This is because some of the differences will be positive and others negative. In the process of addition, deviations from the mean (both positive and negative deviations) will combine to yield a zero total deviation. To correct this, we consider instead the squares of the differences:

$$(x_1 - \mu)^2, (x_2 - \mu)^2, \ldots, (x_N - \mu)^2,$$

which are all ≥ 0. To obtain a measure of deviation from the mean, we multiply each of these expressions by the probability of the corresponding outcome. The number thus obtained is called the *variance of the probability distribution* (or of the associated random variable). That is, the variance of a probability distribution is given by the formula

$$[\text{variance}] = (x_1 - \mu)^2 p_1 + (x_2 - \mu)^2 p_2 + \cdots + (x_N - \mu)^2 p_N. \tag{1}$$

EXAMPLE 1   Compute the variance of the following probability distribution:

Outcome	Probability
0	.1
1	.3
2	.5
3	.1

Solution The mean is given by

$$\mu = 0 \cdot (.1) + 1 \cdot (.3) + 2 \cdot (.5) + 3 \cdot (.1) = 1.6.$$

In the notation of random variables, the calculation of the variance may be summarized as follows:

k	$\Pr(X = k)$	$k - \mu$	$(k - \mu)^2$	$(k - \mu)^2 \Pr(X = k)$
0	.1	$0 - 1.6 = -1.6$	2.56	.256
1	.3	$1 - 1.6 = -.6$	.36	.108
2	.5	$2 - 1.6 = .4$	.16	.080
3	.1	$3 - 1.6 = 1.4$	1.96	.196

$$[\text{variance}] = .256 + .108 + .080 + .196 = .640.$$

Actually, a much more commonly used measure of dispersal about the mean is the *standard deviation*, which is just the square root of the variance:

$$[\text{standard deviation}] = \sqrt{[\text{variance}]}.$$

The most commonly used notation for standard deviation is the Greek letter σ (sigma). Thus, for example, for the probability distribution of Example 1 we have

$$\sigma = \sqrt{[\text{variance}]} = \sqrt{.64} = .8.$$

The reason for using the standard deviation as opposed to the variance is that the former is expressed in the same units of measurement as X, whereas the latter is not.

EXAMPLE 2 Compute the standard deviations for the distributions of sales in car dealerships A and B.

Solution The probability distribution for dealership A is given by:

Weekly sales	Probability
5	$\frac{2}{52}$
6	$\frac{2}{52}$
7	$\frac{13}{52}$
8	$\frac{20}{52}$
9	$\frac{10}{52}$
10	$\frac{4}{52}$
11	$\frac{1}{52}$

The mean of this distribution was seen to be 7.96. Therefore, the variance is given by

$$[\text{variance}] = (5 - 7.96)^2(\tfrac{2}{52}) + (6 - 7.96)^2(\tfrac{2}{52}) + (7 - 7.96)^2(\tfrac{13}{52})$$
$$+ (8 - 7.96)^2(\tfrac{20}{52}) + (9 - 7.96)^2(\tfrac{10}{52}) + (10 - 7.96)^2(\tfrac{4}{52})$$
$$+ (11 - 7.96)^2(\tfrac{1}{52}) \approx 1.42.$$

Therefore, the standard deviation σ_A corresponding to dealership A is given by

$$\sigma_A = \sqrt{[\text{variance}]} = \sqrt{1.42} \approx 1.19.$$

In a similar way we find that the standard deviation σ_B corresponding to dealership B is equal to 2.02. Note that σ_A is smaller than σ_B, indicating that the sales of dealership B are more dispersed about the mean. In practical terms, this means that the sales of dealership B exhibit greater variation than those of dealership A. On the average the sales of dealership B are higher, but those of dealership A exhibit greater consistency.

The variance and standard deviation are used in many sophisticated statistical analyses, which are beyond the scope of this book. For our purposes let us discuss an application of the standard deviation to statistical estimation.

Suppose that we are observing an experiment with numerical outcomes and that the experiment has mean μ and standard deviation σ. We may wish to estimate the proportion of the outcomes which fall within c units of the mean by using the so-called *Chebychev inequality*.

Chebychev Inequality Suppose that a probability distribution with numerical outcomes has expected value μ and standard deviation σ. Then the probability that a randomly chosen outcome lies between $\mu - c$ and $\mu + c$ is at least $1 - (\sigma^2/c^2)$.

A verification of the Chebychev inequality can be found in most elementary statistics texts.

EXAMPLE 3 Suppose that a probability distribution has mean 5 and standard deviation 1. Use the Chebychev inequality to estimate the probability that an outcome lies between 3 and 7.

Solution Here $\mu = 5, \sigma = 1$. Since we wish to estimate the probability of an outcome lying between 3 and 7, we set $c = 2$. Then by the Chebychev inequality the desired probability is at least $1 - (\sigma^2/c^2) = 1 - \frac{1}{4} = .75$. That is, if the experiment is repeated a large number of times, we expect at least 75% of the outcomes to be between 3 and 7.

The Chebychev inequality has many practical applications, one of which is illustrated in the next example.

EXAMPLE 4 Apex Drug Supply Company sells bottles of 100 capsules of penicillin. Owing to the bottling procedure, not every bottle contains exactly 100 capsules. A statistical study reveals that the average number of capsules in a bottle is indeed 100 and the standard deviation is 2. If the company ships 5000 bottles, estimate the number having between 95 and 105 capsules inclusive.

Solution Here our experiment consists of observing the number of capsules in the bottle. We are given that $\mu = 100$, $\sigma = 2$. Therefore, by Chebychev's inequality, the proportion of bottles having between $100 - 5$ and $100 + 5$ capsules should be at least $1 - (2^2/5^2) = \frac{21}{25} = .84$. That is, we expect at least 84% of the 5000 bottles, or 4200 bottles, to have a number of capsules in the desired range.

Note that the estimate provided by Chebychev's inequality is a crude one. In more advanced statistics books you can find sharper estimates. Also, in the case of the normal distribution, we shall provide a much more precise way of estimating the probability of falling within c units of the mean, based on use of a table of areas under the normal curve. (See Section 5.)

PRACTICE PROBLEMS 3

1. (a) Compute the variance of the probability distribution in Table 1.

TABLE 1

Outcome	Probability
21	$\frac{1}{16}$
22	$\frac{1}{8}$
23	$\frac{5}{8}$
24	$\frac{1}{8}$
25	$\frac{1}{16}$

   (b) Using Table 1, find the probability that the outcome is between 22 and 24 inclusive.

2. Refer to the probability distribution of Problem 1. Use the Chebychev inequality to approximate the probability that the outcome is between 22 and 24.

EXERCISES 3

1. Compute the variance of the probability distribution in Table 2(a).

TABLE 2

Outcome	Probability
70	.5
71	.2
72	.1
73	.2

(a)

Outcome	Probability
-1	$\frac{1}{8}$
$-\frac{1}{2}$	$\frac{3}{8}$
0	$\frac{1}{8}$
$\frac{1}{2}$	$\frac{1}{8}$
1	$\frac{2}{8}$

(b)

2. Compute the variance of the probability distribution in Table 2(b).

3. Determine by inspection which one of the probability distributions, Fig. 2(a) or (b), has the greater variance.

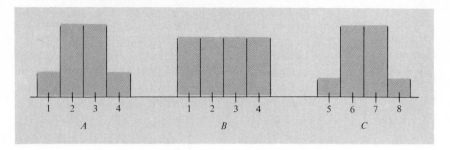

FIGURE 2

4. Determine by inspection which one of the probability distributions, B or C in Fig. 2, has the greater variance.

5. Table 3 gives the probability distribution for the possible returns from two different investments.

(a) Compute the mean and variance for each investment.

(b) Which investment has the higher expected return (i.e., mean)?

(c) Which investment is the less risky (i.e., has lesser variance)?

TABLE 3

Investment A		Investment B	
Return ($ millions)	Probability	Return ($ millions)	Probability
−10	$\frac{1}{5}$	0	.3
20	$\frac{3}{5}$	10	.4
25	$\frac{1}{5}$	30	.3

6. Two golfers recorded their scores for 20 nine-hole rounds of golf. Golfer A's scores were 39, 39, 40, 40, 40, 40, 40, 40, 41, 41, 41, 41, 41, 41, 41, 42, 43, 43, 43, 44. Golfer B's scores were 40, 40, 40, 41, 41, 41, 41, 42, 42, 42, 42, 42, 43, 43, 43, 43, 43, 43, 44, 44.

(a) Compute the mean and variance of the scores of each golfer.

(b) Who is the best golfer?

(c) Who is the most consistent golfer?

7. Table 4 gives the probability distribution for the weekly sales of two businesses.

(a) Compute the mean and variance for each business.

TABLE 4

Business A		Business B	
Sales	Probability	Sales	Probability
100	.1	100	0
101	.2	101	.2
102	.3	102	0
103	0	103	.2
104	0	104	.1
105	.2	105	.2
106	.2	106	.3

(b) Which business has the best sales record?

(c) Which business has the most consistent sales record?

8. Student A received the following course grades during her freshman year: 4, 4, 4, 4, 3, 3, 2, 2, 2, 0. Student B received the following course grades during her freshman year: 4, 4, 4, 4, 4, 4, 3, 1, 1, 1.

(a) Write down the probability distribution tables for each student and compute the means and variances.

(b) Which student had the best grade point average?

(c) Which student was the most consistent?

9. Suppose that a probability distribution has mean 35 and standard deviation 5. Use the Chebychev inequality to estimate the probability that an outcome will lie between

(a) 25 and 45. (b) 20 and 50. (c) 29 and 41.

10. Suppose that a probability distribution has mean 8 and standard deviation .4. Use the Chebychev inequality to estimate the probability that an outcome will lie between

(a) 6 and 10. (b) 7.2 and 8.8. (c) 7.5 and 8.5.

11. For certain types of fluorescent lights the number of hours a bulb will burn before requiring replacement has a mean of 3000 hours and a standard deviation of 250 hours. Suppose that 5000 such bulbs are installed in an office building. Estimate the number that will require replacement between 2000 and 4000 hours from the time of installation.

12. An electronics firm determines that the number of defective transistors in each batch averages 15 with standard deviation 10. Suppose that 100 batches are produced. Estimate the number of batches having between 0 and 30 defective transistors.

13. Suppose that a probability distribution has mean 75 and standard deviation 6. Use the Chebychev inequality to find the value of c for which the probability that the outcome lies between $75 - c$ and $75 + c$ is at least $\frac{7}{16}$.

14. Suppose that a probability distribution has mean 17 and standard deviation .2. Use the Chebychev inequality to find the value of c for which the probability that the outcome lies between $17 - c$ and $17 + c$ is at least $\frac{15}{16}$.

15. The probability distribution for the sum of numbers obtained from tossing a pair of dice is given in Table 5(a).

   (a) Compute the mean and variance of this probability distribution.

   (b) Using the table, give the probability that the number is between 4 and 10 inclusive.

   (c) Use the Chebychev inequality to estimate the probability that the number is between 4 and 10 inclusive.

TABLE 5

Number	Probability
2	$\frac{1}{36}$
3	$\frac{2}{36}$
4	$\frac{3}{36}$
5	$\frac{4}{36}$
6	$\frac{5}{36}$
7	$\frac{6}{36}$
8	$\frac{5}{36}$
9	$\frac{4}{36}$
10	$\frac{3}{36}$
11	$\frac{2}{36}$
12	$\frac{1}{36}$

(a)

Number of "ones"	Probability
0	.112
1	.269
2	.296
3	.197
4	.089
5	.028
6	.007
7	.001
8	.000
9	.000
10	.000
11	.000
12	.000

(b)

16. The probability distribution for the number of "ones" obtained from tossing 12 dice is given in Table 5(b). This probability distribution has mean 2 and standard deviation 1.291 ($\sigma^2 = \frac{5}{3}$).

   (a) Using the table, give the probability that the number of "ones" is between 0 and 4 inclusive.

   (b) Use the Chebychev inequality to estimate the probability that the number of "ones" is between 0 and 4 inclusive.

17. If X is a random variable with mean μ, then the variance of X can also be computed as $E(X^2) - \mu^2$. Redo Example 1 using this formula.

18. If X is a random variable, then the variance of X equals the variance of $X - a$ for any number a. Redo Exercise 1 using this result with $a = 70$.

19. If X is a random variable, then the variance of aX equals a^2 times the variance of X. Verify this result for the random variable in Exercise 2 with $a = 2$.

20. If X is a random variable, then $E(X - a) = E(X) - a$, and $E(aX) = aE(X)$ for any number a. Give intuitive justifications of these results.

SOLUTIONS TO PRACTICE PROBLEMS 3

1. (a)

k	$\Pr(X = k)$	$k - \mu$	$(k - \mu)^2$	$(k - \mu)^2 \Pr(X = k)$
21	$\frac{1}{16}$	-2	4	$\frac{4}{16}$
22	$\frac{1}{8}$	-1	1	$\frac{1}{8}$
23	$\frac{5}{8}$	0	0	0
24	$\frac{1}{8}$	1	1	$\frac{1}{8}$
25	$\frac{1}{16}$	2	4	$\frac{4}{16}$

$$\mu = 21 \cdot \frac{1}{16} + 22 \cdot \frac{1}{8} + 23 \cdot \frac{5}{8} + 24 \cdot \frac{1}{8} + 25 \cdot \frac{1}{16}$$

$$= \frac{21}{16} + \frac{44}{16} + \frac{230}{16} + \frac{48}{16} + \frac{25}{16} = \frac{368}{16} = 23.$$

$$[\text{variance}] = \frac{4}{16} + \frac{1}{8} + 0 + \frac{1}{8} + \frac{4}{16}$$

$$= \frac{2}{8} + \frac{1}{8} + 0 + \frac{1}{8} + \frac{2}{8} = \frac{6}{8} = \frac{3}{4}.$$

(b) $\frac{7}{8}$. The probability that the outcome is between 22 and 24 is $\Pr(22) + \Pr(23) + \Pr(24) = \frac{1}{8} + \frac{5}{8} + \frac{1}{8} = \frac{7}{8}$.

2. Probability $\geq \frac{1}{4}$. Here $\mu = 23$, $\sigma^2 = \frac{3}{4}$, and $c = 1$. By the Chebychev inequality, the probability that the outcome is between $23 - 1$ and $23 + 1$ is at least $1 - (\frac{3}{4}/1^2) = 1 - \frac{3}{4} = \frac{1}{4}$. [From 1(b) we obtained the actual probability of $\frac{7}{8}$, which is much greater than $\frac{1}{4}$. In the next two sections we will study techniques which give better estimates. However, these techniques hold only for special types of probability distributions.]

10.4. The Normal Distribution

Throughout our discussion of probability theory we have been dealing with experiments having a finite number of possible outcomes. Let us now glimpse into the realm of so-called continuous probability by studying one particular class of experiments having infinitely many possible outcomes—namely experiments with *normally distributed outcomes*. For such experiments, as we shall see, the probabilities of events are computed as areas under certain special curves called *normal curves*. It is no exaggeration to say that experiments with normally distributed outcomes are among the most significant in probability theory. Such experiments abound in the world around us. Here are a few examples.

1. Choose an individual at random and observe his (or her) IQ.
2. Choose a 1-day-old infant and observe its weight.
3. Choose an 8-year-old male at random and observe his height.

4. Choose a leaf at random from a particular tree and observe its length.
5. A lumber mill is cutting planks which are supposed to be 8 feet long; choose a plank at random and observe its actual length.

Associated to each of the foregoing experiments is a bell-shaped curve, as shown in Fig. 1. Such a curve is called a *normal curve*. The curve is symmetric about a line drawn through its top point. This line of symmetry indicates the mean value of the corresponding experiment. The mean value is denoted as usual by the Greek letter μ. Thus, for example, if in experiment 4 above the average length of the leaves on the tree is 5 inches, then $\mu = 5$ and the corresponding bell-shaped curve is symmetric about the line $x = 5$.

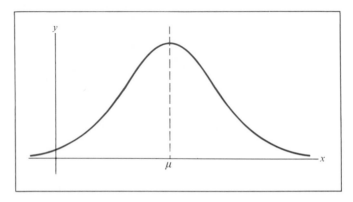

FIGURE 1

The connection between an experiment with normally distributed outcomes and its associated normal curve is this:

The probability that the experimental outcome is between a and b equals the area under the associated normal curve from $x = a$ to $x = b$. (This is the shaded region in Fig. 2.)

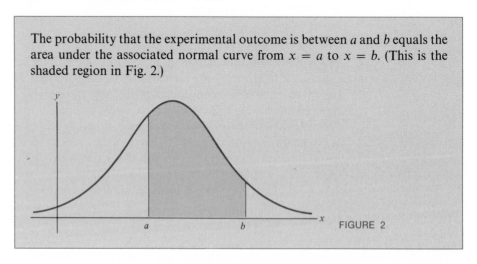

FIGURE 2

EXAMPLE 1 A certain experiment has normally distributed outcomes with mean $\mu = 1$. Shade the region corresponding to the probabilities of the following outcomes.

(a) The outcome lies between 1 and 3.

(b) The outcome lies between 0 and 2.

(c) The outcome is less than .5.

(d) The outcome is greater than 2.

Solution The outcomes are plotted along the x-axis. We then shade the appropriate area under the curve.

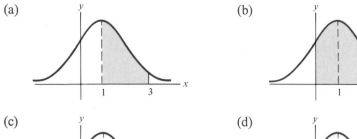

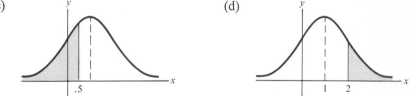

There are many different normal curves with the same mean. For instance, in Fig. 3 we have drawn three normal curves, all with $\mu = 0$. Roughly speaking, the difference between these normal curves is in the width of the center "hump." A sharper hump indicates that the outcomes are more likely to be close to the mean. A flatter hump indicates a greater likelihood for the outcomes to be spread out. As we have seen, the spread of the outcomes about the mean is described by the

FIGURE 3

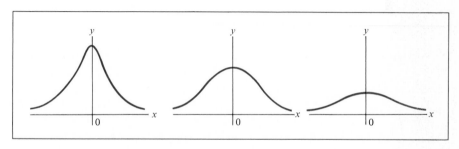

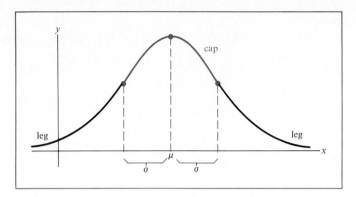

FIGURE 4

standard deviation, denoted by the Greek letter σ. In the case of a normal curve, the standard deviation has a simple geometric meaning: The normal curve "twists" (or, in calculus terminology, "inflects") at a distance σ on either side of the mean (Fig. 4). More specifically, a normal curve may be thought of as made up of two pieces, a "cap," which looks like an upside-down bowl, and a pair of legs which curve in the opposite direction. The places at which the cap and legs are joined are at a distance σ from the mean. Thus, it is clear that the size of σ controls the sharpness of the hump.

A normal curve is completely described by its mean μ and standard deviation σ. In fact, given μ and σ, we may write down the equation of the associated normal curve:

$$y = \frac{1}{\sigma\sqrt{2\pi}} e^{-(1/2)[(x-\mu)/\sigma]^2},$$

where $\pi \approx 3.1416$ and $e \approx 2.7183$. Fortunately, we will not need this rather complicated formula in what follows. But it is only fair to say that all theoretical work on the normal curve ultimately rests on this equation.

For our purposes we will compute areas under normal curves by consulting a table. One might expect that a separate table would be needed for each normal curve, but such is not the case. Only one table is needed: the table corresponding to the so-called *standard normal curve*, which is the one for which $\mu = 0$, $\sigma = 1$. So let us begin our discussion of areas under normal curves by considering the standard normal curve.

Let z be any number and let $A(z)$ denote the area under the standard normal curve to the left of z (Fig. 5). Table 1 gives $A(z)$ for various values of z, with the values of $A(z)$ rounded off to four decimal places.

A more extensive table can be found in Table 1 of the Appendix. It can be shown, using methods beyond the scope of this book, that the total area under the standard normal curve is 1. Using this fact, together with Table 1 above, we can find areas of various types of regions.

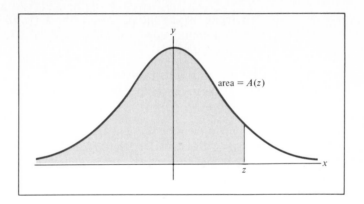

FIGURE 5

TABLE 1

z	A(z)	z	A(z)	z	A(z)
−4.00	.0000	−1.25	.1056	1.50	.9332
−3.75	.0001	−1.00	.1587	1.75	.9599
−3.50	.0002	−.75	.2266	2.00	.9772
−3.25	.0006	−.50	.3085	2.25	.9878
−3.00	.0013	−.25	.4013	2.50	.9938
−2.75	.0030	0	.5000	2.75	.9970
−2.50	.0062	.25	.5987	3.00	.9987
−2.25	.0122	.50	.6915	3.25	.9994
−2.00	.0228	.75	.7734	3.50	.9998
−1.75	.0401	1.00	.8413	3.75	.9999
−1.50	.0668	1.25	.8944	4.00	1.0000

EXAMPLE 2   Use Table 1 to determine the areas of the regions under the standard normal curve pictured in Fig. 6.

FIGURE 6

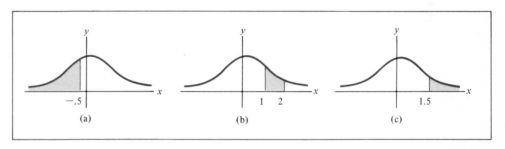

(a)   (b)   (c)

Solution (a) This region is just the portion of the curve to the left of $-.5$. So its area is $A(-.5)$. Looking down the middle pair of columns of the table, we find that $A(-.5) = .3085$.

(b) This region results from beginning with the region to the left of 2 and subtracting the region to the left of 1. We obtain an area of $A(2) - A(1) = .9772 - .8413 = .1359$.

(c) This region can be thought of as the entire region under the curve, with the region to the left of 1.5 removed. Therefore, the area is $1 - A(1.5) = 1 - .9332 = .0668$.

The problem of finding the area of a region under any normal curve can be reduced to finding the area of a region under the standard normal curve. To illustrate the computation procedure, let us consider a numerical example.

EXAMPLE 3 Find the area under the normal curve with $\mu = 3$, $\sigma = 2$ from $x = 1$ to $x = 5$.

Solution We have sketched the described region in Fig. 7. It extends from one standard deviation below the mean to one standard deviation above. Draw the corresponding region under the standard normal curve. That is, draw the region from one standard deviation below to one standard deviation above the mean (Fig. 8). It is

FIGURE 7

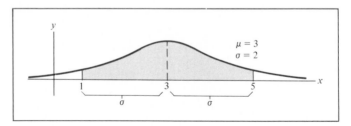

FIGURE 8

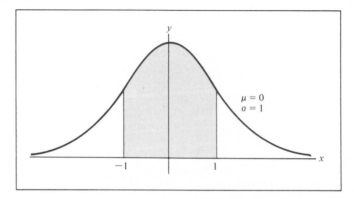

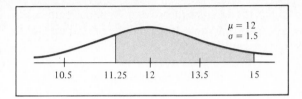

FIGURE 9

a theorem that this new region has the same area as the original one. But the area in Fig. 8 may be computed from Table 1 as $A(1) - A(-1)$. So our desired area is

$$A(1) - A(-1) = .8413 - .1587 = .6826.$$

EXAMPLE 4 Consider the normal curve with $\mu = 12$, $\sigma = 1.5$. Find the area of the region under the curve between $x = 11.25$ and $x = 15$ (Fig. 9).

Solution The number 11.25 is .75 below the mean 12. And .75 is $.75/1.5 = .5$ standard deviations. The number 15 is 3 above the mean. And 3 is $3/1.5 = 2$ standard deviations. Therefore, the region has the same area as the region under the standard normal curve from $-.5$ to 2, which is $A(2) - A(-.5) = .9772 - .3085 = .6687$.

Suppose that a normal curve has mean μ and standard deviation σ. Then the area under the curve from $x = a$ to $x = b$ is

$$A\left(\frac{b - \mu}{\sigma}\right) - A\left(\frac{a - \mu}{\sigma}\right).$$

The numbers $b - \mu$ and $a - \mu$, respectively, measure the distances of b and a from the mean. The numbers $(b - \mu)/\sigma$ and $(a - \mu)/\sigma$ express these distances as multiples of the standard deviation σ. So the area under the normal curve from $x = a$ to $x = b$ is computed by expressing x in terms of standard deviations from the mean and then treating the curve as if it were the standard normal curve.

Let us now use our knowledge of areas under normal curves to calculate probabilities arising in some applied problems.

EXAMPLE 5 Suppose that for a certain population the birth weights of infants in pounds is normally distributed with $\mu = 7.75$ and $\sigma = 1.25$. Find the probability that an infant's birth weight is more than 9 pounds 10 ounces. (*Note:* 9 pounds 10 ounces $= 9\frac{5}{8}$ pounds.)

Solution Since weights are normally distributed, the probability of an infant's birth weight being more than $9\frac{5}{8}$ is given by the area under the appropriate normal curve to the right of $9\frac{5}{8}$—that is, the area shaded in Fig. 10. Since $9\frac{5}{8} = 9.625$, the number $9\frac{5}{8}$ lies $9.625 - 7.75 = 1.875$ units above the mean. In turn, this is $1.875/1.25 = 1.5$ standard deviations. Therefore, the area is $1 - A(1.5) = -.9332 = .0668$. So the probability that an infant weighs more than 9 pounds 10 ounces is .0668.

FIGURE 10

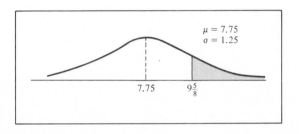

EXAMPLE 6 A wholesale produce dealer finds that the number of boxes of bananas sold each day is normally distributed with $\mu = 1200$ and $\sigma = 100$. Find the probability that the number sold on a particular day is less than 1000.

Solution Since daily sales are normally distributed, the desired probability is the area to the left of 1000 (Fig. 11). The number 1000 is two standard deviations below the mean ($\mu = 1200$, $\sigma = 100$). Therefore, the area is $A(-2) = .0228$. So the probability is .0228 that less than 1000 boxes will be sold.

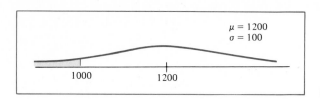

$\mu = 1200$
$\sigma = 100$

1000   1200

FIGURE 11

PRACTICE PROBLEMS 4

1. Refer to Fig. 12(a). Find the value of z for which the area of the shaded region is .0802.

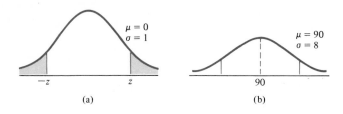

$\mu = 0$
$\sigma = 1$

$-z$ z

(a)

$\mu = 90$
$\sigma = 8$

90

(b)   FIGURE 12

2. Refer to the normal curve in Fig. 12(b). Express the following numbers in terms of standard deviations from the mean.

(a) 90   (b) 82   (c) 94   (d) 104

EXERCISES 4

Use the table for $A(z)$ to find the areas of the shaded regions under the standard normal curve.

1.
1.25

2.
$-.75$ 1

3.
.25

4.
-1 1

5.
.5 1.5

6.
-1

7.
$-.5$ .5

8.
-1.25

In Exercises 9–12, find the value of z for which the area of the shaded region under the standard normal curve is as specified.

9. Area is .0401.

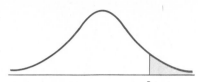

10. Area is .0456.

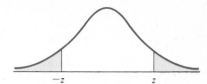

11. Area is .5468.

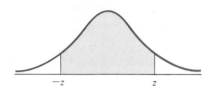

12. Area is .6915.

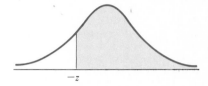

In Exercises 13–16 determine μ and σ by inspection.

13.

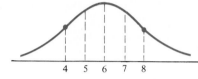

14.

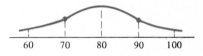

15.

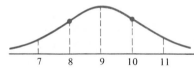

16.

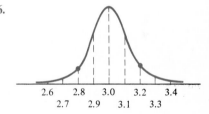

Exercises 17–20 refer to the normal curve with $\mu = 8$, $\sigma = \frac{3}{4}$.

17. Convert 6 into standard deviations from the mean.

18. Convert $9\frac{1}{4}$ into standard deviations from the mean.

19. What value of x corresponds to 10 standard deviations above the mean?

20. What value of x corresponds to 2 standard deviations below the mean?

In Exercises 21–24, find the areas of the shaded regions under the given normal curves.

21.

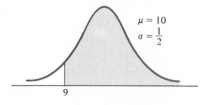

$\mu = 10$
$\sigma = \frac{1}{2}$

22.

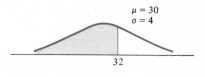

$\mu = 30$
$\sigma = 4$

23.

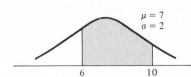

$\mu = 7$
$\sigma = 2$

6   10

24.

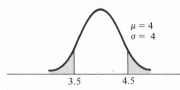

$\mu = 4$
$\sigma = 4$

3.5   4.5

25. Suppose that the height (at the shoulder) of adult African bull bush elephants is normally distributed with $\mu = 3.3$ meters and $\sigma = .2$ meter. The elephant on display at the Smithsonian Institution has height 4 meters and is the largest elephant on record. What is the probability that an adult African bull bush elephant has height 4 meters or more?

26. At a soft-drink bottling plant, the amount of cola put into the bottles is normally distributed with $\mu = 16\frac{3}{4}$ ounces and $\sigma = \frac{1}{2}$. What is the probability that a bottle will contain less than 16 ounces?

27. Bolts produced by a machine are acceptable provided that their length is within the range 5.95 to 6.05 centimeters. Suppose that the lengths of the bolts produced are normally distributed with $\mu = 6$ centimeters and $\sigma = .02$. What is the probability that a bolt will be of an acceptable length?

28. Suppose that IQ scores are normally distributed with $\mu = 100$ and $\sigma = 10$. What percent of the population have IQ scores of 125 or more?

29. The amount of gas sold weekly by a certain gas station is normally distributed with $\mu = 30{,}000$ gallons and $\sigma = 4000$. If the station has 39,000 gallons on hand at the beginning of the week, what is the probability of its running out of gas before the end of the week?

30. Suppose that the lifetimes of a certain type of light bulb are normally distributed with $\mu = 1200$ hours and $\sigma = 160$. Find the probability that a light bulb will burn out in less than 1000 hours.

SOLUTIONS TO PRACTICE PROBLEMS 4

1. 1.75. Owing to the symmetry of normal curves, each piece of the shaded region has area $\frac{1}{2}(.0802) = .0401$. Therefore, $A(-z) = .0401$ and so by Table 1, $-z = -1.75$. Thus, $z = 1.75$.

2. (a) 0. Since 90 *is* the mean, it is 0 standard deviations from the mean.

    (b) -1. Since 82 is $90 - 8$, it is 8 units or 1 standard deviation below the mean.

    (c) .5. Here $94 = 90 + 4$ is 4 units or .5 standard deviation above the mean.

    (d) 1.75. Here $104 = 90 + 14$ is 14 units or $\frac{14}{8} = 1.75$ standard deviations above the mean.

10.5. Binomial Trials

In this section we fix our attention on the simplest experiments: those with just two outcomes. These experiments, called *binomial trials* (or *Bernoulli trials*), occur in many applications. Here are some examples of binomial trials.

1. Flip a coin and observe the outcome, heads or tails.
2. Administer a drug to a sick individual and classify the reaction as "effective" or "ineffective."
3. Manufacture a light bulb and classify it as "nondefective" or "defective."

The outcomes of a binomial trial are usually called "success" and "failure." Of course, the labels "success" and "failure" need have no connection with the usual meanings of these words. For example, in experiment 2 above we might label the outcome "ineffective" as "success" and "effective" as "failure." Throughout our discussion of binomial trials we will always denote the probability of "success" by p and "failure" by q. Since a binomial experiment has only two outcomes, we have $p + q = 1$ or

$$q = 1 - p. \tag{1}$$

Suppose that we consider a particular binomial trial. Let us perform the following experiment. Repeat the binomial trial n times and observe the number of successes which occur. Assume that the n successive trials are independent of one another. The fundamental problem of the theory of binomial trials is to calculate the probabilities of the various outcomes of this experiment. Let X denote the random variable associated with the experiment. Then X denotes the number of successes which occur. For example, if $n = 20$ and $X = 3$, then 3 of the 20 binomial trials yield success and 17 failure. The possible values of X (the possible outcomes of the experiment) are $0, 1, 2, \ldots, n$. We shall let $\Pr(X = k)$ denote the probability that $X = k$, namely the probability that k of the n trials result in success. Moreover, let μ denote the expected value of X. Then we have the following fundamental facts:

$$\Pr(X = k) = \binom{n}{k} p^k q^{n-k} \tag{2}$$

$$\mu = np. \tag{3}$$

Note that the right-hand side of (2) is one of the terms in the binomial expansion of $(p + q)^n$. The derivations of (2) and (3) are given at the end of this section.
Let us now illustrate how the formulas above may be used.

EXAMPLE 1   A plumbing supplies manufacturer produces faucet washers which are packaged in boxes of 300. Quality control studies have shown that 2% of the washers are defective.

(a) What is the probability that a box of washers contains exactly 9 defective washers?

(b) What is the average number of defective washers per box?

Solution (a) Deciding whether a single washer is or is not defective is a binomial trial. Since we wish to consider the number of defective washers in a box, let "success" be the outcome "defective." Then $p = .02$, $q = 1 - .02 = .98$, $n = 300$. The probability that 9 out of 300 washers are defective equals

$$\Pr(X = 9) = \binom{300}{9}(.02)^9(.98)^{291} \approx .07.$$

(b) The average number of defective washers per box is given by the expected value $\mu = np = 300 \cdot (.02) = 6$.

EXAMPLE 2 The recovery rate for a certain cattle disease is 25%. If 40 cattle are afflicted with the disease, what is the probability that exactly 10 will recover?

Solution In this example the binomial trial consists of observing a single cow, with recovery as success. Then $p = .25$, $q = 1 - .25 = .75$, $n = 40$. The probability of 10 successes is

$$\Pr(X = 10) = \binom{40}{10}(.25)^{10}(.75)^{30} \approx .14.$$

The numerical work in Examples 1 and 2 required an electronic calculator with an x^y key [to compute, for example, $(.98)^{291}$]. In other problems the calculations can be stickier. Suppose, for example, that we return to Example 2. What is the probability that 16 or more cattle recover? Using formula (2) to compute the probabilities that 16, 17, ..., 40 cattle recover, the desired probability is

$$\Pr(X = 16) + \Pr(X = 17) + \cdots + \Pr(X = 40)$$

$$= \binom{40}{16}(.25)^{16}(.75)^{24} + \binom{40}{17}(.25)^{17}(.75)^{23} + \cdots + \binom{40}{40}(.25)^{40}(.75)^0.$$

Each of the terms in this sum is difficult to compute. The thought of computing all of them should be sufficient motivation to seek an alternate approach. Fortunately, there exists a reasonably simple method of approximating a sum of this type. Let us illustrate the technique with an example.

Toss a coin 20 times and observe the number of heads. By using formula (2) we can calculate the probability of k heads. The results are displayed in Table 1. The data of Table 1 can be displayed in histogram form, as in Fig. 1.

TABLE 1   Probability of k Heads (to four places)

k	0	1	2	3	4	5	6	7	8	9	10
Probability of k heads	.0000	.0000	.0002	.0011	.0046	.0148	.0370	.0739	.1201	.1602	.1762

k	11	12	13	14	15	16	17	18	19	20
Probability of k heads	.1602	.1201	.0739	.0370	.0148	.0046	.0011	.0002	.0000	.0000

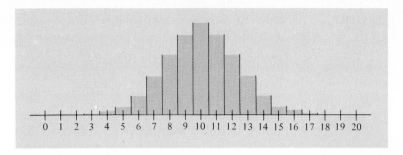

FIGURE 1

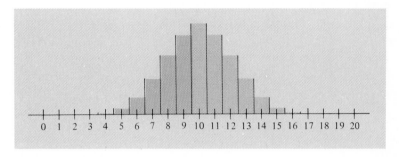

FIGURE 2

As we have seen, various probabilities may now be interpreted as areas. For example, the probability that at most 9 heads occur is equal to the sum of the areas of all the rectangles to the left of the central one (Fig. 2). The shape of the graph in Figs. 1 and 2 suggests that we might approximate it with a normal curve. Indeed we can! In Fig. 3 we have superimposed on the graph of Fig. 1 the normal curve with $\mu = 10$, $\sigma = 2.24$. (See below for the reasons behind our choice of values of μ, σ.) Notice that the normal curve provides a reasonably good fit to the histogram, so the areas of the rectangles can be approximated by areas under the normal curve. Let us see how this approximation works out in practice.

FIGURE 3   Approximation by the normal curve with $\mu = 10$, $\sigma = 2.24$.

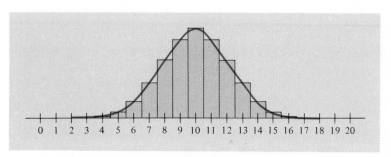

EXAMPLE 3   Suppose that we toss a coin 20 times. Use the normal curve discussed above to calculate the following.

(a) The probability of exactly 10 heads.

(b) The probability of less than 10 heads.

Solution   (a) The probability of exactly 10 heads equals the area of the central rectangle in Fig. 4(a). This rectangle extends from 9.5 to 10.5 on the x-axis. So we approximate it by the area under the appropriate normal curve from 9.5 to 10.5. [See Fig. 4(b).] Using the methods of Section 4, we find this area to be .1742. Notice the good agreement between this value and the actual area of the rectangle, which is .1762.

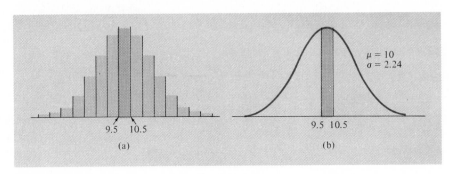

FIGURE 4

(b) The probability of less than 10 heads is the area of the shaded rectangles to the left of 9.5. Therefore, we approximate it by the area under the normal curve to the left of 9.5 (Fig. 5). This area is .4129. The exact probability may be calculated as

$$\Pr(X = 0) + \Pr(X = 1) + \cdots + \Pr(X = 9),$$

where each term of the sum is calculated from formula (2). The resulting probability is .4119, again yielding a close agreement with the value obtained from the approximating normal curve.

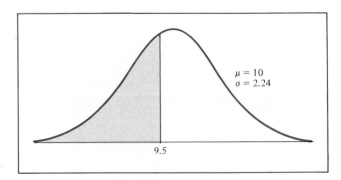

FIGURE 5

The probability distribution for the number of successes in any n binomial trials has a histogram similar to that in Fig. 1. And any such histogram may be approximated by a normal curve. We determine *which* normal curve as follows: Formula (3) asserts that the mean μ of the probability distribution is given by $\mu = np$. Moreover, it can be shown that the standard deviation is $\sigma = \sqrt{npq}$. This suggests that we approximate with a normal curve having mean $\mu = np$ and standard deviation $\sigma = \sqrt{npq}$. This turns out to be a very good approximation in most applications. To summarize, then, we have the following result.

> Suppose that we perform a sequence of n binomial trials with probability of success p and probability of failure q and observe the number of successes. Then the histogram for the resulting probability distribution may be approximated by the normal curve with $\mu = np$ and $\sigma = \sqrt{npq}$.

EXAMPLE 4 Refer to Example 1. What is the probability that more than 10 of the washers in a single box are defective?

Solution We use an approximating normal curve with

$$\mu = np = 300 \cdot (.02) = 6$$

$$\sigma = \sqrt{npq} = \sqrt{300 \cdot (.02) \cdot (.98)} \approx 2.425.$$

The probability that more than 10 washers are defective corresponds to the shaded area in Fig. 6. This shaded area equals .0314. So approximately 3.14% of the boxes have more than 10 defective washers.

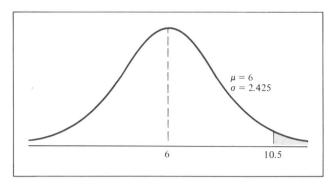

$\mu = 6$
$\sigma = 2.425$

6   10.5

FIGURE 6

Note In Fig. 6 we shaded the region to the right of 10.5 rather than to the right of 10. This gives a better approximation to the corresponding area under the histogram, since the rectangle corresponding to 11 "successes" has its left endpoint at 10.5.

Let us close this section with an application to medical research.

EXAMPLE 5

Consider the cattle disease of Example 2, from which 25% of the cattle recover. A veterinarian discovers a serum to combat the disease. In a test of the serum she observes that 16 of a herd of 40 recover. Suppose that the serum had not been used. What is the likelihood that at least 16 cattle would have recovered?

Solution   In this case $p = .25, q = .75, n = 40$. The approximating normal curve has

$$\mu = np = 40 \cdot (.25) = 10$$

$$\sigma = \sqrt{npq} = \sqrt{40 \cdot (.25) \cdot (.75)} \approx 2.74.$$

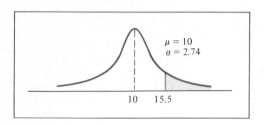

The likelihood that at least 16 cattle recover is approximated by the area shaded in Fig. 7. From our table we find that this area is approximately .0228. The very small value of this probability indicates that it is highly unlikely that the 16 observed recoveries could have occurred by chance. The veterinarian can reasonably conclude that the serum is effective against the disease.

FIGURE 7

Verification of Formulas (2) and (3)

Suppose that the experiment results in the outcome $X = k$. Then there are k successes and $n - k$ failures. Consider the probability of any specific sequence of trials having k successes—say the first k trials are successes and the last $n - k$ are failures. Since the various trials are independent of one another, the probability of k successes followed by $n - k$ failures is

$$\underbrace{p \cdot p \cdot \ldots \cdot p}_{k \text{ times}} \underbrace{q \cdot q \cdot \ldots \cdot q}_{n - k \text{ times}} = p^k q^{n-k}.$$

The number of different sequences having k successes is $\binom{n}{k}$. So the desired probability is

$$\Pr(X = k) = \binom{n}{k} p^k q^{n-k},$$

which is formula (2).

To prove formula (3), note that

$$\mu = 0 \cdot \Pr(X = 0) + 1 \cdot \Pr(X = 1) + \cdots + n \cdot \Pr(X = n)$$

$$= 0 \cdot \binom{n}{0} p^0 q^{n-0} + 1 \cdot \binom{n}{1} p^1 q^{n-1} + \cdots + n \cdot \binom{n}{n} p^n q^{n-n}.$$

Moreover, note that

$$k \binom{n}{k} = k \cdot \frac{n(n-1) \cdot \ldots \cdot (n-k+1)}{k(k-1) \cdot \ldots \cdot 2 \cdot 1} = n \cdot \frac{(n-1) \cdot \ldots \cdot (n-k+1)}{(k-1) \cdot \ldots \cdot 2 \cdot 1}$$

$$= n \cdot \binom{n-1}{k-1} \qquad (k = 1, 2, 3, \ldots, n).$$

Therefore,

$$\mu = 1 \cdot \binom{n}{1} p^1 q^{n-1} + 2 \cdot \binom{n}{2} p^2 q^{n-2} + \cdots + n \cdot \binom{n}{n} p^n q^0$$

$$= n \binom{n-1}{0} p^1 q^{n-1} + n \binom{n-1}{1} p^2 q^{n-2} + \cdots + n \binom{n-1}{n-1} p^n q^0 \qquad \text{[by (4)]}$$

$$= np \left[\binom{n-1}{0} p^0 q^{n-1} + \binom{n-1}{1} pq^{n-2} + \cdots + \binom{n-1}{n-1} p^{n-1} q^0 \right]$$

$$= np(p+q)^{n-1} \qquad \text{(by the binomial theorem)}$$

$$= np \qquad \text{(since } p + q = 1\text{).}$$

PRACTICE PROBLEMS 5

1. A new drug is being tested on laboratory mice. The mice have been given a disease for which the recovery rate is $\frac{1}{2}$.

   (a) In the first experiment the drug is given to 5 of the mice and all 5 recover. Find the probability that the success of this experiment was due to luck. That is, find the probability that 5 out of 5 mice recover in the event that the drug has no effect on the illness.

   (b) In a second experiment the drug is given to 25 mice and 18 recover. Find the probability that 18 or more recover in the event that the drug has no effect on the illness.

2. What conclusions can be drawn from the results in Problem 1?

EXERCISES 5

1. A single die is tossed four times. Find the probability that exactly two of the tosses show a "one."

2. Find the probability of obtaining exactly three heads when tossing a fair coin six times.

3. A salesman determines that the probability of making a sale to a customer is $\frac{1}{4}$. What is the probability of making a sale to three of the next four customers?

4. A basketball player makes free throws with probability .7. What is the probability of making exactly two out of five free throws?

5. Suppose that 60% of the voters in a state intend to vote for a certain candidate. What is the probability that a survey polling five people reveals that two or less intend to vote for that candidate?

6. An exam consists of six "true or false" questions. What is the probability that a person can get five or more correct by just guessing?

In Exercises 7–14, use the normal curve to approximate the probability.

7. An experiment consists of 25 binomial trials, each having probability $\frac{1}{5}$ of success. Use an approximating normal curve to estimate the probability of

   (a) Exactly 5 successes.

   (b) Between 3 and 7 successes inclusive.

   (c) Less than 10 successes.

8. An experiment consists of 18 binomial trials, each having probability $\frac{2}{3}$ of success. Use an approximating normal curve to estimate the probability of

   (a) Exactly 10 successes.

   (b) Between 8 and 16 successes inclusive.

   (c) More than 12 successes.

9. Laboratory mice are given an illness for which the usual recovery rate is $\frac{1}{6}$. A new drug is tested on 20 of the mice, and 8 of them recover. What is the probability that 8 or more would have recovered if the 20 mice had not been given the drug?

10. A person claims to have ESP (extrasensory perception). A coin is tossed 16 times, and each time he is asked to predict in advance whether the coin will land "heads" or "tails." He predicts correctly 75% of the time (i.e., on 12 tosses). What is the probability of being correct 12 or more times by pure guessing?

11. A wine-taster claims that she can usually distinguish between domestic and imported wines. As a test, she is given 100 wines to test and correctly identifies 63 of them. What is the probability that she accomplished that good a record by pure guessing? That is, what is the probability of being correct 63 or more times out of 100 by pure guessing?

12. In American roulette, the probability of winning when betting "red" is $\frac{9}{19}$. What is the probability of being ahead after betting the same amount 90 times?

13. A basketball player makes each free throw with probability $\frac{3}{4}$. What is the probability of making 68 or more shots out of 75 trials?

14. A bookstore determines that two-fifths of the people who come into the store make a purchase. What is the probability that of the 54 people who come into the store during a certain hour, less than 14 make a purchase?

SOLUTIONS TO PRACTICE PROBLEMS 5

1. (a) Giving the drug to a single mouse is a binomial trial with "recovery" as "success" and "death" as "failure." If the drug has no effect, then the probability of success is $\frac{1}{2}$. The probability of five successes in five trials is given by formula (2), with $n = 5$, $p = \frac{1}{2}$, $q = \frac{1}{2}$, $k = 5$.

$$\Pr(X = 5) = \binom{5}{5}(\tfrac{1}{2})^5(\tfrac{1}{2})^0 = (\tfrac{1}{2})^5 = \tfrac{1}{32} = .03125.$$

   (b) As in part (a), this experiment is a binomial experiment with $p = \frac{1}{2}$. However, now $n = 25$. The probability that 18 or more mice recover is

$$\Pr(X = 18) + \Pr(X = 19) + \cdots + \Pr(X = 25).$$

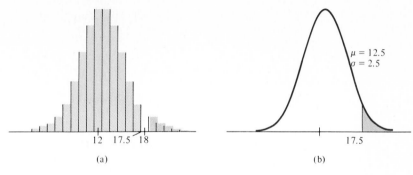

(a)   (b)

FIGURE 8

This probability is the area of the shaded portion of the histogram in Fig. 8(a). The histogram can be approximated by the normal curve with $\mu = 25 \cdot \frac{1}{2} = 12.5$ and $\sigma = \sqrt{25 \cdot \frac{1}{2} \cdot \frac{1}{2}} = \sqrt{\frac{25}{4}} = \frac{5}{2} = 2.5$ [Fig. 8(b)]. Since the shaded portion of the histogram begins at the point 17.5, the desired probability is approximately the area of the shaded region under the normal curve. The number 17.5 is $(17.5 - 12.5)/2.5 = 5/2.5 = 2$ standard deviations to the right of the mean. Hence, the area under the curve is $1 - A(2) = 1 - .9772 = .0228$. Therefore, the probability that 18 or more mice recover is approximately .0228.

2. Both experiments offer convincing evidence that the drug is helpful in treating the illness. The likelihood of obtaining the results by pure chance is slim. The second experiment might be considered more conclusive than the first, since the result has a lower probability of being due to chance.

Chapter 10: CHECKLIST

☐ Frequency distribution
☐ Probability distribution
☐ Histogram
☐ Random variable
☐ Average value, mean, expected value
☐ Variance
☐ Standard deviation
☐ Chebychev inequality
☐ Normal distribution
☐ Standard normal curve
☐ Binomial trial
☐ Use of normal curve to approximate binomial probabilities

Chapter 10: SUPPLEMENTARY EXERCISES

1. An experiment consists of three binomial trials, each having probability $\frac{1}{3}$ of success.

   (a) Determine the probability distribution table for the number of successes.

   (b) Use the table to compute the mean and variance of the probability distribution.

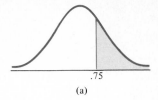

(a)

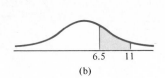

(b)

FIGURE 1

2. Find the area of the shaded region under the standard normal curve shown in Fig. 1(a).

3. Find the area of the shaded region under the normal curve with $\mu = 5, \sigma = 3$ shown in Fig. 1(b).

4. An archer has probability .3 of hitting a certain target. What is the probability of hitting the target exactly two times in four attempts?

5. Suppose that a probability distribution has mean 10 and standard deviation $\frac{1}{3}$. Use the Chebychev inequality to estimate the probability that an outcome will lie between 9 and 11.

6. Table 1 gives the probability distribution of the random variable X. Compute the mean and the variance of the random variable.

TABLE 1

k	$\Pr(X = k)$
0	.2
1	.3
5	.1
10	.4

7. The height of adult males in the United States is normally distributed with $\mu = 5.75$ feet and $\sigma = .2$. What percent of the adult male population has height of 6 feet or greater?

8. An urn contains four red balls and four white balls. An experiment consists of selecting at random a sample of four balls and recording the number of red balls in the sample. Set up the probability distribution and compute its mean and variance.

9. In a certain city two-fifths of the registered voters are women. Out of a group of 54 voters allegedly selected at random for jury duty, 13 are women. A local civil liberties group has charged that the selection procedure discriminated against women. Use the normal curve to estimate the probability of 13 or less women being selected in a truly random selection process.

10. In a complicated production process, $\frac{1}{4}$ of the items produced have to be readjusted. Find the probability that out of a batch of 75 items, between 8 and 22 (inclusive) of the items require readjustment.

11. Figure 2(a) is a normal curve with $\mu = 80$ and $\sigma = 15$. Find the value of z for which the area of the shaded region is .8664.

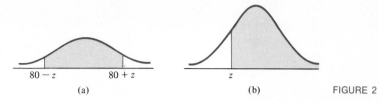

(a)

(b)

FIGURE 2

12. Figure 2(b) is a standard normal curve. Find the value of z for which the area of the shaded region is .7734.

The Derivative

Chapter II

The *derivative* is a mathematical tool used to measure change. To illustrate the sort of change we have in mind, consider the following example. Suppose that a rumor spreads through a town of 10,000 people.* Denote by $N(t)$ the number of people who have heard the rumor after t days. We have tabulated the values for $N(t)$ corresponding to $t = 0, 1, \ldots, 10$ in the chart in Fig. 1. Thus, for example, the rumor starts at $t = 0$ with one person $[N(0) = 1]$; after 1 day, the number

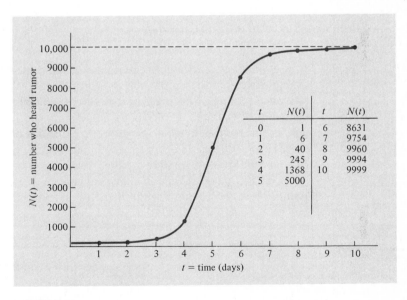

t	$N(t)$	t	$N(t)$
0	1	6	8631
1	6	7	9754
2	40	8	9960
3	245	9	9994
4	1368	10	9999
5	5000		

FIGURE 1

* The graph used in this example is derived from a mathematical model used by sociologists. See J. Coleman, *An Introduction to Mathematical Sociology* (London: Collier-Macmillan Ltd., 1964), p. 43.

having heard the rumor increases to 6; after 2 days, to 40; and so forth. Each day the number of people who have heard the rumor increases. We may describe the increase geometrically by plotting the points corresponding to $t = 0, 1, \ldots, 10$ and drawing a smooth curve through them. (See Fig. 1.)

Let us describe the change that is taking place in this example. Initially, the number of *new* people to hear the rumor each day is rather small (five the first day, 34 the second, 205 the third). As we shall see later, this situation is reflected in the fact that the graph is not very steep at $t = 1, 2, 3$. (It is almost flat.) On the fourth day, 1123 additional people hear the rumor; on the fifth day, 3632. The more rapid spread of the rumor on the fourth and fifth days is reflected in the increasing steepness of the graph at $t = 4$ and $t = 5$. After the fifth day the rumor spreads less rapidly, since there are fewer people to whom it can spread. Correspondingly, after $t = 5$, the graph becomes less steep. Finally, at $t = 9$ and $t = 10$, the graph is practically flat, indicating little change in the number who heard the rumor.

In the example above, there is a correlation between the rate at which $N(t)$ is changing and the steepness of the graph. This illustrates one of the fundamental ideas of calculus, which may be put roughly as follows: Measure rates of change in terms of steepness of graphs. This chapter is devoted to the *derivative*, which provides a numerical measure of the steepness of a curve at a particular point. By studying the derivative we will be able to deal numerically with the rates of change that occur in applied problems.

11.1. The Slope of a Curve at a Point

In Section 2.4 we introduced the notion of the slope of a straight line. In several examples we showed how the steepness property allowed the slope to be interpreted as a rate of change. This suggests that one possible approach to measuring rates of change is to examine the notion of slope more thoroughly. That is precisely the goal of this section, in which we generalize the idea of slope from straight lines to more general curves. In order to accomplish this, we must first discuss the concept of a tangent line to a curve at a point.

We have a clear idea of what is meant by the tangent line to a circle at a point P. It is the straight line that touches the circle at just the one point P. Let us

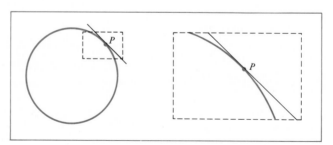

FIGURE 1

focus on the region near P, designated by the dashed rectangle shown in Fig. 1. The enlarged portion of the circle looks almost straight, and the straight line that it resembles is the tangent line. Further enlargements would make the circle near P look even straighter and have an even closer resemblance to the tangent line. In this sense, the tangent line to the circle at the point P is the straight line through P that best approximates the circle near P. In particular, the tangent line at P reflects the steepness of the circle at P. Thus it seems reasonable to define the *slope* of the circle at P to be the slope of the tangent line at P.

Similar reasoning leads us to a suitable definition of slope for an arbitrary curve at a point P. Consider the three curves drawn in Fig. 2. We have drawn an enlarged version of the dashed box around each point P. Notice that the portion of each curve lying in the boxed region looks almost straight. If we further magnify the curve near P, it would appear even straighter. Indeed, if we applied higher and higher magnification, the portion of the curve near P approaches a certain straight line ever more exactly. (See Fig. 3.) This straight line is called the *tangent line to the curve at P*. This line best approximates the curve near P. We define the *slope of a curve at a point P* to be the slope of the tangent line to the curve at P.

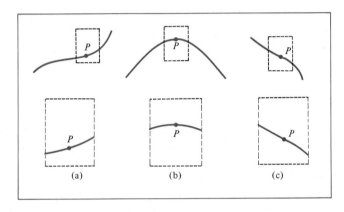

FIGURE 2

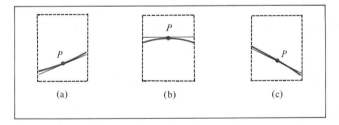

FIGURE 3

The portion of the curve near P can be, at least to within an approximation, replaced by the tangent line. Therefore, the slope of the curve at P—that is, the slope of the tangent line at P—measures the rate of increase or decrease of the curve as it passes through P.

EXAMPLE 1 Suppose that an apartment complex keeps a continuous record of the oil level in the storage tank. The graph for a typical 2-day period appears in Fig. 4. What is the physical significance of the slope of the graph at the point P?

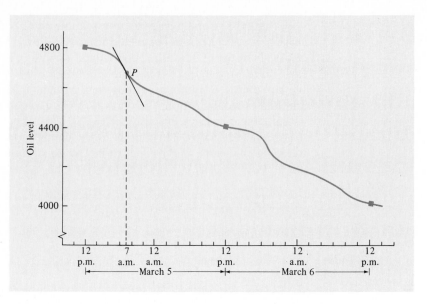

FIGURE 4

Solution The curve near P is closely approximated by its tangent line. So think of the curve as replaced by its tangent line near P. Then the slope at P is just the rate of decrease of the oil at 7 A.M. on March 5.

Notice that during the entire day of March 5 the graph in Fig. 4 seems to be the steepest at 7 A.M. That is, the oil level is falling the fastest at that time. This corresponds to the fact that most people awake around 7 A.M., turn up their thermostats, take showers, and so on. Example 1 provides a typical illustration of the manner in which slopes can be interpreted as rates of change. We shall return to this idea in Section 8.

In calculus, we can usually compute slopes by using formulas. For instance, in Section 3 we shall show that the tangent line to the graph of $y = x^2$ at the point $(1, 1)$ has slope 2; the tangent line at $(3, 9)$ has slope 6; and the tangent line at $(-\frac{5}{2}, \frac{25}{4})$ has slope -5. The various tangent lines are shown in Fig. 5. Notice that the slope at each point is two times the x-coordinate of the point. This is a general fact. In Section 3 we shall derive this simple formula (see Fig. 6.):

$$[\text{slope of the graph of } y = x^2 \text{ at the point } (x, y)] = 2x.$$

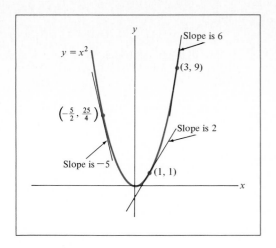

FIGURE 5

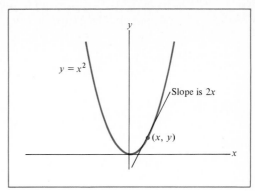

FIGURE 6

EXAMPLE 2 (a) What is the slope of the graph of $y = x^2$ at the point $(\frac{3}{4}, \frac{9}{16})$?

(b) Write the equation of the tangent line to the graph of $y = x^2$ at the point $(\frac{3}{4}, \frac{9}{16})$.

Solution (a) The x-coordinate of $(\frac{3}{4}, \frac{9}{16})$ is $\frac{3}{4}$, so the slope of $y = x^2$ at this point is $2(\frac{3}{4}) = \frac{3}{2}$.

(b) We shall write the equation of the tangent line in the point-slope form. The point is $(\frac{3}{4}, \frac{9}{16})$, and the slope is $\frac{3}{2}$ by part (a). Hence the equation is

$$y - \tfrac{9}{16} = \tfrac{3}{2}(x - \tfrac{3}{4}).$$

PRACTICE PROBLEMS 1

1. Refer to the accompanying graph.

   (a) What is the slope of the curve at $(3, 4)$?

   (b) What is the equation of the tangent line at the point where $x = 3$?

2. What is the equation of the tangent line to the graph of $y = \frac{1}{2}x + 1$ at the point $(4, 3)$?

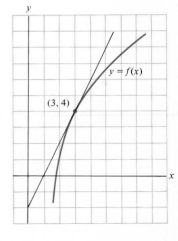

EXERCISES 1

Trace the curves in Exercises 1–6 onto another piece of paper and sketch the tangent line in each case at the designated point P.

1.

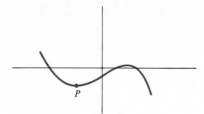

2.

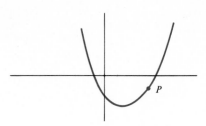

3.

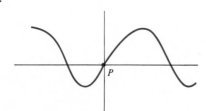

4.

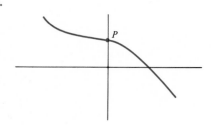

5.

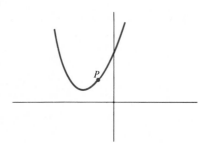

6.

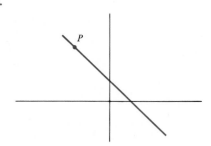

Estimate the slope of each of the following curves at the designated point P.

7.

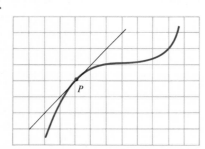

8.

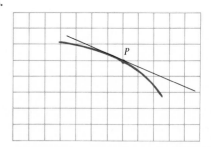

9.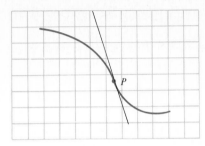

10.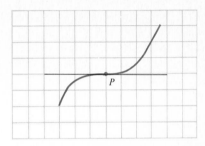

11.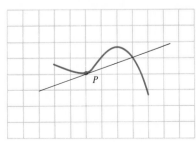

12.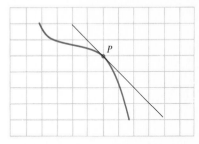

In Exercises 13–15, find the slope of the tangent line to the graph of $y = x^2$ at the point indicated, and then write the corresponding equation of the tangent line.

13. $(-2, 4)$ **14.** $(-.4, .16)$ **15.** $(\frac{4}{3}, \frac{16}{9})$

16. Find the slope of the tangent line to the graph of $y = x^2$ at the point where $x = -\frac{1}{2}$.

17. Write the equation of the tangent line to the graph of $y = x^2$ at the point where $x = 1.5$.

18. Write the equation of the tangent line to the graph of $y = x^2$ at the point where $x = .6$.

19. Find the point on the graph of $y = x^2$ where the curve has slope $\frac{5}{3}$.

20. Find the point on the graph of $y = x^2$ where the curve has slope -4.

21. Find the point on the graph of $y = x^2$ where the tangent line is parallel to the line $x + 2y = 4$.

22. Find the point on the graph of $y = x^2$ where the tangent line is parallel to the line $3x - y = 2$.

In the next section we shall see that the tangent line to the graph of $y = x^3$ at the point (x, y) has slope $3x^2$. Using this result, find the slope of the curve at the following points.

23. $(2, 8)$

24. $(\frac{3}{2}, \frac{27}{8})$

25. $(-\frac{1}{2}, -\frac{1}{8})$

26. Find the slope of the curve $y = x^3$ at the point where $x = \frac{1}{4}$.

27. Write the equation of the line tangent to the graph of $y = x^3$ at the point where $x = -1$.

28. Write the equation of the line tangent to the graph of $y = x^3$ at the point where $x = \frac{1}{2}$.

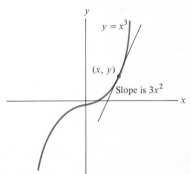

1. (a) The slope of the curve at the point (3, 4) is, by definition, the slope of the tangent line at (3, 4). Note that the point (4, 6) is also on the line. Therefore, the slope is

$$\frac{6 - 4}{4 - 3} = \frac{2}{1} = 2.$$

(b) Use the point-slope formula. The equation of the line passing through the point (3, 4) and having slope 2 is

$$y - 4 = 2(x - 3)$$

or

$$y = 2x - 2.$$

2. The tangent line at (4, 3) is, by definition, the line that best approximates the curve at (4, 3). Since the "curve" in this case is itself a line, the curve and its tangent line at (4, 3) (and at every other point) must be the same. Therefore, the equation is $y = \frac{1}{2}x + 1$.

11.2. The Derivative

Suppose that a curve is the graph of a function $f(x)$. It is usually possible to obtain a formula that gives the slope of the curve $y = f(x)$ at any point. This slope formula is called the *derivative* of $f(x)$ and is written $f'(x)$. For each value of x, $f'(x)$ gives the slope of the curve $y = f(x)$ at the point with first coordinate x.* (See Fig. 1.)

FIGURE 1

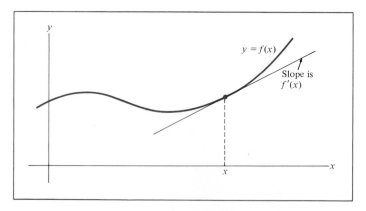

---

* As we shall see, there are are curves which do not have tangent lines at every point. At values of x corresponding to such points, the derivative $f'(x)$ is not defined. For the sake of the current discussion, which is designed to develop an intuitive feeling for the derivative, let us assume that the graph of $f(x)$ has a tangent line for each x in the domain of f.

The process of computing $f'(x)$ for a given function $f(x)$ is called *differentiation*.

As we shall see later, the concept of a derivative occurs frequently in applications. In economics, the derivatives are often described by the adjective "marginal." For instance, if $C(x)$ is a cost function (the cost of producing x units of a commodity), then the derivative $C'(x)$ is called the *marginal cost function*. The derivative $P'(x)$ of a profit function $P(x)$ is called the *marginal profit function*; the derivative of a revenue function is called the *marginal revenue function*; and so on. We shall discuss the economic meaning of these marginal concepts in Section 8.

For the remainder of this section as well as the next few sections we will concentrate on calculating derivatives.

The case of a linear function $f(x) = mx + b$ is particularly simple. The graph of $y = mx + b$ is a straight line L of slope m. The tangent line to L (at any point) is just L itself, and so the slope of the graph is m at every point. (See Fig. 2.) In other words, the value of the derivative $f'(x)$ is always equal to m. We summarize this fact as follows:

$$\text{If } f(x) = mx + b, \text{ then } f'(x) = m. \tag{1}$$

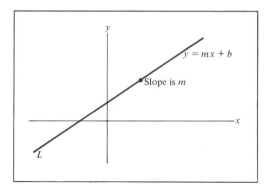

FIGURE 2

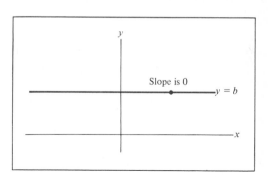

FIGURE 3

Set $m = 0$ in equation (1). Then the function becomes $f(x) = b$, which has the value b for each value of x. The graph is a horizontal line of slope 0, so $f'(x) = 0$ for all x. (See Fig. 3.) Thus we have:

$$\text{The derivative of a constant function } f(x) = b \text{ is } f'(x) = 0. \tag{2}$$

Next, consider the function $f(x) = x^2$. As we stated in Section 2 (and will prove at the end of this section), the slope of the graph of $y = x^2$ at the point

(x, y) is equal to $2x$. That is, the value of the derivative $f'(x)$ is $2x$:

$$\text{If } f(x) = x^2, \text{ then } f'(x) = 2x. \tag{3}$$

In Exercises 23–28 of Section 2 we made use of the fact that the slope of the graph of $y = x^3$ at the point (x, y) is $3x^2$. This can be restated in terms of derivatives as follows:

$$\text{If } f(x) = x^3, \text{ then } f'(x) = 3x^2. \tag{4}$$

We should, at this stage at least, keep the geometric meaning of these formulas clearly in mind. Figure 4 shows the graphs of x^2 and x^3 together with the interpretations of formulas (3) and (4) in terms of slope.

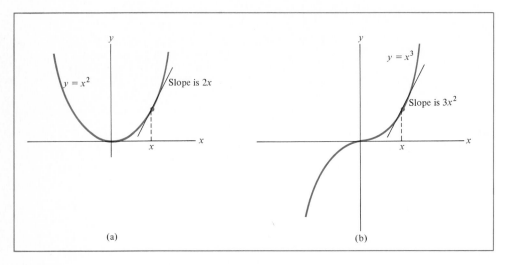

(a)

(b)

FIGURE 4

One of the reasons calculus is so useful is that it provides general techniques which can be easily used to determine derivatives. One such general rule, which contains formulas (3) and (4) as special cases, is the so-called power rule.

Power Rule Let r be any number and let $f(x) = x^r$. Then $f'(x) = rx^{r-1}$.

Indeed, if $r = 2$, then $f(x) = x^2$ and $f'(x) = 2x^{2-1} = 2x$, which is formula (3). If $r = 3$, then $f(x) = x^3$ and $f'(x) = 3x^{3-1} = 3x^2$, which is (4). We shall prove the power rule in Chapter 4. Until then, we shall use it to calculate derivatives.

EXAMPLE 1 Let $f(x) = \sqrt{x}$. What is $f'(x)$?

Solution   Recall that $\sqrt{x} = x^{1/2}$. We may apply the power rule with $r = \frac{1}{2}$.

$$f(x) = x^{1/2}$$

$$f'(x) = \tfrac{1}{2}x^{1/2-1} = \tfrac{1}{2}x^{-1/2}$$

$$= \frac{1}{2} \cdot \frac{1}{x^{1/2}} = \frac{1}{2\sqrt{x}}.$$

Another important special case of the power rule occurs for $r = -1$, corresponding to $f(x) = x^{-1}$. In this case, $f'(x) = (-1)x^{-1-1} = -x^{-2}$. However, since $x^{-1} = 1/x$ and $x^{-2} = 1/x^2$, the power rule for $r = -1$ may also be written as follows:*

$$\text{If } f(x) = \frac{1}{x}, \text{ then } f'(x) = -\frac{1}{x^2} \ (x \neq 0). \tag{5}$$

EXAMPLE 2   Find the slope of the curve $y = 1/x$ at $(2, \frac{1}{2})$.

Solution   Set $f(x) = 1/x$. The point $(2, \frac{1}{2})$ corresponds to $x = 2$, so in order to find the slope at this point, we compute $f'(2)$. From formula (5) we find that

$$f'(2) = -\frac{1}{2^2} = -\frac{1}{4}.$$

Thus the slope of $y = 1/x$ at the point $(2, \frac{1}{2})$ is $-\frac{1}{4}$. (See Fig. 5.)

FIGURE 5

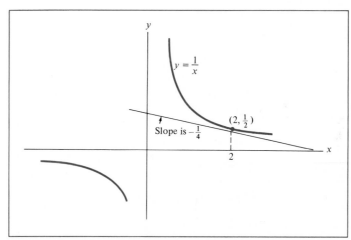

* The formula gives $f'(x)$ for $x \neq 0$. The derivative of $f(x)$ is not defined at $x = 0$ since $f(x)$ itself is not defined there.

Warning Do not confuse $f'(2)$, the value of the derivative at 2, with $f(2)$, the value of the y-coordinate at the point on the graph at which $x = 2$. In Example 2 we have $f'(2) = -\frac{1}{4}$, whereas $f(2) = \frac{1}{2}$. The number $f'(2)$ gives the *slope* of the graph at $x = 2$; the number $f(2)$ gives the *height* of the graph at $x = 2$.

Notation   The operation of forming a derivative $f'(x)$ from a function $f(x)$ is also indicated by the symbol $\dfrac{d}{dx}$ (read: the derivative with respect to x). Thus

$$\frac{d}{dx} f(x) = f'(x).$$

For example,

$$\frac{d}{dx} (x^6) = 6x^5, \qquad \frac{d}{dx} (x^{5/3}) = \frac{5}{3} x^{2/3}, \qquad \frac{d}{dx}\left(\frac{1}{x}\right) = -\frac{1}{x^2}.$$

When working with an equation of the form $y = f(x)$, we often write $\dfrac{dy}{dx}$ as a symbol for the derivative $f'(x)$. For example, if $y = x^6$, we may write

$$\frac{dy}{dx} = 6x^5.$$

The Secant-Line Calculation of the Derivative   So far, we have said nothing about how to derive differentiation formulas, such as (3), (4), or (5). Let us remedy that omission now. The derivative gives the slope of the tangent line, so we must describe a procedure for calculating that slope.*

The fundamental idea for calculating the slope of the tangent line at a point P is to approximate the tangent line very closely by *secant lines*. A secant line at P is a straight line passing through P and a nearby point Q on the curve. (See Fig. 6.) By taking Q very close to P, we can make the slope of the secant line approximate the slope of the tangent line to any desired degree of accuracy. Let us see what this amounts to in terms of calculations.

Suppose that the point P is $(x, f(x))$. Suppose also that Q is h horizontal units away from P. Then Q has x-coordinate $x + h$ and y-coordinate $f(x + h)$. The slope of the secant line through the points $P = (x, f(x))$ and $Q = (x + h, f(x + h))$ is simply

$$[\text{slope of secant line}] = \frac{f(x + h) - f(x)}{(x + h) - x} = \frac{f(x + h) - f(x)}{h}.$$

(See Fig. 7.)

In order to move Q close to P along the curve, we let h approach zero. Then the secant line approaches the tangent line, and so

$$[\text{slope of secant line}] \quad \text{approaches} \quad [\text{slope of tangent line}];$$

---

* The following discussion is designed to develop geometric intuition for the derivative. It will also provide the basis for a formal definition of the derivative in terms of limits in Section 4.

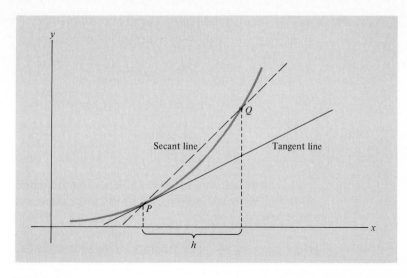

FIGURE 6

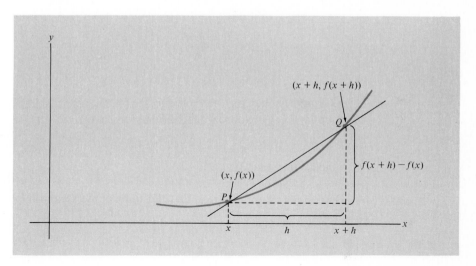

FIGURE 7

that is,

$$\frac{f(x + h) - f(x)}{h} \quad \text{approaches} \quad f'(x).$$

Since we can make the secant line as close to the tangent line as we wish, by taking h sufficiently small, the quantity $[f(x + h) - f(x)]/h$ can be made to approximate $f'(x)$ to any desired degree of accuracy. And we arrive at the following method to compute the derivative $f'(x)$.

To calculate $f'(x)$:

1. First calculate $\dfrac{f(x + h) - f(x)}{h}$ for $h \neq 0$.

2. Then let h approach zero.

3. The quantity $\dfrac{f(x + h) - f(x)}{h}$ will approach $f'(x)$.

Let us use this method to verify the differentiation formulas (3) and (5), which as we saw were special cases of the power rule for $r = 2$ and $r = -1$, respectively.

Verification of the Power Rule for $r = 2$

Here $f(x) = x^2$, so that the slope of the secant line is

$$\frac{f(x + h) - f(x)}{h} = \frac{(x + h)^2 - x^2}{h}.$$

However, by multiplying out, we have $(x + h)^2 = x^2 + 2xh + h^2$, so that

$$\frac{f(x + h) - f(x)}{h} = \frac{x^2 + 2xh + h^2 - x^2}{h} = \frac{(2x + h)h}{h}$$

$$= 2x + h.$$

As h approaches zero (i.e., as the secant line approaches the tangent line), the quantity $2x + h$ approaches $2x$. Thus we have

$$f'(x) = 2x,$$

which is formula (3).

Verification of the Power Rule for $r = -1$

Here $f(x) = x^{-1} = 1/x$, so that the slope of the secant line is

$$\frac{f(x + h) - f(x)}{h} = \frac{1}{h}\left[\frac{1}{x + h} - \frac{1}{x}\right] = \frac{1}{h}\left[\frac{x - (x + h)}{(x + h)x}\right]$$

$$= \frac{1}{h}\left[\frac{-h}{(x + h)x}\right] = -\frac{1}{x(x + h)}.$$

As h approaches zero, $-\dfrac{1}{x(x + h)}$ approaches $-\dfrac{1}{x^2}$. Hence

$$f'(x) = -\frac{1}{x^2},$$

which is formula (5).

Similar arguments can be used to verify the power rule for other values of r. The cases $r = 3$ and $r = \frac{1}{2}$ are outlined in Exercises 47 and 48, respectively.

PRACTICE PROBLEMS 2

1. Consider the curve $y = f(x)$ in the accompanying sketch.

   (a) Find $f(5)$.

   (b) Find $f'(5)$.

2. Let $f(x) = 1/x^4$.

   (a) Find its derivative.

   (b) Find $f'(2)$.

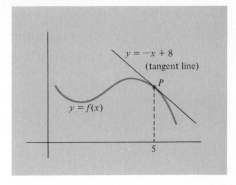

$y = -x + 8$
(tangent line)

P

$y = f(x)$

5

EXERCISES 2

Find the derivatives of the following functions.

1. $f(x) = 2x - 5$ 2. $f(x) = 3 - \frac{1}{2}x$ 3. $f(x) = x^8$

4. $f(x) = x^{75}$ 5. $f(x) = x^{5/2}$ 6. $f(x) = x^{4/3}$

7. $f(x) = \sqrt[3]{x}$ 8. $f(x) = x^{3/4}$ 9. $f(x) = x^{-2}$

10. $f(x) = 5$ 11. $f(x) = x^{-1/4}$ 12. $f(x) = x^{-3}$

13. $f(x) = \frac{3}{4}$ 14. $f(x) = 1/\sqrt[3]{x}$ 15. $f(x) = 1/x^3$

16. $f(x) = 1/x^5$

In Exercises 17–24, find the derivative of $f(x)$ at the designated value of x.

17. $f(x) = x^6$, at $x = -2$ 18. $f(x) = x^3$, at $x = \frac{1}{4}$

19. $f(x) = 1/x$, at $x = 3$ 20. $f(x) = 5x$, at $x = 2$

21. $f(x) = 4 - x$, at $x = 5$ 22. $f(x) = x^{2/3}$, at $x = 1$

23. $f(x) = x^{3/2}$, at $x = 9$ 24. $f(x) = 1/x^2$, at $x = 2$

25. Find the slope of the curve $y = x^4$ at $x = 3$.

26. Find the slope of the curve $y = x^5$ at $x = -2$.

27. Find the slope of the curve $y = \sqrt{x}$ at $x = 9$.

28. Find the slope of the curve $y = x^{-3}$ at $x = 3$.

29. If $f(x) = x^2$, compute $f(-5)$ and $f'(-5)$.

30. If $f(x) = x + 6$, compute $f(3)$ and $f'(3)$.

31. If $f(x) = 1/x^5$, compute $f(2)$ and $f'(2)$.

32. If $f(x) = 1/x^2$, compute $f(5)$ and $f'(5)$.

33. If $f(x) = x^{4/3}$, compute $f(8)$ and $f'(8)$.

34. If $f(x) = x^{3/2}$, compute $f(16)$ and $f'(16)$.

35. Find the slope of the tangent line to the curve $y = x^3$ at the point $(4, 64)$, and write the equation of this line.

36. Find the slope of the tangent line to the curve $y = \sqrt{x}$ at the point $(25, 5)$, and write the equation of this line.

In Exercises 37–44, find the indicated derivative.

37. $\dfrac{d}{dx}(x^8)$

38. $\dfrac{d}{dx}(x^{-3})$

39. $\dfrac{d}{dx}(x^{3/4})$

40. $\dfrac{d}{dx}(x^{-1/3})$

41. $\dfrac{dy}{dx}$ if $y = 1$

42. $\dfrac{dy}{dx}$ if $y = x^{-4}$

43. $\dfrac{dy}{dx}$ if $y = x^{1/5}$

44. $\dfrac{dy}{dx}$ if $y = \dfrac{x-1}{3}$

45. Consider the curve $y = f(x)$ in Fig. 8. Find $f(6)$ and $f'(6)$.

46. Consider the curve $y = f(x)$ in Fig. 9. Find $f(1)$ and $f'(1)$.

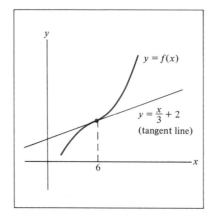

FIGURE 8

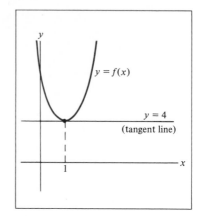

FIGURE 9

47. Use an argument like those used to verify formulas (3) and (5) to show that the derivative of $f(x) = x^3$ is $3x^2$. [*Hint*: Recall that $(x + h)^3 = x^3 + 3x^2h + 3xh^2 + h^3$.]

48. Use an argument like those used to verify formulas (3) and (5) to show that the derivative of $f(x) = \sqrt{x}$ is $1/(2\sqrt{x})$. [*Hint*: After forming the expression for the slope of a secant line, eliminate the square roots from the numerator by multiplying both the numerator and denominator by the quantity $\sqrt{x + h} + \sqrt{x}$.]

SOLUTIONS TO PRACTICE PROBLEMS 2

1. (a) The number $f(5)$ is the y-coordinate of the point P. Since the tangent line passes through P, the coordinates of P satisfy the equation $y = -x + 8$. Since its x-coordinate is 5, its y-coordinate is $-5 + 8 = 3$. Therefore, $f(5) = 3$.

(b) The number $f'(5)$ is the slope of the tangent line at P, which is readily seen to be -1.

2. (a) The function $1/x^4$ can be written as the power function x^{-4}. Here $r = -4$. Therefore,

$$f'(x) = (-4)x^{(-4)-1} = -4x^{-5} = \frac{-4}{x^5}.$$

(b) $f'(2) = -4/2^5 = -4/32 = -\frac{1}{8}.$

11.3. Limits and the Derivative

The notion of a limit is one of the fundamental ideas of calculus. Indeed, any "theoretical" development of calculus rests on an extensive use of the theory of limits. Even in this book, where we have adopted an intuitive viewpoint, limit arguments are used occasionally (although in an informal way). In this section we give a brief introduction to limits and their role in calculus. As we shall see, the limit concept will allow us to define the notion of a derivative independently of our geometric reasoning.

Actually, we have already considered a limit in our discussion of the derivative, although we did not use the term "limit." Using the geometric reasoning of the previous section, we have the following procedure for calculating the derivative of a function $f(x)$ at $x = a$. First calculate the *difference quotient*

$$\frac{f(a + h) - f(a)}{h},$$

where h is a nonzero number. Next, allow h to approach zero by allowing it to assume both positive and negative numbers arbitrarily close to zero but different from zero. In symbols, we write $h \to 0$. The values of the difference quotient then approach the value of the derivative $f'(a)$. We say that the number $f'(a)$ is the *limit* of the difference quotient as h approaches zero, and in symbols we write

$$f'(a) = \lim_{h \to 0} \frac{f(a + h) - f(a)}{h}. \tag{1}$$

As a numerical example of formula (1), consider the case of the derivative of $f(x) = x^2$ at $x = 2$. The difference quotient in this case has the form

$$\frac{f(2 + h) - f(2)}{h} = \frac{(2 + h)^2 - 2^2}{h}.$$

Table 1 gives some typical values of this difference quotient for progressively smaller values of h, both positive and negative. It is clear that the values of the difference quotient are approaching 4 as $h \to 0$. In other words, 4 *is the limit* of the difference quotient as $h \to 0$. Thus

$$\lim_{h \to 0} \frac{(2 + h)^2 - 2^2}{h} = 4.$$

Since the values of the difference quotient approach the derivative $f'(2)$, we conclude that $f'(2) = 4$.

TABLE 1

h	$\dfrac{(2 + h)^2 - 2^2}{h}$	h	$\dfrac{(2 + h)^2 - 2^2}{h}$
1	$\dfrac{(2 + 1)^2 - 2^2}{1} = 5$	-1	$\dfrac{(2 + (+1))^2 - 2^2}{-1} = 3$
.1	$\dfrac{(2 + .1)^2 - 2^2}{.1} = 4.10$	$-.1$	$\dfrac{(2 + (-.1))^2 - 2^2}{-.1} = 3.9$
.01	$\dfrac{(2 + .01)^2 - 2^2}{.01} = 4.01$	$-.01$	$\dfrac{(2 + (-.01))^2 - 2^2}{-.01} = 3.99$
.001	$\dfrac{(2 + .001)^2 - 2^2}{.001} = 4.001$	$-.001$	$\dfrac{(2 + (-.001))^2 - 2^2}{-.001} = 3.999$
.0001	$\dfrac{(2 + .0001)^2 - 2^2}{.0001} = 4.0001$	$-.0001$	$\dfrac{(2 + (-.0001))^2 - 2^2}{-.0001} = 3.9999$

Our discussion of the derivative has been based on an intuitive geometric concept of the tangent line. However, the limit on the right in (1) may be considered independently of its geometric interpretation. In fact, we may use (1) to define $f'(a)$. We say that f is *differentiable* at $x = a$ if $\dfrac{f(a + h) - f(a)}{h}$ approaches some number as $h \to 0$, and we denote this limiting number by $f'(a)$. If the difference quotient $\dfrac{f(a + h) - f(a)}{h}$ does not approach any specific number as $h \to 0$, we say that f is *nondifferentiable* at $x = a$. Essentially all of the functions in this text are differentiable at all points in their domain. A few exceptions are described in Section 5.

To better understand the limit concept used to define the derivative, it will be helpful to look at limits in a more general setting. If we let $g(h) = \dfrac{(2 + h)^2 - 2^2}{h}$, then Table 1 gives values of $g(h)$ as $h \to 0$. These values obviously approach 4. We may express this by writing

$$\lim_{h \to 0} g(h) = 4.$$

The discussion above suggests the following definition: Let $g(x)$ be a function, a a number. We say that the number L *is the limit of $g(x)$ as x approaches a* provided that, as x gets arbitrarily close (but not equal) to a, the values of $g(x)$ approach L. In this case we write

$$\lim_{x \to a} g(x) = L.$$

If, as x approaches a, the values $g(x)$ do *not* approach a specific number, then we say that the limit of $g(x)$ as x approaches a *does not exist*. Let us give some further examples of limits.

EXAMPLE 1 Determine $\lim_{x \to 2} (3x - 5)$.

Solution Let us make a table of values of x approaching 2 and the corresponding values of $3x - 5$:

x	$3x - 5$	x	$3x - 5$
2.1	1.3	1.9	.7
2.01	1.03	1.99	.97
2.001	1.003	1.999	.997
2.0001	1.0003	1.9999	.9997

As x approaches 2, we see that $3x - 5$ approaches 1. In terms of our notation,

$$\lim_{x \to 2} (3x - 5) = 1.$$

EXAMPLE 2 For each of the following functions, determine if $\lim_{x \to 2} g(x)$ exists. (The circles drawn on the graphs are meant to represent breaks in the graph, indicating that the functions under consideration are not defined at $x = 2$.)

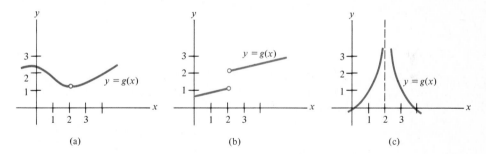

(a) (b) (c)

Solution (a) $\lim_{x \to 2} g(x) = 1$. We can see that as x gets closer and closer to 2, the values of $g(x)$ get closer and closer to 1. This is true for values of x to both the right and the left of 2.

(b) $\lim_{x \to 2} g(x)$ does not exist. As x approaches 2 from the right, $g(x)$ approaches 2. However, as x approaches 2 from the left, $g(x)$ approaches 1. In order for a limit to exist, the function must approach the *same* value from each direction.

(c) $\lim_{x \to 2} g(x)$ does not exist. As x approaches 2, the values of $g(x)$ become larger and larger and do not approach a fixed number.

    The following limit theorems, which we cite without proof, allow us to reduce the computation of limits for combinations of functions to computations of limits involving the constituent functions.

> *Limit Theorems* Suppose that $\lim_{x \to a} f(x)$ and $\lim_{x \to a} g(x)$ both exist. Then we
> have the following results.
>
> I. If k is a constant, then $\lim_{x \to a} k \cdot f(x) = k \cdot \lim_{x \to a} f(x)$.
>
> II. If r is a positive constant, then $\lim_{x \to a} [f(x)]^r = \left[\lim_{x \to a} f(x) \right]^r$.
>
> III. $\lim_{x \to a} [f(x) + g(x)] = \lim_{x \to a} f(x) + \lim_{x \to a} g(x)$.
>
> IV. $\lim_{x \to a} [f(x) - g(x)] = \lim_{x \to a} f(x) - \lim_{x \to a} g(x)$.
>
> V. $\lim_{x \to a} [f(x) \cdot g(x)] = \left[\lim_{x \to a} f(x) \right]\left[\lim_{x \to a} g(x) \right]$.
>
> VI. If $\lim_{x \to a} g(x) \neq 0$, then $\lim_{x \to a} \dfrac{f(x)}{g(x)} = \dfrac{\lim_{x \to a} f(x)}{\lim_{x \to a} g(x)}$.

EXAMPLE 3   Use the limit theorems to compute the following limits.

(a) $\lim_{x \to 2} x^3$ (b) $\lim_{x \to 2} 5x^3$ (c) $\lim_{x \to 2} (5x^3 - 15)$ (d) $\lim_{x \to 2} \sqrt{5x^3 - 15}$

(e) $\lim_{x \to 2} (\sqrt{5x^3 - 15}/x^5)$

Solution   (a) Since $\lim_{x \to 2} x = 2$, we have by Limit Theorem II that

$$\lim_{x \to 2} x^3 = \left(\lim_{x \to 2} x \right)^3 = 2^3 = 8.$$

(b) $\lim_{x \to 2} 5x^3 = 5 \lim_{x \to 2} x^3$ (Limit Theorem I with $k = 5$)

$$= 5 \cdot 8 \qquad [\text{by part (a)}]$$

$$= 40.$$

(c) $\lim_{x \to 2} (5x^3 - 15) = \lim_{x \to 2} 5x^3 - \lim_{x \to 2} 15$ (Limit Theorem IV).

Note that $\lim_{x \to 2} 15 = 15$. This is because the constant function $g(x) = 15$ always
has the value 15, and so its limit as x approaches *any* number is 15. By part (b),
$\lim_{x \to 2} 5x^3 = 40$. Thus

$$\lim_{x \to 2} (5x^3 - 15) = 40 - 15 = 25.$$

(d) $\displaystyle\lim_{x \to 2} \sqrt{5x^3 - 15} = \lim_{x \to 2} (5x^3 - 15)^{1/2}$

$$= \left[\lim_{x \to 2} (5x^3 - 15) \right]^{1/2} \qquad \text{[Limit Theorem II with } r = \tfrac{1}{2},$$
$$f(x) = 5x^3 - 15]$$

$$= 25^{1/2} \qquad\qquad \text{[by part (c)]}$$

$$= 5.$$

(e) The limit of the denominator is $\displaystyle\lim_{x \to 2} x^5$, which is $2^5 = 32$, a nonzero number. So by Limit Theorem VI we have

$$\lim_{x \to 2} \frac{\sqrt{5x^3 - 15}}{x^5} = \frac{\displaystyle\lim_{x \to 2} \sqrt{5x^3 - 15}}{\displaystyle\lim_{x \to 2} x^5}$$

$$= \frac{5}{32} \qquad \text{[by part (d)]}.$$

The following facts, which may be deduced by repeated applications of the various limit theorems, are extremely handy in evaluating limits.

> *Limit of a Polynomial Function* Let $p(x)$ be a polynomial function, a any number. Then
>
> $$\lim_{x \to a} p(x) = p(a).$$

> *Limit of a Rational Function* Let $r(x) = p(x)/q(x)$ be a rational function, where $p(x)$ and $q(x)$ are polynomials. Let a be a number such that $q(a) \neq 0$. Then
>
> $$\lim_{x \to a} r(x) = r(a).$$

In other words, to determine a limit for a polynomial or a rational function, simply evaluate the function at $x = a$, provided, of course, that the function is defined at $x = a$. For instance, we can rework the solution to Example 3(c) as follows:

$$\lim_{x \to 2} (5x^3 - 15) = 5(2)^3 - 15 = 25.$$

Many situations require algebraic simplifications before the limit theorems can be applied.

EXAMPLE 4   Compute the following limits.

(a) $\lim\limits_{x \to 3} \dfrac{x^2 - 9}{x - 3}$

(b) $\lim\limits_{x \to 0} \dfrac{\sqrt{x + 4} - 2}{x}$

Solution   (a) The function $\dfrac{x^2 - 9}{x - 3}$ is not defined when $x = 3$, since $\dfrac{3^2 - 9}{3 - 3} = \dfrac{0}{0}$, which is undefined. That causes no difficulty, since the limit as x approaches 3 depends only on the values of x *near* 3 and excludes consideration of the value at $x = 3$ itself. To evaluate the limit, note that $x^2 - 9 = (x - 3)(x + 3)$. So for $x \neq 3$,

$$\frac{x^2 - 9}{x - 3} = \frac{(x - 3)(x + 3)}{x - 3} = x + 3.$$

As x approaches 3, $x + 3$ approaches 6. Therefore,

$$\lim_{x \to 3} \frac{x^2 - 9}{x - 3} = 6.$$

(b) Since the denominator approaches zero when taking the limit, we may not apply Limit Theorem VI directly. However, if we first apply an algebraic trick, the limit may be evaluated. Multiply numerator and denominator by $\sqrt{x + 4} + 2$.

$$\frac{\sqrt{x + 4} - 2}{x} \cdot \frac{\sqrt{x + 4} + 2}{\sqrt{x + 4} + 2} = \frac{(x + 4) - 4}{x(\sqrt{x + 4} + 2)}$$

$$= \frac{x}{x(\sqrt{x + 4} + 2)}$$

$$= \frac{1}{\sqrt{x + 4} + 2}.$$

Thus

$$\lim_{x \to 0} \frac{\sqrt{x + 4} - 2}{x} = \lim_{x \to 0} \frac{1}{\sqrt{x + 4} + 2}$$

$$= \frac{\lim\limits_{x \to 0} 1}{\lim\limits_{x \to 0} (\sqrt{x + 4} + 2)} \qquad \text{(Limit Theorem VI)}$$

$$= \frac{1}{4}.$$

EXAMPLE 5   Use limits to compute the derivative $f'(5)$ for the following functions.

(a) $f(x) = 15 - x^2$

(b) $f(x) = \dfrac{1}{2x - 3}$ $2x + 3$

$\dfrac{2x+3}{4x^2-9}$

$(2x - 3)^{-1}$

Solution In each case, we must calculate $\lim\limits_{h \to 0} \dfrac{f(5+h) - f(5)}{h}$.

(a)
$$\frac{f(5+h) - f(5)}{h} = \frac{[15 - (5+h)^2] - (15 - 5^2)}{h}$$

$$= \frac{15 - (25 + 10h + h^2) - (15 - 25)}{h}$$

$$= \frac{-10h - h^2}{h} = -10 - h.$$

Therefore, $f'(5) = \lim\limits_{h \to 0} (-10 - h) = -10$.

(b)
$$\frac{f(5+h) - f(5)}{h} = \frac{\dfrac{1}{2(5+h) - 3} - \dfrac{1}{2(5) - 3}}{h}$$

$$= \frac{\dfrac{1}{7 + 2h} - \dfrac{1}{7}}{h} = \frac{\dfrac{7 - (7 + 2h)}{(7 + 2h)7}}{h}$$

$$= \frac{-2h}{(7 + 2h)7 \cdot h} = \frac{-2}{(7 + 2h)7} = \frac{-2}{49 + 14h}.$$

$$f'(5) = \lim_{h \to 0} \frac{-2}{49 + 14h} = -\frac{2}{49}.$$

Remark When computing the limit in the example above, we considered only values of h near zero (and not $h = 0$ itself). Therefore, we were able to freely divide both numerator and denominator by h.

Infinity and Limits   Consider the function $f(x)$ whose graph is sketched in Fig. 1. As x grows large the value of $f(x)$ approaches 2. In this circumstance, we say that 2 is *the limit of $f(x)$ as x approaches infinity.* Infinity is denoted by the symbol ∞.

FIGURE 1

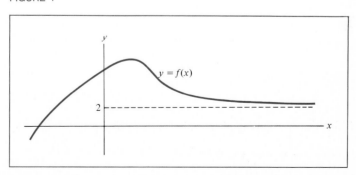

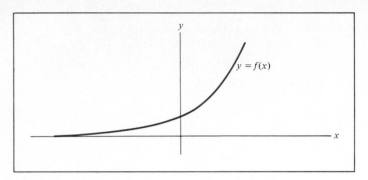

FIGURE 2

The limit statement above is expressed in the following notation:

$$\lim_{x \to \infty} f(x) = 2.$$

In a similar vein, consider the function whose graph is sketched in Fig. 2. As x grows large in the negative direction, the value of $f(x)$ approaches 0. In this circumstance, we say that 0 is *the limit of $f(x)$ as x approaches minus infinity*. In symbols,

$$\lim_{x \to -\infty} f(x) = 0.$$

EXAMPLE 6   Calculate the following limits.

(a) $\displaystyle \lim_{x \to \infty} \frac{1}{x^2 + 1}$

(b) $\displaystyle \lim_{x \to \infty} \frac{x + 1}{x - 1}$

Solution   (a) As x increases without bound, so does $x^2 + 1$. Therefore, $1/(x^2 + 1)$ approaches zero as x approaches ∞.

(b) Both $x + 1$ and $x - 1$ increase without bound as x does. To determine the limit of their quotient, we employ an algebraic trick. Divide both numerator and denominator by x to obtain

$$\lim_{x \to \infty} \frac{x + 1}{x - 1} = \lim_{x \to \infty} \frac{1 + \dfrac{1}{x}}{1 - \dfrac{1}{x}}.$$

As x increases without bound, $1/x$ approaches zero, so that both $1 + (1/x)$ and $1 - (1/x)$ approach 1. Thus the desired limit is $1/1 = 1$.

PRACTICE PROBLEMS 3

Determine which of the following limits exist. Compute the limits that exist.

1. $\displaystyle \lim_{x \to 6} \frac{x^2 - 4x - 12}{x - 6}$

2. $\displaystyle \lim_{x \to 6} \frac{4x + 12}{x - 6}$

EXERCISES 3

For each of the following functions, $g(x)$, determine whether or not $\lim_{x \to 3} g(x)$ exists. If so, give the limit.

1.
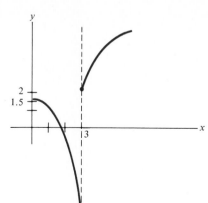

2.
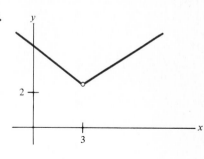

3.
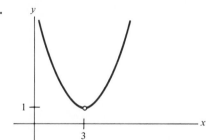

4.
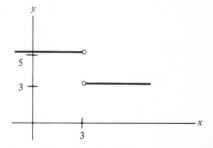

5.
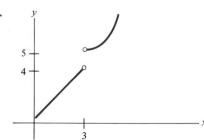

6.
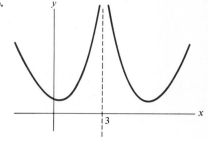

Determine which of the following limits exist. Compute the limits that exist.

7. $\displaystyle\lim_{x \to 1} (1 - 6x)$

8. $\displaystyle\lim_{x \to 2} \frac{x}{x - 2}$

9. $\displaystyle\lim_{x \to 3} \sqrt{x^2 + 16}$

10. $\displaystyle\lim_{x \to 4} (x^3 - 7)$

11. $\displaystyle\lim_{x \to 5} \frac{x^2 + 1}{5 - x}$

12. $\displaystyle\lim_{x \to 6} \left(\sqrt{6x} + 3x - \frac{1}{x} \right)(x^2 - 4)$

13. $\lim\limits_{x \to 7} (x + \sqrt{x} - 6)(x^2 - 2x + 1)$

14. $\lim\limits_{x \to 8} \dfrac{\sqrt{5x - 4} - 1}{3x^2 + 2}$

15. $\lim\limits_{x \to 9} \dfrac{\sqrt{x^2 - 5x - 36}}{8 - 3x}$

16. $\lim\limits_{x \to 10} (2x^2 - 15x - 50)^{20}$

17. $\lim\limits_{x \to 0} \dfrac{x^2 + 3x}{x}$

18. $\lim\limits_{x \to 1} \dfrac{x^2 - 1}{x - 1}$

19. $\lim\limits_{x \to 2} \dfrac{-2x^2 + 4x}{x - 2}$

20. $\lim\limits_{x \to 3} \dfrac{x^2 - x - 6}{x - 3}$

21. $\lim\limits_{x \to 4} \dfrac{x^2 - 16}{4 - x}$

22. $\lim\limits_{x \to 5} \dfrac{2x - 10}{x^2 - 25}$

23. $\lim\limits_{x \to 6} \dfrac{x^2 - 6x}{x^2 - 5x - 6}$

24. $\lim\limits_{x \to 7} \dfrac{x^3 - 2x^2 + 3x}{x^2}$

25. $\lim\limits_{x \to 8} \dfrac{x^2 + 64}{x - 8}$

26. $\lim\limits_{x \to 9} \dfrac{1}{(x - 9)^2}$

27. $\lim\limits_{x \to 0} \dfrac{-2}{\sqrt{x + 16} + 7}$

28. $\lim\limits_{x \to 0} \dfrac{4x}{x(x^2 + 3x + 5)}$

Use limits to compute the following derivatives.

29. $f'(3)$ where $f(x) = x^2 + 1$

30. $f'(2)$ where $f(x) = x^3$

31. $f'(0)$ where $f(x) = x^3 + 3x + 1$

32. $f'(0)$ where $f(x) = x^2 + 2x + 2$

33. $f'(3)$ where $f(x) = \dfrac{1}{2x + 5}$

34. $f'(4)$ where $f(x) = \sqrt{2x - 1}$

35. $f'(2)$ where $f(x) = \sqrt{5 - x}$

36. $f'(3)$ where $f(x) = \dfrac{1}{7 - 2x}$

37. $f'(0)$ where $f(x) = \sqrt{1 - x^2}$

38. $f'(2)$ where $f(x) = (5x - 4)^2$

39. $f'(0)$ where $f(x) = (x + 1)^3$

40. $f'(0)$ where $f(x) = \sqrt{x^2 + x + 1}$

Compute the following limits.

41. $\lim\limits_{x \to \infty} \dfrac{1}{x^2}$

42. $\lim\limits_{x \to -\infty} \dfrac{1}{x^2}$

43. $\lim\limits_{x \to \infty} \dfrac{1}{x - 8}$

44. $\lim\limits_{x \to \infty} \dfrac{1}{3x + 5}$

45. $\lim\limits_{x \to \infty} \dfrac{2x + 1}{x + 2}$

46. $\lim\limits_{x \to \infty} \dfrac{x^2 + x}{x^2 - 1}$

SOLUTIONS TO PRACTICE PROBLEMS 3

1. The function under consideration is a rational function. Since the denominator has value 0 at $x = 6$, we cannot immediately determine the limit by just evaluating the function at $x = 6$. Now also, $\lim\limits_{x \to 6} (x - 6) = 0$. Since the function in the denominator has limit 0, we cannot apply Limit Theorem VI. However, since the definition of limit considers only values of x different from 6, the quotient can be simplified by factoring and canceling.

$$\frac{x^2 - 4x - 12}{x - 6} = \frac{(x + 2)(x - 6)}{(x - 6)} = x + 2 \qquad \text{for } x \neq 6.$$

Now $\lim_{x \to 6} (x + 2) = 8$. Therefore, $\lim_{x \to 6} \dfrac{x^2 - 4x - 12}{x - 6} = 8$.

2. No limit exists. It is easily seen that $\lim_{x \to 6} (4x + 12) = 36$ and $\lim_{x \to 6} (x - 6) = 0$. As x approaches 6, the denominator gets very small and the numerator approaches 36. For example, if $x = 6.00001$, then the numerator is 36.00004 and the denominator is .00001. The quotient is 3,600,004. As x approaches 6 even more closely, the quotient gets arbitrarily large and cannot possibly approach a limit.

11.4. Differentiability and Continuity

In the preceding section we defined differentiability of $f(x)$ at $x = a$ in terms of a limit. If this limit does not exist, then we say that $f(x)$ is *nondifferentiable* at $x = a$. Geometrically, the nondifferentiability of $f(x)$ at $x = a$ can manifest itself in several different ways. First of all, the graph of $f(x)$ could have no tangent line at

FIGURE 1

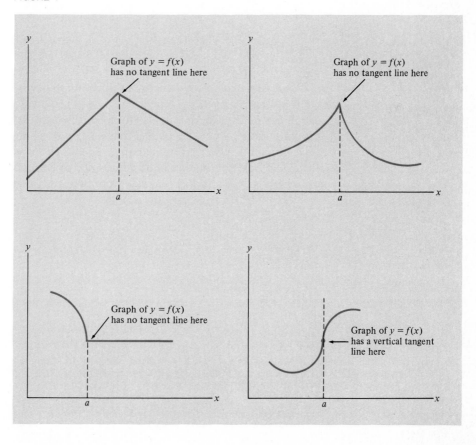

$x = a$. Second, the graph could have a vertical tangent line at $x = a$. (Recall that slope is not defined for vertical lines.) Some of the various geometric possibilities are illustrated in Fig. 1.

The following example illustrates how nondifferentiable functions can arise in practice.

EXAMPLE 1   A railroad company charges $10 per mile to haul a boxcar up to 200 miles and $8 per mile for each mile exceeding 200. In addition, the railroad charges a $1000 handling charge per boxcar. Graph the cost of sending a boxcar x miles.

Solution   If x is at most 200 miles, then the cost $C(x)$ is given by $C(x) = 1000 + 10x$ dollars. The cost for 200 miles is $C(200) = 1000 + 2000 = 3000$ dollars. If x exceeds 200 miles, then the total cost will be

$$C(x) = \underbrace{3000}_{\substack{\text{cost of first} \\ \text{200 miles}}} + \underbrace{8(x - 200)}_{\substack{\text{cost of miles in} \\ \text{excess of 200}}} = 1400 + 8x.$$

Thus

$$C(x) = \begin{cases} 1000 + 10x, & 0 < x \le 200, \\ 1400 + 8x, & x > 200. \end{cases}$$

The graph of $C(x)$ is sketched in Fig. 2. Note that $C(x)$ is not differentiable at $x = 200$.

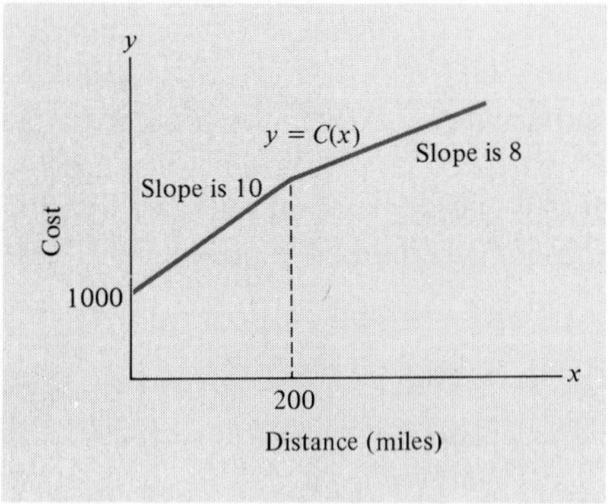

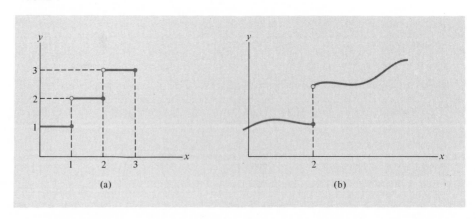

FIGURE 2

Closely related to the concept of differentiability is that of continuity. We say that a function $f(x)$ is *continuous* at $x = a$ provided that, roughly speaking, its graph has no breaks (or gaps) as it passes through the point $(a, f(a))$. That is, $f(x)$ is continuous at $x = a$ provided that we can draw the graph through $(a, f(a))$ without lifting our pencil from the paper. The functions whose graphs are drawn in Figs. 1 and 2 are continuous for all values of x. By contrast, however, the function whose graph is drawn in Fig. 3(a) is not continuous (we say it is *discontinuous*) at

FIGURE 3

(a)

(b)

$x = 1$ and $x = 2$, since the graph has breaks there. Similarly, the function whose graph is drawn in Fig. 3(b) is discontinuous at $x = 2$.

Discontinuous functions can occur in applications, as the following example shows.

EXAMPLE 2 Suppose that a manufacturing plant is capable of producing 15,000 units in one shift of 8 hours. For each shift worked, there is a fixed cost of $2000 (for light, heat, etc.). Suppose that the variable cost (the cost of labor and raw materials) is $2 per unit. Graph the cost $C(x)$ of manufacturing x units.

Solution If $x \leq 15,000$, a single shift will suffice, so that

$$C(x) = 2000 + 2x, \qquad 0 \leq x \leq 15,000.$$

If x is between 15,000 and 30,000, one extra shift will be required, and

$$C(x) = 4000 + 2x, \qquad 15,000 < x \leq 30,000.$$

If x is between 30,000 and 45,000, the plant will need to work three shifts, and

$$C(x) = 6000 + 2x, \qquad 30,000 < x \leq 45,000.$$

The graph of $C(x)$ for $0 \leq x \leq 45,000$ is drawn in Fig. 4. Note that the graph has breaks at two points.

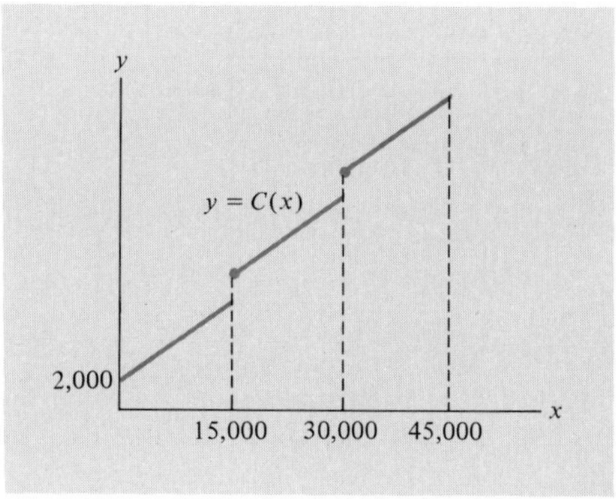

FIGURE 4

The relationship between differentiability and continuity is this:

☐ **Theorem 1** If $f(x)$ is differentiable at $x = a$, then $f(x)$ is continuous at $x = a$.

Note, however, that the converse statement is definitely false: A function may be continuous at $x = a$ but still not be differentiable there. The functions whose graphs are drawn in Fig. 1 provide examples of this phenomenon.

Just as with differentiability, the notion of continuity can be phrased in terms of limits. In order for $f(x)$ to be continuous at $x = a$, the values of $f(x)$ for all x near a must be close to $f(a)$ (otherwise, the graph would have a break at $x = a$). In fact, the closer x is to a, the closer $f(x)$ must be to $f(a)$ (again, in order to avoid a break in the graph). In terms of limits, we must therefore have

$$\lim_{x \to a} f(x) = f(a).$$

Conversely, an intuitive argument shows that if the limit relation above holds, then the graph of $y = f(x)$ has no break at $x = a$. So we may state the following result.

> *Limit Criterion for Continuity*   A function $f(x)$ is continuous at $x = a$ provided the following limit relation holds:
>
> $$\lim_{x \to a} f(x) = f(a). \qquad (1)$$

In order for (1) to hold, three conditions must be fulfilled.

1. $f(x)$ must be defined at $x = a$.
2. $\lim\limits_{x \to a} f(x)$ must exist.
3. The limit $\lim\limits_{x \to a} f(x)$ must have the value $f(a)$.

A function will fail to be continuous at $x = a$ when any one of these conditions fails to hold. The various possibilities are illustrated in the next example.

EXAMPLE 3 Determine whether the functions whose graphs are drawn in Fig. 5 are continuous at $x = 3$. Use the limit criterion.

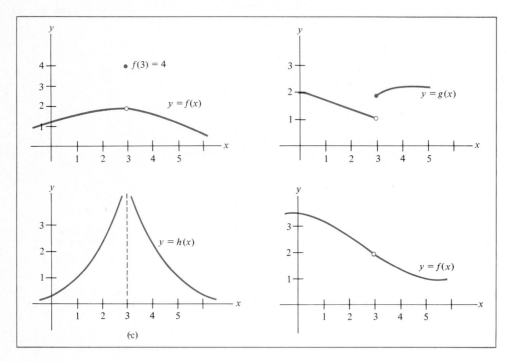

FIGURE 5

Solution (a) Here $\lim\limits_{x \to 3} f(x) = 2$. However, $f(3) = 4$. So

$$\lim_{x \to 3} f(x) \neq f(3)$$

and $f(x)$ is not continuous at $x = 3$. (Geometrically, this is clear. The graph has a break at $x = 3$.)

(b) $\lim\limits_{x \to 3} g(x)$ does not exist, so $g(x)$ is not continuous at $x = 3$.

(c) $\lim\limits_{x \to 3} h(x)$ does not exist, so $h(x)$ is not continuous at $x = 3$.

(d) $f(x)$ is not defined at $x = 3$, so $f(x)$ is not continuous at $x = 3$.

Using our result on the limit of a polynomial function (Section 4), we see that

$$p(x) = a_0 + a_1 x + \cdots + a_n x^n, \qquad a_0, \ldots, a_n \text{ constants,}$$

is continuous at all x. Similarly, a rational function

$$\frac{p(x)}{q(x)}, \qquad p(x), q(x) \text{ polynomials,}$$

is continuous at all x for which $q(x) \neq 0$.

PRACTICE PROBLEMS 4

Let

$$f(x) = \begin{cases} \dfrac{x^2 - x - 6}{x - 3} & \text{for } x \neq 3 \\ 4 & \text{for } x = 3. \end{cases}$$

1. Is $f(x)$ continuous at $x = 3$?
2. Is $f(x)$ differentiable at $x = 3$?

EXERCISES 4

Is the function whose graph is drawn in Fig. 6 continuous at the following values of x?

1. $x = 0$ 2. $x = -3$ 3. $x = 3$

4. $x = .001$ 5. $x = -2$ 6. $x = 2$

Is the function whose graph is drawn in Fig. 6 differentiable at the following values of x?

7. $x = 0$ 8. $x = -3$ 9. $x = 3$

10. $x = .001$ 11. $x = -2$ 12. $x = 2$

Determine whether each of the following functions is continuous and/or differentiable at $x = 1$.

13. $f(x) = x^2 + 8x$ 14. $f(x) = x - \dfrac{1}{x}$

FIGURE 6

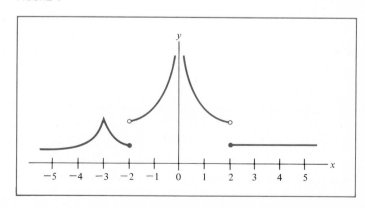

15. $f(x) = \begin{cases} x + 2 & \text{for } -1 \le x \le 1 \\ 3x & \text{for } 1 < x < 5 \end{cases}$

16. $f(x) = \begin{cases} x^2 & \text{for } 1 \le x \le 2 \\ x^3 & \text{for } 0 \le x < 1 \end{cases}$

17. $f(x) = \begin{cases} 2x - 1 & \text{for } 0 \le x \le 1 \\ 1 & \text{for } 1 < x \end{cases}$

18. $f(x) = \begin{cases} x & \text{for } x \ne 1 \\ 2 & \text{for } x = 1 \end{cases}$

19. $f(x) = \begin{cases} \dfrac{1}{x - 1} & \text{for } x \ne 1 \\ 0 & \text{for } x = 1 \end{cases}$

20. $f(x) = \begin{cases} x - 1 & \text{for } 0 \le x < 1 \\ 1 & \text{for } x = 1 \\ 2x - 2 & \text{for } x > 1 \end{cases}$

SOLUTIONS TO PRACTICE PROBLEMS 4

1. The function $f(x)$ is defined at $x = 3$, namely, $f(3) = 4$. When computing $\lim\limits_{x \to 3} f(x)$ we exclude consideration of $x = 3$; therefore, we can simplify the expression for $f(x)$ as follows:

$$f(x) = \frac{x^2 - x - 6}{x - 3} = \frac{(x - 3)(x + 2)}{x - 3} = x + 2.$$

Clearly,

$$\lim_{x \to 3} f(x) = \lim_{x \to 3} (x + 2) = 5.$$

Since $\lim\limits_{x \to 3} f(x) = 5 \ne 4 = f(3)$, $f(x)$ is not continuous at $x = 3$.

2. There is no need to compute any limits in order to answer this question. By Theorem 1, since $f(x)$ is not continuous at $x = 3$, it cannot possibly be differentiable there.

11.5. Some Rules for Differentiation

Three additional rules of differentiation greatly extend the number of functions that we can differentiate.

---

1. **Constant-Multiple Rule**

$$\frac{d}{dx}[k \cdot f(x)] = k \cdot \frac{d}{dx}[f(x)], \qquad k \text{ a constant.}$$

2. **Sum Rule**

$$\frac{d}{dx}[f(x) + g(x)] = \frac{d}{dx}[f(x)] + \frac{d}{dx}[g(x)].$$

3. **General Power Rule**

$$\frac{d}{dx}([g(x)]^r) = r \cdot [g(x)]^{r-1} \cdot \frac{d}{dx}[g(x)].$$

We shall discuss these rules and then prove the first two.

The Constant-Multiple Rule Starting with a function $f(x)$, we can multiply it by a constant number k in order to obtain a new function $k \cdot f(x)$. For instance, if $f(x) = x^2 - 4x + 1$ and $k = 2$, then

$$2f(x) = 2(x^2 - 4x + 1) = 2x^2 - 8x + 2.$$

The constant-multiple rule says that the derivative of the new function $k \cdot f(x)$ is just k times the derivative of the original function.* In other words, when faced with the differentiation of a constant times a function, simply carry along the constant and differentiate the function.

EXAMPLE 1 Calculate.

(a) $\dfrac{d}{dx}(2x^5)$ (b) $\dfrac{d}{dx}\left(\dfrac{x^3}{4}\right)$ (c) $\dfrac{d}{dx}\left(-\dfrac{3}{x}\right)$ (d) $\dfrac{d}{dx}(5\sqrt{x})$.

Solution (a) With $k = 2$ and $f(x) = x^5$, we have

$$\frac{d}{dx}(2 \cdot x^5) = 2 \cdot \frac{d}{dx}(x^5) = 2(5x^4) = 10x^4.$$

(b) Write $\dfrac{x^3}{4}$ in the form $\dfrac{1}{4} \cdot x^3$. Then

$$\frac{d}{dx}\left(\frac{x^3}{4}\right) = \frac{1}{4} \cdot \frac{d}{dx}(x^3) = \frac{1}{4}(3x^2) = \frac{3}{4}x^2.$$

(c) Write $-\dfrac{3}{x}$ in the form $(-3) \cdot \dfrac{1}{x}$. Then

$$\frac{d}{dx}\left(-\frac{3}{x}\right) = (-3) \cdot \frac{d}{dx}\left(\frac{1}{x}\right) = (-3) \cdot \frac{-1}{x^2} = \frac{3}{x^2}.$$

(d) $\dfrac{d}{dx}(5\sqrt{x}) = 5\dfrac{d}{dx}(\sqrt{x}) = 5\dfrac{d}{dx}(x^{1/2}) = \dfrac{5}{2}x^{-1/2}.$

This answer may also be written in the form $\dfrac{5}{2\sqrt{x}}$.

The Sum Rule To differentiate a sum of functions, differentiate each function individually and add the derivatives together.† Another way of saying this is "the derivative of a sum of functions is the sum of the derivatives."

---

* More precisely, the constant-multiple rule asserts that if $f(x)$ is differentiable at $x = a$, then so is the function $k \cdot f(x)$, and the derivative of $k \cdot f(x)$ at $x = a$ may be computed using the given formula.

† More precisely, the sum rule asserts that if both $f(x)$ and $g(x)$ are differentiable at $x = a$, then so is $f(x) + g(x)$, and the derivative (at $x = a$) of the sum is then the sum of the derivatives (at $x = a$).

EXAMPLE 2 Find.

(a) $\dfrac{d}{dx}(x^3 + 5x)$ 　　　　 (b) $\dfrac{d}{dx}\left(x^4 - \dfrac{3}{x^2}\right)$ 　　　　 (c) $\dfrac{d}{dx}(2x^7 - x^5 + 8)$

Solution　(a) Let $f(x) = x^3$ and $g(x) = 5x$. Then

$$\frac{d}{dx}(x^3 + 5x) = \frac{d}{dx}(x^3) + \frac{d}{dx}(5x) = 3x^2 + 5.$$

(b) The sum rule applies to differences as well as sums (see Exercise 36). Indeed, by the sum rule,

$$\frac{d}{dx}\left(x^4 - \frac{3}{x^2}\right) = \frac{d}{dx}(x^4) + \frac{d}{dx}\left(-\frac{3}{x^2}\right) \qquad \text{(sum rule)}$$

$$= \frac{d}{dx}(x^4) - 3\frac{d}{dx}(x^{-2}) \qquad \text{(constant-multiple rule)}$$

$$= 4x^3 - 3(-2x^{-3})$$

$$= 4x^3 + 6x^{-3}.$$

After some practice, one usually omits most or all of the intermediate steps and simply writes

$$\frac{d}{dx}\left(x^4 - \frac{3}{x^2}\right) = 4x^3 + 6x^{-3}.$$

(c) We apply the sum rule repeatedly and use the fact that the derivative of a constant function is 0:

$$\frac{d}{dx}(2x^7 - x^5 + 8) = \frac{d}{dx}(2x^7) - \frac{d}{dx}(x^5) + \frac{d}{dx}(8)$$

$$= 2(7x^6) - 5x^4 + 0$$

$$= 14x^6 - 5x^4.$$

The General Power Rule　Frequently, we will encounter expressions of the form $[g(x)]^r$—for instance, $(x^3 + 5)^2$, where $g(x) = x^3 + 5$ and $r = 2$. The general power rule says that, to differentiate $[g(x)]^r$, we must first treat $g(x)$ as if it were simply an x, form $r[g(x)]^{r-1}$, and then multiply it by a "correction factor" $g'(x)$.* Thus

$$\frac{d}{dx}(x^3 + 5)^2 = 2(x^3 + 5)^1 \cdot \frac{d}{dx}(x^3 + 5)$$

$$= 2(x^3 + 5) \cdot (3x^2)$$

$$= 6x^2(x^3 + 5).$$

---

　　* More precisely, the general power rule asserts that if $g(x)$ is differentiable at $x = a$, and if $[g(x)]^r$ and $[g(x)]^{r-1}$ are both defined at $x = a$, then $[g(x)]^r$ is also differentiable at $x = a$ and its derivative is given by the formula stated.

In this special case it is easy to verify that the general power rule gives the correct answer. We first expand $(x^3 + 5)^2$ and then differentiate.

$$(x^3 + 5)^2 = (x^3 + 5)(x^3 + 5) = x^6 + 10x^3 + 25.$$

From the constant-multiple rule and the sum rule, we have

$$\frac{d}{dx}(x^3 + 5)^2 = \frac{d}{dx}(x^6 + 10x^3 + 25)$$

$$= 6x^5 + 30x^2 + 0$$

$$= 6x^2(x^3 + 5).$$

The two methods give the same answer.

Note that if we set $g(x) = x$ in the general power rule, we recover the power rule. So the general power rule contains the power rule as a special case.

EXAMPLE 3   Differentiate $\sqrt{1 - x^2}$.

Solution
$$\frac{d}{dx}(\sqrt{1 - x^2}) = \frac{d}{dx}((1 - x^2)^{1/2}) = \frac{1}{2}(1 - x^2)^{-1/2} \cdot \frac{d}{dx}(1 - x^2)$$

$$= \frac{1}{2}(1 - x^2)^{-1/2} \cdot (-2x)$$

$$= \frac{-x}{(1 - x^2)^{1/2}} = \frac{-x}{\sqrt{1 - x^2}}.$$

EXAMPLE 4   Differentiate $y = \dfrac{1}{x^3 + 4x}$.

Solution
$$y = \frac{1}{x^3 + 4x} = (x^3 + 4x)^{-1}.$$

$$\frac{dy}{dx} = (-1)(x^3 + 4x)^{-2} \cdot \frac{d}{dx}(x^3 + 4x)$$

$$= \frac{-1}{(x^3 + 4x)^2}(3x^2 + 4)$$

$$= -\frac{3x^2 + 4}{(x^3 + 4x)^2}.$$

Proofs of the Constant-Multiple and Sum Rules   Let us verify both rules when x has the value a. Recall that if $f(x)$ is differentiable at $x = a$, then its derivative is the limit

$$\lim_{h \to 0} \frac{f(a + h) - f(a)}{h}.$$

Constant-Multiple Rule   We assume that $f(x)$ is differentiable at $x = a$. We must prove that $k \cdot f(x)$ is differentiable at $x = a$ and that its derivative there is $k \cdot f'(a)$. This amounts to showing that the limit

$$\lim_{h \to 0} \frac{k \cdot f(a + h) - k \cdot f(a)}{h}$$

exists and has the value $k \cdot f'(a)$. However,

$$\lim_{h \to 0} \frac{k \cdot f(a + h) - k \cdot f(a)}{h} = \lim_{h \to 0} k \left[\frac{f(a + h) - f(a)}{h} \right]$$

$$= k \cdot \lim_{h \to 0} \frac{f(a + h) - f(a)}{h} \qquad \text{(by Limit Theorem I)}$$

$$= k \cdot f'(a) \qquad \text{(since } f(x) \text{ is differentiable at } x = a\text{)},$$

which is what we desired to show.

Sum Rule   We assume that both $f(x)$ and $g(x)$ are differentiable at $x = a$. We must prove that $f(x) + g(x)$ is differentiable at $x = a$ and that its derivative is $f'(a) + g'(a)$. That is, we must show that the limit

$$\lim_{h \to 0} \frac{[f(a + h) + g(a + h)] - [f(a) + g(a)]}{h}$$

exists and equals $f'(a) + g'(a)$. Using Limit Theorem III and the fact that $f(x)$ and $g(x)$ are differentiable at $x = a$, we have

$$\lim_{h \to 0} \frac{[f(a + h) + g(a + h)] - [f(a) + g(a)]}{h}$$

$$= \lim_{h \to 0} \left[\frac{f(a + h) - f(a)}{h} + \frac{g(a + h) - g(a)}{h} \right]$$

$$= \lim_{h \to 0} \frac{f(a + h) - f(a)}{h} + \lim_{h \to 0} \frac{g(a + h) - g(a)}{h}$$

$$= f'(a) + g'(a).$$

The general power rule will be proven as a special case of the chain rule in Chapter 13.

PRACTICE PROBLEMS 5

1.   Find $\dfrac{d}{dx}(x)$.

2.   Differentiate $y = \dfrac{x + (x^5 + 1)^{10}}{3}$.

EXERCISES 5

Differentiate.

1. $y = x^3 + x^2$

2. $y = x^2 + \dfrac{1}{x}$

3. $y = x^2 + 3x - 1$

4. $y = x^3 + 2x + 5$

5. $f(x) = x^5 + \dfrac{1}{x}$

6. $f(x) = x^8 - x$

7. $f(x) = x^4 + x^3 + x$

8. $f(x) = x^5 + x^2 - x$

9. $y = 3x^2$

10. $y = 2x^3$

11. $y = x^3 + 7x^2$

12. $y = -2x$

13. $y = \dfrac{4}{x^2}$

14. $y = 2\sqrt{x}$

15. $y = 3x - \dfrac{1}{x}$

16. $y = -x^2 + 3x + 1$

17. $f(x) = \tfrac{1}{3}x^3 - \tfrac{1}{2}x^2$

18. $f(x) = 100x^{100}$

19. $f(x) = -\dfrac{1}{5x^5}$

20. $f(x) = x^2 - \dfrac{1}{x^2}$

21. $f(x) = 1 - \sqrt{x}$

22. $f(x) = -3x^2 + 7$

23. $f(x) = (3x + 1)^{10}$

24. $f(x) = \dfrac{1}{x^2 + x + 1}$

25. $f(x) = \sqrt{3x^3 + x}$

26. $y = \dfrac{1}{(x^2 - 7)^5}$

27. $y = (2x^2 - x + 4)^6$

28. $y = \sqrt{-2x + 1}$

29. $y = \dfrac{x}{3} + \dfrac{3}{x}$

30. $y = \dfrac{2x - 1}{5}$

31. $y = \dfrac{2}{1 - 5x}$

32. $y = \dfrac{4}{3\sqrt{x}}$

33. $y = \dfrac{1}{1 - x^4}$

34. $y = \left(x^3 + \dfrac{x}{2} + 1\right)^5$

35. $f(x) = \dfrac{4}{\sqrt{x^2 + x}}$

36. $f(x) = \dfrac{6}{x^2 + 2x + 5}$

37. $f(x) = \left(\dfrac{\sqrt{x}}{2} + 1\right)^{3/2}$

38. $f(x) = \left(4 - \dfrac{2}{x}\right)^3$

Find the slope of the graph of $y = f(x)$ at the designated point.

39. $f(x) = 3x^2 - 2x + 1, (1, 2)$

40. $f(x) = x^{10} + 1 + \sqrt{1 - x}, (0, 2)$

41. Find the slope of the tangent line to the curve $y = x^3 + 3x - 8$ at $(2, 6)$.

42. Write the equation of the tangent line to the curve $y = x^3 + 3x - 8$ at $(2, 6)$.

43. Find the slope of the tangent line to the curve $y = (x^2 - 15)^6$ at $x = 4$. Then write the equation of this tangent line.

44. Find the equation of the tangent line to the curve $y = \dfrac{8}{x^2 + x + 2}$ at $x = 2$.

45. Differentiate the function $f(x) = (3x^2 + x - 2)^2$ in two ways.

    (a) Use the general power rule.

    (b) Multiply $3x^2 + x - 2$ by itself and then differentiate the resulting polynomial.

46. Using the sum rule and the constant-multiple rule, show that for any functions $f(x)$ and $g(x)$

$$\frac{d}{dx}[f(x) - g(x)] = \frac{d}{dx}f(x) - \frac{d}{dx}g(x).$$

47. Find m such that the line $y = mx$ is tangent to the curve $y = (3x - 5)^2 + 11$. [*Hint:* The line and the curve must both pass through some point (x, y) and have the same slope at that point.]

48. There are two points on the curve $y = x^2 + 3$ where the tangent line to the curve passes through the point $(1, 0)$. Find these two points.

SOLUTIONS TO PRACTICE PROBLEMS 5

1. The problem asks for the derivative of the function $y = x$, a straight line of slope 1. Therefore, $\dfrac{d}{dx}(x) = 1$. The result can also be obtained from the power rule with $r = 1$. If $f(x) = x^1$, then $\dfrac{d}{dx}(f(x)) = 1 \cdot x^{1-1} = x^0 = 1$.

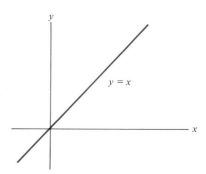

2. All three rules are required to differentiate this function.

$$\frac{dy}{dx} = \frac{d}{dx}\frac{1}{3} \cdot [x + (x^5 + 1)^{10}]$$

$$= \frac{1}{3}\frac{d}{dx}[x + (x^5 + 1)^{10}] \qquad \text{(constant-multiple rule)}$$

$$= \frac{1}{3}\left[\frac{d}{dx}(x) + \frac{d}{dx}(x^5 + 1)^{10}\right] \qquad \text{(sum rule)}$$

$$= \tfrac{1}{3}[1 + 10(x^5 + 1)^9 \cdot (5x^4)] \qquad \text{(general power rule)}$$

$$= \tfrac{1}{3}[1 + 50x^4(x^5 + 1)^9].$$

11.6. More About Derivatives

In many applications it is convenient to use variables other than x and y. One might, for instance, study the function $f(t) = t^2$ instead of writing $f(x) = x^2$. In this case, the notation for the derivative involves t rather than x, but the concept of the derivative as a slope formula is unaffected. (See Fig. 1.) When the independent variable is t instead of x, we write $\dfrac{d}{dt}$ in place of $\dfrac{d}{dx}$. For instance,

$$\frac{d}{dt}(t^3) = 3t^2, \qquad \frac{d}{dt}(2t^2 + 3t) = 4t + 3.$$

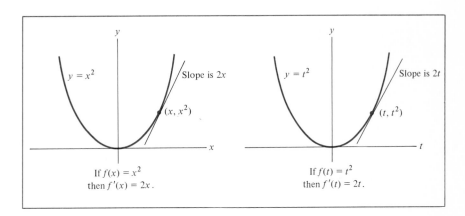

FIGURE 1

Recall that if y is a function of x, say $y = f(x)$, then we may write $\dfrac{dy}{dx}$ in place of $f'(x)$. We sometimes call $\dfrac{dy}{dx}$ "the derivative of y with respect to x." Similarly, if v is a function of t, then the derivative of v with respect to t is written as $\dfrac{dv}{dt}$. For example, if $v = 4t^2$, then $\dfrac{dv}{dt} = 8t$.

Of course, other letters can be used to denote variables. The formulas

$$\frac{d}{dP}(P^3) = 3P^2, \qquad \frac{d}{ds}(s^3) = 3s^2, \qquad \frac{d}{dz}(z^3) = 3z^2$$

are all expressing the same basic fact that the slope formula for the cubic curve $y = x^3$ is given by $3x^2$.

EXAMPLE 1   Compute.

(a) $\dfrac{ds}{dp}$ if $s = 3(p^2 + 5p + 1)^{10}$ (b) $\dfrac{d}{dt}(at^2 + St^{-1} + S^2)$

Solution   (a)
$$\frac{d}{dp}3(p^2 + 5p + 1)^{10} = 30(p^2 + 5p + 1)^9 \cdot \frac{d}{dp}(p^2 + 5p + 1)$$

$$= 30(p^2 + 5p + 1)^9(2p + 5).$$

(b) Although the expression $at^2 + St^{-1} + S^2$ contains several letters, the notation $\frac{d}{dt}$ indicates that all letters except t are to be considered as constants. Hence

$$\frac{d}{dt}(at^2 + St^{-1} + S^2) = \frac{d}{dt}(at^2) + \frac{d}{dt}(St^{-1}) + \frac{d}{dt}(S^2)$$

$$= a \cdot \frac{d}{dt}(t^2) + S \cdot \frac{d}{dt}(t^{-1}) + 0$$

$$= 2at - St^{-2}.$$

$$\left[\text{The derivative } \frac{d}{dt}(S^2) \text{ is zero because } S^2 \text{ is a constant.} \right]$$

The Second Derivative When we differentiate a function $f(x)$, we obtain a new function $f'(x)$ that is a formula for the slope of the curve $y = f(x)$. If we differentiate the function $f'(x)$, we obtain what is called the *second derivative* of $f(x)$, denoted by $f''(x)$. That is,

$$\frac{d}{dx}f'(x) = f''(x).$$

EXAMPLE 2   Find the second derivatives of the following functions.

  (a) $f(x) = x^3 + (1/x)$ (b) $f(x) = 2x + 1$ (c) $f(t) = t^{1/2} + t^{-1/2}$

Solution   (a) $f(x) = x^3 + (1/x) = x^3 + x^{-1}$
$\quad\quad f'(x) = 3x^2 - x^{-2}$
$\quad\quad f''(x) = 6x + 2x^{-3}.$

  (b) $f(x) = 2x + 1$
$\quad\quad f'(x) = 2$ (a constant function whose value is 2)
$\quad\quad f''(x) = 0.$ (The derivative of a constant function is zero.)

  (c) $f(t) = t^{1/2} + t^{-1/2}$
$\quad\quad f'(t) = \frac{1}{2}t^{-1/2} - \frac{1}{2}t^{-3/2}$
$\quad\quad f''(t) = -\frac{1}{4}t^{-3/2} + \frac{3}{4}t^{-5/2}.$

The first derivative of a function $f(x)$ gives the slope of the graph of $f(x)$ at any point. The second derivative of $f(x)$ gives important additional information about the shape of the curve near any point. We shall examine this subject carefully in the next chapter.

Other Notation for Derivatives Unfortunately, the process of differentiation does not have a standardized notation. Consequently, it is important to become familiar with alternative terminology.

  If y is a function of x, say $y = f(x)$, then we may denote the first and second derivatives of this function in several ways.

Prime notation	$\dfrac{d}{dx}$ notation
$f'(x)$	$\dfrac{d}{dx} f(x)$
y'	$\dfrac{dy}{dx}$
$f''(x)$	$\dfrac{d^2}{dx^2} f(x)$
y''	$\dfrac{d^2 y}{dx^2}$

The notation $\dfrac{d^2}{dx^2}$ is purely symbolic. It reminds us that the second derivative is obtained by differentiating $\dfrac{d}{dx} f(x)$; that is,

$$f'(x) = \frac{d}{dx} f(x),$$

$$f''(x) = \frac{d}{dx}\left[\frac{d}{dx} f(x) \right].$$

If we evaluate the derivative $f'(x)$ at a specific value of x, say $x = a$, we get a number $f'(a)$ that gives the slope of the curve $y = f(x)$ at the point $(a, f(a))$. Another way of writing $f'(a)$ is

$$\left. \frac{dy}{dx} \right|_{x=a}.$$

If we have a second derivative $f''(x)$, then its value when $x = a$ is written

$$f''(a) \quad \text{or} \quad \left. \frac{d^2 y}{dx^2} \right|_{x=a}.$$

EXAMPLE 3   If $y = x^4 - 5x^3 + 7$, find $\left. \dfrac{d^2 y}{dx^2} \right|_{x=3}$.

Solution

$$\frac{dy}{dx} = \frac{d}{dx}(x^4 - 5x^3 + 7) = 4x^3 - 15x^2$$

$$\frac{d^2 y}{dx^2} = \frac{d}{dx}(4x^3 - 15x^2) = 12x^2 - 30x$$

$$\left. \frac{d^2 y}{dx^2} \right|_{x=3} = 12(3)^2 - 30(3) = 108 - 90 = 18.$$

EXAMPLE 4 If $s = t^3 - 2t^2 + 3t$, find

$$\left.\frac{ds}{dt}\right|_{t=-2} \quad \text{and} \quad \left.\frac{d^2s}{dt^2}\right|_{t=-2}.$$

Solution

$$\frac{ds}{dt} = \frac{d}{dt}(t^3 - 2t^2 + 3t) = 3t^2 - 4t + 3$$

$$\left.\frac{ds}{dt}\right|_{t=-2} = 3(-2)^2 - 4(-2) + 3 = 12 + 8 + 3 = 23.$$

To find the value of the second derivative at $t = -2$, we must first differentiate $\frac{ds}{dt}$.

$$\frac{d^2s}{dt^2} = \frac{d}{dt}(3t^2 - 4t + 3) = 6t - 4$$

$$\left.\frac{d^2s}{dt^2}\right|_{t=-2} = 6(-2) - 4 = -12 - 4 = -16.$$

PRACTICE PROBLEMS 6

1. Let $f(t) = t + (1/t)$. Find $f''(2)$.

2. Differentiate $g(r) = 2\pi rh$.

EXERCISES 6

Find the first derivatives.

1. $f(t) = (t^2 + 1)^5$

2. $f(P) = P^4 - P^3 + 4P^2 - P$

3. $v = \sqrt{2t - 1}$

4. $g(z) = (z^3 - z + 1)^2$

5. $y = (T^3 + 5T)^{2/3}$

6. $s = \sqrt{t} + \dfrac{1}{\sqrt{t}}$

7. Find $\dfrac{d}{dP}(3P^2 - \frac{1}{2}P + 1)$.

8. Find $\dfrac{d}{dz}(\sqrt{z^2 - 1})$.

9. Find $\dfrac{d}{dt}(a^2t^2 + b^2t + c^2)$.

10. Find $\dfrac{d}{dx}(x^3 + t^3)$.

Find the first and second derivatives.

11. $f(x) = \frac{1}{2}x^2 - 7x + 2$

12. $y = \dfrac{1}{x^2} + 1$

13. $y = \sqrt{x}$

14. $f(t) = t^{100} + t + 1$

15. $f(r) = \pi hr^2 + 2\pi r$

16. $v = t^{3/2} + t$

17. $g(x) = 2 - 5x$

18. $V(r) = \frac{4}{3}\pi r^3$

19. $f(P) = (3P + 1)^5$

20. $u = \dfrac{t^6}{30} - \dfrac{t^4}{12}$

Compute the following.

21. $\dfrac{d}{dx}(2x^2 - 3)\Big|_{x=5}$

22. $\dfrac{d}{dt}(1 - 2t - 3t^2)\Big|_{t=-1}$

23. $\dfrac{d}{dz}(z^2 - 4)^3\Big|_{z=1}$

24. $\dfrac{d}{dT}\left(\dfrac{1}{3T+1}\right)\Big|_{T=2}$

25. $\dfrac{d^2}{dx^2}(3x^3 - x^2 + 7x - 1)\Big|_{x=2}$

26. $\dfrac{d}{dt}\left(\dfrac{dv}{dt}\right)$, where $v = 2t^{-3}$

27. $\dfrac{d}{dP}\left(\dfrac{dy}{dP}\right)$, where $y = \dfrac{k}{2P-1}$

28. $\dfrac{d^2V}{dr^2}\Big|_{r=2}$, where $V = ar^3$

29. $f'(3)$ and $f''(3)$, when $f(x) = \sqrt{10 - 2x}$

30. $g'(2)$ and $g''(2)$, when $g(T) = (3T - 5)^{10}$

31. Suppose a company finds that the revenue R generated by spending x dollars on advertising is given by $R = 1000 + 80x - .02x^2$, for $0 \le x \le 2000$. Find $\dfrac{dR}{dx}\Big|_{x=1500}$

32. A supermarket finds that its average daily volume of business V (in thousands of dollars) and the number of hours t the store is open for business each day are approximately related by the formula

$$V = 20\left(1 - \dfrac{100}{100 + t^2}\right), \qquad 0 \le t \le 24.$$

Find $\dfrac{dV}{dt}\Big|_{t=10}$.

33. The *third derivative* of a function $f(x)$ is the derivative of the second derivative $f''(x)$ and is denoted $f'''(x)$. Compute $f'''(x)$ for the following functions.

(a) $f(x) = x^5 - x^4 + 3x$

(b) $f(x) = 4x^{5/2}$

34. Compute the third derivatives of the following functions.

(a) $f(t) = t^{10}$

(b) $f(z) = \dfrac{1}{z + 5}$

SOLUTIONS TO PRACTICE PROBLEMS 6

1.
$$f(t) = t + t^{-1}$$
$$f'(t) = 1 + (-1)t^{(-1)-1} = 1 - t^{-2}$$
$$f''(t) = -(-2)t^{(-2)-1} = 2t^{-3} = \dfrac{2}{t^3}$$

Therefore, $f''(2) = \dfrac{2}{2^3} = \dfrac{1}{4}$. [*Note:* It is essential to first compute the function $f''(t)$ and *then* evaluate the function at $t = 2$.]

2. The expression $2\pi rh$ contains two numbers, 2 and π, and two letters, r and h. The notation "$g(r)$" tells us that the expression $2\pi rh$ is to be regarded as a function of r. Therefore, h and hence $2\pi h$ is to be treated like a constant, and differentiation is done with respect to

the variable r. That is,

$$g(r) = (2\pi h)r$$
$$g'(r) = 2\pi h.$$

11.7. The Derivative as a Rate of Change

An important interpretation of the derivative of a function is as a rate of change. Consider the following situation, for example. The weight of an animal changes as time passes and we may think of the weight as a function of time, say $W(t)$. Suppose that we measure the weight at some time $t = a$ and at a later time $t = a + h$. Then the change in weight over this time interval is

$$W(a + h) - W(a).$$

The average *rate* of change of weight with respect to time during this time interval is found by dividing the change in weight by the length of the time interval:

$$\left.\begin{array}{l} \text{average rate of change} \\ \text{of } W(t) \text{ with respect to} \\ t, \text{ from } a \text{ to } a + h \end{array}\right\} = \frac{W(a + h) - W(a)}{h}$$

This average rate of change takes into account the change in weight as t increases from $t = a$ to $t = a + h$. In order to study the change in weight $W(t)$ only for t close to a, we must take h very small. But we know that as h becomes small, the quantity $[W(a + h) - W(a)]/h$ approaches the derivative $W'(a)$. Thus

$$\left.\begin{array}{l} \text{average rate of change} \\ \text{of } W(t) \text{ with respect to} \\ t, \text{ from } a \text{ to } a + h \end{array}\right\} \approx W'(a),$$

for h very small.

In this sense, the derivative $W'(a)$ gives the rate of change of $W(t)$ with respect to t precisely at the point when $t = a$.

In geometric terms, the average rate of change of $W(t)$ from a to $a + h$ is the slope of the secant line in Fig. 1. The derivative $W'(a)$ is the slope of the tangent

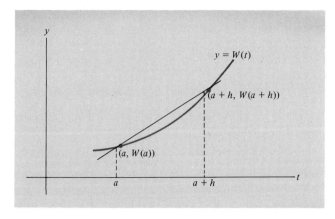

FIGURE 1

line through $(a, W(a))$ and gives the best measure of the rate of change of the weight at time a.

The derivative of any function $f(t)$ may be interpreted as a rate of change, just as we did for the weight function $W(t)$. We state it formally in the following definition.

The derivative $f'(a)$ measures the rate of change of $f(t)$ with respect to t, at the point $t = a$.

The examples and exercises that follow show why it is often convenient to use a derivative to measure a rate of change. Other illustrations will be given later in the book.

EXAMPLE 1 A flu epidemic hits a midwestern town. Public Health officials estimate that the number of persons sick with the flu at time t (measured in days from the beginning of the epidemic) is approximated by $P(t) = 60t^2 - t^3$, provided that $0 \le t \le 40$. At what rate is the flu spreading at time $t = 30$?

Solution The rate at which the flu spreads is given by the rate of change of $P(t)$, measured in people per day. Now

$$P'(t) = 120t - 3t^2$$

$$P'(30) = 120(30) - 3(30)^2 = 900.$$

Thus 30 days after the beginning of the epidemic the flu is spreading at the rate of 900 people per day.

EXAMPLE 2 A common clinical procedure for studying a person's calcium metabolism (the rate at which the body assimilates and uses calcium) is to inject some chemically "labeled" calcium into the bloodstream and then measure how fast this calcium is removed from the blood by the person's bodily processes. Suppose that t days after an injection of calcium, the amount A of the labeled calcium remaining in the blood is $A = t^{-3/2}$ for $t \ge .5$, where A is measured in suitable units.* How fast (in "units of calcium per day") is the body removing calcium from the blood when $t = 1$ day?

Solution The rate of change (per day) of calcium in the blood is given by the derivative

$$\frac{dA}{dt} = -\tfrac{3}{2}t^{-5/2}.$$

When $t = 1$, this rate equals

$$\left.\frac{dA}{dt}\right|_{t=1} = -\tfrac{3}{2}(1)^{-5/2} = -\tfrac{3}{2}.$$

* For a discussion of this mathematical model, see J. Defares, I. Sneddon, and M. Wise, *An Introduction to the Mathematics of Medicine and Biology* (Chicago: Year Book Publishers, Inc., 1973), pp. 609–619.

The amount of calcium in the blood is changing at the rate of $-\frac{3}{2}$ units per day when $t = 1$. The negative sign indicates that the amount of calcium is decreasing rather than increasing.

Approximating the Change in a Function   As we have seen above, when h is near 0

$$\frac{f(a + h) - f(a)}{h} \approx f'(a),$$

where the symbol "$\approx$" means "is approximately equal to." Suppose, for example, that $a = 2$ and $f'(a) = 5$. Then for small values of h, we would have

$$\frac{f(2 + h) - f(2)}{h} \approx 5$$

or

$$f(2 + h) - f(2) \approx 5h.$$

In other words, if x is increased from 2 to $2 + h$, the change in the value of $f(x)$ [namely $f(2 + h) - f(2)$] is approximately $5h$. That is, as x varies near 2, the value of $f(x)$ varies at a rate equal to 5 times the change in x. We may generalize this observation to yield the following important fact.

> If h is small, then
>
> $$f(a + h) - f(a) \approx f'(a) \cdot h. \tag{1}$$

The Marginal Concept in Economics   Suppose a company determines that the cost of producing x units of its product is $C(x)$ dollars. Recall from Section 3 that $C'(x)$ is called the marginal cost function. The value of the derivative of $C(x)$ at $x = a$—that is, $C'(a)$—is called the *marginal cost at production level a*. Since the marginal cost is just a derivative, its value gives the rate at which costs are increasing with respect to the level of production, assuming that production is at level a. The use of marginal cost is illustrated in the next example.

EXAMPLE 3   Suppose that the cost function is $C(x) = .005x^3 - 3x$ and production is proceeding at 1000 units per day.

(a) What is the marginal cost at this production level?

(b) What is the cost of increasing production from 1000 to 1001 units per day?

Solution   (a) The marginal cost at production level 1000 is $C'(1000)$. Thus

$$C'(x) = .015x^2 - 3$$

$$C'(1000) = 14{,}997.$$

(b) The cost of increasing production by one unit, from 1000 to 1001, is $C(1001) - C(1000)$, which equals (rounded to the nearest dollar)

$$[.005(1001)^3 - 3(1001)] - [.005(1000)^3 - 3(1000)] = 5,012,012 - 4,997,000$$

$$= 15,012.$$

Note that the marginal cost approximately equals the cost of increasing production by one unit. This fact is true of marginal cost functions generally.

Velocity and Acceleration   An everyday illustration of rate of change is given by the velocity of a moving object. Suppose that we are driving a car along a straight road and at each time t we let $s(t)$ be our position on the road, measured from some convenient reference point. See Fig. 2, where distances are positive to

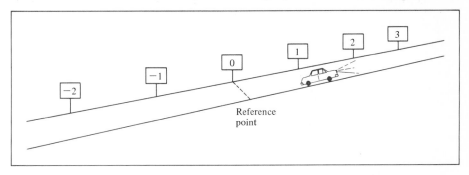

FIGURE 2

the right of the reference point and negative to the left. We call $s(t)$ the *position function* of the car. For the moment we shall assume that we are proceeding in only the positive direction along the road.

At any instant, the car's speedometer tells us how fast we are moving—that is, how fast our position $s(t)$ is changing. To show how the speedometer reading is related to our calculus concept of a derivative, let us examine what is happening at a specific time, say $t = 1$. Consider a short time interval of duration h from $t = 1$ to $t = 1 + h$. Our car will move from position $s(1)$ to position $s(1 + h)$, a distance of $s(1 + h) - s(1)$. Thus the *average velocity from $t = 1$ to $t = 1 + h$ is*

$$\frac{[\text{distance traveled}]}{[\text{time elapsed}]} = \frac{s(1 + h) - s(1)}{h}. \tag{2}$$

If the car is traveling at a steady speed during this time period, then the speedometer reading will equal the average velocity in equation (2).

From our discussion in Section 3 the ratio (2) approaches the derivative $s'(1)$ as h approaches zero. For this reason we call $s'(1)$ *the (instantaneous) velocity at $t = 1$.* This number will agree with the speedometer reading at $t = 1$ because when h is very small, the car's speed will be nearly steady over the time interval from $t = 1$ to $t = 1 + h$, and so the average velocity over this time interval will be nearly the same as the speedometer reading at $t = 1$.

The reasoning used for $t = 1$ holds for an arbitrary t as well. Thus the following definition makes sense:

> If $s(t)$ denotes the position function of an object moving in a straight line, then the velocity $v(t)$ of the object at time t is given by
>
> $$v(t) = s'(t).$$

In our discussion we assumed that the car moved in the positive direction. If the car moves in the opposite direction, the ratio (2) and the limiting value $s'(1)$ will be negative. So we interpret negative velocity as movement in the negative direction along the road.

The derivative of the velocity function $v(t)$ is called the *acceleration* function and is often written as $a(t)$:

> $$a(t) = v'(t).$$

Since $v'(t)$ measures the rate of change of the velocity $v(t)$, this use of the word acceleration agrees with our common usage in connection with automobiles. Note that since $v(t) = s'(t)$, the acceleration is actually the second derivative of the position function $s(t)$:

$$a(t) = s''(t).$$

PRACTICE PROBLEMS 7

When a ball is thrown straight up into the air, it travels along a straight line and its motion can be described in the same manner as the motion of a car. Regard "up" as the positive direction and let $s(t)$ be the height of the ball in feet after t seconds. Suppose that $s(t) = -16t^2 + 128t + 5$.

1. What will be the velocity after 2 seconds?

2. What will be the acceleration after 2 seconds?

3. After how many seconds will the ball attain its greatest height?

EXERCISES 7

1. If $f(t) = t^2 + 3t - 7$, what is the rate of change of $f(t)$ with respect to t when $t = 5$?

2. If $f(t) = 3t + 2 - \dfrac{5}{t}$, what is the rate of change of $f(t)$ with respect to t when $t = 2$?

3. An analysis of the daily output of a factory assembly line shows that about $60t + t^2 - \frac{1}{12}t^3$ units are produced after t hours of work, $0 \le t \le 8$. What is the rate of production (in units per hour) when $t = 2$?

4. Liquid is pouring into a large vat. After t hours, there are $5t - t^{1/2}$ gallons in the vat. At what rate is the liquid flowing into the vat (in gallons per hour) when $t = 4$?

5. Suppose that the weight in grams of a cancerous tumor at time t is $W(t) = .1t^2$, where t is measured in weeks. What is the rate of growth of the tumor (in grams per week) when $t = 5$?

6. After an advertising campaign, the sales of a product often increase and then decrease. Suppose that t days after the end of the advertising, the daily sales are $-3t^2 + 30t + 100$ units. At what rate (in units per day) are the sales changing when $t = 2$?

7. A sewage treatment plant accidentally discharged untreated sewage into a lake for a few days. This temporarily decreased the amount of dissolved oxygen in the lake. Let $f(t)$ be the amount of oxygen in the lake (measured in suitable units) t days after the sewage started flowing into the lake. Experimental data suggest that $f(t)$ is given approximately by

$$f(t) = 500\left[1 - \frac{10}{t + 10} + \frac{100}{(t + 10)^2}\right].$$

Find the rate of change (in units per day) of the oxygen content of the lake at $t = 5$ and at $t = 15$. Is the oxygen content increasing or decreasing when $t = 15$?

8. Suppose that t hours after being placed in a freezer, the temperature of a piece of meat is given by $T(t) = 70 - 12t + 4/(t + 1)$ degrees, where $0 \le t \le 5$. How fast is the temperature falling after 1 hour?

9. A manufacturer estimates that the hourly cost of producing x units of a product on an assembly line is $.1x^3 - 6x^2 + 136x + 200$ dollars.

   (a) Find the marginal cost when the production level is 20 units.

   (b) Show that this marginal cost is approximately the extra cost of raising the production from 20 to 21 units.

10. A market finds that if it prices a certain product so as to sell x units each week, then the revenue received will be approximately $2x - .001x^2$ dollars.

   (a) Find the marginal revenue at a sales level of 600 units.

   (b) Show that this marginal revenue is approximately the extra revenue produced by raising the sales from 600 to 601 units.

11. Suppose that the profit from producing x units of a product is given by $P(x) = .0003x^3 + .01x$.

   (a) Compute the marginal profit at a production level of 100 units.

   (b) Compute the additional profit gained from increasing sales from 100 to 101 units.

12. Suppose that the revenue from producing x units of a product is given by $R(x) = .01x^2 - 3x$.

   (a) Compute the marginal revenue at a production level of 20,000.

   (b) Compute the additional revenue gained by increasing production from 20,000 to 20,001 units.

13. An object moving in a straight line travels $s(t)$ kilometers in t hours, where $s(t) = \frac{1}{2}t^2 + 4t$.

   (a) What is the object's velocity when $t = 3$?

   (b) How far has the object traveled in 3 hours?

14. Suppose that the position of a car at time t is given by $s(t) = 50t - 7/(t + 1)$, where the position is measured in kilometers. Find the velocity and acceleration of the car at $t = 0$.

15. A toy rocket fired straight up into the air has height $s(t) = 160t - 16t^2$ feet after t seconds.

    (a) What is the rocket's initial velocity (when $t = 0$)?

    (b) What is the velocity after 2 seconds?

    (c) What is the acceleration when $t = 3$?

    (d) At what time will the rocket hit the ground?

    (e) At what velocity will the rocket be traveling just as it smashes into the ground?

16. A helicopter is rising straight up in the air. Its distance from the ground t seconds after takeoff is $s(t)$ feet, where $s(t) = t^2 + t$.

    (a) How long will it take for the helicopter to rise 20 feet?

    (b) Find the velocity and the acceleration of the helicopter when it is 20 feet above the ground.

17. During the epidemic in Example 1, is there any day when approximately 1200 people become sick with the flu?

18. Suppose that the cost of producing x units of some product is $\frac{1}{3}x^3 - 5x^2 + 30x + 10$ dollars, while the revenue received from the sale of x units is $39x - x^2$ dollars. At what value(s) of x will the marginal revenue equal the marginal cost?

19. A subway train travels from station A to station B in 2 minutes. Its distance from station A is $s(t) = t^2 - \frac{1}{3}t^3$ kilometers after t minutes. At what time(s) will the train be traveling at the rate of $\frac{1}{2}$ kilometer per minute?

20. Let $f(t)$ be the function in Exercise 7 that gives the amount of oxygen in the lake at time t. Find the time when the amount of oxygen stops decreasing and begins to increase.

21. Let $f(x) = x^3$. Use formula (1) with $f(x) = x^3$ to approximate $f(2.01) - f(2)$.

22. Suppose that the side of a cube is increased from 5 to 5.001 meters. Use the derivative to estimate the increase in volume.

23. Use the derivative to estimate $\sqrt{4.01}$. [*Hint:* Let $f(x) = \sqrt{x}$. Then $f(4 + h) - f(4) \approx f'(4) \cdot (.01).$]

24. Use the derivative to estimate $\sqrt[3]{8.024}$.

25. The monthly payment on a $10,000 3-year car loan is a function $f(r)$ of the interest rate, $r\%$. Now, $f(16) = 351.57$ and $f'(16) = 4.94$. What is the significance of the amount $4.94?

SOLUTIONS TO PRACTICE PROBLEMS 7

1. Since velocity is rate of change of position, $v(t) = s'(t) = -32t + 128$. The velocity after 2 seconds is $v(2) = -32(2) + 128 = 64$. Therefore, after 2 seconds the ball will be traveling upward at a velocity of 64 feet per second.

2. The acceleration is the rate of change of velocity, so $a(t) = v'(t) = -32$. The acceleration is always -32. This constant acceleration is due to the downward (and therefore negative) force of gravity.

3. At the instant at which the ball attains its maximum height it will have zero velocity. Therefore, we can determine this time by setting $v(t) = 0$ and solving for t.

$$-32t + 128 = 0$$
$$t = 4.$$

So the ball attains its maximum height after 4 seconds.

Chapter 11: CHECKLIST

☐ Slope of a curve at a point
☐ The derivative of a constant function is zero
☐ Limit definition of the derivative
☐ Power rule:

$$\frac{d}{dx}(x^r) = rx^{r-1}$$

☐ General power rule:

$$\frac{d}{dx}([g(x)]^r) = r \cdot [g(x)]^{r-1} \cdot \frac{d}{dx}[g(x)]$$

☐ Constant-multiple rule:

$$\frac{d}{dx}[k \cdot f(x)] = k \cdot \frac{d}{dx}[f(x)]$$

☐ Sum rule:

$$\frac{d}{dx}[f(x) + g(x)] = \frac{d}{dx}[f(x)] + \frac{d}{dx}[g(x)]$$

☐ Notation for first and second derivatives
☐ The derivative $f'(a)$ measures the rate of change of $f(t)$ with respect to t at $t = a$
☐ $f(a + h) - f(a) \approx f'(a) \cdot h$
☐ Marginal concept in economics
☐ Position, velocity, and acceleration functions
☐ Differentiable at $x = a$
☐ Continuous at $x = a$
☐ $\lim\limits_{x \to a} f(x)$
☐ $\lim\limits_{x \to \infty} f(x), \ \lim\limits_{x \to -\infty} f(x)$

Chapter 11: SUPPLEMENTARY EXERCISES

Differentiate each of the following.

1. $8x - 1$
2. $-2x + 3$
3. 5
4. -3
5. $9x^2$
6. $x^2 - 3\sqrt{x}$

7. $x + 1 - (2/x)$ **8.** $(x + 2)^2$ **9.** $y = x^7 + x^3$

10. $y = 5x^8$ **11.** $y = 6\sqrt{x}$ **12.** $y = x^7 + 3x^5 + 1$

13. $y = 3/x$ **14.** $y = x^4 - (4/x)$ **15.** $y = (3x^2 - 1)^8$

16. $y = \frac{3}{4}x^{4/3} + \frac{4}{3}x^{3/4}$ **17.** $y = \dfrac{1}{5x - 1}$ **18.** $y = (x^3 + x^2 + 1)^5$

19. $y = \sqrt{x^2 + 1}$ **20.** $y = \dfrac{5}{7x^2 + 1}$

21. $f(x) = 1/\sqrt[4]{x}$ **22.** $f(x) = (2x + 1)^3 - 5(2x + 1)^2$

23. $f(x) = 5$ **24.** $f(x) = \dfrac{5x}{2} - \dfrac{2}{5x}$

25. $f(x) = [x^5 - (x - 1)^5]^{10}$ **26.** $f(t) = t^{10} - 10t^9$

27. $g(t) = 3\sqrt{t} - \dfrac{3}{\sqrt{t}}$ **28.** $h(t) = 3\sqrt{2}$

29. $f(t) = \dfrac{2}{t - 3t^3}$ **30.** $g(P) = 4P^{.7}$

31. $h(x) = \frac{3}{2}x^{3/2} - 6x^{2/3}$ **32.** $f(x) = \sqrt{x + \sqrt{x}}$

33. If $f(t) = 3t^3 - 2t^2$, find $f'(2)$.   **34.** If $V(r) = 15\pi r^2$, find $V'(\frac{1}{3})$.

35. If $g(u) = 3u - 1$, find $g(5)$ and $g'(5)$.   **36.** If $h(x) = -\frac{1}{2}$, find $h(-2)$ and $h'(-2)$.

37. If $f(x) = x^{5/2}$, what is $f''(4)$?   **38.** If $g(t) = \frac{1}{4}(2t - 7)^4$, what is $g''(3)$?

39. Find the slope of the graph of $y = (3x - 1)^3 - 4(3x - 1)^2$ at $x = 0$.

40. Find the slope of the graph of $y = (4 - x)^5$ at $x = 5$.

Compute.

41. $\dfrac{d}{dx}(x^4 - 2x^2)$ **42.** $\dfrac{d}{dt}(t^{5/2} + 2t^{3/2} - t^{1/2})$

43. $\dfrac{d}{dP}(\sqrt{1 - 3P})$ **44.** $\dfrac{d}{dn}(n^{-5})$

45. $\dfrac{d}{dz}(z^3 - 4z^2 + z - 3)\Big|_{z = -2}$ **46.** $\dfrac{d}{dx}(4x - 10)^5\Big|_{x = 3}$

47. $\dfrac{d^2}{dx^2}(5x + 1)^4$ **48.** $\dfrac{d^2}{dt^2}(2\sqrt{t})$

49. $\dfrac{d^2}{dt^2}(t^3 + 2t^2 - t)\Big|_{t = -1}$ **50.** $\dfrac{d^2}{dP^2}(3P + 2)\Big|_{P = 4}$

51. $\dfrac{d^2y}{dx^2}$ where $y = 4x^{3/2}$ **52.** $\dfrac{d}{dt}\left(\dfrac{dy}{dt}\right)$, where $y = \dfrac{1}{3t}$

53. What is the slope of the graph of $f(x) = x^3 - 4x^2 + 6$ at $x = 2$? Write the equation of the line tangent to the graph of $f(x)$ at $x = 2$.

54. What is the slope of the curve $y = 1/(3x - 5)$ at $x = 1$? Write the equation of the line tangent to this curve at $x = 1$.

55. Find the equation of the tangent line to the curve $y = x^2$ at the point $(\frac{3}{2}, \frac{9}{4})$. Sketch the graph of $y = x^2$ and sketch the tangent line at $(\frac{3}{2}, \frac{9}{4})$.

56. Find the equation of the tangent line to the curve $y = x^2$ at the point $(-2, 4)$. Sketch the graph of $y = x^2$ and sketch the tangent line at $(-2, 4)$.

57. Determine the equation of the tangent line to the curve $y = 3x^3 - 5x^2 + x + 3$ at $x = 1$.

58. Determine the equation of the tangent line to the curve $y = (2x^2 - 3x)^3$ at $x = 2$.

59. A helicopter is rising at a rate of 32 feet per second. At a height of 128 feet the pilot drops a pair of binoculars. After t seconds, the binoculars have height $s(t) = -16t^2 + 32t + 128$ feet from the ground. How fast will they be falling when they hit the ground?

60. Each day, the total output of a coal mine after t hours of operation is approximately $40t + t^2 - \frac{1}{15}t^3$ tons, $0 \le t \le 12$. What is the rate of output (in tons of coal per hour) at $t = 5$ hours?

Determine whether the following limits exist. If so, compute the limit.

61. $\lim\limits_{x \to 2} \dfrac{x^2 - 4}{x - 2}$

62. $\lim\limits_{x \to 3} \dfrac{1}{x^2 - 4x + 3}$

63. $\lim\limits_{x \to 4} \dfrac{x - 4}{x^2 - 8x + 16}$

64. $\lim\limits_{x \to 5} \dfrac{x - 5}{x^2 - 7x + 2}$

Use limits to compute the following derivatives.

65. $f'(5)$, where $f(x) = 1/(2x)$

66. $f'(3)$, where $f(x) = x^2 - 2x + 1$

67. If you deposit $100 into a savings account at the end of each month for 2 years, the balance will be a function $f(r)$ of the interest rate, $r\%$. At 7% interest (compounded monthly), $f(7) = 2568.10$ and $f'(7) = 25.06$. Approximately how much additional money would you earn if the bank paid $7\frac{1}{2}\%$ interest?

Applications
of the Derivative

Chapter 12

Calculus techniques can be applied to a wide variety of problems in real life. We consider many examples in this chapter. In each case we construct a function as a "mathematical model" of some problem and then analyze the function and its derivatives in order to gain information about the original problem. Our principal method for analyzing a function will be to sketch its graph. For this reason we devote the first part of the chapter to curve sketching.

12.1. Describing Graphs of Functions

The graph of the function in Fig. 1 is either rising or falling, depending on whether we look at it from left to right or from right to left. To avoid confusion, we shall always follow the accepted practice of reading a graph from left to right. We say that a function $f(x)$ is *increasing* at $x = a$ if, in some small region near the point $(a, f(a))$, the graph is rising as we move from left to right. The function is *decreasing* at $x = a$ if, in some small region near the point $(a, f(a))$, the graph is falling. (See Fig. 2.)

An *extreme point* of a function is a point at which its graph changes from increasing to decreasing, or vice versa. We distinguish the two possibilities in an obvious way. A *maximum point* is a point at which the graph changes from increasing to decreasing; a *minimum point* is a point at which the graph changes from decreasing to increasing (Fig. 3).

The *maximum value* (or *absolute maximum value*) of a function is the largest value that the function assumes on its domain. The *minimum value* (or *absolute minimum value*) of a function is the smallest value that the function assumes on

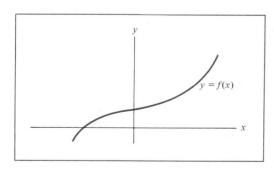

FIGURE 1

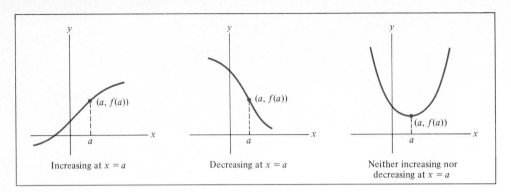

FIGURE 2

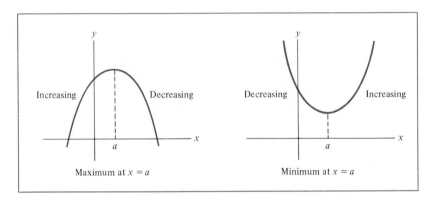

FIGURE 3

its domain. Functions may or may not have maximum and or minimum values. (See Fig. 4.) However, it can be shown that a continuous function whose domain is an interval of the form $a \le x \le b$ has both a maximum value and a minimum value.

Maximum values and minimum values of functions usually occur at maximum points and minimum points, as in Fig. 4(a) and (b). However, they can occur at endpoints of the domain, as in Fig. 4(c). If so, we say that the function has an *endpoint extreme value* (or *endpoint extremum*).

EXAMPLE 1 When a drug is injected intramuscularly (into a muscle), the concentration of the drug in the veins has the time-concentration curve shown in Fig. 5. Describe this graph, using the terms introduced above.

Solution Initially (when $t = 0$), there is no drug in the veins. When the drug is injected into the muscle, it begins to diffuse into the bloodstream. The concentration in the veins increases until it reaches its maximum value at $t = 2$. After this time the concentration begins to decrease, as the body's metabolic processes remove the drug from the blood. Eventually the drug concentration decreases to a level so small that, for all practical purposes, it is zero.

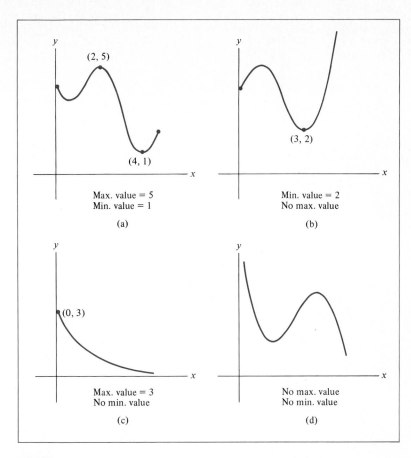

Max. value = 5
Min. value = 1

(a)

Min. value = 2
No max. value

(b)

Max. value = 3
No min. value

(c)

No max. value
No min. value

(d)

FIGURE 4

FIGURE 5

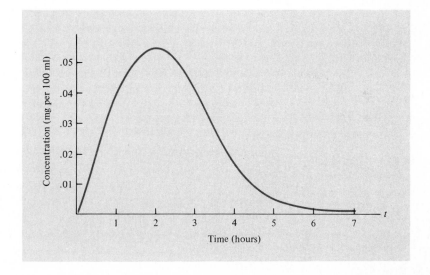

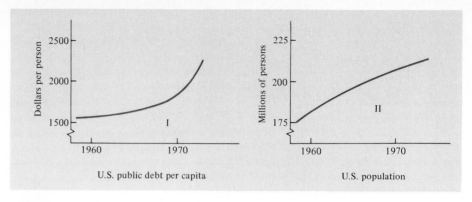

U.S. public debt per capita   U.S. population

FIGURE 6

The graphs in Fig. 6 are both increasing, but there is a fundamental difference in the way they are increasing. Graph I, which describes the U.S. national debt per person, is steeper for 1970 than for 1960. That is, the *slope* of graph I is *increasing* as we move from left to right. On the other hand, the *slope* of graph II is *decreasing* as we move from left to right. Although the U.S. population is rising each year, the rate of increase was not as great in 1970 as it was in 1960.

The difference between the two graphs in Fig. 6 can also be described in geometric terms: Graph I opens up and lies above its tangent line at each point, whereas graph II opens down and lies below its tangent line at each point (Fig. 7).

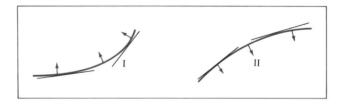

FIGURE 7

We say that a function $f(x)$ is *concave up* at $x = a$ if, in some small region near the point $(a, f(a))$, the graph of $f(x)$ lies above its tangent line. Equivalently, $f(x)$ is concave up at $x = a$ if the slope of the graph increases as we move from left to right through $(a, f(a))$. Graph I is an example of a function that is concave up at each point.

Similarly, we say that a function $f(x)$ is *concave down* at $x = a$ if, in some small region near $(a, f(a))$, the graph of $f(x)$ lies below its tangent line. Equivalently, $f(x)$ is concave down at $x = a$ if the slope of the graph decreases as we move from left to right through $(a, f(a))$. Graph II is concave down at each point.

An *inflection point* of a graph is a point at which the graph changes from concave up to concave down, or vice versa. At such a point, the graph crosses its tangent line (Fig. 8).

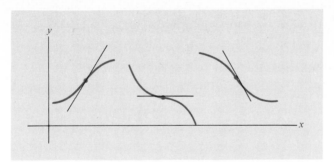

FIGURE 8

EXAMPLE 2   Use the terms defined earlier to describe the graph shown in Fig. 9.

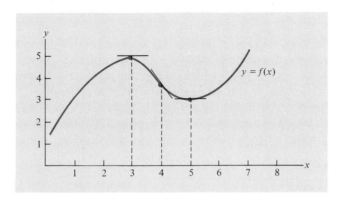

FIGURE 9

Solution   (a) For $x < 3$, $f(x)$ is increasing and concave down.

(b) Maximum point at $x = 3$.

(c) For $3 < x < 4$, $f(x)$ is decreasing and concave down.

(d) Inflection point at $x = 4$.

(e) For $4 < x < 5$, $f(x)$ is decreasing and concave up.

(f) Minimum point at $x = 5$.

(g) For $x > 5$, $f(x)$ is increasing and concave up.

FIGURE 10

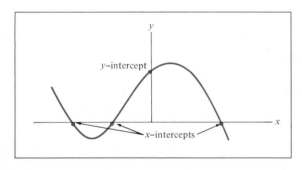

Intercepts, Undefined Points, Asymptotes   A point at which a graph crosses the y-axis is called a *y-intercept*, and a point at which it crosses the x-axis is called an *x-intercept*. The x-coordinate of an x-intercept is sometimes called a "zero" of the function, since the function has the value zero there. (See Fig. 10.)

Some functions are not defined for all values of x. For instance, $f(x) = 1/x$ is not defined for $x = 0$, and $f(x) = \sqrt{x}$ is not defined for $x < 0$. (See Fig. 11.) Many functions that arise in ap-

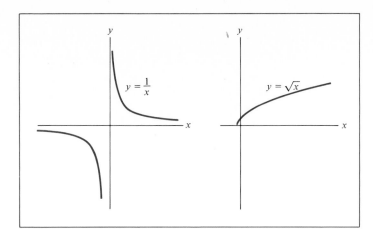

FIGURE 11

plications are defined only for $x \geq 0$. A properly drawn graph should leave no doubt as to the values of x for which the function is defined.

Graphs in applied problems sometimes straighten out and approach some straight line as x gets large (Fig. 12). Such a straight line is called an *asymptote* of the curve. The most common asymptotes are horizontal as in (a) and (b) of Fig. 12. In Example 1 the t-axis is an asymptote of the drug time-concentration curve.

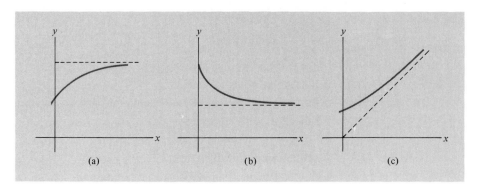

FIGURE 12   Graphs that approach asymptotes as x gets large.

The horizontal asymptotes of a graph may be determined by calculating the limits

$$\lim_{x \to \infty} f(x) \quad \text{and} \quad \lim_{x \to -\infty} f(x).$$

If either limit exists, then the value of the limit determines a horizontal asymptote.

Occasionally, a graph will approach a vertical line as x approaches some fixed value, as in Fig. 13. Such a line is a *vertical asymptote*. Most often, we expect a vertical asymptote at a value x which would result in division by zero in the definition of $f(x)$. For example, $f(x) = 1/(x - 3)$ has a vertical asymptote $x = 3$.

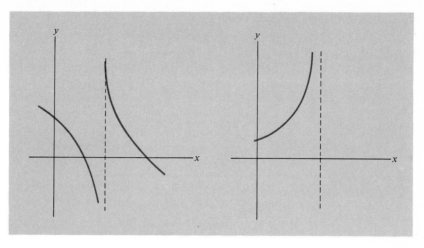

FIGURE 13   Examples of vertical asymptotes.

We now have six categories for describing the graph of a function:

1. Increasing, decreasing, maximum points, minimum points
2. Maximum value, minimum value
3. Concave up, concave down, inflection points
4. x-intercept, y-intercept
5. Undefined points
6. Asymptotes

For us, the first three categories will be the most important. However, the last three categories should not be forgotten.

PRACTICE PROBLEMS 1

1. Does the slope of the curve in Fig. 14 increase or decrease as x increases?

2. At what value of x is the slope of the curve in Fig. 15 minimized?

FIGURE 14

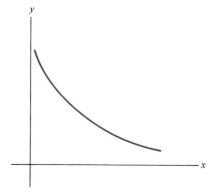

FIGURE 15

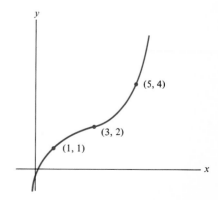

EXERCISES 1

Exercises 1–4 refer to graphs (a)–(f) in Fig. 16.

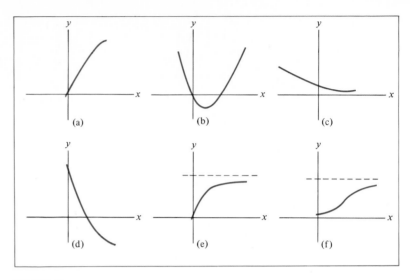

FIGURE 16

1. Which functions are increasing for all x?

2. Which functions are decreasing for all x?

3. Which functions have the property that the slope always increases as x increases?

4. Which functions have the property that the slope always decreases as x increases?

Describe each graph below. Your description should include each of the six categories mentioned above.

5.

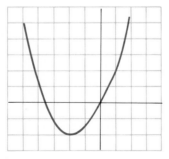

6.

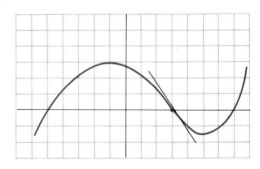

7.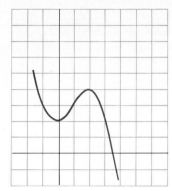

8.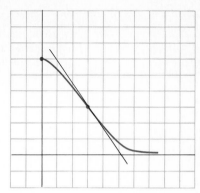

9.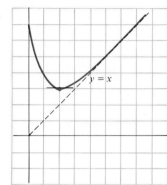

10.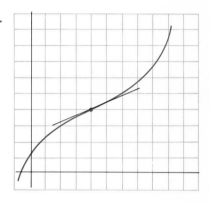

11.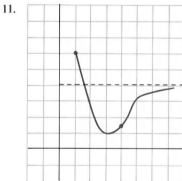

12.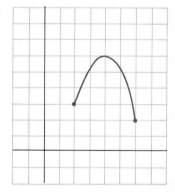

13. Describe the way the *slope* changes as you move along the graph (from left to right) in Exercise 5.

14. Describe the way the *slope* changes on the graph in Exercise 6.

15. Describe the way the *slope* changes on the graph in Exercise 8.

16. Describe the way the *slope* changes on the graph in Exercise 10.

17. Suppose that some organic waste products are dumped into a lake at time $t = 0$, and suppose that the oxygen content of the lake at time t is given by the graph in Fig. 17. Describe the graph in physical terms. Indicate the significance of the inflection point at $t = b$.

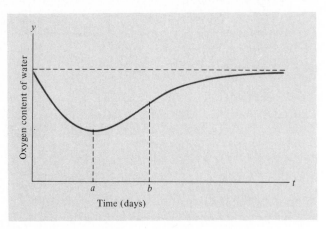

FIGURE 17

18. Let $C(x)$ denote the total cost of manufacturing x units of some product. Then $C(x)$ is an increasing function for all x. For small values of x, the rate of increase of $C(x)$ decreases. (This is because of the savings that are possible with "mass production.") Eventually, however, for large values of x, the cost $C(x)$ increases at an increasing rate. (This happens when production facilities are strained and become less efficient.) Sketch a graph that could represent $C(x)$.

19. The annual world consumption of oil rises each year. Furthermore, the amount of the annual *increase* in oil consumption is also rising each year. Sketch a graph that could represent the annual world consumption of oil.

20. In certain professions the average annual income has been rising at an increasing rate. Let $f(T)$ denote the average annual income at year T for persons in one of these professions and sketch a graph that could represent $f(T)$.

21. Let $s(t)$ be the distance (in feet) traveled by a parachutist after t seconds from the time of opening the chute and suppose that $s(t)$ has the line $y = -15t$ as an asymptote. What does this imply about the velocity of the parachutist? [*Note:* Distance traveled downward is given a negative value.]

22. Let $P(t)$ be the population of a bacteria culture after t days and suppose that $P(t)$ has the line $y = 25,000,000$ as an asymptote. What does this imply about the size of the population?

23. Consider a smooth curve with no undefined points.

(a) If it has two maximum points, must it have a minimum point?

(b) If it has two extreme points, must it have an inflection point?

24. Can a point of a graph be both

    (a) a minimum point and an x-intercept?

    (b) an x-intercept and a y-intercept?

    (c) a maximum point and a minimum point?

    (d) an inflection point and an x-intercept?

In Exercises 25–30, sketch the graph of a function having the given properties.

25. (i) minimum point at $x = 0$

    (ii) inflection point at $x = 3$

    (iii) no maximum point

26. (i) maximum points at $x = 1$ and $x = 5$

    (ii) minimum point at $x = 3$

    (iii) inflection points at $x = 2$ and $x = 4$

27. (i) defined and increasing for all $x \geq 0$

    (ii) inflection point at $x = 5$

    (iii) asymptotic to the line $y = \frac{3}{4}x + 5$

28. (i) inflection point at $x = 1$

    (ii) increasing for all x

29. (i) defined for $0 \leq x \leq 10$

    (ii) maximum point at $x = 3$

    (iii) maximum value at $x = 10$

30. (i) defined for $x \geq 0$

    (ii) minimum value at $x = 0$

    (iii) maximum point at $x = 4$

    (iv) asymptotic to the line $y = (x/2) + 1$

The difference between the pressure inside the lungs and the pressure surrounding the lungs is called the *transmural pressure* (or transmural pressure gradient). The figure at the top of page 490 shows for three different persons how the volume of the lungs is related to the transmural pressure, based on static measurements taken while there is no air flowing through the mouth. (The functional reserve capacity mentioned on the vertical axis is the volume of air in the lungs at the end of a normal expiration.) The rate of change of lung volume with respect to transmural pressure is called the *lung compliance*.

31. If the lungs are less flexible than normal, an increase in pressure will cause a smaller change in lung volume than in a normal lung. In this case, is the compliance relatively high or low?

32. Most lung diseases cause a decrease in lung compliance. However, the compliance of a person with emphysema is higher than normal. Which curve (I or II) in the figure could correspond to a person with emphysema?

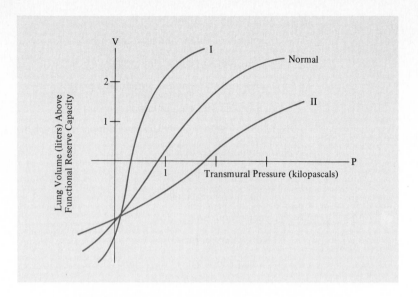

SOLUTIONS TO PRACTICE PROBLEMS 1

1. The curve is concave up, so the slope increases. Even though the curve itself is decreasing, the slope becomes less negative as we move from left to right.

2. At $x = 3$. We have drawn in tangent lines at various points. Note that as we move from left to right, the slopes decrease steadily until the point $(3, 2)$, at which time they start to increase. This is consistent with the fact that the graph is concave down (hence, slopes are decreasing) to the left of $(3, 2)$ and concave up (hence, slopes are increasing) to the right of $(3, 2)$. Extreme values of slopes always occur at inflection points.

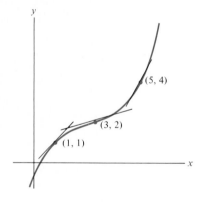

12.2. The First and Second Derivative Rules

We shall now show how properties of the graph of a function $f(x)$ are determined by properties of the derivatives, $f'(x)$ and $f''(x)$. These relationships will provide the key to the curve-sketching and optimization problems discussed in the rest of the chapter.*

---

* Throughout this chapter, we shall assume that we are dealing with functions which are not "too badly behaved." More precisely, it suffices to assume that all of our functions have continuous first and second derivatives in the interval(s) (in x) where we are considering their graphs.

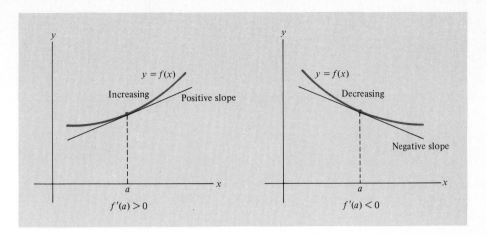

FIGURE 1

We begin with a discussion of the first derivative of a function $f(x)$. Suppose that for some value of x, say $x = a$, the derivative $f'(a)$ is positive. Then the tangent line at $(a, f(a))$ has positive slope and is a rising line (moving from left to right, of course). Since the graph of $f(x)$ near $(a, f(a))$ resembles its tangent line, the function must be increasing at $x = a$. Similarly, when $f'(a) < 0$, the function is decreasing at $x = a$. (See Fig. 1.)

Thus we have the following useful result.

> *First Derivative Rule* If $f'(a) > 0$, then $f(x)$ is increasing at $x = a$. If $f'(a) < 0$, then $f(x)$ is decreasing at $x = a$.

When $f'(a) = 0$, the first derivative rule is not decisive. In this case the function $f(x)$ might be increasing or decreasing, or have an extreme point, at $x = a$.

EXAMPLE 1 Sketch the graph of a function $f(x)$ that has all of the following properties.

(a) $f(3) = 4$

(b) $f'(x) > 0$ for $x < 3$, $f'(3) = 0$ and $f'(x) < 0$ for $x > 3$

Solution The only specific point on the graph is $(3, 4)$ [property (a)]. We plot this point and then use the fact that $f'(3) = 0$ to sketch the tangent line at $x = 3$ (Fig. 2).

From property (b) and the first derivative rule, we know that $f(x)$ must be increasing for x less than 3 and decreasing for x greater than 3. A graph with these properties might look like the curve in Fig. 3.

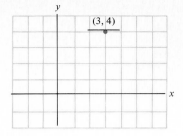

FIGURE 2

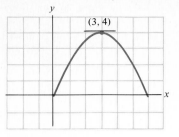

FIGURE 3

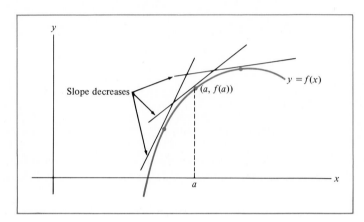

FIGURE 4

The second derivative of a function $f(x)$ gives useful information about the concavity of the graph of $f(x)$. Suppose that $f''(a)$ is negative. Then since $f''(x)$ is the derivative of $f'(x)$, we conclude that $f'(x)$ has a negative derivative at $x = a$. In this case, $f'(x)$ must be a decreasing function at $x = a$; that is, the slope of the graph of $f(x)$ is decreasing as we move from left to right on the graph near $(a, f(a))$. (See Fig. 4.) This means that the graph of $f(x)$ is concave down at $x = a$. A similar analysis shows that if $f''(a)$ is positive, then $f(x)$ is concave up at $x = a$. Thus we have the following rule.

> *Second Derivative Rule* If $f''(a) > 0$, then $f(x)$ is concave up at $x = a$.
> If $f''(a) < 0$, then $f(x)$ is concave down at $x = a$.

When $f''(a) = 0$, the second derivative rule gives no information. In this case, the function might be concave up, concave down, or neither, at $x = a$.

The chart below shows how a graph may combine the properties of increasing, decreasing, concave up, and concave down.

Condition on the derivatives	Description of $f(x)$ at $x = a$	Graph of $y = f(x)$ near $x = a$
1. $f'(a)$ positive $f''(a)$ positive	$f(x)$ increasing $f(x)$ concave up	
2. $f'(a)$ positive $f''(a)$ negative	$f(x)$ increasing $f(x)$ concave down	
3. $f'(a)$ negative $f''(a)$ positive	$f(x)$ decreasing $f(x)$ concave up	
4. $f'(a)$ negative $f''(a)$ negative	$f(x)$ decreasing $f(x)$ concave down	

EXAMPLE 2 Sketch the graph of a function $f(x)$ with all of the following properties.

(a) $(2, 3)$, $(4, 5)$, and $(6, 7)$ are on the graph

(b) $f'(6) = 0$ and $f'(2) = 0$

(c) $f''(x) > 0$ for $x < 4$, $f''(4) = 0$ and $f''(x) < 0$ for $x > 4$

Solution First we plot the three points from property (a) and then sketch two tangent lines, using the information from property (b). (See Fig. 5.) From condition (c) and the second derivative rule, we know that $f(x)$ is concave up for $x < 4$. In particular, $f(x)$ is concave up at $(2, 3)$. Also, $f(x)$ is concave down for $x > 4$ and, in particular, at $(6, 7)$. Note that $f(x)$ must have an inflection point at $x = 4$ because the concavity changes there. We now sketch small portions of the curve near $(2, 3)$ and $(6, 7)$. (See Fig. 6.) We can now complete the sketch (Fig. 7), taking care to make the curve concave up for $x < 4$ and concave down for $x > 4$.

FIGURE 5

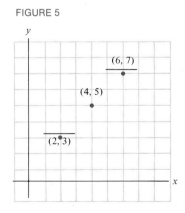

FIGURE 6

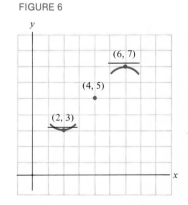

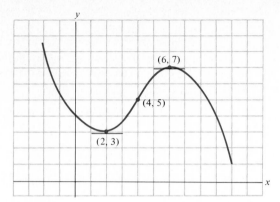

FIGURE 7

PRACTICE PROBLEMS 2

1. Make a good sketch of the function $f(x)$ near the point where $x = 2$, given that $f(2) = 5$, $f'(2) = 1$, and $f''(2) = -3$.

2. The graph of $f(x) = x^3$ is on the right.

   (a) Is the function increasing at $x = 0$?

   (b) Compute $f'(0)$.

   (c) Reconcile your answers to parts (a) and (b) with the first derivative rule.

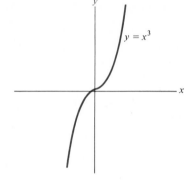

EXERCISES 2

Exercises 1–4 refer to the functions whose graphs are given in Fig. 8.

1. Which functions have a positive first derivative for all x?

2. Which functions have a negative first derivative for all x?

3. Which functions have a positive second derivative for all x?

4. Which functions have a negative second derivative for all x?

In Exercises 5–10, sketch the graph of a function $f(x)$ that has the properties described.

5. $f(2) = 1$; $f'(2) = 0$; concave up for all x.

6. $f(-1) = 0$; $f'(x) < 0$ for $x < -1$, $f'(-1) = 0$ and $f'(x) > 0$ for $x > -1$.

7. $f(3) = 5$; $f'(x) > 0$ for $x < 3$, $f'(3) = 0$ and $f'(x) > 0$ for $x > 3$.

8. $(-2, -1)$ and $(2, 5)$ are on the graph; $f'(-2) = 0$ and $f'(2) = 0$; $f''(x) > 0$ for $x < 0$; $f''(0) = 0$, $f''(x) < 0$ for $x > 0$.

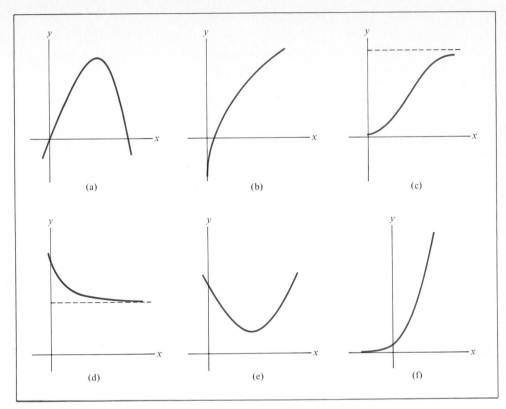

FIGURE 8

9. $(0, 6)$, $(2, 3)$, and $(4, 0)$ are on the graph; $f'(0) = 0$ and $f'(4) = 0$; $f''(x) < 0$ for $x < 2$; $f''(2) = 0$, $f''(x) > 0$ for $x > 2$.

10. $f(x)$ defined only for $x \geq 0$; $(0, 0)$ and $(5, 6)$ are on the graph; $f'(x) > 0$ for $x \geq 0$; $f''(x) < 0$ for $x < 5$; $f''(5) = 0$, $f''(x) > 0$ for $x > 5$.

11. By looking at the first derivative, decide which of the curves in Fig. 9 could *not* be the graph of $f(x) = (3x^2 + 1)^4$ for $x \geq 0$.

FIGURE 9

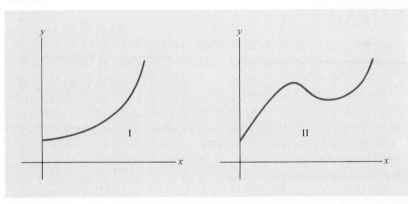

12. By looking at the first derivative, decide which of the curves in Fig. 9 could *not* be the graph of $f(x) = x^3 - 9x^2 + 24x + 1$ for $x \geq 0$. [*Hint:* Factor the formula for $f'(x)$.]

13. By looking at the second derivative, decide which of the curves in Fig. 10 could be the graph of $f(x) = \sqrt{x}$.

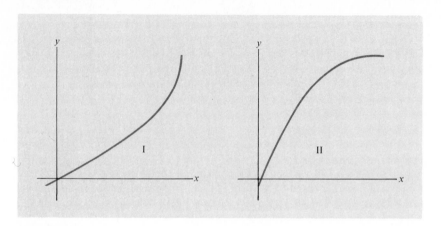

FIGURE 10

14. By looking at the second derivative, decide which of the curves in Fig. 10 could be the graph of $f(x) = x^{5/2}$.

In Exercises 15–20, use the given information to make a good sketch of the function $f(x)$ near $x = 3$.

15. $f(3) = 4, f'(3) = -\frac{1}{2}, f''(3) = 5$

16. $f(3) = -2, f'(3) = 0, f''(3) = 1$

17. $f(3) = 1, f'(3) = 0, f''(3) = 0$, inflection point at $x = 3, f'(x) > 0$ for $x > 3$

18. $f(3) = 4, f'(3) = -\frac{3}{2}, f''(3) = -2$

19. $f(3) = -2, f'(3) = 2, f''(3) = 3$

20. $f(3) = 3, f'(3) = 1, f''(3) = 0$, inflection point at $x = 3, f''(x) < 0$ for $x > 3$

SOLUTIONS TO PRACTICE PROBLEMS 2

1. Since $f(2) = 5$, the point $(2, 5)$ is on the graph [Fig. 11(a)]. Since $f'(2) = 1$, the tangent line at the point $(2, 5)$ has slope 1. Draw in the tangent line [Fig. 11(b)]. Near the point $(2, 5)$ the graph looks approximately like the tangent line. Since $f''(2) = -3$, a negative number, the graph is concave down at the point $(2, 5)$. Now we are ready to sketch the graph [Fig. 11(c)].

2. (a) Yes. The graph is steadily increasing as we pass through the point $(0, 0)$.

   (b) Since $f'(x) = 3x^2, f'(0) = 3 \cdot 0^2 = 0$.

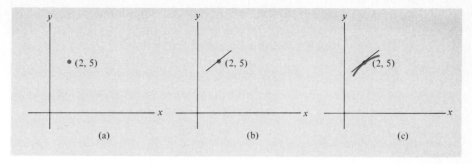

FIGURE 11

(c) There is no contradiction here. The first derivative rule says that if the derivative is positive, the function is increasing. However, it does not say that this is the only condition under which a function is increasing. As we have just seen, sometimes we can have the first derivative zero and the function still increasing.

12.3. Curve Sketching (Introduction)

In this section and the next we develop our ability to sketch the graphs of functions. There are two important reasons for doing so. First, a geometric "picture" of a function is often easier to comprehend than its abstract formula. Second, the material in this section will provide a foundation for the applications in Sections 5 to 7.

A "sketch" of the graph of a function $f(x)$ should convey the general shape of the graph—it should show where $f(x)$ is increasing and decreasing and it should indicate, insofar as possible, where $f(x)$ is concave up and concave down. In addition, one or more key points should be accurately located on the graph. These points usually include extreme points, inflection points, and x- and y-intercepts. Other features of a graph may be important, too, but we shall discuss them as they arise in examples and applications.

Our general approach to curve sketching will involve four main steps:

1. Starting with $f(x)$, we compute $f'(x)$ and $f''(x)$.
2. Next, we locate all maximum and minimum points and make a partial sketch.
3. We study the concavity of $f(x)$ and locate all inflection points.
4. We consider other properties of the graph, such as the intercepts, and complete the sketch.

The first step was the main subject of the preceding chapter. We discuss the second and third steps in this section, and then present several completely worked examples that combine all four steps in the next.

Locating Extreme Points   The tangent line at a maximum or a minimum point of a function $f(x)$ has zero slope; that is, the derivative is zero there. Thus we may state the following useful rule.

> Look for possible extreme points of $f(x)$ by setting $f'(x) = 0$ and solving for x. (1)

Suppose that $f'(a) = 0$. Then $x = a$ is a candidate for an extreme point of $f(x)$. There are several ways to determine if $f(x)$ has a maximum or a minimum (or neither) at $x = a$. The method that works in most cases is described in our first two examples.

EXAMPLE 1   The graph of the quadratic function $f(x) = \frac{1}{4}x^2 - x + 2$ is a parabola and so has one extreme point. Find it and sketch the graph.

Solution   We begin by computing the first and second derivatives of $f(x)$.

$$f(x) = \tfrac{1}{4}x^2 - x + 2$$

$$f'(x) = \tfrac{1}{2}x - 1$$

$$f''(x) = \tfrac{1}{2}.$$

Setting $f'(x) = 0$, we have $\frac{1}{2}x - 1 = 0$, so that $x = 2$. Thus $f'(2) = 0$. Geometrically, this means that the graph of $f(x)$ will have a horizontal tangent line at the point where $x = 2$. To plot this point, we substitute the value 2 for x in the original expression for $f(x)$.

$$f(2) = \tfrac{1}{4}(2)^2 - (2) + 2 = 1.$$

Figure 1 shows the point $(2, 1)$ together with the horizontal tangent line. Is $(2, 1)$ an extreme point? In order to decide, we look at $f''(x)$. Since $f''(2) = \frac{1}{2}$, which is

FIGURE 1

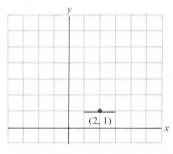

FIGURE 2

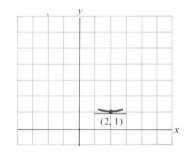

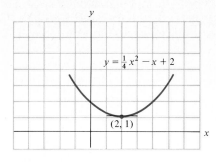

$$y = \tfrac{1}{4}x^2 - x + 2$$

(2, 1)

FIGURE 3

positive, the graph of $f(x)$ is concave up at $x = 2$. So a partial sketch of the graph near $(2, 1)$ should look something like Fig. 2.

We see that $(2, 1)$ is a minimum point. In fact, it is the only extreme point, for there is no other place where the tangent line is horizontal. Since the graph has no other "turning points," it must be decreasing before it gets to $(2, 1)$ and then increasing to the right of $(2, 1)$. Note that since $f''(x)$ is positive (and equal to $\tfrac{1}{2}$) for all x, the graph is concave upward at each point. A completed sketch is given in Fig. 3.

EXAMPLE 2 Locate all possible extreme points on the graph of the function $f(x) = x^3 - 3x^2 + 5$. Check the concavity at these points and use this information to sketch the graph of $f(x)$.

Solution We have

$$f(x) = x^3 - 3x^2 + 5$$
$$f'(x) = 3x^2 - 6x$$
$$f''(x) = 6x - 6.$$

The easiest way to find those values of x for which $f'(x)$ is zero is to factor the expression for $f'(x)$:

$$3x^2 - 6x = 3x(x - 2).$$

From this factorization it is clear that $f'(x)$ will be zero if and only if $x = 0$ or $x = 2$. In other words, the graph will have horizontal tangent lines when $x = 0$ and $x = 2$, and nowhere else.

To plot the points on the graph where $x = 0$ and $x = 2$, we substitute these values back into the original expression for $f(x)$. That is, we compute

$$f(0) = (0)^3 - 3(0)^2 + 5 = 5$$
$$f(2) = (2)^3 - 3(2)^2 + 5 = 1.$$

FIGURE 4

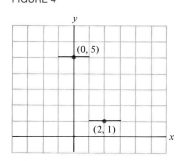

(0, 5)

(2, 1)

Figure 4 shows the points $(0, 5)$ and $(2, 1)$, along with the corresponding tangent lines.

Next, we check the concavity of the graph at these points by evaluating $f''(x)$ at $x = 0$ and $x = 2$:

$$f''(0) = 6(0) - 6 = -6$$
$$f''(2) = 6(2) - 6 = +6.$$

Since $f''(0)$ is negative, the graph is concave down at $x = 0$; since $f''(2)$ is positive, the graph is concave up at $x = 2$. A partial sketch of the graph is given in Fig. 5.

It is clear from Fig. 5 that $(0, 5)$ is a maximum point and $(2, 1)$ is a minimum point. Since they are the only turning points, the graph must be increasing before it gets to $(0, 5)$, decreasing from $(0, 5)$ to $(2, 1)$, and then increasing again to the right of $(2, 1)$. A sketch incorporating these properties appears in Fig. 6.

The facts that we used to sketch Fig. 6 could equally well be used to produce the graph in Fig. 7. Which graph really corresponds to $f(x) = x^3 - 3x^2 + 5$? The answer will be clear when we find the inflection points on the graph of $f(x)$.

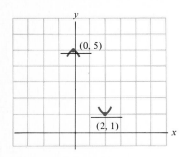

FIGURE 5

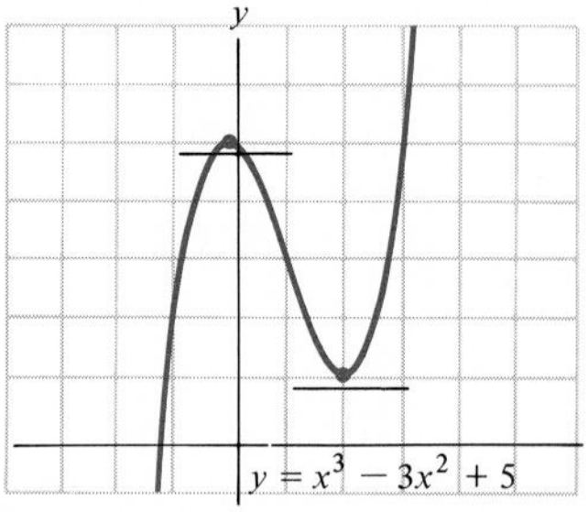

FIGURE 6

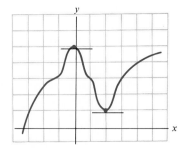

FIGURE 7

Locating Inflection Points   An inflection point of a function $f(x)$ can occur only at a value of x for which $f''(x)$ is zero, because the curve is concave up where $f''(x)$ is positive and concave down where $f''(x)$ is negative. Thus we have the following test.

> Look for possible inflection points by setting $f''(x) = 0$ and solving for x.   (2)

Once we have a value of x where the second derivative is zero, say at $x = b$, we must check the concavity of $f(x)$ at nearby points to see if the concavity really changes at $x = b$.

EXAMPLE 3   Find the inflection points of the function $f(x) = x^3 - 3x^2 + 5$ and explain why the graph in Fig. 6 has the correct shape.

Solution   From Example 2 we have $f''(x) = 6x - 6 = 6(x - 1)$. Clearly, $f''(x) = 0$ if and only if $x = 1$. We will want to plot the corresponding point on the graph, so we

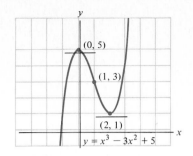

$(0, 5)$

$(1, 3)$

$(2, 1)$

$y = x^3 - 3x^2 + 5$

FIGURE 8

compute

$$f(1) = (1)^3 - 3(1)^2 + 5 = 3.$$

Therefore, the only possible inflection point is $(1, 3)$.

Now look back at Fig. 5, where we indicated the concavity of the graph at the extreme points. Since $f(x)$ is concave down at $(0, 5)$ and concave up at $(2, 1)$, the concavity must reverse somewhere between these points. Hence $(1, 3)$ must be an inflection point. Furthermore, since the concavity of $f(x)$ reverses nowhere else, the concavity at all points to the left of $(1, 3)$ must be the same (i.e., concave down). Similarly, the concavity at all points to the right of $(1, 3)$ must be the same (i.e., concave up). Thus the graph in Fig. 6 has the correct shape. The graph in Fig. 7 has too many "wiggles," caused by frequent changes in concavity; that is, there are too many inflection points. A correct sketch showing the one inflection point at $(1, 3)$ is given in Fig. 8.

EXAMPLE 4   Sketch the graph of $y = -\frac{1}{3}x^3 + 3x^2 - 5x$.

Solution   Let

$$f(x) = -\tfrac{1}{3}x^3 + 3x^2 - 5x.$$

Then

$$f'(x) = -x^2 + 6x - 5$$
$$f''(x) = -2x + 6.$$

We set $f'(x) = 0$ and solve for x:

$$-(x^2 - 6x + 5) = 0$$
$$-(x - 1)(x - 5) = 0$$
$$x = 1 \quad \text{or} \quad x = 5.$$

Substituting these values of x back into $f(x)$, we find that

$$f(1) = -\tfrac{1}{3}(1)^3 + 3(1)^2 - 5(1) = -\tfrac{7}{3}$$
$$f(5) = -\tfrac{1}{3}(5)^3 + 3(5)^2 - 5(5) = \tfrac{25}{3}.$$

The information we have so far is given in Fig. 9(a). The sketch in Fig. 9(b) is obtained by computing

$$f''(1) = -2(1) + 6 = 4$$
$$f''(5) = -2(5) + 6 = -4.$$

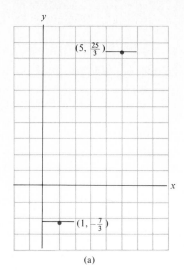
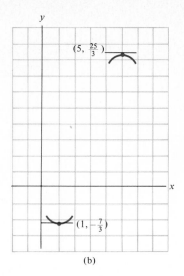

(a)   (b)

FIGURE 9

The curve is concave up at $x = 1$ because $f''(1)$ is positive, and the curve is concave down at $x = 5$ because $f''(5)$ is negative.

Since the concavity reverses somewhere between $x = 0$ and $x = 5$, there must be at least one inflection point. If we set $f''(x) = 0$, we find that

$$-2x + 6 = 0$$

$$x = 3.$$

FIGURE 10

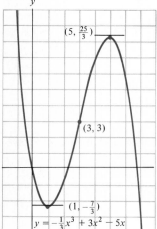

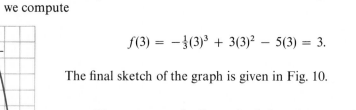

So the inflection point must occur at $x = 3$. In order to plot the inflection point, we compute

$$f(3) = -\tfrac{1}{3}(3)^3 + 3(3)^2 - 5(3) = 3.$$

The final sketch of the graph is given in Fig. 10.

The argument in Example 4 that there must be an inflection point because concavity reverses is valid whenever $f(x)$ is a polynomial. However, it does not always apply to a function whose graph has a break in it. For example, the function $f(x) = 1/x$ is concave down at $x = -1$ and concave up at $x = 1$, but there is no inflection point in between.

A summary of curve-sketching techniques appears at the end of the next section. You may find steps 1, 2, and 3 helpful when working the exercises.

PRACTICE PROBLEMS 3

1. Which of the curves below could possibly be the graph of a function of the form $f(x) = ax^2 + bx + c$, where $a \neq 0$?

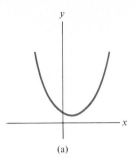

(a)

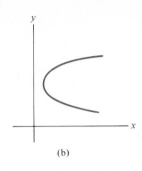

(b)

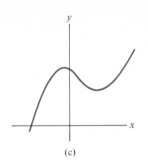

(c)

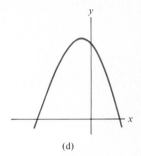

(d)

2. Which of the curves below could possibly be the graph of a function of the form $f(x) = ax^3 + bx^2 + cx + d$, where $a \neq 0$?

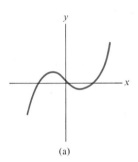

(a)

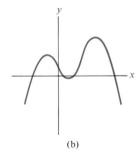

(b)

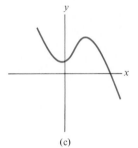

(c)

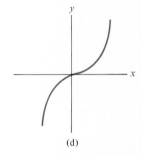

(d)

EXERCISES 3

Each of the graphs of the functions in Exercises 1–8 has one extreme point. Plot this point and check the concavity there. Using only this information, sketch the graph. [As you work the problems, observe that if $f(x) = ax^2 + bx + c$, then $f(x)$ has a minimum point when $a > 0$ and a maximum point when $a < 0$.]

1. $f(x) = 2x^2 - 8$

2. $f(x) = 3x^3 + 6x - 5$

3. $f(x) = \frac{1}{2}x^2 + x - 4$

4. $f(x) = -\frac{1}{2}x^2 + x - 4$

5. $f(x) = 1 + 6x - x^2$

6. $f(x) = 1 + x + x^2$

7. $f(x) = -x^2 - 8x - 10$

8. $f(x) = -3x^2 + 18x - 20$

Each of the graphs of the functions in Exercises 9–16 has one maximum and one minimum point. Plot these two points and check the concavity there. Using only this information, sketch the graph.

9. $f(x) = x^3 + 6x^2 + 9x$

10. $f(x) = \frac{1}{9}x^3 - x^2$

11. $f(x) = x^3 - 12x$

12. $f(x) = -\frac{1}{3}x^3 + 9x - 2$

13. $f(x) = -\frac{1}{9}x^3 + x^2 + 9x$

14. $f(x) = 2x^3 - 15x^2 + 36x - 24$

15. $f(x) = -\frac{1}{3}x^3 + 2x^2 - 12$

16. $f(x) = \frac{1}{3}x^3 + 2x^2 - 5x + \frac{8}{3}$

Sketch the following curves, indicating all extreme points and inflection points.

17. $y = x^3 - 3x + 2$

18. $y = x^3 - 6x^2 + 9x + 3$

19. $y = 1 + 3x^2 - x^3$

20. $y = -x^3 + 12x - 4$

21. $y = \frac{1}{3}x^3 - x^2 - 3x + 5$

22. $y = x^3 + \frac{3}{2}x^2 - 6x + 4$

23. $y = 2x^3 - 3x^2 - 36x + 20$

24. $y = 11 + 9x - 3x^2 - x^3$

25. Let a, b, c be fixed numbers with $a \neq 0$ and let $f(x) = ax^2 + bx + c$. Is it possible for the graph of $f(x)$ to have an inflection point? Explain your answer.

26. Let a, b, c, d be fixed numbers with $a \neq 0$ and let $f(x) = ax^3 + bx^2 + cx + d$. Is it possible for the graph of $f(x)$ to have more than one inflection point? Explain your answer.

The graph of each function below has one extreme point. Find it (giving both x- and y-coordinates) and determine if it is a maximum or a minimum point. Do not include a sketch of the graph of the function.

27. $f(x) = \frac{1}{4}x^2 - 2x + 7$

28. $f(x) = 5 - 12x - 2x^2$

29. $g(x) = 3 + 4x - 2x^2$

30. $g(x) = x^2 + 10x + 10$

31. $f(x) = 5x^2 + x - 3$

32. $f(x) = 30x^2 - 1800x + 29,000$

SOLUTIONS TO PRACTICE PROBLEMS 3

1. Answer: (a) and (d). Curve (b) has the shape of a parabola, but it is not the graph of any function, since vertical lines cross it twice. Curve (c) has two extreme points, but the derivative of $f(x)$ is a linear function which could not be zero for two different values of x.

2. Answer: (a), (c), (d). Curve (b) has three extreme points, but the derivative of $f(x)$ is a quadratic function which could not be zero for three different values of x.

12.4. Curve Sketching (Conclusion)

In Section 3 we discussed the main techniques for curve sketching. Here we add a few finishing touches and examine some slightly more complicated curves.

The more points we plot on a graph, the more accurate the graph becomes. This statement is true even for the simple quadratic and cubic curves in Section 3. Of course, the most important points on a curve are the extreme points and the inflection points. In addition, the x- and y-intercepts often have some intrinsic interest in an applied problem. The y-intercept is $(0, f(0))$. To find the x-intercepts on the graph of $f(x)$, we must find those values of x for which $f(x) = 0$. Since this can be a difficult (or impossible) problem, we shall find x-intercepts only when they are easy to find or when a problem specifically requires us to find them.

When $f(x)$ is a quadratic function, as in Example 1, we can easily compute the x-intercepts (if they exist) either by factoring the expression for $f(x)$ or by using the quadratic formula.

EXAMPLE 1 Sketch the graph of $y = \frac{1}{2}x^2 - 4x + 7$.

Solution Let

$$f(x) = \tfrac{1}{2}x^2 - 4x + 7.$$

Then

$$f'(x) = x - 4$$

$$f''(x) = 1.$$

Since $f'(x) = 0$ only when $x = 4$ and since $f''(4)$ is positive, $f(x)$ must have a minimum point at $x = 4$. The minimum point is $(4, f(4)) = (4, -1)$.

The y-intercept is $(0, f(0)) = (0, 7)$. To find the x-intercepts, we set $f(x) = 0$ and solve for x:

$$\tfrac{1}{2}x^2 - 4x + 7 = 0.$$

The expression for $f(x)$ is not easily factored, so we use the quadratic formula to solve the equation.

$$x = \frac{-(-4) \pm \sqrt{(-4)^2 - 4(\frac{1}{2})(7)}}{2(\frac{1}{2})} = 4 \pm \sqrt{2}.$$

The x-intercepts are $(4 - \sqrt{2}, 0)$ and $(4 + \sqrt{2}, 0)$. To plot these points we use the approximation: $\sqrt{2} \approx 1.4$. (See Fig. 1.)

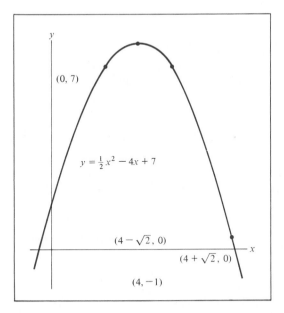

$(0, 7)$

$y = \frac{1}{2}x^2 - 4x + 7$

$(4 - \sqrt{2}, 0)$

$(4 + \sqrt{2}, 0)$

$(4, -1)$

FIGURE 1

EXAMPLE 2 Sketch the graph of $f(x) = \frac{1}{6}x^3 - \frac{3}{2}x^2 + 5x + 1$.

Solution

$$f(x) = \tfrac{1}{6}x^3 - \tfrac{3}{2}x^2 + 5x + 1$$
$$f'(x) = \tfrac{1}{2}x^2 - 3x + 5$$
$$f''(x) = x - 3.$$

Let us set $f'(x) = 0$ and try to solve for x:

$$\tfrac{1}{2}x^2 - 3x + 5 = 0. \tag{1}$$

If we apply the quadratic formula with $a = \tfrac{1}{2}$, $b = -3$, and $c = 5$, we see that $b^2 - 4ac$ is negative, and so there is no solution to (1). In other words, $f'(x)$ is never zero. Thus the graph cannot have extreme points. If we evaluate $f'(x)$ at some x, say $x = 0$, we see that the first derivative is positive, and so $f(x)$ is increasing there. Since the graph of $f(x)$ is a smooth curve with no extreme points and no breaks, $f(x)$ must be increasing for all x. [For if $f(x)$ is increasing at $x = a$ and decreasing at $x = b$, then $f(x)$ has an extreme point between a and b.]

Now let us check the concavity.

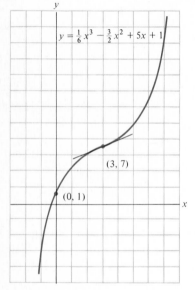

	$f''(x) = x - 3$	*Graph of $f(x)$*
$x < 3$	Negative	Concave down
$x = 3$	Zero	Concavity reverses
$x > 3$	Positive	Concave up

The inflection point is $(3, f(3)) = (3, 7)$. The y-intercept is $(0, f(0)) = (0, 1)$. We omit the x-intercept because it is difficult to solve the cubic equation $\tfrac{1}{6}x^3 - \tfrac{3}{2}x^2 + 5x + 1 = 0$.

The quality of our sketch of the curve will be improved if we first sketch the tangent line at the inflection point. To do this, we need to know the slope of the graph at $(3, 7)$:

$$f'(3) = \tfrac{1}{2}(3)^2 - 3(3) + 5 = \tfrac{1}{2}.$$

We draw a line through $(3, 7)$ with slope $\tfrac{1}{2}$ and then complete the sketch as shown in Fig. 2.

FIGURE 2

The methods used so far will work for most of the functions we shall study. Occasionally, however, $f'(x)$ and $f''(x)$ are both zero at some value of x, say $x = a$, and we cannot tell from the second derivative if the function has a minimum or a maximum at $x = a$. We show how to handle this unusual situation in the next example.

EXAMPLE 3 Sketch the graph of $f(x) = (x - 2)^4 - 1$.

Solution

$$f(x) = (x - 2)^4 - 1$$
$$f'(x) = 4(x - 2)^3$$
$$f''(x) = 12(x - 2)^2.$$

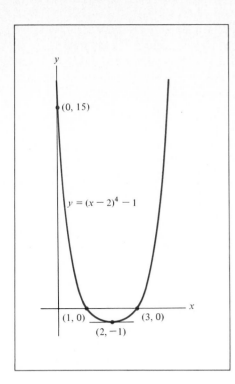

$y = (x - 2)^4 - 1$

(0, 15)

(1, 0)   (3, 0)

(2, −1)

FIGURE 3

Clearly, $f'(x) = 0$ only if $x = 2$. So the curve has a horizontal tangent at $(2, f(2)) = (2, -1)$. Since $f''(2) = 0$, the second derivative rule tells us nothing about the concavity at $x = 2$. Let us look at $f''(x)$ for x near 2. Since $f''(x) = 12(x - 2)^2 > 0$ for $x \neq 2$, the graph is concave up at all points close to $(2, -1)$. In other words, the graph lies above its tangent line at each point near $(2, -1)$. Consequently, the graph cannot drop below its tangent line at $(2, -1)$, and it is concave up at $(2, -1)$.

The y-intercept is $(0, f(0)) = (0, 15)$. To find the x-intercepts, we set $f(x) = 0$ and solve for x:

$$(x - 2)^4 - 1 = 0$$

$$(x - 2)^4 = 1$$

$$x - 2 = 1 \quad \text{or} \quad x - 2 = -1$$

$$x = 3 \quad \text{or} \quad x = 1.$$

(See Fig. 3.)

A Graph with Asymptotes Graphs similar to the one in the next example will arise in several applications later in this chapter.

EXAMPLE 4 Sketch the graph of $f(x) = x + (1/x)$, for $x > 0$.

Solution

$$f(x) = x + \frac{1}{x}$$

$$f'(x) = 1 - \frac{1}{x^2}$$

$$f''(x) = \frac{2}{x^3}.$$

We set $f'(x) = 0$ and solve for x:

$$1 - \frac{1}{x^2} = 0$$

$$1 = \frac{1}{x^2}$$

$$x^2 = 1$$

$$x = 1.$$

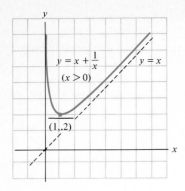

$y = x + \dfrac{1}{x}$
$(x > 0)$

$y = x$

$(1, .2)$

FIGURE 4

(We exclude the case $x = -1$ because we are only considering positive values of x.) The graph has a horizontal tangent at $(1, f(1)) = (1, 2)$. Now, $f''(1) = 2 > 0$, and so the graph is concave up at $x = 1$ and $(1, 2)$ is a minimum point. In fact, $f''(x) = (2/x^3) > 0$ for all positive x, and therefore the graph is concave up at all points.

Before sketching the graph, notice that as x approaches zero (a point at which $f(x)$ is not defined), the term $1/x$ in the formula for $f(x)$ becomes arbitrarily large. Thus $f(x)$ has the y-axis as an asymptote. For large values of x, $f(x) = x + (1/x)$ is only slightly larger than x; that is, the graph of $f(x)$ is slightly above the graph of $y = x$. As x increases, the graph of $f(x)$ has the line $y = x$ as an asymptote. (See Fig. 4.)

Summary of Curve-Sketching Techniques

1. Compute $f'(x)$ and $f''(x)$.

2. Find all maximum and minimum points.
   (a) Set $f'(x) = 0$ and solve for x. Suppose that $f'(a) = 0$.
      (i) If $f''(a) > 0$, the curve is concave up at $x = a$, so $f(x)$ has a minimum point at $x = a$. The minimum point is $(a, f(a))$.
      (ii) If $f''(a) < 0$, the curve is concave down at $x = a$, so $f(x)$ has a maximum point at $x = a$. The maximum point is $(a, f(a))$.
      (iii) If $f''(a) = 0$, examine $f'(x)$ to the left and right of $x = a$ in order to determine if the function changes from increasing to decreasing, or vice versa.
   (b) Make a partial sketch of the graph near each point where $f(x)$ has a horizontal tangent line.

3. Determine the concavity of $f(x)$.
   (a) Set $f''(x) = 0$ and solve for x. Suppose that $f''(b) = 0$. Next, test the concavity for x near to b. If the concavity changes at $x = b$, then $(b, f(b))$ is an inflection point. Otherwise, the concavity at $x = b$ is the same as at other nearby points.

4. Consider other properties of the function and complete the sketch.
   (a) If $f(x)$ is defined at $x = 0$, the y-intercept is $(0, f(0))$.
   (b) Does the partial sketch suggest that there are x-intercepts? If so, they are found by setting $f(x) = 0$ and solving for x. (Solve only in easy cases or when a problem essentially requires you to calculate the x-intercepts.)
   (c) Observe where $f(x)$ is defined. Sometimes the function is given only for restricted values of x. Sometimes the formula for $f(x)$ is meaningless for certain values of x.
   (d) Look for possible asymptotes.
      (i) Examine the formula for $f(x)$. If some terms become insignificant as x gets large and if the rest of the formula gives the equation of a straight line, then that straight line is an asymptote.
      (ii) Suppose that there is some point a such that $f(x)$ is defined for x near a but not at a (e.g., $1/x$ at $x = 0$). If $f(x)$ gets arbitrarily large (in the positive or negative sense) as x approaches a, then the vertical line $x = a$ is an asymptote for the graph.
   (e) Complete the sketch.

PRACTICE PROBLEMS 4

1. Determine whether each of the following functions has an asymptote as x gets large. If so, give the equation of the straight line which is the asymptote.

   (a) $f(x) = (3/x) - 2x + 1$

   (b) $f(x) = \sqrt{x} + x$

   (c) $f(x) = 7 + 1/(2x + 6)$

2. Sketch the graph of $f(x) = 1/(2 - x) + 5$, for $x > 2$.

EXERCISES 4

Find the x-intercepts of the following curves.

1. $y = x^2 - 3x + 1$

2. $y = x^2 + 5x + 5$

3. $y = 2x^2 + 5x + 2$

4. $y = 4 - 2x - x^2$

5. $y = 4x - 4x^2 - 1$

6. $y = 3x^2 + 7x + 2$

7. Show that the function $f(x) = \frac{1}{3}x^3 - 2x^2 + 5x$ has no extreme points.

8. Show that the function $f(x) = 5 - 11x + 6x^2 - \frac{4}{3}x^3$ is always decreasing.

Sketch the graphs of the following functions.

9. $f(x) = x^3 - 2x^2 + 4x - 6$

10. $f(x) = -x^3$

11. $f(x) = x^3 + 3x + 1$

12. $f(x) = 4 - x - x^3$

13. $f(x) = 5 - 13x + 6x^2 - x^3$

14. $f(x) = 2x^3 + x - 2$

15. $f(x) = \frac{4}{3}x^3 - 2x^2 + x$

16. $f(x) = 1 - 4x - 3x^2 - x^3$

17. $f(x) = 1 - 3x + 3x^2 - x^3$

18. $f(x) = x^3 - 6x^2 + 12x - 5$

19. $f(x) = x^4 - 6x^2$

20. $f(x) = 1 + 6x^2 - 3x^4$

21. $f(x) = (x - 3)^4$

22. $f(x) = (x + 2)^4 - 1$

Sketch the graphs of the following functions.

23. $y = \dfrac{1}{x} + 3$

24. $y = \dfrac{1}{x - 1}$

25. $y = \dfrac{1}{x} + \frac{1}{4}x$

26. $y = \dfrac{1}{x} + 9x$

27. $y = \dfrac{9}{x} + x + 1 \quad (x > 0)$

28. $y = \dfrac{12}{x} + 3x + 1 \quad (x > 0)$

29. $y = \dfrac{360}{x} + 10x \quad (x > 0)$

30. $y = \dfrac{40{,}000}{x} + 2500x + 320{,}000 \quad (x > 0)$

31. $y = \dfrac{1}{x-2} + \dfrac{x}{4} \quad (x > 2)$

32. $y = \dfrac{4}{2x-5} + 2x \quad (x > \tfrac{5}{2})$

33. $y = 6\sqrt{x} - x \quad (x > 0)$

34. $y = \dfrac{1}{x} + \dfrac{1}{4-x} \quad (0 < x < 4)$

SOLUTIONS TO PRACTICE PROBLEMS 4

1. Functions with asymptotes as x gets large have the form $f(x) = g(x) + mx + b$, where $g(x)$ approaches zero as x gets large. The function $g(x)$ often looks like c/x or $c/(ax + d)$. The asymptote will be the straight line $y = mx + b$.

   (a) Here $g(x)$ is $3/x$ and the asymptote is $y = -2x + 1$.

   (b) This function has no asymptote as x gets large. Of course, it can be written as $g(x) + mx + b$, where $m = 1$ and $b = 0$. However, $g(x) = \sqrt{x}$ does not approach zero as x gets large.

   (c) Here $g(x)$ is $1/(2x + 6)$ and the asymptote is the line $y = 7$.

2.
$$f(x) = \frac{1}{2-x} + 5 = (2-x)^{-1} + 5, \quad x > 2$$

$$f'(x) = (2-x)^{-2} = \frac{1}{(2-x)^2}$$

$$f''(x) = 2(2-x)^{-3} = \frac{2}{(2-x)^3}$$

The first derivative, $1/(2-x)^2$ is never zero, so there are no extreme points. The second derivative, $2/(2-x)^3$, is also never zero, so there are no inflection points.

   Just to have a point of reference, let us find some point on the graph. For instance, let $x = 3$; then $f(3) = 4$. So the point $(3, 4)$ is on the graph. [See Fig. 5(a).] The first derivative is always positive. Thus the graph is always increasing. The second derivative

FIGURE 5

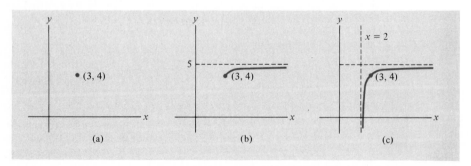

(a)   (b)   (c)

is always negative, since $x > 2$ implies that $2 - x$, and hence $(2 - x)^3$, is negative. There-fore, the graph is concave down. Now for the asymptotes. As x gets large, the term $1/(2 - x)$ approaches zero, so the graph of $f(x)$ has the straight line $y = 5$ as an asymptote [See Fig. 5(b).]

The function also has a vertical asymptote at $x = 2$. For values of x just to the right of 2, $1/(2 - x)$ is negative and very large in magnitude. [See Fig. 5(c).]

12.5. Optimization Problems

One of the most important applications of the derivative concept is to "optimization" problems, in which some quantity must be maximized or minimized. Examples of such problems abound in many areas of life. An airline must decide how many daily flights to schedule between two cities in order to maximize its profits. A doctor wants to find the minimum amount of a drug that will produce a desired response in one of her patients. A manufacturer needs to determine how often to replace certain equipment in order to minimize maintenance and replace-ment costs.

Our purpose in this section is to illustrate how calculus can be used to solve optimization problems. In each example we will find or construct a function that provides a "mathematical model" for the problem. Then, by sketching the graph of this function, we will be able to determine the answer to the original optimization problem by locating the maximum value or the minimum value of the function.

The first two examples are quite simple because the functions to be studied are explicitly given.

EXAMPLE 1 Find the minimum value of the function $f(x) = 2x^3 - 15x^2 + 24x + 19$ for $x \geq 0$.

Solution Using the curve-sketching techniques from Section 3, we obtain the graph in Fig. 1. The lowest point on the graph is the minimum point $(4, 3)$. The minimum *value* of the function $f(x)$ is the y-coordinate of this point—namely, 3.

EXAMPLE 2 Suppose that a ball is thrown straight up into the air and its height after t seconds is $4 + 48t - 16t^2$ feet. Determine how long it will take for the ball to reach its maximum height and determine the maximum height.

Solution Consider the function $f(t) = 4 + 48t - 16t^2$. For each value of t, $f(t)$ is the height of the ball at time t. We want to find the value of t for which $f(t)$ is the greatest. Using the techniques of Section 3, we sketch the graph of $f(t)$. (See Fig. 2.) Note that we may neglect the portions of the graph corresponding to points for which either $t < 0$ or $f(t) < 0$. [A negative value of $f(t)$ would correspond to the ball being underneath the ground.] We see that $f(t)$ is greatest when $t = \frac{3}{2}$. At this

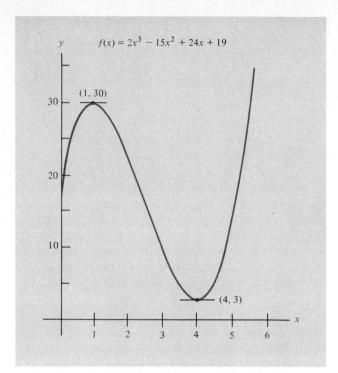

$f(x) = 2x^3 - 15x^2 + 24x + 19$

(1, 30)

(4, 3)

FIGURE 1

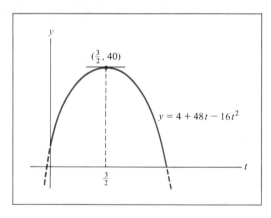

$(\frac{3}{2}, 40)$

$y = 4 + 48t - 16t^2$

$\frac{3}{2}$

FIGURE 2

value of t, the ball attains a height of 40 feet. [Note that the curve in Fig. 2 is the graph of $f(t)$, *not* a picture of the physical path of the ball.]

Answer   The ball reaches its maximum height of 40 feet in 1.5 seconds.

EXAMPLE 3   A man wants to plant a rectangular garden along one side of his house, with a picket fence on the other three sides of the garden. Find the dimensions of the largest garden that can be enclosed using 40 feet of fencing.

Solution   The first step is to make a simple diagram and assign letters to the quantities that may vary. Let us denote the dimensions of the rectangular garden by w and x (Fig. 3). The fencing on three sides must total 40 running feet; that is,

$$2x + w = 40. \tag{1}$$

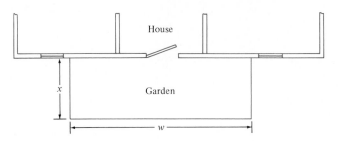

House

Garden

x

w

FIGURE 3

The phrase "largest garden" above indicates that we must maximize the area, A, of the garden. In terms of the variables w and x,

$$A = wx. \tag{2}$$

We now solve equation (1) for w in terms of x:

$$w = 40 - 2x. \tag{3}$$

Substituting this expression for w into equation (2), we have

$$A = (40 - 2x)x = 40x - 2x^2. \tag{4}$$

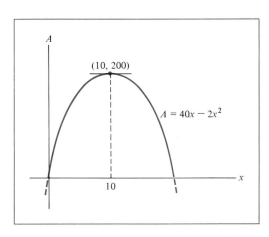

A

$(10, 200)$

$A = 40x - 2x^2$

10

x

We now have a formula for the area A that depends on just one variable, and so we may graph A as a function of x.

Using curve-sketching techniques, we obtain the graph in Fig. 4. We see from the graph that the area is maximized when $x = 10$. (The maximum area is 200 square feet, but this fact is not needed for the problem.) From equation (3) we find that when $x = 10$,

$$w = 40 - 2(10) = 20.$$

FIGURE 4

Answer $w = 20$ feet, $x = 10$ feet.

Equation (1) in Example 3 is called a *constraint equation* because it places a limit or constraint on the way x and w may vary. Equation (2) is called the *objective equation.* It expresses the quantity to be maximized (the area of the garden) in terms of the variables w and x. The key step in the solution was to use the constraint equation to simplify the objective equation—that is, to write down the objective equation in a way that involves only one variable, as in (4).

EXAMPLE 4 The manager of a department store wants to build a 600-square-foot rectangular enclosure on the store's parking lot in order to display some equipment. Three sides of the enclosure will be built of redwood fencing, at a cost of $7 per running foot. The fourth side will be built of cement blocks, at a cost of $14 per running foot. Find the dimensions of the enclosure that will minimize the total cost of the building materials.

Solution Let x be the length of the side built out of cement blocks and let y be the length of an adjacent side, as shown in Fig. 5. Since the area of the enclosure must be 600 square feet, the constraint equation is

$$xy = 600. \tag{5}$$

The phrase "minimizes the total cost . . ." tells us that the objective equation should be a formula giving the total cost of the building materials.

$$[\text{cost of redwood}] = [\text{length of redwood fencing}] \cdot [\text{cost per foot}]$$
$$= (x + 2y) \cdot 7 = 7x + 14y.$$

$$[\text{cost of cement blocks}] = [\text{length of cement wall}] \cdot [\text{cost per foot}]$$
$$= x \cdot 14.$$

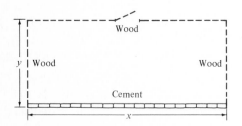

FIGURE 5

If C denotes the total cost of the materials, then

$$C = (7x + 14y) + 14x$$
$$C = 21x + 14y \qquad \text{(objective equation)}. \tag{6}$$

We simplify this objective equation by solving (5) for one of the variables, say y, and substituting into (6):

$$C = 21x + 14\left(\frac{600}{x}\right) = 21x + \frac{8400}{x}.$$

We may now graph C as a function of the one variable x, where $x > 0$ (Fig. 6). (A similar curve was sketched in Example 4 of Section 4.) The minimum cost of $840 occurs where $x = 20$. From equation (5) we find that the corresponding value of y is $\frac{600}{20} = 30$.

Answer $x = 20$ feet, $y = 30$ feet.

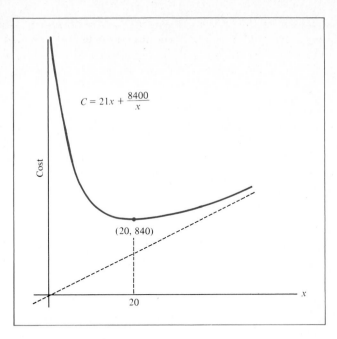

$$C = 21x + \frac{8400}{x}$$

Cost

(20, 840)

20

x

FIGURE 6

EXAMPLE 5   U.S. parcel post regulations state that packages must have length plus girth of no more than 84 inches. Find the dimensions of the cylindrical package of greatest volume that is mailable by parcel post.

Solution   Let l be the length of the package and let r be the radius of the circular end. (See Fig. 7.) The girth equals the circumference of the end—that is, $2\pi r$. Since we want the package to be as large as possible, we must use the entire 84 inches allowable:

$$\text{length} + \text{girth} = 84$$

$$l + 2\pi r = 84 \qquad \text{(constraint equation)}. \tag{7}$$

FIGURE 7

The phrase "greatest volume . . ." tells us that the objective equation should express the volume of the package in terms of the dimensions l and r. Let V denote the volume. Then

$$V = [\text{area of base}] \cdot [\text{length}]$$

$$V = \pi r^2 l \qquad \text{(objective equation)}. \tag{8}$$

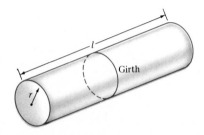

Girth

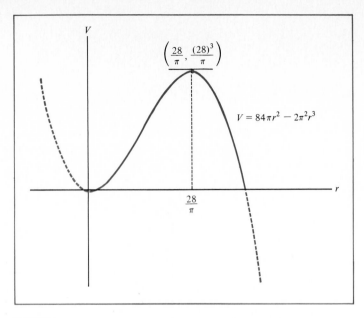

FIGURE 8

We now solve equation (7) for one of the variables, say $l = 84 - 2\pi r$. Substituting this expression into (8), we obtain

$$V = \pi r^2(84 - 2\pi r) = 84\pi r^2 - 2\pi^2 r^3. \tag{9}$$

Let $f(r) = 84\pi r^2 - 2\pi^2 r^3$. Then, for each value of r, $f(r)$ is the volume of the parcel with end radius r that meets the postal regulations. We want to find that value of r for which $f(r)$ is as large as possible.

   Using curve-sketching techniques, we obtain the graph in Fig. 8. [The graph is shown with a dashed curve where r is negative and where $f(r)$ is negative, for these portions of the graph have no physical significance.] We see that the volume is greatest when $r = 28/\pi$. From (7) we find that the corresponding value of l is

$$l = 84 - 2\pi r = 84 - 2\pi\left(\frac{28}{\pi}\right)$$

$$= 84 - 56 = 28.$$

The girth when $r = 28/\pi$ is

$$2\pi r = 2\pi\left(\frac{28}{\pi}\right) = 56.$$

Answer $l = 28$ inches, $r = 28/\pi$ inches, girth $= 56$ inches.

Note Optimization problems often involve geometric formulas. The most common formulas are:

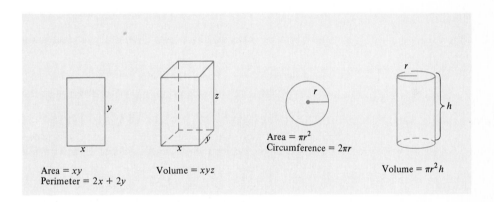

Area = xy
Perimeter = $2x + 2y$

Volume = xyz

Area = πr^2
Circumference = $2\pi r$

Volume = $\pi r^2 h$

PRACTICE PROBLEMS 5

1. A canvas wind shelter for the beach has a back, two square sides, and a top (Fig. 9). Suppose that 96 square feet of canvas are to be used. Find the dimensions of the shelter for which the space inside the shelter (i.e., the volume) will be maximized.

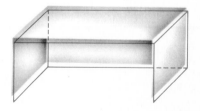

FIGURE 9

2. In Practice Problem 1, what are the objective equation and the constraint equation?

EXERCISES 5

1. Find the maximum value of the function $f(x) = 12x - x^2$.

2. Find the minimum value of the function $g(x) = \frac{1}{2}x^2 - 8x + 35$.

3. Find the minimum value of the function $f(t) = t^3 - 6t^2 + 40$, $t \geq 0$, and give the value of t where this minimum occurs.

4. Find the maximum value of the function $f(s) = 9s^2 - s^3$, $s \geq 0$, and give the value of s where this maximum occurs.

5. For what x does the function $g(x) = 10 + 40x - x^2$ have its maximum value?

6. For what t does the function $f(t) = t^2 - 24t$ have its minimum value?

7. A ball thrown straight up into the air has a height of $-16t^2 + 96t + 6$ feet after t seconds. How high will the ball go?

8. The height of a bullet t seconds after it is fired from a rifle is $-16t^2 + 50t$ meters. How soon will the bullet reach its maximum height?

9. A manufacturer estimates that his profit from the daily production and sale of x power saws will be approximately $-.1x^2 + 50x - 1000$ dollars. How many saws should be produced each day in order to maximize the profit?

10. A manufacturer of costume jewelry estimates that the profit from producing x hundred necklaces per week is $300x - x^3 - 500$ dollars. What is the largest possible weekly profit?

11. Find two positive numbers whose sum is 100 and whose product is as large as possible. [*Note:* Denote the two numbers by x and h; then $x + h = 100$ and the product is $x \cdot h$.]

12. Find two positive numbers whose product is 100 and whose sum is as small as possible.

13. Figure 10(a) shows an open rectangular box with a square base. Find x and h such that the volume is 32 cubic inches and the amount of material needed to construct the box is minimal. [*Note:* The amount of material needed is $x^2 + 4xh$ square inches.]

14. A large soup can is to be designed so that the can will hold 16π cubic inches (about 28 fluid ounces) of soup. [See Fig. 10(b).] Find the values of x and h such that the amount

FIGURE 10

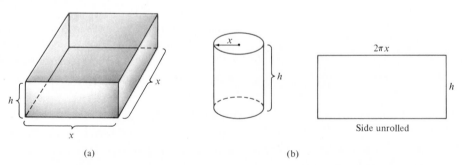

(a)   (b)

of metal needed is as small as possible. [*Note:* The side requires $2\pi xh$ square inches of metal and each end requires πx^2 square inches.]

15. Figure 11(a) shows a "Norman window," which consists of a rectangle capped by a semicircular region. Find the value of x such that the perimeter of the window will be 14 feet and the area of the window will be as large as possible. [*Note:* The perimeter of the window is $2x + 2h + \pi x$.]

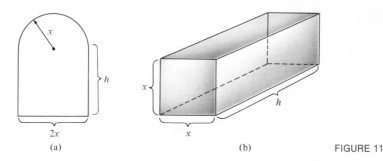

2x

(a)

x

(b)

FIGURE 11

16. Postal requirements specify that parcels must have length plus girth of at most 84 inches. Find the dimensions of the square-ended rectangular package [Fig. 11(b)] of greatest volume that is mailable. [*Note:* The length plus girth is $h + 4x$.]

17. A rectangular garden of area 75 square feet is to be surrounded on three sides by a brick wall costing $10 per foot and on one side by a fence costing $5 per foot. Find the dimensions of the garden such that the cost of the materials is minimized. [*Note:* Letting x be the length of the fence and h be the length of the other dimension, the cost of the materials is $5x + 10(2h + x)$.]

18. A closed rectangular box with square base and a volume of 12 cubic feet is to be constructed using two different types of materials. The top is made of a metal costing $2 per square foot and the remainder of wood costing $1 per square foot. Find the dimensions of the box for which the cost of materials is minimized.

19. Find the dimensions of the closed rectangular box with square base and volume 8000 cubic centimeters that can be constructed using the least amount of material.

20. Find the dimensions of the rectangular garden of greatest area that can be fenced off (all four sides) with 300 meters of fencing.

21. A storage shed is to be built in the shape of a box with a square base. It is to have a volume of 150 cubic feet. The concrete for the base costs $4 per square foot, the material for the roof costs $2 per square foot, and the material for the sides costs $2.50 per square foot. Find the dimensions of the most economical shed.

22. A canvas wind shelter for the beach has a back, two square sides, and a top. Find the dimensions for which the volume will be 250 cubic feet and which requires the least possible amount of canvas.

23. In Example 3 one can solve the constraint equation (1) for x instead of w to get $x = 20 - \frac{1}{2}w$. Substituting this for x in (2), one has $A = xw = (20 - \frac{1}{2}w)w$. Sketch the

graph of the equation $A = 20w - \frac{1}{2}w^2$ and show that the maximum area occurs when $w = 20$ and $x = 10$.

24. In Example 4 one can solve the constraint equation (5) for x instead of y to get $x = 600/y$. Substituting this for x in (6), one has $C = 21x + 14y = (12{,}600/y) + 14y$. Sketch the graph of C as a function of y and show that the minimum cost occurs when $y = 30$ feet.

SOLUTIONS TO PRACTICE PROBLEMS 5

1. Since the sides of the wind shelter are square, we may let x represent the length of each side of the square. The remaining dimension of the wind shelter can be denoted by the letter h. (See Fig. 12.) The volume of the shelter is x^2h, and this is to be maximized. Since we have learned to maximize only functions of a single variable, we must express h in terms of x. We must use the information that 96 feet of canvas are used—that is, $2x^2 + 2xh = 96$. [*Note:* The roof and the back each have area xh, and each end has area x^2.] We now solve this equation for h.

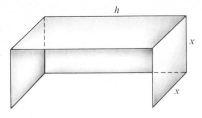

FIGURE 12

$$2x^2 + 2xh = 96$$

$$2xh = 96 - 2x^2$$

$$h = \frac{96}{2x} - \frac{2x^2}{2x} = \frac{48}{x} - x.$$

The volume, V, is

$$x^2h = x^2\left(\frac{48}{x} - x\right) = 48x - x^3.$$

By sketching the graph of $V = 48x - x^3$, we see that V has a maximum value when $x = 4$. Then, $h = \frac{48}{4} - 4 = 12 - 4 = 8$. So each end of the shelter should be a 4-foot-by-4-foot square and the top should be 8 feet long.

2. The objective equation is: $V = x^2h$, since it expresses the volume (the quantity to be maximized) in terms of the variables. The constraint equation is $2x^2 + 2xh = 96$, for it relates the variables to each other; that is, it can be used to express one of the variables in terms of the other.

12.6. Further Optimization Problems

In this section we apply the optimization techniques developed in the preceding section to some practical situations.

EXAMPLE 1   Suppose that, on a certain route, an airline carries 8000 passengers per month, each paying $50. The airline wants to increase the fare. However, the market research department estimates that for each $1 increase in fare, the airline will lose 100 passengers. Determine the price which maximizes the airline's revenue.

Solution   Since the problem calls for setting an optimum price, let x be the price per ticket. The other variable is the number of passengers, which we can denote by n. The goal is to maximize revenue.

$$[\text{revenue}] = [\text{number of passengers}][\text{price per ticket}]$$
$$= n \cdot x.$$

If R denotes revenue, then the objective equation is

$$R = nx.$$

Now the constraint equation is given by

$$[\text{number of passengers}] = [\text{original number}] - [\text{loss due to fare increase}]$$
$$n \qquad = \qquad 8000 \qquad\qquad (x - 50) \cdot 100$$
$$= 13{,}000 - 100x.$$

Therefore,

$$R = nx = (13{,}000 - 100x)x = 13{,}000x - 100x^2.$$

Figure 1 shows that revenue is maximized when the price per ticket is $65.

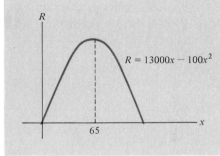

$R = 13000x - 100x^2$

FIGURE 1

EXAMPLE 2   A supermarket manager wants to establish an optimal inventory policy for frozen orange juice. It is estimated that a total of 1200 cases will be sold at a steady rate during the next quarter-year. Since the orange juice requires costly freezer space for storage, she does not want to order all 1200 cases at once. On the other hand, since each delivery incurs additional handling costs, she cannot place too many separate orders. She decides to place several orders of the same size, say each consisting of x cases, equally spaced throughout the quarter-year. Use the data below to determine the optimal reorder quantity x that minimizes total purchasing and inventory costs.

1. The handling cost for each delivery is $75.
2. Inventory cost per quarter is $4x$ dollars. [Here is how such a model arises. Suppose that it costs $8 to store one case for one quarter-year. Since the stock begins at each delivery with x cases and is depleted steadily, there will be an average of $x/2$ cases in stock at any time. Therefore, inventory cost for the quarter will be $(x/2) \cdot 8$ or $4x$ dollars.]

Solution Let r be the number of orders placed during the quarter. Since each order calls for x cases, the constraint equation is

$$r \cdot x = 1200.$$

Now

$$[\text{total cost}] = [\text{handling cost}] + [\text{inventory cost}]$$
$$= 75r + 4x.$$

If C denotes the total cost, then the objective equation is

$$C = 75r + 4x.$$

The constraint equation says that $r = 1200/x$. Substitution into the objective equation yields

$$C = \frac{90,000}{x} + 4x.$$

Figure 2 is the graph of C as a function of x, for $x > 0$. The total cost is at a minimum when $x = 150$. Therefore, the optimum inventory policy is to order 150 cases at a time and to place $1200/150 = 8$ orders during the quarter-year.

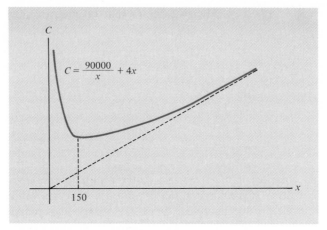

FIGURE 2

EXAMPLE 3 What should the inventory policy of Example 2 be if sales of frozen orange juice increase fourfold (i.e., 4800 cases are sold each quarter), but all other conditions are the same?

Solution The only change in our previous solution is in the constraint equation, which now becomes

$$r \cdot x = 4800.$$

The objective equation is, as before,

$$C = 75r + 4x.$$

Since $r = 4800/x$,

$$C = 75 \cdot \frac{4800}{x} + 4x$$

$$= \frac{360{,}000}{x} + 4x.$$

Now

$$C' = -\frac{360{,}000}{x^2} + 4.$$

Setting $C' = 0$ yields

$$\frac{360{,}000}{x^2} = 4$$

$$90{,}000 = x^2$$

$$x = 300.$$

Therefore, the optimal reorder quantity is 300 cases.

Notice that although the sales increased by a factor of 4, the optimal reorder quantity only increased by a factor of 2 ($=\sqrt{4}$). In general, a store's inventory of an item should be proportional to the square root of the expected sales. (See Exercise 16 for a derivation of this result.) Many stores tend to keep their average inventories at a fixed percentage of sales. For example, each order may contain enough goods to last for 4 or 5 weeks. This policy is likely to create excessive inventories of high-volume items and uncomfortably low inventories of slower-moving items.

EXAMPLE 4   When a person coughs, the trachea (windpipe) contracts. (See Fig. 3.) Let

r_0 = normal radius of the trachea,

r = radius during a cough,

P = increase in air pressure in the trachea during cough,

v = velocity of air through trachea during cough.

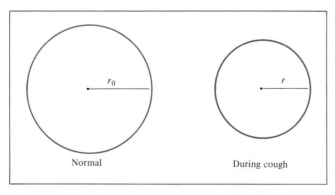

Normal   During cough

FIGURE 3

Use the principles of fluid flow given below to determine how much the trachea should contract in order to create the greatest air velocity—that is, the most effective condition for clearing the lungs and the trachea.

1. $r_0 - r = aP$, for some positive constant a. (Experiment has shown that, during coughing, the decrease in the radius of the trachea is nearly proportional to the increase in the air pressure.)
2. $v = b \cdot P \cdot \pi r^2$, for some positive constant b. (The theory of fluid flow requires that the velocity of the air forced through the trachea is proportional to the product of the increase in the air pressure and the area of a cross section of the trachea.)

Solution   In this problem the constraint equation (1) and the objective equation (2) are given directly. Solving equation (1) for P and substituting this result into equation (2), we have

$$v = b\left(\frac{r_0 - r}{a}\right)\pi r^2 = k(r_0 - r)r^2,$$

where $k = b\pi/a$. To find the radius at which the velocity v is a maximum, we first compute the derivatives:

$$v = k(r_0 r^2 - r^3)$$

$$\frac{dv}{dr} = k(2r_0 r - 3r^2) = kr(2r_0 - 3r)$$

$$\frac{d^2v}{dr^2} = kr(2r_0 - 6r).$$

We see that $\dfrac{dv}{dr} = 0$ when $r = 0$ or when $2r_0 - 3r = 0$; that is, when $r = \frac{2}{3}r_0$. It is easy to see that $\dfrac{d^2v}{dr^2}$ is positive at $r = 0$ and is negative at $r = \frac{2}{3}r_0$. The graph of v as a function of r is drawn in Fig. 4. The air velocity is maximized at $r = \frac{2}{3}r_0$.

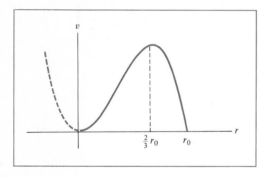

FIGURE 4

When solving optimization problems, we have always made a rough sketch of the graph of the function being optimized before drawing any conclusions. This procedure is necessary, as the next example shows.

EXAMPLE 5   A rancher has 204 meters of fencing from which to build two corrals; one square and the other rectangular with length that is twice the width. Find the dimensions which result in the greatest combined area.

Solution   Let x be the width of the rectangular corral and h the length of each side of the square corral. (See Fig. 5.) Then

$$204 = [\text{perimeter of square}] + [\text{perimeter of rectangle}] = 4h + 6x.$$

FIGURE 5

Let A be the combined area. Then

$$A = [\text{area of square}] + [\text{area of rectangle}]$$
$$= h^2 + 2x^2.$$

Since the perimeter of the rectangle cannot exceed 204, we must have $0 \le 6x \le 204$ or $0 \le x \le 34$. Solving the constraint equation for h and substituting into the objective equation leads to the function graphed in Fig. 6. The graph reveals that the area is minimized when $x = 18$. However, the problem asks for the *maximum* possible area. From Fig. 6 we see that this occurs at the endpoint where $x = 0$. Therefore, the rancher should build only the square corral, with $h = 204/4 = 51$ meters. In this example, the objective function has an endpoint extremum. Namely, the maximum value occurs at the endpoint $x = 0$.

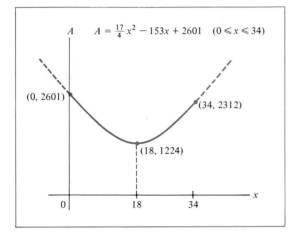

FIGURE 6

PRACTICE PROBLEMS 6

1. An apple orchard produces a profit of \$40 a tree when planted with 1000 trees. Because of overcrowding, the profit per tree (for each tree in the orchard) is reduced by 2 cents for each additional tree planted. How many trees should be planted in order to maximize the total profit from the orchard?

2. In the inventory problem of Example 2, suppose that the sales of frozen orange juice increase ninefold; that is, 10,800 cases are sold each quarter. What is now the optimal reorder quantity?

EXERCISES 6

1. Find the dimensions of the rectangular garden of area 100 square meters for which the amount of fencing needed to surround the garden is as small as possible.

2. A certain airline requires that rectangular packages carried on an airplane by passengers be such that the sum of the three dimensions is at most 120 centimeters. Find the dimensions of the square-ended rectangular package of greatest volume that meets this requirement.

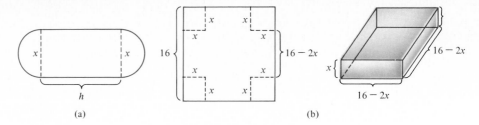

(a)   (b)

FIGURE 7

3. An athletic field [Fig. 7(a)] consists of a rectangular region with a semicircular region at each end. The perimeter will be used for a 440-yard track. Find the value of x for which the area of the rectangular region is as large as possible.

4. An open rectangular box is to be constructed by cutting square corners out of a 16 × 16-inch piece of cardboard and folding up the flaps. [See Fig. 7(b).] Find the value for x for which the volume of the box will be as large as possible.

5. A long rectangular sheet of metal 30 inches wide is to be made into a gutter by turning up strips vertically along the two sides. How many inches should be turned up on each side so as to maximize the amount of water that the gutter can carry?

6. Design an open rectangular box with square ends, having volume 36 cubic inches, that minimizes the amount of material required for construction.

7. A supermarket is to be designed in the form of a rectangle whose area is 12,000 square feet. The front of the building will be mostly glass and will cost $70 per running foot for materials. The other three sides will be constructed of brick and cement block, at a cost of $50 per running foot. Ignore all other costs (labor, cost of foundation and roof, etc.) and find the dimensions of the building that will minimize the cost of the materials for the four sides of the building.

8. A manufacturer can produce x units of a certain commodity at a cost of $C(x) = .01x^3 - .04x^2 + 2x + 100$ dollars and receives a revenue of $R(x) = 50x - .04x^2$ dollars. How many units should be produced in order to maximize the profit? [*Note:* The profit from producing x units is $P(x) = R(x) - C(x)$.]

9. A rectangular corral of 54 square meters is to be fenced off and then divided by a fence into two sections. (See Fig. 8.) Find the dimensions of the corral so that the amount of fencing required is minimized.

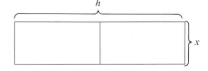

FIGURE 8

10. Referring to Exercise 9, suppose that the cost of the fencing for the boundary is $5 per meter and the dividing fence costs $2 per meter. Find the dimensions of the corral that minimize the cost of the fencing.

11. A swimming club offers memberships at the rate of $200, provided that a minimum of 100 people join. For each member in excess of 100, the membership fee will be reduced by $1 per person (for every member). At most 160 memberships will be sold. How many memberships should the club try to sell in order to maximize its revenue?

12. In the planning of a sidewalk cafe, it is estimated that if there are 12 tables, the daily profit will be $10 per table. Because of overcrowding, for each additional table the profit per table (for every table in the cafe) will be reduced by $.50. How many tables should be provided in order to maximize the profit from the cafe?

13. A certain toll road averages 36,000 cars per day when charging $1 per car. A survey concludes that increasing the toll will result in 300 fewer cars for each cent of increase. What toll should be charged in order to maximize the revenue?

14. A small orchard yields 25 bushels of fruit per tree when planted with 40 trees. Because of overcrowding, the yield per tree (for each tree in the orchard) is reduced by $\frac{1}{2}$ bushel for each additional tree that is planted. How many trees should be planted in order to maximize the total yield of the orchard?

15. A publishing company sells 400,000 copies of a certain book each year. Ordering the entire amount printed at the beginning of the year ties up valuable storage space. However, running off the copies in several partial runs throughout the year results in added costs for setting up each printing run. Using the data below, find the production run size, x, that minimizes the total setting up and storage costs.

    (i) Setting up each production run costs $1000.

    (ii) The storage cost is $x/4$, where x is the size of each production run. (There will be an average of $x/2$ books in storage during the year and storage costs are 50 cents per book per year.)

16. A store manager wants to establish an optimal inventory policy for an item. Sales are expected to be at a steady rate and should total Q items sold during the year. Each time an order is placed, a handling cost of h dollars is incurred. If the size of each order is x, then storage costs for the year will be $(x/2) \cdot s$ dollars. (That is, there will be an average of $x/2$ items in storage during the year, and the cost of storage is s dollars per item per year.) Show that total costs for handling and storage are minimized when each order calls for $\sqrt{2hQ/s}$ items.

17. Foggy Optics, Inc., makes laboratory microscopes and similar items. Setting up each production run costs $2500. The expense of holding one microscope in inventory for 1 year is $50. Suppose that the company expects to sell 1600 microscopes at a fairly uniform rate throughout the year. Determine the production run size that will minimize the company's expenses.

18. The Great American Tire Co. expects to sell 600,000 tires of a particular size and grade during the next year. Sales tend to be roughly the same from month to month. Setting up each production run costs the company an additional $15,000. The inventory expense for one tire is $5 per year. What size production runs should be scheduled for the coming year in order to minimize the overall cost of producing these tires?

19. Let $f(t)$ be the amount of oxygen in a lake t days after sewage is dumped into the lake, and suppose that $f(t)$ is given approximately by

$$f(t) = 500\left[1 - \frac{10}{t + 10} + \frac{100}{(t + 10)^2}\right].$$

At what time is the oxygen content increasing the fastest?

20. The daily output of a coal mine after t hours of operation is approximately $40t + t^2 - \frac{1}{15}t^3$ tons, $0 \le t \le 12$. Find the maximum rate of output (in tons of coal per hour).

21. Consider a parabolic arch whose shape may be represented by the graph of $y = 9 - x^2$, where the base of the arch lies on the x-axis from $x = -3$ to $x = 3$. Find the dimensions of the rectangular window of maximum area that can be constructed inside the arch.

22. Advertising for a certain product is terminated, and t weeks later the weekly sales are $f(t)$ cases, where $f(t) = 1000(t + 8)^{-1} - 4000(t + 8)^{-2}$. At what time is the weekly sales amount falling the fastest?

SOLUTIONS TO PRACTICE PROBLEMS 6

1. Since the question asks for the optimum "number of trees," let x be the number of trees to be planted. The other quantity that varies is "profit per tree." So let p be the profit per tree. The objective is to maximize total profit, call it T. Then

$$[\text{total profit}] = [\text{profit per tree}] \cdot [\text{number of trees}]$$

$$T = p \cdot x.$$

Now,

$$[\text{profit per tree}] = [\text{original profit per tree}] - [\text{loss per tree due to increase}]$$

$$p = 40 - (x - 1000)(.02)$$

$$p = 60 - .02x.$$

Therefore,

$$T = p \cdot x = (60 - .02x)x = 60x - .02x^2.$$

By computing first and second derivatives and sketching the graph, we easily find that total profit is maximum when $x = 1500$. Therefore, 1500 trees should be planted.

2. This problem can be solved in the same manner that Example 3 was solved. However, the comment made at the end of Example 3 indicates that the optimal reorder quantity should increase by a factor of 3, since $3 = \sqrt{9}$. Therefore, the optimal reorder quantity is $3 \cdot 150 = 450$ cases.

12.7. Applications of Calculus to Business and Economics

In recent years economic decision making has become more and more mathematically oriented. Faced with huge masses of statistical data, depending on hundreds or even thousands of different variables, business analysts and economists have increasingly turned to mathematical methods to help them describe what is happening, predict the effects of various policy alternatives, and choose reasonable courses of action from the myriad possibilities. Among the mathematical methods employed is calculus. In this section we illustrate just a few of the many applications of calculus to business and economics. All our applications will center around what economists call *the theory of the firm*. In other words, we study the activity of a business (or possibly a whole industry) and restrict our analysis to a time period during which background conditions (such as supplies of raw materials, wage rates, taxes) are fairly constant. We then show how calculus can help the management of such a firm make vital production decisions.

Management, whether or not it knows calculus, utilizes many functions of the sort we have been considering. Examples of such functions are

$C(x)$ = cost of producing x units of the product,

$R(x)$ = revenue generated by producing x units of the product,

$P(x) = R(x) - C(x)$ = the profit (or loss) generated by producing x units of the product.

Note that the functions $C(x)$, $R(x)$, $P(x)$ often are defined only for nonnegative integers—that is, for $x = 0, 1, 2, 3, \ldots$. The reason is that it does not make sense to speak about the cost of producing -1 cars or the revenue generated by selling 3.62 refrigerators. Thus each of these functions may give rise to a set of discrete points on a graph, as in Fig. 1(a). In studying these functions, however, economists usually draw a smooth curve through the points and assume that $C(x)$ is actually defined for all positive x. Of course, we must often interpret answers to problems in light of the fact that x is, in most cases, a nonnegative integer.

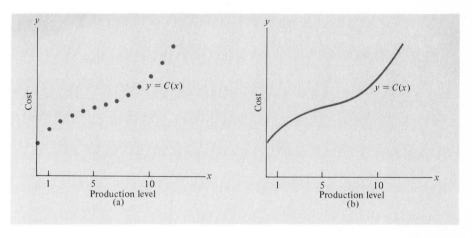

FIGURE 1

Cost Functions   If we assume that a cost function $C(x)$ has a smooth graph as in Fig. 1(b), we can use the tools of calculus to study it. A typical cost function is analyzed in Example 1.

EXAMPLE 1   Suppose that the cost function for a manufacturer is given by $C(x) = (10^{-6})x^3 - .003x^2 + 5x + 1000$ dollars.

(a) Describe the behavior of the marginal cost.

(b) Sketch the graph of $C(x)$.

Solution   The first two derivatives of $C(x)$ are given by

$$C'(x) = (3 \cdot 10^{-6})x^2 - .006x + 5$$

$$C''(x) = (6 \cdot 10^{-6})x - .006.$$

Let us sketch the marginal cost $C'(x)$ first. From the behavior of $C'(x)$, we will be able to graph $C(x)$. The marginal cost function $y = (3 \cdot 10^{-6})x^2 - .006x + 5$ has as its graph a parabola that opens upward. Since $y' = C''(x) = .000006(x - 1000)$, we see that the parabola has a horizontal tangent at $x = 1000$. So the minimum value of y occurs at $x = 1000$. The corresponding y-coordinate is

$$(3 \cdot 10^{-6})(1000)^2 - .006 \cdot (1000) + 5 = 3 - 6 + 5 = 2.$$

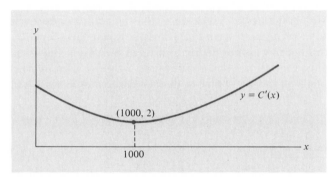

FIGURE 2

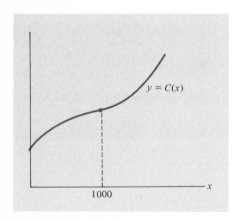

FIGURE 3

The graph of $y = C'(x)$ is shown in Fig. 2. Consequently, at first the marginal cost decreases. It reaches a minimum of 2 at production level 1000 and increases thereafter. This answers part (a). Let us now graph $C(x)$. Since the graph shown in Fig. 2 is the graph of the derivative of $C(x)$, we see that $C'(x)$ is never zero, so that there are no extreme points. Since $C'(x)$ is always positive, $C(x)$ is always increasing (as any cost curve should be). Moreover, since $C'(x)$ decreases for x less than 1000 and increases for x greater than 1000, we see that $C(x)$ is concave down for x less than 1000, concave up for x greater than 1000, and has an inflection point at $x = 1000$. The graph of $C(x)$ is drawn in Fig. 3. Note that the inflection point of $C(x)$ occurs at the value of x for which marginal cost is a minimum.

Actually, most marginal cost functions have the same general shape as the marginal cost curve of Example 1. For when x is small, production of additional units is subject to economies of production, which lower unit costs. Thus, for x small, marginal cost decreases. However, increased production eventually leads to overtime, use of less efficient, older plants, and competition for scarce raw materials. As a result, the cost of additional units will increase for very large x. So we see that $C'(x)$ initially decreases and then increases.

Revenue Functions   In general, a business is concerned not only with its costs, but also with its revenues. Recall that if $R(x)$ is the revenue received from the sale of x units of some commodity, then the derivative $R'(x)$ is called the *marginal revenue*. Economists use this to measure the rate of increase in revenue per unit increase in sales.

   If x units of a product are sold at a price p per unit, then the total revenue $R(x)$ is given by

$$R(x) = x \cdot p.$$

If a firm is small and is in competition with many other companies, its sales have little effect on the market price. Then since the price is constant as far as the one firm is concerned, the marginal revenue $R'(x)$ equals the price p [that is, $R'(x)$ is the amount that the firm receives from the sale of one additional unit]. In this case, the revenue function will have a graph as in Fig. 4.

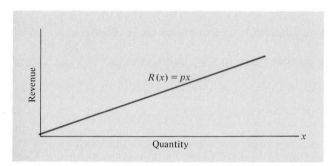

FIGURE 4

   An interesting problem arises when a single firm is the only supplier of a certain product or service—that is, when the firm has a "monopoly." Consumers will buy large amounts of the commodity if the price per unit is low and less if the price is raised. For each quantity x, let $f(x)$ be the highest price per unit that can be set in order to sell all x units to consumers. Since selling greater quantities requires a lowering of the price, $f(x)$ will be a decreasing function. Figure 5 shows a typical "demand curve" that relates the quantity demanded, x, to the price $p = f(x)$.

   The *demand equation* $p = f(x)$ determines the total revenue function. If the firm wants to sell x units, the highest price it can set is $f(x)$ dollars per unit, and so the total revenue from the sale of x units is

$$R(x) = x \cdot p = x \cdot f(x).$$

   The concept of a demand curve applies to an entire industry (with many producers) as well as to a single monopolistic firm. In this case, many producers offer the same product for sale. If x denotes the total output of the industry, then $f(x)$ is the market price per unit of output and $x \cdot f(x)$ is the total revenue earned from the sale of the x units.

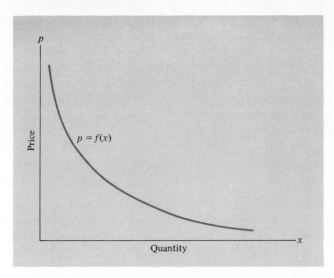

FIGURE 5

EXAMPLE 2   The demand equation for a certain product is $p = 6 - \frac{1}{2}x$. Find the level of production that results in maximum revenue.

Solution   The revenue function $R(x)$, in this case, is

$$R(x) = x \cdot p = x(6 - \tfrac{1}{2}x) = 6x - \tfrac{1}{2}x^2.$$

The marginal revenue is given by

$$R'(x) = 6 - x.$$

The graph of $R(x)$ is a parabola that opens downward (Fig. 6). It has a horizontal tangent precisely at those x for which $R'(x) = 0$—that is, for those x at which marginal revenue is 0. The only such x is $x = 6$. The corresponding value of revenue is

$$R(6) = 6 \cdot 6 - \tfrac{1}{2}(6)^2 = 18.$$

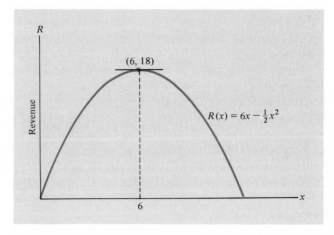

FIGURE 6

Thus the rate of production resulting in maximum revenue is $x = 6$, which results in total revenue of 18.

EXAMPLE 3 The WMA Bus Lines offers sightseeing tours of Washington, D.C. One of the tours, priced at $7 per person, had an average demand of about 1000 customers per week. When the price was lowered to $6, the weekly demand jumped to about 1200 customers. Assuming that the demand equation is linear, find the tour price that should be charged per person in order to maximize the total revenue each week.

Solution First we must find the demand equation. Let x be the number of customers per week and let p be the price of a tour ticket. Then $(x, p) = (1000, 7)$ and $(x, p) = (1200, 6)$ are on the demand curve (Fig. 7). Using the point-slope formula for the line through these two points, we have

$$p - 7 = \frac{7 - 6}{1000 - 1200} \cdot (x - 1000)$$

$$= -\frac{1}{200}(x - 1000)$$

$$= -\frac{1}{200}x + 5,$$

so

$$p = 12 - \frac{1}{200}x. \tag{1}$$

From equation (1) we obtain the revenue function

$$R(x) = x(12 - \tfrac{1}{200}x) = 12x - \tfrac{1}{200}x^2.$$

The marginal revenue is

$$R'(x) = 12 - \tfrac{1}{100}x = -\tfrac{1}{100}$$

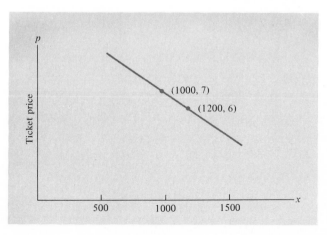

FIGURE 7

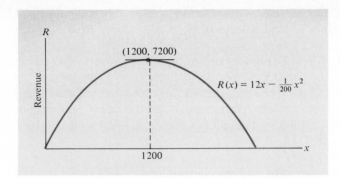

FIGURE 8

Using $R(x)$ and $R'(x)$, we can sketch the graph of $R(x)$ (Fig. 8). The maximum revenue occurs when the marginal revenue is zero—that is, when $x = 1200$. The price corresponding to this number of customers is found from the demand equation (1),

$$p = 12 - \tfrac{1}{200}(1200) = 6.$$

Thus the price of $6 is most likely to bring in the greatest revenue per week.

Profit Functions Once we know the cost function $C(x)$ and the revenue function $R(x)$, we can compute the profit function $P(x)$ from

$$P(x) = R(x) - C(x).$$

EXAMPLE 4 Suppose that the demand equation for a monopolist is $p = 100 - .01x$ and the cost function is $C(x) = 50x + 10{,}000$. Find the value of x that maximizes the profit and determine the corresponding price and total profit for this level of production. (See Fig. 9.)

Solution The total revenue function is

$$R(x) = x \cdot p = x(100 - .01x) = 100x - .01x^2.$$

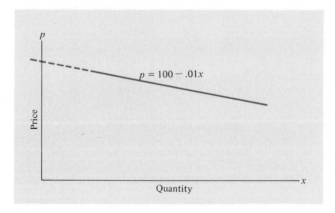

FIGURE 9

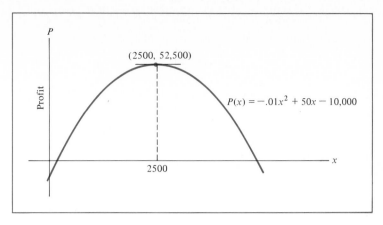

FIGURE 10

Hence the profit function is

$$P(x) = R(x) - C(x)$$
$$= 100x - .01x^2 - (50x + 10,000)$$
$$= -.01x^2 + 50x - 10,000.$$

The graph of this function is a parabola that opens downward. (See Fig. 10.) Its highest point will be where the curve has zero slope—that is, where the marginal profit $P'(x)$ is zero. Now

$$P'(x) = -.02x + 50 = -.02(x - 2500).$$

So $P'(x) = 0$ when $x = 2500$. The profit for this level of production is

$$P(2500) = -.01(2500)^2 + 50(2500) - 10,000$$
$$= 52,500.$$

Finally, we return to the demand equation to find the highest price that can be charged per unit in order to sell all 2500 units:

$$p = 100 - .01(2500)$$
$$= 100 - 25 = 75.$$

Answer   Produce 2500 units and sell them at $75 per unit. The profit will be $52,500.

EXAMPLE 5   Rework Example 4 under the condition that the government imposes an excise tax of $10 per unit.

Solution   For each unit sold, the manufacturer will have to pay $10 to the government. In other words, $10x$ dollars is added to the cost of producing and selling x units. The cost function now will be

$$C(x) = (50x + 10,000) + 10x = 60x + 10,000.$$

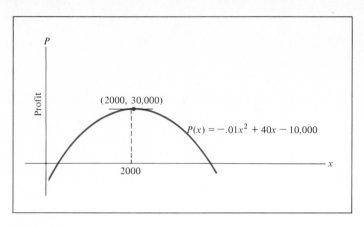

FIGURE 11

The demand equation is unchanged by this tax, so the revenue function is still

$$R(x) = 100x - .01x^2.$$

Proceeding as before, we have

$$P(x) = R(x) - C(x)$$
$$= 100x - .01x^2 - (60x + 10,000)$$
$$= -.01x^2 + 40x - 10,000.$$
$$P'(x) = -.02x + 40 = -.02(x - 2000).$$

The graph of $P(x)$ is still a parabola that opens downward, and the highest point is where $P'(x) = 0$—that is, where $x = 2000$. (See Fig. 11.) The corresponding profit is

$$P(2000) = -.01(2000)^2 + 40(2000) - 10,000$$
$$= 30,000.$$

From the demand equation, $p = 100 - .01x$, we find the price that corresponds to $x = 2000$:

$$p = 100 - .01(2000) = 80.$$

Answer Produce 2000 units and sell them at $80 per unit. The profit will be $30,000.

Notice in Example 5 that the optimal price is raised from $75 to $80. If the monopolist wishes to maximize profits, he should pass only half the $10 tax on to the consumer. The monopolist cannot avoid the fact that profits will be substantially lowered by the imposition of the tax. This is one reason why industries lobby against taxation.

Setting Production Levels Suppose that a firm has cost function $C(x)$ and revenue function $R(x)$. In a free enterprise economy the firm will set production x

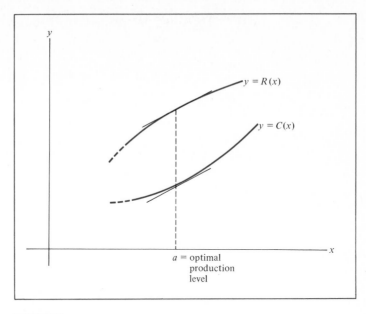

FIGURE 12

in such a way as to maximize its profit function

$$P(x) = R(x) - C(x).$$

We have seen that if $P(x)$ has a maximum at $x = a$, then $P'(a) = 0$. In other words, since $P'(x) = R'(x) - C'(x)$, we see that

$$R'(a) - C'(a) = 0$$

$$R'(a) = C'(a).$$

Thus profit is maximized at a production level for which marginal revenue equals marginal cost. (See Fig. 12.)

PRACTICE PROBLEMS 7

1. Rework Example 4 by finding the production level at which marginal revenue equals marginal cost.

2. Rework Example 4 under the condition that the fixed cost is increased from $10,000 to $15,000.

EXERCISES 7

1. Given the cost function $C(x) = x^3 - 6x^2 + 13x + 15$, find the minimum marginal cost.

2. Suppose that a total cost function is $C(x) = .0001x^3 - .06x^2 + 12x + 100$. Is the marginal cost increasing, decreasing, or not changing at $x = 100$? Find the minimum marginal cost.

3. The revenue function for a one-product firm is

$$R(x) = 200 - \frac{1600}{x + 8} - x.$$

Find the value of x that results in maximum revenue.

4. The revenue function for a particular product is $R(x) = x(4 - .0001x)$. Find the largest possible revenue.

5. A one-product firm estimates that its daily total cost function (in suitable units) is $C(x) = x^3 - 6x^2 + 13x + 15$ and its total revenue function is $R(x) = 28x$. Find the value of x that maximizes the daily profit.

6. A small tie shop sells men's ties for $3.50 each. The daily cost function is estimated to be $C(x)$ dollars, where x is the number of ties sold on a typical day and $C(x) = .0006x^3 - .03x^2 + 2x + 80$. Find the value of x that will maximize the store's daily profit.

7. The demand equation for a certain commodity is $p = \frac{1}{12}x^2 - 10x + 300, 0 \le x \le 60$. Find the value of x and the corresponding price p that maximize the revenue.

8. The demand equation for a product is $p = 2 - .001x$. Find the value of x and the corresponding price p that maximize the revenue.

9. Some years ago it was estimated that the demand for steel approximately satisfied the equation $p = 256 - 50x$, and the total cost of producing x units of steel was $C(x) = 182 + 56x$. (The quantity x was measured in millions of tons and the price and total cost were measured in millions of dollars.) Determine the level of production and the corresponding price that maximize the profits.

10. Consider a rectangle in the xy-plane, with corners at $(0, 0), (a, 0), (0, b),$ and (a, b). Suppose that (a, b) lies on the graph of the equation $y = 30 - x$. Find a and b such that the area of the rectangle is maximized. What economic interpretation can be given to your answer if the equation $y = 30 - x$ represents a demand curve and y is the price corresponding to the demand x?

11. Until recently hamburgers at the city sports arena cost $1 each. The food concessionaire sold an average of 10,000 hamburgers on a game night. When the price was raised to $1.20, hamburger sales dropped off to an average of 8000 per night.

(a) Assuming a linear demand curve, find the price of a hamburger that will maximize the nightly hamburger revenue.

(b) Suppose that the concessionaire has fixed costs of $1000 per night and the variable cost is $.30 per hamburger. Find the price of a hamburger that will maximize the nightly hamburger profit.

12. The average ticket price for a concert at the opera house was $32. The 4000-seat auditorium was filled for nearly every performance. When the ticket price was raised to $34 attendance declined to an average of 3800 persons per performance. Salaries, electricity, and maintenance expenses total $60,000 per performance. Costs that vary with the size of the audience are negligible. What should the average ticket price be in order to maximize the profit for the opera house? (Assume a linear demand curve.)

13. The monthly demand equation for an electric utility company is estimated to be

$$p = 60 - (10^{-5})x,$$

where p is measured in dollars and x is measured in thousands of kilowatt-hours. The utility has fixed costs of $7,000,000 per month and variable costs of $30 per thousand kilowatt-hours of electricity generated, so that the cost function is

$$C(x) = 7 \cdot 10^6 + 30x.$$

(a) Find the value of x and the corresponding price for a thousand kilowatt-hours that maximize the utility's profit.

(b) Suppose that rising fuel costs increase the utility's variable costs from $30 to $40 so that its new cost function is

$$C_1(x) = 7 \cdot 10^6 + 40x.$$

Should the utility pass all this increase of $10 per thousand kilowatt-hours on to consumers? Explain your answer.

14. The demand equation for a monopolist is $p = 200 - 3x$, and the cost function is $C(x) = 75 + 80x - x^2, 0 \le x \le 40$.

(a) Determine the value of x and the corresponding price that maximizes the profit.

(b) Suppose that the government imposes a tax on the monopolist of $4 per unit quantity produced. Determine the new price that maximizes the profit.

(c) Suppose that the government imposes a tax of T dollars per unit quantity produced, so that the new cost function is

$$C(x) = 75 + (80 + T)x - x^2, \qquad 0 \le x \le 40.$$

Determine the new value of x that maximizes the monopolist's profit as a function of T. Assuming that the monopolist cuts back production to this level, express the tax revenues received by the government as a function of T. Finally, determine the value of T that will maximize the tax revenue received by the government.

SOLUTIONS TO PRACTICE PROBLEMS 7

1. The revenue function is $R(x) = 100x - .01x^2$, so the marginal revenue function is $R'(x) = 100 - .02x$. The cost function is $C(x) = 50x + 10,000$, so the marginal cost function is $C'(x) = 50$. Let us now equate the two marginal functions and solve for x.

$$R'(x) = C'(x)$$

$$100 - .02x = 50$$

$$-.02x = -50$$

$$x = \frac{-50}{-.02} = \frac{5000}{2} = 2500.$$

Of course, we obtain the same level of production as before.

2. If the fixed cost is increased from $10,000 to $15,000, the new cost function will be $C(x) = 50x + 15,000$ but the marginal cost function will still be $C'(x) = 50$. Therefore, the solution will be the same: 2500 units should be produced and sold at $75 per unit. [Increases in fixed costs should not necessarily be passed on to the consumer if the objective is to maximize the profit.]

Chapter 12: CHECKLIST

- ☐ Increasing, decreasing
- ☐ Extreme point, maximum point, minimum point
- ☐ Maximum value, minimum value
- ☐ Concave up, concave down
- ☐ Inflection point
- ☐ x-intercept, y-intercept
- ☐ Asymptote
- ☐ First derivative rule
- ☐ Second derivative rule
- ☐ How to look for possible extreme points
- ☐ How to decide whether an extreme point is a maximum or minimum point
- ☐ How to look for possible inflection points
- ☐ Summary of curve-sketching techniques
- ☐ Objective equation
- ☐ Constraint equation
- ☐ Suggestions for solving an optimization problem

Chapter 12: SUPPLEMENTARY EXERCISES

Properties of various functions are described below. In each case draw some conclusion about the graph of the function.

1. $f(1) = 2, f'(1) > 0$

2. $g(1) = 5, g'(1) = -1$

3. $h'(3) = 4, h''(3) = 1$

4. $F'(2) = -1, F''(2) < 0$

5. $G(10) = 2, G'(10) = 0, G''(10) > 0$

6. $f(4) = -2, f'(4) > 0, f''(4) = -1$

7. $g(5) = -1, g'(5) = -2, g''(5) = 0$

8. $H(0) = 0, H'(0) = 0, H''(0) = 1$

9. $F(-2) = 0, F'(-2) = 0, F''(-2) = -1$

10. $h(-3) = 4, h'(-3) = 1, h''(-3) = 0$

Sketch the following parabolas. Include their x- and y-intercepts.

11. $y = 3 - x^2$

12. $y = 7 + 6x - x^2$

13. $y = x^2 + 3x - 10$

14. $y = 4 + 3x - x^2$

15. $y = -2x^2 + 10x - 10$

16. $y = x^2 - 9x + 19$

17. $y = x^2 + 3x + 2$

18. $y = -x^2 + 8x - 13$

19. $y = -x^2 + 20x - 90$

20. $y = 2x^2 + x - 1$

Sketch the following curves.

21. $y = 2x^3 + 3x^2 + 1$

22. $y = x^3 - \frac{3}{2}x^2 - 6x$

23. $y = x^3 - 3x^2 + 3x - 2$

24. $y = 100 + 36x - 6x^2 - x^3$

25. $y = \frac{11}{3} + 3x - x^2 - \frac{1}{3}x^3$

26. $y = x^3 - 3x^2 - 9x + 7$

27. $y = -\frac{1}{3}x^3 - 2x^2 - 5x$ 28. $y = x^3 - 6x^2 - 15x + 50$

29. $y = x^4 - 2x^2$ 30. $y = x^4 - 4x^3$

31. Let $f(x) = (x^2 + 2)^{3/2}$. Show that the graph of $f(x)$ has a possible extreme point at $x = 0$.

32. Show that the function $f(x) = (2x^2 + 3)^{3/2}$ is decreasing for $x < 0$ and increasing for $x > 0$.

33. Let $f(x)$ be a function whose *derivative* is

$$f'(x) = \frac{1}{1 + x^2}.$$

Note that $f'(x)$ is always positive. Show that the graph of $f(x)$ definitely has an inflection point at $x = 0$.

34. Let $f(x)$ be a function whose *derivative* is

$$f'(x) = \sqrt{5x^2 + 1}.$$

Show that the graph of $f(x)$ definitely has an inflection point at $x = 0$.

35. A closed rectangular box is to be constructed such that the base has a length twice as long as its width. Suppose that the volume must be 9 cubic feet. Find the dimensions of the box that will minimize the total surface area and show that this minimum surface area is 27 square feet.

36. A closed rectangular box is to be constructed such that the base has a length twice as long as its width. Suppose that the total surface area must be 27 square feet. Find the dimensions of the box that will maximize the volume and show that this maximum volume is 9 cubic feet.

37. The strength S of a rectangular beam is proportional to its width x and the square of its depth y, so that $S = kxy^2$, where k is a positive constant (Fig. 13). Find the dimensions of the strongest beam that can be cut from a circular log of diameter 75 centimeters.

FIGURE 13

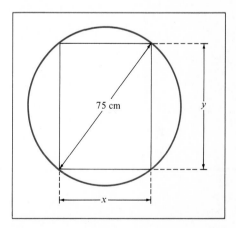

75 cm

y

x

38. A fabric store purchases scissors for $5 a pair and resells them for $9. It costs $10 to place an order with the manufacturer, and the yearly inventory cost per pair of scissors is $.80. If the store sells about 900 pairs of scissors each year at a relatively constant rate, how many scissors should be ordered at one time?

39. A California distributor of sporting equipment expects to receive orders during the coming year for 100,000 cans of tennis balls. Yearly inventory costs per can are about $.50, and the cost of placing an order with the manufacturer is $10. Assuming a fairly constant demand for the tennis balls, determine the optimal reorder quantity for the distributor.

40. Suppose that the demand equation for a monopolist is $p = 150 - .02x$ and the cost function is $C(x) = 10x + 300$. Find the value of x that maximizes the profit.

Chapter 13

Techniques of Differentiation

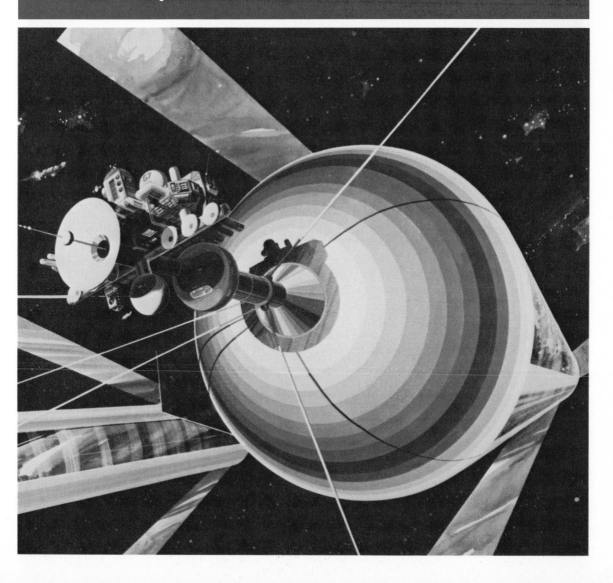

We have seen that the derivative is useful in many applications. However, our ability to differentiate functions is somewhat limited. For example, we cannot yet readily differentiate

$$(x^{3/2} + 1)\sqrt{1 - x^2}, \qquad \frac{x}{x^2 + 1}, \qquad \frac{x^3}{3x^2 - 3x + 1}.$$

In this chapter we develop differentiation techniques that apply to functions like those given above. Two new rules are the *product rule* and the *quotient rule*. In Section 2 we extend the general power rule into a powerful formula called the *chain rule*.

13.1 The Product and Quotient Rules

We observed in our discussion of the sum rule for derivatives that the derivative of the sum of two differentiable functions is the sum of the derivatives. Unfortunately, however, the derivative of the product $f(x)g(x)$ is *not* the product of the derivatives. Rather, the derivative of a product is determined from the following rule.

Product Rule

$$\frac{d}{dx}\left[f(x)g(x)\right] = f(x)g'(x) + g(x)f'(x).$$

The derivative of the product of two functions is the first function times the derivative of the second plus the second function times the derivative of the first. At the end of the section we show why this statement is true.

EXAMPLE 1   Show that the product rule works for the case $f(x) = x^2$, $g(x) = x^3$.

Solution   Since $x^2 \cdot x^3 = x^5$, we know that

$$\frac{d}{dx}[x^2 \cdot x^3] = \frac{d}{dx}[x^5] = 5x^4.$$

On the other hand, using the product rule,

$$\frac{d}{dx}(x^2 \cdot x^3) = x^2 \frac{d}{dx}(x^3) + x^3 \frac{d}{dx}(x^2)$$

$$= x^2(3x^2) + x^3(2x)$$

$$= 3x^4 + 2x^4 = 5x^4.$$

Thus the product rule gives the correct answer.

EXAMPLE 2   Differentiate the product $(2x^3 - 5x)(3x + 1)$.

Solution   Let $f(x) = 2x^3 - 5x$ and $g(x) = 3x + 1$. Then

$$\frac{d}{dx}[(2x^3 - 5x)(3x + 1)] = (2x^3 - 5x) \cdot \frac{d}{dx}(3x + 1) + (3x + 1) \cdot \frac{d}{dx}(2x^3 - 5x)$$

$$= (2x^3 - 5x)(3) + (3x + 1)(6x^2 - 5)$$

$$= 6x^3 - 15x + 18x^3 - 15x + 6x^2 - 5$$

$$= 24x^3 + 6x^2 - 30x - 5.$$

EXAMPLE 3   Apply the product rule to $y = g(x) \cdot g(x)$.

Solution

$$\frac{d}{dx}[g(x) \cdot g(x)] = g(x) \cdot g'(x) + g(x) \cdot g'(x)$$

$$= 2g(x)g'(x).$$

This answer is the same as that given by the general power rule:

$$\frac{d}{dx}[g(x) \cdot g(x)] = \frac{d}{dx}[g(x)]^2 = 2g(x)g'(x).$$

EXAMPLE 4   Differentiate the function $(x^3 - 1)(x^2 + 1)^5$.

Solution   Let $f(x) = x^3 - 1$ and $g(x) = (x^2 + 1)^5$. When we compute $g'(x)$ we will need the general power rule. Using the product rule first, we find that

$$\frac{d}{dx}[(x^3 - 1)(x^2 + 1)^5] = (x^3 - 1)\frac{d}{dx}(x^2 + 1)^5 + (x^2 + 1)^5 \frac{d}{dx}(x^3 - 1)$$

$$= (x^3 - 1) \cdot 5(x^2 + 1)^4(2x) + (x^2 + 1)^5 \cdot (3x^2)$$

$$= 10x(x^3 - 1)(x^2 + 1)^4 + 3x^2(x^2 + 1)^5.$$

The Quotient Rule Another useful formula for differentiating functions is the quotient rule.

Quotient Rule

$$\frac{d}{dx}\left[\frac{f(x)}{g(x)}\right] = \frac{g(x)f'(x) - f(x)g'(x)}{[g(x)]^2}.$$

One must be careful to remember the order of the terms in this formula because of the minus sign in the numerator.

EXAMPLE 5 Differentiate $\dfrac{x^2 + 1}{x^3 + 5}$.

Solution Let $f(x) = x^2 + 1$ and $g(x) = x^3 + 5$.

$$\frac{d}{dx}\left(\frac{x^2 + 1}{x^3 + 5}\right) = \frac{(x^3 + 5)\dfrac{d}{dx}(x^2 + 1) - (x^2 + 1)\dfrac{d}{dx}(x^3 + 5)}{(x^3 + 5)^2}$$

$$= \frac{(x^3 + 5)(2x) - (x^2 + 1)(3x^2)}{(x^3 + 5)^2}$$

$$= \frac{-x^4 - 3x^2 + 10x}{(x^3 + 5)^2}.$$

EXAMPLE 6 Suppose that the total cost of manufacturing x units of a certain product is given by the function $C(x)$. Then the *average cost per unit*, AC, is defined by

$$AC = \frac{C(x)}{x}.$$

Recall that the *marginal cost*, MC, is defined by

$$MC = C'(x).$$

Show that at the level of production where the average cost is at a minimum, the average cost equals the marginal cost.

Solution In practice, the marginal cost and average cost curves will have the general shapes shown in Fig. 1. So the minimum point on the average cost curve will occur when $\dfrac{d}{dx}(AC) = 0$. To compute the derivative, we need the quotient rule,

$$\frac{d}{dx}(AC) = \frac{d}{dx}\left[\frac{C(x)}{x}\right] = \frac{x \cdot C'(x) - C(x)}{x^2}.$$

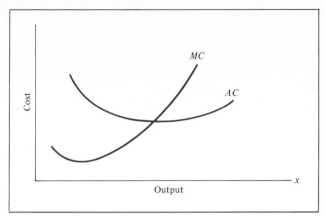

FIGURE 1　Marginal cost and average cost functions.

Setting the derivative equal to zero and multiplying by x^2, we obtain

$$0 = x \cdot C'(x) - C(x)$$

$$C(x) = x \cdot C'(x)$$

$$\frac{C(x)}{x} = C'(x)$$

$$AC = MC.$$

Thus when the output x is chosen so that the average cost is minimized, the average cost equals the marginal cost.

Verification of the Product and Quotient Rules

Verification of the Product Rule　From our discussion of limits we compute the derivative of $f(x)g(x)$ at $x = a$ as the limit

$$\frac{d}{dx}\left[f(x)g(x)\right]\Big|_{x=a} = \lim_{h \to 0} \frac{f(a+h)g(a+h) - f(a)g(a)}{h}.$$

Let us add and subtract the quantity $f(a)g(a+h)$ in the numerator. After factoring and applying Limit Theorem III, we obtain

$$\lim_{h \to 0} \frac{\left[f(a+h)g(a+h) - f(a)g(a+h)\right] + \left[f(a)g(a+h) - f(a)g(a)\right]}{h}$$

$$= \lim_{h \to 0} g(a+h) \cdot \frac{f(a+h) - f(a)}{h} + \lim_{h \to 0} f(a) \cdot \frac{g(a+h) - g(a)}{h}.$$

And this expression may be rewritten by Limit Theorem V as

$$\lim_{h \to 0} g(a+h) \cdot \lim_{h \to 0} \frac{f(a+h) - f(a)}{h} + \lim_{h \to 0} f(a) \cdot \lim_{h \to 0} \frac{g(a+h) - g(a)}{h}.$$

Note, however, that since $g(x)$ is differentiable at $x = a$, it is continuous there, so that $\lim_{h \to 0} g(a+h) = g(a)$. Therefore, the expression above equals

$$g(a)f'(a) + f(a)g'(a).$$

That is, we have proved that

$$\frac{d}{dx}[f(x)g(x)]\Big|_{x=a} = g(a)f'(a) + f(a)g'(a),$$

which is the product rule. An alternative verification of the product rule not involving limit arguments is outlined in Exercise 45.

Verification of the Quotient Rule   From the general power rule, we know that

$$\frac{d}{dx}\left[\frac{1}{g(x)}\right] = \frac{d}{dx}[g(x)]^{-1} = (-1)[g(x)]^{-2} \cdot g'(x).$$

We can now derive the quotient rule from the product rule.

$$\frac{d}{dx}\left[\frac{f(x)}{g(x)}\right] = \frac{d}{dx}\left[\frac{1}{g(x)} \cdot f(x)\right]$$

$$= \frac{1}{g(x)} \cdot f'(x) + f(x) \cdot \frac{d}{dx}\left[\frac{1}{g(x)}\right]$$

$$= \frac{g(x)f'(x)}{[g(x)]^2} + f(x) \cdot (-1)[g(x)]^{-2} \cdot g'(x)$$

$$= \frac{g(x)f'(x) - f(x)g'(x)}{[g(x)]^2}.$$

PRACTICE PROBLEMS 1

1. Consider the function $y = (\sqrt{x} + 1)x$.

   (a) Differentiate y by the product rule.

   (b) Multiply out first and then differentiate.

2. Differentiate $y = x/(x^4 - x^3 + 1)$.

EXERCISES 1

Differentiate the following functions.

1. $(x + 1)(x^3 + 5x + 2)$

2. $(2x - 1)(x^2 - 3)$

3. $(3x^2 - x + 2)(2x^2 - 1)$

4. $(x^3 + x^2 + x)(5x - 4)$

5. $(x^4 + 1)(3x + 5)$

6. $(10x^2 - 2x + 1)(x^2 + 7x - 8)$

7. $(x^2 - 3x + 1)\left(x + \dfrac{2}{x}\right)$

8. $\left(4 - \dfrac{1}{x^2}\right)(3x^3 - 5x + 8)$

9. $(2x - 7)(x - 1)^5$

10. $(x^2 + 4)(x + 3)^6$

11. $(x^2 + 3)(x^2 - 3)^{10}$

12. $(x + 1)^2(x - 3)^3$

13. $(2x + 9)^{-2}(5 - 3x)$

14. $(1 - x^2)(x^2 - 4x - 1)^{-4}$

15. $\sqrt{x}(3x^2 - 1)^3$

16. $(5x - 1)^4\left(\sqrt{x} + \dfrac{1}{\sqrt{x}}\right)$

17. $(1 - 4x)\sqrt{x^2 - 1}$

18. $(x^{3/2} + 2x^{1/2})(4x^{5/2} - x^{3/2})$

19. $\dfrac{x^2 - 1}{x^2 + 1}$

20. $\dfrac{x^2 + x + 1}{5 - x}$

21. $\dfrac{1}{5x^2 + 2x + 5}$

22. $\dfrac{2x - 1}{x}$

23. $\dfrac{\sqrt{x}}{x + 1}$

24. $\dfrac{3x^2 + 5x + 2}{x^2 - 3}$

25. $\dfrac{x}{(x^2 + 1)^2}$

26. $\dfrac{4 + x - x^2}{1 + \sqrt{x}}$

27. $\dfrac{x - 2}{x + 2}$

28. $\dfrac{x - 1}{(2x + 1)^2}$

29. $\dfrac{3 - x}{\sqrt{4x + x^2}}$

30. $\dfrac{1 + \sqrt{2x}}{1 - \sqrt{2x}}$

31. $\dfrac{(x^5 + 1)^3}{(3x^2 + 1)^2}$

32. $\left(\dfrac{4x - 1}{3x + 1}\right)^3$

33. Find the equation of the tangent line to the curve $y = (x - 2)^5(x + 1)^2$ at the point $(3, 16)$.

34. Find the equation of the tangent line to the curve $y = (x + 1)/(x - 1)$ at the point $(2, 3)$.

35. A sugar refinery can produce x tons of sugar per week at a weekly cost of $\frac{1}{10}x^2 + 5x + 2250$ dollars. Find the level of production for which the average cost is at a minimum and show that the average cost equals the marginal cost at that level of production.

36. A cigar manufacturer produces x cases of cigars per day at a daily cost of $50x(x + 200)/(x + 100)$ dollars. Show that his cost increases and his average cost decreases as output increases.

37. An open rectangular box is 3 feet long and has a surface area of 16 cubic feet. Find the dimensions of the box for which the volume is as large as possible.

38. A closed rectangular box is to be constructed with one side 1 meter long. The material for the top costs \$20 per square meter, and the material for the sides and bottom costs \$10 per square meter. Find the dimensions of the box with largest possible volume that can be built at a cost of \$240 for materials.

39. (*A Voting Model*) Let x be the proportion of the total popular vote a Democratic candidate for president receives in a U.S. national election. Political scientists have observed that a good estimate of the proportion of seats in the House of Representatives going to Democratic candidates is given by

$$f(x) = \frac{x^3}{3x^2 - 3x + 1}, \qquad 0 \le x \le 1.$$

This formula is referred to as the "cube law" by some political scientists, primarily because $f(x)$ may also be written as

$$f(x) = \frac{x^3}{x^3 + (1 - x)^3}.$$

(Note that $1 - x$ is approximately the proportion of votes received by the Republicans.)

(a) Are $f(0)$, $f(\frac{1}{2})$, and $f(1)$ what you would expect them to be?

(b) Show that $f'(\frac{1}{2}) = 3$.

(c) It can be shown (by the quotient rule) that

$$f''(x) = \frac{6x(x - 1)(2x - 1)}{(3x^2 - 3x + 1)^3}.$$

What does this tell you about $f(x)$?

(d) Sketch the graph of $y = f(x)$ for $0 \le x \le 1$.

40. Let $f(x) = 1/x$ and $g(x) = x^3$.

(a) Show that the product rule yields the correct derivative of $(1/x) \cdot x^3 = x^2$.

(b) Compute the product $f'(x)g'(x)$ and note that it is *not* the derivative of $f(x)g(x)$.

41. The derivative of $(x^3 - 4x)/x$ is obviously $2x$ for $x \ne 0$, because $(x^3 - 4x)/x = x^2 - 4$ for $x \ne 0$. Verify that the quotient rule gives the same derivative.

42. Let $f(x)$, $g(x)$, $h(x)$ be differentiable functions. Find a formula for the derivative of $f(x)g(x)h(x)$.

43. Suppose that $f(x)$ is a function whose derivative is $f'(x) = 1/x$. Find the derivative of $xf(x) - x$.

44. Suppose that $f(x)$ is a function whose derivative is $f'(x) = \dfrac{1}{1 + x^2}$. Find the derivative

of $\dfrac{f(x)}{1 + x^2}$.

45. (*Alternative Verification of the Product Rule*) Apply the special case of the general power

rule $\dfrac{d}{dx}[h(x)]^2 = 2h(x)h'(x)$ and the identity $fg = \frac{1}{4}[(f + g)^2 - (f - g)^2]$ to prove

the product rule.

SOLUTIONS TO PRACTICE PROBLEMS 1

1. (a) Apply the product rule to $y = (\sqrt{x} + 1)x$ with

$$f(x) = \sqrt{x} + 1 = x^{1/2} + 1$$

$$g(x) = x$$

$$\frac{dy}{dx} = (x^{1/2} + 1) \cdot 1 + x \cdot \tfrac{1}{2}x^{-1/2}$$

$$= x^{1/2} + 1 + \tfrac{1}{2}x^{1/2}$$

$$= \tfrac{3}{2}\sqrt{x} + 1.$$

(b) $y = (\sqrt{x} + 1)x = (x^{1/2} + 1)x = x^{3/2} + x.$

$$\frac{dy}{dx} = \tfrac{3}{2}x^{1/2} + \tfrac{3}{2}\sqrt{x} + 1.$$

Comparing parts (a) and (b), we note that sometimes it is helpful to simplify the function before differentiating.

2. Apply the quotient rule to $y = x/(x^4 - x^3 + 1)$.

$$\frac{dy}{dx} = \frac{(x^4 - x^3 + 1) \cdot 1 - x(4x^3 - 3x^2)}{(x^4 - x^3 + 1)^2}$$

$$= \frac{x^4 - x^3 + 1 - 4x^4 + 3x^3}{(x^4 - x^3 + 1)^2}$$

$$= \frac{-3x^4 + 2x^3 + 1}{(x^4 - x^3 + 1)^2}.$$

13.2 The Chain Rule and the General Power Rule

A useful way of combining functions $f(x)$ and $g(x)$ is to replace every occurrence of the variable x in $f(x)$ by the function $g(x)$. The resulting function is called the *composition* (or *composite*) of $f(x)$ and $g(x)$ and is denoted $f(g(x))$.

EXAMPLE 1   Let $f(x) = x^2$, $g(x) = x^3 - 3x + 1$. What is $f(g(x))$?

Solution   Replace each occurrence of x in $f(x)$ by $g(x)$ to obtain

$$f(g(x)) = [g(x)]^2 = (x^3 - 3x + 1)^2.$$

EXAMPLE 2   Let $f(x) = \dfrac{x - 1}{x + 1}$, $g(x) = x^3$. What is $f(g(x))$?

Solution   Replace each occurrence of x in $f(x)$ by $g(x)$ to obtain

$$f(g(x)) = \frac{g(x) - 1}{g(x) + 1}$$

$$= \frac{x^3 - 1}{x^3 + 1}.$$

EXAMPLE 3   Let $f(x) = \dfrac{1}{x}$, $g(x) = x^2 + 3x + 1$. What are $f(g(x))$ and $g(f(x))$?

Solution

$$f(g(x)) = \frac{1}{g(x)} = \frac{1}{x^2 + 3x + 1}.$$

$$g(f(x)) = [f(x)]^2 + 3[f(x)] + 1$$

$$= \left(\frac{1}{x}\right)^2 + 3 \cdot \left(\frac{1}{x}\right) + 1$$

$$= \frac{1}{x^2} + \frac{3}{x} + 1.$$

Not only is it important to be able to compute $f(g(x))$ when $f(x)$ and $g(x)$ are given, but it is also important (see below) to be able to recognize a given function as the composite of two simpler functions.

EXAMPLE 4   Let $h(x) = (x^5 + 9x + 7)^{18}$, $k(x) = \sqrt{3x^2 + 1}$. Write $h(x)$ and $k(x)$ as composites of simpler functions.

Solution   (a) Let $f(x) = x^{18}$, $g(x) = x^5 + 9x + 7$. Then $f(g(x)) = (x^5 + 9x + 7)^{18} = h(x)$.
(b) Let $f(x) = \sqrt{x}$, $g(x) = 3x^2 + 1$. Then $f(g(x)) = \sqrt{3x^2 + 1} = k(x)$.

A function of the form $[g(x)]^r$ is a composite $f(g(x))$, where $f(x) = x^r$. Moreover, we have already given a rule for differentiating this function—namely

$$\frac{d}{dx}[g(x)]^r = r[g(x)]^{r-1} \cdot g'(x). \tag{1}$$

This rule is a special case of the *chain rule*, which describes how to differentiate a composite $f(g(x))$. This rule enables us to calculate the derivative of $f(g(x))$ in terms of the derivatives of $f(x)$ and $g(x)$.

> *The Chain Rule*   To differentiate $f(g(x))$, first differentiate $f(x)$ and substitute $g(x)$ for x in the result. Then multiply by the derivative of $g(x)$. Symbolically,
>
> $$\frac{d}{dx}[f(g(x))] = f'(g(x)) \cdot g'(x).$$

EXAMPLE 5   Use the chain rule to compute the derivative of $h(x) = (x^5 + 9x + 7)^{18}$.

Solution   We saw in Example 4(a) that $h(x)$ can be represented as the composite $f(g(x))$, where $f(x) = x^{18}$, $g(x) = x^5 + 9x + 7$. Therefore, we apply the chain rule to compute $h'(x)$.

$$f'(x) = 18x^{17}, \qquad g'(x) = 5x^4 + 9$$

$$f'(g(x)) = 18(x^5 + 9x + 7)^{17}$$

$$f'(g(x)) \cdot g'(x) = 18(x^5 + 9x + 7)^{17} \cdot (5x^4 + 9).$$

Therefore,

$$h'(x) = 18(x^5 + 9x + 7)^{17} \cdot (5x^4 + 9).$$

Note that this is the same answer as derived using the general power rule.

EXAMPLE 6 Use the chain rule to compute the derivative of

$$h(x) = \frac{(x^3 + 1)^5 - 1}{(x^3 + 1)^5 + 1}.$$

Solution Note that $h(x) = f(g(x))$, where

$$f(x) = \frac{x^5 - 1}{x^5 + 1}, \qquad g(x) = x^3 + 1.$$

By the quotient rule,

$$f'(x) = \frac{(x^5 + 1) \cdot 5x^4 - (x^5 - 1) \cdot 5x^4}{(x^5 + 1)^2} = \frac{10x^4}{(x^5 + 1)^2}.$$

Therefore,

$$f'(g(x)) = \frac{10(x^3 + 1)^4}{[(x^3 + 1)^5 + 1]^2}.$$

$$g'(x) = 3x^2$$

$$h'(x) = f'(g(x)) \cdot g'(x)$$

$$= \frac{10(x^3 + 1)^4}{[(x^3 + 1)^5 + 1]^2} \cdot 3x^2$$

$$= \frac{30x^2(x^3 + 1)^4}{[(x^3 + 1)^5 + 1]^2}.$$

Note that this answer can also be derived using the quotient rule together with the general power rule.

Here is a good mnemonic device for remembering the chain rule. Suppose that we wish to differentiate a function of the form $y = f(g(x))$. If we set $u = g(x)$, we can view y as both a function of x and a function of u. Then the chain rule can be stated:

$$\frac{dy}{dx} = \frac{dy}{du} \cdot \frac{du}{dx}.$$

That is, to compute the derivative of y with respect to x, we multiply the derivative of y with respect to u, by the derivative of u with respect to x. [Of course, we express $\dfrac{dy}{du}$ as a function of x by setting $u = g(x)$.] You may remember this formulation of the chain rule by thinking of the derivative as a fraction and the equation above arising from "canceling" the du's.

At this stage, the chain rule does not enable us to differentiate any functions which we could not already differentiate via the general power rule. However, as we shall see in our discussion of the exponential and logarithmic functions, the chain rule does indeed expand the repertoire of functions we can differentiate.

When the power rule and the general power rule were introduced in Chapter 10, we promised to justify these rules later in the text. The general power rule can be deduced from the chain rule and the ordinary power rule. The chain rule is justified below.

Verification of the Chain Rule Suppose that $f(x)$ and $g(x)$ are differentiable, and let $x = a$ be a number in the domain of $f(g(x))$. Since every differentiable function is continuous, we have

$$\lim_{h \to 0} g(a + h) = g(a),$$

which implies that

$$\lim_{h \to 0} \left[g(a + h) - g(a) \right] = 0. \tag{2}$$

Now $g(a)$ is a number in the domain of f, and the limit definition of the derivative gives us

$$f'(g(a)) = \lim_{k \to 0} \frac{f(g(a) + k) - f(g(a))}{k}. \tag{3}$$

Let $k = g(a + h) - g(a)$. By equation (2), k approaches zero as h approaches zero. Also, $g(a + h) = g(a) + k$. Therefore, (3) may be rewritten in the form

$$f'(g(a)) = \lim_{h \to 0} \frac{f(g(a + h)) - f(g(a))}{g(a + h) - g(a)}. \tag{4}$$

[Strictly speaking, we must assume that the denominator in (4) is never zero. This assumption may be avoided by a somewhat different and more technical argument which we omit.] Finally, we show that the function $f(g(x))$ has a derivative at $x = a$. We use the limit definition of the derivative, Limit Theorem V, and (4) above.

$$\frac{d}{dx} \left[f(g(x)) \right] \bigg|_{x=a} = \lim_{h \to 0} \frac{f(g(a + h)) - f(g(a))}{h}$$

$$= \lim_{h \to 0} \left[\frac{f(g(a + h)) - f(g(a))}{g(a + h) - g(a)} \cdot \frac{g(a + h) - g(a)}{h} \right]$$

$$= \lim_{h \to 0} \frac{f(g(a + h)) - f(g(a))}{g(a + h) - g(a)} \cdot \lim_{h \to 0} \frac{g(a + h) - g(a)}{h}$$

$$= f'(g(a)) \cdot g'(a).$$

PRACTICE PROBLEMS 2

Consider the function $h(x) = (2x^3 - 5)^5 + (2x^3 - 5)^4$.

1. Write $h(x)$ as a composite function, $f(g(x))$.

2. Compute $f'(x)$ and $f'(g(x))$.

3. Use the chain rule to differentiate $h(x)$.

EXERCISES 2

Compute $f(g(x))$, where $f(x)$ and $g(x)$ are the following.

1. $f(x) = \dfrac{1}{x + 1}$, $g(x) = x^3$

2. $f(x) = \dfrac{1}{x^2 - 1}$, $g(x) = \dfrac{1}{x}$

3. $f(x) = \sqrt{x + 1}$, $g(x) = \dfrac{1}{x^2 - 1}$

4. $f(x) = \sqrt{x}$, $g(x) = \dfrac{x - 1}{x + 1}$

5. $f(x) = x^2$, $g(x) = x^2$

6. $f(x) = \sqrt{x}$, $g(x) = \sqrt[3]{x}$

Write the following functions as composites of two functions.

7. $\sqrt{x^2 + 3x - 1}$

8. $(2x - 3)^3 + (2x - 3)^2 + \dfrac{1}{2x - 3}$

9. $\sqrt{\dfrac{x - 1}{x + 1}}$

10. $\dfrac{\sqrt{x} - 1}{\sqrt{x} + 1}$

11. $(x^3 + 9x - 2)^5$

12. $(9x^2 + 2x - 5)^7$

13. $\dfrac{x^2 - 1}{x^2 + 1}$

14. $\dfrac{x^3 + 3x^6}{1 + x^9}$

15. $\sqrt{x^{1/3} + 1}$

16. $(\sqrt{x})^{1/3}$

17. $(x^3 + 1)^2 + 1$

18. $(x^2 + 1)^3 + 1$

19. $x^2(x^4 + 1)^2$

20. $(x^2 + 1)(x^4 + 1)^3$

Differentiate the following functions using the product rule, quotient rule, and chain rule.

21. $(x^2 + 1)^{15}$

22. $(x^4 + x^2 + 1)^{10}$

23. $\sqrt{1 + x^4}$

24. $(2x^3 + 3x - 1)^{-7}$

25. $\left[\dfrac{2}{x} + 5(3x - 1)^4 \right]^3$

26. $[(2x^2 + 1)^3 + 5]^4$

27. $(x^3 - 1)(x^2 + 1)^4$

28. $(2x^2 - 4)^5(5x + 1)^3$

29. $(4 - x)^3(4 + x)^3$

30. $(2x - 1)^{1/4}(2x + 1)^{3/4}$

31. $\sqrt{2 + \sqrt{1 + x^2}}$

32. $\left(\dfrac{x - 1}{x + 1} \right)^3$

33. $\dfrac{x}{\sqrt{1 + x^2}}$

34. $x^2\sqrt{x^2 - 9}$

35. $\dfrac{1 + \sqrt{2x}}{1 - \sqrt{2x}}$

36. $\dfrac{(x^5 + 1)^3}{(3x^2 + 1)^4}$

37. Sketch the graph of $y = 4x/(x + 1)^2, (x > -1)$.

38. Sketch the graph of $y = 2/(1 + x^2)$.

39. Suppose that $f(x)$ and $g(x)$ are differentiable functions. Find $g(x)$ if you know that

$$\frac{d}{dx} f(g(x)) = 3x^2 \cdot f'(x^3 + 1).$$

40. Suppose that $f(x)$ and $g(x)$ are differentiable functions. Find $g(x)$ if you know that $f'(x) = 1/x$ and

$$\frac{d}{dx} f(g(x)) = \frac{2x + 5}{x^2 + 5x - 4}.$$

SOLUTIONS TO PRACTICE PROBLEMS 2

1. Let $f(x) = x^5 + x^4$ and $g(x) = 2x^3 - 5$.

2. $f'(x) = 5x^4 + 4x^3, f'(g(x)) = 5(2x^3 - 5)^4 + 4(2x^3 - 5)^3$.

3. We have $g'(x) = 6x^2$. Then, from the chain rule and the result of Problem 2, we have

$$h'(x) = f'(g(x))g'(x) = [5(2x^3 - 5)^4 + 4(2x^3 - 5)^3](6x^2).$$

13.3 Implicit Differentiation and Related Rates

In this section we apply the chain rule to implicit differentiation and to related rates problems.

Implicit Differentiation   If a function is defined directly by a formula, such as

$$y = (x^2 + 1)^3,$$

then we say that the function is defined *explicitly*. However, in many applications, functions are not defined explicitly but are specified only by an equation involving x and y. An example of such an equation is

$$x^3 + y^8 - y = 9. \tag{1}$$

We say that this equation determines y *implicitly* as a function of x. Roughly speaking, to obtain y explicitly as a function of x, it would be necessary to solve for y in terms of x. However, as the equation above illustrates, this may not be a simple task.

Using the technique of implicit differentiation, it is possible to calculate derivatives of implicitly defined functions without ever determining the functions explicitly! The basic idea is to differentiate all terms in an equation with respect to x, treating y as a function of x, say $y = f(x)$. Let us illustrate this technique for equation (1) above. The first term x^3 has derivative $3x^2$, as usual. The term y^8, which we think of as $(f(x))^8$, must be differentiated using the general power rule.

$$\frac{d}{dx}(f(x))^8 = 8(f(x))^7 f'(x)$$

$$= 8y^7 y'.$$

Finally, the derivatives of the terms y and 9 are, respectively, y' and 0. Thus differentiation of the original equation gives us the relationship

$$3x^2 + 8y^7 y' - y' = 0.$$

Let us solve this equation for y'. We place all terms involving y' on one side of the equation, factor out the y', and then solve for y'.

$$8y^7 y' - y' = -3x^2$$

$$(8y^7 - 1)y' = -3x^2$$

$$y' = \frac{-3x^2}{8y^7 - 1}.$$

The last equation gives us a formula for the derivative y'. Note, however, that this formula involves not only x, but y as well. So to evaluate y', it is necessary to have values for both x and y. This is the price we pay for our original function being given implicitly. However, it is a rather small price. The original equation cannot be solved for y in terms of x in any elementary fashion. Nevertheless, we may calculate the derivative of y.

EXAMPLE 1 (a) Use implicit differentiation to determine y' if y is defined implicitly by the equation $x^2 y + xy - 3x = 5$.

(b) Use the formula of part (a) to evaluate y' when $x = 1$, $y = 4$.

Solution (a) We differentiate the equation term by term, taking care to differentiate $x^2 y$ and xy using the product rule and to treat y as a function of x.

$$x^2 y' + 2xy + xy' + y - 3 = 0.$$

We now solve for y' in terms of x and y.

$$x^2 y' + xy' = 3 - y - 2xy$$

$$(x^2 + x)y' = 3 - y - 2xy$$

$$y' = \frac{3 - y - 2xy}{x^2 + x}.$$

(b) When $x = 1$, $y = 4$, we have

$$y' = \frac{3 - 4 - 2(1)(4)}{1^2 + 1} = -\frac{9}{2}.$$

EXAMPLE 2 Use implicit differentiation to calculate $\dfrac{dy}{dx}$ when y and x are related by the equation

$$x^5 + y^5 = 2x^2.$$

Solution

$$5x^4 + 5y^4 \frac{dy}{dx} = 4x$$

$$5y^4 \frac{dy}{dx} = 4x - 5x^4$$

$$\frac{dy}{dx} = \frac{4x - 5x^4}{5y^4}.$$

It so happens that the equation in Example 2 can be solved explicitly for y in terms of x. Namely, we have

$$y^5 = 2x^2 - x^5$$
$$y = (2x^2 - x^5)^{1/5}.$$

We may differentiate this formula by the general power rule, obtaining

$$\frac{dy}{dx} = \frac{1}{5}(2x^2 - x^5)^{-4/5}(4x - 5x^4).$$

In some respects, this formula for $\dfrac{dy}{dx}$ is more cumbersome than the formula obtained in Example 2.

Related Rates In implicit differentiation, we differentiate an equation involving x and y, where y is viewed as a function of x. However, in many applications, x and y are related by an equation, but both are functions of a third variable t (which often represents time). Then by differentiating the equation connecting x and y we may derive a relationship between the derivatives $\dfrac{dx}{dt}$ and $\dfrac{dy}{dt}$.

EXAMPLE 3 Suppose that x and y are both functions of t and are related by the equation

$$x^2 + 5xy = 1.$$

(a) Compute $\dfrac{dy}{dt}$ in terms of $\dfrac{dx}{dt}$, x and y.

(b) Suppose that at a certain time $x = 4$, $y = -\frac{3}{4}$, and $\dfrac{dx}{dt} = 80$. Calculate $\dfrac{dy}{dt}$.

Solution (a) We differentiate the given equation with respect to t. The derivative of x^2 is then $2x \dfrac{dx}{dt}$ (by the chain rule, with x regarded as a function of t), and the derivative of $5xy$ is $5x\dfrac{5y}{dt} + 5y\dfrac{dx}{dt}$ (by the product rule). Thus we have

$$2x\frac{dx}{dt} + 5x\frac{dy}{dx} + 5y\frac{dx}{dt} = 0$$

$$5x\frac{dy}{dt} = -2x\frac{dx}{dt} - 5y\frac{dx}{dt}$$

$$= -(2x + 5y)\frac{dx}{dt}.$$

Finally,

$$\frac{dy}{dt} = -\frac{2x + 5y}{5x} \cdot \frac{dx}{dt}.$$

(b) If $x = 4$, $y = -\frac{3}{4}$, and $\dfrac{dx}{dt} = 80$, then

$$\frac{dy}{dt} = -\frac{2(4) + 5(-\frac{3}{4})}{5(4)} \cdot 80$$

$$= -17.$$

EXAMPLE 4 A corporation manufactures manual typewriters. Its research department estimates that the demand for its product t years from now will be x hundred thousand typewriters per year, where

$$x = \frac{30}{t + 2} + 1.$$

The profit P (in hundreds of thousands of dollars per year) from the sale of manual typewriters is related to the quantity sold by the equation

$$2P - 3x^2 = -5.$$

At what rate will profits be changing 3 years from now?

Solution Both P and x are functions of t. We are asked to calculate $\dfrac{dP}{dt}$ when $t = 3$. From the equation relating P and x we see that

$$2\frac{dP}{dt} - 3 \cdot 2x \cdot \frac{dx}{dt} = 0$$

$$\frac{dP}{dt} = 3x\frac{dx}{dt}.$$

However,

$$\frac{dx}{dt} = -\frac{30}{(t + 2)^2},$$

so that

$$\frac{dP}{dt} = 3x\left[-\frac{30}{(t+2)^2}\right]$$

$$= -\frac{90x}{(t+2)^2}.$$

In particular, when $t = 3$, we have $x = 30/(t+2) + 1 = 30/5 + 1 = 7$, and

$$\frac{dP}{dt} = -\frac{90(7)}{5^2}$$

$$= -25.2.$$

In other words, 3 years from now, profits will be declining at the rate of 25.2 hundred thousand dollars per year, or 2.52 million dollars per year.

EXAMPLE 5 A corporate planning department estimates that the demand q (in thousands of units per year) for a certain type of computer terminal is related to their price by the demand equation $q = 100 - .8p$. Suppose that technological advances are causing the price to fall at a rate of $200 per year. The current price of a terminal is $1000. At what rate are revenues changing?

Solution The revenue R is related to the quantity sold by the equation

$$[\text{revenue}] = [\text{quantity}] \cdot [\text{price}]$$

$$R = q \cdot p$$

$$= (100 - .8p)p$$

$$= 100p - .8p^2.$$

Therefore, the rate at which revenue is changing equals

$$\frac{dR}{dt} = 100\frac{dp}{dt} - .8 \cdot \left(2p\frac{dp}{dt}\right)$$

$$= 100\frac{dp}{dt} - 1.6p\frac{dp}{dt}.$$

We are given that $p = 1000$ and $\dfrac{dp}{dt} = -200$. Therefore,

$$\frac{dR}{dt} = 100(-200) - 1.6(1000)(-200)$$

$$= -20{,}000 + 320{,}000$$

$$= 300{,}000.$$

Therefore, revenue is increasing at the rate of $300,000 per year. (Can you explain why revenue is increasing even though the price is falling?)

EXAMPLE 6 An epidemiologist finds that a certain tropical disease is carried by mosquitoes. He relates the number of reported cases N of the disease to the number of mosquitoes M observed in a certain marsh, where the mosquito population is estimated using a standardized experiment. He finds that N and M are related by the equation

$$N = .0048M^{2/3}.$$

A program is initiated to destroy the breeding grounds of the mosquito. At one point the mosquito population is 8,000,000, and the program is estimated to be destroying 1,000,000 mosquitoes per day. At what rate is the number of reported cases of the disease changing?

Solution We are asked to calculate $\dfrac{dN}{dt}$. From the given equation, we have

$$\frac{dN}{dt} = .0032M^{-1/3}\frac{dM}{dt}.$$

We are given that $M = 8,000,000$ and $\dfrac{dM}{dt} = -1,000,000$. Thus

$$\frac{dN}{dt} = .0032(8,000,000)^{-1/3}(-1,000,000)$$

$$= .0032\left(\frac{1}{200}\right)(-1,000,000)$$

$$= -16.$$

That is, the number of reported cases of the disease is decreasing at the rate of 16 cases per day.

PRACTICE PROBLEMS 3

Suppose that x and y are related by the equation

$$x^2y^2 - 4x = 3y^4.$$

1. Use implicit differentiation to find a formula for $\dfrac{dy}{dx}$.

2. Find $\dfrac{dy}{dx}$ for $x = 3$, $y = -2$.

EXERCISES 3

In Exercises 1–12, suppose that x and y are related by the given equation and use implicit differentiation to determine $\dfrac{dy}{dx}$.

1. $x^2 + y^2 = 1$ 2. $x^3 + y^3 - 6 = 0$ 3. $y^5 - 3x^2 = x$

4. $x^4 + (y + 3)^4 = x^2$ **5.** $xy = 5$ **6.** $xy^3 = 2$

7. $xy^{-2} = 1$ **8.** $x(y + 2)^5 = 8$ **9.** $(x + 1)^2(y - 1)^2 = 1$

10. $x^3y^2 - 4x^2 = 1$ **11.** $y^2 - y = x^2$ **12.** $x + y - xy = 0$

In Exercises 13–18, suppose that x and y are related by the given equation. Use implicit differentiation to calculate the value of $\dfrac{dy}{dx}$ for the given values of x and y.

13. $4y^3 - x^2 = -5; x = 3, y = 1$ **14.** $xy = 9; x = -3, y = -3$

15. $xy + y^3 = 14; x = 3, y = 2$ **16.** $y^2 - 3xy = -5; x = 2, y = 1$

17. $x^2y + y^2x = 6; x = 1, y = -3$ **18.** $x^3 + y^3 - 2xy = 0; x = 1, y = 1$

In Exercises 19–24, suppose that x and y are both functions of t and are related by the given equation. Use implicit differentiation to determine $\dfrac{dy}{dt}$ in terms of x, y, and $\dfrac{dx}{dt}$.

19. $x^4 + y^4 = 1$ **20.** $y^4 - x^2 = 1$ **21.** $3xy - 3x^2 = 4$

22. $x^4y^4 = 10$ **23.** $x^2 + 2xy = y^3$ **24.** $x^2y^2 = 2y^3 + 1$

25. A manufacturer of microcomputers estimates that t months from now it will sell x thousand units of its main line of microcomputers per month, where $x = .05t^2 + 2t + 5$. Because of economies of scale, the profit P from manufacturing and selling x thousand units is estimated to be $P = .001x^2 + .1x - .25$ million dollars. Calculate the rate at which the profit will be increasing 5 months from now.

26. Ecologists estimate that the average level of carbon monoxide in the air above a certain city will be $1 + .4x + .0001x^2$ ppm (parts per million) when the population is x thousand persons. The population of the city is estimated to be $x = 750 + 25t + .1t^2$ thousand people t years from the present.

(a) Find the rate of change of carbon monoxide with respect to the population of the city.

(b) How fast (with respect to time) is the carbon monoxide level changing at time $t = 2$?

27. A town library estimates that when the town's population is x thousand persons, approximately y thousand books will be checked out of the library during 1 year, where x and y are related by the equation $y^3 - 8000x^2 = 0$. Currently, the population is 27 thousand persons, and the library's book circulation is 180 thousand books per year. At what rate will the library's book circulation rise if the population is increasing at the rate of 1.2 thousand persons per year?

28. Suppose that in a certain city the wholesale price p of oranges (in dollars per crate) and the daily supply x (in thousands of crates) are related by the equation $px + 7x + 8p = 384$. If there are 4000 crates available today at a price of $25 per crate, and if the supply is decreasing at the rate of 300 crates (i.e., .3 thousand) crates per day, at what rate is the price changing?

29. The volume V of a sphere of radius r is $V = \frac{4}{3}\pi r^3$. When helium is pumped into a spherical balloon, both the radius and volume change with respect to time.

(a) Express $\dfrac{dr}{dt}$ in terms of $\dfrac{dV}{dt}$.

(b) Suppose that the volume of the balloon is increasing at the rate of 500 cm^3 per minute. At what rate is the radius of the balloon changing when the radius is 10 cm?

30. A rectangular aquarium with a 50-cm by 80-cm base is being filled with water at the rate of 5000 cm^3 per minute. At what rate is the height of the water changing when the height is 15 cm?

SOLUTIONS TO PRACTICE PROBLEMS 3

1. We differentiate term by term, regarding y as a function of x.

$$\frac{d}{dx}(x^2y^2) - \frac{d}{dx}(4x) = \frac{d}{dx}(3y^4)$$

$$\overbrace{x^2 \cdot 2y\frac{dy}{dx} + y^2 \cdot 2x} - 4 = 12y^3\frac{dy}{dx}.$$

Next, we move the terms involving $\dfrac{dy}{dx}$ to the left side of the equation and move the other terms to the right side. Then we can factor out $\dfrac{dy}{dx}$ and solve for $\dfrac{dy}{dx}$.

$$2x^2y\frac{dy}{dx} - 12y^3\frac{dy}{dx} = 4 - 2xy^2$$

$$(2x^2y - 12y^3)\frac{dy}{dx} = 4 - 2xy^2$$

$$\frac{dy}{dx} = \frac{4 - 2xy^2}{2x^2y - 12y^3} = \frac{2 - xy^2}{x^2y - 6y^3}.$$

2. Substituting $x = 3$ and $y = -2$, we have

$$\frac{dy}{dx} = \frac{2 - (3)(-2)^2}{3^2(-2) - 6(-2)^3} = \frac{-10}{30} = -\frac{1}{3}.$$

Chapter 13: CHECKLIST

☐ $\dfrac{d}{dx}[f(x) \cdot g(x)] = f(x)g'(x) + g(x)f'(x)$

☐ $\dfrac{d}{dx}\left[\dfrac{f(x)}{g(x)}\right] = \dfrac{g(x)f'(x) - f(x)g'(x)}{[g(x)]^2}$

☐ $\dfrac{d}{dx}[f(g(x))] = f'(g(x)) \cdot g'(x)$

☐ Implicit differentiation
☐ Related rates

Chapter 13: SUPPLEMENTARY EXERCISES

Differentiate the following functions.

1. $\sqrt{x}(x^3 + 1)$

2. $\sqrt{x + 2}(x^2 + 5x + 1)$

3. $\dfrac{\sqrt{x}}{\sqrt{x + 1}}$

4. $(x^2 - 1)^5(x^3 + 5)^4$

5. $(x^2 + 5x + 9)(x^3 - 2)^{10}$

6. $(\sqrt{x} - 1)^3(\sqrt[3]{x} - 2)^2$

7. $\dfrac{(x + 2)^2 - (x - 2)^2}{x^2 + 2}$

8. $\dfrac{x - \sqrt{x}}{x + \sqrt{x}}$

9. $(x + 2 + (x + 2)^2)^2$

10. $(\sqrt{x} + \sqrt{x + 1})^2$

In Exercises 11–16, find formulas for $f(g(x))$ and $f'(g(x))$, where $g(x)$ represents an arbitrary differentiable function.

11. $f(x) = x^2 + x + 3$

12. $f(x) = x^3 - 8x$

13. $f(x) = \sqrt{x}(x + 1)$

14. $f(x) = \dfrac{x}{x + 4}$

15. $f(x) = (3x - 1)^5$

16. $f(x) = (5 - \tfrac{1}{2}x)^6$

In Exercises 17–22, find a formula for $\dfrac{d}{dx}[f(g(x))]$, where $g(x)$ represents an arbitrary differentiable function.

17. $f(x) = \dfrac{5}{x^3}$

18. $f(x) = \sqrt{x}$

19. $f(x) = (4 - x)^{3/4}$

20. $f(x) = (2x + 3)^{5/3}$

21. $f(x) = \dfrac{2}{\sqrt{x + 1}}$

22. $f(x) = \dfrac{8}{(x - 1)^2}$

In Exercises 23–25, find a formula for $\dfrac{d}{dx}[f(g(x))]$, where $f(x)$ is a function such that $f'(x) = 1/(x^2 + 1)$.

23. $g(x) = x^3$

24. $g(x) = \dfrac{1}{x}$

25. $g(x) = x^2 + 1$

In Exercises 26–28, find a formula for $\dfrac{d}{dx}[f(g(x))]$, where $f(x)$ is a function such that $f'(x) = x\sqrt{1 - x^2}$.

26. $g(x) = x^2$

27. $g(x) = \sqrt{x}$

28. $g(x) = x^{3/2}$

29. Suppose that $y = f(x)$ is a function with the property that the slope of the tangent line at each point (x, y) on its graph is $1/y$. In addition, suppose that $(1, 2)$ is a point on the graph. What is the slope of the tangent line to the curve $y = f(x^3)$ at $(1, 2)$?

30. Rework Exercise 29 with $1/y$ replaced by $1/x$.

31. Animal physiologists have determined experimentally that the weight W (in kilograms) and the surface area S (in square meters) of a typical horse are related by the empirical

equation $S = 0.1W^{2/3}$. How fast is the surface area of a horse increasing at a time when the horse weighs 350 kg and is gaining weight at the rate of 200 kg per year? [*Hint:* Use the chain rule.]

32. When oxygen is compressed adiabatically (i.e., in such a way that no heat enters or leaves the gas), the pressure P (in torr) and the volume V (in liters) are related by the equation $P = 7.6/V^{1.4}$. Suppose that at some moment the volume is 2 liters, the pressure is 2.88 torr, and the volume is decreasing at the rate of .6 liter per minute. How fast is the pressure increasing?

In Exercises 33–38, x and y are related by the given equation. Use implicit differentiation to calculate the value of $\dfrac{dy}{dx}$ for the given values of x and y.

33. $x^2y^2 = 9; x = 1, y = 3$

34. $xy^4 = 48; x = 3, y = 2$

35. $x^2 - xy^3 = 20; x = 5, y = 1$

36. $xy^2 - x^3 = 10; x = 2, y = 3$

37. $x^3y^3 - x^2y^2 = 0; x = 5, y = \frac{1}{5}$

38. $(x + y)^3 = x^3 + y^3; x = 1, y = -1$

39. Suppose that a kitchen appliance company's monthly sales and advertising expenses are approximately related by the equation $xy - 6x + 20y = 0$, where x is thousands of dollars spent on advertising and y is thousands of dishwashers sold. Currently, the company is spending 10 thousand dollars on advertising and is selling 2 thousand dishwashers each month. If the company plans to increase monthly advertising expenditures at the rate of $1.5 thousand per month, how fast will sales rise? Use implicit differentiation to answer the question.

The Exponential and Natural Logarithm Functions

Chapter 14

When an investment grows steadily at 15% per year, the rate of growth of the investment at any time is proportional to the value of the investment at that time. When a bacteria culture grows in a laboratory dish, the rate of growth of the culture at any moment is proportional to the total number of bacteria in the dish at that moment. These situations are examples of what is called *exponential growth*. A pile of radioactive uranium U^{235} decays at a rate that at each moment is proportional to the amount of U^{235} present. This decay of uranium (and of radioactive elements in general) is called *exponential decay*. Both exponential growth and exponential decay can be described and studied in terms of exponential functions and the natural logarithm function. The properties of these functions are investigated in this chapter. Subsequently, we shall explore a wide range of applications, in fields such as business, biology, archeology, public health, and pyschology.

14.1 Exponential Functions

Throughout this section b will denote a positive number. The function

$$f(x) = b^x$$

is called an *exponential function*, because the variable x is in the exponent. The number b is called the *base* of the exponential function. In Section 1.3 we reviewed the definition of b^x for various values of b and x (although we used the letter r there instead of x). For instance, if $f(x)$ is the exponential function with base 2,

$$f(x) = 2^x,$$

then

$$f(0) = 2^0 = 1, \quad f(1) = 2^1 = 2, \quad f(4) = 2^4 = 2 \cdot 2 \cdot 2 \cdot 2 = 16,$$

and

$$f(-1) = 2^{-1} = \tfrac{1}{2}, \qquad f(\tfrac{1}{2}) = 2^{1/2} = \sqrt{2}, \qquad f(\tfrac{3}{5}) = (2^{1/5})^3 = (\sqrt[5]{2})^3.$$

Actually, in Section 1.3, we only defined b^x for rational (i.e., integer or fractional) values of x. For other values of x (such as $\sqrt{3}$ or π), it is possible to define b^x by first approximating x with rational numbers and then applying a limiting process. We shall omit the details and simply assume henceforth that b^x can be defined for all numbers x in such a way that the usual laws of exponents remain valid.

Let us state the laws of exponents for reference.

(i) $b^x \cdot b^y = b^{x+y}$	(iv) $(b^y)^x = b^{xy}$
(ii) $b^{-x} = \dfrac{b}{b^x}$	(v) $a^x b^x = (ab)^x$
(iii) $\dfrac{b^x}{b^y} = b^x \cdot b^{-y} = b^{x-y}$	(vi) $\dfrac{a^x}{b^x} = \left(\dfrac{a}{b}\right)^x$

Property (iv) may be used to change the appearance of an exponential function. For instance, the function $f(x) = 8^x$ may also be written as $f(x) = (2^3)^x = 2^{3x}$, and $g(x) = (\tfrac{1}{9})^x$ may be written as $g(x) = (1/3^2)^x = (3^{-2})^x = 3^{-2x}$.

EXAMPLE 1 Use properties of exponents to write the following functions in the form 2^{kx} for a suitable constant k.

(a) $4^{5x/2}$ (b) $(2^{4x} \cdot 2^{-x})^{1/2}$ (c) $8^{x/3} \cdot 16^{3x/4}$ (d) $\dfrac{10^x}{5^x}$

Solution (a) First express the base 4 as a power of 2, and then use Property (iv):

$$4^{5x/2} = (2^2)^{5x/2} = 2^{2(5x/2)} = 2^{5x}.$$

(b) Use Property (i) first to simplify the quantity inside the parentheses, and then use Property (iv):

$$(2^{4x} \cdot 2^{-x})^{1/2} = (2^{4x-x})^{1/2} = (2^{3x})^{1/2} = 2^{(3/2)x}.$$

(c) First express the bases 8 and 16 as powers of 2, and then use (iv) and (i):

$$8^{x/3} \cdot 16^{3x/4} = (2^3)^{x/3} \cdot (2^4)^{3x/4} = 2^x \cdot 2^{3x} = 2^{4x}.$$

(d) Use (v) to change the numerator 10^x, and then cancel the common term 5^x:

$$\frac{10^x}{5^x} = \frac{(2 \cdot 5)^x}{5^x} = \frac{2^x \cdot 5^x}{5^x} = 2^x.$$

An alternative method is to use Property (vi):

$$\frac{10^x}{5^x} = \left(\frac{10}{5}\right)^x = 2^x.$$

x	2^x
-3	.125
-2	.25
-1	.5
0	1.0
1	2.0
2	4.0
3	8.0

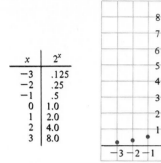

FIGURE 1

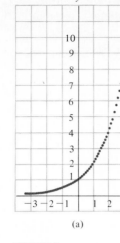

(a)

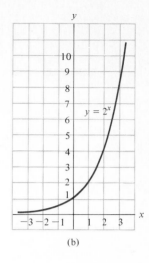

(b)

FIGURE 2

Let us now study the graph of the exponential function $y = b^x$ for various values of b. We begin with the special case $b = 2$.

We have tabulated the values of 2^x for $x = 0, \pm 1, \pm 2, \pm 3$ and plotted these values in Fig. 1. Other intermediate values of 2^x for $x = \pm .1, \pm .2, \pm .3, \ldots$, may be obtained from tables or from a calculator with a y^x key. [See Fig. 2(a).] By passing a smooth curve through these points, we obtain the graph of $y = 2^x$, in Fig. 2(b).

In the same manner, we have sketched the graph of $y = 3^x$ (Fig. 3). The graphs of $y = 2^x$ and $y = 3^x$ have the same basic shape. Also note that they both pass through the point $(0, 1)$ (because $2^0 = 1$, $3^0 = 1$).

In Fig. 4 we have sketched the graphs of several more exponential functions. Notice that the graph of $y = 5^x$ has a large slope at $x = 0$, since the graph at

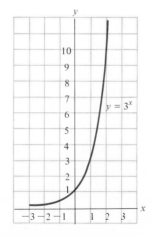

FIGURE 3

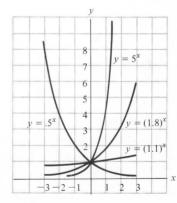

FIGURE 4

$x = 0$ is quite steep; however, the graph of $y = (1.1)^x$ is nearly horizontal at $x = 0$, and hence the slope is close to zero.

EXAMPLE 2 Let $f(x) = 3^{5x}$. Determine all x for which $f(x) = 27$.

Solution Since $27 = 3^3$, we must determine all x for which

$$3^{5x} = 3^3.$$

Equating exponents, we have

$$5x = 3$$

$$x = \tfrac{3}{5}.$$

In general, for $b > 1$, the equation $b^r = b^s$ implies that $r = s$. This is because the graph of $y = b^x$ has the same basic shape as $y = 2^x$ and $y = 3^x$. Similarly, when $0 < b < 1$, the equation $b^r = b^s$ implies that $r = s$, because the graph of $y = b^x$ resembles the graph of $y = (\tfrac{1}{3})^x$ and is always decreasing.

There is no need at this point to become familiar with the graphs of the functions b^x. We have shown a few graphs merely to make the reader more comfortable with the concept of an exponential function. The main purpose of this section has been to review properties of exponents in a context that is appropriate for our future work.

PRACTICE PROBLEMS 1

1. Can a function such as $f(x) = 5^{3x}$ be written in the form $f(x) = b^x$? If so, what is b?
2. Solve the equation $7 \cdot 2^{6-3x} = 28$.

EXERCISES 1

Write each function in Exercises 1–14 in the form 2^{kx} or 3^{kx}, for a suitable constant k.

1. $4^x, (\sqrt{3})^x, (\tfrac{1}{9})^x$

2. $27^x, (\sqrt[3]{2})^x, (\tfrac{1}{8})^x$

3. $8^{2x/3}, 9^{3x/2}, 16^{-3x/4}$

4. $9^{-x/2}, 8^{4x/3}, 27^{-2x/3}$

5. $(\tfrac{1}{4})^{2x}, (\tfrac{1}{8})^{-3x}, (\tfrac{1}{81})^{x/2}$

6. $(\tfrac{1}{9})^{2x}, (\tfrac{1}{27})^{x/3}, (\tfrac{1}{16})^{-x/2}$

7. $2^{3x} \cdot 2^{-5x/2}, 3^{2x} \cdot (\tfrac{1}{3})^{2x/3}$

8. $2^{5x/4} \cdot (\tfrac{1}{2})^x, 3^{-2x} \cdot 3^{5x/2}$

9. $(2^{-3x} \cdot 2^{-2x})^{2/5}, (9^{1/2} \cdot 9^4)^{x/9}$

10. $(3^{-x} \cdot 3^{x/5})^5, (16^{1/4} \cdot 16^{-3/4})^{3x}$

11. $\dfrac{3^{4x}}{3^{2x}}, \dfrac{2^{5x+1}}{2 \cdot 2^{-x}}, \dfrac{9^{-x}}{27^{-x/3}}$

12. $\dfrac{2^x}{6^x}, \dfrac{3^{-5x}}{3^{-2x}}, \dfrac{16^x}{8^{-x}}$

13. $6^x \cdot 3^{-x}, \dfrac{15^x}{5^x}, \dfrac{12^x}{2^{2x}}$ **14.** $7^{-x} \cdot 14^x, \dfrac{2^x}{6^x}, \dfrac{32^x}{18^x}$

From a table we have $2^{1/2} \approx 1.414$, $2^{1/10} \approx 1.072$, $(2.7)^{1/2} \approx 1.643$, and $(2.7)^{1/10} \approx 1.104$. (These figures are accurate to three decimal places.) Compute the following numbers, rounding off your answer to two decimal places, if necessary.

15. (a) 2^x for $x = 3$ (b) 2^x for $x = -3$

   (c) 2^x for $x = \frac{5}{2}$ $\left[Hint: \frac{5}{2} = 2 + \frac{1}{2}.\right]$ (d) 2^x for $x = 4.1$

   (e) 2^x for $x = .2$ (f) 2^x for $x = .9$ $[Hint: .9 = 1 - .1.]$

   (g) 2^x for $x = -2.5$ $[Hint: -2.5 = -3 + .5.]$

   (h) 2^x for $x = -3.9$

16. (a) $(2.7)^x$ for $x = .2$ (b) $(2.7)^x$ for $x = 1.5$ (c) $(2.7)^x$ for $x = 0$

   (d) $(2.7)^x$ for $x = -1$ (e) $(2.7)^x$ for $x = 1.1$ (f) $(2.7)^x$ for $x = .6$

Solve the following equations for x.

17. $5^{2x} = 5^2$ **18.** $10^{-x} = 10^2$

19. $(2.5)^{2x+1} = (2.5)^5$ **20.** $(3.2)^{x-3} = (3.2)^5$

21. $10^{1-x} = 100$ **22.** $2^{4-x} = 8$

23. $3(2.7)^{5x} = 8.1$ **24.** $4(2.7)^{2x-1} = 10.8$

25. $(2^{x+1} \cdot 2^{-3})^2 = 2$ **26.** $(3^{2x} \cdot 3^2)^4 = 3$

27. $2^{3x} = 4 \cdot 2^{5x}$ **28.** $3^{5x} \cdot 3^x - 3 = 0$

29. $(1 + x)2^{-x} - 5 \cdot 2^{-x} = 0$ **30.** $(2 - 3x)5^x + 4 \cdot 5^x = 0$

The expressions in Exercises 31–38 may be factored as shown. Find the missing factors.

31. $2^{3+h} = 2^3 (\quad)$ **32.** $5^{2+h} = 25 (\quad)$

33. $2^{x+h} - 2^x = 2^x (\quad)$ **34.** $5^{x+h} + 5^x = 5^x (\quad)$

35. $3^{x/2} + 3^{-x/2} = 3^{-x/2} (\quad)$ **36.** $5^{7x/2} - 5^{x/2} = \sqrt{5^x} (\quad)$

37. $3^{10x} - 1 = (3^{5x} - 1) (\quad)$ **38.** $2^{6x} - 2^x = (2^{3x} - 2^{x/2}) (\quad)$

SOLUTIONS TO PRACTICE PROBLEMS 1

1. If $5^{3x} = b^x$, then when $x = 1$, $5^{3(1)} = b^1$, which says that $b = 125$. This value of b certainly works, because

$$5^{3x} = (5^3)^x = 125^x.$$

2. Divide both sides of the equation by 7. We then obtain

$$2^{6-3x} = 4.$$

Now 4 can be written as 2^2. So we have

$$2^{6-3x} = 2^2.$$

Equate exponents.

$$6 - 3x = 2$$

$$4 = 3x$$

$$x = \tfrac{4}{3}.$$

14.2 The Exponential Function e^x

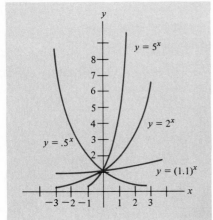

FIGURE 1

Let us begin by examining the graphs of the exponential functions shown in Fig. 1. They all pass through $(0, 1)$, but with different slopes there. Notice that the graph of 5^x is quite steep at $x = 0$, while the graph of $(1.1)^x$ is nearly horizontal at $x = 0$. It turns out that at $x = 0$, the graph of 2^x has a slope of approximately .69, while the graph of 3^x has a slope of approximately 1.1.

Evidently, there is a particular value of the base b, between 2 and 3, where the graph of b^x has slope *exactly* 1 at $x = 0$. We denote this special value of b by the letter e, and we call

$$f(x) = e^x$$

the exponential function. The number e is an important constant of nature that has been calculated to thousands of decimal places. To 10 significant digits, we have $e = 2.718281828$. For our purposes, it is usually sufficient to think of e as "approximately 2.7."

Our goal in this section is to find a formula for the derivative of e^x. It turns out that the calculations for e^x and 2^x are very similar. Since many people are more comfortable working with 2^x rather than e^x, we shall first analyze the graph of 2^x. Then we shall draw the appropriate conclusions about the graph of e^x.

Before computing the slope of $y = 2^x$ at an arbitrary x, let us consider the special case $x = 0$. Denote the slope at $x = 0$ by m. We shall use the secant-line approximation of the derivative to approximate m. We proceed by constructing a secant line in Fig. 2. The slope of the secant line through $(0, 1)$ and $(h, 2^h)$ is $\dfrac{2^h - 1}{h}$. As h approaches zero, the slope of the secant line approaches the slope of $y = 2^x$ at $x = 0$. That is,

$$m = \lim_{h \to 0} \frac{2^h - 1}{h}. \tag{1}$$

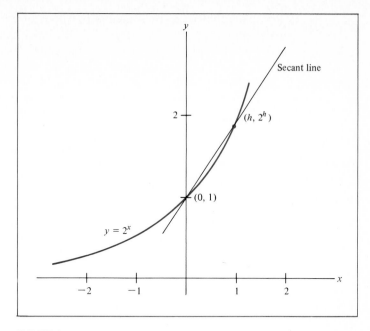

FIGURE 2

We can estimate the value of m by taking h smaller and smaller. When $h = .1$, we have $2^h \approx 1.072$ (from a table of values of 2^x), and

$$\frac{2^h - 1}{h} \approx \frac{.072}{.1} = .72.$$

When $h = .01$, we have $2^h \approx 1.00696$, and

$$\frac{2^h - 1}{h} \approx \frac{.00696}{.01} \approx .696.$$

When $h = .001$, $2^h \approx 1.0006934$, so that

$$\frac{2^h - 1}{h} \approx \frac{.0006934}{.001} \approx .693.$$

Thus it is reasonable to conclude that $m \approx .69$.* Since m equals the slope of $y = 2^x$ at $x = 0$, we have

$$m = \frac{d}{dx}(2^x)\bigg|_{x=0} \approx .69. \tag{2}$$

---

* To 10 decimal places, m is .6931471806.

Now that we have estimated the slope of $y = 2^x$ at $x = 0$, let us compute the slope for an arbitrary value of x. We construct a secant line through $(x, 2^x)$ and a nearby point $(x + h, 2^{x+h})$ on the graph. The slope of the secant line is

$$\frac{2^{x+h} - 2^x}{h}. \tag{3}$$

By law of exponents (i), we have $2^{x+h} - 2^x = 2^x(2^h - 1)$ so that, by (1), we see that

$$\lim_{h \to 0} \frac{2^{x+h} - 2^x}{h} = \lim_{h \to 0} 2^x \frac{2^h - 1}{h} = 2^x \lim_{h \to 0} \frac{2^h - 1}{h} = m2^x. \tag{4}$$

However, the slope of the secant (3) approaches the derivative of 2^x as h approaches zero. Consequently, we have

$$\frac{d}{dx}(2^x) = m2^x, \qquad \text{where } m = \frac{d}{dx}(2^x)\Big|_{x=0}. \tag{5}$$

EXAMPLE 1 Calculate (a) $\dfrac{d}{dx}(2^x)\Big|_{x=3}$ and (b) $\dfrac{d}{dx}(2^x)\Big|_{x=-1}$.

Solution (a) $\dfrac{d}{dx}(2^x)\Big|_{x=3} = m \cdot 2^3 = 8m \approx 8(.69) = 5.52.$

(b) $\dfrac{d}{dx}(2^x)\Big|_{x=-1} = m \cdot 2^{-1} = .5m \approx .5(.69) = .345.$

The calculations just carried out for $y = 2^x$ can be carried out for $y = b^x$, where b is any positive number. Equation (5) will read exactly the same except that 2 will be replaced by b. Thus we have the following formula for the derivative of the function $f(x) = b^x$:

$$\frac{d}{dx}(b^x) = mb^x, \qquad \text{where } m = \frac{d}{dx}(b^x)\Big|_{x=0}. \tag{6}$$

Our calculations showed that if $b = 2$, then $m \approx .69$. If $b = 3$, then it turns out that $m \approx 1.1$. (See Exercise 1.) Obviously, the derivative formula in (6) is simplest when $m = 1$, that is, when the graph of b^x has slope 1 at $x = 0$. As we said earlier, this special value of b is denoted by the letter e. Thus the number e has the property that

$$\frac{d}{dx}(e^x)\Big|_{x=0} = 1 \tag{7}$$

and

$$\frac{d}{dx}(e^x) = 1 \cdot e^x = e^x. \tag{8}$$

The graphical interpretation of (7) is that the curve $y = e^x$ has slope 1 at $x = 0$. The graphical interpretation of (8) is that the slope of the curve $y = e^x$ at an arbitrary value of x is exactly equal to the value of the function e^x at that point. (See Fig. 3.)

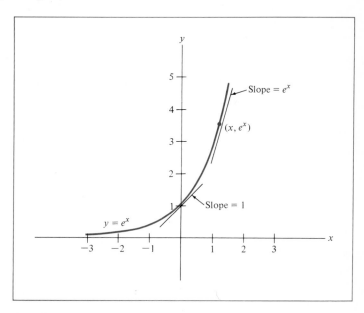

FIGURE 3

The function e^x is the same type of function as 2^x and 3^x except that taking derivatives of e^x is much easier. For this reason, functions based on e^x are used in almost all applications that require an exponential-type function to describe a physical phenomenon. Extensive tables are available from which to determine the value of e^x for a wide range of values of x. Also, scientific calculators provide for calculating e^x at the push of a button.

PRACTICE PROBLEMS 2

In the following problems, use the number 20 as the (approximate) value of e^3.

1. Find the equation of the tangent line to the graph of $y = e^x$ at $x = 3$.

2. Solve the following equation for x:

$$4e^{6x} = 80.$$

EXERCISES 2

1. Use the table below to show that

$$\frac{d}{dx}(3^x)\Big|_{x=0} \approx 1.1.$$

That is, calculate the slope

$$\frac{3^h - 1}{h}$$

of the secant line passing through the point $(0, 1)$ and $(h, 3^h)$. Take $h = .1, .01,$ and $.001$.

x	3^x
0	1.00000
.001	1.00110
.010	1.01105
.100	1.11612

2. Use the table below to show that

$$\frac{d}{dx}(2.7)^x\Big|_{x=0} \approx .99.$$

That is, calculate

$$\frac{(2.7)^h - 1}{h}$$

for $h = .1, .01,$ and $.001$.

x	$(2.7)^x$
0	1.00000
.001	1.00099
.010	1.00998
.100	1.10443

3. Consider the secant line on the graph of e^x passing through $(0, 1)$ and (h, e^h). Its slope is

$$\frac{e^h - 1}{h}.$$

Compute this quantity for $h = .01, .005,$ and $.001$.

x	e^x
0	1.00000
.001	1.00100
.005	1.00501
.010	1.01005

4. Use (8) and a familiar rule for differentiation to find

$$\frac{d}{dx}(5e^x).$$

5. Use (8) and a familiar rule for differentiation to find

$$\frac{d}{dx}(e^x)^{10}.$$

6. Use the fact that $e^{2+x} = e^2 \cdot e^x$ to find

$$\frac{d}{dx}(e^{2+x}).$$

[Remember that e^2 is just a constant—approximately $(2.7)^2$.]

7. Use the fact that $e^{4x} = (e^x)^4$ to find

$$\frac{d}{dx}(e^{4x}).$$

8. Find $\dfrac{d}{dx}(e^x + x^2)$.

Simplify.

9. $e^{2x}(1 + e^{3x})$ **10.** $(e^x)^2$ **11.** $e^{1-x} \cdot e^{2x}$

12. $\dfrac{5e^{3x}}{e^x}$ **13.** $\dfrac{1}{e^{-2x}}$ **14.** $e^3 \cdot e^{x+1}$

Use Table 8 of the Appendix to determine the following numbers:

15. e^2 **16.** $e^{-1.30}$ **17.** $e^{-.5}$ **18.** $e^{3/2}$

Solve the following equations for x.

19. $e^{5x} = e^{20}$ **20.** $e^{1-x} = e^2$

21. $e^{x^2-2x} = e^8$ **22.** $e^{-x} = 1$

Differentiate the following functions.

23. xe^x **24.** $\dfrac{e^x}{x}$ **25.** $\dfrac{e^x}{1 + e^x}$

26. $(1 + x^2)e^x$ **27.** $(1 + 5e^x)^4$ **28.** $(xe^x - 1)^{-3}$

SOLUTIONS TO PRACTICE PROBLEMS 2

1. When $x = 3$, $y = e^3 = 20$. So the point $(3, 20)$ is on the tangent line. Since $\dfrac{d}{dx}(e^x) = e^x$, the slope of the tangent line is e^3 or 20. Therefore, the equation of the tangent line in point-slope form is $y - 20 = 20(x - 3)$.

This problem is similar to the Practice Problem 2 of Section 1. First divide both sides of the equation by 4.

$$e^{6x} = 20.$$

The idea is to express 20 as a power of e and then equate exponents.

$$e^{6x} = e^3$$

$$6x = 3$$

$$x = \tfrac{1}{2}.$$

14.3 Differentiation of Exponential Functions

We have shown that $\dfrac{d}{dx}(e^x) = e^x$. Using this fact and the chain rule, we may differentiate functions of the form $e^{g(x)}$, where $g(x)$ is any differentiable function. This is because $e^{g(x)}$ is the composite of two functions. Indeed, if $f(x) = e^x$, then

$$e^{g(x)} = f(g(x)).$$

Thus, by the chain rule, we have

$$\frac{d}{dx}(e^{g(x)}) = f'(g(x))g'(x)$$

$$= f(g(x))g'(x) \qquad [\text{since } f'(x) = f(x)]$$

$$= e^{g(x)}g'(x).$$

So we have the following result:

> *Chain Rule for Exponential Functions* Let $g(x)$ be any differentiable function. Then
>
> $$\frac{d}{dx}(e^{g(x)}) = e^{g(x)}g'(x).$$

(1)

EXAMPLE 1   Differentiate e^{x^2+1}.

Solution   Here $g(x) = x^2 + 1$, $g'(x) = 2x$, so

$$\frac{d}{dx}(e^{x^2+1}) = e^{x^2+1} \cdot 2x$$

$$= 2xe^{x^2+1}.$$

EXAMPLE 2   Differentiate $e^{3x^2 - (1/x)}$.

Solution

$$\frac{d}{dx}(e^{3x^2-(1/x)}) = e^{3x^2-(1/x)} \cdot \frac{d}{dx}\left(3x^2 - \frac{1}{x}\right)$$

$$= e^{3x^2-(1/x)}\left(6x + \frac{1}{x^2}\right).$$

EXAMPLE 3 Differentiate e^{5x}.

Solution

$$\frac{d}{dx}(e^{5x}) = e^{5x} \cdot \frac{d}{dx}(5x) = e^{5x} \cdot 5 = 5e^{5x}.$$

Using a computation similar to that used in Example 3, we may differentiate e^{kx} for any constant k. (In Example 3 we have $k = 5$.) The result is the following useful formula.

$$\frac{d}{dx}(e^{kx}) = ke^{kx}. \tag{2}$$

Many applications involve exponential functions of the form $y = Ce^{kx}$, where C and k are constants. In the next example we differentiate such functions.

EXAMPLE 4 Differentiate the following exponential functions.

(a) $3e^{5x}$

(b) $3e^{kx}$, where k is a constant

(c) Ce^{kx}, where C and k are constants

Solution (a) $\dfrac{d}{dx}(3e^{5x}) = 3\dfrac{d}{dx}(e^{5x}) = 3 \cdot 5e^{5x} = 15e^{5x}.$

(b) $\dfrac{d}{dx}(3e^{kx}) = 3\dfrac{d}{dx}(e^{kx})$

$$= 3 \cdot ke^{kx} \qquad [\text{by (2)}]$$
$$= 3ke^{kx}.$$

(c) $\dfrac{d}{dx}(Ce^{kx}) = C\dfrac{d}{dx}(e^{kx}) = Cke^{kx}.$

The result of part (c) may be summarized in an extremely useful fashion as follows: Suppose that we let $y = Ce^{kx}$. By part (c) we have

$$y' = Cke^{kx}$$
$$= k \cdot (Ce^{kx})$$
$$= ky.$$

In other words, the derivative of the function Ce^{kx} is k times the function itself. Let us record this fact.

> Let C, k be any constants and let $y = Ce^{kx}$. Then y satisfies the equation
> $$y' = ky.$$

The equation $y' = ky$ expresses a relationship between the function y and its derivative y'. Any equation expressing a relationship between a function y and one or more of its derivatives is called a *differential equation*.

Very often an applied problem will involve a function $y = f(x)$ which satisfies the differential equation $y' = ky$. It can be shown that y must then necessarily be an exponential function of the form Ce^{kx}. That is, we have the following result.

> Suppose that $y = f(x)$ satisfies the differential equation
> $$y' = ky.$$
> Then y is an exponential function of the form
> $$y = Ce^{kx}, \qquad C \text{ a constant.}$$

(3)

We shall verify this result in Exercise 47.

EXAMPLE 5 Determine all functions $y = f(x)$ such that $y' = -.2y$.

Solution The equation $y' = -.2y$ has the form $y' = ky$ with $k = -.2$. Therefore, any solution of the equation has the form
$$y = Ce^{-.2x},$$
where C is a constant.

EXAMPLE 6 Determine all functions $y = f(x)$ such that $y' = y/2$ and $f(0) = 4$.

Solution The equation $y' = y/2$ has the form $y' = ky$ with $k = \frac{1}{2}$. Therefore,
$$f(x) = Ce^{(1/2)x}$$
for some constant C. We also require that $f(0) = 4$. That is,
$$4 = f(0) = Ce^{(1/2) \cdot 0} = Ce^0 = C.$$
So $C = 4$ and
$$f(x) = 4e^{(1/2)x}.$$

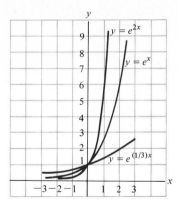

FIGURE 1

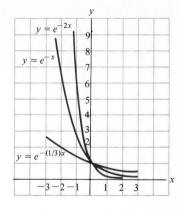

FIGURE 2

The Functions e^{kx} Exponential functions of the form e^{kx} occur in many applications. Figure 1 shows the graphs of several functions of this type when k is a positive number. These curves $y = e^{kx}$, k positive, have several properties in common:

1. $(0, 1)$ is on the graph.
2. The graph lies strictly above the x-axis (e^{kx} is never zero).
3. The x-axis is an asymptote as x becomes large negatively.
4. The graph is always increasing and concave up.

When k is negative, the graph of $y = e^{kx}$ is decreasing. (See Fig. 2.) Note the following properties of the curves $y = e^{kx}$, k negative:

1. $(0, 1)$ is on the graph.
2. The graph lies strictly above the x-axis.
3. The x-axis is an asymptote as x becomes large positively.
4. The graph is always decreasing and concave up.

The Functions b^x If b is a positive number, then the function b^x may be written in the form e^{kx} for some k. For example, take $b = 2$. From Fig. 3 of the preceding section it is clear that there is some value of x such that $e^x = 2$. Call this value k, so that $e^k = 2$. Then

$$2^x = (e^k)^x = e^{kx}$$

for all x. In general, if b is any positive number, there is a value of x, say $x = k$, such that $e^k = b$. In this case, $b^x = (e^k)^x = e^{kx}$. Thus all the curves $y = b^x$ discussed in Section 2 can be written in the form $y = e^{kx}$. This is one reason why we have focused on exponential functions with base e instead of studying 2^x, 3^x, and so on.

PRACTICE PROBLEMS 3

1. Differentiate $[e^{-3x}(1 + e^{6x})]^{12}$.

2. Determine all functions $y = f(x)$ such that $y' = -y/20$, $f(0) = 2$.

EXERCISES 3

Differentiate the following.

1. $y = e^{-x}$

2. $f(x) = e^{10x}$

3. $f(x) = 5e^x$

4. $y = \dfrac{e^x + e^{-x}}{2}$

5. $f(t) = e^{t^2}$

6. $f(t) = e^{-2t}$

7. $f(x) = \dfrac{e^x - e^{-x}}{2}$

8. $f(x) = 2e^{1-x}$

9. $y = e^{-2x} - 2x$

10. $f(x) = \frac{1}{10}e^{-x^2/2}$

11. $g(x) = (e^x + e^{-x})^3$

12. $y = (e^{-x})^2$

13. $y = \frac{1}{3}e^{3-2x}$

14. $g(x) = e^{1/x}$

15. $f(t) = e^t(e^{2t} - e^{-t})$

16. $f(t) = \dfrac{e^t + e^{-t}}{e^t}$

[*Hint:* In Exercises 15 and 16, simplify $f(t)$ before differentiating.]

17. $y = e^{x^3 + x - (1/x)}$

18. $y = (e^{x^2} + x^2)^5$

19. $f(x) = (2x + 1 - e^{2x+1})^4$

20. $f(x) = e^{1/(3x-7)}$

21. $x^3 e^{x^2}$

22. $x^2 e^{-3x}$

23. e^{x^3}/x

24. $e^{(x-1)/x}$

25. $(x + 1)e^{-x+2}$

26. $x\sqrt{2 + e^x}$

27. $\left(\dfrac{1}{x} + 3\right)e^x$

28. $\dfrac{e^{-3x}}{1 - 3x}$

29. $\dfrac{e^x - 1}{e^x + 1}$

30. $\dfrac{xe^x - 3}{x + 1}$

In Exercises 31–40, find all values of x such that the function has a possible maximum or minimum point. Use the second derivative test to determine the nature of the function at these points. (Recall that e^x is positive for all x.)

31. $f(x) = (1 + x)e^{-x/2}$

32. $f(x) = (1 - x)e^{-x/2}$

33. $f(x) = \dfrac{3 - 2x}{e^{x/4}}$

34. $f(x) = \dfrac{4x - 3}{e^{x/2}}$

35. $f(x) = (8 - 2x)e^{x+5}$

36. $f(x) = (4x - 1)e^{3x-2}$

37. $f(x) = \dfrac{(x-1)^2}{e^x}$

38. $f(x) = (x+3)^2 e^x$

39. $f(x) = (x+5)^2 e^{2x-1}$

40. $f(x) = \dfrac{(x-3)^2}{e^{2x}}$

41. Let a and b be positive numbers: A curve whose equation is $y = e^{-ae^{-bx}}$ is called a *Gompertz growth curve*. These curves are used in biology to describe certain types of population growth. Compute the derivative of $y = e^{-2e^{-.01x}}$.

42. Find $\dfrac{dy}{dx}$ if $y = e^{-(1/10)e^{-x/2}}$.

43. Determine all solutions of the differential equation

$$y' = -4y.$$

44. Determine all solutions of the differential equation

$$y' = \tfrac{1}{3}y.$$

45. Determine all functions $y = f(x)$ such that

$$y' = -.5y \quad \text{and} \quad f(0) = 1.$$

46. Determine all functions $y = f(x)$ such that

$$y' = 3y \quad \text{and} \quad f(0) = \tfrac{1}{2}.$$

47. Verify the result (3). [*Hint:* Let $g(x) = f(x)e^{-kx}$. Show that $g'(x) = 0$.] You may assume that only a constant function has a zero derivative.

48. Let $f(x)$ be a function with the property that $f'(x) = 1/x$. Let $g(x) = f(e^x)$, and compute $g'(x)$.

Graph the following functions.

49. $y = e^{-x^2}$

50. $y = xe^{-x}$ for $x \geq 0$

SOLUTIONS TO PRACTICE PROBLEMS 3

1. We must use the general power rule. However, this is most easily done if we first use the laws of exponents to simplify the function inside the brackets.

$$e^{-3x}(1 + e^{6x}) = e^{-3x} + e^{-3x} \cdot e^{6x}$$
$$= e^{-3x} + e^{3x}.$$

Now

$$\frac{d}{dx}[e^{-3x} + e^{3x}]^{12} = 12 \cdot [e^{-3x} + e^{3x}]^{11} \cdot (-3e^{-3x} + 3e^{3x})$$

$$= 36 \cdot [e^{-3x} + e^{3x}]^{11} \cdot (-e^{-3x} + e^{3x}).$$

2. The differential equation $y' = -y/20$ is of the type $y' = ky$, where $k = -\frac{1}{20}$. Therefore, any solution has the form $f(x) = Ce^{-(1/20)x}$. Now $f(0) = Ce^{-(1/20)\cdot 0} = Ce^0 = C$, so that $f(0) = 2$ when $C = 2$. Therefore, the desired function is $f(x) = 2e^{-(1/20)x}$.

14.4 The Natural Logarithm Function

As a preparation for the definition of the natural logarithm, we shall make a geometrical digression. In Fig. 1 we have plotted several pairs of points. Observe how they are related to the line $y = x$.

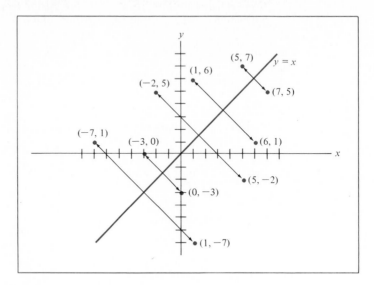

FIGURE 1

The points $(5, 7)$ and $(7, 5)$, for example, are the same distance from the line $y = x$. If we were to plot the point $(5, 7)$ with wet ink and then fold the page along the line $y = x$, the ink blot would produce a second blot at the point $(7, 5)$. If we think of the line $y = x$ as a mirror, then $(7, 5)$ is the mirror image of $(5, 7)$. We say that $(7, 5)$ is the *reflection* of $(5, 7)$ through the line $y = x$. Similarly, $(5, 7)$ is the reflection of $(7, 5)$ through the line $y = x$.

Now let us consider all points lying on the graph of the exponential function $y = e^x$ [see Fig. 2(a)]. If we reflect each such point through the line $y = x$, we obtain a new graph [see Fig. 2(b)]. For each positive x, there is exactly one value of y such that (x, y) is on the new graph. We call this value of y the *natural logarithm of* x, denoted ln x. Thus the reflection of the graph of $y = e^x$ through the line $y = x$ is the graph of the natural logarithm function $y = \ln x$.

We may deduce some properties of the natural logarithm function from an inspection of its graph.

1. The point $(1, 0)$ is on the graph of $y = \ln x$ (because $(0, 1)$ is on the graph of $y = e^x$). In other words,

$$\ln 1 = 0. \tag{1}$$

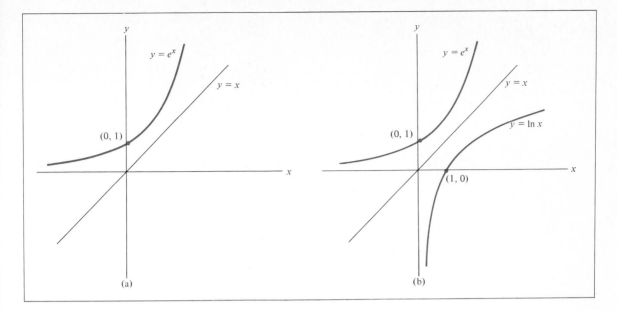

FIGURE 2

2. ln x is defined only for positive values of x.
3. ln x is negative for x between 0 and 1.
4. ln x is positive for x greater than 1.
5. ln x is an increasing function.

Let us study the relationship between the natural logarithm and exponential functions more closely. From the way in which the graph of ln x was obtained we know that (a, b) is on the graph of ln x if and only if (b, a) is on the graph of e^x. However, a typical point on the graph of ln x is of the form $(a, \ln a)$, $a > 0$. So for any positive value of a, the point $(\ln a, a)$ is on the graph of e^x. That is,

$$e^{\ln a} = a.$$

Since a was an arbitrary positive number, we have the following important relationship between the natural logarithm and exponential functions.

$$e^{\ln x} = x \qquad \text{for } x > 0. \tag{2}$$

Equation (2) can be put into verbal form.

For each positive number x, ln x is that exponent to which we must raise e in order to get x.

If b is any number, then e^b is positive and hence $\ln(e^b)$ makes sense. What is $\ln(e^b)$? Since (b, e^b) is on the graph of e^x, we know that (e^b, b) must be on the graph of $\ln x$. That is, $\ln(e^b) = b$. Thus we have shown that

$$\ln(e^x) = x \qquad \text{for any } x. \tag{3}$$

The identities (2) and (3) express the fact that the natural logarithm is the *inverse* of the exponential function. For instance, if we take a number x and compute e^x, then, by (3), we can undo the effect of the exponentiation by taking the natural logarithm; that is, the logarithm of e^x equals the original number x. Similarly, if we take a positive number x and compute $\ln x$, then, by (2), we can undo the effect of the logarithm by raising e to the $\ln x$ power; that is, $e^{\ln x}$ equals the original number x.

Scientific calculators have an "$\ln x$" key that will compute the natural logarithm of a number to as many as ten significant figures. For instance, entering the number 2 into the calculator and pressing the $\ln x$ key, one obtains $\ln 2 = .6931471806$ (to 10 significant figures). If a scientific calculator is unavailable, one may use a table of logarithms, such as Table 9 of the Appendix, which gives the values of $\ln x$ to five significant figures. For instance, this table shows that $\ln .8 = -.22314$ (to five significant figures).

The relationships (2) and (3) between e^x and $\ln x$ may be used to solve equations, as the next examples show.

EXAMPLE 1   Solve the equation $5e^{x-3} = 4$ for x.

Solution   First divide each side by 5,

$$e^{x-3} = .8.$$

Taking the logarithm of each side and using (3), we have

$$\ln(e^{x-3}) = \ln .8$$

$$x - 3 = \ln .8$$

$$x = 3 + \ln .8.$$

[If desired, the numerical value of x can be obtained by using a scientific calculator or a natural logarithm table, namely, $x = 3 - .22314 = 2.77686$ (to five decimal places).]

EXAMPLE 2   Solve the equation $2 \ln x + 7 = 0$ for x.

Solution

$$2 \ln x = -7$$

$$\ln x = -3.5$$

$$e^{\ln x} = e^{-3.5}$$

$$x = e^{-3.5} \qquad [\text{by (2)}].$$

Other Exponential and Logarithm Functions   In our discussion of the exponential function, we mentioned that all exponential functions of the form b^x, where b is a fixed positive number, can be expressed in terms of *the* exponential function e^x. Now we can be quite explicit. For since $b = e^{\ln b}$, we see that

$$b^x = (e^{\ln b})^x = e^{(\ln b)x}.$$

Hence we have shown that

$$b^x = e^{kx}, \qquad \text{where } k = \ln b.$$

The natural logarithm function is sometimes called the *logarithm to the base e*, for it is the inverse of the exponential function e^x. If we reflect the graph of the function $y = 2^x$ through the line $y = x$, we obtain the graph of a function called the *logarithm to the base* 2, denoted by $\log_2 x$. Similarly, if we reflect the graph of $y = 10^x$ through the line $y = x$, we obtain the graph of a function called the *logarithm to the base* 10, denoted by $\log_{10} x$. (See Fig. 3.)

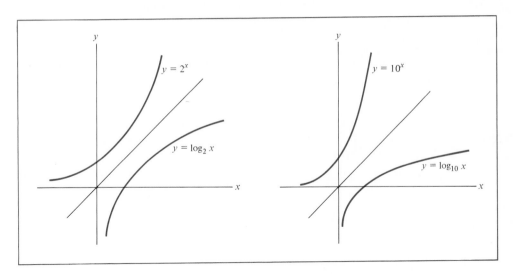

FIGURE 3

Logarithms to the base 10 are sometimes called *common* logarithms. Common logarithms are usually introduced into algebra courses for the purpose of simplifying certain arithmetic calculations. However, with the advent of the modern digital computer and the widespread availability of pocket electronic calculators, the need for common logarithms has diminished considerably. It can be shown that

$$\log_{10} x = \frac{1}{\ln 10} \cdot \ln x,$$

so that $\log_{10} x$ is simply a constant multiple of $\ln x$. However, we shall not need this fact.

The natural logarithm function is used in calculus because differentiation and integration formulas are simpler than for $\log_{10} x$ or $\log_2 x$, and so on. (Recall that we prefer the function e^x over the functions 10^x and 2^x for the same reason.) Also, $\ln x$ arises "naturally" in the process of solving certain differential equations that describe various growth processes.

PRACTICE PROBLEMS 4

1. Find $\ln e$.

2. Solve $e^{-3x} = 2$ using the natural logarithm function.

EXERCISES 4

1. Find $\ln(1/e)$.   2. Find $\ln(\sqrt{e})$.

3. If $e^{-x} = 1.7$, write x in terms of the natural logarithm.

4. If $e^x = 3.5$, write x in terms of the natural logarithm.

5. If $\ln x = 2.2$, write x using the exponential function.

6. If $\ln x = -5.7$, write x using the exponential function.

Simplify the following expressions.

7. $\ln e^2$
8. $e^{\ln 1.37}$
9. $e^{e^{\ln 1}}$
10. $\ln(e^{.73 \ln e})$
11. $e^{5 \ln 1}$
12. $\ln(\ln e)$

Solve the following equations for x.

13. $e^{2x} = 5$
14. $e^{3x-1} = 4$
15. $\ln(4 - x) = \frac{1}{2}$
16. $\ln 3x = 2$
17. $\ln x^2 = 6$
18. $e^{x^2} = 7$
19. $6e^{-.00012x} = 3$
20. $2 - \ln x = 0$
21. $\ln 5x = \ln 3$
22. $\ln(x^2 - 3) = 0$
23. $\ln(\ln 2x) = 0$
24. $3 \ln x = 8$
25. $2e^{x/3} - 9 = 0$
26. $4 - 3e^{x+6} = 0$
27. $300e^{.2x} = 1800$
28. $750e^{-.4x} = 375$
29. $e^{5x} \cdot e^{\ln 5} = 2$
30. $e^{x^2 - 5x + 6} = 1$
31. $4e^x \cdot e^{-2x} = 6$
32. $(e^x)^2 \cdot e^{2-3x} = 4$

In Exercises 33–36, find the coordinates of each extreme point of the given function and determine if the point is a maximum point or a minimum point.

33. $f(x) = e^{-x} + 3x$
34. $f(x) = 5x - 2e^x$
35. $f(x) = \frac{1}{3}e^{2x} - x + \frac{1}{2}\ln\frac{3}{2}$
36. $f(x) = 5 - \frac{1}{2}x - e^{-3x}$

37. When a drug or vitamin is administered intramuscularly (into a muscle), the concentration in the blood at time t after injection can be approximated by a function of the form $f(t) = c(e^{-k_1 t} - e^{-k_2 t})$. Sketch the graph of the function $f(t) = 5(e^{-.01t} - e^{-.51t})$ for $t \geq 0$ and discuss why this graph could be a reasonable model for the concentration of a drug injected intramuscularly.

38. Under certain geographic conditions, the wind velocity v at a height x centimeters above the ground is given by $v = K \ln(x/x_0)$, where K is a positive constant (depending on the air density, average wind velocity, etc.), and x_0 is a roughness parameter (depending on the roughness of the vegetation on the ground).* Suppose that $x_0 = .7$ cm (a value that applies to lawn grass 3 cm high) and $K = 300$ cm/sec.

(a) At what height above the ground is the wind velocity zero?

(b) At what height is the wind velocity 1200 cm/sec?

39. Use the tables of values of e^x and $\ln x$ to estimate $(1.6)^{10}$.

40. Find k such that $2^x = 2^{kx}$ for all x.

SOLUTIONS TO PRACTICE PROBLEMS 4

1. Answer: 1. The number $\ln e$ is that exponent to which e must be raised in order to obtain e.

2. Take the logarithm of each side and use (3) to simplify the left side:

$$\ln e^{-3x} = \ln 2$$

$$-3x = \ln 2$$

$$x = -\frac{\ln 2}{3}.$$

14.5 The Derivative of ln x

Let us now compute the derivative of $\ln x$ for $x > 0$. Since $e^{\ln x} = x$, we have

$$\frac{d}{dx}(e^{\ln x}) = \frac{d}{dx}(x) = 1. \tag{1}$$

On the other hand, if we differentiate $e^{\ln x}$ by the chain rule, we find that

$$\frac{d}{dx}(e^{\ln x}) = e^{\ln x} \cdot \frac{d}{dx}(\ln x) = x \cdot \frac{d}{dx}(\ln x), \tag{2}$$

where the last equality used the fact that $e^{\ln x} = x$. By combining equations (1) (2) we obtain

$$x \cdot \frac{d}{dx}(\ln x) = 1.$$

---

* G. Cox, B. Collier, A. Johnson, and P. Miller, *Dynamic Ecology* (Englewood Cliffs, N.J.: Prentice-Hall, Inc., 1973), pp. 113–115.

In other words,

$$\frac{d}{dx}(\ln x) = \frac{1}{x}, \qquad x > 0. \tag{3}$$

By combining this differentiation formula with the chain rule, product rule, and quotient rule, we may differentiate many functions involving ln x.

EXAMPLE 1 Differentiate.

(a) $(\ln x)^5$ (b) $x \ln x$ (c) $\ln(x^3 + 5x^2 + 8)$

Solution (a) By the general power rule,

$$\frac{d}{dx}(\ln x)^5 = 5(\ln x)^4 \cdot \frac{d}{dx}(\ln x)$$

$$= 5(\ln x)^4 \cdot \frac{1}{x}$$

$$= \frac{5(\ln x)^4}{x}.$$

(b) By the product rule,

$$\frac{d}{dx}(x \ln x) = x \cdot \frac{d}{dx}(\ln x) + (\ln x) \cdot 1$$

$$= x \cdot \frac{1}{x} + \ln x$$

$$= 1 + \ln x.$$

(c) By the chain rule,

$$\frac{d}{dx}\ln(x^3 + 5x^2 + 8) = \frac{1}{x^3 + 5x^2 + 8} \cdot \frac{d}{dx}(x^3 + 5x^2 + 8)$$

$$= \frac{3x^2 + 10x}{x^3 + 5x^2 + 8}.$$

Let $g(x)$ be any differentiable function. For any value of x for which $g(x)$ is positive, the function $\ln(g(x))$ is defined. For such a value of x, the derivative is given by the chain rule as

$$\frac{d}{dx}[\ln g(x)] = \frac{1}{g(x)} \cdot \frac{d}{dx}g(x)$$

$$= \frac{g'(x)}{g(x)}.$$

Example 1(c) illustrates a special case of this formula.

EXAMPLE 2   The function $f(x) = (\ln x)/x$ has an extreme point for some $x > 0$. Find the point and determine whether it is a maximum or a minimum point.

Solution   By the quotient rule,

$$f'(x) = \frac{x \cdot \dfrac{1}{x} - \ln x \cdot 1}{x^2} = \frac{1 - \ln x}{x^2}$$

$$f''(x) = \frac{x^2 \cdot \left(-\dfrac{1}{x}\right) - (1 - \ln x)(2x)}{x^4} = \frac{2 \ln x - 3}{x^3}.$$

If we set $f'(x) = 0$, then

$$1 - \ln x = 0$$
$$\ln x = 1$$
$$e^{\ln x} = e^1 = e$$
$$x = e.$$

Therefore, the only possible extreme point is at $x = e$. When $x = e$, $f(e) = (\ln e)/e = 1/e$. Furthermore,

$$f''(e) = \frac{2 \ln e - 3}{e^3} = -\frac{1}{e^3} < 0,$$

which implies that the graph of $f(x)$ is concave down at $x = e$. Therefore, $(e, 1/e)$ is a maximum point of the graph of $f(x)$.

The next example introduces a function that will be needed later when we study integration.

EXAMPLE 3   The function $\ln|x|$ is defined for all nonzero values of x. Its graph is sketched in Fig. 1. Compute the derivative of $\ln|x|$.

Solution   If x is positive, then $|x| = x$, so

$$\frac{d}{dx} \ln|x| = \frac{d}{dx} \ln x = \frac{1}{x}.$$

FIGURE 1

If x is negative, then $|x| = -x$; and, by the chain rule,

$$\frac{d}{dx} \ln|x| = \frac{d}{dx} \ln(-x)$$

$$= \frac{1}{-x} \cdot \frac{d}{dx}(-x)$$

$$= \frac{1}{-x} \cdot (-1) = \frac{1}{x}.$$

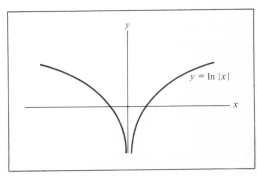
$y = \ln|x|$

Therefore, we have established the following useful fact.

$$\frac{d}{dx} \ln|x| = \frac{1}{x}, \qquad x \neq 0.$$

PRACTICE PROBLEMS 5

Differentiate.

1. $f(x) = \dfrac{1}{\ln(x^4 + 5)}$

2. $f(x) = \ln(\ln x)$.

EXERCISES 5

Differentiate the following functions.

1. $\ln 2x$

2. $\ln x^2$

3. $\ln(x + 5)$

4. $x^2 \ln x$

5. $\dfrac{1}{x} \ln(x + 1)$

6. $\sqrt{\ln x}$

7. $e^{\ln x + x}$

8. $\ln\left(\dfrac{x}{x - 3}\right)$

9. $4 + \ln\left(\dfrac{x}{2}\right)$

10. $\ln \sqrt{x}$

11. $(\ln x)^2 + \ln x$

12. $\ln(x^3 + 2x + 1)$

13. $\ln(kx), k$ constant

14. $\dfrac{x}{\ln x}$

15. $\dfrac{x}{(\ln x)^2}$

16. $(\ln x)e^{-x}$

17. $e^{2x} \ln x$

18. $(\ln x + 1)^3$

19. $\ln(e^{5x} + 1)$

20. $\ln(e^{e^x})$

Find.

21. $\dfrac{d}{dt}(t^2 \ln 4)$

22. $\dfrac{d^2}{dx^2} \ln(1 + x^2)$

23. $\dfrac{d^2}{dt^2} (\ln t)^3$

24. Find the slope of the graph of $y = \ln |x|$ at $x = 3$ and $x = -3$.

25. Write the equation of the tangent line to the graph of $y = \ln(x^2 + e)$ at $x = 0$.

26. The function $f(x) = (\ln x + 1)/x$ has an extreme point for $x > 0$. Find the coordinates of the point. Is it a maximum point?

27. The function $f(x) = (\ln x)/\sqrt{x}$ has an extreme point for $x > 0$. Find the coordinates of the point. Is it a maximum point?

28. The function $f(x) = x/(\ln x + x)$ has an extreme point for $x > 1$. Find the coordinates of the point. Is it a minimum point?

29. Sketch the graph of the function $y = \ln x + (1/x) - \frac{1}{2}$.

30. Sketch the graph of $y = 1 + \ln(x^2 - 6x + 10)$.

31. If a cost function is $C(x) = (\ln x)/(40 - 3x)$, find the marginal cost when $x = 10$.

32. Suppose that the demand equation for a certain commodity is $p = 45/(\ln x)$. Determine the marginal revenue function for this commodity, and compute the marginal revenue when $x = 20$.

33. Suppose that the total revenue function for a manufacturer is $R(x) = 300 \ln(x + 1)$, so that the sale of x units of a product brings in about $R(x)$ dollars. Suppose also that the total cost of producing x units is $C(x)$ dollars, where $C(x) = 2x$. Find the value of x at which the profit function $R(x) - C(x)$ will be maximized. Show that the profit function has a maximum and not a minimum point at this value of x.

34. Evaluate $\displaystyle\lim_{h \to 0} \frac{\ln(7 + h) - \ln 7}{h}$.

35. Find the maximum area of a rectangle in the first quadrant with one corner at the origin, two sides on the coordinate axes, and one corner on the graph of $y = -\ln x$.

SOLUTIONS TO PRACTICE PROBLEMS 5

1. Here $f(x) = [\ln(x^4 + 5)]^{-1}$. By the chain rule,

$$f'(x) = -[\ln(x^4 - 5)]^{-2} \cdot \frac{d}{dx} \ln(x^4 + 5)$$

$$= -[\ln(x^4 + 5)]^{-2} \cdot \frac{4x^3}{x^4 + 5}.$$

2. $f'(x) = \dfrac{d}{dx} \ln(\ln x) = \dfrac{1}{\ln x} \cdot \dfrac{d}{dx} \ln x$

$$= \frac{1}{\ln x} \cdot \frac{1}{x} = \frac{1}{x \ln x}.$$

Chapter 14: CHECKLIST

☐ $b^x \cdot b^y = b^{x+y}$

☐ $b^{-x} = 1/b^x$

☐ $b^x/b^y = b^{x-y}$

☐ $(b^y)^x = b^{xy}$

☐ $a^x b^x = (ab)^x$

☐ $a^x/b^x = (a/b)^x$

☐ $b^0 = 1$

☐ Definition of e and e^x

☐ $\dfrac{d}{dx}(e^{kx}) = ke^{kx}$

☐ $\dfrac{d}{dx}(e^{g(x)}) = e^{g(x)}g'(x)$

☐ Graph of e^{kx}, k positive

- $\square$ Graph of e^{kx}, k negative
- $\square$ If $y = f(x)$ satisfies $y' = ky$, then $y = Ce^{kx}$ for some constant C.
- $\square$ Reflection in the line $y = x$
- $\square$ Definition of $\ln x$
- $\square$ $\ln 1 = 0$
- $\square$ $\ln e = 1$
- $\square$ $e^{\ln x} = x$, $x > 0$
- $\square$ $\ln e^x = x$
- $\square$ $\dfrac{d}{dx}(\ln x) = \dfrac{1}{x}$

Chapter 14: SUPPLEMENTARY EXERCISES

Calculate the following.

1. $27^{4/3}$
2. $4^{1.5}$
3. 5^{-2}
4. $16^{-.25}$
5. $(2^{5/7})^{14/5}$
6. $8^{1/2} \cdot 2^{1/2}$
7. $9^{5/2}/9^{3/2}$
8. $4^2 \cdot 4^3$

Simplify the following.

9. $(e^{x^2})^3$
10. $e^{5x} \cdot e^{2x}$
11. e^{3x}/e^x
12. $2^x \cdot 3^x$
13. $(e^{8x} + 7e^{-2x})e^{3x}$
14. $\dfrac{e^{5x/2} - e^{3x}}{\sqrt{e^x}}$

Solve the following equations for x.

15. $e^{-3x} = e^{-12}$
16. $e^{x^2-x} = e^2$
17. $(e^x \cdot e^2)^3 = e^{-9}$
18. $e^{-5x} \cdot e^4 = e$

Differentiate the following functions.

19. $10e^{7x}$
20. $e^{\sqrt{x}}$
21. xe^{x^2}
22. $\dfrac{e^x + 1}{x - 1}$
23. $e^{(e^x)}$
24. $(\sqrt{x} + 1)e^{-2x}$
25. $\dfrac{x^2 - x + 5}{e^{3x} + 3}$
26. x^e

27. Determine all solutions of the differential equation $y' = -y$.

28. Determine all functions $y = f(x)$ such that $y' = -1.5y$ and $f(0) = 2000$.

29. Determine all solutions of the differential equation $y' = 1.5y$ and $f(0) = 2$.

30. Determine all solutions of the differential equation $y' = \frac{1}{3}y$.

Graph the following functions.

31. $e^{-x} + x$
32. $e^x - x$
33. $e^{-(1/2)x^2}$
34. $100(x - 2)e^{-x}$ for $x \geq 2$

Simplify the following expressions.

35. $\ln e^{5x}$
36. $e^{2 \ln 2}$
37. $e^{-5 \ln 1}$

38. $\left[e^{\ln x}\right]^2$ **39.** $e^{(\ln 5)/2}$ **40.** $e^{\ln(x^2)}$

Solve the following equations for t.

41. $3e^{2t} = 15$ **42.** $3e^{t/2} - 12 = 0$ **43.** $2 \ln t = 5$

44. $2e^{-.3t} = 1$ **45.** $5 \ln(t - 2) = 8$ **46.** $\ln(\ln 3t) = 0$

Differentiate the following functions.

47. $\ln(5x - 7)$ **48.** $\ln(9x)$ **49.** $(\ln x)^2$

50. $(x \ln x)^3$ **51.** $\ln(x^6 + 3x^4 + 1)$ **52.** $\dfrac{x}{\ln x}$

53. $e^{4x} \ln x$ **54.** $e^{x + \ln x}$ **55.** $\ln(\ln \sqrt{x})$

56. $1/\ln x$ **57.** $x \ln x - x$ **58.** $\ln(x^2 + e^x)$

Sketch the following curves.

59. $y = x - \ln x$ **60.** $y = \ln(x^2 + 1)$

61. $y = (\ln x)^2$ **62.** $y = (\ln x)^3$

Applications of the Exponential and Natural Logarithm Functions

Chapter 15

Earlier, we introduced the exponential function e^x and the natural logarithm function $\ln x$ and studied their most important properties. From the way we introduced these functions, it is by no means clear that they have any substantial connection with the physical world. However, as this chapter will demonstrate, the exponential and natural logarithm functions intrude into the study of many physical problems, often in a very curious and unexpected way.

Here the most significant fact that we require is that the exponential function is uniquely characterized by its differential equation. In other words, we will constantly make use of the following fact, which was stated previously.

The function* $y = Ce^{kt}$ satisfies the differential equation
$$y' = ky.$$
Conversely, if $y = f(t)$ satisfies the differential equation, then $y = Ce^{kt}$ for some constant C.

If $f(t) = Ce^{kt}$, then, by setting $t = 0$, we have
$$f(0) = Ce^0 = C.$$
Therefore, C is the value of $f(t)$ at $t = 0$.

15.1. Exponential Growth and Decay

In biology, chemistry, and economics it is often necessary to study the behavior of a quantity that is increasing as time passes. If, at every instant, the rate of increase of the quantity is proportional to the quantity at that instant, then we say

---

* Note that we use the variable t instead of x. The reason is that, in most applications, the variable of our exponential function is time. The variable t will be used throughout this chapter.

that the quantity is *growing exponentially* or is *exhibiting exponential growth*. A simple example of exponential growth is exhibited by the growth of bacteria in a culture. Under ideal laboratory conditions a bacteria culture grows at a rate proportional to the number of bacteria present. It does so because the growth of the culture is accounted for by the division of the bacteria. The more bacteria there are at a given instant, the greater the possibilities for division and hence the more rapid is the rate of growth.

Let us study the growth of a bacteria culture as a typical example of exponential growth. Suppose that $P(t)$ denotes the number of bacteria in a certain culture at time t. The rate of growth of the culture at time t is $P'(t)$. We assume that this rate of growth is proportional to the size of the culture at time t, so that

$$P'(t) = kP(t), \tag{1}$$

where k is a positive constant of proportionality. If we let $y = P(t)$, then (1) can be written as

$$y' = ky.$$

Therefore, from our discussion at the beginning of this chapter we see that

$$y = P(t) = P_0 e^{kt}, \tag{2}$$

where P_0 is the number of bacteria in the culture at time $t = 0$. The number k is called the *growth constant*.

EXAMPLE 1   Suppose that a certain bacteria culture grows at a rate proportional to its size. At time $t = 0$, approximately 20,000 bacteria are present. In 5 hours there are 400,000 bacteria. Determine a function that expresses the size of the culture as a function of time, measured in hours.

Solution   Let $P(t)$ be the number of bacteria present at time t. By assumption, $P(t)$ satisfies a differential equation of the form $y' = ky$, so $P(t)$ has the form

$$P(t) = P_0 e^{kt},$$

where the constants P_0 and k must be determined. The value of P_0 and k can be obtained from the data that give the population size at two different times. We are told that

$$P(0) = 20{,}000, \qquad P(5) = 400{,}000. \tag{3}$$

The first condition immediately implies that $P_0 = 20{,}000$, so

$$P(t) = 20{,}000 e^{kt}.$$

Using the second condition in (3), we have

$$20{,}000 e^{k \cdot 5} = P(5) = 400{,}000$$

$$e^{5k} = 20$$

$$5k = \ln 20 \tag{4}$$

$$k = \frac{\ln 20}{5} \approx .60*.$$

---

* Here and elsewhere in this chapter, we have used Table 9 of the Appendix and have carried out calculations to two significant figures.

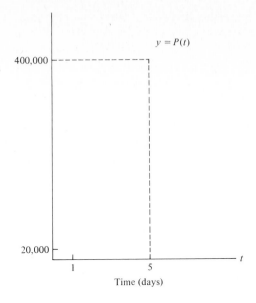

FIGURE 1 Graph of $y = 20{,}000e^{.6t}$.

So we may take

$$P(t) = 20{,}000e^{.6t}.$$

This function is a mathematical model of the growth of the bacteria culture. (See Fig. 1.)

EXAMPLE 2 Suppose that a colony of fruit flies is growing according to the exponential law $P(t) = P_0e^{kt}$ and suppose that the size of the colony doubles in 12 days. Determine the growth constant k.

Solution We do not know the initial size of the population at $t = 0$. However, we are told that $P(12) = 2P(0)$; that is,

$$P_0e^{k \cdot 12} = 2P_0$$

$$e^{12k} = 2$$

$$12k = \ln 2$$

$$k = \tfrac{1}{12} \ln 2 \approx .058.$$

Notice that the initial size P_0 of the population was not given in Example 2. We were able to determine the growth constant because we were told the amount of time required for the colony to double in size. Thus the growth constant does not depend on the initial size of the population. This property is characteristic of exponential growth.

EXAMPLE 3 Suppose that the initial size of the colony in Example 2 was 300. At what time will the colony contain 1800 fruit flies?

Solution   From Example 2 we have $P(t) = P_0 e^{.058t}$. Since $P(0) = 300$, we conclude that

$$P(t) = 300e^{.058t}.$$

Now that we have the explicit formula for the size of the colony, we can set $P(t) = 1800$ and solve for t:

$$300e^{.058t} = 1800$$

$$e^{.058t} = 6$$

$$.058t = \ln 6$$

$$t = \frac{\ln 6}{.058} \approx 31 \text{ days.}$$

The table below shows the growth of the colony in Example 3. Notice that 1800 is exactly halfway between 1200 ($t = 24$) and 2400 ($t = 36$). It is incorrect to guess that the population will equal 1800 when t is halfway between $t = 24$ and $t = 36$—that is, when $t = 30$. We saw in Example 3 that it takes approximately 31 days for the colony to reach 1800 fruit flies.

Population size	Day
300	0
600	12
1,200	24
2,400	36
4,800	48
9,600	60
19,200	72
⋮	⋮

Exponential Decay   An example of negative exponential growth, or *exponential decay*, is given by the disintegration of a radioactive element such as uranium 235. It is known that, at any instant, the rate at which a radioactive substance is decaying is proportional to the amount of the substance that has not yet disintegrated. If $P(t)$ is the quantity present at time t, then $P'(t)$ is the rate of decay. Of course, $P'(t)$ must be negative, since $P(t)$ is decreasing. Thus we may write $P'(t) = kP(t)$ for some negative constant k. To emphasize the fact that the constant is negative, k is often replaced by $-\lambda$*, where λ is a positive constant. Then $P(t)$ satisfies the differential equation

$$P'(t) = -\lambda P(t). \tag{5}$$

The general solution of (5) has the form

$$P(t) = P_0 e^{-\lambda t}$$

for some positive number P_0. We call such a function an *exponential decay function*. The constant λ is called the *decay constant*.

---

* λ is the Greek lowercase letter lambda.

EXAMPLE 4   The decay constant for strontium 90 is $\lambda = .0244$, where the time is measured in years. How long will it take for a quantity P_0 of strontium 90 to decay to one-half its original size?

Solution   We have

$$P(t) = P_0 e^{-.0244t}.$$

Next, set $P(t)$ equal to $\frac{1}{2}P_0$ and solve for t:

$$P_0 e^{-.0244t} = \tfrac{1}{2}P_0$$

$$e^{-.0244t} = \tfrac{1}{2} = .5$$

$$-.0244t = \ln .5$$

$$t = \frac{\ln .5}{-.0244} \approx 28 \text{ years.}$$

The *half-life* of a radioactive element is the length of time required for a given quantity of that element to decay to one-half its original size. Thus strontium 90 has a half-life of about 28 years. (See Fig. 2.) Notice from Example 4 that the half-life does not depend on the initial amount P_0.

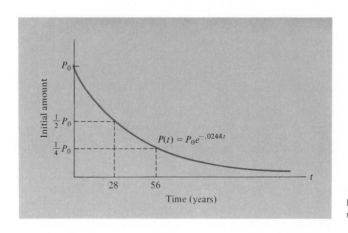

FIGURE 2   Half-life of radioactive strontium 90.

EXAMPLE 5   Radioactive carbon 14 has a half-life of about 5730 years. Find its decay constant.

Solution   If P_0 denotes the initial amount of carbon 14, then the amount after t years will be

$$P(t) = P_0 e^{-\lambda t}.$$

After 5730 years, $P(t)$ will equal $\frac{1}{2}P_0$. That is,

$$P_0 e^{-\lambda(5730)} = P(5730) = \tfrac{1}{2}P_0 = .5P_0.$$

Solving for λ gives

$$e^{-5730\lambda} = .5$$

$$-5730\lambda = \ln .5$$

$$\lambda = \frac{\ln .5}{-5730} \approx .00012.$$

One of the problems connected with aboveground nuclear explosions is the radioactive debris that falls on plants and grass, thereby contaminating the food supply of animals. Strontium 90 is one of the most dangerous components of "fallout" because it has a relatively long half-life and because it is chemically similar to calcium and is absorbed into the bone structure of animals (and humans) who eat contaminated food. Iodine 131 is also produced by nuclear explosions, but it presents less of a hazard because it has a half-life of 8 days.

EXAMPLE 6   If dairy cows eat hay containing too much iodine 131, their milk will be unfit to drink. Suppose that some hay contains 10 times the maximum allowable level of iodine 131. How many days should the hay be stored before it is fed to dairy cows?

Solution   Let P_0 be the amount of iodine 131 present in the hay. Then the amount at time t is $P(t) = P_0 e^{-\lambda t}$ (t in days). The half-life of iodine 131 is 8 days, so

$$P_0 e^{-8\lambda} = .5P_0$$

$$e^{-8\lambda} = .5$$

$$-8\lambda = \ln .5$$

$$\lambda = \frac{\ln .5}{-8} \approx .087,$$

and

$$P(t) = P_0 e^{-.087t}.$$

Now that we have the formula for $P(t)$, we want to find t such that $P(t) = \frac{1}{10}P_0$. We have

$$P_0 e^{-.087t} = .1P_0,$$

so

$$e^{-.087t} = .1$$

$$-.087t = \ln .1$$

$$t = \frac{\ln .1}{-.087} \approx 26 \text{ days.}$$

Radiocarbon Dating   Knowledge about radioactive decay is valuable to social scientists who want to estimate the age of objects belonging to ancient civilizations. Several different substances are useful for radioactive-dating techniques; the most common is radiocarbon, C^{14}. Carbon 14 is produced in the upper atmosphere when cosmic rays react with atmospheric nitrogen. Because the C^{14} eventually decays, the concentration of C^{14} cannot rise above certain levels. An equilibrium is reached where C^{14} is produced at the same rate as it decays. Scientists usually assume that the total amount of C^{14} in the biosphere has remained constant over the past 50,000 years. Consequently, it is assumed that the *ratio* of C^{14} to ordinary nonradioactive carbon 12, C^{12}, has been constant during this same period. (The ratio is about one part C^{14} to 10^{12} parts of C^{12}.) Both C^{14} and C^{12} are in the atmosphere in the form of carbon dioxide. All living vegetation and most forms of animal life contain C^{14} and C^{12} in the same proportion as the atmosphere. The reason is that plants absorb carbon dioxide through photosynthesis. The C^{14} and C^{12} in plants is distributed through the various food chains to almost all animal life.

When an organism dies, it stops replacing its carbon, and therefore the amount of C^{14} begins to decrease through radioactive decay. (The C^{12} in the dead organism remains constant.) At a later date, the ratio of C^{14} to C^{12} can be measured in order to determine when the organism died. (See Fig. 3.)

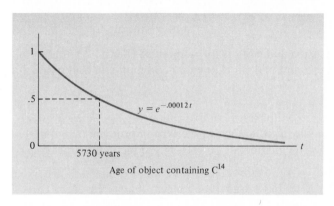

Age of object containing C^{14}

FIGURE 3 C^{14}–C^{12} ratio compared to the ratio in living plants.

EXAMPLE 7   A parchment fragment was discovered that had a C^{14}–C^{12} ratio of about .8 times the ratio found today in living matter. Estimate the age of the parchment.

Solution   We assume that the original C^{14}–C^{12} ratio in the parchment was the same as the ratio in living organisms today. Consequently, about eight-tenths of the original C^{14} remains. From Example 5 we obtain the formula for the amount of C^{14} present t years after the parchment was made from an animal skin:

$$P(t) = P_0 e^{-.00012t},$$

where P_0 = initial amount. We want to find t such that $P(t) = .8P_0$.

$$P_0 e^{-.00012t} = .8P_0$$

$$e^{-.00012t} = .8$$

$$-.00012t = \ln .8$$

$$t = \frac{\ln .8}{-.00012} \approx 1900 \text{ years old.}$$

A Sales Decay Curve   Marketing studies* have demonstrated that if advertising and other promotion of a particular product are stopped and if other market conditions remain fairly constant, then, at any time t, the sales of that product will be declining at a rate proportional to the amount of current sales at t. (See Fig. 4.) If S_0 is the number of sales in the last month during which advertising

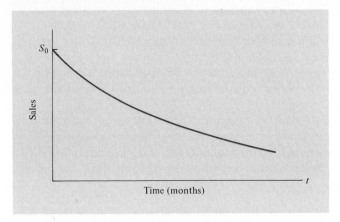

FIGURE 4
Exponential decay
of sales.

occurred and if $S(t)$ is the number of sales in the tth month following the cessation of promotional effort, then a good mathematical model for $S(t)$ is

$$S(t) = S_0 e^{-\lambda t},$$

where λ is a positive number called the *sales decay constant*. The value of λ depends on many factors, such as the type of product, the number of years of prior advertising, the number of competing products, and other characteristics of the market.

The Time Constant   Consider an exponential decay function $y = Ce^{-\lambda t}$. In Fig. 5, we have drawn the tangent line to the decay curve when $t = 0$. The slope

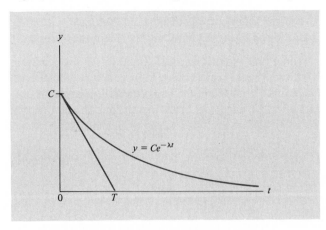

$$y = Ce^{-\lambda t}$$

FIGURE 5   The time constant T in exponential decay: $T = 1/\lambda$.

    * M. Vidale and H. Wolfe, "An Operations-Research Study of Sales Response to Advertising," *Operations Research*, 5 (1957), 370–381. Reprinted in F. Bass et al., *Mathematical Models and Methods in Marketing* (Homewood, Ill.: Richard D. Irwin, Inc., 1961).

there is the initial rate of decay. If the decay process were to continue at this rate, the decay curve would follow the tangent line and y would be zero at some time T. This time is called the *time constant* of the decay curve. It can be shown (see Exercise 29) that $T = 1/\lambda$ for the curve $y = Ce^{-\lambda t}$. Thus $\lambda = 1/T$, and the decay curve may be written in the form

$$y = Ce^{-t/T}.$$

If one has experimental data that tend to lie along an exponential decay curve, then the numerical constants for the curve may be obtained from Fig. 5. First, sketch the curve and estimate the y-intercept, C. Then sketch an approximate tangent line and from this estimate the time constant, T. This procedure is sometimes used in biology and medicine.

PRACTICE PROBLEMS 1

1. (a) Solve the differential equation $P'(t) = -.6P(t)$, $P(0) = 50$.

   (b) Solve the differential equation $P'(t) = kP(t)$, $P(0) = 4000$, where k is some constant.

   (c) Find the value of k in part (b) for which $P(2) = 100P(0)$.

2. Under ideal conditions a colony of *Escherichia coli* bacteria can grow by a factor of 100 every 2 hours. If initially 4000 bacteria are present, how long will it take before there are 1,000,000 bacteria?

EXERCISES 1

1. Let $P(t)$ be the size of a certain insect population after t days, and suppose that $P(t)$ satisfies the differential equation

$$P'(t) = .07P(t), \qquad P(0) = 400.$$

   Find the formula for $P(t)$.

2. Let $P(t)$ be the number of bacteria present in a culture after t minutes, and suppose that $P(t)$ satisfies the differential equation

$$P'(t) = .55P(t).$$

   Find the formula for $P(t)$ if initially there are approximately 10,000 bacteria present.

3. Suppose that after t hours there are $P(t)$ cells present in a culture, where $P(t) = 5000e^{.2t}$.

   (a) How many cells were present initially?

   (b) When will 20,000 cells be present?

4. The size of a certain insect population is given by $P(t) = 300e^{.01t}$, where t is measured in days. At what time will the population equal 600? 1200?

5. Determine the growth constant of a population that is growing at a rate proportional to its size, where the population doubles in size every 40 days.

6. Determine the growth constant of a population that is growing at a rate proportional to its size, where the population triples in size every 5 hours.

7. The world's population was 4 billion on January 1, 1974 and was expected to reach 5 billion on January 1, 1985. Assume that at any time the population grows at a rate proportional to the population at that time.

   (a) Find the formula for $P(t)$, the world's population t years after January 1, 1974.

   (b) What will the population be on January 1, 2000?

   (c) In what year will the world's population reach 6 billion?

8. Mexico City is expected to become the most heavily populated city in the world by the end of this century. At the beginning of 1980, 14.5 million people lived in the metropolitan area of Mexico City, and the population was growing exponentially with growth constant .044. (Part of the growth is due to immigration.)

   (a) If this trend continues, how large will the population be in the year 2000?

   (b) In what year will the current population have doubled?

9. The growth rate of a certain bacteria culture is proportional to its size. If the bacteria culture doubles in size every 20 minutes, how long will it take for the culture to increase 12-fold?

10. The growth rate of a certain cell culture is proportional to its size. In 10 hours a population of 1 million cells grew to 9 million. How large will the cell culture be after 15 hours?

11. The weight in grams after t years $P(t)$ of a certain radioactive substance satisfies the differential equation

$$P'(t) = -.08P(t), \qquad P(0) = 30.$$

   (a) Find the formula for $P(t)$.

   (b) What is $P(10)$?

12. Five milligrams of a drug is injected into a patient, and the amount of drug present t hours after the injection satisfies the differential equation

$$P'(t) = -.09P(t).$$

   (a) Find the formula for $P(t)$.

   (b) Determine the amount of drug present 20 hours after the injection.

13. One hundred grams of a radioactive substance with decay constant .01 is buried in the ground. Assume that time is measured in years.

   (a) Give the formula for the amount remaining after t years.

   (b) How much will remain after 30 years?

   (c) What is the half-life of this radioactive substance?

14. The decay constant for cesium 137 is .023 when time is measured in years. Find the half-life of cesium 137.

15. Radioactive cobalt 60 has a half-life of 5.3 years.

    (a) Find the decay constant of cobalt 60.

    (b) If the initial amount of cobalt 60 is 10 grams, how much will be present after 2 years?

16. Five grams of a certain radioactive material decays to 3 grams in 1 year. After how many years will just 1 gram remain?

17. A 4500-year-old wooden chest was found in the tomb of the twenty-fifth century B.C. Chaldean king Meskalumdug of Ur. What percentage of the original C^{14} would you expect to find in the wooden chest? (Recall that the decay constant for C^{14} is .00012.)

18. In 1947, a cave with beautiful prehistoric wall paintings was discovered in Lascaux, France. Some charcoal found in the cave contained 20% of the C^{14} expected in living trees. How old are the Lascaux cave paintings?

19. Sandals woven from strands of tree bark were found recently in Fort Rock Creek Cave in Oregon. The bark contained 34% of the level of C^{14} found in living bark. Approximately how old are the sandals?

20. Many scientists believe there have been four ice ages in the past 1 million years. Before the technique of carbon dating was known, geologists erroneously believed that the retreat of the Fourth Ice Age began about 25,000 years ago. In 1950, logs from ancient spruce trees were found under glacial debris near Two Creeks, Wisconsin. Geologists determined that these trees had been crushed by the advance of ice during the Fourth Ice Age. Wood from the spruce trees contained 27% of the level of C^{14} found in living trees. Approximately how long ago did the Fourth Ice Age actually occur?

21. Let $f(t)$ be the value of the dollar t years after January 1, 1982, where the value is in terms of the purchasing power on January 1, 1982, and $f(0) = 1.00$. Suppose that the rate of decrease of $f(t)$ at any time t is proportional to $f(t)$. Give the formula for $f(t)$ if by January 1, 1984, the dollar had lost 15% of its purchasing power.

22. The consumer price index (CPI) gives a measure of the prices of commodities commonly purchased by consumers. For example, an increase of 10% in the CPI corresponds to a 10% average increase in the prices of consumer goods. Let $f(t)$ be the CPI at time t, where time is years since January 1, 1980. Suppose that $f(t)$ satisfies the differential equation

    $$f'(t) = .12f(t), \qquad f(0) = 100.$$

    (This means that at each time t, the CPI is rising at an annual rate of 12%, and the index is set equal to 100 on January 1, 1980.) How many years will it take for the CPI to double?

23. An island in the Pacific Ocean is contaminated by fallout from a nuclear explosion. If the strontium 90 is 100 times the level that scientists believe is "safe," how many years will it take for the island to once again be "safe" for human habitation? The half-life of strontium 90 is 28 years.

24. A common infection of the urinary tract in humans is caused by the bacterium *E. coli*. The infection is generally noticed when the bacteria colony reaches a population of about 10^8. The colony doubles in size about every 20 minutes. When a full bladder is emptied, about 90% of the bacteria are eliminated. Suppose that at the beginning of a certain time period, a person's bladder and urinary tract contain 10^8 *E. coli* bacteria. During an interval of T minutes the person drinks enough liquid to fill the bladder. Find the value of T such that if the bladder is emptied after T minutes, about 10^8 bacteria will still

remain. [*Note:* The average bladder holds about 1 liter of urine. It is seldom possible to eliminate an *E. coli* infection by diuresis without drugs—such as by drinking large amounts of water.]

25. By 1974 the United States had an estimated 80 million gallons of radioactive products from nuclear power plants and other nuclear reactors. These waste products were stored in various sorts of containers (made of such materials as stainless steel and cement), and the containers were buried in the ground and the ocean. Scientists feel that the waste products must be prevented from contaminating the rest of the earth until more than 99.99% of the radioactivity is gone (i.e., until the level is less than 0.0001 times the original level). If a storage cylinder contains waste products whose half-life is 1500 years, how many years must the container survive without leaking? [*Note:* Some of the containers are already leaking.]

26. In 1950 the world's population required 1×10^9 hectares* of arable land for food growth and in 1980 2×10^9 hectares were required. Current population trends indicate that if $A(t)$ denotes the amount of land needed t years after 1950, then

$$\frac{dA}{dt} = kA$$

for some constant k.

(a) Derive a formula for $A(t)$.

(b) The total amount of arable land on the earth's surface is estimated at 3.2×10^9 hectares. In what year will the earth exhaust its supply of land for growing food? [Data based on the Club of Rome's report, *The Limits to Growth* by D. H. Meadows, D. L. Meadows, J. Randers, and W. Behrens III (New York: Universe Books, 1972).]

27. A drought in the African veldt causes the death of much of the animal population. A typical herd of wildebeests suffers a death rate proportional to its size. The herd numbers 500 at the onset of the drought and only 200 remain 4 months later.

(a) Find a formula for the herd's population at time t months.

(b) How long will it take for the herd to diminish to one-tenth its original size?

28. In a laboratory experiment, 8 units of a sulfate were injected into a dog. After 50 minutes, only 4 units remained in the dog. Let $f(t)$ be the amount of sulfate present after t minutes. At any time, the rate of change of $f(t)$ is proportional to the value of $f(t)$. Find the formula for $f(t)$.

29. Let T be the time constant of the curve $y = Ce^{-\lambda t}$ as defined in Fig. 5. Show that $T = 1/\lambda$. [*Hint:* Express the slope of the tangent line in Fig. 5 in terms of C and T. Then set this slope equal to the slope of the curve $y = Ce^{-\lambda t}$ at $t = 0$.]

30. Suppose that a person is given an injection of 300 milligrams of penicillin at time $t = 0$, and let $f(t)$ be the amount (in milligrams) of penicillin present in the person's bloodstream t hours after the injection. Then the amount of penicillin present decays exponentially, and a typical formula for $f(t)$ is $f(t) = 300e^{-(2/3)t}$.

(a) What is the initial rate of decay of the penicillin?

(b) What is the time constant for this decay curve $y = f(t)$? (See the discussion accompanying Fig. 5.)

---

* A hectare equals 2.471 acres.

SOLUTIONS TO PRACTICE PROBLEMS 1

1. (a) **Answer:** $P(t) = 50e^{-.6t}$. Differential equations of the type $y' = ky$ have as their solution $P(t) = Ce^{kt}$, where C is $P(0)$.

    (b) **Answer:** $P(t) = 4000e^{kt}$. This problem is like the previous one except that the constant is not specified. Additional information is needed if one wants to determine a specific value for k.

    (c) **Answer:** $P(t) = 4000e^{2.3t}$. From the solution to part (b) we know that $P(t) = 4000e^{kt}$. We are given that $P(2) = 100P(0) = 100(4000) = 400{,}000$. So

    $$P(2) = 4000e^{k(2)} = 400{,}000$$

    $$e^{2k} = 100$$

    $$2k = \ln 100$$

    $$k = \frac{\ln 100}{2} \approx 2.3.$$

2. Let $P(t)$ be the number of bacteria present after t hours. We must first find an expression for $P(t)$ and then determine the value of t for which $P(t) = 1{,}000{,}000$. From the discussion at the beginning of the section we know that $P'(t) = k \cdot P(t)$. Also, we are given that $P(2)$ (the population after 2 hours) is $100P(0)$ (100 times the initial population). From Problem 1(c) we have an expression for $P(t)$:

    $$P(t) = 4000e^{2.3t}.$$

    Now we must solve $P(t) = 1{,}000{,}000$ for t.

    $$4000e^{2.3t} = 1{,}000{,}000$$

    $$e^{2.3t} = 250$$

    $$2.3t = \ln 250$$

    $$t = \frac{\ln 250}{2.3} \approx 2.4.$$

    Therefore, after 2.4 hours there will be 1,000,000 bacteria.

15.2. Compound Interest

When money is deposited in a savings account, interest is paid at stated intervals. If this interest is added to the account and thereafter earns interest itself, then the interest is called *compound interest*. The original amount deposited is called the *principal amount*. The principal amount plus the compound interest is called the *compound amount*. The interval between interest payments is referred to as the *interest period*. In formulas for compound interest, the interest rate is expressed as a decimal rather than a percent. Thus 6% is written as .06.

If $1000 is deposited at 6% annual interest, compounded annually, the compound amount at the end of the first year will be

$$A_1 = \underset{\text{principal}}{1000} + \underset{\text{interest}}{1000(.06)} = 1000(1 + .06).$$

At the end of the second year the compound amount will be

$$A_2 = \underset{\substack{\text{compound} \\ \text{amount}}}{A_1} + \underset{\text{interest}}{A_1(.06)} = A_1(1 + .06)$$

$$= [1000(1 + .06)](1 + .06) = 1000(1 + .06)^2.$$

At the end of 3 years

$$A_3 = A_2 + A_2(.06) = A_2(1 + .06)$$
$$= [1000(1 + .06)^2](1 + .06) = 1000(1 + .06)^3.$$

After n years the compound amount will be

$$A = 1000(1 + .06)^n.$$

In this example the interest period was 1 year. The important point to note, however, is that at the end of each interest period the amount on deposit grew by a factor of $(1 + .06)$. In general, if the interest rate is i instead of .06, the compound amount will grow by a factor of $(1 + i)$ at the end of each interest period.

Suppose that a principal amount P is invested at a compound interest rate i per interest period, for a total of n interest periods. Then the compound amount A at the end of the nth period will be

$$A = P(1 + i)^n. \tag{1}$$

EXAMPLE 1 Suppose that $5000 is invested at 8% per year, with interest compounded annually. What is the compound amount after 3 years?

Solution Substituting $P = 5000$, $i = .08$, and $n = 3$ into formula (1), we have

$$A = 5000(1 + .08)^3 = 5000(1.08)^3$$
$$= 5000(1.259712) = 6298.56 \text{ dollars.}$$

It is common practice to state the interest rate as a percent per year ("per annum"), even though each interest period is often shorter than 1 year. If the annual rate is r and if interest is paid and compounded m times per year, then the interest rate i for each period is given by

$$[\text{rate per period}] \quad i = \frac{r}{m} = \frac{[\text{annual interest rate}]}{[\text{periods per year}]}.$$

Many banks pay interest quarterly. If the stated annual rate is 5%, then $i = .05/4 = .0125$.

If interest is compounded for t years, with m interest periods each year, there will be a total of mt interest periods. If in formula (1) we replace n by mt and replace i by r/m, we obtain the following formula for the compound amount:

$$A = P\left(1 + \frac{r}{m}\right)^{mt},$$

where P = principal amount,
$\quad r$ = interest rate per annum,
$\quad m$ = number of interest periods per year,
$\quad t$ = number of years. $\qquad$ (2)

EXAMPLE 2 Suppose that $1000 is deposited in a savings account that pays 6% per annum, compounded quarterly. If no additional deposits or withdrawals are made, how much will be in the account at the end of 1 year?

Solution We use (2) with $P = 1000$, $r = .06$, $m = 4$, and $t = 1$.

$$A = 1000\left(1 + \frac{.06}{4}\right)^4 = 1000(1.015)^4$$

$$= 1000(1.06136355) \approx 1061.36 \text{ dollars.}$$

Note that the $1000 in Example 2 earned a total of $61.36 in (compound) interest. This is 6.136% of $1000. Savings institutions sometimes advertise this rate as the *effective* annual interest rate. That is, the savings institutions mean that *if* they paid interest only once a year, they would have to pay a rate of 6.136% in order to produce the same earnings as their 6% rate compounded quarterly. The stated annual rate of 6% is often called the *nominal rate*.

The effective annual rate can be increased by compounding the interest more often. Some savings institutions compound interest monthly or even daily.

EXAMPLE 3 Suppose that the interest in Example 2 were compounded monthly. How much would be in the account at the end of 1 year? What about the case when 6% annual interest is compounded daily?

Solution For monthly compounding, $m = 12$. From (2) we have

$$A = 1000\left(1 + \frac{.06}{12}\right)^{12} = 1000(1.005)^{12}$$

$$\approx 1000(1.06167781) \approx 1061.68 \text{ dollars.}$$

The effective rate in this case is 6.168%.

A "bank year" usually consists of 360 days (in order to simplify calculations). So, for daily compounding, we take $m = 360$. Then

$$A = 1000\left(1 + \frac{.06}{360}\right)^{360} \approx 1000(1.00016667)^{360}$$

$$\approx 1000(1.06183133) \approx 1061.83 \text{ dollars.}$$

With daily compounding, the effective rate is 6.183%.

What would happen if the interest in Example 3 were compounded more often than once a day? Would the total interest be much more than $61.83 if the interest were compounded every hour? Every minute? To answer these questions, we shall connect the notion of compound interest to the exponential function. Recall that the compound amount A is given by

$$A = P\left(1 + \frac{r}{m}\right)^{mt} = P\left(1 + \frac{r}{m}\right)^{(m/r)\cdot rt}.$$

If we set $h = r/m$, then $1/h = m/r$, and

$$A = P(1 + h)^{(1/h)\cdot rt}.$$

As the frequency of compounding is increased, m gets large and $h = r/m$ approaches 0. To determine what happens to the compound amount, we must therefore examine the limit

$$\lim_{h \to 0} P(1 + h)^{(1/h)\cdot rt}.$$

The following remarkable fact is proved in the appendix at the end of this section:

$$\lim_{h \to 0} (1 + h)^{1/h} = e.$$

Using this fact together with two limit theorems, we have

$$\lim_{h \to 0} P(1 + h)^{(1/h)rt} = P\left[\lim_{h \to 0} (1 + h)^{1/h}\right]^{rt} = Pe^{rt}.$$

These calculations show that the compound amount calculated from the formula $P\left(1 + \frac{r}{m}\right)^{mt}$ gets closer to Pe^{rt} as the number m of interest periods per year is increased. When the formula

$$A = Pe^{rt} \tag{3}$$

is used to calculate the compound amount, we say that the interest is *compounded continuously*.

We can now answer the question posed following Example 3. Suppose that $1000 is deposited for 1 year in an account paying 6% per annum. Then $P = 1000$, $r = .06$, and $t = 1$. If the interest is compounded continuously, the compound amount at the end of 1 year will be

$$1000e^{.06} \approx 1061.84 \text{ dollars.}$$

Recall from Example 3 that daily compounding of the 6% interest produced $1061.83. Consequently, more frequent compounding (such as every hour or every second) will produce at most 1 cent more.

In recent years, when banks have wanted to offer the maximum effective rate of interest permitted by law, many banks and savings institutions have advertised savings accounts that pay interest compounded continuously. However, as we have seen, the effect of compounding continuously is practically the same as compounding daily, unless the principal amount P is quite large.

When interest is compounded continuously, the compound amount $A(t)$ is an exponential function of the number of years t that interest is earned, $A(t) = Pe^{rt}$. Hence $A(t)$ satisfies the differential equation

$$\frac{dA}{dt} = rA.$$

The rate of growth of the compound amount is proportional to the amount of money present. Since the growth comes from the interest, we conclude that under continuous compounding, interest is earned continuously at a rate of growth proportional to the amount of money present.

The formula $A = Pe^{rt}$ contains four variables. (Remember that the letter e here represents a specific constant, $e = 2.718\ldots$.) In a typical problem, we are given values for three of these variables and must solve for the remaining variable.

EXAMPLE 4   How long is required for an investment of $1000 to double if the interest is 10%, compounded continuously?

Solution   Here $P = 1000$ and $r = .10$. For each time t, the value of the investment is $Pe^{rt} = 1000e^{.10t}$. We must find t such that this value is $2000. So we set

$$2000 = 1000e^{.10t}$$

and solve for t. We divide both sides by 1000 and then take logarithms of both sides to obtain

$$2 = e^{.10t}$$

$$\ln 2 = .10t,$$

and

$$t = 10 \ln 2 \approx 6.9 \text{ years.}$$

Remark The calculations in Example 4 would be essentially unchanged after the first step if the initial amount of the investment were changed from $1000 to any arbitrary amount P. When this investment doubles, the compound amount will be $2P$. So one sets $2P = Pe^{.10t}$ and solves for t as we did above, to conclude that any amount doubles in about 6.9 years.

If P dollars are invested today, the formula $A = Pe^{rt}$ gives the value of this investment after t years (assuming continuously compounded interest). We say that P is the *present value* of the amount A to be received in t years. If we solve for P in terms of A, we obtain

$$P = Ae^{-rt}. \tag{4}$$

The concept of the present value of money is an important theoretical tool in business and economics. Problems involving depreciation of equipment, for example, may be analyzed by calculus techniques when the present value of money is computed from (4) using continuously compounded interest.

EXAMPLE 5 Find the present value of $5000 to be received in 2 years if money can be invested at 12% compounded continuously.

Solution Use (4) with $A = 5000$, $r = .12$, and $t = 2$.

$$P = 5000e^{-(.12)(2)} = 5000e^{-.24}$$
$$\approx 5000(0.78663) = 3933.15 \text{ dollars.}$$

APPENDIX A Limit Formula for e

For $h \neq 0$, we have

$$\ln(1 + h)^{1/h} = 1/h \ln(1 + h).$$

Taking the exponential of both sides, we find that

$$(1 + h)^{1/h} = e^{(1/h)\ln(1+h)}.$$

Since the exponential function is continuous,

$$\lim_{h \to 0} (1 + h)^{1/h} = e^{\left[\lim_{h \to 0} (1/h)\ln(1 + h)\right]}. \tag{5}$$

To examine the limit inside the exponential function, we note that $\ln 1 = 0$, and hence

$$\lim_{h \to 0} (1/h)\ln(1 + h) = \lim_{h \to 0} \frac{\ln(1 + h) - \ln 1}{h}.$$

The limit on the right is a difference quotient of the type used to compute a derivative. In fact,

$$\lim_{h \to 0} \frac{\ln(1 + h) - \ln 1}{h} = \frac{d}{dx} \ln x \Big|_{x=1}$$

$$= \frac{1}{x}\Big|_{x=1} = 1.$$

Thus the limit inside the exponential function in (5) is 1. That is,

$$\lim_{h \to 0} (1 + h)^{1/h} = e^{[1]} = e.$$

PRACTICE PROBLEMS 2

1. One thousand dollars is to be invested in a bank for 4 years. Would 8% interest compounded semiannually be better than $7\frac{3}{4}$% interest compounded continuously?

2. A building was bought for $150,000 and sold 10 years later for $400,000. What interest rate (compounded continuously) was earned on the investment?

EXERCISES 2

1. Suppose that $1000 is deposited in a savings account at 10% interest compounded annually. What is the compound amount after 2 years?

2. Five thousand dollars is deposited in a savings account at 6% interest compounded monthly. Give the formula that describes the compound amount after 4 years.

3. Ten thousand dollars is invested at 8% interest compounded quarterly. Give the formula that describes the value of the investment after 3 years.

4. What is the effective annual rate of interest of a savings account paying 8% interest compounded semiannually?

5. One thousand dollars is invested at 14% interest compounded continuously. Compute the value of the investment at the end of 6 years.

6. A painting was purchased in 1983 for $100,000. If it appreciates at 12% compounded continuously, how much will it be worth in 1990?

7. Five hundred dollars is deposited in a savings account paying 7% interest compounded daily. *Estimate* the balance in the account at the end of 3 years.

8. Ten thousand dollars is deposited into a money market fund paying 18% interest compounded continuously. How much interest will be earned during the first half-year if this rate of 18% does not change?

9. An office building, built in 1975 at a cost of 10 million dollars, was appraised at 25 million dollars in 1983. At what annual rate of interest (compounded continuously) did the building appreciate?

10. One thousand dollars is deposited in a savings account at 6% interest compounded continuously. How many years are required for the balance in the account to reach $2500?

11. Ten thousand dollars is to be invested in a highly speculative venture for 1 year. Would you rather receive 40% interest compounded semiannually or 39% interest compounded continuously?

12. A lot purchased in 1964 for $5000 was appraised at $60,000 in 1983. If the lot continues to appreciate at the same rate, when will it be worth $100,000?

13. Ten thousand dollars is invested at 15% interest compounded continuously. When will the investment be worth $38,000?

14. How many years are required for an investment to double in value if it is appreciating at the rate of 13% compounded continuously?

15. A farm purchased in 1970 for $1,000,000 was valued at $3,000,000 in 1980. If the farm continues to appreciate at the same rate (with continuous compounding), when will it be worth $10,000,000?

16. Find the present value of $1000 payable at the end of 3 years, if money may be invested at 14% with interest compounded continuously.

17. Find the present value of $2000 to be received in 10 years, if money may be invested at 15% with interest compounded continuously.

18. A parcel of land bought in 1975 for $10,000 was worth $16,000 in 1980. If the land continues to appreciate at this rate, in what year will it be worth $45,000?

19. One hundred dollars is deposited in a savings account at 7% interest compounded continuously. What is the effective annual rate of return?

20. In a certain town, property values tripled from 1973 to 1984. If this trend continues, when will property values be at five times their 1973 level? (Use an exponential model for the property value at time t.)

21. How much money must you invest now at 12% interest compounded continuously, in order to have $10,000 at the end of 5 years?

22. Investment A is currently worth $70,200 and is growing at the rate of 13% per year compounded continuously. Investment B is currently worth $60,000 and is growing at the rate of 14% per year compounded continuously. After how many years will the two investments have the same value?

23. Suppose that the present value of $1000 to be received in 2 years is $559.90. What rate of interest, compounded continuously, was used to compute this present value?

24. Two thousand dollars is deposited in a savings account at 5% interest compounded continuously. Let $f(t)$ be the compound amount after t years. Find and interpret $f'(2)$.

SOLUTIONS TO PRACTICE PROBLEMS 2

1. Let us compute the balance after 4 years for each type of interest.

    8% compounded semiannually: Use formula (2). Here $P = 1000$, $r = .08$, $m = 2$ (semiannually means that there are two interest periods per year), and $t = 4$. Therefore,

    $$A = 1000\left(1 + \frac{.08}{2}\right)^{2 \cdot 4}$$

    $$= 1000(1.04)^8 = 1368.57.$$

    $7\frac{3}{4}\%$ compounded continuously: Use the formula $A = Pe^{rt}$, where $P = 1000$, $r = .0775$, and $t = 4$. Then

    $$A = 1000e^{(.0775) \cdot 4}$$

    $$= 1000e^{.31} = 1363.43.$$

    Therefore, 8% compounded semiannually is best.

2. If the \$150,000 had been compounded continuously for 10 years at interest rate r, the balance would be $150{,}000e^{r \cdot 10}$. The question asks: For what value of r will the balance be 400,000? We need just solve an equation for r.

    $$150{,}000e^{r \cdot 10} = 400{,}000$$

    $$e^{r \cdot 10} \approx 2.67$$

    $$r \cdot 10 = \ln 2.67$$

    $$r = \frac{\ln 2.67}{10} \approx .098.$$

    Therefore, the investment earned 9.8% interest per year.

15.3. Applications of the Natural Logarithm Function to Economics

In this section we consider two applications of the natural logarithm to the field of economics. Our first application is concerned with relative rates of change and the second with elasticity of demand.

Relative Rates of Change   The *logarithmic derivative* of a function $f(t)$ is defined by the equation

$$\frac{d}{dt} \ln f(t) = \frac{f'(t)}{f(t)}. \tag{1}$$

The quantity on either side of equation (1) is often called the *relative rate of change of $f(t)$ per unit change of t*. Indeed, this quantity compares the rate of change of $f(t)$ [namely $f'(t)$] with $f(t)$ itself. The *percentage rate of change* is the relative rate of change of $f(t)$ expressed as a percentage.

A simple example will illustrate these concepts. Suppose that $f(t)$ denotes the average price per pound of sirloin steak at time t and $g(t)$ denotes the average price of a new car (of a given make and model) at time t, where $f(t)$ and $g(t)$ are given in dollars and time is measured in years. Then the ordinary derivatives $f'(t)$ and $g'(t)$ may be interpreted as the rate of change of the price of a pound of sirloin steak and of a new car, respectively, where both are measured in dollars per year. Suppose that, at a given time t_0, we have $f(t_0) = \$5.25$ and $g(t_0) = \$12,000$. Moreover, suppose that $f'(t_0) = \$.75$ and $g'(t_0) = \$1500$. Then at time t_0 the price per pound of steak is increasing at a rate of $\$.75$ per year, while the price of a new car is increasing at a rate of $\$1500$ per year. Which price is increasing more quickly? It is not meaningful to say that the car price is increasing faster simply because $\$1500$ is larger than $\$.75$. We must take into account the vast difference between the actual cost of a car and the cost of steak. The usual basis of comparison of price increases is the percentage rate of increase. In other words, at $t = t_0$, the price of sirloin steak is increasing at the percentage rate

$$\frac{f'(t_0)}{f(t_0)} = \frac{.75}{5.25} \approx .143 = 14.3\%$$

per year, but at the same time the price of a new car is increasing at the percentage rate

$$\frac{g'(t_0)}{g(t_0)} = \frac{1500}{12,000} \approx .125 = 12.5\%$$

per year. Thus the price of sirloin steak is increasing at a faster percentage rate than the price of a new car.

Economists often use relative rates of change (or percentage rates of change) when discussing the growth of various economic quantities, such as national income or national debt, because such rates of change can be meaningfully compared.

EXAMPLE 1 Suppose that the Gross National Product of the United States at time t (measured in years from January 1, 1984) is predicted by a certain school of economists to be

$$f(t) = 3.4 + .04t + .13e^{-t},$$

where the Gross National Product is measured in trillions of dollars. What is the percentage rate of growth (or decline) of the economy at $t = 0$ and $t = 1$?

Solution Since

$$f'(t) = .04 - .13e^{-t},$$

we see that

$$\frac{f'(0)}{f(0)} = \frac{.04 - .13}{3.4 + .13} = -\frac{.09}{3.53} \approx -2.6\%.$$

$$\frac{f'(1)}{f(1)} = \frac{.04 - .13e^{-1}}{3.4 + .04 + .13e^{-1}} = -\frac{.00782}{3.4878} \approx -.2\%.$$

So on January 1, 1984, the economy is predicted to contract at a relative rate of 2.6% per year; on January 1, 1985, the economy is predicted to be still contracting but only at a relative rate of .2% per year.

EXAMPLE 2 Suppose that the value in dollars of a certain business investment at time t may be approximated empirically by the function $f(t) = 750,000e^{.6\sqrt{t}}$. Use a logarithmic derivative to describe how fast the value of the investment is increasing when $t = 5$ years.

Solution We have

$$\frac{f'(t)}{f(t)} = \frac{d}{dt} \ln f(t) = \frac{d}{dt} (\ln 750,000 + \ln e^{.6\sqrt{t}})$$

$$= \frac{d}{dt} (\ln 750,000 + .6\sqrt{t})$$

$$= (.6)\left(\frac{1}{2}\right)t^{-1/2} = \frac{.3}{\sqrt{t}}.$$

When $t = 5$,

$$\frac{f'(5)}{f(5)} = \frac{.3}{\sqrt{5}} \approx .1345 = 13.4\%.$$

Thus, when $t = 5$ years, the value of the investment is increasing at the relative rate of 13.4% per year.

In certain mathematical models, it is assumed that for a limited period of time, the percentage rate of change of a particular function is constant. The following example shows that such a function must be an exponential function.

EXAMPLE 3 Suppose that the function $f(t)$ has a constant relative rate of change k. Show that $f(t) = Ce^{kt}$ for some constant C.

Solution We are given that

$$\frac{d}{dt} \ln f(t) = k.$$

That is,

$$\frac{f'(t)}{f(t)} = k.$$

Hence $f'(t) = kf(t)$. But this is just the differential equation satisfied by the exponential function. Therefore, we must have $f(t) = Ce^{kt}$ for some constant C.

Elasticity of Demand   In Section 12.7 we considered demand equations for monopolists and for entire industries. Recall that a demand equation expresses, for each quantity q to be produced, the market price which will generate a demand of exactly q. For instance, the demand equation

$$p = 150 - .01x \tag{2}$$

says that in order to sell x units, the price must be set at $150 - .01x$ dollars. To be specific: In order to sell 6000 units, the price must be set at $150 - .01(6000) = \$90$ per unit.

Equation (2) may be solved for x in terms of p to yield

$$x = 100(150 - p). \tag{3}$$

This last equation gives quantity in terms of price. If we let the letter q represent quantity, equation (3) becomes

$$q = 100(150 - p). \tag{3$'$}$$

This equation is of the form $q = f(p)$, where in this case $f(p)$ is the function $f(p) = 100(150 - p)$. In what follows it will be convenient to always write our demand functions so that the quantity q is expressed as a function $f(p)$ of the price p.

Usually, raising the price of a commodity lowers demand. Therefore, the typical demand function $q = f(p)$ is decreasing and has a negative slope everywhere.

A demand function $q = f(p)$ relates the quantity demanded to the price. Therefore, the derivative $f'(p)$ compares the change in quantity demanded with the change in price. By way of contrast, the concept of elasticity is designed to compare the *relative* rate of change of the quantity demanded with the *relative* rate of change of price.

Let us be more explicit. Consider a particular demand function $q = f(p)$ and a particular price p. Then at this price, the ratio of the relative rates of change of the quantity demanded and the price is given by

$$\frac{[\text{relative rate of change of quantity}]}{[\text{relative rate of change of price}]} = \frac{\dfrac{d}{dp}\ln f(p)}{\dfrac{d}{dp}\ln p}$$

$$= \frac{f'(p)/f(p)}{1/p}$$

$$= \frac{pf'(p)}{f(p)}.$$

Since $f'(p)$ is always negative for a typical demand function, the quantity $pf'(p)/f(p)$ will be negative for all values of p. For convenience, economists prefer to work with positive numbers and therefore the *elasticity of demand* is taken to be this quantity multiplied by -1.

The elasticity of demand $E(p)$ at price p for the demand function $q = f(p)$ is defined to be

$$E(p) = \frac{-pf'(p)}{f(p)}.$$

EXAMPLE 4 Suppose that the demand function for a certain metal is $q = 100 - 2p$, where p is the price per pound and q is the quantity demanded (in millions of pounds).

(a) What quantity can be sold at $30 per pound?

(b) Determine the function $E(p)$.

(c) Determine and interpret the elasticity of demand at $p = 30$.

(d) Determine and interpret the elasticity of demand at $p = 20$.

Solution (a) In this case, $q = f(p)$, where $f(p) = 100 - 2p$. When $p = 30$, we have $q = f(30) = 100 - 2(30) = 40$. Therefore, 40 million pounds of the metal can be sold. We also say that the *demand* is 40 million pounds.

(b)
$$E(p) = \frac{-pf'(p)}{f(p)}$$

$$= \frac{-p(-2)}{100 - 2p}$$

$$= \frac{2p}{100 - 2p}.$$

(c) The elasticity of demand at price $p = 30$ is $E(30)$.

$$E(30) = \frac{2(30)}{100 - 2(30)}$$

$$= \frac{60}{40}$$

$$= \frac{3}{2}.$$

When the price is set at $30 per pound, a small increase in price will result in a relative rate of decrease in quantity demanded of about $\frac{3}{2}$ times the relative rate of increase in price. For example, if the price is increased from $30 by 1%, then the quantity demanded will decrease by about 1.5%.

(d) When $p = 20$, we have

$$E(20) = \frac{2(20)}{100 - 2(20)} = \frac{40}{60} = \frac{2}{3}.$$

When the price is set at $20 per pound, a small increase in price will result in a relative rate of decrease in quantity demanded of only $\frac{2}{3}$ of the relative rate of increase of price. For example, if the price is increased from $20 by 1%, the quantity demanded will decrease by $\frac{2}{3}$ of 1%.

Economists say that demand is *elastic* at price p_0 if $E(p_0) > 1$ and *inelastic* at price p_0 if $E(p_0) < 1$. In Example 4, the demand for the metal is elastic at $30 per pound and inelastic at $20 per pound.

The significance of the concept of elasticity may perhaps best be appreciated by studying how revenue, $R(p)$, responds to changes in price. Recall that

$$[\text{revenue}] = [\text{quantity}] \cdot [\text{price per unit}],$$

that is,

$$R(p) = f(p) \cdot p.$$

If we differentiate $R(p)$ using the product rule, we find that

$$R'(p) = \frac{d}{dp}[f(p) \cdot p] = f(p) \cdot 1 + p \cdot f'(p)$$

$$= f(p)\left[1 + \frac{pf'(p)}{f(p)}\right]$$

$$= f(p)[1 - E(p)]. \tag{4}$$

Now suppose that demand is elastic at some price p_0. Then $E(p_0) > 1$ and $1 - E(p_0)$ is negative. Since $f(p)$ is always positive, we see from (4) that $R'(p_0)$ is negative. Therefore, by the first derivative rule, $R(p)$ is decreasing at p_0. So an increase in price will result in a decrease in revenue, and a decrease in price will result in an increase in revenue. On the other hand, if demand is inelastic at p_0, then $1 - E(p_0)$ will be positive and hence $R'(p_0)$ will be positive. In this case an increase in price will result in an increase in revenue, and a decrease in price will result in an decrease in revenue. We may summarize this as follows:

> The change in revenue is in the opposite direction of the change in price when demand is elastic and in the same direction when demand is inelastic.

PRACTICE PROBLEMS 3

The current toll for the use of a certain toll road is $2.50. A study conducted by the state highway department determined that with a toll of p dollars, q cars will use the road each day, where $q = 60,000e^{-.5p}$.

1. Compute the elasticity of demand at $p = 2.5$.

2. Is demand elastic or inelastic at $p = 2.5$?

3. If the state increases the toll slightly, will the revenue increase or decrease?

EXERCISES 3

Determine the percentage rate of change of the functions at the points indicated.

1. $f(t) = t^2$ at $t = 10$ and $t = 50$

2. $f(t) = t^{10}$ at $t = 10$ and $t = 50$

3. $f(x) = e^{.3x}$ at $x = 10$ and $x = 20$

4. $f(x) = e^{-.05x}$ at $x = 1$ and $x = 10$

5. $f(t) = e^{.3t^2}$ at $t = 1$ and $t = 5$

6. $G(s) = e^{-.05s^2}$ at $s = 1$ and $s = 10$

7. $f(p) = 1/(p + 2)$ at $p = 2$ and $p = 8$

8. $g(p) = 5/(2p + 3)$ at $p = 1$ and $p = 11$

9. Suppose that the annual sales S (in dollars) of a company may be approximated empirically by the formula

$$S = 50{,}000\sqrt{e^{\sqrt{t}}},$$

where t is the number of years beyond some fixed reference date. Use a logarithmic derivative to determine the percentage rate of growth of sales at $t = 4$.

10. Suppose that the price of wheat per bushel at time t (in months) is approximated by

$$f(t) = 4 + .001t + .01e^{-t}.$$

What is the percentage rate of change of $f(t)$ at $t = 0$? $t = 1$? $t = 2$?

11. Suppose that an investment grows at a constant 12% rate per year. In how many years will the value of the investment double?

12. Suppose that the value of a piece of property is growing at a constant $r\%$ rate per year and that the value doubles in 3 years. Find r.

For each demand function, find $E(p)$ and determine if demand is elastic or inelastic (or neither) at the indicated price.

13. $q = 700 - 5p$, $p = 80$

14. $q = 600e^{-.2p}$, $p = 10$

15. $q = 400(116 - p^2)$, $p = 6$

16. $q = (77/p^2) + 3$, $p = 1$

17. $q = p^2 e^{-(p+3)}$, $p = 4$

18. $q = 700/(p + 5)$, $p = 15$

19. Currently, 1800 people ride a certain commuter train each day and pay $4 for a ticket. The number of people q willing to ride the train at price p is $q = 600(5 - \sqrt{p})$. The railroad would like to increase its revenue.

   (a) Is demand elastic or inelastic at $p = 4$?

   (b) Should the price of a ticket be raised or lowered?

20. A company can sell $q = 9000/(p + 60) - 50$ radios at a price of p dollars per radio. The current price is $30.

   (a) Is demand elastic or inelastic at $p = 30$?

   (b) If the price is lowered slightly, will revenue increase or decrease?

21. A movie theater has a seating capacity of 3000 people. The number of people attending a show at price p dollars per ticket is $q = (18{,}000/p) - 1500$. Currently, the price is $6 per ticket.

(a) Is demand elastic or inelastic at $p = 6$?

(b) If the price is lowered, will revenue increase or decrease?

22. A subway charges 65 cents per person and has 10,000 riders each day. The demand function for the subway is $q = 2000\sqrt{90 - p}$.

(a) Is demand elastic or inelastic at $p = 65$?

(b) Should the price of a ride be raised or lowered in order to increase the amount of money taken in by the subway?

23. A country which is the major supplier of a certain commodity wishes to improve its balance of trade position by lowering the price of the commodity. The demand function is $q = 1000/p^2$.

(a) Compute $E(p)$.

(b) Will the country succeed in raising its revenue?

24. Show that any demand function of the form $q = a/p^m$ has constant elasticity m.

A cost function $C(x)$ gives the total cost of producing x units of a product. The *elasticity of cost at quantity* x is defined to be

$$E_c(x) = \frac{\dfrac{d}{dx} \ln C(x)}{\dfrac{d}{dx} \ln x}.$$

25. Show that $E_c(x) = x \cdot C'(x)/C(x)$.

26. Show that E_c is equal to the marginal cost divided by the average cost.

27. Let $C(x) = (1/10)x^2 + 5x + 300$. Show that $E_c(50) < 1$. (Hence when producing 50 units, a small relative increase in production results in an even smaller relative increase in total cost. Also, the average cost of producing 50 units is greater than the marginal cost at $x = 50$.)

28. Let $C(x) = 1000e^{.02x}$. Determine and simplify the formula for $E_c(x)$. Show that $E_c(60) > 1$ and interpret this result.

SOLUTIONS TO PRACTICE PROBLEMS 3

1. The demand function is $f(p) = 60,000e^{-.5p}$.

$$f'(p) = -30,000e^{-.5p}.$$

$$E(p) = \frac{-pf'(p)}{f(p)} = \frac{-p(-30,000)e^{-.5p}}{60,000e^{-.5p}} = \frac{p}{2}.$$

$$E(2.5) = \frac{2.5}{2} = 1.25.$$

2. The demand is elastic, because $E(2.5) > 1$.

3. Since demand is elastic at \$2.50, a slight change in price causes revenue to change in the *opposite* direction. Hence revenue will decrease.

15.4. Further Exponential Models

A skydiver, on jumping out of an airplane, falls at an increasing rate. However, the wind rushing past the skydiver's body creates an upward force that begins to counterbalance the downward force of gravity. This air friction finally becomes so great that the skydiver's velocity reaches a limiting speed called the *terminal velocity*. If we let $v(t)$ be the downward velocity of the skydiver after t seconds of free fall, then a good mathematical model for $v(t)$ is given by

$$v(t) = M(1 - e^{-kt}), \tag{1}$$

where M is the terminal velocity and k is some positive constant (Fig. 1). When t is close to zero, e^{-kt} is close to one and the velocity is small. As t increases, e^{-kt} becomes small and so $v(t)$ approaches M.

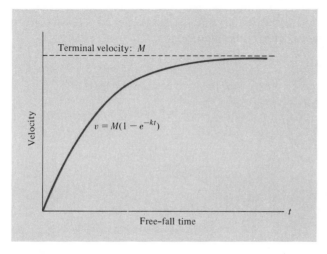

FIGURE 1

EXAMPLE 1 Show that the velocity given in (1) satisfies the differential equation

$$\frac{dv}{dt} = k[M - v(t)], \qquad v(0) = 0. \tag{2}$$

Solution From (1) we have $v(t) = M - Me^{-kt}$. Then

$$\frac{dv}{dt} = Mke^{-kt}.$$

However,

$$k[M - v(t)] = k[M - (M - Me^{-kt})] = kMe^{-kt},$$

so that the differential equation $\dfrac{dv}{dt} = k[M - v(t)]$ holds. Also,

$$v(0) = M - Me^0 = M - M = 0.$$

The differential equation (2) says that the rate of change in v is proportional to the difference between the terminal velocity M and the actual velocity v. It is not difficult to show that the only solution of (2) is given by the formula in (1).

The two equations (1) and (2) arise as mathematical models in a variety of situations. Some of these applications are described below.

The Learning Curve   Psychologists have found that in many learning situations a person's rate of learning is rapid at first and then slows down. Finally, as the task is mastered, the person's level of performance reaches a level above which it is almost physically impossible to rise. For example, within reasonable limits, each person seems to have a certain maximum capacity for memorizing a list of nonsense syllables. Suppose that a subject can memorize M syllables in a row if given sufficient time, say an hour, to study the list but cannot memorize $M + 1$ syllables in a row even if allowed several hours of study. By giving the subject different lists of syllables and varying lengths of time to study the lists, the psychologist can determine an empirical relationship between the number of nonsense syllables memorized accurately and the number of minutes of study time. It turns out that a good model for this situation is

$$y = M(1 - e^{-kt}) \tag{3}$$

for some appropriate positive constant k. (See Fig. 2.)

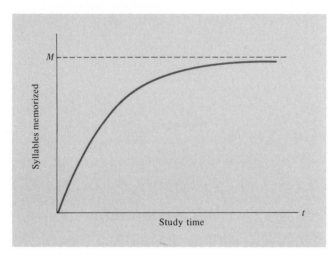

FIGURE 2   Learning curve, $y = M(1 - e^{-kt})$.

The *slope* of this learning curve at time t is approximately the number of additional syllables that can be memorized if the subject is given one more minute of study time. Thus the slope is a measure of the *rate of learning*. The differential equation satisfied by the function in (3) is

$$y' = k(M - y), \qquad f(0) = 0.$$

This equation says that if the subject is given a list of M nonsense syllables, then the rate of memorization is proportional to the number of syllables remaining to be memorized.

Diffusion of Information by Mass Media   Sociologists have found that the differential equation (2) provides a good model for the way information is spread (or "diffused") through a population when the information is being propagated constantly by mass media, such as television or magazines.* Given a fixed population P, let $f(t)$ be the number of people who have already heard a certain piece of information by time t. Then $P - f(t)$ is the number who have not yet heard the information. Also, $f'(t)$ is the rate of increase of the number of people who have heard the news (the "rate of diffusion" of the information). If the information is being publicized often by some mass media, then it is likely that the number of *newly informed* people per unit time is proportional to the number of people who have not yet heard the news. Therefore,

$$f'(t) = k[P - f(t)].$$

Assume that $f(0) = 0$ (i.e., there was a time $t = 0$ when nobody had heard the news). Then the remark following Example 1 shows that

$$f(t) = P(1 - e^{-kt}). \tag{4}$$

(See Fig. 3.)

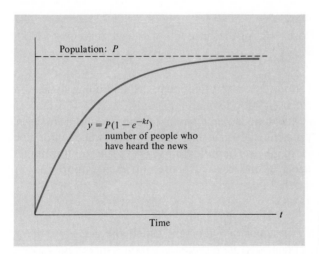

$y = P(1 - e^{-kt})$
number of people who
have heard the news

FIGURE 3   Diffusion of information by mass media.

EXAMPLE 2   Suppose that a certain piece of news (such as the resignation of a public official) is broadcast frequently by radio and television stations. Also suppose that one-half of the residents of a city have heard the news within 4 hours of its initial release. Use the exponential model (4) to estimate when 90% of the residents will have heard the news.

Solution   We must find the value of k in (4). If P is the number of residents, then the number who have heard the news in four hours is given by (4) with $t = 4$. By assumption,

---

* J. Coleman, *Introduction to Mathematical Sociology* (New York: The Free Press, 1964), p. 43.

this number is half the population. So

$$\tfrac{1}{2}P = P(1 - e^{-k \cdot 4})$$
$$.5 = 1 - e^{-4k}$$
$$e^{-4k} = 1 - .5 = .5.$$

Solving for k, we find that $k \approx .17$. So the model for this particular situation is

$$f(t) = P(1 - e^{-.17t}).$$

Now we want to find t such that $f(t) = .90P$. We solve for t:

$$.90P = P(1 - e^{-.17t})$$
$$.90 = 1 - e^{-.17t}$$
$$e^{-.17t} = 1 - .90 = .10$$
$$-.17t = \ln .10$$
$$t = \frac{\ln .10}{-.17} \approx 14.$$

Therefore, 90% of the residents will hear the news within 14 hours of its initial release.

Intravenous Infusion of Glucose   The human body both manufactures and uses glucose ("blood sugar"). Usually, there is a balance in these two processes, so that the bloodstream has a certain "equilibrium level" of glucose. Suppose that a patient is given a single intravenous injection of glucose and let $A(t)$ be the amount of glucose (in milligrams) above the equilibrium level. Then the body will start using up the excess glucose at a rate proportional to the amount of excess glucose; that is,

$$A'(t) = -\lambda A(t), \tag{5}$$

where λ is a positive constant called the *velocity constant of elimination*. This constant depends on how fast the patient's metabolic processes eliminate the excess glucose from the blood. Equation (5) describes a simple exponential decay process.

Now suppose that, instead of a single shot, the patient receives a continuous intravenous infusion of glucose. A bottle of glucose solution is suspended above the patient, and a small tube carries the glucose down to a needle that runs into a vein. In this case, there are two influences on the amount of excess glucose in the blood: the glucose being added steadily from the bottle and the glucose being removed from the blood by metabolic processes. Let r be the rate of infusion of glucose (often from 10 to 100 milligrams per minute). If the body did not remove any glucose, the excess glucose would increase at a constant rate of r milligrams per minute; that is,

$$A'(t) = r. \tag{6}$$

Taking into account the two influences on $A'(t)$ described by (5) and (6), we can write

$$A'(t) = r - \lambda A(t). \tag{7}$$

If we let $M = r/\lambda$, then

$$A'(t) = \lambda(M - A(t)).$$

It can be shown that a solution of this differential equation is given by

$$A(t) = M(1 - e^{-\lambda t}) = \frac{r}{\lambda}(1 - e^{-\lambda t}). \tag{8}$$

Reasoning as in Example 1, we conclude that the amount of excess glucose rises until it reaches a stable level. (See Fig. 4).

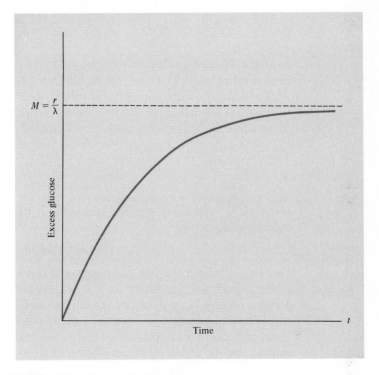

FIGURE 4   Continuous infusion of glucose.

$$y = \frac{r}{\lambda}(1 - e^{-\lambda t}).$$

The Logistic Growth Curve   The model for simple exponential growth discussed in Section 1 is adequate for describing the growth of many types of populations, but obviously a population cannot increase exponentially forever. The simple exponential growth model becomes inapplicable when the environment begins to inhibit the growth of the population. The logistic growth curve is an important exponential model that takes into account some of the effects of the environment on a population (Fig. 5). For small values of t, the curve has the same basic shape

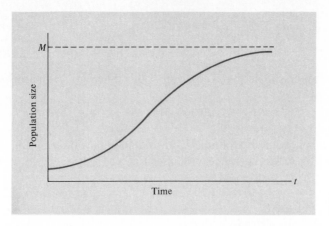

FIGURE 5
Logistic growth.

as an exponential growth curve. Then when the population begins to suffer from overcrowding or lack of food, the growth rate (the slope of the population curve) begins to slow down. Eventually, the growth rate tapers off to zero as the population reaches the maximum size that the environment will support. This latter part of the curve resembles the growth curves studied earlier in this section.

The equation for logistic growth has the general form

$$y = \frac{M}{1 + Be^{-Mkt}}, \tag{9}$$

where B, M, and k are positive constants. It can be shown that y satisfies the differential equation

$$y' = ky(M - y). \tag{10}$$

The factor y reflects the fact that the growth rate (y') depends in part on the size y of the population. The factor $M - y$ reflects the fact that the growth rate also depends on how close y is to the maximum level M.

The logistic curve is often used to fit experimental data that lie along an "S-shaped" curve. Examples are given by the growth of a fish population in a lake and the growth of a fruit fly population in a laboratory container. Also, certain enzyme reactions in animals follow a logistic law. One of the earliest applications of the logistic curve occurred in about 1840 when the Belgian sociologist P. Verhulst fit a logistic curve to six U.S. census figures, 1790 to 1840, and predicted the U.S. population for 1940. His prediction missed by less than 1 million persons (an error of about 1%).

EXAMPLE 3   Suppose that a lake is stocked with 100 fish. After 3 months there are 250 fish. A study of the ecology of the lake predicts that the lake can support 1000 fish. Find a formula for the number $P(t)$ of fish in the lake t months after it has been stocked.

Solution The limiting population M is 1000. Therefore, we have

$$P(t) = \frac{1000}{1 + Be^{-1000kt}}.$$

At $t = 0$ there are 100 fish, so that

$$100 = P(0) = \frac{1000}{1 + Be^0} = \frac{1000}{1 + B}.$$

Thus $1 + B = 10$, or $B = 9$. Finally, since $P(3) = 250$, we have

$$250 = \frac{1000}{1 + 9e^{-3000k}}$$

$$1 + 9e^{-3000k} = 4$$

$$e^{-3000k} = \tfrac{1}{3}.$$

$$-3000k = \ln \tfrac{1}{3}$$

$$k \approx .00037.$$

Therefore,

$$P(t) = \frac{1000}{1 + 9e^{-.37t}}.$$

Several theoretical justifications can be given for using (9) and (10) in situations where the environment prevents a population from exceeding a certain size. A discussion of this topic may be found in *Mathematical Models and Applications* by D. Maki and M. Thompson (Englewood Cliffs, N.J.: Prentice-Hall, Inc., 1973), pp. 312–317.

An Epidemic Model It will be instructive to actually "build" a mathematical model. Our example concerns the spread of a highly contagious disease. We begin by making several simplifying assumptions:

1. The population is a fixed number P and each member of the population is susceptible to the disease.
2. The duration of the disease is long, so that no cures occur during the time period under study.
3. All infected individuals are contagious and circulate freely among the population.
4. During each unit time period (such as 1 day or 1 week) each infected person makes c contacts, and each contact with an uninfected person results in transmission of the disease.

Consider a short period of time from t to $t + h$. Each infected person makes $c \cdot h$ contacts. How many of these contacts are with uninfected persons? If $f(t)$ is the number of infected persons at time t, then $P - f(t)$ is the number of uninfected persons, and $[P - f(t)]/P$ is the fraction of the population that is uninfected. Thus, of the $c \cdot h$ contacts made,

$$\left[\frac{P - f(t)}{P}\right] \cdot c \cdot h$$

will be with uninfected persons. This is the number of new infections produced by one infected person during the time period of length h. The total number of *new* infections during this period is

$$f(t)\left[\frac{P - f(t)}{P}\right]ch.$$

But this number must equal $f(t + h) - f(t)$, where $f(t + h)$ is the total number of infected persons at time $t + h$. So

$$f(t + h) - f(t) = f(t)\left[\frac{P - f(t)}{P}\right]ch.$$

Dividing by h, the length of the time period, we obtain the average number of new infections per unit time (during the small time period):

$$\frac{f(t + h) - f(t)}{h} = \frac{c}{P}f(t)[P - f(t)].$$

If we let h approach zero and let y stand for $f(t)$, the left-hand side approaches the rate of change in the number of infected persons and we derive the following equation:

$$\frac{dy}{dt} = \frac{c}{P}y(P - y). \tag{11}$$

This is the same type of equation as that used in (10) for logistic growth, although the two situations leading to this model appear to be quite dissimilar.

Comparing (11) with (10), we see that the number of infected individuals at time t is described by a logistic curve with $M = P$ and $k = c/P$. Therefore, by (9), we can write

$$f(t) = \frac{P}{1 + Be^{-ct}}.$$

B and c can be determined from the characteristics of the epidemic. (See Example 4 below.)

The logistic curve has an inflection point at that value of t for which $f(t) = P/2$. The position of this inflection point has great significance for applications of the logistic curve. From inspecting a graph of the logistic curve, we see that the inflection point is the point at which the curve has greatest slope. In other words, the inflection point corresponds to the instant of fastest growth of the logistic curve. This means, for example, that in the foregoing epidemic model the disease is spreading with the greatest rapidity precisely when half the population is infected. Any attempt at disease control (through immunization, for example) must strive to reduce the incidence of the disease to as low a point as possible, but in any case at least below the inflection point at $P/2$, at which point the epidemic is spreading fastest.

EXAMPLE 4 The Public Health Service monitors the spread of an epidemic of a particularly long-lasting strain of flu in a city of 500,000 people. At the beginning of the first week of monitoring, 200 cases have been reported; during the first week 300 new cases are reported. Estimate the number of infected individuals after 6 weeks.

Solution Here $P = 500{,}000$. If $f(t)$ denotes the number of cases at the end of t weeks, then

$$f(t) = \frac{P}{1 + Be^{-ct}}$$

$$= \frac{500{,}000}{1 + Be^{-ct}}.$$

Moreover, $f(0) = 200$, so that

$$200 = \frac{500{,}000}{1 + Be^0} = \frac{500{,}000}{1 + B},$$

and $B = 2499$. Consequently, since $f(1) = 300 + 200 = 500$, we have

$$500 = f(1) = \frac{500{,}000}{1 + 2499e^{-c}},$$

so that $e^{-c} \approx .4$ and $c \approx .92$. Finally,

$$f(t) = \frac{500{,}000}{1 + 2499e^{-.92t}}$$

and

$$f(6) = \frac{500{,}000}{1 + 2499e^{-.92(6)}} \approx 45{,}000.$$

After 6 weeks, about 45,000 individuals are infected.

This epidemic model is used by sociologists (where it is still called an epidemic model) to describe the spread of a rumor. In economics the model is used to describe the diffusion of knowledge about a product. An "infected person" represents an individual who possesses knowledge of the product. In both cases, it is assumed that the members of the population are themselves primarily responsible for the spread of the rumor or knowledge of the product. This situation is in contrast to the model described earlier where information was spread through a population by external sources, such as radio and television.

There are several limitations to this epidemic model. Each of the four simplifying assumptions made at the outset is unrealistic in varying degrees. More complicated models can be constructed that rectify one or more of these defects, but they require more advanced mathematical tools.

The Exponential Function in Lung Physiology   Let us conclude this section by deriving a useful model for the pressure in a person's lungs when the air is allowed to escape passively from the lungs with no use of the person's muscles. Let V be the volume of air in the lungs and let P be the relative pressure in the lungs when compared with the pressure in the mouth. The *total compliance* (of the respiratory system) is defined to be the derivative $\dfrac{dV}{dP}$. For normal values of V and P we may assume that the total compliance is a positive constant, say C. That is,

$$\frac{dV}{dP} = C. \tag{12}$$

We shall assume that the airflow during the passive respiration is smooth and not turbulent. Then Poiseuille's law of fluid flow says that the rate of change of volume as a function of time (i.e., the rate of air flow) satisfies

$$\frac{dV}{dt} = -\frac{P}{R}, \tag{13}$$

where R is a (positive) constant called the airway resistance. Under these conditions, we may derive a formula for P as a function of time. Since the volume is a function of the pressure, and the pressure is in turn a function of time, we may use the chain rule to write

$$\frac{dV}{dt} = \frac{dV}{dP} \cdot \frac{dP}{dt}.$$

From (12) and (13),

$$-\frac{P}{R} = C \cdot \frac{dP}{dt},$$

so

$$\frac{dP}{dt} = -\frac{1}{RC} \cdot P.$$

From this differential equation we conclude that P must be an exponential function of t. In fact,

$$P = P_0 e^{kt},$$

where P_0 is the initial pressure at time $t = 0$ and $k = -1/RC$. This relation between k and the product RC is useful to lung specialists, because they can experimentally compute k and the compliance C, and then use the formula $k = -1/RC$ to determine the airway resistance R.

PRACTICE PROBLEMS 4

1. A sociological study* was made to examine the process by which doctors decide to adopt a new drug. The doctors were divided into two groups. The doctors in group A had little interaction with other doctors and so received most of their information via mass media. The doctors in group B had extensive interaction with other doctors and so received most of their information via word of mouth. For each group, let $f(t)$ be the number who have learned about a new drug after t months. Examine the appropriate differential equations to explain why the two graphs were of the types shown below.

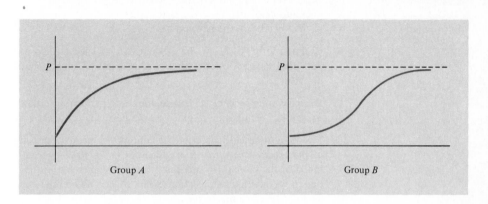

Group A   Group B

EXERCISES 4

1. Consider the function $f(x) = 5(1 - e^{-2x})$, $x \geq 0$.

   (a) Show that $f(x)$ is increasing and concave down for all $x \geq 0$.

   (b) Explain why $f(x)$ approaches 5 as x gets large.

   (c) Sketch the graph of $f(x)$, $x \geq 0$.

2. Consider the function $g(x) = 10 - 10e^{-.1x}$, $x \geq 0$.

   (a) Show that $g(x)$ is increasing and concave down for $x \geq 0$.

   (b) Explain why $g(x)$ approaches 10 as x gets large.

   (c) Sketch the graph of $g(x)$, $x \geq 0$.

3. Suppose that $y = 2(1 - e^{-x})$. Compute y' and show that $y' = 2 - y$.

4. Suppose that $y = 5(1 - e^{-2x})$. Compute y' and show that $y' = 10 - 2y$.

5. Suppose that $f(x) = 3(1 - e^{-10x})$. Show that $y = f(x)$ satisfies the differential equation

$$y' = 10(3 - y), \qquad f(0) = 0.$$

---

* James S. Coleman, Elihu Katz, and Herbert Menzel, "The Diffusion of an Innovation Among Physicians," *Sociometry*, 20 (1957), 253–270.

6. (*Ebbinghaus Model for Forgetting*) Suppose that a student learns a certain amount of material for some class. Let $f(t)$ denote the percentage of the material that the student can recall t weeks later. The psychologist Ebbinghaus found that this percent retention can be modeled by a function of the form

$$f(t) = (100 - a)e^{-\lambda t} + a,$$

where λ and a are positive constants and $0 < a < 100$. Sketch the graph of the function $f(t) = 85e^{-.5t} + 15, t \geq 0$.

7. When a grand jury indicted the mayor of a certain town for accepting bribes, the newspaper, radio, and television immediately began to publicize the news. Within an hour one-quarter of the citizens heard about the indictment. Estimate when three-quarters of the town heard the news.

8. Examine formula (8) for the amount $A(t)$ of excess glucose in the bloodstream of a patient at time t. Describe what would happen if the rate r of infusion of glucose were doubled.

9. Describe an experiment that a doctor could perform in order to determine the velocity constant of elimination of glucose for a particular patient.

10. Physiologists usually describe the continuous intravenous infusion of glucose in terms of the excess *concentration* of glucose, $C(t) = A(t)/V$, where V is the total volume of blood in the patient. In this case, the rate of increase in the concentration of glucose due to the continuous injection is r/V. Find a differential equation that gives a model for the rate of change of the excess concentration of glucose.

SOLUTIONS TO PRACTICE PROBLEMS 4

1. The difference between transmission of information via mass media and via word of mouth is that in the second case the rate of transmission depends not only on the number of people who have not yet received the information, but also on the number of people who know the information and therefore are capable of spreading it. Therefore, for group A, $f'(t) = k[P - f(t)]$, and for group B, $f'(t) = kf(t)[P - f(t)]$. Note that the spread of information by word of mouth follows the same pattern as the spread of an epidemic.

Chapter 15: CHECKLIST

☐ $y' = ky$ has the solution $y = Ce^{kt}$
☐ Exponential growth
☐ Exponential decay
☐ Half-life of a radioactive element
☐ Continuous compounding of interest
☐ Present value of money
☐ Percentage rate of change
☐ Elasticity of demand
☐ $y = M(1 - e^{-kt})$
☐ $y = M/(1 + Be^{-Mkt})$ (logistic growth)

Chapter 15: SUPPLEMENTARY EXERCISES

1. The atmospheric pressure (measured in inches of mercury) at height x miles above sea level $P(x)$ satisfies the differential equation $P'(x) = -.2P(x)$. Find the formula for $P(t)$ if the atmospheric pressure at sea level is 29.92.

2. The herring gull population in North America has been doubling every 13 years since 1900. Give a differential equation satisfied by $P(t)$, the population t years after 1900.

3. Find the present value of $10,000 payable at the end of 5 years if money may be invested at 12% with interest compounded continuously.

4. One thousand dollars is deposited in a savings account at 10% interest compounded continuously. How many years are required for the balance in the account to reach $3000?

5. The half-life of the radioactive element tritium is 12 years. Find its decay constant.

6. A piece of charcoal found at Stonehenge contained 63% of the level of C^{14} found in living trees. Approximately how old is the charcoal?

7. From 1970 to 1980, the population of Texas grew from 11.2 million to 14.2 million.

    (a) Give the formula for the population t years after 1970.

    (b) If this growth rate continues, how large will the population be in 1990?

    (c) In what year will the population reach 19 million?

8. A stock portfolio increased in value from $100,000 to $117,000 in 2 years. What rate of interest, compounded continuously, did this investment earn?

9. An investor initially invests $10,000 in a speculative venture. Suppose that the investment earns 20% interest compounded continuously for 5 years and 6% interest compounded continuously for 5 years thereafter.

    (a) How much does the $10,000 grow to after 10 years?

    (b) Suppose that the investor has the alternative of an investment paying 14% interest compounded continuously. Which investment is superior over a 10-year period, and by how much?

10. Two different bacteria colonies are growing near a pool of stagnant water. Suppose that the first colony initially has 1000 bacteria and doubles every 21 minutes. The second colony has 710,000 bacteria and doubles every 33 minutes. How much time will elapse before the first colony becomes as large as the second?

11. Find the percentage rate of change of the function $f(t) = 50e^{.2t^2}$ at $t = 10$.

12. Find $E(p)$ for the demand function $q = 4000 - 40p^2$, and determine if demand is elastic or inelastic at $p = 5$.

13. Suppose that for a certain demand function, $E(8) = 1.5$. If the price is increased to $8.16, estimate the percentage decrease in the quantity demanded. Will the revenue increase or decrease?

14. Find the percentage rate of change of the function $f(p) = \dfrac{1}{3p + 1}$ at $p = 1$.

15. A company can sell $q = 1000p^2 e^{-.02(p+5)}$ calculators at a price of p dollars per calculator. The current price is \$200. If the price is decreased, will the revenue increase or decrease?

16. Consider a demand function of the form $q = ae^{-bp}$, where a and b are positive numbers. Find $E(p)$ and show that the elasticity equals 1 when $p = 1/b$.

17. Refer to Practice Problems 4. Out of 100 doctors in group A, none knew about the drug at time $t = 0$, but 66 of them were familiar with the drug after 13 months. Find the formula for $f(t)$.

18. The growth of the yellow nutsedge weed is described by a formula $f(t)$ of type (9) in Section 4. A typical weed has length 8 centimeters after 9 days, length 48 centimeters after 25 days and reaches length 55 centimeters at maturity. Find the formula for $f(t)$.

Integration

Chapter 16

The derivative is one of the two fundamental concepts of calculus. The other is the *integral*, which we introduce in this chapter.

16.1. Antidifferentiation

We have developed several techniques for calculating the derivative $F'(x)$ of a function $F(x)$. In many applications, however, it is necessary to proceed in reverse. We are given the derivative $F'(x)$ and must determine the function $F(x)$. The process of determining $F(x)$ from $F'(x)$ is called *antidifferentiation*. The next example gives a typical application involving antidifferentiation.

EXAMPLE 1 For a number of years, the annual worldwide rate of oil consumption was growing exponentially with a growth constant of about .07. At the beginning of 1970, the rate was about 16.1 billion barrels of oil per year. Let $R(t)$ denote the rate of oil consumption at time t, where t is the number of years since the beginning of 1970. Then a reasonable model for $R(t)$ is given by

$$R(t) = 16.1e^{.07t}. \tag{1}$$

Use this formula for $R(t)$ to estimate the total amount of oil that will be consumed from 1970 to 1990.

Solution Let $T(t)$ be the total amount of oil consumed from time 0 (1970) until time t. We wish to calculate $T(20)$, the amount of oil consumed from 1970 to 1990. We do this by first determining a formula for $T(t)$. Since $T(t)$ is the total oil consumed, the derivative $T'(t)$ is the *rate* of oil consumption, namely $R(t)$. Thus, although we do not yet have a formula for $T(t)$, we do know that

$$T'(t) = R(t).$$

Thus the problem of determining a formula for $T(t)$ has been reduced to a problem of antidifferentiation: Find a function whose derivative is $R(t)$. We shall solve this particular problem after developing some techniques for solving antidifferentiation problems in general.

Suppose that $f(x)$ is a given function and $F(x)$ is a function having $f(x)$ as its derivative—that is, $F'(x) = f(x)$. We call $F(x)$ an *antiderivative* of $f(x)$.

EXAMPLE 2 Find an antiderivative of $f(x) = x^2$.

Solution One such function is $F(x) = \frac{1}{3}x^3$, since

$$F'(x) = \frac{1}{3} \cdot 3x^2 = x^2.$$

Another antiderivative is $F(x) = \frac{1}{3}x^3 + 2$, since

$$\frac{d}{dx}\left(\frac{1}{3}x^3 + 2\right) = \frac{1}{3} \cdot 3x^2 + 0 = x^2.$$

In fact, if C is any constant, the function $F(x) = \frac{1}{3}x^3 + C$ is also an antiderivative of x^2, since

$$\frac{d}{dx}\left(\frac{1}{3}x^3 + C\right) = \frac{1}{3} \cdot 3x^2 + 0 = x^2.$$

(The derivative of a constant function is zero.)

EXAMPLE 3 Find an antiderivative of the function $f(x) = 2x - (1/x^2)$.

Solution Since

$$\frac{d}{dx}(x^2) = 2x \quad \text{and} \quad \frac{d}{dx}\left(\frac{1}{x}\right) = -\frac{1}{x^2},$$

we see that one antiderivative of $f(x)$ is given by

$$F(x) = x^2 + \frac{1}{x}.$$

However, any function of the form $x^2 + (1/x) + C$, C a constant, will do, since

$$\frac{d}{dx}\left(x^2 + \frac{1}{x} + C\right) = 2x - \frac{1}{x^2} + 0 = 2x - \frac{1}{x^2}.$$

Using the same reasoning as in Examples 2 and 3, we see that if $F(x)$ is an antiderivative of $f(x)$, then so is $F(x) + C$, where C is any constant. Thus if we know one antiderivative $F(x)$ of a function $f(x)$, we can write down an infinite number by adding all possible constants C to $F(x)$. It turns out that in this way we obtain all antidervatives of $f(x)$. That is, we have the following fundamental result.

> **Theorem I** If $F_1(x)$ and $F_2(x)$ are two antiderivatives of the same function $f(x)$, then $F_1(x)$ and $F_2(x)$ differ by a constant. In other words, there is a constant C such that
>
> $$F_2(x) = F_1(x) + C.$$

Our verification of this theorem will be based on the following fact, which is important in its own right.

> **Theorem II** If $F'(x) = 0$ for all x, then $F(x) = C$ for some constant C.

Verification of Theorem II If $F'(x) = 0$ for all x, then the curve $y = F(x)$ has slope equal to zero at every point. Thus the tangent line to $y = F(x)$ at any point is horizontal, which implies that the graph of $y = F(x)$ is a horizontal line. (Try to draw the graph of a function with horizontal tangent everywhere. There is no choice but to keep your pencil moving on a constant, horizontal line!) If the horizontal line is $y = C$, then $F(x) = C$ for all x.

Verification of Theorem I If $F_1(x)$ and $F_2(x)$ are two antiderivatives of $f(x)$, then the function $F(x) = F_2(x) - F_1(x)$ has derivative

$$F'(x) = F'_2(x) - F'_1(x)$$

$$= f(x) - f(x)$$

$$= 0.$$

So, by Theorem II, we know that $F(x) = C$ for some constant C. In other words, $F_2(x) - F_1(x) = C$, so that

$$F_2(x) = F_1(x) + C,$$

which is Theorem I.

Using Theorem I, we can find *all* antiderivatives of a given function once we know one antiderivative. For instance, since one antiderivative of x^2 is $\frac{1}{3}x^3$ (Example 2), all antiderivatives of x^2 have the form $\frac{1}{3}x^3 + C$, where C is a constant.

Suppose that $f(x)$ is a function whose antiderivatives are $F(x) + C$. The standard way to express this fact is to write

$$\int f(x)\,dx = F(x) + C.$$

The symbol $\int$ is called an *integral sign*. The entire notation $\int f(x)\,dx$ is called an *indefinite integral* and stands for antidifferentiation of the function $f(x)$. We always record the variable of interest by writing it, prefaced by the letter d. For example, if the variable of interest is t rather than x, then we write $\int f(t)\,dt$ for the antiderivative of $f(t)$.

EXAMPLE 4 Determine.

(a) $\int x^r dx, \quad r \neq -1$

(b) $\int e^{kx} dx, \quad k \neq 0$

Solution (a) By the constant multiple and power rules,

$$\frac{d}{dx}\left(\frac{1}{r+1} x^{r+1}\right) = \frac{1}{r+1} \cdot \frac{d}{dx} x^{r+1} = \frac{1}{r+1} \cdot (r+1)x^r = x^r.$$

Thus $x^{r+1}/(r+1)$ is an antiderivative of x^r. Letting C represent any constant, we have

$$\int x^r dx = \frac{1}{r+1} x^{r+1} + C, \qquad r \neq -1. \qquad (2)$$

(b) An antiderivative of e^{kx} is e^{kx}/k, since

$$\frac{d}{dx}\left(\frac{1}{k} e^{kx}\right) = \frac{1}{k} \cdot \frac{d}{dx} e^{kx} = \frac{1}{k}(ke^{kx}) = e^{kx}.$$

Hence

$$\int e^{kx} dx = \frac{1}{k} e^{kx} + C, \qquad k \neq 0. \qquad (3)$$

Formula (2) does not give an antiderivative of x^{-1} because $1/(r+1)$ is undefined for $r = -1$. However, we know that for $x \neq 0$, the derivative of $\ln|x|$ is $1/x$. Hence $\ln|x|$ is an antiderivative of $1/x$, and we have

$$\int \frac{1}{x} dx = \ln|x| + C, \qquad x \neq 0. \qquad (4)$$

Formulas (2), (3), and (4) each followed by "reversing" a familiar differentiation rule. In a similar fashion, one may use the sum rule and constant-multiple rule for derivatives to obtain corresponding rules for antiderivatives:

$$\int [f(x) + g(x)] dx = \int f(x) dx + \int g(x) dx \qquad (5)$$

$$\int kf(x) dx = k \int f(x) dx, \qquad k \text{ a constant.} \qquad (6)$$

In words, (5) says that a sum of functions may be antidifferentiated term by term, and (6) says that a constant multiple may be moved through the integral sign.

EXAMPLE 5 Compute:

$$\int \left(x^{-3} + 7e^{5x} + \frac{4}{x} \right) dx.$$

Solution Using the rules given above, we have

$$\int \left(x^{-3} + 7e^{5x} + \frac{4}{x} \right) dx = \int x^{-3} dx + \int 7e^{5x} dx + \int \frac{4}{x} dx$$

$$= \int x^{-3} dx + 7 \int e^{5x} dx + 4 \int \frac{1}{x} dx$$

$$= \frac{1}{-2} x^{-2} + 7 \left(\frac{1}{5} e^{5x} \right) + 4 \ln|x| + C$$

$$= -\frac{1}{2} x^{-2} + \frac{7}{5} e^{5x} + 4 \ln|x| + C.$$

After some practice, most of the steps shown in the solution of Example 5 can be omitted.

A function $f(x)$ has infinitely many different antiderivatives, corresponding to the various choices of the constant C. In applications, it is often necessary to satisfy an additional condition, which then determines a specific value for C. As an illustration of this technique, consider the next example.

EXAMPLE 6 Find the antiderivative $F(x)$ of $3x^2 - 5$ for which $F(0) = 2$.

Solution One antiderivative of $3x^2 - 5$ is $x^3 - 5x$. Therefore, by Theorem I, the antiderivatives are precisely the functions

$$F(x) = x^3 - 5x + C, \qquad C \text{ a constant.}$$

If $F(0) = 2$, then

$$2 = F(0) = 0^3 - 5 \cdot 0 + C = C,$$

so that $C = 2$, and the unique antiderivative satisfying the given condition is

$$F(x) = x^3 - 5x + 2.$$

Having introduced the basics of antidifferentiation, let us now solve the oil-consumption problem.

Solution of Example 1 (Continued) The rate of oil consumption at time t is $R(t) = 16.1e^{.07t}$ billion barrels per year. Moreover, we observed that the total consumption $T(t)$, from time 0 to time t, is an antiderivative of $R(t)$. Using (3) and (6), we have

$$T(t) = \int 16.1e^{.07t} dt = \frac{16.1}{.07} e^{.07t} + C = 230e^{.07t} + C,$$

where C is a constant. However, in our particular example, $T(0) = 0$, since $T(0)$ is the amount of oil used from time 0 to time 0. Therefore, the constant C must satisfy

$$0 = T(0) = 230e^{.07(0)} + C = 230 + C,$$

$$C = -230.$$

Therefore,

$$T(t) = 230e^{.07t} - 230 = 230(e^{.07t} - 1).$$

The total oil that will be consumed from 1970 to 1990 is

$$T(20) = 230(e^{.07(20)} - 1) \approx 703 \text{ billion barrels.}$$

Antidifferentiation can be used to solve a variety of applied problems, of which the next two examples are typical.

EXAMPLE 7   A rocket is fired vertically into the air. Its velocity at t seconds after liftoff is $v(t) = 20t + 50$ meters per second. How far will the rocket travel during the first 100 seconds?

Solution   If $s(t)$ denotes the height (in meters) of the rocket at time t seconds after liftoff, then $s'(t)$ is the rate (in meters per second) at which the height is changing at time t, which is just $v(t)$. In other words, $s(t)$ is an antiderivative of $v(t)$. Thus

$$s(t) = \int v(t)\, dt = \int (20t + 50)\, dt$$

$$= 10t^2 + 50t + C,$$

where C is a constant. At time 0 the height of the rocket is 0, so $s(0) = 0$ and

$$0 = s(0) = 10(0)^2 + 50(0) + C = C$$

$$C = 0$$

$$s(t) = 10t^2 + 50t.$$

At $t = 100$, the height of the rocket is

$$s(100) = 10(100)^2 + 50(100) = 105{,}000 \text{ meters.}$$

EXAMPLE 8   A factory's marginal cost function is $\frac{1}{100}x^2 - 2x + 120$, where x denotes the number of units produced per day. The factory has fixed costs of $1000 per day. Find the cost of producing x units per day.

Solution   Let $C(x)$ be the cost of producing x units per day. The derivative $C'(x)$ is just the marginal cost function. In other words, $C(x)$ is an antiderivative of the marginal cost function. Thus

$$C(x) = \int (\tfrac{1}{100}x^2 - 2x + 120)\, dx$$

$$= \tfrac{1}{300}x^3 - x^2 + 120x + C$$

for some constant C. However, the fixed costs are just the costs experienced when producing 0 units. That is, the fixed costs equal $C(0)$. So the given data imply that $C(0) = 1000$. Thus

$$1000 = C(0) = \tfrac{1}{300}(0)^3 - (0)^2 - 120(0) + C$$

$$C = 1000.$$

Therefore, the cost function $C(x)$ is given by

$$C(x) = \tfrac{1}{300}x^3 - x^2 + 120x + 1000.$$

PRACTICE PROBLEMS 1

1. Determine.

   (a) $\int (x^3 + 4x)\,dx$ (b) $\int t^{7/2}\,dt$

2. Let $R(x)$ be the revenue received from the sale of x units of a product. The marginal revenue function is $5 - 3x$. Find $R(x)$. [Note that the revenue is 0 if no units are sold. That is, $R(0) = 0$.]

EXERCISES 1

Find all antiderivatives of each of the following functions.

1. $f(x) = x$ 2. $f(x) = 9x^8$ 3. $f(x) = e^{3x}$

4. $f(x) = e^{-3x}$ 5. $f(x) = 3$ 6. $f(x) = -4x$

In Exercises 7–22, find the value of k that makes the antidifferentiation formula true. [*Note:* You can check your answer without looking in the answer section. How?]

7. $\int x^{-5}\,dx = kx^{-4} + C$ 8. $\int x^{1/3}\,dx = kx^{4/3} + C$

9. $\int \sqrt{x}\,dx = kx^{3/2} + C$ 10. $\int \dfrac{6}{x^3}\,dx = \dfrac{k}{x^2} + C$

11. $\int \dfrac{10}{t^6}\,dt = kt^{-5} + C$ 12. $\int \dfrac{3}{\sqrt{t}}\,dt = k\sqrt{t} + C$

13. $\int 5e^{-2t}\,dt = ke^{-2t} + C$ 14. $\int 3e^{t/10}\,dt = ke^{t/10} + C$

15. $\int 2e^{4x-1}\,dx = ke^{4x-1} + C$ 16. $\int \dfrac{4}{e^{3x+1}}\,dx = \dfrac{k}{e^{3x+1}} + C$

17. $\int (x-7)^{-2}\,dx = k(x-7)^{-1} + C$ 18. $\int \sqrt{x+1}\,dx = k(x+1)^{3/2} + C$

19. $\int (x+4)^{-1}\,dx = k\ln|x+4| + C$ 20. $\int \dfrac{5}{(x-8)^4}\,dx = \dfrac{k}{(x-8)^3} + C$

21. $\int (3x+2)^4\,dx = k(3x+2)^5 + C$ 22. $\int (2x-1)^3\,dx = k(2x-1)^4 + C$

Determine the following.

23. $\int (x^2 - x - 1)\,dx$

24. $\int (x^3 + 6x^2 - x)\,dx$

25. $\int \left(\dfrac{2}{\sqrt{x}} - 3\sqrt{x}\right)dx$

26. $\int \left[\dfrac{\sqrt{t}}{4} - 4(t - 3)^{-2}\right]dt$

27. $\int \left(4 - 5e^{-5t} + \dfrac{e^{2t}}{3}\right)dt$

28. $\int (e^2 + 3t^2 - 2e^{3t})\,dt$

Find all functions $f(t)$ with the following property.

29. $f'(t) = t^{3/2}$

30. $f'(t) = \dfrac{4}{6 + t}$

31. $f'(t) = 0$

32. $f'(t) = t^2 - 5t - 7$

Find all functions $f(x)$ with the following properties.

33. $f'(x) = x,\ f(0) = 3$

34. $f'(x) = 8x^{1/3},\ f(1) = 4$

35. $f'(x) = \sqrt{x} + 1,\ f(4) = 0$

36. $f'(x) = x^2 + \sqrt{x},\ f(1) = 3$

37. $f'(x) = 2/x,\ f(1) = 2$

38. $f'(x) = 3,\ f(2) = 7$

39. A ball is thrown upward from a height of 256 feet above the ground, with an initial velocity of 96 feet per second. From physics it is known that the velocity at time t is $96 - 32t$ feet per second.

   (a) Find $s(t)$, the function giving the height of the ball at time t.

   (b) How long will it take for the ball to reach the ground?

   (c) How high will the ball go?

40. A rock is dropped from the top of a 400-foot cliff. Its velocity at time t seconds is $v(t) = -32t$ feet per second.

   (a) Find $s(t)$, the height of the rock above the ground at time t.

   (b) How long will it take to reach the ground?

   (c) What will be its velocity when it hits the ground?

41. Let $P(t)$ be the total output of a factory assembly line after t hours of work. Suppose that the rate of production at time t is $60 + 2t - \frac{1}{4}t^2$ units per hour. Find the formula for $P(t)$. [*Hint:* The rate of production is $P'(t)$ and $P(0) = 0$.]

42. After t hours of operation a coal mine is producing coal at the rate of $40 + 2t - \frac{1}{5}t^2$ tons of coal per hour. Find a formula for the total output of the coal mine after t hours of operation.

43. A package of frozen strawberries is taken from a freezer at $-5°C$ into a room at $20°C$. At time t the average temperature of the strawberries is increasing at the rate of $10e^{-.4t}$ degrees Celsius per hour. Find the temperature of the strawberries at time t.

44. A flu epidemic hits a town. Let $P(t)$ be the number of persons sick with the flu at time t, where time is measured in days from the beginning of the epidemic and $P(0) = 100$. Suppose that after t days the flu is spreading at the rate of $120t - 3t^2$ people per day. Find the formula for $P(t)$.

45. A small tie shop finds that at a sales level of x ties per day its marginal profit is $MP(x)$ dollars per tie, where $MP(x) = 1.30 + .06x - .0018x^2$. Also, the shop will lose \$95 per day at a sales level of $x = 0$. Find the profit from operating the shop at a sales level of x ties per day.

46. A soap manufacturer estimates that its marginal cost of producing soap powder is $.2x + 1$ hundred dollars per ton at a production level of x tons per day. Fixed costs are \$200 per day. Find the cost of producing x tons of soap powder per day.

47. The United States has been consuming iron ore at the rate of $R(t)$ million metric tons per year at time t, where $t = 0$ corresponds to 1980 and $R(t) = 94e^{.016t}$. Find a formula for the total U.S. consumption of iron ore from 1980 until time t.

48. The rate of production of natural and manufactured gas in the United States has been $R(t)$ quadrillion Btu per year at time t, with $t = 0$ corresponding to 1976 and $R(t) = 20e^{.02t}$. Find a formula for the total U.S. production of natural and manufactured gas from 1976 until time t.

SOLUTIONS TO PRACTICE PROBLEMS 1

1. (a) $\int (x^3 + 4x) dx = \frac{1}{4}x^4 + 2x^2 + C$

(b) $\int t^{7/2} dt = \dfrac{1}{\frac{9}{2}} t^{9/2} + C = \frac{2}{9}t^{9/2} + C$

2. $R(x)$ is an antiderivative of the marginal revenue function. So

$$R(x) = \int (5 - 3x) dx = 5x - \tfrac{3}{2}x^2 + C$$

for some constant C. Since $R(0) = 0$,

$$0 = R(0) = 5(0) - \tfrac{3}{2}(0)^2 + C = C$$

$$C = 0.$$

Thus $R(x) = 5x - \tfrac{3}{2}x^2$.

16.2. Definite Integrals

In order to motivate the definite integral, let us return for a few moments to Example 1 of Section 1. Recall that we were given a formula $R(t)$ for the annual rate of oil consumption at time t. From this formula we determined the function $T(t)$, which gives the total oil consumed from time 0 to time t. Recall that $t = 0$ corresponds to 1970 and that $T(t)$ is an antiderivative of $R(t)$.

Consider the following related question: How much oil was consumed between 1975 and 1980? This is easy to determine using the function $T(t)$. The year 1975 corresponds to $t = 5$, and 1980 corresponds to $t = 10$. So the oil consumed between 1975 and 1980 is

$$T(10) - T(5). \tag{1}$$

[Indeed, $T(10)$ gives the oil consumed between 1970 and 1980 and $T(5)$ gives the oil consumed during the first five of those years.] The quantity $T(10) - T(5)$ is called the *net change* of $T(t)$ over the interval $t = 5$ to $t = 10$. More generally, if $F(x)$ is any function and a, b are numbers, then the *net change of $F(x)$ over the interval $x = a$ to $x = b$* is the number

$$F(b) - F(a).$$

The quantity $F(b) - F(a)$ is often abbreviated by the symbol $F(x)|_a^b$. For example, the net change (1) can be written in this notation as $T(t)|_5^{10}$.

Since $T(t)$ is an antiderivative of $R(t)$, the quantity (1) is the net change of an antiderivative of $R(t)$ over a certain interval. This suggests the following definition.

Definition of the Definite Integral Let $f(x)$ be a function and a, b be numbers. Then the definite integral of $f(x)$ over the interval from $x = a$ to $x = b$, denoted $\int_a^b f(x)\, dx$, is the net change of an antiderivative of $f(x)$ over that interval. Thus, if $F(x)$ is an antiderivative of $f(x)$, we have

$$\int_a^b f(x)\, dx = F(x)\Big|_a^b = F(b) - F(a).$$

For example, the quantity (1) can be expressed as the definite integral

$$\int_5^{10} R(t)\, dt.$$

EXAMPLE 1   Evaluate.

(a) $\displaystyle \int_2^5 3x^2\, dx$ (b) $\displaystyle \int_{-1}^2 e^{3x}\, dx$

Solution   (a) An antiderivative of $3x^2$ is x^3. So

$$\int_2^5 3x^2\, dx = x^3\Big|_2^5 = 5^3 - 2^3 = 117.$$

(b) An antiderivative of e^{3x} is $\frac{1}{3}e^{3x}$. Therefore,

$$\int_{-1}^2 e^{3x}\, dx = \tfrac{1}{3}e^{3x}\Big|_{-1}^2 = \tfrac{1}{3}e^{3\cdot 2} - \tfrac{1}{3}e^{3\cdot(-1)} = \tfrac{1}{3}e^6 - \tfrac{1}{3}e^{-3}.$$

In computing the definite integral we may use *any* antiderivative of $f(x)$. The value of the definite integral does not depend on the choice of antiderivative. For instance, consider the definite integral $\int_2^5 3x^2\, dx$. The antiderivatives of $3x^2$ are all of the form $x^3 + C$, where C is any constant. Moreover,

$$(x^3 + C)\Big|_2^5 = (5^3 + C) - (2^3 + C) = 5^3 - 2^3 = 117.$$

So the value of the definite integral does not depend on the choice of C. The same argument works for a general function $f(x)$ in place of $3x^2$.

EXAMPLE 2 Refer to the oil-consumption data of Example 1, Section 1. How much oil will be consumed from 1978 to 1985?

Solution As we saw above, the total consumption during a given time interval can be represented as a definite integral over that interval. Since $t = 8$ corresponds to 1978 and $t = 15$ to 1985, the desired quantity equals

$$\int_8^{15} R(t)\,dt = \int_8^{15} 16.1e^{.07t}\,dt.$$

Since an antiderivative of $16.1e^{.07t}$ is $230e^{.07t}$, the definite integral above equals

$$230e^{.07t}\Big|_8^{15} = 230e^{.07(15)} - 230e^{.07(8)}$$

$$\approx 255 \text{ billion barrels.}$$

EXAMPLE 3 Suppose that the velocity $v(t)$ of a rocket t seconds after liftoff is $v(t) = 20t + 50$ meters per second. Determine the distance it travels during the time interval from $t = 60$ to $t = 70$.

Solution Let $s(t)$ denote the height of the rocket at time t. The desired distance is the net change $s(70) - s(60)$. Since $s(t)$ is an antiderivative of $v(t)$, this net change is a definite integral:

$$s(70) - s(60) = \int_{60}^{70} v(t)\,dt$$

$$= \int_{60}^{70} (20t + 50)\,dt$$

$$= (10t^2 + 50t)\Big|_{60}^{70}$$

$$= 13{,}500 \text{ meters.}$$

Thus the rocket travels 13,500 meters during the time interval from $t = 60$ to $t = 70$.

EXAMPLE 4 The marginal cost function for a certain factory is $\frac{1}{100}x^2 - 2x + 120$ dollars per unit, where x is the total daily production. Find the net increase in cost if production is raised from 100 to 105 units per day.

Solution Let $C(x)$ denote the cost of producing x units per day. Then $C'(x) = \frac{1}{100}x^2 - 2x + 120$. Hence

$$C(105) - C(100) = \int_{100}^{105} C'(x)\,dx$$

$$= \int_{100}^{105} \left(\frac{1}{100}x^2 - 2x + 120 \right) dx$$

$$= (\tfrac{1}{300}x^3 - x^2 + 120x)\Big|_{100}^{105}$$

$$= 5433.75 - 5333.33 = 100.42.$$

Thus, if the factory is currently producing 100 units per day, it will cost an additional $100.42 to produce 5 additional units per day.

One of the most important applications of the definite integral is to the calculation of areas under curves.

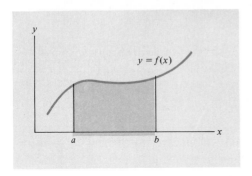

PROBLEM Let $f(x)$ be a function that is continuous and nonnegative for all x between a and b, where $a < b$. Determine the area of the region lying beneath the curve $y = f(x)$ and above the x-axis, from $x = a$ to $x = b$.

The region referred to is the shaded region in Fig. 1. We see that the region is bounded on the left by the vertical line $x = a$ and on the right by the vertical line $x = b$. We shall refer to the area of this region simply as the *area under the curve* $y = f(x)$ *from* $x = a$ *to* $x = b$.

This problem has a rather simple (and surprising) solution.

FIGURE 1

The Fundamental Theorem of Calculus Let $f(x)$ be continuous, and nonnegative for x between a and b. Then the area under the curve $y = f(x)$ from $x = a$ to $x = b$ is given by the definite integral

$$\int_a^b f(x)\, dx.$$

At first, this theorem may not seem very fundamental. (It may not even seem very interesting!) However, it allows us to accomplish several impressive goals. First of all, it permits us to apply everything we know about antidifferentiation to the calculation of areas. Although it is not at all apparent that antidifferentiation and calculation of areas are related problems, they indeed are, and the fundamental theorem of calculus provides the relationship. A second consequence of the fundamental theorem is that it allows us to interpret various quantities involving antiderivatives in terms of areas. Such interpretations are useful in many applications.

We shall explain why this theorem is true after we examine two examples of its application.

EXAMPLE 5 Find the area under the curve $y = x$ from $x = 1$ to $x = 2$.

Solution The region whose area we must calculate is the shaded region in Fig. 2(a). Here $f(x) = x$, $a = 1$, and $b = 2$. The value of the definite integral is then

$$\int_1^2 x\, dx = \tfrac{1}{2}x^2 \Big|_1^2 = \tfrac{1}{2} \cdot 2^2 - \tfrac{1}{2} \cdot 1^2 = \tfrac{3}{2}.$$

Therefore, the area is $\tfrac{3}{2}$.

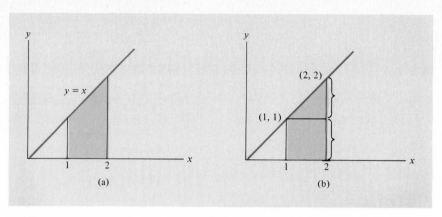

FIGURE 2

In Example 5 it is also possible to compute the area by elementary geometry. Let us break up the region into a triangle and a square, as pictured in Fig. 2(b). The area of the triangle is $\frac{1}{2} \cdot (\text{base}) \cdot (\text{height}) = \frac{1}{2} \cdot 1 \cdot 1 = \frac{1}{2}$. The area of the square is $1 \cdot 1 = 1$. Thus the area of the region is $\frac{1}{2} + 1 = \frac{3}{2}$, the same result that we obtained by using the definite integral.

EXAMPLE 6   Compute the area under the curve $y = x^2 - 3x + 5$ from $x = 1$ to $x = 4$.

Solution   Using our curve-sketching techniques ($y' = 2x - 3$, $y' = 0$ when $x = \frac{3}{2}$), we obtain the graph shown in Fig. 3. Since the function $f(x) = x^2 - 3x + 5$ is non-negative for all x between 1 and 4, the area under the curve is given by the integral

$$\int_1^4 (x^2 - 3x + 5)\,dx = \left(\frac{x^3}{3} - \frac{3}{2}x^2 + 5x\right)\Bigg|_1^4$$

$$= \left(\frac{64}{3} - \frac{48}{2} + 20\right) - \left(\frac{1}{3} - \frac{3}{2} + 5\right) = 13\tfrac{1}{2}.$$

FIGURE 3

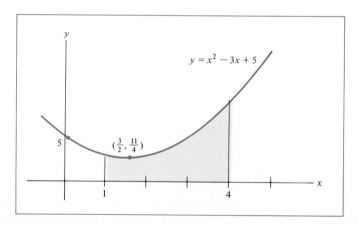

Historically, calculus centered about two problems: finding the slope of a curve at a point (differentiation) and finding the area of a region under the graph of a function. The fundamental theorem of calculus is so named because it connects these two problems. Computing an area can be accomplished by finding an antiderivative, the reverse process of differentiation.

To explain why the Fundamental Theorem of Calculus is true, we need the following theorem.

Theorem III Let $f(x)$ be a continuous nonnegative function for $a \leq x \leq b$. Let $A(x)$ be the area under the graph of the function from a to the number x. (See Fig. 4.) Then $A(x)$ is an antiderivative of $f(x)$.

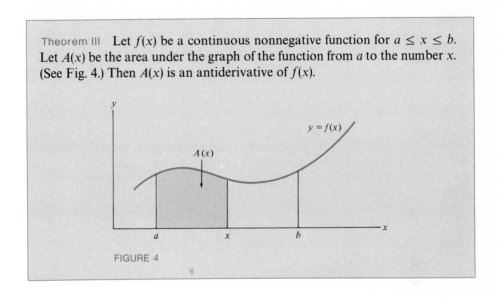

FIGURE 4

Verification of the Fundamental Theorem of Calculus

Note that the "area function" $A(x)$ in Theorem III has the property that $A(a)$ is zero and $A(b)$ equals the area under the curve $y = f(x)$ from a to b. Therefore, we may conclude from Theorem III that

$$\int_a^b f(x)\,dx = A(b) - A(a) = A(b) - 0$$

$$= [\text{area under } y = f(x) \text{ from } a \text{ to } b].$$

This is the Fundamental Theorem of Calculus.

It is not difficult to explain why Theorem III is true. Let h be a small positive number. Then $A(x + h) - A(x)$ is the area of the shaded region in Fig. 5. This shaded region is approximately a rectangle of width h, height $f(x)$, and area $h \cdot f(x)$. Thus

$$A(x + h) - A(x) \approx h \cdot f(x),$$

where the approximation becomes better as h approaches zero. Dividing by h, we have

$$\frac{A(x + h) - A(x)}{h} \approx f(x).$$

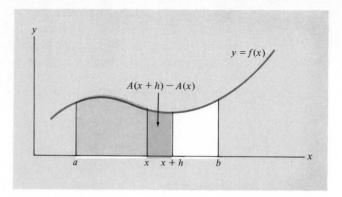

FIGURE 5

Since the approximation becomes more exact as h approaches zero, the quotient must approach $f(x)$. However, the limit definition of the derivative tells us that the quotient approaches $A'(x)$ as h approaches zero. Therefore, we have $A'(x) = f(x)$. Since x represented any number between a and b, this shows that $A(x)$ is an antiderivative of $f(x)$.

Note The verification above applies only for $f(x)$ nonnegative for x between a and b. However, this restriction on $f(x)$ is easily removed.

PRACTICE PROBLEMS 2

1. Find the area under the curve $y = e^{x/2}$ from $x = -3$ to $x = 2$.

2. Let $C(x)$ be the cost of producing x units of a certain commodity. A common notation for the marginal cost function is $MC(x)$ (or just MC). What economic interpretation can be given to the area of the shaded region in Fig. 6?

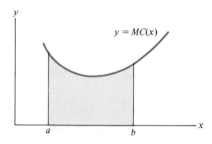

FIGURE 6

EXERCISES 2

Calculate the following definite integrals.

1. $\int_{-1}^{1} x\, dx$

2. $\int_{4}^{5} e^2\, dx$

3. $\int_{1}^{2} 5\, dx$

4. $\int_{-1}^{-1/2} \frac{1}{x^2}\, dx$

5. $\int_{1}^{2} 8x^3\, dx$

6. $\int_{0}^{1} e^{x/3}\, dx$

7. $\int_{0}^{1} 4e^{-3x}\, dx$

8. $\int_{1}^{3} \frac{5}{x}\, dx$

9. $\int_{1}^{4} 3\sqrt{x}\, dx$

10. $\int_{1}^{8} 2x^{1/3}\, dx$

11. $\int_{0}^{5} e^{\cdot 2t}\, dt$

12. $\int_{0}^{1} \frac{4}{e^{3t}}\, dt$

13. $\int_{3}^{6} x^{-1}\, dx$

14. $\int_{1}^{3} (5t - 1)^3\, dt$

15. $\int_{-1}^{1} \frac{4}{(t + 2)^3}\, dt$

16. $\int_{-3}^{0} \sqrt{25 + 3t}\, dt$

17. $\int_{2}^{3} (5 - 2t)^4\, dt$

18. $\int_{4}^{9} \frac{3}{t - 2}\, dt$

19. $\int_{0}^{3} (x^3 + x - 7)\, dx$

20. $\int_{-5}^{5} (e^{x/10} - x^2 - 1)\, dx$

21. $\int_{2}^{4} \left(x^2 + \frac{2}{x^2} - \frac{1}{x + 5}\right) dx$

22. $\int_{1}^{2} (4x^3 + 3x^{-4} - 5)\, dx$

23. Suppose that the velocity of a car at time t is $40 + 8/(t + 1)^2$ kilometers per hour. How far does the car travel from $t = 1$ to $t = 9$ hours?

24. A helicopter is rising straight up in the air. Its velocity at time t is $2t + 1$ feet per second for the first 5 seconds. How high does it rise in that time period?

25. After t hours of operation an assembly line is assembling power lawn mowers at the rate of $21 - \frac{1}{5}t$ mowers per hour. How many mowers are assembled during the time from $t = 2$ to $t = 6$ hours?

26. Some food is placed in a freezer. After t hours the temperature of the food is dropping at a rate of $12 + 4/(t + 3)^2$ degrees Fahrenheit per hour. How many degrees will the temperature of the food change during the first 2 hours?

27. Suppose that the marginal cost function of a handbag manufacturer is $\frac{3}{32}x^2 - x + 200$ dollars per unit at production level x (where x is measured in units of 100 handbags). Find the total cost of producing 6 additional units if 2 units are currently being produced.

28. If $C(x)$ is the cost of producing x units of some commodity, then $C(0)$ represents the fixed costs. Suppose that the marginal cost function is $C'(x) = .06x^2 - 8x + 250$ dollars per unit. Find the total cost of producing the first 100 units, not counting the fixed costs.

29. Let $P(x)$ denote the profit earned from the sale of x units of some commodity. Then $P'(x)$ is the marginal profit. Let a and b be two values of x, with $a < b$, and give an economic interpretation to the area under the marginal profit curve $y = P'(x)$ from $x = a$ to $x = b$.

30. Suppose that the marginal profit function for a company is $100 + 50x - 3x^2$. Find the total profit earned from the sale of 3 additional units if 5 units are currently being produced.

Find the area under each of the given curves.

31. $y = 4x$; $x = 2$ to $x = 3$

32. $y = 3x^2$; $x = -1$ to $x = 1$

33. $y = e^{x/2}; x = 0$ to $x = 1$ **34.** $y = \sqrt{x}; x = 0$ to $x = 4$

35. $y = (x - 3)^4; x = 1$ to $x = 4$ **36.** $y = e^{3x}; x = -\frac{1}{3}$ to $x = 0$

Find the area under each of the given curves by antidifferentiation and use elementary geometry to check the answer.

37. $y = 5; x = -1$ to $x = 2$ **38.** $y = x + 1; x = 2$ to $x = 4$

39. $y = 2x; x = 0$ to $x = 3$ **40.** $y = 2x + 1; x = 0$ to $x = 3$

41. Verify the formula

$$\ln b = \int_1^b \frac{1}{t}\, dt, \qquad b > 0.$$

42. For $x > 1$, show how $\ln x$ may be interpreted as the area under an appropriate curve $y = f(x)$. (This interpretation of $\ln x$ is used in some texts as the *definition* of the natural logarithm function.)

43. For each positive number x, let $A(x)$ be the area under the curve $y = x^2 + 1$ from 0 to x. Find $A'(3)$.

44. For each number $x > 2$, let $A(x)$ be the area under the curve $y = x^3$ from 2 to x. Find $A'(6)$.

SOLUTIONS TO PRACTICE PROBLEMS 2

1. The desired area is

$$\int_{-3}^{2} e^{x/2}\, dx = 2e^{x/2}\Big|_{-3}^{2} = 2e - 2e^{-3/2} \approx 4.99.$$

2. The area shaded is the area under the curve MC from $x = a$ to $x = b$. This area can be computed as the definite integral $\int_a^b MC(x)\, dx$. Since $C(x)$ is an antiderivative of $MC(x)$,

$$\int_a^b MC(x)\, dx = C(x)\Big|_a^b = C(b) - C(a).$$

So the shaded area equals the additional cost of increasing production from a units to b units.

16.3. Areas in the xy-Plane

In Section 2 we introduced the concept of the definite integral and showed how the definite integral is used to compute the area of the region lying above the x-axis and below the graph of the nonnegative function $y = f(x)$. In this section we show how to use the definite integral to compute areas of more general regions.

Three simple but important properties of the integral will be used repeatedly.

Let $f(x)$ and $g(x)$ be functions and a, b, k any constants. Then

$$\int_a^b f(x)\,dx + \int_a^b g(x)\,dx = \int_a^b [f(x) + g(x)]\,dx \qquad (1)$$

$$\int_a^b f(x)\,dx - \int_a^b g(x)\,dx = \int_a^b [f(x) - g(x)]\,dx \qquad (2)$$

$$\int_a^b kf(x)\,dx = k\int_a^b f(x)\,dx. \qquad (3)$$

Verification of (1) Let $F(x)$ and $G(x)$ be antiderivatives of $f(x)$ and $g(x)$, respectively. Then $F(x) + G(x)$ is an antiderivative of $f(x) + g(x)$, so

$$\int_a^b [f(x) + g(x)]\,dx = [F(x) + G(x)]\Big|_a^b$$

$$= [F(b) + G(b)] - [F(a) + G(a)]$$

$$= [F(b) - F(a)] + [G(b) - G(a)]$$

$$= \int_a^b f(x)\,dx + \int_a^b g(x)\,dx.$$

The verifications of (2) and (3) are similar and use the facts that $F(x) - G(x)$ is an antiderivative of $f(x) - g(x)$ and $kF(x)$ is an antiderivative of $kf(x)$.

Let us now consider regions that are bounded both above and below by graphs of functions. Referring to Fig. 1, we would like to find a simple expression for the area of the shaded region under the graph of $y = f(x)$ and above the graph of $y = g(x)$ from $x = a$ to $x = b$. It is the region under the graph of $f(x)$ with the region under the graph of $g(x)$ taken away. Therefore,

[area of shaded region] = [area under $f(x)$] − [area under $g(x)$]

$$= \int_a^b f(x)\,dx - \int_a^b g(x)\,dx$$

$$= \int_a^b [f(x) - g(x)]\,dx \qquad \text{[by property (2)].}$$

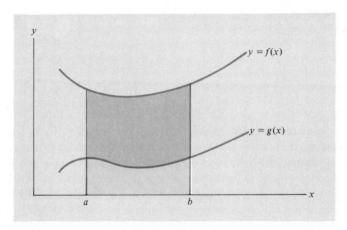

FIGURE 1

> *Area Between Two Curves* If $y = f(x)$ lies above $y = g(x)$ from $x = a$ to $x = b$, the area of the region between $f(x)$ and $g(x)$ from $x = a$ to $x = b$ is
> $$\int_a^b [f(x) - g(x)]\,dx.$$

EXAMPLE 1 Find the area of the region between $y = 2x^2 - 4x + 6$ and $y = -x^2 + 2x + 1$ from $x = 1$ to $x = 2$.

Solution Upon sketching the two graphs (Fig. 2), we see that $f(x) = 2x^2 - 4x + 6$ lies above $g(x) = -x^2 + 2x + 1$ for $1 \le x \le 2$. Therefore, our formula gives the area of the shaded region as

$$\int_1^2 [(2x^2 - 4x + 6) - (-x^2 + 2x + 1)]\,dx = \int_1^2 (3x^2 - 6x + 5)\,dx$$

$$= (x^3 - 3x^2 + 5x)\Big|_1^2$$

$$= 6 - 3 = 3.$$

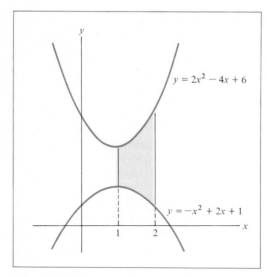

$y = 2x^2 - 4x + 6$

$y = -x^2 + 2x + 1$

FIGURE 2

EXAMPLE 2 Find the area of the region between $y = x^2$ and $y = (x - 2)^2 = x^2 - 4x + 4$ from $x = 0$ to $x = 3$.

Solution Upon sketching the graphs (Fig. 3), we see that the two graphs cross; and by setting $x^2 = x^2 - 4x + 4$, we find that they cross when $x = 1$. Thus one graph does not always lie above the other from $x = 0$ to $x = 3$, so that we cannot directly apply our rule for finding the area between two curves. However, the difficulty is easily surmounted if we break the region into two parts, namely the area from $x = 0$ to $x = 1$ and the area from $x = 1$ to $x = 3$. For from $x = 0$

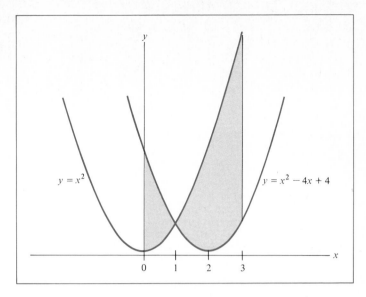

FIGURE 3

to $x = 1$, $y = x^2 - 4x + 4$ is on top; from $x = 1$ to $x = 3$, $y = x^2$ is on top. Consequently,

$$[\text{area from } x = 0 \text{ to } x = 1] = \int_0^1 [(x^2 - 4x + 4) - (x^2)]\, dx$$

$$= \int_0^1 (-4x + 4)\, dx$$

$$= (-2x^2 + 4x)\Big|_0^1$$

$$= 2 - 0 = 2.$$

$$[\text{area from } x = 1 \text{ to } x = 3] = \int_1^3 [(x^2) - (x^2 - 4x + 4)]\, dx$$

$$= \int_1^3 (4x - 4)\, dx$$

$$= (2x^2 - 4x)\Big|_1^3$$

$$= 6 - (-2) = 8.$$

Thus the total area is $2 + 8 = 10$.

In our derivation of the formula for the area between two curves, we examined functions that are nonnegative. However, the statement of the rule does not contain this stipulation, and rightfully so. Consider the case where $f(x)$ and $g(x)$ are not always positive. Let us determine the area of the shaded region in Fig. 4(a). Select some constant c such that the graphs of the functions $f(x) + c$ and $g(x) + c$ lie completely above the x-axis [Fig. 4(b)]. The region between them will have the same area as the original region. Using the rule as applied to

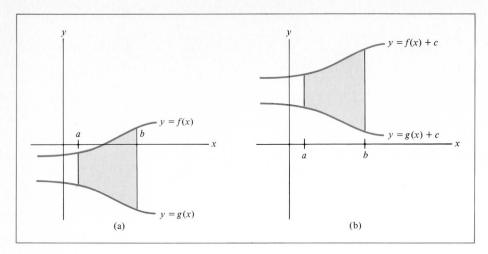

FIGURE 4

nonnegative functions, we have

$$[\text{area of the region}] = \int_a^b \left[(f(x) + c) - (g(x) + c) \right] dx$$

$$= \int_a^b \left[f(x) - g(x) \right] dx.$$

Therefore, we see that our rule is valid for any functions $f(x)$ and $g(x)$ as long as the graph of $f(x)$ lies above the graph of $g(x)$ for all x from $x = a$ to $x = b$.

EXAMPLE 3   Set up the integral that gives the area between the curves $y = x^2 - 2x$ and $y = -e^x$ from $x = -1$ to $x = 2$.

Solution   Since $y = x^2 - 2x$ lies above $y = -e^x$ (Fig. 5), the rule for finding the area between two curves can be applied directly. The area between the curves is

$$\int_{-1}^2 (x^2 - 2x + e^x) \, dx.$$

Sometimes we are asked to find the area between two curves without being given the values of a and b. In these cases there is a region that is completely enclosed by the two curves. As the next examples illustrate, we must first find the points of intersection of the two curves in order to obtain the values of a and b. In such problems careful curve sketching is especially important.

EXAMPLE 4   Set up the integral that gives the area bounded by the curves $y = x^2 + 2x + 3$ and $y = 2x + 4$.

Solution   The two curves are sketched in Fig. 6, and the region bounded by them is shaded. In order to find the points of intersection, we set $x^2 + 2x + 3 = 2x + 4$ and solve for x. We obtain $x^2 = 1$ or $x = -1$ and $x = +1$. When $x = -1$, $2x + 4 = 2(-1) + 4 = 2$. When $x = 1$, $2x + 4 = 2(1) + 4 = 6$. Thus the curves intersect at the points $(1, 6)$ and $(-1, 2)$.

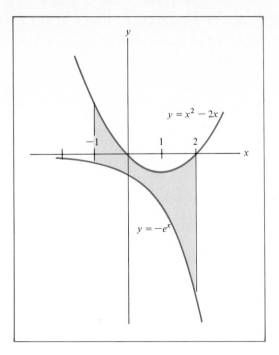

FIGURE 5

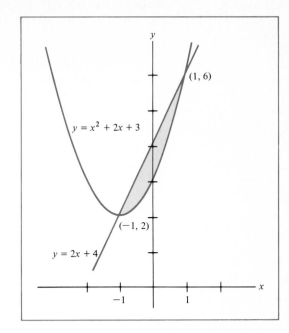

FIGURE 6

Since $y = 2x + 4$ lies above $y = x^2 + 2x + 3$ from $x = -1$ to $x = 1$, the area between the curves is given by

$$\int_{-1}^{1} \left[(2x + 4) - (x^2 + 2x + 3) \right] dx = \int_{-1}^{1} (1 - x^2) \, dx.$$

EXAMPLE 5 Set up the integral that gives the area bounded by the two curves $y = 2x^2$ and $y = x^3 - 3x$.

FIGURE 7

Solution First we make a rough sketch of the two curves as in Fig. 7. The curves intersect where $x^3 - 3x = 2x^2$ or $x^3 - 2x^2 - 3x = 0$. Note that

$$x^3 - 2x^2 - 3x$$
$$= x(x^2 - 2x - 3)$$
$$= x(x - 3)(x + 1).$$

So the solutions to $x^3 - 2x^2 - 3x = 0$ are $x = 0, 3, -1$, and the curves intersect at $(-1, 2)$, $(0, 0)$, and $(3, 18)$.

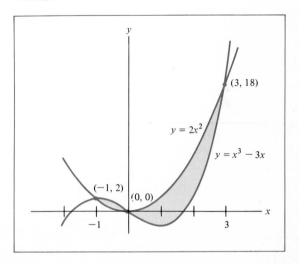

From $x = -1$ to $x = 0$, the curve $y = x^3 - 3x$ lies above $y = 2x^2$. But from $x = 0$ to $x = 3$, the reverse is true. Thus the area between the curves is given by

$$\int_{-1}^{0} (x^3 - 3x - 2x^2)\,dx + \int_{0}^{3} (2x^2 - x^3 + 3x)\,dx.$$

EXAMPLE 6 Beginning in 1974, with the advent of dramatically higher oil prices, the exponential rate of growth of world oil consumption slowed down from a growth constant of 7% to a growth constant of 4% per year. A fairly good model for the annual rate of oil consumption since 1974 is given by

$$R_1(t) = 21.3e^{.04(t-4)}, \qquad t \geq 4,$$

where $t = 0$ corresponds to 1970. Determine the total amount of oil saved between 1976 and 1980 by not consuming oil at the rate predicted by the model of Example 1, Section 1, namely

$$R(t) = 16.1e^{.07t}, \qquad t \geq 0.$$

Solution If oil consumption had continued to grow as it did prior to 1974, then the total oil consumed between 1976 and 1980 would have been

$$\int_{6}^{10} R(t)\,dt. \tag{4}$$

However, taking into account the slower increase in the rate of oil consumption since 1974, we find that the total oil consumed between 1976 and 1980 was approximately

$$\int_{6}^{10} R_1(t)\,dt. \tag{5}$$

The integrals in (4) and (5) may be interpreted as the areas under the curves $y = R(t)$ and $y = R_1(t)$, respectively, from $t = 6$ to $t = 10$. (See Fig. 8.) By

FIGURE 8

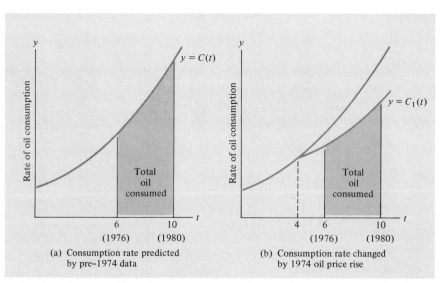

(a) Consumption rate predicted by pre–1974 data

(b) Consumption rate changed by 1974 oil price rise

superimposing the two curves we see that the area between them from $t = 6$ to $t = 10$ represents the total oil that was saved by consuming oil at the rate given by $R_1(t)$ instead of $R(t)$. (See Fig. 9.) The area between the two curves equals

$$\int_6^{10} [R(t) - R_1(t)] \, dt = \int_6^{10} [16.1e^{.07t} - 21.3e^{.04(t-4)}] \, dt$$

$$= \left(\frac{16.1}{.07} e^{.07t} - \frac{21.3}{.04} e^{.04(t-4)} \right) \Bigg|_6^{10}$$

$$\approx 13.02.$$

Thus about 13 billion barrels of oil were saved between 1976 and 1980.

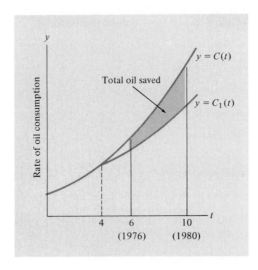

FIGURE 9

PRACTICE PROBLEMS 3

1. Find the area between the curves $y = x + 3$ and $y = x^2 + x - 13$ from $x = 1$ to $x = 3$.

2. A company plans to increase its production from 10 to 15 units per day. The present marginal cost function is $MC_1(x) = x^2 - 20x + 108$. By redesigning the production process and purchasing new equipment the company can change the marginal cost function to $MC_2(x) = \frac{1}{2}x^2 - 12x + 75$. Determine the area between the graphs of the two marginal cost curves from $x = 10$ to $x = 15$. Interpret this area in economic terms.

EXERCISES 3

Find the area of the region between the curves.

1. $y = 2x^2$ and $y = 8$ (a horizontal line) from $x = -2$ to $x = 2$

2. $y = 13 - 3x^2$ and $y = 1$ from $x = -2$ to $x = 2$

3. $y = x^2 - 6x + 12$ and $y = 1$ from $x = 0$ to $x = 4$

4. $y = x(2 - x)$ and $y = 4$ from $x = 0$ to $x = 2$

5. $y = x^2$ and $y = (x - 4)^2$ from $x = 0$ to $x = 3$

6. $y = x^2$ and $y = (x + 3)^2$ from $x = -3$ to $x = 0$

7. $y = 3x^2$ and $y = -3x^2$ from $x = -1$ to $x = 2$

8. $y = e^{2x}$ and $y = -e^{2x}$ from $x = -1$ to $x = 1$

Find the area of the region bounded by the curves.

9. $y = x^2 + x$ and $y = 3 - x$

10. $y = 3x - x^2$ and $y = 4 - 2x$

11. $y = -x^2 + 6x - 5$ and $y = 2x - 5$

12. $y = 2x^2 + x - 7$ and $y = x + 1$

13. Find the area bounded by $y = x^2 - 3x$ and the x-axis

   (a) between $x = 0$ and $x = 3$.

   (b) between $x = 0$ and $x = 4$.

   (c) between $x = -2$ and $x = 3$.

14. Find the area bounded by $y = x^2$ and $y = 1/x^2$

   (a) between $x = 1$ and $x = 4$.

   (b) between $x = \frac{1}{2}$ and $x = 4$.

15. Find the area bounded by $y = 1/x^2$, $y = x$, and $y = 8x$, for $x \geq 0$. (Make a careful sketch.)

16. Find the area bounded by $y = 1/x$, $y = 4x$, and $y = \frac{1}{2}x$, for $x \geq 0$.

17. Find the area bounded by $y = 12/x$, $y = \frac{3}{2}\sqrt{x}$, and $y = \frac{1}{3}x$, for $0 \leq x \leq 6$.

18. Find the area of the region in the first quadrant of the xy-plane that is bounded by $y = 12 - x^2$, $y = 4x$, and $y = x$.

19. Refer to the oil-consumption data in Example 1, Section 1. Suppose that in 1970 the growth constant for the annual rate of oil consumption had been held to .04. What effect would this action have had on oil consumption from 1970 to 1974?

20. The marginal profit function for a certain company is $MP_1(x) = -x^2 + 14x - 24$. The company expects the daily production level to rise from $x = 6$ to $x = 8$ units. The management is considering a plan that would have the effect of changing the marginal profit to $MP_2(x) = -x^2 + 12x - 20$. Should the company adopt the plan? Determine the area between the graphs of the two marginal cost functions from $x = 6$ to $x = 8$. Interpret this area in economic terms.

SOLUTIONS TO PRACTICE PROBLEMS 3

1. First graph the two curves, as shown in the accompanying figure. The curve $y = x + 3$ lies on top. So the area between the curves is

$$\int_1^3 [(x + 3) - (x^2 + x - 13)] \, dx = \int_1^3 [-x^2 + 16] \, dx$$

$$= (-\tfrac{1}{3}x^3 + 16x)\Big|_1^3$$

$$= 23\tfrac{1}{3}.$$

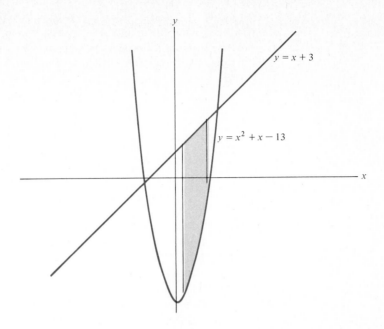

2. Graphing the two marginal cost functions yields the results shown in the accompanying figure. So the area between the curves equals

$$\int_{10}^{15} \left[MC_1(x) - MC_2(x) \right] dx = \int_{10}^{15} (x^2 - 20x + 108) - (\tfrac{1}{2}x^2 - 12x + 75) \right] dx$$

$$= \int_{10}^{15} \left[\tfrac{1}{2}x^2 - 8x + 33 \right] dx$$

$$= \left(\tfrac{1}{6}x^3 - 4x^2 + 33x \right) \Big|_{10}^{15}$$

$$= 60\tfrac{5}{6}.$$

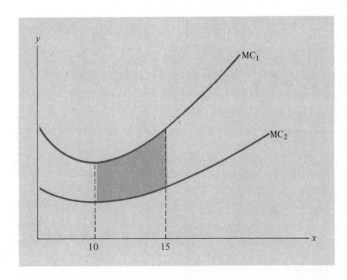

The area under a marginal cost curve is total cost. So the area between the curves represents the difference in the cost of increasing production from 10 to 15 units using the old production process versus the new production process. Using the new production process allows the company to perform the increased production for $60\frac{5}{6}$ dollars less than using the old process.

16.4. Riemann Sums

We have seen many applied problems which require us to calculate a definite integral $\int_a^b f(x)\,dx$. So far, our only technique for doing so requires us to determine an antiderivative of $f(x)$. However, as we shall now demonstrate, it is possible to numerically approximate a definite integral without first determining an antiderivative. The numerical approximation is in terms of so-called *Riemann sums*.

Let us begin by defining Riemann sums for a continuous function $f(x)$ over the interval from $x = a$ to $x = b$. Divide the interval on the x-axis from $x = a$ to $x = b$ into n equal subintervals (Fig. 1). (Here n may be any positive integer.) Since the entire interval is of length $b - a$, each subinterval has length $(1/n)$th of the total, or $(b - a)/n$. For brevity, let us set $\Delta x = (b - a)/n$. In each subinterval choose a point (any point will do). Label these points $x_1, x_2, x_3, \ldots, x_n$. (See Fig. 1.) Now form the sum

$$f(x_1)\Delta x + f(x_2)\Delta x + \cdots + f(x_n)\Delta x.$$

Such a sum is called a *Riemann sum* for the function $f(x)$ over the interval from $x = a$ to $x = b$.

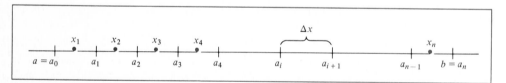

FIGURE 1

The significance of Riemann sums is that they approximate the definite integral $\int_a^b f(x)\,dx$. More precisely, we can state the following fact.

Suppose that the interval from $x = a$ to $x = b$ is divided into n equal subintervals of length $\Delta x = (b - a)/n$. Let x_i be any point of the ith subinterval. Then

$$\int_a^b f(x)\,dx \approx f(x_1)\Delta x + f(x_2)\Delta x + \cdots + f(x_n)\Delta x, \tag{1}$$

where the approximation becomes exact as n gets large.

We shall justify (1) only for the case when $f(x)$ is nonnegative for x between a and b. [Our argument can easily be modified to cover general $f(x)$.] Note that by the Fundamental Theorem of Calculus the left side of (1) equals the area under the curve $y = f(x)$ from $x = a$ to $x = b$.

Let us approximate this area by the area enclosed by n rectangles, as in Fig. 2. The first rectangle has height $f(x_1)$ and width Δx, so its area is $f(x_1)\Delta x$. The second rectangle has height $f(x_2)$ and width Δx, so its area is $f(x_2)\Delta x$. And so on. Each term on the right side of (1) corresponds to the area of one of the rectangles in Fig. 2. The sum of the areas of these rectangles closely approximates the area under the curve $y = f(x)$ from $x = a$ to $x = b$. This explains why (1) is true. Moreover, as n increases, the width of the rectangles decreases and the rectangular area approximates the area under the curve more closely. This completes the justification.

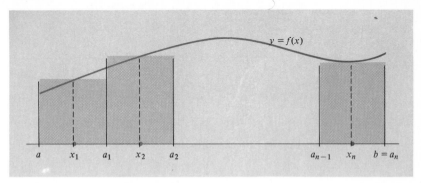

FIGURE 2

The Riemann sum approximation (1) is used in some texts to *define* the definite integral. Indeed, it is possible to prove that if f is continuous for $a \le x \le b$, then as Δx approaches 0, the limit of the Riemann sum is equal to the definite integral. Thus it is possible to define the definite integral as a limit of a Riemann sum. In this approach, our version of the fundamental theorem of calculus is a direct consequence of the definition of the definite integral. However, then the relationship of the integral with antidifferentiation must be proved.

The Riemann sum approximation may be used to explain the notation

$$\int f(x)\,dx$$

which is used for the integral. The symbol $\int$ is an old-fashioned letter "s" and stands for the Latin word "summum" (meaning sum). And, indeed, by (1) the integral is (approximately) a sum of quantities of the form $f(x)\Delta x$. Sir Isaac Newton, one of the inventors of calculus, denoted passage to the limit by changing the Δx to dx and therefore decided to write the integral as $\int f(x)\,dx$.

The approximation (1) may be used in a number of ways. First of all, we may apply it to estimate definite integrals. For this purpose, it is often convenient to choose $x_1, x_2, \ldots, x_n$ to be the midpoints of their respective subintervals, in which case approximation (1) is called the *rectangle rule*.

EXAMPLE 1 Approximate $\int_1^3 x^2\,dx$, using the rectangle rule with $n = 5$.

Note We know, of course, that

$$\int_1^3 x^2\,dx = \tfrac{1}{3}x^3 \Big|_1^3 = 9 - \tfrac{1}{3} = 8.3333\ldots,$$

so the rectangle rule is not really necessary here. But this example will give us some assurance that the rectangle rule works. Also, it will give us some idea of the accuracy we obtain for a given amount of calculation.

Solution We have $\Delta x = (b - a)/n = (3 - 1)/5 = 2/5 = .4$. The endpoints of the five subintervals begin at $a = 1$ and are spaced .4 unit apart. The first midpoint of a subinterval is at $a + \dfrac{\Delta x}{2} = 1 + .2 = 1.2$. The midpoints are also spaced .4 unit apart.

Therefore, $x_1 = 1.2$, $x_2 = 1.6$, $x_3 = 2.0$, $x_4 = 2.4$, $x_5 = 2.8$, and

$$\int_1^3 x^2\,dx \approx (1.2)^2(.4) + (1.6)^2(.4) + (2.0)^2(.4) + (2.4)^2(.4) + (2.8)^2(.4)$$

$$= [1.44 + 2.56 + 4 + 5.76 + 7.84](.4)$$

$$= 8.64.$$

The error of this approximation is less than .30667.

Approximate techniques of integration become manageable for most examples by using one of the currently available calculators. (The examples in this book were done that way.) However, for very large n (such as $n = 100$), even the minicalculator must make way for a high-speed digital computer or a programmable calculator.

The fundamental approximation (1) can be applied in a manner completely different from that described above. Namely it may be used to express certain quantities as definite integrals. The principle behind such applications of (1) is as follows: Suppose that we are interested in finding the value of a certain quantity, call it Q. We observe that we can obtain an approximation of Q by subdividing an interval from $x = a$ to $x = b$ into n equal subintervals and forming the sum $f(x_1)\Delta x + \cdots + f(x_n)\Delta x$, where $f(x)$ is some function (not given originally) and where x_i, Δx are as before. Also, we observe that this approximation to Q becomes exact as n gets large. We then invoke our fundamental formula to conclude that $Q = \int_a^b f(x)\,dx$.

Examples 2–5 illustrate such applications of (1).

EXAMPLE 2 Revolve the region of Fig. 3(a) about the x-axis so that it sweeps out a solid [Fig. 3(b)]. Derive a formula for the volume of this *solid of revolution*.

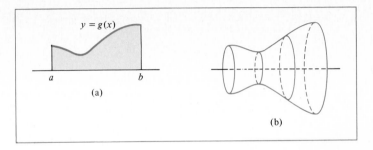

FIGURE 3

Solution   Let us break the x-axis between a and b into a large number n of equal subintervals, each of length $\Delta x = (b - a)/n$. Using each subinterval as a base, we can divide the region into strips (see Fig. 4).

Let x_i be a point in the ith subinterval. Then the volume swept out by revolving the ith strip is approximately the same as the volume of the cylinder swept out by revolving the rectangle of height $g(x_i)$ and base Δx around the x-axis (Fig. 5). The volume of the cylinder is

$$[\text{area of circular side}] \cdot [\text{width}] = \pi[g(x_i)]^2 \cdot \Delta x.$$

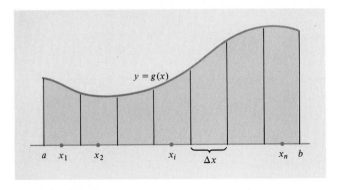

FIGURE 4

FIGURE 5

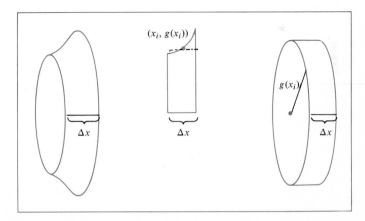

The total volume swept out by all the strips is approximated by the total volume swept out by the rectangles, which is

$$\text{volume} \approx \pi[g(x_1)]^2 \, \Delta x + \pi[g(x_2)]^2 \, \Delta x + \cdots + \pi[g(x_n)]^2 \, \Delta x.$$

As n gets larger and larger, this approximation becomes arbitrarily close to the true volume. The expression on the right is a Riemann sum with $f(x) = \pi[g(x)]^2$. Therefore, we conclude that

$$\text{volume} = \int_a^b \pi[g(x)]^2 \, dx.$$

EXAMPLE 3 Find the volume of the solid of revolution obtained by revolving the region of Fig. 6 about the x-axis.

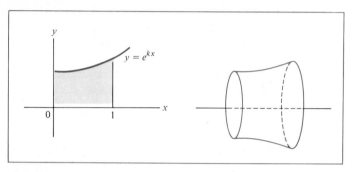

FIGURE 6

Solution Using the formula derived in Example 2, we have

$$\text{volume} = \int_0^1 \pi(e^{kx})^2 \, dx$$

$$= \int_0^1 \pi e^{2kx} \, dx = \frac{\pi}{2k} e^{2kx} \Big|_0^1 = \frac{\pi}{2k}(e^{2k} - 1).$$

EXAMPLE 4 Find the volume of a right circular cone of radius r and height h.

Solution The cone [Fig. 7(a)] is the solid of revolution swept out when the shaded region in Fig. 7(b) is revolved about the x-axis. Using the formula developed in Example 2,

FIGURE 7

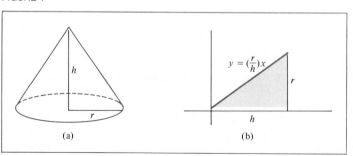

(a)   (b)

the volume of the cone is

$$\int_0^h \pi \left(\frac{r}{h}x\right)^2 dx = \frac{\pi r^2}{h^2} \int_0^h x^2 \, dx = \frac{\pi r^2}{h^2} \left.\frac{x^3}{3}\right|_0^h = \frac{\pi r^2 h}{3}.$$

The next example provides an application of Riemann sums to economics.

EXAMPLE 5 (*Consumers' Surplus*) Using a demand curve of economics, derive a formula showing the amount that consumers benefit from an open system that has no price discrimination.

Solution Figure 8(a) is a *demand curve* for a commodity. It is determined by complex economic factors and gives a relationship between the quantity sold and the unit price of a commodity. Specifically, it says that, in order to sell x units, the price must be set at $f(x)$ dollars per unit. Since, for most commodities, selling larger quantities requires a lowering of the price, demand functions are usually decreasing. Interactions between supply and demand determine the amount of a quantity available. Let A designate the amount of the commodity currently available and $B = f(A)$ the current selling price.

  Divide the interval from 0 to A into n subintervals, each of length $\Delta x = (A - 0)/n$, and take x_i to be the right-hand endpoint of the ith interval. Consider the first subinterval, from 0 to x_1. [See Fig. 8(b).] Suppose that only x_1 units had been available. Then the price per unit could have been set at $f(x_1)$ dollars and these x_1 units sold. Of course, at this price we could not have sold any more units. However, those people who paid $f(x_1)$ dollars had a great demand for the commodity. It is extremely valuable to them, and there is no advantage to substituting another commodity at that price. They are actually paying what the commodity is worth to them. In theory, then, the first x_1 units of the commodity could be sold to these people at $f(x_1)$ dollars per unit, yielding (price per unit) · (number of units) $= f(x_1) \cdot (x_1) = f(x_1) \cdot \Delta x$ dollars.

  After selling the first x_1 units, suppose that more units become available, so that now a total of x_2 units have been produced. Setting the price at $f(x_2)$, the remaining $x_2 - x_1 = \Delta x$ units can be sold, yielding $f(x_2) \cdot \Delta x$ dollars. Here, again,

FIGURE 8 Consumers' surplus.

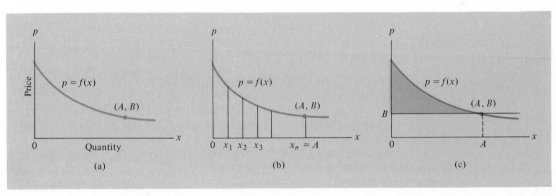

(a) (b) (c)

the second group of buyers would have paid as much for the commodity as it is worth to them. Continuing this process of price discrimination, the amount of money paid by the consumers would be

$$f(x_1)\Delta x + f(x_2)\Delta x + \cdots + f(x_n)\Delta x.$$

Taking n large, we see that this amount approaches $\int_0^A f(x)\,dx$.

Of course, in our open system, everyone pays the same price, B, and so the total amount paid is [price per unit] · [number of units] = $B \cdot A$. The amount of money saved by the consumers is

$$\int_0^A f(x)\,dx - B \cdot A = \int_0^A [f(x) - B]\,dx$$

and is called the *consumers' surplus*. Referring to Fig. 8(c), we see that the consumers' surplus is the area of the shaded region. It gives a numerical value to one benefit of a modern efficient economy.

The next example shows how the definite integral can be used to approximate the sum of a large number of terms.

EXAMPLE 6 Suppose that money is deposited daily into a saving account at an annual rate of $1000. The account pays 6% interest compounded continuously. Approximate the amount of money in the account at the end of 5 years.

Solution Divide the time interval from 0 to 5 years into daily subintervals. Each subinterval is then of duration $\Delta t = \frac{1}{365}$ years. Let $t_1, t_2, \ldots, t_n$ be points chosen from these subintervals. Since we deposit money at an annual rate of $1000, the amount deposited during one of the subintervals is $1000\,\Delta t$ dollars. If this amount is deposited at time t_i, the $1000\,\Delta t$ dollars will earn interest for the remaining $5 - t_i$ years. The total amount resulting from this one deposit at time t_i is then

$$1000\,\Delta t e^{.06(5 - t_i)}.$$

Add up the effects of the deposits at times $t_1, t_2, \ldots, t_n$ to arrive at the total balance in the account:

$$A = 1000e^{.06(5 - t_1)}\Delta t + 1000e^{.06(5 - t_2)}\Delta t + \cdots + 1000e^{.06(5 - t_n)}\Delta t.$$

This sum is a Riemann sum for the integral

$$\int_0^5 1000e^{.06(5 - t)}\,dt = \frac{1000}{-.06} e^{.06(5 - t)}\bigg|_0^5$$

$$= \frac{1000}{-.06} (1 - e^{.3})$$

$$\approx 5831.$$

That is, the approximate balance in the account at the end of 5 years is $5831.

Average Value of a Function Let $f(x)$ be a function. The definite integral may be used to define the *average value of $f(x)$ from $x = a$ to $x = b$*. To calculate the average of a collection of numbers $y_1, y_2, \ldots, y_n$, we add up the numbers and divide by n to obtain

$$\frac{y_1 + y_2 + \cdots + y_n}{n}.$$

To determine the average value of $f(x)$ from $x = a$ to $x = b$ we proceed similarly. Choose n values of x, say $x_1, x_2, \ldots, x_n$, and calculate the corresponding function values $f(x_1), f(x_2), \ldots, f(x_n)$. The average of these values is

$$\frac{f(x_1) + f(x_2) + \cdots + f(x_n)}{n}. \tag{2}$$

If the points $x_1, x_2, \ldots, x_n$ are "evenly" spread out from a to b, we would expect this sum to closely approximate the average value of $f(x)$ from $x = a$ to $x = b$. In fact, as n becomes large the average (2) should approximate the average value of $f(x)$ to any arbitrary degree of accuracy. To guarantee that the points $x_1, x_2, \ldots, x_n$ are "evenly" spread out from a to b, let us divide the interval from $x = a$ to $x = b$ into n subintervals of equal length $\Delta x = (b - a)/n$ (Fig. 9). Then choose x_1 from the first subinterval, x_2 from the second, and so forth. Thus, we see that for n large,

[average value of $f(x)$ from $x = a$ to $x = b$]

$$\approx f(x_1) \cdot \frac{1}{n} + f(x_2) \cdot \frac{1}{n} + \cdots + f(x_n) \cdot \frac{1}{n}$$

$$= \frac{1}{b - a} \left[f(x_1) \cdot \frac{b - a}{n} + f(x_2) \cdot \frac{b - a}{n} + \cdots + f(x_n) \cdot \frac{b - a}{n} \right]$$

$$= \frac{1}{b - a} \left[f(x_1) \Delta x + f(x_2) \Delta x + \cdots + f(x_n) \Delta x \right]$$

$$\approx \frac{1}{b - a} \int_a^b f(x)\,dx,$$

where the approximation becomes exact as n becomes large. Thus we have motivated the following important definition:

The *average value* of $f(x)$ from $x = a$ to $x = b$ is defined as the quantity

$$\frac{1}{b - a} \int_a^b f(x)\,dx. \tag{3}$$

FIGURE 9

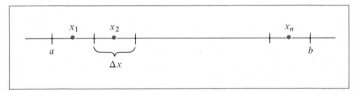

EXAMPLE 7   Find the average value of $f(x) = \sqrt{x}$ over the interval from 0 to 9.

Solution   Using (2) with $a = 0$ and $b = 9$, the average value of $f(x) = \sqrt{x}$ from $x = 0$ to $x = 9$ is

$$\frac{1}{9 - 0} \int_0^9 \sqrt{x}\, dx.$$

Since $\sqrt{x} = x^{1/2}$, an antiderivative of $\sqrt{x}$ is $\frac{2}{3}x^{3/2}$. Therefore,

$$\frac{1}{9} \int_0^9 \sqrt{x}\, dx = \frac{1}{9}\left(\frac{2}{3}x^{3/2}\Big|_0^9\right) = \frac{1}{9}\left(\frac{2}{3} \cdot 9^{3/2} - 0\right) = \frac{1}{9}\left(\frac{2}{3} \cdot 27\right) = 2,$$

so that the average value of $\sqrt{x}$ over the interval from 0 to 9 is 2. The area of the shaded region is the same as the area of the pictured rectangle (Fig. 10).

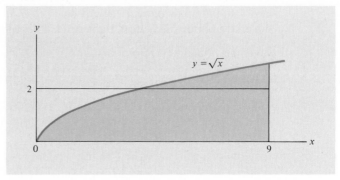

FIGURE 10

EXAMPLE 8   Suppose that the current world population is 4.2×10^9 and the population t years from now is given by the exponential growth law

$$P(t) = (4.2 \times 10^9)e^{.023t}.$$

Find the average population of the earth during the next 30 years. (This number is important in long-range planning for agricultural production and for the allocation of goods and services.)

Solution   The average value of the population $P(t)$ from $t = 0$ to $t = 30$ is

$$\frac{1}{30 - 0} \int_0^{30} P(t)\, dt = \frac{1}{30} \int_0^{30} (4.2 \times 10^9)e^{.023t}\, dt$$

$$= \left(\frac{1}{30}\right)\frac{4.2 \times 10^9}{.023}e^{.023t}\Big|_0^{30}$$

$$= \frac{4.2 \times 10^9}{.69}(e^{.69} - 1)$$

$$\approx 6 \times 10^9.$$

PRACTICE PROBLEMS 4

1. Approximate $\int_1^3 \frac{1}{x} dx$ using the rectangle rule with $n = 4$.

2. Suppose that the interval $0 \leq x \leq 2$ is divided into 100 subintervals, each of width Δx. Show that the sum,

$$[3(\tfrac{1}{100})^2 - \tfrac{1}{100}]\Delta x + [3(\tfrac{3}{100})^2 - \tfrac{3}{100}]\Delta x$$
$$+ [3(\tfrac{5}{100})^2 - \tfrac{5}{100}]\Delta x + \cdots + [3(\tfrac{199}{100})^2 - \tfrac{199}{100}]\Delta x,$$

is close to 6.

EXERCISES 4

Approximate the following integrals by the rectangle rule using the given value of n. Also, find the exact value of the integral. State your answers to five decimal places.

1. $\int_0^1 x^3 dx; n = 2, 4$

2. $\int_1^2 \frac{1}{(4x - 3)^2} dx; n = 4$

3. $\int_1^2 e^{-x} dx; n = 5$

4. $\int_0^4 (x^2 - 5) dx; n = 2, 4$

5. $\int_4^6 \frac{1}{x - 3} dx; n = 5$

6. $\int_1^4 \sqrt{x - 1} dx; n = 6$

Find the consumers' surplus for each of the following demand curves at the given sales level x.

7. $p = 3 - (x/10); x = 20$

8. $p = \dfrac{x^2}{200} - x + 50; x = 20$

9. $p = -.01x + 5; x = 100$

10. $p = \sqrt{16 - .02x}; x = 350$

Figure 11 shows a supply curve for a commodity. It gives the relationship between the selling price of the commodity and the quantity that producers will manufacture. At a higher selling price, a greater quantity will be produced. Therefore, the curve is increasing. If (A, B) is a point on the curve, then, in order to stimulate the production of A units of the commodity, the price per unit must be B dollars. Of course, some producers will be willing to produce the commodity even with a lower selling price. Since everyone receives the same price in an open efficient economy, most producers are receiving more than their minimal required price. The excess is called the *producers' surplus*. Using an argument analogous to that of the *consumers' surplus*, one can show that the total producers' surplus when the price is B is the area of the shaded region in Fig. 11. Find the producers' surplus for each of the following supply curves at the given sales level x.

11. $p = .01x + 3; x = 200$

12. $p = (x^2/9) + 1; x = 3$

13. $p = (x/2) + 7; x = 10$

14. $p = 1 + \tfrac{1}{2}\sqrt{x}; x = 36$

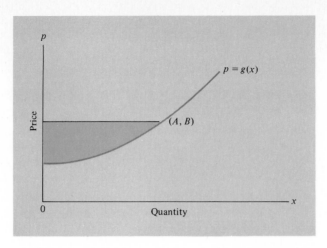

FIGURE 11   Producers' surplus.

For a particular commodity, the quantity produced and the unit price are given by the coordinates of the point where the supply and demand curves intersect. For each pair of supply and demand curves, determine the point of intersection (A, B) and the consumers' and producers' surplus. (See Fig. 12.)

15. Demand curve: $p = 12 - (x/50)$; supply curve: $p = (x/20) + 5$.

16. Demand curve: $p = \sqrt{25 - .1x}$; supply curve: $p = \sqrt{.1x + 9} - 2$.

17. Suppose that money is deposited daily into a savings account at an annual rate of $1000. If the account pays 5% interest compounded continuously, estimate the balance in the account at the end of 3 years.

18. Suppose that money is deposited daily into a savings account at an annual rate of $2000. If the account pays 6% interest compounded continuously, approximately how much will be in the account at the end of 2 years?

19. Suppose that money is deposited daily into a savings account at the rate of P dollars per year. The savings account pays 6% interest compounded continuously. Estimate the balance at the end of 4 years.

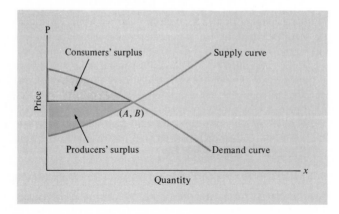

FIGURE 12

20. Suppose that money is deposited daily into a savings account at the rate of P dollars per year for T years. The account pays 5% interest compounded continuously. Find a definite integral that gives the approximate balance at the end of the T years.

Find the average value of $f(x)$ over the interval from $x = a$ to $x = b$, where:

21. $f(x) = x^2; a = 0, b = 3$
22. $f(x) = e^{x/3}; a = 0, b = 3$

23. $f(x) = x^3; a = -1, b = 1$
24. $f(x) = 5; a = 1, b = 10$

25. $f(x) = 1/x^2; a = \frac{1}{4}, b = \frac{1}{2}$
26. $f(x) = 2x - 6; a = 2, b = 4$

27. During a certain 12-hour period the temperature at time t (measured in hours from the start of the period) was $47 + 4t - \frac{1}{3}t^2$ degrees. What was the average temperature during that period?

28. Assuming that a country's population is now 3 million and growing exponentially with growth constant .02, what will be the average population during the next 50 years?

29. One hundred grams of radioactive radium having a half-life of 1690 years is placed in a concrete vault. What will be the average amount of radium in the vault during the next 1000 years?

30. One hundred dollars is deposited in the bank at 5% interest compounded continuously. What will be the average value of the money in the account during the next 20 years?

31. Suppose the interval $0 \leq x \leq 3$ is divided into 100 subintervals of width Δx. Let $x_1, \ldots,$ x_{100} be points in these subintervals. Suppose that in some application one needs to estimate the sum

$$(3 - x_1)^2 \, \Delta x + (3 - x_2)^2 \, \Delta x + \cdots + (3 - x_{100})^2 \, \Delta x.$$

Show that this sum is close to 9.

32. Suppose that the interval $0 \leq t \leq 3$ is divided into 1000 subintervals of width Δt. Let $t_1, t_2, \ldots, t_{1000}$ denote the right endpoints of these subintervals. Suppose that in some problem one needs to estimate the sum.

$$5000e^{-.1t_1} \, \Delta t + 5000e^{-.1t_2} \, \Delta t + \cdots + 5000e^{-.1t_{1000}} \, \Delta t.$$

Show that this sum is close to 13,000. [*Note:* A sum such as this would arise if one wanted to compute the present value of a continuous stream of income of $5000 per year for 3 years, with interest compounded continuously at 10%.]

33. Suppose that the interval $0 \leq x \leq 1$ is divided into 100 subintervals of width $\Delta x = \frac{1}{100}$. Show that the sum

$$\left[2 \cdot \tfrac{1}{100} + \left(\tfrac{1}{100}\right)^3\right] \Delta x + \left[2 \cdot \tfrac{2}{100} + \left(\tfrac{2}{100}\right)^3\right] \Delta x$$
$$+ \left[2 \cdot \tfrac{3}{100} + \left(\tfrac{3}{100}\right)^3\right] \Delta x + \cdots + \left[2 \cdot \tfrac{100}{100} + \left(\tfrac{100}{100}\right)^3\right] \Delta x$$

is close to $\frac{5}{4}$.

34. Suppose that water is flowing into a tank at a rate of $r(t)$ gallons per hour, where the rate depends on the time t according to the formula

$$r(t) = 20 - 4t, \qquad 0 \leq t \leq 5.$$

(a) Consider a brief period of time, say from t_1 to t_2. The length of this time period is $\Delta t = t_2 - t_1$. During this period the rate of flow does not change much and is approximately $20 - 4t_1$ (the rate at the beginning of the brief time interval). Approximately how much water flows into the tank during the time from t_1 to t_2?

(b) Explain why the total amount of water added to the tank during the time interval from $t = 0$ to $t = 5$ is given by $\int_0^5 r(t)\,dt$.

35. Suppose that the world consumes oil at the rate of $R_1(t)$ billion barrels per year at time t, where $t = 0$ corresponds to 1970 and $R_1(t) = 21.3e^{.04(t-4)}$ for $t \geq 4$. Use the rectangle rule with $n = 4$ to approximate $\int_6^{10} R_1(t)\,dt$. What interpretation can you give to each of the four terms in the approximation?

36. Find the average rate of oil consumption between 1976 and 1980, where the rate $R_1(t)$ is given in Exercise 35.

Find the volume of the solid of revolution generated by revolving about the x-axis the region under each of the following curves.

37. $y = \sqrt{r^2 - x^2}$ from $x = -r$ to $x = r$ (generates a sphere of radius r)

38. $y = r$ from $x = 0$ to $x = h$ (generates a cylinder)

39. $y = x^2$ from $x = 1$ to $x = 2$

40. $y = \dfrac{1}{x}$ from $x = 1$ to $x = 100$

41. $y = \sqrt{x}$ from $x = 0$ to $x = 4$ (The solid generated is called a *paraboloid*.)

42. $y = 2x - x^2$ from $x = 0$ to $x = 2$

43. $y = e^{-x}$ from $x = 0$ to $x = r$

44. $y = 2x + 1$ from $x = 0$ to $x = 1$ (The solid generated is called a *truncated cone*.)

SOLUTIONS TO PRACTICE PROBLEMS 4

1. Here $a = 1$, $b = 3$, $n = 4$, so $\Delta x = (3 - 1)/4 = \frac{1}{2}$. Moreover, by drawing the interval from $x = 1$ to $x = 3$, we see that $x_1 = 1.25$, $x_2 = 1.75$, $x_3 = 2.25$, $x_4 = 2.75$ (see Fig. 13). Thus

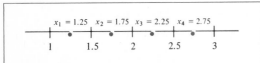

$$\int_1^3 \frac{1}{x}\,dx \approx \left[\frac{1}{1.25} + \frac{1}{1.75} + \frac{1}{2.25} + \frac{1}{2.75}\right]\cdot\frac{1}{2}$$
$$\approx 1.0898.$$

FIGURE 13

We note that

$$\int_1^3 \frac{1}{x}\,dx = \ln x \Big|_1^3 = \ln 3.$$

To four decimal places, $\ln 3 = 1.0986$, so the error from the rectangle rule is .0088. When $n = 10$, the error is .0015.

2. Here $\Delta x = (2 - 0)/100 = .02$. The subdivision is shown in Fig. 14, where the respective midpoints of the subintervals are seen to be $x_1 = .01$, $x_2 = .03$, $x_3 = .05, \ldots, x_{100} = 1.99$.

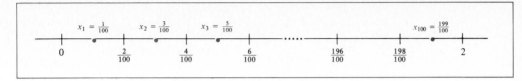

FIGURE 14

Therefore, the sum in question is

$$[3x_1^2 - x_1]\Delta x + [3x_2^2 - x_2]\Delta x + [3x_3^2 - x_3]\Delta x + \cdots + [3x_{100}^2 - x_{100}]\Delta x,$$

a Riemann sum for the function $f(x) = 3x^2 - x$ over the interval from $x = 0$ to $x = 2$. By (1), this sum is approximately equal to $\int_0^2 (3x^2 - x)\,dx$. Now,

$$\int_0^2 (3x^2 - x)\,dx = \left(x^3 - \frac{x^2}{2} \right)\Big|_0^2 = 6.$$

16.5. Techniques of Integration

This section presents two techniques of integration that will greatly expand our ability to find antiderivatives. Like the other rules of integration discussed in Section 1, these new methods are based on corresponding rules of differentiation: the chain rule and the product rule.

Integration by Substitution Many integrals that appear complex have a simple form that becomes evident after making an appropriate change of variable. Integration by substitution is a technique for making such changes of variables.

Let $f(x)$ and $g(x)$ be two given functions and let $F(x)$ be an antiderivative of $f(x)$. The chain rule asserts that

$$\frac{d}{dx}[F(g(x))] = F'(g(x))g'(x)$$

$$= f(g(x))g'(x) \qquad [\text{since } F'(x) = f(x)].$$

Turning this formula into an integration formula, we have

$$\int f(g(x))g'(x)\,dx = F(g(x)) + C, \tag{1}$$

where C is any constant. One way to use this formula is illustrated in the following example.

EXAMPLE 1 Determine $\int (x^2 + 1)^3 \cdot 2x\,dx$.

Solution If we set $f(x) = x^3$, $g(x) = x^2 + 1$, then $f(g(x)) = (x^2 + 1)^3$ and $g'(x) = 2x$. Therefore, we can apply formula (1). An antiderivative $F(x)$ of $f(x)$ is given by

$$F(x) = \tfrac{1}{4}x^4,$$

so that, by formula (1), we have

$$\int (x^2 + 1)^3 \cdot 2x \, dx = F(g(x)) + C$$
$$= \tfrac{1}{4}(x^2 + 1)^4 + C.$$

Formula (1) can be elevated from the status of a sometimes useful formula to a technique of integration by the introduction of a simple mnemonic device. Suppose that we are faced with integrating a function of the form $f(g(x))g'(x)$. Of course, we know the answer from formula (1). However, let us proceed somewhat differently. Replace the expression $g(x)$ by a new variable u and replace $g'(x) \, dx$ by du. Such a substitution has the advantage that it reduces the generally complex expression $f(g(x))$ to the simpler form $f(u)$. In terms of u, the integration problem may be written

$$\int f(g(x))g'(x) \, dx = \int f(u) \, du.$$

However, the integral on the right is easy to evaluate, since

$$\int f(u) \, du = F(u) + C.$$

Since $u = g(x)$, we then obtain

$$\int f(g(x))g'(x) \, dx = F(u) + C = F(g(x)) + C,$$

which is the correct answer by (1). Remember, however, that replacing $g'(x) \, dx$ by du only has status as a correct mathematical statement because doing so leads to the correct answers. We do not, in this book, seek to explain in any deeper way what this replacement means.

Let us rework Example 1 using this method.

Second Solution of Example 1 Set $u = x^2 + 1$. Then $du = \dfrac{d}{dx}(x^2 + 1) \, dx = 2x \, dx$, and

$$\int (x^2 + 1)^3 \cdot 2x \, dx = \int u^3 \, du$$
$$= \tfrac{1}{4}u^4 + C$$
$$= \tfrac{1}{4}(x^2 + 1)^4 + C \qquad (\text{since } u = x^2 + 1).$$

EXAMPLE 2 Evaluate $\int 2xe^{x^2} \, dx$.

Solution Let $u = x^2$, so that $du = \dfrac{d}{dx}(x^2) \, dx = 2x \, dx$. Therefore,

$$\int 2xe^{x^2} \, dx = \int e^{x^2} \cdot 2x \, dx \ .$$
$$= \int e^u \, du$$
$$= e^u + C$$
$$= e^{x^2} + C.$$

From Examples 1 and 2 we can deduce the following method for integration of functions of the form $f'(g(x))g'(x)$.

Integration by Substitution

1. Define a new variable $u = g(x)$, where $g(x)$ is chosen in such a way that, when written in terms of u, the integrand is simpler than when written in terms of x.

2. Transform the integral with respect to x into an integral with respect to u by replacing $g(x)$ everywhere by u and $g'(x)\,dx$ by du.

3. Integrate the resulting function of u.

4. Rewrite the answer in terms of x by replacing u by $g(x)$.

Let us try a few more examples.

EXAMPLE 3 Evaluate $\int 3x^2 \sqrt{x^3 + 1}\,dx$.

Solution The first problem facing us is to find an appropriate substitution that will simplify the integral. An immediate possibility is offered by setting $u = x^3 + 1$. Then $\sqrt{x^3 + 1}$ will become $\sqrt{u}$, a significant simplification. If $u = x^3 + 1$, then $du = \dfrac{d}{dx}(x^3 + 1)\,dx = 3x^2\,dx$, so that

$$\int 3x^2 \sqrt{x^3 + 1}\,dx = \int \sqrt{u}\,du$$
$$= \tfrac{2}{3}u^{3/2} + C$$
$$= \tfrac{2}{3}(x^3 + 1)^{3/2} + C.$$

Knowing the correct substitution to make is a skill that develops through practice. Basically, we look for an occurrence of function composition, $f(g(x))$, where $f(x)$ is a function that we know how to integrate and where $g'(x)$ also appears in the integrand. Sometimes $g'(x)$ does not appear exactly but can be obtained by multiplying by a constant. Such a shortcoming is easily remedied, as illustrated in Examples 4 and 5.

EXAMPLE 4 Find $\int x^2 e^{x^3}\,dx$.

Solution Let $u = x^3$; then $du = 3x^2\,dx$. A simple trick allows us to introduce the needed factor of 3 into the integrand. Note that

$$\int x^2 e^{x^3}\,dx = \int \frac{1}{3} \cdot 3x^2 e^{x^3}\,dx = \frac{1}{3}\int 3x^2 e^{x^3}\,dx.$$

Substituting, we obtain

$$\int x^2 e^{x^3} \, dx = \frac{1}{3} \int e^u \, du$$

$$= \frac{1}{3} e^u + C$$

$$= \frac{1}{3} e^{x^3} + C \qquad (\text{since } u = x^3).$$

EXAMPLE 5   Find $\displaystyle\int \frac{x}{x^2 + 1} \, dx.$

Solution   If $u = x^2 + 1$, then $du = 2x \, dx$, and

$$\int \frac{x}{x^2 + 1} \, dx = \int \frac{1}{2} \cdot \frac{1}{x^2 + 1} \cdot 2x \, dx$$

$$= \frac{1}{2} \int \frac{1}{x^2 + 1} \cdot 2x \, dx = \frac{1}{2} \int \frac{1}{u} \, du$$

$$= \tfrac{1}{2} \ln|u| + C = \tfrac{1}{2} \ln|x^2 + 1| + C.$$

Integration by Parts   Integration by parts is a technique of integration that exchanges a given integral for another (hopefully less complicated) integral. Let $f(x)$ and $g(x)$ be given functions and let $G(x)$ be an antiderivative of $g(x)$. The product rule asserts that

$$\frac{d}{dx}\left[f(x)G(x)\right] = f(x)G'(x) + G(x)f'(x)$$

$$= f(x)g(x) + f'(x)G(x) \qquad \left[\text{since } G'(x) = g(x)\right].$$

Therefore,

$$f(x)G(x) = \int f(x)g(x) \, dx + \int f'(x)G(x) \, dx.$$

This last formula may be rewritten in the following more useful form.

$$\int f(x)g(x) \, dx = f(x)G(x) - \int f'(x)G(x) \, dx. \tag{2}$$

Equation (2) is the basis of *integration by parts*, one of the most important techniques of integration. The next three examples illustrate this technique.

EXAMPLE 6   Evaluate $\int x e^x \, dx.$

Solution Set $f(x) = x$, $g(x) = e^x$. Then $f'(x) = 1$, $G(x) = e^x$, and (2) yields

$$\int x e^x \, dx = x e^x - \int 1 \cdot e^x \, dx$$

$$= x e^x - e^x + C.$$

The following principles underlie Example 6 and also illustrate general features of situations to which integration by parts may be applied:

1. The integrand is the product of two functions $f(x)$ and $g(x)$.
2. It is easy to compute $f'(x)$ and $G(x)$. That is, we can differentiate $f(x)$ and integrate $g(x)$.
3. The integral $\int f'(x)G(x) \, dx$ can be calculated.

Let us consider another example in order to see how these three principles work.

EXAMPLE 7 Evaluate $\int x(x + 5)^8 \, dx$.

Solution Our calculations can be set up as follows:

$$f(x) = x, \qquad g(x) = (x + 5)^8$$

$$f'(x) = 1, \qquad G(x) = \frac{1}{9}(x + 5)^9.$$

Then

$$\int x(x + 5)^8 \, dx = x \cdot \frac{1}{9}(x + 5)^9 - \int 1 \cdot \frac{1}{9}(x + 5)^9 \, dx$$

$$= \frac{1}{9} x(x + 5)^9 - \frac{1}{9} \int (x + 5)^9 \, dx$$

$$= \frac{1}{9} x(x + 5)^9 - \frac{1}{9} \cdot \frac{1}{10}(x + 5)^{10} + C$$

$$= \frac{1}{9} x(x + 5)^9 - \frac{1}{90}(x + 5)^{10} + C.$$

We were led to try integration by parts because our integrand is the product of two functions. We were led to choose $f(x) = x$ [and not $(x + 5)^9$] because $f'(x) = 1$, so that the factor x in the integrand is made to disappear, thereby simplifying the integral.

EXAMPLE 8 Evaluate $\int x^2 \ln x \, dx$.

Solution Set

$$f(x) = \ln x, \qquad g(x) = x^2$$

$$f'(x) = \frac{1}{x}, \qquad G(x) = \frac{x^3}{3}.$$

Then

$$\int x^2 \ln x \, dx = \frac{x^3}{3} \ln x - \int \frac{1}{x} \cdot \frac{x^3}{3} \, dx$$

$$= \frac{x^3}{3} \ln x - \frac{1}{3} \int x^2 \, dx$$

$$= \frac{x^3}{3} \ln x - \frac{1}{9} x^3 + C.$$

Let us illustrate how an integration technique, such as integration by parts, can arise in a fairly simple business situation. Recall from our discussion of compound interest that the present value of A dollars to be received t years from now is given by

$$P = Ae^{-rt},$$

where r is a specified annual rate of interest, with interest compounded continuously. Now suppose that a sum of money is to be received in a series of frequent payments over a period of time instead of in one lump sum at the end of the period. Such a series of payments is often viewed as a "continuous stream of income." If $K(t)$ is the annual rate of income at time t, and if the income is to be received over the next T years, then the *present value P of the stream of income* at interest rate r is defined by the integral

$$P = \int_0^T K(t)e^{-rt} \, dt. \tag{3}$$

A Riemann sum argument can be used to show that it takes a fund of P dollars available today in order to create a continuous stream of income of $K(t)$ dollars (annual rate) at time t.

The concept of the present value of a continuous stream of income is an important tool in management decision processes involving the selection or replacement of equipment. Even when $K(t)$ is a simple function, the evaluation of the integral in (3) usually requires special techniques such as integration by parts, as we see in the following example.

EXAMPLE 9 A printing company estimates that the rate of revenue generated by one of its printing presses at time t will be $80 - 2t$ thousand dollars per year. Find the present value of this continuous stream of income over the next 4 years at a 10% interest rate.

Solution We use (3) with $K(t) = 80 - 2t$, $T = 4$, and $r = .1$. Our first task is to find an antiderivative of $K(t)e^{-rt} = (80 - 2t)e^{-.1t}$. We integrate by parts, with $f(t) = 80 - 2t$, $g(t) = e^{-.1t}$, and $G(t) = -10e^{-.1t}$.

$$\int (80 - 2t)e^{-.1t} \, dt = (80 - 2t)(-10e^{-.1t}) - \int 20e^{-.1t} \, dt$$

$$= -800e^{-.1t} + 20te^{-.1t} + 200e^{-.1t} + C$$

$$= 20te^{-.1t} - 600e^{-.1t} + C.$$

Then

$$P = \int_0^4 (80 - 2t)e^{-.1t}\, dt = (20te^{-.1t} - 600e^{-.1t})\Big|_0^4$$

$$= (80e^{-.4} - 600e^{-.4}) - (0 - 600e^0)$$
$$= 600 - 520e^{-.4}$$
$$\approx 251.$$

Thus the present value of the machine's earnings is approximately 251 thousand dollars.

PRACTICE PROBLEMS 5

Calculate the following integrals.

1. $\displaystyle\int \frac{e^{5x} + x^4}{(e^{5x} + x^5)^2}\, dx$ [*Hint:* Let $u = e^{5x} + x^5$.]

2. $\displaystyle\int \frac{2x - 1}{(x + 4)^{1/3}}\, dx$ [*Hint:* Let $f(x) = 2x - 1$, $g(x) = (x + 4)^{-1/3}$.]

EXERCISES 5

Calculate each of the following indefinite integrals.

1. $\displaystyle\int 2x(x^2 + 4)^5\, dx$.

2. $\displaystyle\int 3x^2(x^3 + 1)^2\, dx$

3. $\displaystyle\int (x^2 - 5x)^3(2x - 5)\, dx$

4. $\displaystyle\int 2x\sqrt{x^2 + 3}\, dx$

5. $\displaystyle\int 5e^{5x-3}\, dx$

6. $\displaystyle\int 2xe^{-x^2}\, dx$

7. $\displaystyle\int \frac{3x^2}{x^3 - 1}\, dx$

8. $\displaystyle\int \frac{2x + 1}{(x^2 + x + 3)^6}\, dx$

Calculate the following integrals, making the indicated substitutions.

9. $\displaystyle\int \frac{x^2}{\sqrt{x^3 - 1}}\, dx; u = x^3 - 1$

10. $\displaystyle\int \frac{1}{\sqrt{2x + 1}}\, dx; u = 2x + 1$

11. $\displaystyle\int \frac{e^{1/x}}{x^2}\, dx; u = \frac{1}{x}$

12. $\displaystyle\int \frac{e^{3x}}{e^{3x} + 1}\, dx; u = e^{3x} + 1$

13. $\displaystyle\int \frac{x^2 + 1}{x^3 + 3x + 2}\, dx; u = x^3 + 3x + 2$

14. $\displaystyle\int \frac{\ln x}{x}\, dx; u = \ln x$

Determine the following integrals by making an appropriate substitution.

15. $\displaystyle\int (x^5 - 2x + 1)^{10}(5x^4 - 2)\,dx$

16. $\displaystyle\int (x + 1)e^{x^2 + 2x + 4}\,dx$

17. $\displaystyle\int \frac{3}{2x - 4}\,dx$

18. $\displaystyle\int \frac{e^{\sqrt{x}}}{\sqrt{x}}\,dx$

19. $\displaystyle\int \frac{x}{e^{x^2}}\,dx$

20. $\displaystyle\int \frac{3x - x^3}{x^4 - 6x^2 + 5}\,dx$

Use integration by parts to determine the following integrals.

21. $\displaystyle\int xe^{5x}\,dx$

22. $\displaystyle\int xe^{-x/2}\,dx$

23. $\displaystyle\int x(2x + 1)^4\,dx$

24. $\displaystyle\int (x + 1)e^x\,dx$

25. $\displaystyle\int x\sqrt{x + 1}\,dx$

26. $\displaystyle\int x(x + 5)^{-3}\,dx$

27. $\displaystyle\int \frac{3x}{e^x}\,dx$

28. $\displaystyle\int x^3 \ln x\,dx$

29. $\displaystyle\int \ln x\,dx$

30. $\displaystyle\int \frac{x}{\sqrt{3 + 2x}}\,dx$

$[\text{Let } f(x) = \ln x,\ g(x) = 1.]$

$[\text{Let } f(x) = x,\ g(x) = (3 + 2x)^{-1/2}.]$

Determine the following indefinite integrals.

31. $\displaystyle\int xe^{2x}\,dx$

32. $\displaystyle\int x\sqrt{x^2 - 1}\,dx$

33. $\displaystyle\int xe^{x^2}\,dx$

34. $\displaystyle\int x\sqrt{x - 1}\,dx$

35. $\displaystyle\int \frac{2x - 1}{\sqrt{3x - 3}}\,dx$

36. $\displaystyle\int \frac{2x - 1}{3x^2 - 3x + 1}\,dx$

37. $\displaystyle\int (4x + 3)(6x^2 + 9x)^{-7}\,dx$

38. $\displaystyle\int (4x + 3)(6x + 9)^{-7}\,dx$

39. $\displaystyle\int \frac{\ln x}{\sqrt{x}}\,dx$

40. $\displaystyle\int \frac{x + 4}{e^{4x}}\,dx$

41. $\displaystyle\int \frac{1}{x \ln 5x}\,dx$

42. $\displaystyle\int x \ln 5x\,dx$

43. Suppose that an investment produces a continuous stream of income at the rate of $300t - 500$ dollars per year at time t. Find the present value of the investment income over the next 5 years, using a 10% interest rate.

44. Recompute the present value of the stream of income in Exercise 43 over 5 years using a 5% interest rate.

SOLUTIONS TO PRACTICE PROBLEMS 5

1. Let $u = e^{5x} + x^5$, $du = (5e^{5x} + 5x^4)\,dx$, and note that $5e^{5x} + 5x^4 = 5(e^{5x} + x^4)$. Then

$$\int \frac{e^{5x} + x^4}{(e^{5x} + x^5)^2}\,dx = \int \frac{1}{5} \cdot \frac{5e^{5x} + 5x^4}{(e^{5x} + x^5)^2}\,dx$$

$$= \frac{1}{5} \int \frac{1}{(e^{5x} + x^5)^2} \cdot (5e^{5x} + 5x^4)\,dx$$

$$= \frac{1}{5} \int \frac{1}{u^2}\,du$$

$$= \frac{1}{5}(-u^{-1}) + C$$

$$= -\frac{1}{5}(e^{5x} + x^5)^{-1} + C.$$

2. Use integration by parts with $f(x) = 2x - 1$, $g(x) = (x + 4)^{-1/3}$, and $G(x) = \frac{3}{2}(x + 4)^{2/3}$.

$$\int \frac{2x - 1}{(x + 4)^{1/3}}\,dx = \int (2x - 1)(x + 4)^{-1/3}\,dx$$

$$= (2x - 1) \cdot \tfrac{3}{2}(x + 4)^{2/3} - \int 2 \cdot \tfrac{3}{2}(x + 4)^{2/3}\,dx$$

$$= \tfrac{3}{2}(2x - 1)(x + 4)^{2/3} - 3\int (x + 4)^{2/3}\,dx$$

$$= \tfrac{3}{2}(2x - 1)(x + 4)^{2/3} - \tfrac{9}{5}(x + 4)^{5/3} + C.$$

16.6. Improper Integrals

In applications of calculus, especially to statistics, it is often necessary to consider the area of a region that extends infinitely far to the right or left along the x-axis. We have drawn several such regions in Fig. 1. The areas of such "infinite" regions may be computed using *improper integrals*.

FIGURE 1

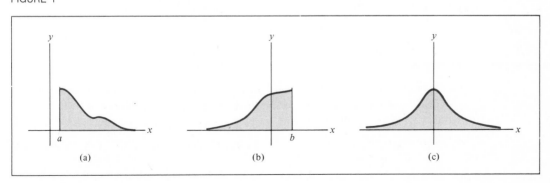

(a) (b) (c)

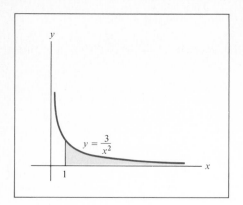

FIGURE 2

In order to motivate the idea of an improper integral, let us attempt to calculate the area under the curve $y = 3/x^2$ to the right of $x = 1$ (Fig. 2).

First, we shall compute the area under the graph of this function from $x = 1$ to $x = b$, where b is some number greater than 1. [See Fig. 3(a).] Then we shall examine how the area increases as we let b get larger, as in Fig. 3(b) and (c). The area from 1 to b is given by

$$\int_1^b \frac{3}{x^2}\, dx = -\frac{3}{x}\bigg|_1^b = \left(-\frac{3}{b}\right) - \left(-\frac{3}{1}\right) = 3 - \frac{3}{b}.$$

When b is large, $3/b$ is small and the integral nearly equals 3. That is, the area under the curve from 1 to b nearly equals 3. (See Table 1.) In fact, the area gets arbitrarily close to 3 as b gets larger. Thus it is reasonable to say that the region under the curve $y = 3/x^2$ for $x \geq 1$ has area 3.

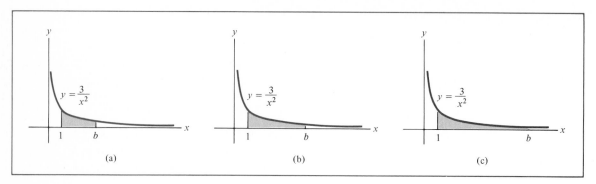

(a)   (b)   (c)

FIGURE 3

TABLE 1

b	$Area = \int_1^b \dfrac{3}{x^2}\, dx = 3 - \dfrac{3}{b}$
10	2.7000
100	2.9700
1,000	2.9970
10,000	2.9997

Recall from Chapter 11 that we write $b \to \infty$ as a shorthand for "b gets arbitrarily large, without bound." Then, to express the fact that the value of $\int_1^b \frac{3}{x^2}\, dx$ approaches 3 as $b \to \infty$, we write

$$\int_1^\infty \frac{3}{x^2}\, dx = 3.$$

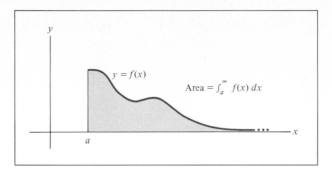

FIGURE 4

We call $\int_1^\infty \dfrac{3}{x^2}\,dx$ an *improper* integral because the upper limit of the integral is ∞ (infinity) rather than a finite number.

□ **Definition** Let a be fixed and suppose that $f(x)$ is a nonnegative function for all $x \geq a$. If $\displaystyle\lim_{b \to \infty} \int_a^b f(x)\,dx = L$, then we define

$$\int_a^\infty f(x)\,dx = \lim_{b \to \infty} \int_a^b f(x)\,dx = L.$$

We say that the improper integral $\int_a^\infty f(x)\,dx$ is *convergent* and that the region under the curve $y = f(x)$ for $x \geq a$ has area L. (See Fig. 4.)

It is possible to consider improper integrals in which $f(x)$ is both positive and negative. However, we shall consider only nonnegative functions, since this is the case occurring in most applications.

EXAMPLE 1 Find the area under the curve $y = e^{-x}$ for $x \geq 0$ (Fig. 5).

Solution We must calculate the improper integral

$$\int_0^\infty e^{-x}\,dx.$$

We take $b > 0$ and compute

FIGURE 5

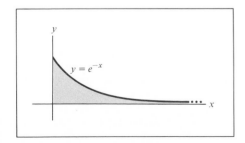

$$\int_0^b e^{-x}\,dx = -e^{-x}\Big|_0^b = (-e^{-b}) - (-e^0)$$

$$= 1 - e^{-b} = 1 - \frac{1}{e^b}.$$

We now consider the limit as $b \to \infty$ and note that $1/e^b$ approaches zero. Thus

$$\int_0^\infty e^{-x}\,dx = \lim_{b \to \infty} \int_0^b e^{-x}\,dx = \lim_{b \to \infty}\left(1 - \frac{1}{e^b}\right) = 1.$$

Therefore, the region in Fig. 5 has area 1.

EXAMPLE 2 Evaluate the improper integral $\int_7^\infty \dfrac{1}{(x-5)^2} \, dx$.

Solution
$$\int_7^b \frac{1}{(x-5)^2} \, dx = -\frac{1}{x-5}\Big|_7^b = -\frac{1}{b-5} - \left(-\frac{1}{7-5}\right) = \frac{1}{2} - \frac{1}{b-5}.$$

As $b \to \infty$, the fraction $1/(b-5)$ approaches zero, so
$$\int_7^\infty \frac{1}{(x-5)^2} \, dx = \lim_{b \to \infty} \int_7^b \frac{1}{(x-5)^2} \, dx = \lim_{b \to \infty} \left(\frac{1}{2} - \frac{1}{b-5}\right) = \frac{1}{2}.$$

Not every improper integral is convergent. If the value of $\int_a^b f(x)\,dx$ does not have a limit as $b \to \infty$, then we cannot assign any numerical value to $\int_a^\infty f(x)\,dx$, and we say that the improper integral $\int_a^\infty f(x)\,dx$ is *divergent*.

EXAMPLE 3 Show that $\int_1^\infty \dfrac{1}{\sqrt{x}} \, dx$ is divergent.

Solution For $b > 1$ we have
$$\int_1^b \frac{1}{\sqrt{x}} \, dx = 2\sqrt{x}\,\Big|_1^b = 2\sqrt{b} - 2. \tag{1}$$

As $b \to \infty$, the quantity $2\sqrt{b} - 2$ increases without bound. That is, $2\sqrt{b} - 2$ can be made larger than any specific number. Therefore, $\int_1^\infty \dfrac{1}{\sqrt{x}} \, dx$ has no limit as $b \to \infty$, so $\int_1^\infty \dfrac{1}{\sqrt{x}} \, dx$ is divergent.

In some cases it is necessary to consider improper integrals of the form
$$\int_{-\infty}^b f(x)\,dx.$$

Let b be fixed and examine the value of $\int_a^b f(x)\,dx$ as $a \to -\infty$; that is, as a moves arbitrarily far to the left on the number line. If $\lim\limits_{a \to -\infty} \int_a^b f(x)\,dx = L$, we say that the improper integral $\int_{-\infty}^b f(x)\,dx$ is *convergent* and we write
$$\int_{-\infty}^b f(x)\,dx = L.$$

Otherwise, we say that the improper integral is *divergent*. An integral of the form $\int_{-\infty}^b f(x)\,dx$ may be used to compute the area of a region like that shown in Fig. 1(b).

EXAMPLE 4 Determine if $\int_{-\infty}^0 e^{5x}\,dx$ is convergent. If convergent, find its value.

Solution

$$\int_{-\infty}^{0} e^{5x}\, dx = \lim_{a \to -\infty} \int_{a}^{0} e^{5x}\, dx$$

$$= \lim_{a \to -\infty} \tfrac{1}{5}e^{5x}\Big|_{a}^{0}$$

$$= \lim_{a \to -\infty} (\tfrac{1}{5} - \tfrac{1}{5}e^{5a}).$$

As $a \to -\infty$, e^{5a} approaches 0 so that $\tfrac{1}{5} - \tfrac{1}{5}e^{5a}$ approaches $\tfrac{1}{5}$. Thus the improper integral converges and has value $\tfrac{1}{5}$.

Areas of regions that extend infinitely far to the left *and* right, such as the region in Fig. 1(c), are calculated using improper integrals of the form

$$\int_{-\infty}^{\infty} f(x)\, dx.$$

We define such an integral to have the value

$$\int_{-\infty}^{0} f(x)\, dx + \int_{0}^{\infty} f(x)\, dx,$$

provided that both of the latter improper integrals are convergent.

An important area that arises in probability theory is the area under the so-called *normal curve*, whose equation is

$$y = \frac{1}{\sqrt{2\pi}} e^{-x^2/2}.$$

(See Fig. 6.) It is of fundamental importance for probability theory that this area is 1. In terms of an improper integral, this fact may be written

$$\int_{-\infty}^{\infty} \frac{1}{\sqrt{2\pi}} e^{-x^2/2}\, dx = 1.$$

The proof of this result is beyond the scope of this book.

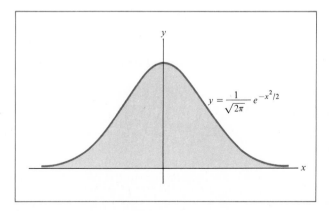

$$y = \frac{1}{\sqrt{2\pi}} e^{-x^2/2}$$

FIGURE 6

PRACTICE PROBLEMS 6

1. Does $1 - 2(1 - 3b)^{-4}$ approach a limit as $b \to \infty$?

2. Evaluate $\int_1^\infty \dfrac{x^2}{x^3 + 8} \, dx$.

3. Evaluate $\int_{-\infty}^{-2} \dfrac{1}{x^4} \, dx$.

EXERCISES 6

In Exercises 1–14, determine if the given expression approaches a limit as $b \to \infty$, and find that number when it exists.

1. $\dfrac{5}{b}$

2. b^2

3. $-3e^{2b}$

4. $\dfrac{1}{b} + \dfrac{1}{3}$

5. $\dfrac{1}{4} - \dfrac{1}{b^2} d$

6. $\frac{1}{2}\sqrt{b}$

7. $2 - (b + 1)^{-1/2}$

8. $\dfrac{2}{b} - \dfrac{3}{b^{3/2}}$

9. $5 - \dfrac{1}{b - 1}$

10. $5(b^2 + 3)^{-1}$

11. $6 - 3b^{-2}$

12. $e^{-b/2} + 5$

13. $2(1 - e^{-3b})$

14. $4(1 - b^{-3/4})$

15. Find the area under the graph of $y = 1/x^2$ for $x \geq 2$.

16. Find the area under the graph of $y = (x + 1)^{-2}$ for $x \geq 0$.

17. Find the area under the graph of $y = e^{-x/2}$ for $x \geq 0$.

18. Find the area under the graph of $y = 4e^{-4x}$ for $x \geq 0$.

19. Find the area under the graph of $y = (x + 1)^{-3/2}$ for $x \geq 3$.

20. Find the area under the graph of $y = (3x + 1)^{-2}$ for $x \geq 1$.

21. Show that the region under the graph of $y = x^{-3/4}$ for $x \geq 1$ cannot be assigned any finite number as its area.

22. Show that the region under the graph of $y = (x - 1)^{-1/3}$ for $x \geq 2$ cannot be assigned any finite number as its area.

Evaluate the following improper integrals whenever they are convergent.

23. $\int_1^\infty \dfrac{1}{x^3} \, dx$

24. $\int_1^\infty \dfrac{2}{x^{3/2}} \, dx$

25. $\int_0^\infty e^{2x} \, dx$

26. $\int_0^\infty x^2 + 1 \, dx$

27. $\int_0^\infty \dfrac{1}{(2x + 3)^2} \, dx$

28. $\int_0^\infty e^{-3x} \, dx$

29. $\int_2^\infty \dfrac{1}{(x - 1)^{5/2}} \, dx$

30. $\int_2^\infty e^{2-x} \, dx$

31. $\int_0^\infty .01e^{-.01x} \, dx$

32. $\int_0^\infty \dfrac{4}{(2x + 1)^3} \, dx$

33. $\int_0^\infty 6e^{1-3x} \, dx$

34. $\int_1^\infty e^{-.2x} \, dx$

35. $\displaystyle\int_3^\infty \frac{x^2}{\sqrt{x^3-1}}\,dx$ **36.** $\displaystyle\int_2^\infty \frac{1}{x\ln x}\,dx$ **37.** $\displaystyle\int_0^\infty xe^{-x^2}\,dx$

38. $\displaystyle\int_0^\infty \frac{x}{x^2+1}\,dx$ **39.** $\displaystyle\int_0^\infty 2x(x^2+1)^{-3/2}\,dx$ **40.** $\displaystyle\int_1^\infty (5x+1)^{-4}\,dx$

41. $\displaystyle\int_{-\infty}^0 e^{4x}\,dx$ **42.** $\displaystyle\int_{-\infty}^0 \frac{8}{(x-5)^2}\,dx$ **43.** $\displaystyle\int_{-\infty}^0 \frac{6}{(1-3x)^2}\,dx$

44. $\displaystyle\int_{-\infty}^0 \frac{1}{\sqrt{4-x}}\,dx$ **45.** $\displaystyle\int_0^\infty \frac{e^{-x}}{(e^{-x}+2)^2}\,dx$ **46.** $\displaystyle\int_{-\infty}^\infty \frac{e^{-x}}{(e^{-x}+2)^2}\,dx$

47. If $k > 0$, show that $\displaystyle\int_0^\infty ke^{-kx}\,dx = 1$.   **48.** If $k > 0$, show that $\displaystyle\int_1^\infty \frac{k}{x^{k+1}}\,dx = 1$.

49. The *capital value* of an asset such as a machine is sometimes defined as the present value of all future net earnings of the asset. (See Section 5.) The actual lifetime of the asset may not be known, and since some assets may last indefinitely, the capital value of the asset may be written in the form $\int_0^\infty K(t)e^{-rt}\,dt$, where $K(t)$ is the annual rate of earnings produced by the asset at time t, and where r is the annual rate of interest, compounded continuously. Find the capital value of an asset that generates income at the rate of \$5000 per year, assuming an interest rate of 10%.

50. Find the capital value of an asset that at time t is producing income at the rate of $6000e^{.04t}$ dollars per year, assuming an interest rate of 16%.

SOLUTIONS TO PRACTICE PROBLEMS 6

1. The expression $1 - 2(1 - 3b)^{-4}$ may also be written in the form

$$1 - \frac{2}{(1-3b)^4}.$$

When b is large, $(1 - 3b)^4$ is very large, and so $2/(1 - 3b)^4$ is very small. Thus $1 - 2(1 - 3b)^{-4}$ approaches 1 as $b \to \infty$.

2. The first step is to find an antiderivative of $x^2/(x^3 + 8)$. Using the substitution $u = x^3 + 8$, $du = 3x^2\,dx$, we obtain

$$\int \frac{x^2}{x^3+8}\,dx = \frac{1}{3}\int \frac{1}{u}\,du = \frac{1}{3}\ln|u| + C = \frac{1}{3}\ln|x^3+8| + C.$$

Now,

$$\int_1^b \frac{x^2}{x^3+8}\,dx = \frac{1}{3}\ln|x^3+8|\,\Big|_1^b = \frac{1}{3}\ln(b^3+8) - \frac{1}{3}\ln 9.$$

Finally, we examine what happens as $b \to \infty$. Certainly, $b^3 + 8$ gets arbitrarily large, and so $\ln(b^3 + 8)$ must also get arbitrarily large. Hence

$$\int_1^b \frac{x^2}{x^3+8}\,dx$$

has no limit as $b \to \infty$, so the improper integral

$$\int_1^\infty \frac{x^2}{x^3 + 8}\,dx$$

is divergent.

3.

$$\int_a^{-2} \frac{1}{x^4}\,dx = \int_a^{-2} x^{-4}\,dx = \left.\frac{x^{-3}}{-3}\right|_a^{-2} = \left.\frac{1}{-3x^3}\right|_a^{-2}$$

$$= \frac{1}{-3(-2)^3} - \left(\frac{1}{-3 \cdot a^3}\right)$$

$$= \frac{1}{24} + \frac{1}{3a^3}.$$

$$\int_{-\infty}^{-2} \frac{1}{x^4}\,dx = \lim_{a \to -\infty} \int_a^{-2} \frac{1}{x^4}\,dx = \lim_{a \to -\infty} \left(\frac{1}{24} - \frac{1}{3a^3}\right) = \frac{1}{24}.$$

16.7. Applications of Calculus to Probability

Consider a cell population that is growing vigorously. Suppose that when a cell is 3 days old it divides and forms two new "daughter" cells. If the population is sufficiently large, it will contain cells of many different ages between 0 and 3, and it will turn out that the proportion of cells of various ages remains constant. That is, if a and b are any two numbers between 0 and 3, with $a < b$, then the proportion of cells whose ages lie between a and b will be essentially constant from one moment to the next, even though individual cells are aging and new cells are being formed all the time. In fact, biologists have found that under the ideal circumstances described, the proportion of cells whose ages are between a and b is given by the area under the graph of the function $f(x) = 2ke^{-kx}$ from $x = a$ to $x = b$, where $k = (\ln 2)/3$.* (See Fig. 1.)

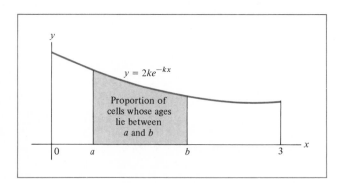

FIGURE 1   Age Distribution of Cells

* See J. R. Cook and T. W. James, "Age Distribution of Cells in Logarithmically Growing Cell Populations," in *Synchrony in Cell Division and Growth*, Erik Zeuthen, ed. (New York: John Wiley & Sons, 1964), pp. 485–495.

Now consider an experiment where we select a cell at random from the population and observe its age, X. Then the probability or likelihood* that X lies between a and b is given by the area under the graph of $f(x) = 2ke^{-kx}$ from a to b, as in Fig. 1. Let us denote this probability by $\Pr(a \leq X \leq b)$. Using the fact that the area under the graph of $f(x)$ is given by a definite integral, we have

$$\Pr(a \leq X \leq b) = \int_a^b f(x)\,dx = \int_a^b 2ke^{-kx}\,dx. \tag{1}$$

The function $f(x)$ that determines the probability in (1) for each a and b is called the *probability density function* of X (or of the experiment whose outcome is X).

The general situation we wish to describe in this section concerns an experiment whose outcome is a number X in a certain interval, say between A and B. For the cell population above, $A = 0$ and $B = 3$. Another typical experiment might consist of choosing a decimal X at random between $A = 5$ and $B = 6$. Or, one could select a random telephone call at some telephone switchboard and observe its duration, X. If we have no way of knowing how long a call might last, then X might be any nonnegative number. In this case it is convenient to say that X lies between 0 and ∞ and to take $A = 0$ and $B = \infty$. A similar situation arises in reliability studies where one measures the lifetime X of a transistor selected at random from a manufacturer's production line. Again, the possible values of X lie between 0 and ∞.

When we are dealing with experiments such as those described above, many questions can be reduced to calculating the probability that the outcome X lies between two specified numbers, say a and b. This probability, $\Pr(a \leq X \leq b)$, is a measure of the likelihood that an outcome of the experiment will lie between a and b. If the experiment is repeated many times, then the proportion of times X has a value between a and b will be close to $\Pr(a \leq X \leq b)$. In experiments of practical interest, it is often possible to find a density function $f(x)$ such that

$$\Pr(a \leq X \leq b) = \int_a^b f(x)\,dx, \tag{2}$$

for all a and b in the range of possible values of X.

Any function $f(x)$ with the following two properties is said to be a (*probability*) *density function:*

I. $f(x) \geq 0,\ A \leq x \leq B$
II. $\int_A^B f(x)\,dx = 1$

The additional condition (2) above is what relates a density function to the outcome X of a specific experiment. Graphically, properties I and II mean that for x between A and B, the graph of $f(x)$ must lie above or on the x-axis and the area under the graph must equal 1. Property II simply says that there is probability 1 (certainty) that X has a value between A and B. Of course, if $B = \infty$, then the integral in property II is an improper integral.

---

*For our purposes, it is sufficient to think of probability in the following intuitive terms. Suppose that an experiment with observed outcome X is repeated very often. Then the probability $\Pr(a \leq X \leq b)$ is given (approximately) as the fraction of repetitions in which X was between a and b.

EXAMPLE 1 Consider the cell population described earlier. Let $f(x) = 2ke^{-kx}$, where $k = (\ln 2)/3$. Show that $f(x)$ is indeed a probability density function on the interval from $x = 0$ to $x = 3$.

Solution Clearly $f(x) \geq 0$, since the exponential function is never negative. Thus property I is satisfied. For property II, we check that

$$\int_0^3 f(x)\,dx = \int_0^3 2ke^{-kx}\,dx = -2e^{-kx}\Big|_0^3 = -2e^{-k \cdot 3} + 2e^0$$

$$= 2 - 2e^{-[(\ln 2)/3]3} = 2 - 2e^{-\ln 2}$$

$$= 2 - 2(e^{\ln 2})^{-1} = 2 - 2(2)^{-1} = 2 - 1 = 1.$$

The simplest probability density function is that which assumes a constant value for $A \leq x \leq B$. In order for property II to hold, this constant value must be $1/(B - A)$; that is,

$$f(x) = \frac{1}{B - A}, \quad A \leq x \leq B.$$

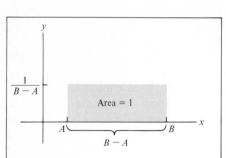

FIGURE 2 A Uniform Density Function.

For in this case the area under the graph of $f(x)$ in Fig. 2 is 1. Such a probability density function is said to be *uniform*. Experiments having uniform density functions generalize those experiments with a finite number of equally likely outcomes.

EXAMPLE 2 Suppose that a subway train leaves the station every 15 minutes. A person who arrives at a random time during the day must wait between 0 and 15 minutes for the next train. What are the chances the person must wait at least 10 minutes?

Solution Let X be the number of minutes the person must wait. The statement that the person arrives at a "random" time is usually interpreted to mean that X has a uniform probability density function. Since the possible values of X lie between 0 and 15, we take $f(x) = \frac{1}{15}$. Then the probability that the person waits at least 10 minutes is given by

$$\Pr(10 \leq X \leq 15) = \int_{10}^{15} \frac{1}{15}\,dx = \frac{1}{15}x\Big|_{10}^{15} = \frac{15}{15} - \frac{10}{15} = \frac{1}{3}.$$

(See Fig. 3.)

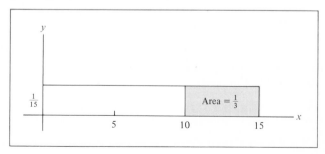

FIGURE 8

As we mentioned earlier, for some experiments the possible outcomes are the numbers between 0 and ∞. In such a case the probability density function $f(x)$ is defined for all $x \geq 0$, and property II is written as

$$\int_0^\infty f(x)\,dx = 1.$$

The most important function of this type has the form $f(x) = \lambda e^{-\lambda x}$, where λ is a positive constant. Just as in Example 1 of Section 6, one easily verifies that

$$\int_0^\infty \lambda e^{-\lambda x}\,dx = 1.$$

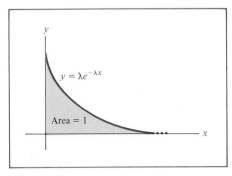

(See Fig. 4.) If the outcome X of an experiment has such a probability density function, the experiment is said to be *exponential* (or *exponentially distributed*). It can be shown that the constant λ may be interpreted as

$$\lambda = \frac{1}{a}, \qquad \text{where } a = \text{average value of } X.$$

Typical uses of exponential probability density functions are given in the next two examples.

FIGURE 4   Exponential Density Function.

EXAMPLE 3   Experiment has shown that the lifetime of a light bulb is exponentially distributed. Let X be the lifetime of a light bulb selected at random from the production line of a light bulb manufacturer. For simplicity, let us measure the lifetime in years, rather than hours, and suppose that the average light bulb produced by the manufacturer burns out in $\frac{1}{4}$ year of continuous use.

(a) What proportion of the light bulbs will burn out within $\frac{1}{2}$ year?

(b) What proportion will continue to burn for at least 1 year?

Solution   The average value of X is $\frac{1}{4}$, so we let $\lambda = 4$ and $f(x) = 4e^{-4x}$.

(a) The proportion of light bulbs where X is less than or equal to $\frac{1}{2}$ is

$$\Pr(0 \leq X \leq \tfrac{1}{2}) = \int_0^{1/2} 4e^{-4x}\,dx = -e^{-4x}\Big|_0^{1/2} = -e^{-2} + 1$$

$$= 1 - .13534 = .86466.$$

(b) The proportion of light bulbs that do not burn out for at least 1 year is

$$\Pr(1 \leq X < \infty) = \int_1^\infty 4e^{-4x}\,dx.$$

To evaluate this improper integral, we compute

$$\int_1^b 4e^{-4x}\,dx = -e^{-4x}\Big|_1^b = -e^{-4b} + e^{-4} \to e^{-4}$$

as $b \to \infty$. Hence $\Pr(1 \leq X \leq \infty) = e^{-4} = .01832$.

EXAMPLE 4 A company makes a survey of the duration of telephone calls made by its employees. It finds that the lengths of calls are exponentially distributed, with the average call lasting 5 minutes. What is the probability that a randomly chosen call will last between 5 and 10 minutes?

Solution Let X be the length of the call. Since the average value of X is 5, we take $\lambda = \frac{1}{5} = .2$ and $f(x) = .2e^{-.2x}$. The desired probability is

$$\Pr(5 \le X \le 10) = \int_5^{10} .2e^{-.2x}\,dx = -e^{-.2x}\Big|_5^{10}$$
$$= -e^{-2} + e^{-1}$$
$$= -.13534 + .36788 = .23254.$$

PRACTICE PROBLEMS 7

1. An experimenter determines that the probability density function for a certain experiment with outcomes between 0 and 1 is given by $f(x) = x$. Why might you doubt his conclusion?

2. An experiment with outcomes between 0 and 1 has probability density function $f(x) = 4x^3$. What is the probability of an outcome larger than $\frac{1}{2}$?

EXERCISES 7

1. An experiment has the probability density function $f(x) = 6(x - x^2)$ and outcomes lying between 0 and 1. Determine the probability that an outcome

   (a) lies between $\frac{1}{4}$ and $\frac{1}{2}$
   (b) lies between 0 and $\frac{1}{3}$
   (c) is at least $\frac{1}{4}$
   (d) is at most $\frac{3}{4}$

2. Suppose that the outcome X of an experiment lies between 0 and 4, and the probability function for X is $f(x) = \frac{1}{8}x$. Find.

   (a) $\Pr(X \le 1)$
   (b) $\Pr(2 \le X \le 2.5)$
   (c) $\Pr(3.5 \le X)$

3. If $f(x) = kx^2$, determine the value of k that makes $f(x)$ a probability density function on $0 \le x \le 2$.

4. If $f(x) = k/\sqrt{x}$, determine the value of k that makes $f(x)$ a probability density function on $1 \le x \le 4$.

5. Suppose that the outcomes X of an experiment lie between 0 and ∞, and X has an exponential density function $f(x) = 2e^{-2x}$. Find.

   (a) $\Pr(X \le .1)$ (b) $\Pr(.1 \le X \le .5)$ (c) $\Pr(1 \le X)$ (d) the average value of X

6. Suppose that the outcomes X of an experiment are exponentially distributed, with density function $f(x) = .25e^{-.25x}$. Find.

   (a) $\Pr(1 \le X \le 2)$ (b) $\Pr(X \le 3)$ (c) $\Pr(4 \le X)$ (d) the average value of X

7. An automated machine produces an automobile part every 3 minutes. An inspector arrives at a random time and must wait X minutes for a part.

(a) Find the probability density function for X.

(b) Find the probability that the inspector must wait at least 1 minute.

(c) Find the probability that the inspector must wait no more than 1 minute.

8. The annual income of the households in a certain community ranges between 5 and 25 thousand dollars. Let X represent the annual income (in thousands of dollars) of a household chosen at random in this community, and suppose that the probability density function for X is $f(x) = kx$, $5 \le x \le 25$.

(a) Find the value of k that makes $f(x)$ a density function.

(b) Find the fraction of the households that have an annual income between 5 and 10 thousand dollars.

(c) What fraction of the households have an income exceeding $20,000?

9. Suppose that in a certain farming region, and in a certain year, the number X of bushels of wheat produced on a given acre has a probability density function $f(x) = (x - 30)/50$, $30 \le x \le 40$.

(a) What is the probability that an acre selected at random produced less than 35 bushels of wheat?

(b) If the farming region had 20,000 acres of wheat, how many acres produced less than 35 bushels of wheat?

10. The parent corporation for a franchised chain of fast-food restaurants claims that the fraction X of their new restaurants that make a profit during their first year of operation has the probability density $f(x) = 12x^2 - 12x^3$, $0 \le x \le 1$.

(a) What is the likelihood that less than 40% of the restaurants opened this year will make a profit during their first year of operation?

(b) What is the likelihood that more than 50% of the restaurants will make a profit during their first year of operation?

11. Suppose that at a certain supermarket, the amount of time X one must wait at the express lane has the probability density function $f(x) = \frac{11}{10}(x + 1)^{-2}$, $0 \le x \le 10$. Find the probability of having to wait less than 4 minutes at the express lane.

12. Suppose that in a certain cell population, cells divide every 10 days and the age X of a cell selected at random has the probability density function $f(x) = 2ke^{-kx}$, $0 \le x \le 10$, where $k = (.1) \ln 2$.

(a) Find the probability that a cell is at most 5 days old.

(b) Upon examining a slide, a microbiologist finds that 10% of the cells are undergoing mitosis (a change in the cell leading to division). Compute the length of time required for mitosis; that is, find the number M such that

$$\int_{10-M}^{10} 2ke^{-kx}\, dx = .10.$$

13. At a certain gas station, it takes an average of 4 minutes to get serviced. Suppose that the service time X for a car has an exponential probability density function.

(a) What fraction of the cars are serviced within 2 minutes?

(b) What is the probability that a car will have to wait at least 4 minutes?

14. The emergency flasher on an automobile is guaranteed for the first 12,000 miles that the car is driven. On the average, the flashers last about 50,000 miles. Let X be the time of failure of the flasher (measured in thousands of miles), and suppose X has an exponential probability density function. What percentage of the emergency flashers will have to be replaced during the warranty period?

15. Let X be the number of seconds between successive cars at a toll booth on the Ohio Turnpike on a typical Saturday afternoon. It can be shown that X has an exponential density function. If the average interarrival time is 2 seconds, find the probability that X is at least 3 seconds.

16. Let X be the relief time (in minutes) of an arthritic patient who has taken an analgesic for pain. Suppose that a certain analgesic provides relief within 4 minutes for 75% of a large group of patients, and suppose the density function for X is $f(x) = ke^{-kx}$. (This model has been used by some medical researchers.) Then one estimates that $\Pr(X \le 4) = .75$. Use this estimate to find an approximate value for k. [*Hint:* First show that $\Pr(X \le 4) = 1 - e^{-4k}$.]

SOLUTIONS TO PRACTICE PROBLEMS 7

1. Property II is not satisfied:

$$\int_0^1 f(x)\,dx = \int_0^1 x\,dx = \left.\frac{x^2}{2}\right|_0^1 = \frac{1}{2}.$$

Property II says that this integral should be 1.

2. $\int_{1/2}^1 4x^3\,dx = x^4\Big|_{1/2}^1 = 1 - \frac{1}{16} = \frac{15}{16}.$

Chapter 16: CHECKLIST

☐ Antiderivative

☐ Indefinite integral

☐ $\int x^r\,dx = \dfrac{1}{r+1}x^{r+1} + C, \qquad r \ne -1$

☐ $\int e^{kx}\,dx = \dfrac{1}{k}e^{kx} + C, \qquad k \ne 0$

☐ $\int \dfrac{1}{x}\,dx = \ln|x| + C, \qquad x \ne 0$

☐ $\int [f(x) + g(x)]\,dx = \int f(x)\,dx + \int g(x)\,dx$

- $\int kf(x)\,dx = k\int f(x)\,dx$

- Net change: $F(x)\big|_a^b$

- $\int_a^b f(x)\,dx = F(b) - F(a),\ F'(x) = f(x)$

- Fundamental Theorem of Calculus

- Area between two curves: $\int_a^b [f(x) - g(x)]\,dx$

- Riemann sum approximation to $\int_a^b f(x)\,dx$

- Rectangle rule

- Average value of $f(x)$: $\dfrac{1}{b-a}\int_a^b f(x)\,dx$

- $\int f(g(x))g'(x)\,dx = \int f(u)\,du, \qquad u = g(x)$

- $\int f(x)g(x)\,dx = f(x)G(x) - \int f'(x)G(x)\,dx, \qquad G'(x) = g(x)$

- Improper integral (convergent and divergent)
- Probability density function
- Uniform density function
- Exponential density function

Chapter 16: SUPPLEMENTARY EXERCISES

Calculate each of the following indefinite integrals.

1. $\displaystyle\int e^{-x/2}\,dx$

2. $\displaystyle\int \frac{5}{\sqrt{x-7}}\,dx$

3. $\displaystyle\int \frac{x^2 - 1}{(x^3 - 3x + 2)^2}\,dx$

4. $\displaystyle\int x^4 e^{-x^5}\,dx$

5. $\displaystyle\int (2x + 3)^7\,dx$

6. $\displaystyle\int \left(9 - 4e^x + \frac{1}{x^4}\right)dx$

7. $\displaystyle\int (e^x + 4)^3 \cdot e^x\,dx$

8. $\displaystyle\int \frac{x - e^{-2x}}{x^2 + e^{-2x}}\,dx$

9. $\displaystyle\int (2x + 1)e^{-x/2}\,dx$

10. $\displaystyle\int \frac{5x}{\sqrt{x-7}}\,dx$

11. $\displaystyle\int \frac{x^2 - 1}{x^3 - 3x + 2}\,dx$

12. $\displaystyle\int x^3 e^{x^2}\,dx$

13. $\displaystyle\int x(2x + 3)^7\,dx$

14. $\displaystyle\int x^{-2}\ln x\,dx$

15. Find the function $f(x)$ for which $f'(x) = (x - 5)^2,\ f(8) = 2$.

16. Find the function $f(x)$ for which $f'(x) = e^{-5x}$, $f(0) = 1$.

Calculate the following definite integrals.

17. $\int_1^4 \frac{1}{x^2} \, dx$

18. $\int_3^6 e^{2-(x/3)} \, dx$

19. $\int_{-1}^1 x^4 e^{x^5-1} \, dx$

20. $\int_0^5 (5 + 3x)^{-1} \, dx$

21. $\int_0^1 \frac{e^x - e^{-x}}{e^x + e^{-x}} \, dx$

22. $\int_2^3 \frac{x-1}{(x^2 - 2x + 1)^2} \, dx$

23. Find the area under the curve $y = 1 + \sqrt{x}$ from $x = 1$ to $x = 9$.

24. Find the area under the curve $y = (3x - 2)^{-3}$ from $x = 1$ to $x = 2$.

25. Find the area of the region bounded by the curves $y = 16 - x^2$ and $y = 10 - x$.

26. Find the area of the region bounded by the curves $y = x^3 - 3x + 1$ and $y = x + 1$.

27. Find the area of the region between the curves $y = 5x + (1/x)$ and $y = 2x + (1/x)$ from $x = 2$ to $x = 3$.

28. Find the area of the region between the curves $y = 2x^2 + x$ and $y = x^2 + 2$ from $x = 0$ to $x = 2$.

29. Use the rectangle rule with $n = 2$ to approximate $\int_2^4 \frac{1}{x+2} \, dx$. Then find the exact value to five decimal places.

30. Use the rectangle rule with $n = 5$ to approximate $\int_0^1 e^{x^2} \, dx$.

31. Find the consumers' surplus for the demand curve $p = \sqrt{25 - .04x}$ at the sales level $x = 400$.

32. Three thousand dollars is deposited in the bank at 6% interest compounded continuously. What will be the average value of the money in the account during the next 10 years?

33. Find the average value of $f(x) = 1/x^3$ from $x = \frac{1}{3}$ to $x = \frac{1}{2}$.

34. Suppose that the interval $0 \le x \le 1$ is divided into 100 subintervals of width $\Delta x = .01$. Show that the sum $[3e^{-.01}]\Delta x + [3e^{-.02}]\Delta x + [3e^{-.03}]\Delta x + \cdots + [3e^{-1}]\Delta x$ is close to $3(1 - e^{-1})$.

35. An airplane tire plant finds that its marginal cost of producing tires is $.04x + 150$ dollars at a production level of x tires per day. If fixed costs are $500 per day, find the cost of producing x tires per day.

36. Suppose the marginal revenue function for a company is $400 - 3x^2$. Find the additional revenue received from doubling production if currently 10 units are being produced.

37. Let $y = f(t)$ be a function such that $y' = kty$ for some constant k. Show that $y = Ce^{kt^2/2}$ for some constant C. [*Hint:* Use the product rule to evaluate $\frac{d}{dt} [f(t)e^{-kt^2/2}]$, and then apply Theorem II of Section 1.]

38. The rate of production of copper ore in North America has been $2.6e^{.034t}$ million metric tons per year at time t, where $t = 0$ corresponds to 1975. Find a formula for the total production from 1980 to time t.

39. The known U.S. natural gas reserves are decreasing at the rate of $R(t)$ quadrillion Btu per year at time t, where $t = 0$ corresponds to 1980 and $R(t) = 22.3e^{-.14t}$.

(a) If the known reserves in 1980 are 159 quadrillion Btu, what will be the reserves in 1986, assuming no new gas sources are discovered?

(b) Compute $\int_0^\infty R(t)\,dt$ and interpret this quantity.

40. The United States has been consuming sulfur at the rate of $R(t)$ million long tons per year at time t, where $t = 0$ corresponds to 1980 and $R(t) = 12.4e^{.035t}$.

(a) How much sulfur will be consumed in the United States between 1981 and 1985?

(b) How much sulfur was consumed in the United States between 1975 and 1980?

41. A drug is injected into a patient at the rate of $f(t)$ cubic centimeters per minute at time t. What does the area under the graph of $y = f(t)$ from $t = 0$ to $t = 4$ represent?

42. Find the area under the graph of $y = e^x/(1 + e^x)^2$ for $x \geq 0$.

Evaluate the following improper integrals whenever they are convergent.

43. $\int_0^\infty e^{6-3x}\,dx$

44. $\int_1^\infty x^{-2/3}\,dx$

45. $\int_{-\infty}^0 \dfrac{8}{(5-2x)^3}\,dx$

46. $\int_0^\infty x^2 e^{-x^3}\,dx$

47. Show that the region under the graph of $y = x^{-1/4}$ for $x \geq 1$ cannot be assigned any finite number as its area.

48. Can the region under the graph of $y = 1/x$ for $x \geq 1$ be assigned a finite number as its area?

49. Suppose that the average lifetime of an electronic component is 72 months, and the lifetimes are exponentially distributed.

(a) Find the probability that a component lasts for more than 24 months.

(b) The *reliability function* $r(t)$ gives the probability that a component will last for more than t months. Compute $r(t)$ in this case.

50. Verify that for any number A, the function $f(x) = e^{A-x}$, $x \geq A$, is a probability density function.

Functions
of Several Variables

Chapter 17

17.1. Examples of Functions of Several Variables

In Chapters 11 through 16 we were concerned with calculus for functions of one variable. In the present chapter we show briefly how calculus can be extended to functions of several variables.

A function $f(x, y)$ of the two variables x and y is a rule that associates to each pair of values for the variables a number. Examples of functions of two variables are

$$f(x, y) = x^2 + 2y$$

$$g(x, y) = e^x(x + y).$$

An example of a function of three variables is

$$f(x, y, z) = 5xyz.$$

It is possible to consider functions of any number of variables.

EXAMPLE 1 Let $f(x, y) = xy + 5y - x^2$.

(a) Calculate $f(2, 3)$. That is, calculate the value of $f(x, y)$ when $x = 2$ and $y = 3$.

(b) Calculate $f(1, 0)$ and $f(0, 1)$.

(c) Calculate $f(x, y)$ when $(x, y) = (-4, 6)$—that is, when $x = -4$ and $y = 6$.

Solution (a) $f(2, 3) = 2 \cdot 3 + 5 \cdot 3 - (2)^2 = 17$.

(b) $f(1, 0) = 1 \cdot 0 + 5 \cdot 0 - 1^2 = -1$.
$f(0, 1) = 0 \cdot 1 + 5 \cdot 1 - 0^2 = 5$.

(c) $f(-4, 6) = (-4) \cdot 6 + 5 \cdot 6 - (-4)^2 = -10$.

A function $f(x, y)$ of two variables may be graphed in a manner analogous to that for functions of one variable. It is necessary to use a three-dimensional coordinate system, where each point is identified by three coordinates (x, y, z). For each choice of x, y, the graph of $f(x, y)$ includes the point $(x, y, f(x, y))$. The graph of $f(x, y)$ is thus a surface in three-dimensional space. (See Fig. 1.) We will

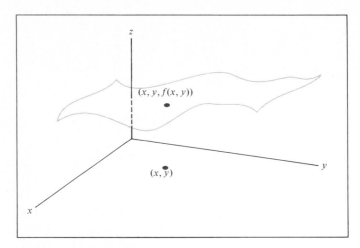

FIGURE 1   Graph of $f(x, y)$.

not employ graphical analysis to study $f(x, y)$ for two reasons. First, it is fairly difficult to draw good graphs in three-dimensional space. This fact makes graphical analysis inconvenient for those not artistically inclined. (See Fig. 2.*) Second, and more significantly, however, we will not use graphical analysis because it is a tool that exists only for functions of one or two variables. It is not possible to graph a function of three or more variables.

Throughout this book we have shown how functions of one variable arise in applications. Functions of several variables are equally important. In Example 2 we present an application to architectural design and in Example 3 an application to economics. These two examples will recur throughout the chapter to illustrate the various techniques that will be developed.

In designing a building, it is necessary to know, at least approximately, how much heat the building loses per day. The heat loss affects many aspects of the design, such as the size of the heating plant, the size and location of duct work, and so on. A building loses heat through its sides, roof, and floor. How much heat is lost will generally differ for each face of the building and will depend on such factors as insulation, materials used in construction, exposure (north, south, east, or west), and climate. It is possible to estimate how much heat is lost per square foot of each face. Using this data, one can construct a heat-loss function as in the following example.

---

* These graphs were drawn by Norton Starr at the University of Waterloo Computing Centre.

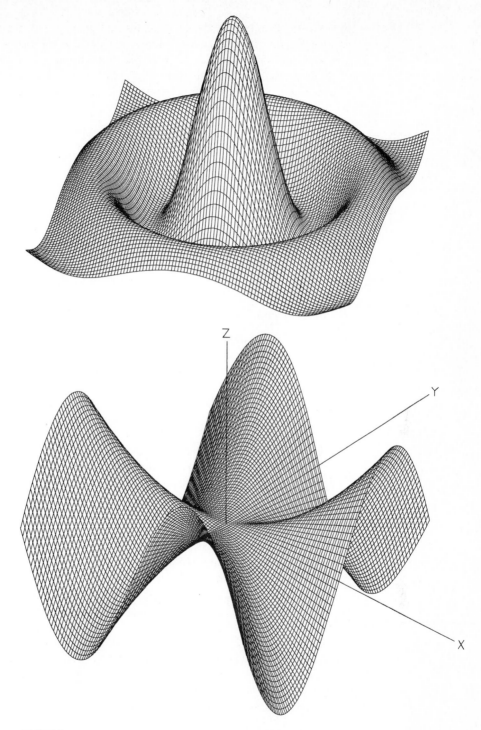

FIGURE 2

EXAMPLE 2 A rectangular industrial building of dimensions x, y, and z is shown in Fig. 3(a). In Fig. 3(b) we give the amount of heat lost per day by each side of the building, measured in suitable units of heat per square foot. Let $f(x, y, z)$ be the total daily heat loss for such a building.

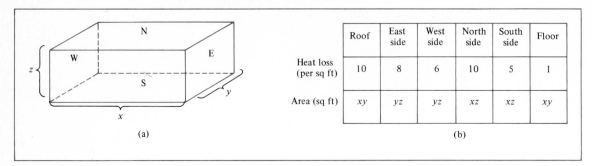

	Roof	East side	West side	North side	South side	Floor
Heat loss (per sq ft)	10	8	6	10	5	1
Area (sq ft)	xy	yz	yz	xz	xz	xy

(a) (b)

FIGURE 3

(a) Find a formula for $f(x, y, z)$.

(b) Find the total daily heat loss if the building has length 100 feet, width 70 feet, and height 50 feet.

Solution (a) The total heat loss is the sum of the amount of heat loss through each side of the building. The heat loss through the roof is

[heat loss per square foot of roof] · [area of roof in square feet] $= 10xy$.

Similarly, the heat loss through the east side is $8yz$. Continuing in this way, we see that the total daily heat loss is

$$f(x, y, z) = 10xy + 8yz + 6yz + 10xz + 5xz + 1 \cdot xy.$$

We collect terms to obtain

$$f(x, y, z) = 11xy + 14yz + 15xz.$$

(b) The amount of heat loss when $x = 100$, $y = 70$, and $z = 50$ is given by $f(100, 70, 50)$, which equals

$$f(100, 70, 50) = 11(100)(70) + 14(70)(50) + 15(100)(50)$$

$$= 77{,}000 + 49{,}000 + 75{,}000 = 201{,}000.$$

In Section 3 we will determine the dimensions x, y, z that minimize the heat loss for a building of specified volume.

The costs of a manufacturing process can generally be classified as one of two types: cost of labor and cost of capital. The meaning of the cost of labor is clear. By the cost of capital, we mean the cost of buildings, tools, machines, and similar items used in the production process. A manufacturer usually has some control over the relative portions of labor and capital utilized in his production process. He can completely automate production so that labor is at a minimum, or he can utilize mostly labor and little capital. Suppose that x units of labor and y units of capital are used.* Let $f(x, y)$ denote the number of units of finished product that are manufactured. Economists have found that $f(x, y)$ is often a function of the form

$$f(x, y) = Cx^A y^{1-A},$$

where A and C are constants, $0 < A < 1$. Such a function is called a *Cobb-Douglas production function.*

EXAMPLE 3   (*Production in a Firm*)   Suppose that during a certain time period the number of units of goods produced when utilizing x units of labor and y units of capital is $f(x, y) = 60x^{3/4}y^{1/4}$.

(a) How many units of goods will be produced by using 81 units of labor and 16 units of capital?

(b) Show that whenever the amounts of labor and capital being used are doubled, so is the production. (Economists say that the production function has "constant returns to scale.")

Solution   (a) $f(81, 16) = 60(81)^{3/4} \cdot (16)^{1/4} = 60 \cdot 27 \cdot 2 = 3240$. There will be 3240 units of goods produced.

(b) Utilization of a units of labor and b units of capital results in the production of $f(a, b) = 60a^{3/4}b^{1/4}$ units of goods. Utilizing $2a$ and $2b$ units of labor and capital, respectively, results in $f(2a, 2b)$ units produced. Set $x = 2a$ and $y = 2b$. Then we see that

$$\begin{aligned}
f(2a, 2b) &= 60(2a)^{3/4}(2b)^{1/4} \\
&= 60 \cdot 2^{3/4} \cdot a^{3/4} \cdot 2^{1/4} \cdot b^{1/4} \\
&= 60 \cdot 2^{(3/4 + 1/4)} \cdot a^{3/4}b^{1/4} \\
&= 2^1 \cdot 60a^{3/4}b^{1/4} \\
&= 2f(a, b).
\end{aligned}$$

Level Curves   It is possible graphically to depict a function $f(x, y)$ of two variables using a family of curves called level curves. Let c be any number. Then the graph of the equation $f(x, y) = c$ is a curve in the xy-plane called the *level curve of height c*. This curve describes all points of height c on the graph of the

---

* Economists normally use L and K, respectively, for labor and capital. However, for simplicity, we use x and y.

function $f(x, y)$. As c varies, we have a family of level curves indicating the sets of points on which $f(x, y)$ assumes various values c. In Fig. 4, we have drawn the graph and various level curves for the function $f(x, y) = x^2 + y^2$.

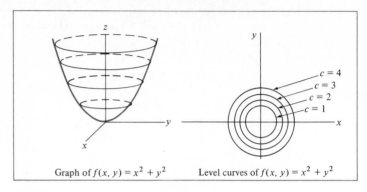

Graph of $f(x, y) = x^2 + y^2$ Level curves of $f(x, y) = x^2 + y^2$

FIGURE 4

Level curves often have interesting physical interpretations. For example, let (x, y) specify the coordinates of a point on the earth (x = latitude, y = longitude) and $f(x, y)$ the current temperature at location (x, y). Then the level curves of the function $f(x, y)$ indicate the locations having equal temperatures. Such curves are called *isotherms*. In Fig. 5 we have drawn several typical isotherms.

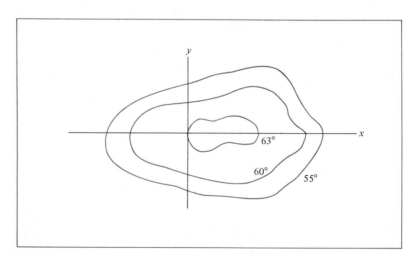

FIGURE 5

EXAMPLE 4   Determine the level curve at height 600 for the production function $f(x, y) = 60x^{3/4}y^{1/4}$ of Example 3.

Solution   The level curve is the graph of $f(x, y) = 600$, or

$$60x^{3/4}y^{1/4} = 600$$

$$y^{1/4} = \frac{10}{x^{3/4}}$$

$$y = \frac{10{,}000}{x^3}.$$

Of course, since x and y represent quantities of labor and capital, they must both be positive. We have sketched the graph of the level curve in Fig. 6. The points

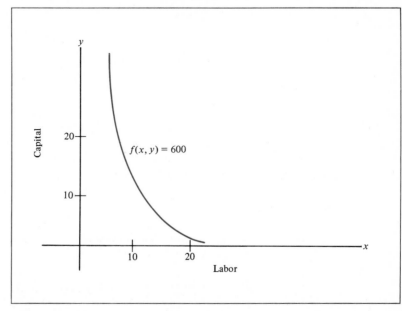

FIGURE 6

on the curve are precisely those combinations of capital and labor which yield 600 units of production. Since all points on the curve yield the same amount of production, the level curve is also called an *isoquant* ("iso" means equal, "quant" means "amount").

PRACTICE PROBLEMS 1

1. Let $f(x, y, z) = x^2 + y/(x - z) - 4$. Compute $f(3, 5, 2)$.

2. Suppose that in a certain country the daily demand for coffee is given by $f(p_1, p_2) = 16p_1/p_2$ thousand pounds, where p_1 and p_2 are the respective prices of tea and coffee per pound. Compute and interpret $f(3, 4)$.

EXERCISES 1

1. Let $f(x, y) = x^2 + 8y$. Compute $f(1, 0)$, $f(0, 1)$, and $f(3, 2)$.

2. Let $g(x, y) = 3xe^y$. Compute $g(2, 1)$, $g(1, 0)$, and $g(0, 0)$.

3. Let $f(L, K) = 3\sqrt{LK}$. Compute $f(0, 1)$, $f(3, 12)$, and $f(a, b)$.

4. Let $f(p, q) = pe^{q/p}$. Compute $f(1, 0)$, $f(3, 12)$, and $f(a, b)$.

5. Let $f(x, y, z) = x/(y - z)$. Compute $f(2, 3, 4)$ and $f(7, 46, 44)$.

6. Let $f(x, y, z) = x^2 e^{\sqrt{y^2 + z^2}}$. Compute $f(1, 0, 1)$ and $f(5, 2, 3)$.

7. Let $f(x, y) = xy$. Show that $f(2 + h, 3) - f(2, 3) = 3h$.

8. Let $f(x, y) = xy$. Show that $f(2, 3 + K) - f(2, 3) = 2K$.

9. Let $f(x, y) = \dfrac{x^2 + 3xy + 3y^2}{x + y}$. Show that $f(2a, 2b) = 2f(a, b)$.

10. Let $f(x, y) = 75x^A y^{1-A}$, where $0 < A < 1$. Show that $f(2a, 2b) = 2f(a, b)$.

11. The present value of A dollars to be paid t years in the future (assuming a 5% continuous interest rate) is $P(A, t) = Ae^{-.05t}$. Find and interpret $P(100, 13.8)$.

12. Refer to Example 3. Suppose that labor costs $100 per unit and capital costs $200 per unit. Express as a function of two variables, $C(x, y)$, the cost of utilizing x units of labor and y units of capital.

Draw the level curves of heights 0, 1, and 2 for each of the following functions.

13. $f(x, y) = 2x + y$

14. $f(x, y) = -x^2 + y$

15. Suppose that a topographic map is viewed as the graph of a certain function $f(x, y)$. What are the level curves?

16. A certain production process uses labor and capital. If the quantities of these commodities are x and y, respectively, then the total cost is $100x + 200y$ dollars. Draw the level curves of height 800, 600, and 8000 for this function. Explain the significance of these curves. (Economists frequently refer to these lines as *budget lines* or *isocost lines*.)

SOLUTIONS TO PRACTICE PROBLEMS 1

1. Substitute 3 for x, 5 for y, and 2 for z.

$$f(3, 5, 2) = 3^2 + \frac{5}{3 - 2} - 4 = 10.$$

2. To compute $f(3, 4)$, substitute 3 for p_1 and 4 for p_2 into $f(p_1, p_2) = 16p_1/p_2$. Thus

$$f(3, 4) = 16 \cdot \tfrac{3}{4} = 12.$$

Therefore, if the price of tea is $3 per pound and the price of coffee is $4 per pound, then 12,000 pounds of coffee will be sold each day. (Notice that as the price of coffee increases, the demand decreases.)

17.2. Partial Derivatives

In Chapter 11 we introduced the notion of a derivative to measure the rate at which a function $f(x)$ is changing with respect to changes in the variable x. Let us now study the analog of the derivative for functions of two (or more) variables.

Let $f(x, y)$ be a function of the two variables x and y. Since we want to know how $f(x, y)$ changes both with respect to changes in the variable x and changes in the variable y, we shall define two derivatives of $f(x, y)$ (to be called "partial derivatives"), one with respect to each of the variables. The *partial derivative of* $f(x, y)$ *with respect to* x, written $\dfrac{\partial f}{\partial x}$, is the derivative of $f(x, y)$, where y is treated as a constant and $f(x, y)$ is considered as a function of x alone. The *partial derivative of* $f(x, y)$ *with respect to* y, written $\dfrac{\partial f}{\partial y}$, is the derivative of $f(x, y)$, where x is treated as a constant.

EXAMPLE 1 Let $f(x, y) = 5x^3y^2$. Compute $\dfrac{\partial f}{\partial x}$ and $\dfrac{\partial f}{\partial y}$.

Solution To compute $\dfrac{\partial f}{\partial x}$, we think of $f(x, y)$ written as

$$f(x, y) = [5y^2]x^3,$$

where the brackets emphasize that $5y^2$ is to be treated as a constant. Therefore, when differentiating with respect to x, $f(x, y)$ is just a constant times x^3. Recall that if k is any constant, then

$$\frac{d}{dx}(kx^3) = 3 \cdot k \cdot x^2.$$

Thus

$$\frac{\partial f}{\partial x} = 3 \cdot [5y^2] \cdot x^2 = 15x^2y^2.$$

After some practice, it is unnecessary to place the y^2 in front of the x^3 before differentiating.

Now, in order to compute $\dfrac{\partial f}{\partial y}$, we think of

$$f(x, y) = [5x^3]y^2.$$

When differentiating with respect to y, $f(x, y)$ is simply a constant (namely $5x^3$) times y^2. Hence

$$\frac{\partial f}{\partial y} = 2 \cdot [5x^3] \cdot y = 10x^3y.$$

EXAMPLE 2 Let $f(x, y) = 3x^2 + 2xy + 5y$. Compute $\dfrac{\partial f}{\partial x}$ and $\dfrac{\partial f}{\partial y}$.

Solution To compute $\dfrac{\partial f}{\partial x}$, we think of

$$f(x, y) = 3x^2 + [2y]x + [5y].$$

Now we differentiate $f(x, y)$ as if it were a quadratic polynomial in x:

$$\frac{\partial f}{\partial x} = 6x + [2y] + 0 = 6x + 2y.$$

Note that $5y$ is treated as a constant when differentiating with respect to x, so the partial derivative of $5y$ with respect to x is zero.

To compute $\frac{\partial f}{\partial y}$, we think of

$$f(x, y) = [3x^2] + [2x]y + 5y.$$

Then

$$\frac{\partial f}{\partial y} = 0 + [2x] + 5 = 2x + 5.$$

Note that $3x^2$ is treated as a constant when differentiating with respect to y, so the partial derivative of $3x^2$ with respect to y is zero.

EXAMPLE 3   Compute $\frac{\partial f}{\partial x}$ and $\frac{\partial f}{\partial y}$ for each of the following.

(a) $f(x, y) = (4x + 3y - 5)^8$

(b) $f(x, y) = e^{xy^2}$

(c) $f(x, y) = y/(x + 3y)$

Solution   (a) To compute $\frac{\partial f}{\partial x}$, we think of

$$f(x, y) = (4x + [3y - 5])^8.$$

By the general power rule,

$$\frac{\partial f}{\partial x} = 8 \cdot (4x + [3y - 5])^7 \cdot 4 = 32(4x + 3y - 5)^7.$$

Here we used the fact that the derivative of $4x + 3y - 5$ with respect to x is just 4.

To compute $\frac{\partial f}{\partial y}$, we think of

$$f(x, y) = ([4x] + 3y - 5)^8.$$

Then

$$\frac{\partial f}{\partial y} = 8 \cdot ([4x] + 3y - 5)^7 \cdot 3 = 24(4x + 3y - 5)^7.$$

(b) To compute $\frac{\partial f}{\partial x}$, we observe that

$$f(x, y) = e^{x[y^2]},$$

so that

$$\frac{\partial f}{\partial x} = [y^2]e^{x[y^2]} = y^2 e^{xy^2}.$$

To compute $\dfrac{\partial f}{\partial y}$, we think of

$$f(x, y) = e^{[x]y^2}.$$

Thus

$$\frac{\partial f}{\partial y} = e^{[x]y^2} \cdot 2[x]y = 2xye^{xy^2}.$$

(c) To compute $\dfrac{\partial f}{\partial x}$, we use the general power rule to differentiate $[y](x + [3y])^{-1}$ with respect to x:

$$\frac{\partial f}{\partial x} = (-1) \cdot [y](x + [3y])^{-2} \cdot 1 = -\frac{y}{(x + 3y)^2}.$$

To compute $\dfrac{\partial f}{\partial y}$, we use the quotient rule to differentiate

$$f(x, y) = \frac{y}{[x] + 3y}$$

with respect to y. We find that

$$\frac{\partial f}{\partial y} = \frac{([x] + 3y) \cdot 1 - y \cdot 3}{([x] + 3y)^2} = \frac{x}{(x + 3y)^2}.$$

The use of brackets to highlight constants is helpful initially in order to compute partial derivatives. From now on we shall merely form a mental picture of those terms to be treated as constants and dispense with brackets.

A partial derivative of a function of several variables is also a function of several variables and hence can be evaluated at specific values of the variables. We write

$$\frac{\partial f}{\partial x}(a, b)$$

for $\dfrac{\partial f}{\partial x}$ evaluated at $x = a, y = b$. Similarly,

$$\frac{\partial f}{\partial y}(a, b)$$

denotes the function $\dfrac{\partial f}{\partial y}$ evaluated at $x = a, y = b$.

EXAMPLE 4 Let $f(x, y) = 3x^2 + 2xy + 5y$.

(a) Calculate $\dfrac{\partial f}{\partial x}(1, 4)$.

(b) Evaluate $\dfrac{\partial f}{\partial y}$ at $(x, y) = (1, 4)$.

Solution (a) $\dfrac{\partial f}{\partial x} = 6x + 2y, \dfrac{\partial f}{\partial x}(1, 4) = 6 \cdot 1 + 2 \cdot 4 = 14.$

(b) $\dfrac{\partial f}{\partial y} = 2x + 5, \dfrac{\partial f}{\partial y}(1, 4) = 2 \cdot 1 + 5 = 7.$

Since $\dfrac{\partial f}{\partial x}$ is simply the ordinary derivative with y held constant, $\dfrac{\partial f}{\partial x}$ gives the rate of change of $f(x, y)$ with respect to x for y held constant. In other words, keeping y constant and increasing x by one (small) unit produces a change in $f(x, y)$ that is approximately given by $\dfrac{\partial f}{\partial x}$. An analogous interpretation holds for $\dfrac{\partial f}{\partial y}$.

EXAMPLE 5 Interpret the partial derivatives of $f(x, y) = 3x^2 + 2xy + 5y$ calculated in Example 4.

Solution We showed in Example 4 that

$$\frac{\partial f}{\partial x}(1, 4) = 14, \qquad \frac{\partial f}{\partial y}(1, 4) = 7.$$

The fact that $\dfrac{\partial f}{\partial x}(1, 4) = 14$ means that if y is kept constant at 4 and x is allowed to vary near 1, then $f(x, y)$ changes at a rate equal to 14 times the change in x. That is, if x increases by one small unit, then $f(x, y)$ increases by approximately 14 units. If x increases by h units (h small), then $f(x, y)$ increases approximately $14 \cdot h$ units. That is, we have

$$f(1 + h, 4) - f(1, 4) \approx 14 \cdot h.$$

Similarly, the fact that $\dfrac{\partial f}{\partial y}(1, 4) = 7$ means that if we keep x constant at 1 and let y vary near 4, then $f(x, y)$ changes at a rate equal to seven times the change in y. So for a small value of k, we have

$$f(1, 4 + k) - f(1, 4) \approx 7 \cdot k.$$

We can generalize the interpretations of $\dfrac{\partial f}{\partial x}$ and $\dfrac{\partial f}{\partial y}$ given in Example 5 to yield the following general fact.

Let $f(x, y)$ be a function of two variables. Then if h and k are small, we have

$$f(a + h, b) - f(a, b) \approx \frac{\partial f}{\partial x}(a, b) \cdot h$$

$$f(a, b + k) - f(a, b) \approx \frac{\partial f}{\partial y}(a, b) \cdot k.$$

Partial derivatives can be computed for functions of any number of variables. When taking the partial with respect to one variable, we treat the other variables as constants.

EXAMPLE 6   Let $f(x, y, z) = x^2yz - 3z$.

(a) Compute $\frac{\partial f}{\partial x}, \frac{\partial f}{\partial y}$, and $\frac{\partial f}{\partial z}$.   (b) Calculate $\frac{\partial f}{\partial z}(2, 3, 1)$.

Solution   (a) $\frac{\partial f}{\partial x} = 2xyz$, $\frac{\partial f}{\partial y} = x^2z$, $\frac{\partial f}{\partial z} = x^2y - 3$.

(b) $\frac{\partial f}{\partial z}(2, 3, 1) = 2^2 \cdot 3 - 3 = 12 - 3 = 9$.

EXAMPLE 7   Let $f(x, y, z)$ be the heat-loss function computed in Example 2 of Section 1. That is, $f(x, y, z) = 11xy + 14yz + 15xz$. Calculate and interpret $\frac{\partial f}{\partial x}(10, 7, 5)$.

Solution   We have

$$\frac{\partial f}{\partial x} = 11y + 15z$$

$$\frac{\partial f}{\partial x}(10, 7, 5) = 11 \cdot 7 + 15 \cdot 5 = 77 + 75 = 152.$$

The quantity $\frac{\partial f}{\partial x}$ is commonly referred to as the *marginal heat loss with respect to change in x*. Specifically, if x is changed from 10 by h units (where h is small) and the values of y and z remain fixed at 7 and 5, then the amount of heat loss will change by approximately $152 \cdot h$ units.

EXAMPLE 8   (*Production*)   Consider the production function $f(x, y) = 60x^{3/4}y^{1/4}$, which gives the number of units of goods produced when utilizing x units of labor and y units of capital.

(a) Find $\frac{\partial f}{\partial x}$ and $\frac{\partial f}{\partial y}$.

(b) Evaluate $\dfrac{\partial f}{\partial x}$ and $\dfrac{\partial f}{\partial y}$ at $x = 81$, $y = 16$.

(c) Interpret the numbers computed in part (b).

Solution (a) $\dfrac{\partial f}{\partial x} = 60 \cdot \dfrac{3}{4} x^{-1/4} \cdot y^{1/4} = 45 x^{-1/4} y^{1/4} = 45 \dfrac{y^{1/4}}{x^{1/4}}$

$\dfrac{\partial f}{\partial y} = 60 \cdot \dfrac{1}{4} x^{3/4} y^{-3/4} = 15 x^{3/4} y^{-3/4} = 15 \dfrac{x^{3/4}}{y^{3/4}}.$

(b) $\dfrac{\partial f}{\partial x}(81, 16) = 45 \cdot \dfrac{16^{1/4}}{81^{1/4}} = 45 \cdot \dfrac{2}{3} = 30$

$\dfrac{\partial f}{\partial y}(81, 16) = 15 \cdot \dfrac{(81)^{3/4}}{(16)^{3/4}} = 15 \cdot \dfrac{27}{8} = \dfrac{405}{8} = 50\tfrac{5}{8}.$

(c) The quantities $\dfrac{\partial f}{\partial x}$ and $\dfrac{\partial f}{\partial y}$ are referred to as the *marginal productivity of labor* and the *marginal productivity of capital*. If the amount of capital is held fixed at $y = 16$ and the amount of labor increases by 1 unit, then the quantity of goods produced will increase by approximately 30 units. Similarly, an increase in capital of 1 unit (with labor fixed at 81) results in an increase in production of approximately $50\tfrac{5}{8}$ units of goods.

Just as we formed second derivatives in the case of one variable, we can form second partial derivatives of a function $f(x, y)$ of two variables. Since $\dfrac{\partial f}{\partial x}$ is a function of x and y, we can differentiate it with respect to x or y. The partial derivative of $\dfrac{\partial f}{\partial x}$ with respect to x is denoted by $\dfrac{\partial^2 f}{\partial x^2}$. The partial derivative of $\dfrac{\partial f}{\partial x}$ with respect to y is denoted by $\dfrac{\partial^2 f}{\partial y\, \partial x}$. Similarly, the partial derivative of the function $\dfrac{\partial f}{\partial y}$ with respect to x is denoted by $\dfrac{\partial^2 f}{\partial x\, \partial y}$, and the partial derivative of $\dfrac{\partial f}{\partial y}$ with respect to y is denoted by $\dfrac{\partial^2 f}{\partial y^2}$. Almost all functions $f(x, y)$ encountered in applications [and all functions $f(x, y)$ in this text] have the property that

$$\frac{\partial^2 f}{\partial y\, \partial x} = \frac{\partial^2 f}{\partial x\, \partial y}.$$

EXAMPLE 9 Let $f(x, y) = x^2 + 3xy + 2y^2$. Calculate $\dfrac{\partial^2 f}{\partial x^2}, \dfrac{\partial^2 f}{\partial y^2}, \dfrac{\partial^2 f}{\partial x\, \partial y}$, and $\dfrac{\partial^2 f}{\partial y\, \partial x}$.

Solution   First we compute $\dfrac{\partial f}{\partial x}$ and $\dfrac{\partial f}{\partial y}$.

$$\frac{\partial f}{\partial x} = 2x + 3y, \qquad \frac{\partial f}{\partial y} = 3x + 4y.$$

To compute $\dfrac{\partial^2 f}{\partial x^2}$, we differentiate $\dfrac{\partial f}{\partial x}$ with respect to x:

$$\frac{\partial^2 f}{\partial x^2} = 2.$$

Similarly, to compute $\dfrac{\partial^2 f}{\partial y^2}$, we differentiate $\dfrac{\partial f}{\partial y}$ with respect to y:

$$\frac{\partial^2 f}{\partial y^2} = 4.$$

To compute $\dfrac{\partial^2 f}{\partial x\,\partial y}$, we differentiate $\dfrac{\partial f}{\partial y}$ with respect to x:

$$\frac{\partial^2 f}{\partial x\,\partial y} = 3.$$

Finally, to compute $\dfrac{\partial^2 f}{\partial y\,\partial x}$, we differentiate $\dfrac{\partial f}{\partial x}$ with respect to y:

$$\frac{\partial^2 f}{\partial y\,\partial x} = 3.$$

PRACTICE PROBLEMS 2

1. The number of TV sets sold per week by an appliance store is given by a function of two variables, $f(x, y)$, where x is the price per TV set and y is the amount of money spent weekly on advertising. Suppose that the current price is $400 per set and that currently $2000 per week is being spent for advertising.

    (a) Would you expect $\dfrac{\partial f}{\partial x}(400, 2000)$ to be positive or negative?

    (b) Would you expect $\dfrac{\partial f}{\partial y}(400, 2000)$ to be positive or negative?

2. The monthly mortgage payment for a house is a function of two variables, $f(A, r)$, where A is the amount of the mortgage and the interest rate is $r\%$. For a 30-year mortgage, $f(92{,}000, 14) = 1090.08$ and $\dfrac{\partial f}{\partial x}(92{,}000, 14) = 72.82$. What is the significance of the number 72.82?

EXERCISES 2

Find $\dfrac{\partial f}{\partial x}$ and $\dfrac{\partial f}{\partial y}$ for each of the following functions.

1. $f(x, y) = 5xy$

2. $f(x, y) = 3x^2 + 2y + 1$

3. $f(x, y) = 2x^2 e^y$

4. $f(x, y) = x + e^{xy}$

5. $f(x, y) = \dfrac{y^2}{x}$

6. $f(x, y) = \dfrac{x}{1 + e^y}$

7. $f(x, y) = (2x - y + 5)^2$

8. $f(x, y) = (9x^2 y + 3x)^{12}$

9. $f(x, y) = x^2 e^{3x} \ln y$

10. $f(x, y) = (x - \ln y)e^{xy}$

11. $f(x, y) = \dfrac{x - y}{x + y}$

12. $f(x, y) = \dfrac{2xy}{e^x}$

13. Let $f(L, K) = 3\sqrt{LK}$. Compute $\dfrac{\partial f}{\partial L}$.

14. Let $f(p, q) = e^{q/p}$. Compute $\dfrac{\partial f}{\partial q}$ and $\dfrac{\partial f}{\partial p}$.

15. Let $f(x, y, z) = (1 + x^2 y)/z$. Compute $\dfrac{\partial f}{\partial x}, \dfrac{\partial f}{\partial y}$, and $\dfrac{\partial f}{\partial z}$.

16. Let $f(x, y, z) = x^2 y + 3yz - z^2$. Compute $\dfrac{\partial f}{\partial x}, \dfrac{\partial f}{\partial y}$, and $\dfrac{\partial f}{\partial z}$.

17. Let $f(x, y, z) = xze^{yz}$. Find $\dfrac{\partial f}{\partial x}, \dfrac{\partial f}{\partial y}$, and $\dfrac{\partial f}{\partial z}$.

18. Let $f(x, y, z) = ze^{z/xy}$. Find $\dfrac{\partial f}{\partial x}, \dfrac{\partial f}{\partial y}$, and $\dfrac{\partial f}{\partial z}$.

19. Let $f(x, y) = x^2 + 2xy + y^2 + 3x + 5y$. Compute $\dfrac{\partial f}{\partial x}(2, -3)$ and $\dfrac{\partial f}{\partial y}(2, -3)$.

20. Let $f(x, y) = xye^{2x-y}$. Evaluate $\dfrac{\partial f}{\partial x}$ and $\dfrac{\partial f}{\partial y}$ at $(x, y) = (1, 2)$.

21. Let $f(x, y, z) = xy^2 z + 5$. Evaluate $\dfrac{\partial f}{\partial y}$ at $(x, y, z) = (2, -1, 3)$.

22. Let $f(x, y, z) = \dfrac{x}{y - z}$. Compute $\dfrac{\partial f}{\partial y}(2, -1, 3)$.

23. Let $f(x, y) = x^3 y + 2xy^2$. Find $\dfrac{\partial^2 f}{\partial x^2}, \dfrac{\partial^2 f}{\partial y^2}, \dfrac{\partial^2 f}{\partial x \, \partial y}$, and $\dfrac{\partial^2 f}{\partial y \, \partial x}$.

24. Let $f(x, y) = xe^y + x^4 y + y^3$. Find $\dfrac{\partial^2 f}{\partial x^2}, \dfrac{\partial^2 f}{\partial y^2}, \dfrac{\partial^2 f}{\partial x \, \partial y}$, and $\dfrac{\partial^2 f}{\partial y \, \partial x}$.

25. A farmer can produce $f(x, y) = 200\sqrt{6x^2 + y^2}$ units of produce by utilizing x units of labor and y units of capital. (The capital is used to rent or purchase land, materials, and equipment.)

   (a) Calculate the marginal productivities of labor and capital when $x = 10$ and $y = 5$.

   (b) Use the result of part (a) to determine the approximate effect on production of utilizing 5 units of capital but cutting back to $9\frac{1}{2}$ units of labor.

26. The productivity of a country is given by $f(x, y) = 300x^{2/3}y^{1/3}$, where x and y are the amounts of labor and capital.

   (a) Compute the marginal productivities of labor and capital when $x = 125$ and $y = 64$.

   (b) What would be the approximate effect of utilizing 125 units of labor but cutting back to 62 units of capital?

27. In a certain suburban community commuters have the choice of getting into the city by bus or train. The demand for these modes of transportation varies with their cost. Let $f(p_1, p_2)$ be the number of people who will take the bus when p_1 is the price of the bus ride and p_2 is the price of the train ride. Explain why $\dfrac{\partial f}{\partial p_1} < 0$ and $\dfrac{\partial f}{\partial p_2} > 0$.

28. The demand for a certain gas-guzzling car is given by $f(p_1, p_2)$, where p_1 is the price of the car and p_2 is the price of gasoline. Explain why $\dfrac{\partial f}{\partial p_1} < 0$ and $\dfrac{\partial f}{\partial p_2} < 0$.

29. Using data collected from 1929–1941, Richard Stone* determined that the yearly quantity Q of beer consumed in the United Kingdom was approximately given by the formula $Q = f(m, p, r, s)$, where

$$f(m, p, r, s) = (1.058)m^{.136}p^{-.727}r^{.914}s^{.816}$$

and m is the aggregate real income (personal income after direct taxes, adjusted for retail price changes), p is the average retail price of the commodity (in this case, beer), r is the average retail price level of all other consumer goods and services, and s is a measure of the strength of the beer. Determine which partial derivatives are positive and which are negative and give interpretations. (For example, since $\dfrac{\partial f}{\partial r} > 0$, people buy more beer when the prices of other goods increase and the other factors remain constant.)

30. Richard Stone (see Exercise 29) determined that the yearly consumption of food in the United States was given by

$$f(m, p, r) = (2.186)m^{.595}p^{-.543}r^{.922}.$$

Determine which partial derivatives are positive and which are negative and give interpretations of these facts.

31. The volume (V) of a certain amount of a gas is determined by the temperature (T) and the pressure (P) by the formula, $V = .08(T/P)$. Calculate and interpret $\dfrac{\partial V}{\partial P}$ and $\dfrac{\partial V}{\partial T}$ when $P = 20$, $T = 300$.

---

* Richard Stone, "The Analysis of Market Demand," *Journal of the Royal Statistical Society*, CVIII (1945), 286–391.

32. For the production function, $f(x, y) = 60x^{3/4}y^{1/4}$, considered in Example 8, think of $f(x, y)$ as the revenue when utilizing x units of labor and y units of capital. Under actual operating conditions, say $x = a$ and $y = b$, $\dfrac{\partial f}{\partial x}(a, b)$ is referred to as the *wage per unit of labor* and $\dfrac{\partial f}{\partial y}(a, b)$ is referred to as the *wage per unit of capital*. Show that

$$f(a, b) = a \cdot \left[\frac{\partial f}{\partial x}(a, b) \right] + b \cdot \left[\frac{\partial f}{\partial y}(a, b) \right].$$

(This equation shows how the revenue is distributed between labor and capital.)

33. Compute $\dfrac{\partial^2 f}{\partial x^2}$ where $f(x, y) = 60x^{3/4}y^{1/4}$, a production function (where x is units of labor). Explain why $\dfrac{\partial^2 f}{\partial x^2}$ is always negative.

34. Compute $\dfrac{\partial^2 f}{\partial y^2}$ where $f(x, y) = 60x^{3/4}y^{1/4}$, a production function (where y is units of capital). Explain why $\dfrac{\partial^2 f}{\partial y^2}$ is always negative.

35. Let $f(x, y) = 3x^2 + 2xy + 5y$, as in Example 5. Show that

$$f(1 + h, 4) - f(1, 4) = 14h + 3h^2.$$

Thus the error in approximating $f(1 + h, 4) - f(1, 4)$ by $14h$ is $3h^2$. (If $h = .01$, for instance, the error is only .0003.)

36. Physicians, particularly pediatricians, sometimes need to know the body surface area of a patient. For instance, the surface area is used to adjust the results of certain tests of kidney performance. Tables are available that give the approximate body surface A in square meters of a person who weighs W kilograms and is H centimeters tall. The following empirical formula* is also used:

$$A = .007W^{.425}H^{.725}.$$

Evaluate $\dfrac{\partial A}{\partial W}$ and $\dfrac{\partial A}{\partial H}$ when $W = 54$, $H = 165$, and give a physical interpretation of your answers. You may use the approximations: $(54)^{.425} \approx 5.4$, $(54)^{-.575} \approx .10$, $(165)^{.725} \approx 40.5$, $(165)^{-.275} \approx .25$.

SOLUTIONS TO PRACTICE PROBLEMS 2

1. (a) Negative. $\dfrac{\partial f}{\partial x}(400, 2000)$ is approximately the change in sales due to a \$1 increase in x (price). Since raising prices lowers sales, we would expect $\dfrac{\partial f}{\partial x}(400, 2000)$ to be negative.

---

* See J. Routh, *Mathematical Preparation for Laboratory Technicians* (Philadelphia: W. B. Saunders Co., 1971), p. 92.

(b) Positive. $\dfrac{\partial f}{\partial y}$ (400, 2000) is approximately the change in sales due to a \$1 increase in advertising. Since spending more money on advertising brings in more customers, we would expect sales to increase; that is, $\dfrac{\partial f}{\partial y}$ (400, 2000) is most likely positive.

2. If the interest rate is raised from 14% to 15%, then the monthly payment will increase by about \$72.82. [An increase to $14\frac{1}{2}\%$ causes an increase in the monthly payment of $\frac{1}{2} \cdot (72.82)$ or \$36.41, and so on.]

17.3. Maxima and Minima of Functions of Several Variables

Previously, we studied how to determine the maxima and minima of functions of a single variable. Let us extend that discussion to functions of several variables.

If $f(x, y)$ is a function of two variables, then we say that $f(x, y)$ has a *maximum* when $x = a$, $y = b$ if $f(x, y)$ is at most equal to $f(a, b)$ whenever x is near a and y is near b. Geometrically, the graph of $f(x, y)$ has a peak at the point (a, b). [See Fig. 1(a).] Similarly, we say that $f(x, y)$ has a *minimum* when $x = a$, $y = b$

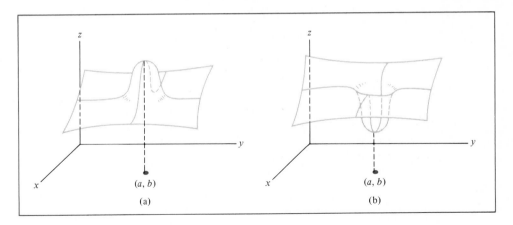

FIGURE 1   Maximum and minimum points.

if $f(x, y)$ is at least equal to $f(a, b)$ whenever x is near a and y is near b. Geometrically, the graph of $f(x, y)$ has a pit with bottom at the point (a, b). [See Fig. 1(b).]

Suppose that $f(x, y)$ has a maximum at $(x, y) = (a, b)$. Then, keeping y constant at b, $f(x, y)$ becomes a function of the variable x with a maximum at $x = a$. Therefore, its derivative with respect to x is zero at $x = a$. That is,

$$\frac{\partial f}{\partial x}(a, b) = 0.$$

Similarly, keeping x constant at a, $f(x, y)$ becomes a function of the variable y with a maximum at $y = b$. Therefore, its derivative with respect to y is zero at

$y = b$. That is,

$$\frac{\partial f}{\partial y}(a, b) = 0.$$

Similar considerations apply if $f(x, y)$ has a minimum at $(x, y) = (a, b)$. Thus we have the following test for extrema in two variables.

First-Derivative Test for Extrema If $f(x, y)$ has either a maximum or a minimum at $(x, y) = (a, b)$, then

$$\frac{\partial f}{\partial x}(a, b) = 0$$

and

$$\frac{\partial f}{\partial y}(a, b) = 0.$$

This result is quite helpful in finding maximum and minimum points, especially when additional information about the function is known.

EXAMPLE 1 The function $f(x, y) = 3x^2 - 4xy + 3y^2 + 8x - 17y + 5$ has the graph pictured in Fig. 2. Find the point (x, y) at which $f(x, y)$ attains its minimum.

Solution We look for those values of x and y at which both partial derivatives are zero. The partial derivatives are

$$\frac{\partial f}{\partial x} = 6x - 4y + 8$$

$$\frac{\partial f}{\partial y} = -4x + 6y - 17.$$

Setting $\dfrac{\partial f}{\partial x} = 0$ and $\dfrac{\partial f}{\partial y} = 0$, we obtain

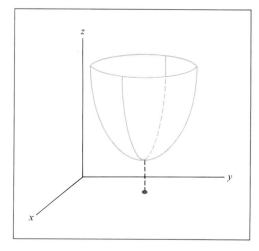

FIGURE 2 Graph of $f(x, y) = 3x^2 - 4xy + 3y^2 + 8x - 17y + 5$.

$$6x - 4y + 8 = 0 \qquad \text{or} \qquad y = \frac{6x + 8}{4}$$

$$-4x + 6y - 17 = 0 \qquad \text{or} \qquad y = \frac{4x + 17}{6}.$$

By equating these two expressions for y, we have

$$\frac{6x + 8}{4} = \frac{4x + 17}{6}.$$

Cross-multiplying, we see that

$$36x + 48 = 16x + 68$$

$$20x = 20$$

$$x = 1.$$

When we substitute this value for x into our first equation for y in terms of x, we obtain

$$y = \frac{6x + 8}{4} = \frac{6 \cdot 1 + 8}{4} = \frac{7}{2}.$$

If $f(x, y)$ has a minimum, it must occur where $\dfrac{\partial f}{\partial x} = 0$ and $\dfrac{\partial f}{\partial y} = 0$. We have determined that the partial derivatives are zero only when $x = 1$, $y = \frac{7}{2}$. From Fig. 2 we know that $f(x, y)$ has a minimum, so it must be at $(x, y) = (1, \frac{7}{2})$.

EXAMPLE 2 (*Price Discrimination*)   A monopolist markets his product in two countries and can charge different amounts in each country. Let x be the number of units to be sold in the first country and y the number of units to be sold in the second country. Due to the laws of demand, the monopolist must set the price at $97 - (x/10)$ dollars in the first country and $83 - (y/20)$ dollars in the second country in order to sell all the units. The cost of producing these units is $20{,}000 + 3(x + y)$. Find the values of x and y that maximize the profit.

Solution Let $f(x, y)$ be the profit derived from selling x units in the first country and y in the second. Then

$$f(x, y) = [\text{revenue from first country}] + [\text{revenue from second country}] - [\text{cost}]$$

$$= \left(97 - \frac{x}{10}\right)x + \left(83 - \frac{y}{20}\right)y - [20{,}000 + 3(x + y)]$$

$$= 97x - \frac{x^2}{10} + 83y - \frac{y^2}{20} - 20{,}000 - 3x - 3y$$

$$= 94x - \frac{x^2}{10} + 80y - \frac{y^2}{20} - 20{,}000.$$

To find where $f(x, y)$ has its maximum value, we look for those values of x and y at which both partial derivatives are zero.

$$\frac{\partial f}{\partial x} = 94 - \frac{x}{5}$$

$$\frac{\partial f}{\partial y} = 80 - \frac{y}{10}.$$

We set $\dfrac{\partial f}{\partial x} = 0$ and $\dfrac{\partial f}{\partial y} = 0$ to obtain

$$94 - \frac{x}{5} = 0 \quad \text{or} \quad x = 470$$

$$80 - \frac{y}{10} = 0 \quad \text{or} \quad y = 800.$$

Therefore, the firm should adjust its prices to levels where it will sell 470 units in the first country and 800 units in the second country.

EXAMPLE 3 Suppose that we want to design a rectangular building having volume 147,840 cubic feet. Assuming that the daily loss of heat is given by

$$11xy + 14yz + 15xz,$$

where x, y, and z are, respectively, the length, width, and height of the building, find the dimensions of the building for which the daily heat loss is minimal.

Solution We must minimize the function

$$11xy + 14yz + 15xz, \tag{1}$$

where x, y, z satisfy the constraint equation

$$xyz = 147,840.$$

For simplicity, let us denote 147,840 by V. Then $xyz = V$, so that $z = V/xy$. We substitute this expression for z into the objective function (1) to obtain a heat-loss function $g(x, y)$ of two variables—namely

$$g(x, y) = 11xy + 14y\frac{V}{xy} + 15x\frac{V}{xy}$$

$$= 11xy + \frac{14V}{x} + \frac{15V}{y}.$$

To minimize this function, we first compute the partial derivatives with respect to x and y; then we equate them to zero.

$$\frac{\partial g}{\partial x} = 11y - \frac{14V}{x^2} = 0$$

$$\frac{\partial g}{\partial y} = 11x - \frac{15V}{y^2} = 0.$$

These two equations yield

$$y = \frac{14V}{11x^2} \tag{2}$$

$$11xy^2 = 15V. \tag{3}$$

If we substitute the value of y from (2) into (3), we see that

$$11x\left(\frac{14V}{11x^2}\right)^2 = 15V$$

$$\frac{14^2V^2}{11x^3} = 15V$$

$$x^3 = \frac{14^2 \cdot V^2}{11 \cdot 15 \cdot V} = \frac{14^2 \cdot V}{11 \cdot 15}$$

$$= \frac{14^2 \cdot 147,840}{11 \cdot 15}$$

$$= 175,616.$$

Therefore, we see that (using a calculator or a table of cube roots)

$$x = 56.$$

From equation (2) we find that

$$y = \frac{14 \cdot V}{11x^2} = \frac{14 \cdot 147,840}{11 \cdot 56^2} = 60.$$

Finally,

$$z = \frac{V}{xy} = \frac{147,840}{56 \cdot 60} = 44.$$

Thus the building should be 56 feet long, 60 feet wide, and 44 feet high in order to minimize the heat loss.*

When considering a function of two variables, we find points (x, y) at which $f(x, y)$ has a potential maximum or minimum by setting $\frac{\partial f}{\partial x}$ and $\frac{\partial f}{\partial y}$ equal to zero and solving for x and y. However, if we are given no additional information about $f(x, y)$, it may be difficult to determine whether we have found a maximum or a minimum (or neither). In the case of functions of one variable, we studied concavity and deduced the second-derivative test. There is an analog of the second-derivative test for functions of two variables, but it is much more complicated than the one-variable test. We state it without proof.

---

* For further discussion of this heat-loss problem, as well as other examples of optimization in architectural design, see L. March, "Elementary Models of Built Forms," Chapter 3 in *Urban Space and Structures*, L. Martin and L. March, eds. (Cambridge: Cambridge University Press, 1972).

Second-Derivative Test for Maxima and Minima Suppose that $f(x, y)$ is a function and (a, b) is a point at which

$$\frac{\partial f}{\partial x}(a, b) = 0 \quad \text{and} \quad \frac{\partial f}{\partial y}(a, b) = 0,$$

and let

$$D(x, y) = \frac{\partial^2 f}{\partial x^2} \cdot \frac{\partial^2 f}{\partial y^2} - \left(\frac{\partial^2 f}{\partial x \, \partial y} \right)^2.$$

1. If

$$D(a, b) > 0 \quad \text{and} \quad \frac{\partial^2 f}{\partial x^2}(a, b) > 0,$$

then $f(x, y)$ has a minimum at (a, b).

2. If

$$D(a, b) > 0 \quad \text{and} \quad \frac{\partial^2 f}{\partial x^2}(a, b) < 0,$$

then $f(x, y)$ has a maximum at (a, b).

3. If

$$D(a, b) < 0,$$

then $f(x, y)$ has neither a maximum nor a minimum at (a, b).

4. If $D(a, b) = 0$, then no conclusion can be drawn from this test.

The saddle-shaped graph in Fig. 3 illustrates a function $f(x, y)$ for which $D(a, b) < 0$. Both partial derivatives are zero at $(x, y) = (a, b)$ and yet the function has neither a maximum nor a minimum there. (Observe that the function has a maximum with respect to x when y is held constant and a minimum with respect to y when x is held constant.)

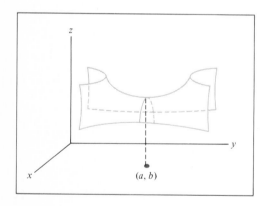

FIGURE 3

EXAMPLE 4 Let $f(x, y) = y^3 - x^2 + 6x - 12y + 5$. Find all possible maximum and minimum points of $f(x, y)$. Use the second-derivative test to determine the nature of each such point.

Solution Since

$$\frac{\partial f}{\partial x} = -2x + 6, \qquad \frac{\partial f}{\partial y} = 3y^2 - 12,$$

we find that $f(x, y)$ has a potential extreme point when

$$-2x + 6 = 0$$
$$3y^2 - 12 = 0.$$

The first equation implies that $x = 3$. From the second equation we have

$$3y^2 = 12$$
$$y^2 = 4$$
$$y = \pm 2.$$

Thus $\dfrac{\partial f}{\partial x}$ and $\dfrac{\partial f}{\partial y}$ are both zero when $(x, y) = (3, 2)$ and when $(x, y) = (3, -2)$. To apply the second-derivative test, we compute

$$\frac{\partial^2 f}{\partial x^2} = -2, \qquad \frac{\partial^2 f}{\partial y^2} = 6y, \qquad \frac{\partial^2 f}{\partial x \, \partial y} = 0,$$

and

$$D(x, y) = \frac{\partial^2 f}{\partial x^2} \cdot \frac{\partial^2 f}{\partial y^2} - \left(\frac{\partial^2 f}{\partial x \, \partial y} \right)^2 = (-2)(6y) - 0 = -12y. \qquad (4)$$

Since $D(3, 2) = -24$ is negative, case 3 of the second derivative test says that $f(x, y)$ has neither a maximum nor a minimum at $(3, 2)$. Next, note that

$$D(3, -2) = 24 > 0, \qquad \frac{\partial^2 f}{\partial x^2}(3, -2) = -2 < 0.$$

Thus, by case 2 of the second-derivative test, the function $f(x, y)$ has a maximum at $(3, -2)$.

In this section we have restricted ourselves to functions of two variables, but the case of three or more variables is handled in a similar fashion. For instance, here is the first-derivative test for a function of three variables.

If $f(x, y, z)$ has a maximum or a minimum at $(x, y, z) = (a, b, c)$, then

$$\frac{\partial f}{\partial x}(a, b, c) = 0$$

$$\frac{\partial f}{\partial y}(a, b, c) = 0$$

$$\frac{\partial f}{\partial z}(a, b, c) = 0.$$

PRACTICE PROBLEMS 3

1. Find all points (x, y) where $f(x, y) = x^3 - 3xy + \frac{1}{2}y^2 + 8$ has a possible maximum or minimum.

2. Apply the second-derivative test to the function $g(x, y)$ of Example 3 to confirm that a minimum point actually occurs when $x = 56$ and $y = 60$.

EXERCISES 3

Find all points (x, y) where $f(x, y)$ has a possible maximum or minimum.

1. $f(x, y) = x^2 - 3y^2 + 4x + 6y + 8$

2. $f(x, y) = \frac{1}{2}x^2 + y^2 - 3x + 2y - 5$

3. $f(x, y) = x^2 - 5xy + 6y^2 + 3x - 2y + 4$

4. $f(x, y) = -3x^2 + 7xy - 4y^2 + x + y$

5. $f(x, y) = x^3 + y^2 - 3x + 6y$

6. $f(x, y) = x^2 - y^3 + 5x + 12y + 1$

7. $f(x, y) = \frac{1}{3}x^3 - 2y^3 - 5x + 6y - 5$

8. $f(x, y) = x^4 - 8xy + 2y^2 - 3$

9. The function $f(x, y) = 2x + 3y + 9 - x^2 - xy - y^2$ has a maximum at some point (x, y). Find the values of x and y where this maximum occurs.

10. The function $f(x, y) = \frac{1}{2}x^2 + 2xy + 3y^2 - x + 2y$ has a minimum at some point (x, y). Find the values of x and y where this minimum occurs.

Find all points (x, y) where $f(x, y)$ has a possible maximum or minimum. Then use the second derivative test to determine, if possible, the nature of $f(x, y)$ at each of these points. If the second derivative test is inconclusive, so state.

11. $f(x, y) = x^2 - 2xy + 4y^2$

12. $f(x, y) = 2x^2 + 3xy + 5y^2$

13. $f(x, y) = -2x^2 + 2xy - y^2 + 4x - 6y + 5$

14. $f(x, y) = -x^2 - 8xy - y^2$

15. $f(x, y) = x^2 + 2xy + 5y^2 + 2x + 10y - 3$

16. $f(x, y) = x^2 - 2xy + 3y^2 + 4x - 16y + 22$

17. $f(x, y) = x^3 - y^2 - 3x + 4y$

18. $f(x, y) = x^3 - 2xy + 4y$

19. $f(x, y) = 2x^2 + y^3 - x - 12y + 7$

20. $f(x, y) = x^2 + 4xy + 2y^4$

21. Find the possible values of x, y, z at which

$$f(x, y, z) = 2x^2 + 3y^2 + z^2 - 2x - y - z$$

assumes its minimum value.

22. Find the possible values of x, y, z at which

$$f(x, y, z) = 5 + 8x - 4y + x^2 + y^2 + z^2$$

assumes its minimum value.

23. U.S. postal rules require that the length plus the girth of a package cannot exceed 84 inches in order to be mailed. Find the dimensions of the rectangular package of greatest volume that can be mailed. [*Note:* From Fig. 4 we see that $84 = $ (length) + (girth) = $l + (2x + 2y)$ and volume = xyl.]

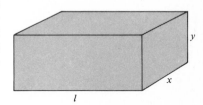

FIGURE 4

24. Find the dimensions of the rectangular box of least surface area that has a volume of 100 cubic inches. [*Note:* From Fig. 4 we see that the surface area = $2xy + 2xl + 2yl$ and volume = $100 = xyl$.]

25. A company manufactures and sells two products, call them I and II, that sell for $10 and $9 per unit, respectively. The cost of producing x units of product I and y units of product II is

$$400 + 2x + 3y + .01(3x^2 + xy + 3y^2).$$

Find the values of x and y that maximize the company's profit. [*Note:* Profit = (revenue) − (cost).]

26. A monopolist manufactures and sells two competing products, call them I and II, that cost $30 and $20 per unit, respectively, to produce. The revenue from marketing x units of product I and y units of product II is $98x + 112y - .04xy - .1x^2 - .2y^2$. Find the values of x and y that maximize the monopolist's profits.

27. A company manufactures and sells two products, call them I and II, that sell for p_1 and p_{II} per unit, respectively. Let $C(x, y)$ be the cost of producing x units of product I and y units of product II. Show that if the company's profit is maximized when $x = a$, $y = b$, then

$$\frac{\partial C}{\partial x}(a, b) = p_1 \quad \text{and} \quad \frac{\partial C}{\partial y}(a, b) = p_{II}.$$

28. A monopolist manufactures and sells two competing products, call them I and II, that cost p_1 and p_{II} per unit, respectively, to produce. Let $R(x, y)$ be the revenue from marketing x units of product I and y units of product II. Show that if the monopolist's profit is maximized when $x = a$, $y = b$, then

$$\frac{\partial R}{\partial x}(a, b) = p_1 \quad \text{and} \quad \frac{\partial R}{\partial y}(a, b) = p_{II}.$$

SOLUTIONS TO PRACTICE PROBLEMS 3

1. Compute the first partial derivatives of $f(x, y)$ and solve the system of equations that results from setting the partials equal to zero.

$$\frac{\partial f}{\partial x} = 3x^2 - 3y = 0$$

$$\frac{\partial f}{\partial y} = -3x + y = 0.$$

Solve each equation for y in terms of x.

$$\begin{cases} y = x^2 \\ y = 3x. \end{cases}$$

Equate expressions for y and solve for x.

$$x^2 = 3x$$

$$x^2 - 3x = 0$$

$$x(x - 3) = 0$$

$$x = 0 \quad \text{or} \quad x = 3,$$

When $x = 0$, $y = 0^2 = 0$. When $x = 3$, $y = 3^2 = 9$. Therefore, the possible maximum or minimum points are $(0, 0)$ and $(3, 9)$.

2. We have

$$g(x, y) = 11xy + \frac{14V}{x} + \frac{15V}{y},$$

$$\frac{\partial g}{\partial x} = 11y - \frac{14V}{x^2}, \quad \text{and} \quad \frac{\partial g}{\partial y} = 11x - \frac{15V}{y^2}.$$

Now,

$$\frac{\partial^2 g}{\partial x^2} = \frac{28V}{x^3}, \quad \frac{\partial^2 g}{\partial y^2} = \frac{30V}{y^3}, \quad \text{and} \quad \frac{\partial^2 g}{\partial x \, \partial y} = 11.$$

Therefore,

$$D(x, y) = \frac{28V}{x^3} \cdot \frac{30V}{y^3} - (11)^2$$

$$D(56, 60) = \frac{28(147{,}840)}{(56)^3} \cdot \frac{30(147{,}840)}{(60)^3} - 121$$

$$= 484 - 121 = 363 > 0,$$

and

$$\frac{\partial^2 g}{\partial x^2}(56, 60) = \frac{28(147{,}840)}{(56)^3} > 0.$$

It follows that $g(x, y)$ has a minimum at $x = 56$, $y = 60$.

17.4. Lagrange Multipliers and Constrained Optimization

We have seen a number of optimization problems in which we were required to minimize (or maximize) an objective function where the variables were subject to a constraint equation. For instance, in Example 4 of Section 12.5, we minimized the cost of a rectangular enclosure by minimizing the objective function $21x + 14y$, where x and y were subject to the constraint equation $600 - xy = 0$. In the preceding section (Example 3) we minimized the daily heat loss from a building by minimizing the objective function $11xy + 14yz + 15xz$, subject to the constraint equation $147{,}840 - xyz = 0$.

In this section we introduce a powerful technique for solving problems of this type. Let us begin with the following general problem, which involves two variables.

> **PROBLEM**   Let $f(x, y)$ and $g(x, y)$ be functions of two variables. Find values of x and y that maximize (or minimize) the objective function $f(x, y)$ and that also satisfy the constraint equation $g(x, y) = 0$.

Of course, if we can solve the equation $g(x, y) = 0$ for one variable in terms of the other and substitute the resulting expression into $f(x, y)$, we arrive at a function of a single variable that can be maximized (or minimized) by using the methods of Chapter 12. However, this technique can be unsatisfactory for two reasons. First, it may be difficult to solve the equation $g(x, y) = 0$ for x or for y. For example, if $g(x, y) = x^4 + 5x^3y + 7x^2y^3 + y^5 - 17 = 0$, then it is difficult to write y as a function of x or x as a function of y. Second, even if $g(x, y) = 0$ can be solved for one variable in terms of the other, substitution of the result into $f(x, y)$ may yield a complicated function.

One clever idea for handling the preceding problem was discovered by the eighteenth-century mathematician Lagrange, and the technique that he pioneered today bears his name—the method of *Lagrange multipliers*. The basic idea of this method is to replace $f(x, y)$ by an auxiliary function of three variables $F(x, y, \lambda)$, defined as

$$F(x, y, \lambda) = f(x, y) + \lambda g(x, y).$$

The new variable λ (lambda) is called a *Lagrange multiplier* and always multiplies the constraint function $g(x, y)$. The following theorem is stated without proof.

> **Theorem**   Suppose that, subject to the constraint $g(x, y) = 0$, the function $f(x, y)$ has a maximum or a minimum at $(x, y) = (a, b)$. Then there is a value of λ, say $\lambda = c$, such that the partial derivatives of $F(x, y, \lambda)$ all equal zero at $(x, y, \lambda) = (a, b, c)$.

The theorem implies that if we locate all points (x, y, λ) where the partial derivatives of $F(x, y, \lambda)$ are all zero, then among the corresponding points (x, y) we will find all possible places where $f(x, y)$ may have a constrained maximum or minimum. Thus the first step in the method of Lagrange multipliers is to set the partial derivatives of $F(x, y, \lambda)$ equal to zero and solve for x, y, and λ:

$$\frac{\partial F}{\partial x} = 0 \tag{L-1}$$

$$\frac{\partial F}{\partial y} = 0 \tag{L-2}$$

$$\frac{\partial F}{\partial \lambda} = 0. \tag{L-3}$$

From the definition of $F(x, y, \lambda)$, we see that $\dfrac{\partial F}{\partial \lambda} = g(x, y)$. Thus the third equation (L-3) is just the original constraint equation $g(x, y) = 0$. So when we find a point (x, y, λ) that satisfies (L-1), (L-2), and (L-3), the coordinates x and y will automatically satisfy the constraint equation.

Let us see how this method works in practice.

EXAMPLE 1   Minimize $x^2 + y^2$, subject to the constraint $2x + 3y - 4 = 0$.

Solution   Here we have $f(x, y) = x^2 + y^2$, $g(x, y) = 2x + 3y - 4$, and

$$F(x, y, \lambda) = x^2 + y^2 + \lambda(2x + 3y - 4).$$

Equations (L-1) to (L-3) read

$$\frac{\partial F}{\partial x} = 2x + 2\lambda = 0 \tag{1}$$

$$\frac{\partial F}{\partial y} = 2y + 3\lambda = 0 \tag{2}$$

$$\frac{\partial F}{\partial \lambda} = 2x + 3y - 4 = 0. \tag{3}$$

Let us solve the first two equations for λ:

$$\lambda = -x \tag{4}$$

$$\lambda = -\tfrac{2}{3}y.$$

If we equate these two expressions for λ, we obtain

$$-x = -\tfrac{2}{3}y$$

$$x = \tfrac{2}{3}y. \tag{5}$$

Then, substituting this expression for x into the equation (3), we have

$$2(\tfrac{2}{3}y) + 3y - 4 = 0$$

$$\tfrac{13}{3}y = 4$$

$$y = \tfrac{12}{13}.$$

Using this value for y, equations (4) and (5) give us the value of x and λ:

$$x = \tfrac{2}{3}y = \tfrac{2}{3}(\tfrac{12}{13}) = \tfrac{8}{13}$$

$$\lambda = -x = -\tfrac{8}{13}.$$

Therefore, the partial derivatives of $F(x, y, \lambda)$ are zero when $x = \tfrac{8}{13}$, $y = \tfrac{12}{13}$, and $\lambda = -\tfrac{8}{13}$. So the minimum value of $x^2 + y^2$, subject to the constraint $2x + 3y - 4 = 0$, is

$$(\tfrac{8}{13})^2 + (\tfrac{12}{13})^2 = \tfrac{16}{13}.$$

The preceding technique for solving three equations in the three variables x, y, λ can usually be applied to solve Lagrange multiplier problems. Here is the basic procedure.

1. Solve (L-1) and (L-2) for λ in terms of x and y; then equate the resulting expressions for λ.
2. Solve the resulting equation for one of the variables.
3. Substitute the expression so derived into the equation (L-3) and solve the resulting equation of one variable.
4. Use the one known variable and the equations of steps 1 and 2 to determine the other two variables.

EXAMPLE 2   Using Lagrange multipliers, minimize $21x + 14y$, subject to the constraint $600 - xy = 0$, where x and y are restricted to positive values. (This problem arose in Example 4 of Section 12.5, where $21x + 14y$ was the cost of building a 600 square foot enclosure having dimensions x and y.)

Solution   We have $f(x, y) = 21x + 14y$, $g(x, y) = 600 - xy$, and

$$F(x, y, \lambda) = 21x + 14y + \lambda(600 - xy).$$

The equations (L-1) to (L-3), in this case, are

$$\frac{\partial F}{\partial x} = 21 - \lambda y = 0$$

$$\frac{\partial F}{\partial y} = 14 - \lambda x = 0$$

$$\frac{\partial F}{\partial \lambda} = 600 - xy = 0.$$

From the first two equations we see that

$$\lambda = \frac{21}{y} = \frac{14}{x} \qquad \text{(step 1)}.$$

Therefore,

$$21x = 14y$$

and

$$x = \tfrac{2}{3}y \qquad \text{(step 2)}.$$

Substituting this expression for x into the third equation, we derive

$$600 - (\tfrac{2}{3}y)y = 0$$

$$y^2 = \tfrac{3}{2} \cdot 600 = 900$$

$$y = \pm 30 \qquad \text{(step 3)}.$$

We discard the case $y = -30$ because we are interested only in positive values of x and y. Using $y = 30$, we find that

$$\left.\begin{array}{l} x = \tfrac{2}{3}(30) = 20 \\ \lambda = \tfrac{14}{20} = \tfrac{7}{10}. \end{array}\right\} \qquad \text{(step 4)}$$

So the minimum value of $21x + 14y$ with x and y subject to the constraint occurs when $x = 20$, $y = 30$, and $\lambda = \tfrac{7}{10}$. And the minimum value of $21x + 14y$, subject to the constraint $600 - xy = 0$, is

$$21 \cdot (20) + 14 \cdot (30) = 840.$$

EXAMPLE 3   (*Production*)   Suppose that x units of labor and y units of capital can produce $f(x, y) = 60x^{3/4}y^{1/4}$ units of a certain product. Also suppose that each unit of labor costs \$100, whereas each unit of capital costs \$200. Assume that \$30,000 is available to spend on production. How many units of labor and how many of capital should be utilized in order to maximize production?

Solution   The cost of x units of labor and y units of capital equals $100x + 200y$. Therefore, since we want to use all the available money (\$30,000), we must satisfy the constraint equation

$$100x + 200y = 30,000$$

or

$$g(x, y) = 30,000 - 100x - 200y = 0.$$

Our objective function is $f(x, y) = 60x^{3/4}y^{1/4}$. In this case, we have

$$F(x, y, \lambda) = 60x^{3/4}y^{1/4} + \lambda(30,000 - 100x - 200y).$$

The equations (L-1) to (L-3) read

$$\frac{\partial F}{\partial x} = 45x^{-1/4}y^{1/4} - 100\lambda = 0 \qquad \text{(L-1)}$$

$$\frac{\partial F}{\partial y} = 15x^{3/4}y^{-3/4} - 200\lambda = 0 \qquad \text{(L-2)}$$

$$\frac{\partial F}{\partial \lambda} = 30{,}000 - 100x - 200y = 0. \qquad \text{(L-3)}$$

By solving the first two equations for λ, we see that

$$\lambda = \tfrac{45}{100}x^{-1/4}y^{1/4} = \tfrac{9}{20}x^{-1/4}y^{1/4}$$

$$\lambda = \tfrac{15}{200}x^{3/4}y^{-3/4} = \tfrac{3}{40}x^{3/4}y^{-3/4}.$$

Therefore, we must have

$$\tfrac{9}{20}x^{-1/4}y^{1/4} = \tfrac{3}{40}x^{3/4}y^{-3/4}.$$

To solve for y in terms of x, let us multiply both sides of this equation by $x^{1/4}y^{3/4}$:

$$\tfrac{9}{20}y = \tfrac{3}{40}x$$

or

$$y = \tfrac{1}{6}x.$$

Inserting this result in (L-3), we find that

$$100x + 200(\tfrac{1}{6}x) = 30{,}000$$

$$\frac{400x}{3} = 30{,}000$$

$$x = 225.$$

Hence

$$y = \tfrac{225}{6} = 37.5.$$

So maximum production is achieved by using 225 units of labor and 37.5 units of capital.

In Example 3 it turns out that, at the optimum value of x and y,

$$\lambda = \tfrac{9}{20}x^{-1/4}y^{1/4} = \tfrac{9}{20}(225)^{-1/4}(37.5)^{1/4} \approx .2875$$

$$\frac{\partial f}{\partial x} = 45x^{-1/4}y^{1/4} = 45(225)^{-1/4}(37.5)^{1/4} \qquad \text{(6)}$$

$$\frac{\partial f}{\partial y} = 15x^{3/4}y^{-3/4} = 15(225)^{3/4}(37.5)^{-3/4}. \qquad \text{(7)}$$

It can be shown that the Lagrange multiplier λ can be interpreted as the marginal productivity of money. That is, if one additional dollar is available, then approximately .2875 additional units of the product can be produced.

Recall that the partial derivatives $\dfrac{\partial f}{\partial x}$ and $\dfrac{\partial f}{\partial y}$ are called the marginal productivity of labor and capital, respectively. From (6) and (7) we have

$$\frac{[\text{marginal productivity of labor}]}{[\text{marginal productivity of capital}]} = \frac{45(225)^{-1/4}(37.5)^{1/4}}{15(225)^{3/4}(37.5)^{-3/4}}$$

$$= \frac{45}{15}(225)^{-1}(37.5)^{1}$$

$$= \frac{3(37.5)}{225} = \frac{37.5}{75} = \frac{1}{2}.$$

On the other hand,

$$\frac{[\text{cost per unit of labor}]}{[\text{cost per unit of capital}]} = \frac{100}{200} = \frac{1}{2}.$$

This result illustrates the following law of economics. *If labor and capital are at their optimal levels, then the ratio of their marginal productivities equals the ratio of their unit costs.*

The method of Lagrange multipliers generalizes to functions of any number of variables. For instance, we can maximize $f(x, y, z)$, subject to the constraint equation $g(x, y, z) = 0$, by forming the Lagrange function

$$F(x, y, z, \lambda) = f(x, y, z) + \lambda g(x, y, z).$$

The analogs of equations (L-1) to (L-3) are

$$\frac{\partial F}{\partial x} = 0$$

$$\frac{\partial F}{\partial y} = 0$$

$$\frac{\partial F}{\partial z} = 0$$

$$\frac{\partial F}{\partial \lambda} = 0.$$

Let us now show how we can solve the heat-loss problem of Section 3 by using this method.

EXAMPLE 4   Use Lagrange multipliers to find the values of x, y, z that minimize the objective function

$$f(x, y, z) = 11xy + 14yz + 15xz,$$

subject to the constraint

$$xyz = 147{,}840.$$

Solution The Lagrange function is

$$F(x, y, z, \lambda) = 11xy + 14yz + 15xz + \lambda(147{,}840 - xyz).$$

The conditions for a minimum are

$$\frac{\partial F}{\partial x} = 11y + 15z - \lambda yz = 0$$

$$\frac{\partial F}{\partial y} = 11x + 14z - \lambda xz = 0$$

$$\frac{\partial F}{\partial z} = 14y + 15x - \lambda xy = 0$$

$$\frac{\partial F}{\partial \lambda} = 147{,}840 - xyz = 0. \tag{8}$$

From the first three equations we have

$$\left.\begin{aligned}
\lambda &= \frac{11y + 15z}{yz} = \frac{11}{z} + \frac{15}{y} \\[2mm]
\lambda &= \frac{11x + 14z}{xz} = \frac{11}{z} + \frac{14}{x} \\[2mm]
\lambda &= \frac{14y + 15x}{xy} = \frac{14}{x} + \frac{15}{y}
\end{aligned}\right\}. \tag{9}$$

Let us equate the first two expressions for λ:

$$\frac{11}{z} + \frac{15}{y} = \frac{11}{z} + \frac{14}{x}$$

$$\frac{15}{y} = \frac{14}{x}$$

$$x = \frac{14}{15}y.$$

Next, we equate the second and third expressions for λ in (9):

$$\frac{11}{z} + \frac{14}{x} = \frac{14}{x} + \frac{15}{y}$$

$$\frac{11}{z} = \frac{15}{y}$$

$$z = \frac{11}{15}y.$$

We now substitute the expressions for x and z into the constraint equation (8) and obtain

$$\tfrac{14}{15}y \cdot y \cdot \tfrac{11}{15}y = 147{,}840$$

$$y^3 = \frac{(147{,}840)(15)^2}{(14)(11)} = 216{,}000$$

$$y = 60.$$

From this, we find that

$$x = \tfrac{14}{15}(60) = 56 \qquad \text{and} \qquad z = \tfrac{11}{15}(60) = 44.$$

We conclude that the heat loss is minimized when $x = 56$, $y = 60$, and $z = 44$.

In the solution of Example 4, we found that at the optimal values of x, y, and z,

$$\frac{14}{x} = \frac{15}{y} = \frac{11}{z}.$$

Referring to Example 2 of Section 1, we see that 14 is the combined heat loss through the east and west sides of the building, 15 is the heat loss through the north and south sides of the building, and 11 is the heat loss through the floor and roof. Thus we have that under optimal conditions

$$\frac{[\text{heat loss through east and west sides}]}{[\text{distance between east and west sides}]} = \frac{[\text{heat loss through north and south sides}]}{[\text{distance between north and south sides}]}$$

$$= \frac{[\text{heat loss through floor and roof}]}{[\text{distance between floor and roof}]}.$$

This is a principle of optimal design: minimal heat loss occurs when the distance between each pair of opposite sides is some fixed constant times the heat loss from the pair of sides.

The value of λ in Example 4 corresponding to the optimal values of x, y, and z is

$$\lambda = \frac{11}{z} + \frac{15}{y} = \frac{11}{44} + \frac{15}{60} = \frac{1}{2}.$$

One can show that the Lagrange multiplier λ is the marginal heat loss with respect to volume. That is, if a building of volume slightly more than 147,840 cubic feet is optimally designed, then $\tfrac{1}{2}$ unit of additional heat will be lost for each additional cubic foot of volume.

PRACTICE PROBLEMS 4

1. Let $F(x, y, \lambda) = 2x + 3y + \lambda(90 - 6x^{1/3}y^{2/3})$. Find $\dfrac{\partial F}{\partial x}$.

2. Refer to Exercise 23 of Section 3. What is the function $F(x, y, \lambda)$ when the exercise is solved using the method of Lagrange multipliers?

EXERCISES 4

Solve the following exercises by the method of Lagrange multipliers.

1. Minimize the function $x^2 + 3y^2 + 10$, subject to the constraint $8 - x - y = 0$.

2. Maximize the function $x^2 - y^2$, subject to the constraint $2x + y - 3 = 0$.

3. Maximize $x^2 + xy - 3y^2$, subject to the constraint $2 - x - 2y = 0$.

4. Minimize $\frac{1}{2}x^2 - 3xy + y^2 + \frac{1}{2}$, subject to the constraint $3x - y - 1 = 0$.

5. Find the values of x, y that maximize the function
    $$-2x^2 - 2xy - \tfrac{3}{2}y^2 + x + 2y,$$
    subject to the constraint $x + y - \frac{5}{2} = 0$.

6. Find the values of x, y that minimize the function
    $$x^2 + xy + y^2 - 2x - 5y,$$
    subject to the constraint $1 - x + y = 0$.

7. Find the values of x, y, z that maximize the function
    $$3x + 5y + z - x^2 - y^2 - z^2,$$
    subject to the constraint $6 - x - y - z = 0$.

8. Find the values of x, y, z that minimize the function
    $$x^2 + y^2 + z^2 - 3x - 5y - z,$$
    subject to the constraint $20 - 2x - y - z = 0$.

9. The material for a rectangular box costs $2 per square foot for the top and $1 per square foot for the sides and bottom. Using Lagrange multipliers, find the dimensions for which the volume of the box is 12 cubic feet and the cost of the materials is minimized. [Referring to Fig. 2(a), the cost will be $3xy + 2xz + 2yz$.]

10. Use Lagrange multipliers to find the three positive numbers whose sum is 15 and whose product is as large as possible.

11. Find the dimensions of the rectangle of maximum area that can be inscribed in the unit circle. [See Fig. 1(a).] That is, find the values of x, y that maximize $4xy$, subject to the constraint $x^2 + y^2 = 1$.

FIGURE 1

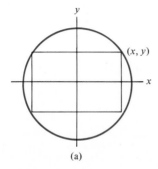

(a)

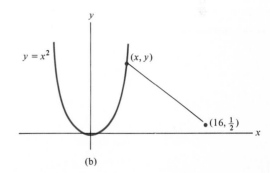

(b)

12. Find the point on the parabola $y = x^2$ that has minimal distance from the point $(16, \frac{1}{2})$. [See Fig. 1(b).] [*Suggestion:* If d denotes the distance from (x, y) to $(16, \frac{1}{2})$, then $d^2 = (x - 16)^2 + (y - \frac{1}{2})^2$. If d^2 is minimized, then d will be minimized. Thus it suffices to minimize $(x - 16)^2 + (y - \frac{1}{2})^2$, subject to the constraint $y - x^2 = 0$.]

13. Find the dimensions of an open rectangular glass tank of volume 32 cubic feet for which the amount of material needed to construct the tank is minimized. [See Fig. 2(a).] That is, find the values of x, y, z that minimize $xy + 2xz + 2yz$, subject to the constraint $32 - xyz = 0$.

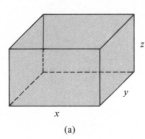

(a)

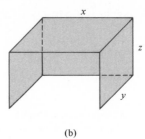

(b)   FIGURE 2

14. A shelter for use at the beach has a back, two sides, and a top made of canvas. [See Fig. 2(b).] Find the dimensions that maximize the volume and require 96 square feet of canvas. That is, find the values of x, y, z that maximize xyz, subject to the constraint $xz + xy + 2yz - 96 = 0$.

15. The production function for a firm is $f(x, y) = 64x^{3/4}y^{1/4}$, where x and y are the number of units of labor and capital utilized. Suppose that labor costs \$96 per unit and capital costs \$162 per unit and that the firm decides to produce 3456 units of goods.

   (a) Determine the amounts of labor and capital that should be utilized in order to minimize the cost. That is, find the values of x, y that minimize $96x + 162y$, subject to the constraint $3456 - 64x^{3/4}y^{1/4} = 0$.

   (b) Find the value of λ at the optimal level of production.

   (c) Show that, at the optimal level of production, we have

   $$\frac{[\text{marginal productivity of labor}]}{[\text{marginal productivity of capital}]} = \frac{[\text{unit price of labor}]}{[\text{unit price of capital}]}.$$

16. The production function for a firm is $f(x, y) = 1000\sqrt{6x^2 + y^2}$, where x and y are, respectively, the number of units of labor and capital utilized. Suppose that labor costs \$480 per unit and capital costs \$40 per unit and that the firm has \$5000 to spend.

   (a) Determine the amounts of labor and capital that should be utilized in order to maximize production.

   (b) Find the value of λ.

   (c) Repeat part (c) of Exercise 15 for $f(x, y) = 1000\sqrt{6x^2 + y^2}$.

17. Suppose that a firm makes two products A and B that use the same raw materials. Given a fixed amount of raw materials and a fixed amount of manpower, the firm must decide how much of its resources should be allocated to the production of A and how much to

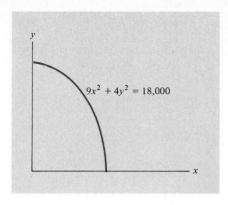

FIGURE 3

B. If x units of A and y units of B are produced, suppose that x and y must satisfy

$$9x^2 + 4y^2 = 18{,}000.$$

The graph of this equation (for $x \geqslant 0$, $y \geqslant 0$) is called a *production possibilities curve* (Fig. 3). A point (x, y) on this curve represents a *production schedule* for the firm, committing it to produce x units of A and y units of B. The reason for the relationship between x and y involves the limitations on personnel and raw materials available to the firm. Suppose that each unit of A yields a $3 profit, whereas each unit of B yields a $4 profit. Then the profit of the firm is

$$P(x, y) = 3x + 4y.$$

Find the production schedule that maximizes the profit function $P(x, y)$.

18. Consider the monopolist of Example 2, Section 3, who sells his goods in two countries. Suppose that he must set the same price in each country. That is, $97 - (x/10) = 83 - (y/20)$. Find the values of x and y that maximize profits under this new restriction.

19. Let $f(x, y)$ be any production function where x represents labor (costing $\$a$ per unit) and y represents capital (costing $\$b$ per unit). Assuming that $\$c$ is available, show that, at the values of x, y that maximize production,

$$\frac{\dfrac{\partial f}{\partial x}}{\dfrac{\partial f}{\partial y}} = \frac{a}{b}.$$

[*Note:* Let $F(x, y, \lambda) = f(x, y) + \lambda(c - ax - by)$. The result follows from (L-1) and (L-2).]

20. By applying the result in Exercise 19 to the production function $f(x, y) = kx^{\alpha}y^{\beta}$, show that, for the values of x, y that maximize production, we have

$$\frac{y}{x} = \frac{a\beta}{b\alpha}.$$

(This tells us that the ratio of capital to labor does not depend on the amount of money available nor on the level of production but only on the numbers a, b, α, and β.)

SOLUTIONS TO PRACTICE PROBLEMS 4

1. The function can be written as

$$F(x, y, \lambda) = 2x + 3y + \lambda \cdot 90 - \lambda \cdot 6x^{1/3}y^{2/3}.$$

When differentiating with respect to x, both y and λ should be treated as constants (so $\lambda \cdot 90$ and $\lambda \cdot 6$ are also regarded as constants).

$$\frac{\partial F}{\partial x} = 2 - \lambda \cdot 6 \cdot \tfrac{1}{3}x^{-2/3} \cdot y^{2/3}$$

$$= 2 - \lambda \cdot 2 \cdot x^{-2/3}y^{2/3}.$$

[*Note:* It is not necessary to write out the multiplication by λ as we did. Most people just do this mentally and then differentiate.]

2. The quantity to be maximized is the volume xyl. The constraint is that length plus girth is 84. This translates to $84 = l + 2x + 2y$ or $84 - l - 2x - 2y = 0$. Therefore,

$$F(x, y, l, \lambda) = xyl + \lambda(84 - l - 2x - 2y).$$

Chapter 17: CHECKLIST

☐ Level curve
☐ Partial derivative

☐ $\dfrac{\partial f}{\partial x}(a, b), \qquad \dfrac{\partial f}{\partial y}(a, b)$

☐ Second partial derivatives
☐ Maxima and minima in several variables
☐ Second-derivative test in two variables
☐ Method of Lagrange multipliers:
  objective function
  constraint equation

Chapter 17: SUPPLEMENTARY EXERCISES

1. Let $f(x, y) = x\sqrt{y}/(1 + x)$. Compute $f(2, 9), f(5, 1)$, and $f(0, 0)$.

2. Let $f(x, y, z) = x^2 e^{y/z}$. Compute $f(-1, 0, 1), f(1, 3, 3)$, and $f(5, -2, 2)$.

3. If A dollars is deposited in a bank (assuming 6% continuous interest rate), the amount in the account after t years is $f(A, t) = Ae^{.06t}$. Find and interpret $f(10, 11.5)$.

4. Let $f(x, y, \lambda) = xy + \lambda(5 - x - y)$. Find $f(1, 2, 3)$.

5. Let $f(x, y) = 3x^2 + xy + 5y^2$. Find $\dfrac{\partial f}{\partial x}$ and $\dfrac{\partial f}{\partial y}$.

6. Let $f(x, y) = 3x - \tfrac{1}{2}y^4 + 1$. Find $\dfrac{\partial f}{\partial x}$ and $\dfrac{\partial f}{\partial y}$.

7. Let $f(x, y) = e^{x/y}$. Find $\dfrac{\partial f}{\partial x}$ and $\dfrac{\partial f}{\partial y}$.

8. Let $f(x, y) = x/(x - 2y)$. Find $\dfrac{\partial f}{\partial x}$ and $\dfrac{\partial f}{\partial y}$.

9. Let $f(x, y, z) = x^3 - yz^2$. Find $\dfrac{\partial f}{\partial x}, \dfrac{\partial f}{\partial y}$ and $\dfrac{\partial f}{\partial z}$.

10. Let $f(x, y, \lambda) = xy + \lambda(5 - x - y)$. Find $\dfrac{\partial f}{\partial x}, \dfrac{\partial f}{\partial y}$, and $\dfrac{\partial f}{\partial \lambda}$.

11. Let $f(x, y) = x^3y + 8$. Compute $\dfrac{\partial f}{\partial x}(1, 2)$ and $\dfrac{\partial f}{\partial y}(1, 2)$.

12. Let $f(x, y, z) = (x + y)z$. Evaluate $\dfrac{\partial f}{\partial y}$ at $(x, y, z) = (2, 3, 4)$.

13. Let $f(x, y) = x^5 - 2x^3y + \dfrac{1}{2}y^4$. Find $\dfrac{\partial^2 f}{\partial x^2}, \dfrac{\partial^2 f}{\partial y^2}, \dfrac{\partial^2 f}{\partial x\,\partial y}$, and $\dfrac{\partial^2 f}{\partial y\,\partial x}$.

14. Let $f(x, y) = 2x^3 + x^2y - y^2$. Compute $\dfrac{\partial^2 f}{\partial x^2}, \dfrac{\partial^2 f}{\partial y^2}$, and $\dfrac{\partial^2 f}{\partial x\,\partial y}$ at $(x, y) = (1, 2)$.

15. A dealer in a certain brand of electronic calculator finds that (within certain limits) the number of calculators he can sell is given by $f(p, t) = -p + 6t - .02pt$, where p is the price of the calculator and t is the number of dollars spent on advertising. Compute $\dfrac{\partial f}{\partial p}(25, 10{,}000)$ and $\dfrac{\partial f}{\partial t}(25, 10{,}000)$ and interpret these numbers.

16. Suppose that the crime rate in a certain city can be approximated by a function $f(x, y, z)$, where x is the unemployment rate, y is the amount of social services available, and z is the size of the police force. Explain why $\dfrac{\partial f}{\partial x} > 0, \dfrac{\partial f}{\partial y} < 0$, and $\dfrac{\partial f}{\partial z} < 0$.

In Exercises 17–20, find all points (x, y) where $f(x, y)$ has a possible maximum or minimum.

17. $f(x, y) = -x^2 + 2y^2 + 6x - 8y + 5$.

18. $f(x, y) = x^2 + 3xy - y^2 - x - 8y + 4$.

19. $f(x, y) = x^3 + 3x^2 + 3y^2 - 6y + 7$.

20. $f(x, y) = \frac{1}{2}x^2 + 4xy + y^3 + 8y^2 + 3x + 2$.

In Exercises 21–23, find all points (x, y) where $f(x, y)$ has a possible maximum or minimum. Then use the second-derivative test to determine, if possible, the nature of $f(x, y)$ at each of these points. If the second-derivative test is inconclusive, so state.

21. $f(x, y) = x^2 + 3xy + 4y^2 - 13x - 30y + 12$.

22. $f(x, y) = 7x^2 - 5xy + y^2 + x - y + 6$.

23. $f(x, y) = x^3 + y^2 - 3x - 8y + 12$.

24. Find the values of x, y, z at which

$$f(x, y, z) = x^2 + 4y^2 + 5z^2 - 6x + 8y + 3$$

assumes its minimum value.

25. Using the method of Lagrange multipliers, maximize the function $3x^2 + 2xy - y^2$, subject to the constraint $5 - 2x - y = 0$.

26. Using the method of Lagrange multipliers, find the values of x, y that minimize the function $-x^2 - 3xy - \frac{1}{2}y^2 + y + 10$, subject to the constraint $10 - x - y = 0$.

27. Using the method of Lagrange multipliers, find the values of x, y, z that minimize the function $3x^2 + 2y^2 + z^2 + 4x + y + 3z$, subject to the constraint $4 - x - y - z = 0$.

28. Using the method of Lagrange multipliers, find the dimensions of the rectangular box of volume 1000 cubic inches for which the sum of the dimensions is minimized.

29. A person wants to plant a rectangular garden along one side of a house and put a fence on the other three sides of the garden. (See Fig. 1). Using the method of Lagrange multipliers, find the dimensions of the garden of greatest area that can be enclosed by using 40 feet of fencing.

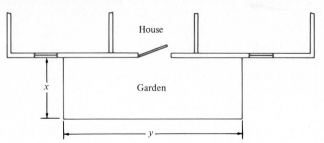

FIGURE 1

30. The present value of y dollars after x years at 15% continuous interest is $f(x, y) = ye^{-.15x}$. Sketch some sample level curves. (Economists call this collection of level curves a "discount system.")

Appendix Tables

TABLE 1 Areas Under the Standard Normal Curve

z	A(z)	z	A(z)	z	A(z)	z	A(z)	z	A(z)
−3.50	.0002	−2.00	.0228	−.50	.3085	1.00	.8413	2.50	.9938
−3.45	.0003	−1.95	0256	−.45	.3264	1.05	.8531	2.55	.9946
−3.40	.0003	−1.90	.0287	−.40	.3446	1.10	.8643	2.60	.9953
−3.35	.0004	−1.85	.0322	−.35	.3632	1.15	.8749	2.65	.9960
−3.30	.0005	−1.80	.0359	−.30	.3821	1.20	.8849	2.70	.9965
−3.25	.0006	−1.75	.0401	−.25	.4013	1.25	.8944	2.75	.9970
−3.20	.0007	−1.70	.0446	−.20	.4207	1.30	.9032	2.80	.9974
−3.15	.0008	−1.65	.0495	−.15	.4404	1.35	.9115	2.85	.9978
−3.10	.0010	−1.60	.0548	−.10	.4602	1.40	.9192	2.90	.9981
−3.05	.0011	−1.55	.0606	−.05	.4801	1.45	.9265	2.95	.9984
−3.00	.0013	−1.50	.0668	.00	.5000	1.50	.9332	3.00	.9987
−2.95	.0016	−1.45	.0735	.05	.5199	1.55	.9394	3.05	.9989
−2.90	.0019	−1.40	.0808	.10	.5398	1.60	.9452	3.10	.9990
−2.85	.0022	−1.35	.0885	.15	.5596	1.65	.9505	3.15	.9992
−2.80	.0026	−1.30	.0968	.20	.5793	1.70	.9554	3.20	.9993
−2.75	.0030	−1.25	.1056	.25	.5987	1.75	.9599	3.25	.9994
−2.70	.0035	−1.20	.1151	.30	.6179	1.80	.9641	3.30	.9995
−2.65	.0040	−1.15	.1251	.35	.6368	1.85	.9678	3.35	.9996
−2.60	.0047	−1.10	.1357	.40	.6554	1.90	.9713	3.40	.9997
−2.55	.0054	−1.05	.1469	.45	.6736	1.95	.9744	3.45	.9997
−2.50	.0062	−1.00	.1587	.50	.6915	2.00	.9772	3.50	.9998
−2.45	.0071	−.95	.1711	.55	.7088	2.05	.9798		
−2.40	.0082	−.90	.1841	.60	.7257	2.10	.9821		
−2.35	.0094	−.85	.1977	.65	.7422	2.15	.9842		
−2.30	.0107	−.80	.2119	.70	.7580	2.20	.9861		
−2.25	.0122	−.75	.2266	.75	.7734	2.25	.9878		
−2.20	.0139	−.70	.2420	.80	.7881	2.30	.9893		
−2.15	.0158	−.65	.2578	.85	.8023	2.35	.9906		
−2.10	.0179	−.60	.2743	.90	.8159	2.40	.9918		
−2.05	.0202	−.55	.2912	.95	.8289	2.45	.9929		

TABLE 2 $(1 + i)^n$ Compound Amount of \$1 Invested for n Interest Periods at Interest Rate i per Period

n	$\frac{1}{50}\%$	$\frac{1}{2}\%$	1%	2%	6%
1	1.000200000	1.00500000	1.01000000	1.02000000	1.06000000
2	1.000400040	1.01002500	1.02010000	1.04040000	1.12360000
3	1.000600120	1.01507513	1.03030100	1.06120800	1.19101600
4	1.000800240	1.02015050	1.04060401	1.08243216	1.26247696
5	1.001000400	1.02525125	1.05101005	1.10408080	1.33822558
6	1.001200600	1.03037751	1.06152015	1.12616242	1.41851911
7	1.001400840	1.03552940	1.07213535	1.14868567	1.50363026
8	1.001601120	1.04070704	1.08285671	1.17165938	1.59384807
9	1.001601447	1.04591058	1.09368527	1.19509257	1.68947896
10	1.002001806	1.05114013	1.10462213	1.21899442	1.79084770
11	1.002202201	1.05639583	1.11566835	1.24337431	1.89829856
12	1.002402642	1.06167781	1.12682503	1.26824179	2.01219647
13	1.002603122	1.06698620	1.13809328	1.29360663	2.13292826
14	1.002802643	1.07232113	1.14947421	1.31947876	2.26090396
15	1.003004204	1.07768274	1.16096896	1.34586834	2.39655819
16	1.003204804	1.08307115	1.17257864	1.37278571	2.54035168
17	1.007405445	1.08848651	1.18430443	1.40024142	2.69277279
18	1.0035051	1.09392894	1.19614748	1.42824625	2.85433915
19	1.003806848	1.09939858	1.20810895	1.45681117	3.02559950
20	1.004007609	1.10489558	1.22019004	1.48594740	3.20713547
21	1.004208411	1.11042006	1.23239194	1.51566634	3.39956360
22	1.004409252	1.11597216	1.24471586	1.54597967	3.60353742
23	1.004610134	1.12155202	1.25716302	1.57689926	3.81974966
24	1.004811056	1.12715978	1.26973465	1.60843725	4.40893464
25	1.005012018	1.13279558	1.28243200	1.64060599	4.29187072
26	1.005213021	1.13845955	1.29525631	1.67341811	4.54938296
27	1.005414063	1.14415185	1.30820888	1.70688648	4.82234594
28	1.005615146	1.14987261	1.32129097	1.74102421	5.11168670
29	1.005816269	1.15562197	1.33450388	1.77584469	5.41838790
30	1.006017433	1.16140008	1.34784892	1.81136158	5.74349117
36	1.007225257	1.19668052	1.43076878	2.03988734	8.14725200
48	1.009645259	1.27048916	1.61222608	2.58707039	16.39387173
52	1.010453217	1.29609015	1.67768892	2.80032819	20.69688534
60	1.012071075	1.34885015	1.81669670	3.28103079	32.98769085
104	1.021015784	1.67984969	2.81464012	7.84183795	428.36106292
120	1.024287860	1.81939673	3.30038689	10.76516303	1,088.18774784
180	1.036652115	2.45409356	5.99580198	35.32083136	35,896.80101597
240	1.049165620	3.31020448	10.89255365	115.88873515	
300	1.061830136	4.46496981	19.78846626	380.23450806	
360	1.074647608	6.02257521	35.94964133	1,247.56112775	
365	1.075722685	6.17465278	37.78343433	1,377.40829197	

TABLE 3 $\dfrac{1}{(1+i)^n}$ Present Value of $1. Principal That Will Accumulate
to $1 in n Interest Periods at a Compound Rate of i per Period

			i		
n	1/50%	1/2%	1%	2%	6%
1	.99980004	.99502488	.99009901	.98039216	.94339623
2	.99960012	.99007450	.98029605	.96116878	.88999644
3	.99940024	.98514876	.97059015	.94232233	.83961928
4	.99920040	.98024752	.96098034	.92384543	.79209366
5	.99900060	.97537067	.95146569	.90573081	.74725817
6	.99880084	.97051808	.94204524	.88797138	.70496054
7	.99860112	.96568963	.93271805	.87056018	.66505711
8	.99840144	.96088520	.92348322	.85349037	.62741237
9	.99820180	.95610468	.91433982	.83675527	.59189846
10	.99800220	.95134794	.90528695	.82034830	.55839478
11	.99780264	.94661487	.89632372	.80426304	.52678753
12	.99760312	.94190534	.88744923	.78849318	.49696936
13	.99740364	.93721924	.87866260	.77303253	.46883902
14	.99720420	.93255646	.86996297	.75787502	.44230096
15	.99700479	.92791688	.86134947	.74301473	.41726506
16	.99680543	.92330037	.85282126	.72844581	.39364628
17	.99660611	.91870684	.84437749	.71416256	.37136442
18	.99640683	.91413616	.83601731	.70015937	.35034379
19	.99620759	.90958822	.82773992	.68643076	.33051301
20	.99600839	.90506290	.81954447	.67297133	.31180473
21	.99580923	.90056010	.81143017	.65977582	.29415540
22	.99561010	.89607971	.80339621	.64683904	.27750510
23	.99541102	.89162160	.79544179	.63415592	.26179726
24	.99521198	.88718567	.78756613	.62172149	.24697855
25	.99501298	.88277181	.77976844	.60953087	.23299863
26	.99481401	.87837991	.77204796	.59757928	.21981003
27	.99461509	87400986	.76440392	.58586204	.20736795
28	.99441621	.86966155	.75683557	.57437455	.19563014
29	.99421736	.86533488	.74934215	.56311231	.18455674
30	.99401856	.86102973	.74192292	.55207089	.17411013
36	.99282657	.83564492	.69892495	.49022315	.12274077
48	.99044688	.78709841	.62026041	.38653761	.06099840
52	.98965492	.77155127	.59605806	.35710100	.04831645
60	.98807290	.74137220	.55044962	.30478227	.03031434
104	.97941686	.59529136	.35528521	.12752113	.00233448
120	.97628805	.54963273	.30299478	.09289223	.00091896
180	.96464377	.40748243	.16678336	.02831190	.00002786
240	.95313836	.30209614	.09180584	.00862897	.00000084
300	.94177018	.22396568	.05053449	.00262996	.00000003
360	.93053759	.16604193	.02781669	.00080156	.00000000
365	.92960762	.16195243	.02646663	.00072600	.00000000

TABLE 4 $s_{\overline{n}|i}$ Future Value of an Ordinary Annuity of n $1 Payments Each, Immediately
After the Last Payment at Compound Interest Rate of i per Period

n	1/2%	1%	1.5%	2%	6%
1	1.00000000	1.00000000	1.00000000	1.00000000	1.00000000
2	2.00500000	2.01000000	2.01500000	2.02000000	2.06000000
3	3.01502500	3.03010000	3.04522500	3.06040000	3.18360000
4	4.03010013	4.06040100	4.09090337	4.12160800	4.37461600
5	5.05025063	5.10100501	5.15226693	5.20404016	5.63709296
6	6.07550188	6.15201506	6.22955093	6.30812096	6.97531854
7	7.10587939	7.21353521	7.32299419	7.43428338	8.39383765
8	8.14140879	8.28567056	8.43283911	8.58296905	9.89746791
9	9.18211583	9.36852727	9.55933169	9.75462843	11.49131598
10	10.22802641	10.46221254	10.70272167	10.94972100	13.18079494
11	11.27916654	11.56683467	11.86326249	12.16871542	14.97164264
12	12.33556237	12.68250301	13.04121143	13.41208973	16.86994120
13	13.39724018	13.80932804	14.23682960	14.68033152	18.88213767
14	14.46422639	14.94742132	15.45038205	15.97393815	21.01506593
15	15.53654752	16.09689554	16.68213778	17.29341692	23.27596988
16	16.61423026	17.25786449	17.93236984	18.63928525	25.67252808
17	17.69730141	18.43044314	19.20135539	20.01207096	28.21287976
18	18.78578791	19.61474757	20.48937572	21.41231238	30.90565255
19	19.87971685	20.81089504	21.79671636	22.84055863	33.75999170
20	20.97911544	22.01900399	23.12366710	24.29736980	36.78559120
21	22.08401101	23.23919403	24.47052211	25.78331719	39.99272668
22	23.19443107	24.47158598	25.83757994	27.29898354	43.39229028
23	24.31040322	25.71630183	27.22514364	28.84496321	46.99582769
24	25.43195524	26.97346485	28.63352080	30.42186247	50.81557735
25	26.55911502	28.24319950	30.06302361	32.03029972	54.86451200
26	27.69191059	29.52563150	31.51396896	33.67090572	59.15638272
27	28.83037015	30.82088781	32.98667850	35.34432383	63.70576568
28	29.97452200	32.12909669	34.48147867	37.05121031	68.52811162
29	31.12439461	33.45038766	35.99870085	38.79223451	73.63979832
30	32.28001658	34.78489153	37.53868137	40.56807921	79.05818622
36	39.33610496	43.07687836	47.27596921	51.99436719	119.12086666
48	54.09783222	61.22260777	69.56521929	79.35351927	256.56452882
52	59.21803075	67.76889215	77.92489152	90.01640927	328.28142239
60	69.77003051	81.66966986	96.21465171	114.05153942	533.12818089
104	135.96993732	181.46401172	246.93411381	342.09189731	7,122.68438195
120	163.87934681	230.03868946	331.28819149	488.25815171	18,119.79579725
180	290.81871245	499.58019754	905.62451261	1,716.04156785	
240	462.04089516	989.25536539	2,308.85437027	5,744.43675765	
300	692.99396243	1,878.84662619	5,737.25330834	18,961.72540308	
360	1,004.51504245	3,494.96413277	14,113.58539279	62,328.05638744	
365	1,034.93055669	3,678.34343329	15,209.49204803	68,820.41459830	

TABLE 5 $\dfrac{1}{s_{\overline{n}|i}}$ Rent per Period for an Ordinary Annuity of n Payments, Compound Interest Rate i per Period, and Future Value \$1

			i		
n	1/2%	1%	1.5%	2%	6%
1	1.00000000	1.00000000	1.00000000	1.00000000	1.00000000
2	.49875312	.49751244	.49627792	.49504950	.48543689
3	.33167221	.33002211	.32838296	.32675467	.31410981
4	.24813279	.24628109	.24444479	.24262375	.22859149
5	.19800997	.19603980	.19408932	.19215839	.17739640
6	.16459546	.16254837	.16052521	.15852581	.14336263
7	.14072854	.13862828	.13655616	.13451196	.11913502
8	.12282886	.12069029	.11858402	.11650980	.10103594
9	.10890736	.10674036	.10460982	.10251544	.08702224
10	.09777057	.09558208	.09343418	.09132653	.07586796
11	.08865903	.08645408	.08429384	.08217794	.06679294
12	.08106643	.07884879	.07667999	.07455960	.05927703
13	.07464224	.07241482	.07024036	.06811835	.05296011
14	.06913609	.06690117	.06472332	.06260197	.04758491
15	.06436436	.06212378	.05994436	.05782547	.04296276
16	.06018937	.05794460	.05576508	.05365013	.03895214
17	.05650579	.05425806	.05207966	.04996984	.03544480
18	.05323173	.05098205	.04880578	.04670210	.03235654
19	.05030253	.04805175	.04587847	.04378177	.02962086
20	.04766645	.04541531	.04324574	.04115672	.02718456
21	.04528163	.04303075	.04086550	.03878477	.02500455
22	.04311380	.04086372	.03870332	.03663140	.02304557
23	.04113465	.03888584	.03673075	.03466810	.02127848
24	.03932061	.03707347	.03492410	.03287110	.01967900
25	.03765186	.03540675	.03326345	.03122044	.01822672
26	.03611163	.03386888	.03173196	02969923	.01690435
27	.03468565	.03244553	.03031527	.02829309	.01569717
28	.03336167	.03112444	.02900108	.02698967	.01459255
29	.03212914	.02989502	.02777878	.02577836	.01357961
30	.03097892	.02874811	.02663919	.02464992	.01264891
36	.02542194	.02321431	.02115240	.01923285	.00839483
48	.01848503	.01633384	.01437500	.01260184	.00389765
52	.01688675	.01475603	.01283287	.01110909	.00304617
60	.01433280	.01224445	.01039343	.00876797	.00187572
104	.00735457	.00551073	.00404966	.00292319	.00014040
120	.00610205	.00434709	.00301852	.00204810	.00005519
180	.00343857	.00200168	.00110421	.00058274	.00000167
240	.00216431	.00101086	.00043312	.00017408	.00000005
300	.00144301	.00053224	.00017430	.00005274	.00000000
360	.00099551	.00028613	.00007085	.00001604	.00000000
365	.00096625	.00027186	.00006575	.00001453	.00000000

TABLE 6 $a_{\overline{n}|i}$ Present Value of an Ordinary Annuity of n Payments of $1 One Period
Before the First Payment with Interest Compounded at i per Period

			i		
n	1/2%	1%	1.5%	2%	6%
1	.99502488	.99009901	.98522167	.98039216	.94339623
2	1.98509938	1.97039506	1.95588342	1.94156094	1.83339267
3	2.97024814	2.94098521	2.91220042	2.88388327	2.67301195
4	3.95049566	3.90196555	3.85438465	3.80772870	3.46510561
5	4.92586633	4.85343124	4.78264497	4.71345951	4.21236379
6	5.89638441	5.79547647	5.69718717	5.60143089	4.91732433
7	6.86207404	6.72819453	6.59821396	6.47199107	5.58238144
8	7.82295924	7.65167775	7.48592508	7.32548144	6.20979381
9	8.77906392	8.56601758	8.36051732	8.16223671	6.80169227
10	9.73041186	9.47130453	9.22218455	8.98258501	7.36008705
11	10.67702673	10.36762825	10.07111779	9.78684805	7.88687458
12	11.61893207	11.25507747	10.90750521	10.57534122	8.38384394
13	12.55615131	12.13374007	11.73153222	11.34837375	8.85268296
14	13.48870777	13.00370304	12.54338150	12.10624877	9.29498393
15	14.41662465	13.86505252	13.34323301	12.84926350	9.71224899
16	15.33992502	14.71787378	14.13126405	13.57770931	10.10589527
17	16.25863186	15.56225127	14.90764931	14.29187188	10.47725969
18	17.17276802	16.39826858	15.67256089	14.99203125	10.82760348
19	18.08235624	17.22600850	16.42616837	15.67846201	11.15811649
20	18.98741915	18.04555297	17.16863879	16.35143334	11.46992122
21	19.88797925	18.85698313	17.90013673	17.01120916	11.76407662
22	20.78405896	19.66037934	18.62082437	17.65804820	12.04158172
23	21.67568055	20.45582113	19.33086145	18.29220412	12.30337898
24	22.56286622	21.24338726	20.03040537	18.91392560	12.55035753
25	23.44563803	22.02315570	20.71961120	19.52345647	12.78335616
26	24.32401794	22.79520366	21.39863172	20.12103576	13.00316619
27	25.19802780	23.55960759	22.06761746	20.70689780	13.21053414
28	26.06768936	24.31644316	22.72671671	21.28127236	13.40616428
29	26.93302423	25.06578530	23.37607558	21.84438466	13.59072102
30	27.79405397	25.80770822	24.01583801	22.39645555	13.76483115
36	32.87101624	30.10750504	27.66068431	25.48884248	14.62098713
48	42.58031778	37.97395949	34.04255365	30.67311957	15.65002661
52	45.68974664	40.39419423	35.92874185	32.14494992	15.86139252
60	51.72556075	44.95503841	39.38026889	34.76088668	16.16142771
104	80.94172854	64.47147918	52.49436634	43.62394373	16.62775868
120	90.07345333	69.70052203	55.49845411	45.35538850	16.65135068
180	118.50351467	83.32166399	62.09556231	48.58440478	16.66620237
240	139.58077168	90.81941635	64.79573209	49.56855168	16.66665259
300	155.20686401	94.94655125	65.90090069	49.86850220	16.66666624
360	166.79161439	97.21833108	66.35324174	49.95992180	16.66666665
365	167.60951473	97.35333747	66.37572674	49.96369994	16.66666666

TABLE 7 $\dfrac{1}{a_{\overline{n}|i}}$ Rent per Period for an Ordinary Annuity of *n* Payments Whose
Present Value is $1 with Interest Compounded at *i* per Period

			i		
n	1/2%	1%	1.5%	2%	6%
1	1.00500000	1.01000000	1.01500000	1.02000000	1.06000000
2	.50375312	.50751244	.51127792	.51504950	.54543689
3	.33667221	.34002211	.34338296	.34675467	.37410981
4	.25313279	.25628109	.25944479	.26262375	.28859149
5	.20300997	.20603980	.20908932	.21215839	.23739640
6	.16959546	.17254837	.17552521	.17852581	.20336263
7	.14572854	.14862828	.15155616	.15451196	.17913502
8	.12782886	.13069029	.13358402	.13650980	.16103594
9	.11390736	.11674036	.11960982	.12251544	.14702224
10	.10277057	.10558208	.10843418	.11132653	.13586796
11	.09365903	.09645408	.09929384	.10217794	.12679294
12	.08606643	.08884879	.09167999	.09455960	.11927703
13	.07964224	.08241482	.08524036	.08811835	.11296011
14	.07413609	.07690117	.07972332	.08260197	.10758491
15	.06936436	.07212378	.07494436	.07782547	.10296276
16	.06518937	.06794460	.07076508	.07365013	.09895214
17	.06150579	.06425806	.06707966	.06996984	.09544480
18	.05823173	.06098205	.06380578	.06670210	.09235654
19	.05530253	.05805175	.06087847	.06378177	.08962086
20	.05266645	.05541531	.05824574	.06115672	.08718456
21	.05028163	.05303075	.05586550	.05878477	.08500455
22	.04811380	.05086372	.05370332	.05663140	.08304557
23	.04613465	.04888584	.05173075	.05466810	.08127848
24	.04432061	.04707347	.04992410	.05287110	.07967900
25	.04265186	.04540675	.04826345	.05122044	.07822672
26	.04111163	.04386888	.04673196	.04969923	.07690435
27	.03968565	.04244553	.04531527	.04829309	.07569717
28	.03836167	.04112444	.04400108	.04698967	.07459255
29	.03712914	.03989502	.04277878	.04577836	.07357961
30	.03597892	.03874811	.04163919	.04464992	.07264891
36	.03042194	.03321431	.03615240	.03923285	.06839483
48	.02348503	.02633384	.02937500	.03260184	.06389765
52	.02188675	.02475603	.02783287	.03110909	.06304617
60	.01933280	.02224445	.02539343	.02876797	.06187572
104	.01235457	.01551073	.01904966	.02292319	.06014040
120	.01110205	.01434709	.01801852	.02204810	.06005519
180	.00843857	.01200168	.01610421	.02058274	.06000167
240	.00716431	.01101086	.01543312	.02017408	.06000005
300	.00644301	.01053224	.01517430	.02005274	.06000000
360	.00599551	.01028613	.01507085	.02001604	.06000000
365	.00596625	.01027186	.01506575	.02001453	.06000000

TABLE 8   The Exponential Function

x	e^x	e^{-x}	x	e^x	e^{-x}	x	e^x	e^{-x}
.00	1.00000	1.00000	**.40**	1.49182	.67032	**.80**	2.22554	.44933
.01	1.01005	.99005	.41	1.50682	.66365	.81	2.24791	.44486
.02	1.02020	.98020	.42	1.52196	.65705	.82	2.27050	.44043
.03	1.03045	.97045	.43	1.53726	.65051	.83	2.29332	.43605
.04	1.04081	.96079	.44	1.55271	.64404	.84	2.31637	.43171
.05	1.05127	.95123	.45	1.56831	.63763	.85	2.33965	.42741
.06	1.06184	.94176	.46	1.58407	.63128	.86	2.36316	.42316
.07	1.07251	.93239	.47	1.59999	.62500	.87	2.38691	.41895
.08	1.08329	.92312	.48	1.61607	.61878	.88	2.41090	.41478
.09	1.09417	.91393	.49	1.63232	.61263	.89	2.43513	.41066
.10	1.10517	.90484	**.50**	1.64872	.60653	**.90**	2.45960	.40657
.11	1.11628	.89583	.51	1.66529	.60050	.91	2.48432	.40252
.12	1.12750	.88692	.52	1.68203	.59452	.92	2.50929	.39852
.13	1.13883	.87810	.53	1.69893	.58860	.93	2.53451	.39455
.14	1.15027	.86936	.54	1.71601	.58275	.94	2.55998	.39063
.15	1.16183	.86071	.55	1.73325	.57695	.95	2.58571	.38674
.16	1.17351	.85214	.56	1.75067	.57121	.96	2.61170	.38289
.17	1.18530	.84366	.57	1.76827	.56553	.97	2.63794	.37908
.18	1.19722	.83527	.58	1.78604	.55990	.98	2.66446	.37531
.19	1.20925	.82696	.59	1.80399	.55433	.99	2.69123	.37158
.20	1.22140	.81873	**.60**	1.82212	.54881	**1.00**	2.71828	.36788
.21	1.23368	.81058	.61	1.84043	.54335	1.01	2.74560	.36422
.22	1.24608	.80252	.62	1.85893	.53794	1.02	2.77319	.36059
.23	1.25860	.79453	.63	1.87761	.53259	1.03	2.80107	.35701
.24	1.27125	.78663	.64	1.89648	.52729	1.04	2.82922	.35345
.25	1.28403	.77880	.65	1.91554	.52205	1.05	2.85765	.34994
.26	1.29693	.77105	.66	1.93479	.51685	1.06	2.88637	.34646
.27	1.30996	.76338	.67	1.95424	.51171	1.07	2.91538	.34301
.28	1.32313	.75578	.68	1.97388	.50662	1.08	2.94468	.33960
.29	1.33643	.74826	.69	1.99372	.50158	1.09	2.97427	.33622
.30	1.34986	.74082	**.70**	2.01375	.49659	**1.10**	3.00417	.33287
.31	1.36343	.73345	.71	2.03399	.49164	1.11	3.03436	.32956
.32	1.37713	.72615	.72	2.05443	.48675	1.12	3.06485	.32628
.33	1.39097	.71892	.73	2.07508	.48191	1.13	3.09566	.32303
.34	1.40495	.71177	.74	2.09594	.47711	1.14	3.12677	.31982
.35	1.41907	.70469	.75	2.11700	.47237	1.15	3.15819	.31664
.36	1.43333	.69768	.76	2.13828	.46767	1.16	3.18993	.31349
.37	1.44773	.69073	.77	2.15977	.46301	1.17	3.22199	.31037
.38	1.46228	.68386	.78	2.18147	.45841	1.18	3.25437	.30728
.39	1.47698	.67706	.79	2.20340	.45384	1.19	3.28708	.30422

TABLE 8 The Exponential Function (*continued*)

x	e^x	e^{-x}	x	e^x	e^{-x}	x	e^x	e^{-x}
1.20	3.32012	.30119	**1.60**	4.95303	.20190	**4.0**	54.598	.01832
1.21	3.35348	.29820	1.61	5.00281	.19989	4.1	60.340	.01657
1.22	3.38719	.29523	1.62	5.05309	.19790	4.2	66.686	.01500
1.23	3.42123	.29229	1.63	5.10387	.19593	4.3	73.700	.01357
1.24	3.45561	.28938	1.64	5.15517	.19398	4.4	81.451	.01228
1.25	3.49034	.28650	1.65	5.20698	.19205	4.5	90.017	.01111
1.26	3.52542	.28365	1.66	5.25931	.19014	4.6	99.484	.01005
1.27	3.56085	.28083	1.67	5.31217	.18825	4.7	109.947	.00910
1.28	3.59664	.27804	1.68	5.36556	.18637	4.8	121.510	.00823
1.29	3.63279	.27527	1.69	5.41948	.18452	4.9	134.290	.00745
1.30	3.66930	.27253	**1.70**	5.47395	.18268	**5.0**	148.41	.00674
1.31	3.70617	.26982	1.71	5.52896	.18087	5.1	164.02	.00610
1.32	3.74342	.26714	1.72	5.58453	.17907	5.2	181.27	.00552
1.33	3.78104	.26448	1.73	5.64065	.17728	5.3	200.34	.00499
1.34	3.81904	.26185	1.74	5.69734	.17552	5.4	221.41	.00452
1.35	3.85743	.25924	1.75	5.75460	.17377	5.5	244.69	.00409
1.36	3.89619	.25666	1.80	6.04965	.16530	5.6	270.43	.00370
1.37	3.93535	.25411	1.85	6.35982	.15724	5.7	298.87	.00335
1.38	3.97490	.25158	1.90	6.68589	.14957	5.8	330.30	.00303
1.39	4.01485	.24908	1.95	7.02869	.14227	5.9	365.04	.00274
1.40	4.05520	.24660	**2.0**	7.3891	.13534	**6.0**	403.43	.00248
1.41	4.09596	.24414	2.1	8.1662	.12246	6.1	445.86	.00224
1.42	4.13712	.24171	2.2	9.0250	.11080	6.2	492.75	.00203
1.43	4.17870	.23931	2.3	9.9742	.10026	6.3	544.57	.00184
1.44	4.22070	.23693	2.4	11.0232	.09072	6.4	601.85	.00166
1.45	4.26311	.23457	2.5	12.1825	.08208	6.5	665.14	.00150
1.46	4.30596	.23224	2.6	13.4637	.07427	6.6	735.10	.00136
1.47	4.34924	.22993	2.7	14.8797	.06721	6.7	812.41	.00123
1.48	4.39295	.22764	2.8	16.4446	.06081	6.8	897.85	.00111
1.49	4.43710	.22537	2.9	18.1741	.05502	6.9	992.27	.00101
1.50	4.48169	.22313	**3.0**	20.086	.04979	**7.0**	1096.6	.00091
1.51	4.52673	.22091	3.1	22.198	.04505	7.5	1808.0	.00055
1.52	4.57223	.21871	3.2	24.533	.04076	8.0	2981.0	.00034
1.53	4.61818	.21654	3.3	27.113	.03688	8.5	4914.8	.00020
1.54	4.66459	.21438	3.4	29.964	.03337	9.0	8103.1	.00012
1.55	4.71147	.21225	3.5	33.115	.03020	9.5	13360	.00007
1.56	4.75882	.21014	3.6	36.598	.02732	10.0	22026	.00005
1.57	4.80665	.20805	3.7	40.447	.02472	10.5	36316	.00003
1.58	4.85496	.20598	3.8	44.701	.02237	11.0	59874	.00002
1.59	4.90375	.20393	3.9	49.402	.02024	11.5	98716	.00001

TABLE 9   The Natural Logarithm Function

x	$\ln x$	x	$\ln x$	x	$\ln x$	x	$\ln x$
		0.40	-0.91629	0.80	-0.22314	1.20	0.18232
0.01	-4.60517	0.41	-0.89160	0.81	-0.21072	1.21	0.19062
0.02	-3.91202	0.42	-0.86750	0.82	-0.19845	1.22	0.19885
0.03	-3.50656	0.43	-0.84397	0.83	-0.18633	1.23	0.20701
0.04	-3.21888	0.44	-0.82098	0.84	-0.17435	1.24	0.21511
0.05	-2.99573	0.45	-0.79851	0.85	-0.16252	1.25	0.22314
0.06	-2.81341	0.46	-0.77653	0.86	-0.15082	1.26	0.23111
0.07	-2.65926	0.47	-0.75502	0.87	-0.13926	1.27	0.23902
0.08	-2.52573	0.48	-0.73397	0.88	-0.12783	1.28	0.24686
0.09	-2.40795	0.49	-0.71335	0.89	-0.11653	1.29	0.25464
0.10	-2.30259	0.50	-0.69315	0.90	-0.10536	1.30	0.26236
0.11	-2.20727	0.51	-0.67334	0.91	-0.09431	1.31	0.27003
0.12	-2.12026	0.52	-0.65393	0.92	-0.08338	1.32	0.27763
0.13	-2.04022	0.53	-0.63488	0.93	-0.07257	1.33	0.28518
0.14	-1.96611	0.54	-0.61619	0.94	-0.06188	1.34	0.29267
0.15	-1.89712	0.55	-0.59784	0.95	-0.05129	1.35	0.30010
0.16	-1.83258	0.56	-0.57982	0.96	-0.04082	1.36	0.30748
0.17	-1.77196	0.57	-0.56212	0.97	-0.03046	1.37	0.31481
0.18	-1.71480	0.58	-0.54473	0.98	-0.02020	1.38	0.32208
0.19	-1.66073	0.59	-0.52763	0.99	-0.01005	1.39	0.32930
0.20	-1.60944	0.60	-0.51083	1.00	0.00000	1.40	0.33647
0.21	-1.56065	0.61	-0.49430	1.01	0.00995	1.41	0.34359
0.22	-1.51413	0.62	-0.47804	1.02	0.01980	1.42	0.35066
0.23	-1.46968	0.63	-0.46204	1.03	0.02956	1.43	0.35767
0.24	-1.42712	0.64	-0.44629	1.04	0.03922	1.44	0.36464
0.25	-1.38629	0.65	-0.43078	1.05	0.04879	1.45	0.37156
0.26	-1.34707	0.66	-0.41552	1.06	0.05827	1.46	0.37844
0.27	-1.30933	0.67	-0.40048	1.07	0.06766	1.47	0.38526
0.28	-1.27297	0.68	-0.38566	1.08	0.07696	1.48	0.39204
0.29	-1.23787	0.69	-0.37106	1.09	0.08618	1.49	0.39878
0.30	-1.20397	0.70	-0.35667	1.10	0.09531	1.50	0.40547
0.31	-1.17118	0.71	-0.34249	1.11	0.10436	1.51	0.41211
0.32	-1.13943	0.72	-0.32850	1.12	0.11333	1.52	0.41871
0.33	-1.10866	0.73	-0.31471	1.13	0.12222	1.53	0.42527
0.34	-1.07881	0.74	-0.30111	1.14	0.13103	1.54	0.43178
0.35	-1.04982	0.75	-0.28768	1.15	0.13976	1.55	0.43825
0.36	-1.02165	0.76	-0.27444	1.16	0.14842	1.56	0.44469
0.37	-0.99425	0.77	-0.26136	1.17	0.15700	1.57	0.45108
0.38	-0.96758	0.78	-0.24846	1.18	0.16551	1.58	0.45742
0.39	-0.94161	0.79	-0.23572	1.19	0.17395	1.59	0.46373

TABLE 9   The Natural Logarithm Function (*continued*)

x	$\ln x$	x	$\ln x$	x	$\ln x$	x	$\ln x$
1.60	0.47000	2.00	0.69315	6.00	1.79176	10.0	2.30259
1.61	0.47623	2.10	0.74194	6.10	1.80829	11.0	2.39790
1.62	0.48243	2.20	0.78846	6.20	1.82455	12.0	2.48491
1.63	0.48858	2.30	0.83291	6.30	1.84055	13.0	2.56495
1.64	0.49470	2.40	0.87547	6.40	1.85630	14.0	2.63906
1.65	0.50078	2.50	0.91629	6.50	1.87180	15.0	2.70805
1.66	0.50682	2.60	0.95551	6.60	1.88707	16.0	2.77259
1.67	0.51282	2.70	0.99325	6.70	1.90211	17.0	2.83321
1.68	0.51879	2.80	1.02962	6.80	1.91692	18.0	2.89037
1.69	0.52473	2.90	1.06471	6.90	1.93152	19.0	2.94444
1.70	0.53063	3.00	1.09861	7.00	1.94591	20.0	2.99573
1.71	0.53649	3.10	1.13140	7.10	1.96009	21.0	3.04452
1.72	0.54232	3.20	1.16315	7.20	1.97408	22.0	3.09104
1.73	0.54812	3.30	1.19392	7.30	1.98787	23.0	3.13549
1.74	0.55389	3.40	1.22378	7.40	2.00148	24.0	3.17805
1.75	0.55962	3.50	1.25276	7.50	2.01490	25.0	3.21888
1.76	0.56531	3.60	1.28093	7.60	2.02815	26.0	3.25810
1.77	0.57098	3.70	1.30833	7.70	2.04122	27.0	3.29584
1.78	0.57661	3.80	1.33500	7.80	2.05412	28.0	3.33220
1.79	0.58222	3.90	1.36098	7.90	2.06686	29.0	3.36730
1.80	0.58779	4.00	1.38629	8.00	2.07944	30.0	3.40120
1.81	0.59333	4.10	1.41099	8.10	2.09186	31.0	3.43399
1.82	0.59884	4.20	1.43508	8.20	2.10413	32.0	3.46574
1.83	0.60432	4.30	1.45862	8.30	2.11626	33.0	3.49651
1.84	0.60977	4.40	1.48160	8.40	2.12823	34.0	3.52636
1.85	0.61519	4.50	1.50408	8.50	2.14007	35.0	3.55535
1.86	0.62058	4.60	1.52606	8.60	2.15176	36.0	3.58352
1.87	0.62594	4.70	1.54756	8.70	2.16332	37.0	3.61092
1.88	0.63127	4.80	1.56862	8.80	2.17475	38.0	3.63759
1.89	0.63658	4.90	1.58924	8.90	2.18605	39.0	3.66356
1.90	0.64185	5.00	1.60944	9.00	2.19722	40.0	3.68888
1.91	0.64710	5.10	1.62924	9.10	2.20827	41.0	3.71357
1.92	0.65233	5.20	1.64866	9.20	2.21920	42.0	3.73767
1.93	0.65752	5.30	1.66771	9.30	2.23001	43.0	3.76120
1.94	0.66269	5.40	1.68640	9.40	2.24071	44.0	3.78419
1.95	0.66783	5.50	1.70475	9.50	2.25129	45.0	3.80666
1.96	0.67294	5.60	1.72277	9.60	2.26176	46.0	3.82864
1.97	0.67803	5.70	1.74047	9.70	2.27213	47.0	3.85015
1.98	0.68310	5.80	1.75786	9.80	2.28238	48.0	3.87120
1.99	0.68813	5.90	1.77495	9.90	2.29253	49.0	3.89182

TABLE 9   The Natural Logarithm Function (*continued*)

x	$\ln x$	x	$\ln x$	x	$\ln x$	x	$\ln x$
50.0	3.91202	90.0	4.49981	400.	5.99146	800.	6.68461
51.0	3.93183	91.0	4.51086	410.	6.01616	810.	6.69703
52.0	3.95124	92.0	4.52179	420.	6.04025	820.	6.70930
53.0	3.97029	93.0	4.53260	430.	6.06379	830.	6.72143
54.0	3.98898	94.0	4.54329	440.	6.08677	840.	6.73340
55.0	4.00733	95.0	4.55388	450.	6.10925	850.	6.74524
56.0	4.02535	96.0	4.56435	460.	6.13123	860.	6.75693
57.0	4.04305	97.0	4.57471	470.	6.15273	870.	6.76849
58.0	4.06044	98.0	4.58497	480.	6.17379	880.	6.77992
59.0	4.07754	99.0	4.59512	490.	6.19441	890.	6.79122
60.0	4.09434	100.	4.60517	500.	6.21461	900.	6.80239
61.0	4.11087	110.	4.70048	510.	6.23441	910.	6.81344
62.0	4.12713	120.	4.78749	520.	6.25383	920.	6.82437
63.0	4.14313	130.	4.86753	530.	6.27288	930.	6.83518
64.0	4.15888	140.	4.94164	540.	6.29157	940.	6.84588
65.0	4.17439	150.	5.01064	550.	6.30992	950.	6.85646
66.0	4.18965	160.	5.07517	560.	6.32794	960.	6.86693
67.0	4.20469	170.	5.13580	570.	6.34564	970.	6.87730
68.0	4.21951	180.	5.19296	580.	6.36303	980.	6.88755
69.0	4.23411	190.	5.24702	590.	6.38012	990.	6.89770
70.0	4.24850	200.	5.29832	600.	6.39693	1000.	6.90776
71.0	4.26268	210.	5.34711	610.	6.41346	—	—
72.0	4.27667	220.	5.39363	620.	6.42972	—	—
73.0	4.29046	230.	5.43808	630.	6.44572	—	—
74.0	4.30407	240.	5.48064	640.	6.46147	—	—
75.0	4.31749	250.	5.52146	650.	6.47697	—	—
76.0	4.33073	260.	5.56068	660.	6.49224	—	—
77.0	4.34381	270.	5.59842	670.	6.50728	—	—
78.0	4.35671	280.	5.63479	680.	6.52209	—	—
79.0	4.36945	290.	5.66988	690.	6.53669	—	—
80.0	4.38203	300.	5.70378	700.	6.55108	—	—
81.0	4.39445	310.	5.73657	710.	6.56526	—	—
82.0	4.40672	320.	5.76832	720.	6.57925	—	—
83.0	4.41884	330.	5.79909	730.	6.59304	—	—
84.0	4.43082	340.	5.82895	740.	6.60665	—	—
85.0	4.44265	350.	5.85793	750.	6.62007	—	—
86.0	4.45435	360.	5.88610	760.	6.63332	—	—
87.0	4.46591	370.	5.91350	770.	6.64639	—	—
88.0	4.47734	380.	5.94017	780.	6.65929	—	—
89.0	4.48864	390.	5.96615	790.	6.67203	—	—

Answers to Odd-Numbered Exercises

CHAPTER 1

1. $\frac{7}{2}$ 3. $\frac{7}{10}$ 5. $\frac{-1}{1}$ 7. $2\frac{13}{10}$ 9. $\frac{1}{20}$ 11. $\frac{1}{200}$ 13. 8   15. 3.6   17. -15 19. 10   21. 3   23. -5 25. 5   27. $-.6$

29. 1   31. -3 33. 1   35. -16 37. 0   39. 2   41. $-\frac{3}{40}$ 43. 12   45. $<$ 47. $>$ 49. $<$ 51. $>$ 53. $=$
55. $=$ 57. $a \geq 0$ 59. $b \geq a$ 61. $a \geq b$ 63. $a \leq 0$ 65. $8 < 12$ 67. $5 < 6$ 69. $2 < 7$ 71. $-3 > -7$
73. $2 \geq -6$

1. $\frac{16}{3}$ 3. $\frac{1}{2}$ 5. 3   7. 3   9. $\frac{9}{4}$ 11. 20   13. $\frac{14}{3}$ 15. 4   17. $-\frac{3}{2}$ 19. $-\frac{5}{4}$ 21. 1   23. 1   25. $x > 4$ 27. $x \geq -\frac{1}{2}$
29. $x > 12$ 31. $x \leq 15$ 33. $x \leq 5$ 35. $x \geq \frac{17}{3}$ 37. $x \geq \frac{1}{5}$

1. 27   3. 1   5. .0001   7. -16 9. 4   11. .01   13. $\frac{1}{6}$ 15. 100   17. 16   19. 125   21. 1   23. 4   25. $\frac{1}{2}$

27. 1000   29. 10   31. 6   33. 16   35. 18   37. $\frac{4}{9}$ 39. 7   41. $x^6 y^6$ 43. $x^3 y^3$ 45. $\frac{1}{\sqrt{x}}$ 47. $\frac{x^{12}}{y^6}$ 49. $x^{12} y^{20}$

51. $x^2 y^6$ 53. $16x^4$ 55. x^2 57. $\frac{1}{x^7}$ 59. x 61. $\frac{27x^6}{8y^3}$ 63. $2\sqrt{x}$ 65. $\frac{1}{8x^6}$ 67. $\frac{1}{32x^2}$ 69. $9x^3$ 71. $x - 1$

73. $1 + 6\sqrt{x}$ 77. 16   79. $\frac{1}{4}$ 81. 8   83. $\frac{1}{32}$

1. 5   3. $-\frac{2}{3}$ 5. $8y$ 7. $\frac{1}{3}xy$ 9. $-3xy$ 11. $\frac{2}{y}$ 13. $6x^2 y$ 15. $9x^2$ 17. $8x^2 + 6x + 7$ 19. $-1 + 2x - x^2$

21. $3x + 5y + 3$ 23. $xy + x - y$ 25. $6x^2 + 8x$ 27. $3x^2 + 10x$ 29. $xy + 3y + 2x + 6$ 31. $x^2 + x - 20$
33. $x^2 + 8x + 16$ 35. $8x^2 + 2x - 15$ 37. $9x^2 + 6x + 1$ 39. $4x^2 - 3x$ 41. $x^2 + 3x$ 43. $3r^2 - 2r$
45. $(x + 5)(x + 3)$ 47. $(x - 4)(x + 4)$ 49. $3(x + 2)^2$ 51. $-2(x + 5)(x - 3)$ 53. $x(3 - x)$
55. $-2x(x + \sqrt{3})(x - \sqrt{3})$ 57. $x(x + 3)^2$ 59. $\frac{1}{2}(x + 2)(x - 1)$ 61. $(x + 1)(x + 2)$

EXERCISES 1.5, page 22

1. $\frac{3}{4}$ **3.** 2 **5.** $\frac{5}{6}x$ **7.** $5x^2$ **9.** $x - 1$ **11.** $\dfrac{3}{3x + 1}$ **13.** $\dfrac{5}{6xy}$ **15.** $\dfrac{-1}{x + 3}$ **17.** $\dfrac{2x^2}{3x + 6}$ **19.** $\dfrac{x + 5}{x - 5}$ **21.** $\frac{13}{15}$ **23.** $\frac{14}{5}$

25. $-\frac{25}{108}$ **27.** $\dfrac{x + 1}{x^2 + 2x}$ **29.** $\dfrac{4x}{x - 3}$ **31.** $\dfrac{x^2 + 3x + 6}{x^2 + 2x}$ **33.** $\dfrac{1 - 2x}{x^2 + x}$ **35.** $\dfrac{2 + 2x - x^2}{(x^2 - 1)(x + 2)}$ **37.** $\dfrac{3}{x + 1}$ **39.** $\dfrac{x^2 + 1}{3}$

41. $\dfrac{2}{3x - 15}$ **43.** $\dfrac{1}{x(x - 1)}$ **45.** $\dfrac{4x^2 + 3x}{2x + 2}$ **47.** $\dfrac{5x^2 + 7x + 2}{x^3 + x^2 - x}$ **49.** $2 + \dfrac{x^2 - 3x - 10}{x^2 + 5x}$

EXERCISES 1.6, page 28

1. 5, 3 **3.** 4 **5.** 0, 17 **7.** 1, -2 **9.** 4, -4 **11.** 5, -4 **13.** 3, -6 **15.** 5, -1 **17.** 2, -4 **19.** 2, $\frac{3}{2}$ **21.** $\frac{3}{2}$
23. No solution **25.** 1, $-\frac{1}{5}$ **27.** 5, 4 **29.** $2 \pm \frac{1}{3}\sqrt{6}$ **31.** 3, -2 **33.** 3, -4 **35.** $(3 \pm \sqrt{5})/2$ **37.** $4 \pm \sqrt{5}$
39. 6, -6, 0 **41.** 4, 5 **43.** $(3 \pm \sqrt{5})/2$ **45.** 0, $-\frac{1}{2}$ **47.** No solution

EXERCISES 1.7, page 34

1. $\log 10{,}000 = 4$ **3.** $\log .5 = -.311$ **5.** $\log_2 8 = 3$ **7.** $\log_8 2 = \frac{1}{3}$ **9.** $\log_2 .125 = -3$ **11.** $\log 31.62 = \frac{3}{2}$
13. $10^{-1} = \frac{1}{10}$ **15.** $10^{.3979} = 2.5$ **17.** $2^4 = 16$ **19.** $36^{1/2} = 6$ **21.** $8^0 = 1$ **23.** $3^{1/4} = 1.316$ **25.** -2 **27.** 5
29. 1.5 **31.** $\frac{1}{2}$ **33.** 8 **35.** 5 **37.** $1 - x$ **39.** $x^2 + 1$ **41.** $1 + t$ **43.** $-.1549$ **45.** .19084 **47.** 333 **49.** 7
51. $1 \pm \sqrt{2}$ **53.** 6 **55.** 1.292 **57.** 1.199 **59.** .9208 **61.** \$2000, 14 years

EXERCISES 1.8, page 38

1. $\log_2 12$ **3.** $\log \frac{5}{2}$ **5.** $\log_2 1250$ **7.** $\log_5 \frac{1}{2}$ **9.** $\log xy^2$ **11.** $\log_2 x^3 y^4$ **13.** $\log x^{-3/2}$ **15.** $\log_2 x^2$
17. $2 \log x + \log y$ **19.** $\frac{1}{2}\log x + \frac{1}{2}\log y$ **21.** $\frac{1}{2}\log x - \log y$ **23.** 2, -5 **25.** 999 **27.** 10,000

29. $y = \dfrac{1}{1000 - x}$ **31.** $y = x - 10$ **33.** $y = \frac{1}{100}(x + 1)$ **35.** 1.4771 **37.** .3891 **39.** .4260

CHAPTER 1: SUPPLEMENTARY EXERCISES, page 40

1. $\frac{1}{8}$ **3.** $x \geq 3$ **5.** $x \leq -1$ **7.** 27 **9.** $(x + 1)^2$ **11.** x **13.** $4x$ **15.** $\dfrac{x}{x - 1}$ **17.** $\dfrac{2x + 15}{x + 5}$ **19.** $-3(x - 4)(x + 2)$

21. $\frac{5}{4}$ **23.** $\frac{1}{2}$ **25.** 2, -7 **27.** $(3 \pm \sqrt{5})/2$ **29.** $1 \pm \frac{1}{2}\sqrt{3}$ **31.** 1, 7 **33.** 3 **35.** 12 **37.** $10^{.8751} = 7.5$ **39.** $3 - x$

41. $\log x^3$

CHAPTER 2

EXERCISES 2.1, page 49

1, 3, 5.
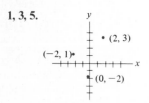

7.
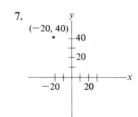

9. $m = 5, b = 8$

11. $m = 0, b = 3$ **13.** $y = -2x + 3$ **15.** $x = \frac{5}{3}$ **17.** $y = \frac{6}{5}x - 6$ **19.** $(2, 0), (0, 8)$

21. $(7, 0)$, none   **23.**

25.

27.

29. $y = 0$

EXERCISES 2.2, page 55

1. False   **3.** True   **5.** $x \geq 4$ **7.** $x \geq 3$ **9.** $y \leq -2x + 5$ **11.** $y \geq 15x - 18$ **13.** $x \geq -\frac{3}{4}$ **15.** True
17. False   **19.** True   **21.** True   **23.**

25.

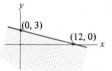

27.

29.

31.

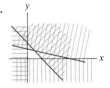

33.

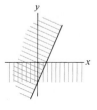

35.

37.

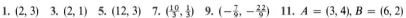

39. In   **41.** Not in

EXERCISES 2.3, page 62

1. $(2, 3)$ **3.** $(2, 1)$ **5.** $(12, 3)$ **7.** $(\frac{10}{3}, \frac{1}{3})$ **9.** $(-\frac{7}{9}, -\frac{22}{9})$ **11.** $A = (3, 4), B = (6, 2)$
13. $A = (0, 0), B = (2, 4), C = (5, 5\frac{1}{2}), D = (5, 0)$ **15.**

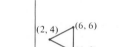

17.

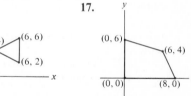

19.

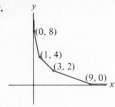

21. $(120, 2100)$ **23.** (a) $19,500$ (b) 5 cents   **25.** $\$3.48$

EXERCISES 2.4, page 72

1. $\frac{2}{3}$ **3.** 5 **5.** $\frac{5}{4}$ **7.** $\frac{5}{4}$ **9.**

11.

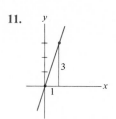

13. $y = -2x + 7$

15. $y = -2x + 4$ **17.** $y = \frac{1}{4}x + \frac{3}{2}$ **19.** $y = -x$ **21.** $y = 3$ **23.** $(0, 3)$
25. Each unit sold yields a commission of $\$5$. In addition, she receives $\$60$ per week base pay.
27. If the monopolist wants to sell one more unit of goods, then the price per unit must be lowered by 2 cents. No one
is willing to pay $\$7$ (or more) for a unit of goods.
29. $17,600$ gallons   **31.** $(0, 30,000)$; on Jan. 1 the tank contained $30,000$ gallons of oil
33.

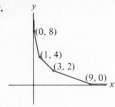

35. $(0, 2.3)$; $\$2.3$ million is the amount of cash reserves on July 1.

37. $\$1.85$ million   **39.** $y = 3x - 1$ **41.** $y = x + 1$ **43.** $y = -7x + 35$ **45.** $y = 4$ **47.** $y - 1 = \frac{1}{2}(x - 2)$
49. $y = -2x$ **51.** $5; 1; -1$ **53.** $-\frac{5}{4}; -\frac{3}{2}; -\frac{3}{4}$ **55.** l_1

CHAPTER 2: SUPPLEMENTARY EXERCISES, page 76

1. $x = 0$ **3.** $(2, -\frac{4}{3})$ **5.** $y = -\frac{1}{2}x + 5$ **7.** Yes   **9.** $y = \frac{1}{5}x + 13$ **11.** $(5, 0)$ **13.** $x = 7, y = 10$ **15.** $(0, 7)$
17. 450 **19.** $\$210; 1700$

CHAPTER 3

EXERCISES 3.1, page 86

1. $0, 10, 0, 70$ **3.** $0, 0, -\frac{9}{8}, a^3 + a^2 - a - 1$ **5.** $\frac{1}{3}, 3, \dfrac{a + 1}{a + 2}$ **7.** $a^2 - 1, a^2 + 2a$ **9.** (a) 1980 sales, (b) 60

11. $x \neq 1, 2$ **13.** $x < 3$ **15.** Function   **17.** Not a function   **19.** Not a function   **21.** 1 **23.** 3 **25.** Positive
27. Positive   **29.** $-1, 5, 9$ **31.** $.03$ **33.** $.04$ **35.** No   **37.** Yes   **39.** $(a + 1)^3$ **41.** $1, 3, 4$ **43.** $\pi, 3, 12$

45. $f(x) = \begin{cases} .06x & \text{for } 50 \le x \le 300 \\ .02x + 12 & \text{for } 300 < x \le 600 \\ .015x + 15 & \text{for } 600 < x \end{cases}$

EXERCISES 3.2, page 94

1.    **3.** 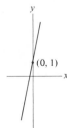   **5.** 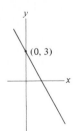   **7.** $(-\frac{1}{3}, 0), (0, 3)$

9. y-intercept $(0, 5)$ **11.** $(12, 0), (0, 3)$ **13.** (a) $K = \frac{1}{250}, V = \frac{1}{50}$ (b) $\left(-\frac{1}{K}, 0\right), \left(0, \frac{1}{V}\right)$ **15.** (a) $42 (b) $12 + .15x$

17. $150 + 135n, n = $ number of days   **19.** $a = 3, b = -4, c = 0$ **21.** $a = -2, b = 3, c = 1$
23. $a = -1, b = 0, c = 1$ **25.** $\frac{1}{8}$ **27.** 2   **29.** 4   **31.** $\frac{1}{5}$ **33.** 2.5   **35.** 10^{-2}

37. 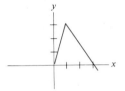   **39.**   **41.**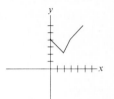

EXERCISES 3.3, page 100

1. $x^2 + 13x$ **3.** $9x^3 + 36x^2$ **5.** $\frac{t + 4}{9}, t \ne 0$ **7.** $\left(\frac{x}{1 - x}\right)^6$ **9.** $\left(\frac{x}{1 - x}\right)^3 - 5\left(\frac{x}{1 - x}\right)^2 + 1$

11. $\frac{t^3 - 5t^2 + 1}{5t^2 - t^3}$ **13.** $3x^2h + 3xh^2 + h^3$ **15.** $4 - 2t - h$ **17.** (a) $C(A(t)) = 3000 + 1600t - 40t^2$ (b) $6040

19. 5 units up   **21.** 2 units to the right   **23.** $y = 4(x - 5)^2 + 4$ **25.** $y = (x + 1)^3 - 3(x + 1) + 2$

27. $A = 15x - 6x^2$ **29.** $A = \frac{1}{5}x + \frac{1}{12x}$ **31.** $A = \frac{5}{2}x^2$ **33.** $A = 5x^2 + 4x + 2$ **35.** $A = x - \frac{1}{2}x^3$

EXERCISES 3.4, page 107

1. Narrower   **3.** Wider   **5.** Narrower   **7.** Down   **9.** Up   **11.** Down   **13.** $(0, 0)$ **15.** $(3, 4)$ **17.** $(-\frac{1}{2}, 2)$
19. $(1, 2)$ **21.** $(\frac{3}{2}, -2)$ **23.** $(-3, -\frac{9}{2})$ **25.**   **27.**

29. $(\frac{2}{3}, 1)$

31. 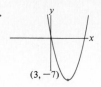 $(3, -7)$

33. $(-2, -9)$

35. $(3, 7)$

CHAPTER 3: SUPPLEMENTARY EXERCISES, page 109

1. $2, 27\frac{1}{3}, -2, -2\frac{1}{8}, \dfrac{5\sqrt{2}}{2}$ **3.** $a^2 - 4a + 2$ **5.** $x \neq 0, -3$ **7.** All x **9.** Yes **11.** $x^2 + x - 1$ **13.** $x^{5/2} - 2x^{3/2}$

15. $x^{3/2} - 2x^{1/2}$ **17.** $\dfrac{x^2 - x + 1}{x^2 - 1}$ **19.** $-\dfrac{3x^2 + 1}{3x^2 + 4x + 1}$ **21.** $\dfrac{-3x^2 + 9x - 10}{3x^2 - 5x - 8}$ **23.** $\dfrac{1}{x^4} - \dfrac{2}{x^2} + 4$ **25.** $(\sqrt{x} - 1)^2$

27. $\dfrac{1}{(\sqrt{x} - 1)^2} - \dfrac{2}{\sqrt{x} - 1} + 4$ **29.** $27, 32, 4$ **31.** $301 + 10t + .04t^2$ **33.** $A = x^2 - \dfrac{24}{x}$ **35.** $A = 84\pi r^2 - 2\pi^2 r^3$

37. $y = \dfrac{1}{x + 2} + x$ **39.** 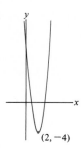 $(2, -4)$

41. 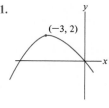 $(-3, 2)$

43. $(4, 5)$

CHAPTER 4

EXERCISES 4.1, page 118

1. (a) $i = .01, n = 24$ (b) $i = .02, n = 20$ (c) $i = .05, n = 40$
3. (a) $i = .06, n = 4, P = \$500, F = \$631.24$ (b) $i = .02, n = 120, P = \$800, F = \$8612.13$
(c) $i = .06, n = 19, P = \$2974.62, F = \$9000$
5. $\$1127.16$ **7.** $\$5053.45$ **9.** $\$42,918.71$ **11.** $\$8874.49$ **13.** (a) 12% (b) $\$10.20, \$1030.30$ (c) $\$12.57, \$1269.73$
15. $\$1700$ in 9 years **17.** 30% compounded annually **19.** $\$7884.93$
21. (a) $r = .07, n = \frac{1}{2}, P = \$500, A = \$517.50$ (b) $r = .08, n = 2, P = \$500, A = \$580$
23. $\$1270$ **25.** $\$2000$ **27.** 50% **29.** 10 years **31.** $P = A/(1 + rn)$ **33.** $\$126.25, 26.25\%$ **35.** 12.36%
37. $a + (a^2/4)$ **39.** $(1 + a/n)^n - 1$

EXERCISES 4.2, page 127

1. (a) 31 (b) 11 (c) $\frac{31}{16}$ (d) $\frac{93}{16}$
3. (a) $i = .005, n = 120, R = \$50, F = \$8193.97$ (b) $i = .06, n = 20, R = \$1767, F = \$65,000$
5. $\$8166.97$ **7.** $\$4698.97$ **9.** $\$12,648.91$ **11.** $\$877.91$ **13.** $\$1000$ at end of each month for 10 years **15.** $\$7239.33$
17. $\$17,584.61$ **19.** $\$10,000$ **21.** $\$4339.35$ **23.** $\$59,189.85$ **25.** (a) $\$1$ (b) $i \cdot s_{\overline{n}|i}$ **29.** $\$3623.52$
31. (a) $\$1269.67$ (b) $\$65,134.88$

EXERCISES 4.3, page 135

1. $\$287.68$ **3.** $\$11,469.92$ **5.** (a) $\$583.31$ (b) $\$16.69$ (c) $\$58,314.31$ (d) $\$26,973.02$ (e) $\$4188.65$ (f) $\$269.73$
7. $\$9000$ **9.** $\$14,042.36$ **13.** $\$2001.60$ **15.** $\$65,900$ **17.** $\$2039.89$

19. (a) \$183,927.96 (b) \$24,989.92 (c) \$105,266.63   **21.** \$2.5 billion
23. Yes. The sinking fund will have a surplus of \$12.26 million.

CHAPTER 4: SUPPLEMENTARY EXERCISES, page 138

1. \$347.77   **3.** \$26,541.30   **5.** 10% compounded annually   **7.** (a) \$2400.34 (b) \$167,304.95   **9.** \$15,149.74
11. \$293.42   **13.** \$1200.17   **15.** \$349,496.41   **17.** Investment A   **19.** \$146,861.85   **21.** \$151,843.34   **23.** 265,720

CHAPTER 5

EXERCISES 5.1, page 150

1. $\begin{cases} x - 6y = 4 \\ 5x + 4y = 1 \end{cases}$ **3.** $\begin{cases} x + 2y = 3 \\ 14y = 16 \end{cases}$ **5.** $\begin{cases} x - 2y + z = 0 \\ y - 2z = 4 \\ 9y - z = 5 \end{cases}$ **7.** $\begin{bmatrix} 1 & 0 & | & 5 \\ 0 & 1 & | & 4 \end{bmatrix}$ **9.** $[2] + 2[1]$
$\quad 2[1]$ $\qquad\qquad [2] + 5[1]$ $\qquad\qquad\qquad [3] + (-4)[1]$ $\qquad\quad [1] + (\frac{1}{2})[2]$

11. $[1] + (-2)[2]$ **13.** Interchange rows 1 and 2 or rows 1 and 3.   **15.** $[1] + (-3)[3]$ **17.** $x = -1, y = 1$
19. $x = -\frac{8}{7}, y = -\frac{9}{7}, z = -\frac{3}{7}$ **21.** $x = -1, y = 1$ **23.** $x = 1, y = 2, z = -1$
25. $x = \$25,000, y = \$50,000, z = \$25,000$

EXERCISES 5.2, page 158

1. $\begin{bmatrix} 1 & -2 & 3 \\ 0 & 13 & -8 \end{bmatrix}$ **3.** $\begin{bmatrix} 9 & -1 & 0 & -7 \\ -\frac{1}{2} & \frac{1}{2} & 1 & 3 \\ 5 & -1 & 0 & -3 \end{bmatrix}$ **5.** $\begin{bmatrix} 1 & \frac{3}{2} \\ 0 & -9 \\ 0 & \frac{7}{2} \end{bmatrix}$ **7.** $\begin{bmatrix} 4 & 3 & 0 \\ 1 & 1 & 0 \\ \frac{1}{6} & \frac{1}{2} & 1 \end{bmatrix}$ **9.** $y =$ any value, $x = 3 + 2y$

11. $x = 1, y = 2$ **13.** No solution   **15.** $z =$ any value, $y = 5, x = -6 - z$ **17.** No solution
19. $z =$ any value, $w =$ any value, $x = 2z + w, y = 5 - 3w$
21. $z =$ any value, $x = 300 - z, y = 100 - z$. Of course, to be realistic, we must have $0 \le z \le 100$.

EXERCISES 5.3, page 170

1. 2×3 **3.** 1×3, row matrix   **5.** 1×1, square matrix   **7.** $\begin{bmatrix} 9 & 3 \\ 7 & -1 \end{bmatrix}$ **9.** $\begin{bmatrix} 1 & 3 \\ 1 & 2 \\ 4 & -2 \end{bmatrix}$ **11.** $[11]$ **13.** $[10]$

15. 3×5 **17.** Not defined   **19.** 3×1 **21.** $\begin{bmatrix} 6 & 17 \\ 6 & 10 \end{bmatrix}$ **23.** $\begin{bmatrix} 21 \\ -4 \\ 8 \end{bmatrix}$ **25.** $\begin{bmatrix} 5 & 6 \\ 7 & 8 \end{bmatrix}$ **27.** $\begin{bmatrix} .48 & .39 \\ .52 & .61 \end{bmatrix}$

29. $\begin{bmatrix} 25 & 17 & 2 \\ 3 & -1 & 2 \\ 1 & 1 & 4 \end{bmatrix}$ **31.** $\begin{cases} 2x + 3y = 6 \\ 4x + 5y = 7 \end{cases}$ **33.** $\begin{cases} x + 2y + 3z = 10 \\ 4x + 5y + 6z = 11 \\ 7x + 8y + 9z = 12 \end{cases}$ **35.** $\begin{bmatrix} 3 & 2 \\ 7 & -1 \end{bmatrix}\begin{bmatrix} x \\ y \end{bmatrix} = \begin{bmatrix} -1 \\ 2 \end{bmatrix}$

37. $\begin{bmatrix} 1 & -2 & 3 \\ 0 & 1 & 1 \\ 0 & 0 & 1 \end{bmatrix}\begin{bmatrix} x \\ y \\ z \end{bmatrix} = \begin{bmatrix} 5 \\ 6 \\ 2 \end{bmatrix}$ **39.** $\begin{bmatrix} 4 & 24 \\ 20 & 24 \end{bmatrix}$

43. Number voting Democratic = 10,100; number voting Republican = 7900.
45. (a) Shift 1: \$1000, shift 2: \$1050, shift 3: \$600;   (b) Carpenters: \$1000, bricklayers: \$1050, plumbers: \$600.

EXERCISES 5.4, page 182

1. $x = 2, y = 0$ **3.** $\begin{bmatrix} 1 & -2 \\ -3 & 7 \end{bmatrix}$ **5.** $\begin{bmatrix} 1 & -1 \\ -2.5 & 3 \end{bmatrix}$ **7.** $\begin{bmatrix} 1.6 & -.4 \\ -.6 & 1.4 \end{bmatrix}$ **9.** $[\frac{1}{3}]$ **11.** $x = 4, y = -\frac{1}{2}$

13. $x = 32, y = -6$

15. (a) $\begin{bmatrix} .8 & .3 \\ .2 & .7 \end{bmatrix} \begin{bmatrix} x \\ y \end{bmatrix} = \begin{bmatrix} m \\ s \end{bmatrix}$ **(b)** $\begin{bmatrix} x \\ y \end{bmatrix} = \begin{bmatrix} 1.4 & -.6 \\ -.4 & 1.6 \end{bmatrix} \begin{bmatrix} m \\ s \end{bmatrix}$ **(c)** $110,000; 40,000$ **(d)** $130,000; 20,000$

17. $x = 9, y = -2, z = -2$ **19.** $x = 1, y = 5, z = -4, w = 9$

EXERCISES 5.5, page 188

1. $\begin{bmatrix} -2 & 3 \\ 5 & -7 \end{bmatrix}$ **3.** $\begin{bmatrix} \frac{1}{19} & \frac{3}{19} \\ \frac{3}{76} & -\frac{5}{38} \end{bmatrix}$ **5.** No inverse **7.** $\begin{bmatrix} -1 & 2 & -4 \\ 1 & -1 & 3 \\ 0 & 0 & 1 \end{bmatrix}$ **9.** No inverse **11.** $\begin{bmatrix} -5 & 6 & 0 & 0 \\ 1 & -1 & 0 & 0 \\ 0 & 0 & -\frac{1}{46} & \frac{1}{46} \\ 0 & 0 & \frac{25}{46} & -\frac{1}{23} \end{bmatrix}$

13. $x = 2, y = -3, z = 2$ **15.** $x = 4, y = -4, z = 3, w = -1$

EXERCISES 5.6, page 195

1. Coal: $8.84 billion, steel: $3.73 billion, electricity: $9.90 billion
3. Computers: $354 million, semiconductors: $172 million **7.** Plastics: $955,000; industrial equipment: $590,000

CHAPTER 5: SUPPLEMENTARY EXERCISES, page 197

1. $\begin{bmatrix} 1 & -2 & \frac{1}{3} \\ 0 & 8 & \frac{16}{3} \end{bmatrix}$ **3.** $x = 4, y = 5$ **5.** $x = -1, y = \frac{2}{3}, z = \frac{1}{3}$ **7.** $z = $ any value, $x = 1 - 3z, y = 4z, w = 5$

9. $\begin{bmatrix} 5 \\ 3 \\ 7 \end{bmatrix}$ **11.** $x = -2, y = 3$ **13.** $\begin{bmatrix} -1 & 3 \\ \frac{1}{2} & -1 \end{bmatrix}$ **15.** Industry I: 20 units, Industry II: 20 units

CHAPTER 6

EXERCISES 6.1, page 205

1. Yes **3.** No
5. (a)

	A	B	Truck capacity
Volume	4	3	300
Weight	100	200	10,000
Earnings	13	9	

(b) $4x + 3y \le 300$
$100x + 200y \le 10,000$

(c) $y \le 2x, x \ge 0, y \ge 0$

(d) $13x + 9y$ **(e)**

EXERCISES 6.2, page 213

1. $x = 20, y = 0$ **3.** $x = 4, y = 2$ **5.** 75 crates of cargo A, 0 crates of cargo B **7.** Produce 16 chairs and 0 sofas

9.
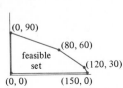
The minimum value is 40 and occurs at the point (4, 3).

11.
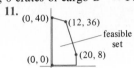
The maximum value is 6600 and occurs at the point (12, 36).

13.
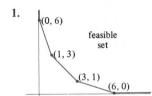
80 homes of first type, 60 homes of second type.

15.
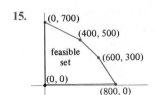
400 cans of Fruit Delight, 500 cans of Heavenly Punch.

EXERCISES 6.3, page 224

1.
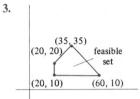
1 can of brand A, 3 cans of brand B.

3.
35 crates of oranges, 35 crates of grapefruit, 30 crates of avocados.

5.
In Detroit make 100 cars, 300 trucks. In Cleveland make 500 cars and 0 trucks.

7. (a) $y = -\frac{3}{2}x + \frac{c}{14}$ (b) Up (c) B

9.
9000 gallons of gasoline, 5000 gallons of jet fuel, 5000 gallons of diesel fuel.

CHAPTER 6: SUPPLEMENTARY EXERCISES, page 228

1.
10 planes of type A, 3 planes of type B

3.
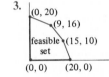
9 hardtops, 16 sports cars

5.

60 elementary, 8 intermediate, 4 advanced.

(51, 17)
(48, 16) — feasible set
(60, 8)

CHAPTER 7

EXERCISES 7.1, page 236

1.
$$\begin{cases} 20x + 30y + u & = 3500 \\ 50x + 10y \quad + v & = 5000 \\ -8x - 13y \qquad + M = 0 \end{cases}$$

Find a solution to the linear system for which $x \geq 0,\, y \geq 0,\, u \geq 0,\, v \geq 0$ and M is as large as possible.

3.
$$\begin{cases} x + y + z + u & = 100 \\ 3x \quad + z \quad + v & = 200 \\ 5x + 10y \qquad + w & = 100 \\ -x - 2y + 3z \qquad + M = 0 \end{cases}$$

Find a solution to the linear system for which $x \geq 0,\, y \geq 0,\, z \geq 0,\, u \geq 0,\, v \geq 0,\, w \geq 0$ and M is as large as possible.

5.
$$\begin{array}{ccccc} x & y & u & v & M \\ \begin{bmatrix} 20 & 30 & 1 & 0 & 0 & | & 3500 \\ 50 & 10 & 0 & 1 & 0 & | & 5000 \\ -8 & -13 & 0 & 0 & 1 & | & 0 \end{bmatrix} \end{array}$$

$x = 0,\, y = 0,\, u = 3500,\, v = 5000,\, M = 0$

7.
$$\begin{array}{ccccccc} x & y & z & u & v & w & M \\ \begin{bmatrix} 1 & 1 & 1 & 1 & 0 & 0 & 0 & | & 100 \\ 3 & 0 & 1 & 0 & 1 & 0 & 0 & | & 200 \\ 5 & 10 & 0 & 0 & 0 & 1 & 0 & | & 100 \\ -1 & -2 & 3 & 0 & 0 & 0 & 1 & | & 0 \end{bmatrix} \end{array}$$

$x = 0,\, y = 0,\, z = 0,\, u = 100,\, v = 200,\, w = 100,\, M = 0$

9. $x = 15,\, y = 0,\, u = 10,\, v = 0,\, M = 20$ **11.** $x = 10,\, y = 0,\, z = 15,\, u = 23,\, v = 0,\, w = 0,\, M = -11$

13. (a)
$$\begin{array}{ccccc} x & y & u & v & M \\ \begin{bmatrix} 1 & \frac{3}{2} & \frac{1}{2} & 0 & 0 & | & 6 \\ 0 & -\frac{1}{2} & -\frac{1}{2} & 1 & 0 & | & 4 \\ 0 & -5 & 5 & 0 & 1 & | & 60 \end{bmatrix} \end{array}$$

$x = 6,\, y = 0,\, u = 0,\, v = 4,\, M = 60$

(b)
$$\begin{array}{ccccc} x & y & u & v & M \\ \begin{bmatrix} \frac{2}{3} & 1 & \frac{1}{3} & 0 & 0 & | & 4 \\ \frac{1}{3} & 0 & -\frac{1}{3} & 1 & 0 & | & 6 \\ \frac{10}{3} & 0 & \frac{20}{3} & 0 & 1 & | & 80 \end{bmatrix} \end{array}$$

$x = 0,\, y = 4,\, u = 0,\, v = 6,\, M = 80$

(c)
$$\begin{array}{ccccc} x & y & u & v & M \\ \begin{bmatrix} 0 & 1 & 1 & -2 & 0 & | & -8 \\ 1 & 1 & 0 & 1 & 0 & | & 10 \\ 0 & -10 & 0 & 10 & 1 & | & 100 \end{bmatrix} \end{array}$$

$x = 10,\, y = 0,\, u = -8,\, v = 0,\, M = 100$

(d)
$$\begin{array}{ccccc} x & y & u & v & M \\ \begin{bmatrix} -1 & 0 & 1 & -3 & 0 & | & -18 \\ 1 & 1 & 0 & 1 & 0 & | & 10 \\ 10 & 0 & 0 & 20 & 1 & | & 200 \end{bmatrix} \end{array}$$

$x = 0,\, y = 10,\, u = -18,\, v = 0,\, M = 200$

15. (d)

EXERCISES 7.2, page 247

1. (a) 3 (b)
$$\begin{array}{ccccc} x & y & u & v & M \\ \begin{bmatrix} \frac{16}{3} & 0 & 1 & -\frac{2}{3} & 0 & | & 6 \\ \frac{1}{3} & 1 & 0 & \frac{1}{3} & 0 & | & 2 \\ 0 & 0 & 0 & 4 & 1 & | & 24 \end{bmatrix} \end{array}$$
(c) $x = 0,\, y = 2,\, u = 6,\, v = 0,\, M = 24$

3. (a) 10 (b)
$$\begin{array}{ccccc|c} x & y & u & v & M \\ -13 & 0 & 1 & -\frac{6}{5} & 0 & 6 \\ \frac{3}{2} & 1 & 0 & \frac{1}{10} & 0 & \frac{1}{2} \\ 7 & 0 & 0 & \frac{1}{5} & 1 & 1 \end{array}$$
(c) $x = 0, y = \frac{1}{2}, u = 6, v = 0, M = 1$

5. $x = 0, y = 5, M = 15$ 7. $x = 12, y = 20, M = 88$ 9. $x = 0, y = \frac{19}{3}, z = 5, M = 44$
11. $x = 0, y = 30, M = 90$ 13. $x = 50, y = 100, M = 1300$ 15. 98 chairs, 4 sofas, 21 tables; $10,640 profit
17. 75 small sofas, 15 large sofas, 30 chairs; max. profit $= 6900$ 19. $x = 100, y = 50$; max. value $= 45,000$

EXERCISES 7.3, page 258

1. $156; x = \frac{3}{5}, y = \frac{22}{5}$ 3. $8; x = \frac{5}{2}, y = \frac{1}{2}$ 5. $59; x = 3, y = 5$
7. 1 serving of food A, 3 servings of food B; cost is $7.50 9. 100 brand A, 50 brand B, 450 brand C

EXERCISES 7.4, page 267

1. Minimize $80u + 76v$ subject to the constraints

$$\begin{cases} 5u + 3v \geq 4 \\ u + 2v \geq 2 \\ u \geq 0, \quad v \geq 0. \end{cases}$$

3. Maximize $u + 2v + w$ subject to the constraints

$$\begin{cases} u - v + 2w \leq 10 \\ 2u + v + 3w \leq 12 \\ u \geq 0, \quad v \geq 0, \quad w \geq 0. \end{cases}$$

5. Maximize $-7u + 10v$ subject to the constraints

$$\begin{cases} -2u + 8v \leq 3 \\ 4u + v \leq 5 \\ 6u + 9v \leq 1 \\ u \geq 0, \quad v \geq 0. \end{cases}$$

7. $x = 12, y = 20, M = 88; u = \frac{2}{7}, v = \frac{6}{7}, M = 88$ 9. $x = 0, y = 2, M = 24; u = 0, v = 12, M = 24$
11. Maximize $3u + 5v$ subject to the constraints

$$\begin{cases} u + 2v \leq 3 \\ v \leq 1 \\ u \geq 0, \quad v \leq 0 \end{cases}$$

Solutions: $x = \frac{5}{2}, y = \frac{1}{2}, M = 8; u = 1, v = 1, M = 8$
13. Minimize $6u + 9v + 12w$ subject to

$$\begin{cases} u + 3v \geq 10 \\ -2u + w \geq 12 \\ v + 3w \geq 10 \\ u \geq 0, \quad v \geq 0, \quad w \geq 0 \end{cases}$$

Solution: $x = 27, y = 3, z = 12, M = 174; u = 0, v = \frac{10}{3}, w = 12, M = 174$
15. $1410 17. $1430 19. 62 cents 21. Nothing

23. Suppose that we had the opportunity to fill the trucks with other cargo using a charge system of u dollars per cubic foot and v dollars per pound. Then the minimum acceptable profit per truck is determined by minimizing $300u + 10,000v$ subject to $4u + 100v \geq 13, 3u + 200v \geq 9, u \geq 0, v \geq 0$.

25. $A = \begin{bmatrix} 5 & 1 \\ 3 & 2 \end{bmatrix}$, $B = \begin{bmatrix} 80 \\ 76 \end{bmatrix}$, $C = [4 \;\; 2]$, $X = \begin{bmatrix} x \\ y \end{bmatrix}$. Maximize CX subject to $AX \leq B, X \geq \mathbf{0}$. Dual: $U = [u \;\; v]$. Minimize UB subject to $UA \geq C, U \geq \mathbf{0}$.

27. $A = \begin{bmatrix} 1 & 2 \\ -1 & 1 \\ 2 & 3 \end{bmatrix}$, $B = \begin{bmatrix} 1 \\ 2 \\ 1 \end{bmatrix}$, $C = [10 \;\; 12]$, $X = \begin{bmatrix} x \\ y \end{bmatrix}$, $U = [u \;\; v \;\; w]$. Minimize CX subject to $AX \geq B, X \geq \mathbf{0}$. Maximize UB subject to $UA \leq C, U \geq \mathbf{0}$.

29. $A = \begin{bmatrix} -2 & 4 & 6 \\ 8 & 9 & 1 \end{bmatrix}$, $B = \begin{bmatrix} -7 \\ 10 \end{bmatrix}$, $C = [3 \;\; 5 \;\; 1]$, $X = \begin{bmatrix} x \\ y \\ z \end{bmatrix}$, $U = [u \;\; v]$. Minimize CX subject to $AX \geq B, X \geq \mathbf{0}$. Maximize UB subject to $UA \leq C, U \geq \mathbf{0}$.

CHAPTER 7: SUPPLEMENTARY EXERCISES, page 270

1. $x = 2, y = 3, M = 18$ **3.** $x = 4, y = 5, M = 23$ **5.** $6; x = 5, y = 1$ **7.** $110; x = 4, y = 1$
9. $x = 1, y = 6, z = 8, M = 884$
11. Minimize $14u + 9v + 24w$ subject to the constraints.

$$\begin{cases} u + v + 3w \geq 2 \\ 2u + v + 2w \geq 3 \\ u \geq 0, \quad v \geq 0, \quad w \geq 0. \end{cases}$$

13. $x = 4, y = 5, M = 23; u = 1, v = 1, w = 0, M = 23$.

15. $A = \begin{bmatrix} 1 & 2 \\ 1 & 1 \\ 3 & 2 \end{bmatrix}$, $B = \begin{bmatrix} 14 \\ 9 \\ 24 \end{bmatrix}$, $C = [2 \;\; 3]$, $X = \begin{bmatrix} x \\ y \end{bmatrix}$. Maximize CX subject to $AX \leq B, X \geq \mathbf{0}$. Dual: $U = [u \;\; v \;\; w]$. Minimize UB subject to $UA \leq C, U \geq \mathbf{0}$.

CHAPTER 8

EXERCISES 8.1, page 279

1. (a) $\{5, 6, 7\}$ (b) $\{1, 2, 3, 4, 5, 7\}$ (c) $\{1, 3\}$ (d) $\{5, 7\}$ **3.** (a) $\{a, b, c, d, e, f\}$ (b) $\{c\}$ (c) $\varnothing$ **5.** $\{1, 2\}, \{1\}, \{2\}, \varnothing$
7. (a) {all male college students who like football} (b) {all female college students} (c) {all female college students who dislike football} (d) {all college students who are either male or like football}
9. (a) $\{1976, 1975, 1967, 1963, 1958, 1951, 1950\}$ (b) $\{1976, 1975, 1967, 1963, 1961, 1958, 1955, 1954, 1951, 1950\}$
(c) $\{1976, 1975, 1967, 1963, 1958, 1951, 1950\}$ (d) $\{1961, 1955, 1954\}$ (e) $\varnothing$
11. From 1950 to 1977, whenever the Standard and Poor's Index increased by 2 or more percent during the first five days of a year, it always increased by at least 16% for that year.
13. (a) $\{d, f\}$ (b) $\{a, b, c, e, f\}$ (c) $\varnothing$ (d) $\{a, c\}$ (e) $\{e\}$ (f) $\{a, c, e, f\}$ **15.** S **17.** U **19.** $\varnothing$ **21.** $L \cup T$
23. $P \cap L$ **25.** $P \cap L \cap T$ **27.** S' **29.** $S \cup D \cup A$ **31.** $(A \cap S)' \cap D$

EXERCISES 8.2, page 286

1. 7 **3.** 0 **5.** 11 **7.** S is a subset of T **9.** 14 million **11.** 19,000

13.

15.

17.
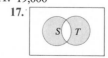

19.
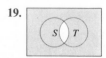

21. **23.** **25.** **27.**

29. **31.** **33.** **35.**

37. $S' \cup T'$ **39.** $S \cap T'$ **41.** U **43.** S' **45.** $R \cap T$ **47.** $T \cap S \cap R'$

EXERCISES 8.3, page 293

1. 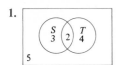 **3.** **5.** **7.**

9. 25 **11.** 2, 5 **13.** 30 **15.** 4 **17.** 35

EXERCISES 8.4, page 298

1. 15 **3.** 676 **5.** 380 **7.** 20 **9.** 120 **11.** 64 **13.** 6840 **15.** 870 **17.** 24 **19.** 32 **21.** 360,000

EXERCISES 8.5, page 304

1. 12 **3.** 120 **5.** 120 **7.** 5 **9.** 720 **11.** 36 **13.** 24 **15.** 35 **17.** 15,600 **19.** $C(100, 15)$ **21.** $C(52, 5)$

EXERCISES 8.6, page 310

1. (a) 64 (b) 20 (c) 22 (d) 57 **3.** 126 **5.** (a) 120 (b) 56 (c) 64 **7.** $C(100, 25) \cdot C(75, 40)$ **9.** 60
11. $C(50, 15) \cdot 2^{15}$ **13.** 24 **15.** 3744

EXERCISES 8.7, page 317

1. 15 **3.** 1 **5.** 816 **7.** 1 **9.** $x^7 + 7x^6y + 21x^5y^2 + 35x^4y^3 + 35x^3y^4 + 21x^2y^5 + 7xy^6 + y^7$
11. $x^{10}, 10x^9y, 45x^8y^2$ **13.** 64 **15.** 16 **17.** 64 **19.** 5040 **21.** (a) 14,520 (b) $C(12, 5) \cdot 2^5 = 25,344$

CHAPTER 8: SUPPLEMENTARY EXERCISES, page 319

1. $\{a, b\}, \{a\}, \{b\}, \varnothing$ **3.** 120 **5.** **7.** 840 **9.** 35 **11.** 0 **13.** 126 **15.** 550

CHAPTER 9

EXERCISES 9.2, page 330

1. (a) {RS, RT, RU, RV, ST, SU, SV, TU, TV, UV} (b) {RS, RT, RU, RV} (c) {TU, TV, UV}
3. (a) {HH, HT, TH, TT} (b) {HH, HT}

5. (a) {(Urn I, red),(urn I, white), (urn II, red), (urn II, white)} (b) {(Urn I, red), (urn I, white)}
7. (a) {all positive numbers} (b) "More than 5 but less than 8 minutes," $\varnothing$, "5 minutes or less," "8 minutes or more," "5 minutes or less," "less than 4 minutes," S
9. (a) No (b) Yes   **11.** $S, \{a, b\}, \{a, c\}, \{b, c\}, \{a\}, \{b\}, \{c\}, \varnothing$ **13.** Yes
15. (a) $\{0, 1, 2, \ldots, 10\}$ (b) $\{6, 7, 8, 9, 10\}$ **17.** (a) No (b) Yes (c) Yes
19. The set of nonnegative integers   **21.** The set of nonnegative numbers

EXERCISES 9.3, page 340

1. (a) $\dfrac{46,277}{774,746}$ (b) $\dfrac{48,132}{774,746}$ **3.** (a) $\frac{5}{36}$ (b) $\frac{1}{6}$ **5.** $\frac{1}{19}$ **7.** $\frac{1}{6}$ **9.** (a) .7 (b) .7   **11.** (a) $\frac{10}{11}$ (b) $\frac{1}{3}$ (c) $\frac{4}{9}$ **13.** (a) .7 (b) .2
15. (a) .7 (b) 7000   **17.**

Failures in:	Prob.
Month 1	.05
Month 2	.05
Month 3	.10
Month 4	.05
Month 5	.05
Month 6	.02
No failures in months 1–6	.68

EXERCISES 9.4, page 347

1. $\frac{5}{11}$ **3.** $\frac{25}{42}$ **5.** $1 - \dfrac{30 \cdot 29 \cdot 28 \cdot 27}{30^4} = .188$ **7.** $\frac{1}{2}$ **9.** (a) .25 (b) .75 (c) .8   **11.** 0   **13.** $\frac{1}{11}$ **15.** $\frac{2}{5}$
17. (a) $\frac{15}{28}$ (b) $\frac{15}{56}$ (c) $\frac{9}{56}$ (d) $\frac{9}{14}$ **19.** $\frac{5}{6}$ **21.** $\frac{1}{3}$

EXERCISES 9.5, page 358

1. $\frac{1}{3}, \frac{1}{5}$ **3.** $\frac{4}{7}$ **5.** No   **7.** .36   **9.** $(.99)^5(.98)^5(.975)^3$ **11.** $\frac{3}{4}, \frac{1}{2}$ **13.** (a) $\frac{2}{5}$ (b) $\frac{3}{5}$ (c) $\frac{3}{4}$ **15.** $\frac{1}{4}$ **17.** .94   **21.** .2   **23.** $\frac{2}{3}$

EXERCISES 9.6, page 365

1.   **3.**   **5.** .08   **7.** .295

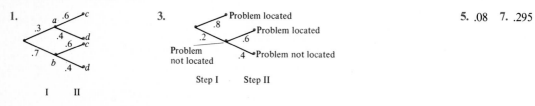

9. $\frac{7}{12}$ **11.** $\frac{4}{5}$ **13.** (a) .026 (b) $\frac{9}{13}$ **15.** .00473   **17.** $1 - (.9999)^3$

EXERCISES 9.7, page 370

1. $\frac{40}{265}$ **3.** $\frac{3}{7}$ **5.** .075

CHAPTER 9: SUPPLEMENTARY EXERCISES, page 373

1. $\frac{31}{32}$ **3.** $\frac{4}{9}$ **5.** $\frac{1}{5}$ **7.** (a) $\frac{1}{12}$ (b) $\frac{1}{2}$ **9.** $\frac{1}{21}$ **11.** No   **13.** $\frac{1}{3}$

CHAPTER 10

1.

Grade	Prob.
0	.08
1	.12
2	.40
3	.24
4	.16

3.

No. Calls	Prob.
20	.05
21	.05
22	.00
23	.10
24	.30
25	.20
26	.00
27	.15
28	.10
29	.05

5.

No. Heads	Prob.
0	$\frac{1}{8}$
1	$\frac{3}{8}$
2	$\frac{3}{8}$
3	$\frac{1}{8}$

7.

No. Shots	Prob.
1	$\frac{1}{3}$
2	$\frac{2}{9}$
3	$\frac{4}{27}$
4	$\frac{8}{27}$

9.

Earnings	Prob.
$5	$\frac{1}{15}$
$1	$\frac{8}{15}$
−$1	$\frac{2}{5}$

11. .6

13.

k	$\Pr(X^2 = k)$
0	.1
1	.2
4	.3
9	.2
16	.2

15.

k	$\Pr(X - 1 = k)$
−1	.1
0	.2
1	.3
2	.2
3	.2

17.

k	$\Pr(\frac{1}{5} Y = k)$
1	.3
2	.4
3	.1
4	.1
5	.1

19.

k	$\Pr((X + 1)^2 = k)$
1	.1
4	.2
9	.3
16	.2
25	.2

1. 2.35   **3.** (a) 2.9 (b)

Grade	Prob.
4	.3
3	.4
2	.2
1	.1

(c) 2.9   **5.** A

7.

Earnings	Prob.
$35	$\frac{1}{38}$
−$1	$\frac{37}{38}$

, $\mu = -5.26$ cents

9.

Earnings	Prob.
$-50¢$	$\frac{1}{3}$
$0¢$	$\frac{4}{15}$
$50¢$	$\frac{1}{5}$
$\$1.00$	$\frac{2}{15}$
$\$1.50$	$\frac{1}{15}$

$, \mu = 16.67$ cents   **11.** $\$1000$ **13.** $\frac{161}{36}$

EXERCISES 10.3, page 399

1. 1.4   **3.** B   **5.** (a) $\mu_A = 15$, $\sigma_A^2 = 160$, $\mu_B = 13$, $\sigma_B^2 = 141$ (b) A (c) B
7. (a) $\mu_A = 103$, $\sigma_A^2 = 4.6$, $\mu_B = 104$, $\sigma_B^2 = 3.4$ (b) B (c) B   **9.** (a) $\geq \frac{3}{4}$ (b) $\geq \frac{8}{9}$ (c) $\geq \frac{11}{36}$ **11.** ≥ 4688 **13.** 8
15. (a) 7, $\frac{35}{6}$ (b) $\frac{5}{6}$ (c) $\geq \frac{19}{54}$
17.

k	$\Pr(X^2 = k)$	$k \cdot \Pr(X^2 = k)$
0	.1	0
1	.3	.3
4	.5	2
9	.1	.9
		$3.2 = E(X^2)$

$$E(X^2) - \mu^2 = 3.2 - (1.6)^2$$
$$= 3.2 - 2.56$$
$$= .64$$

19.

k	$\Pr(2X = k)$	$(k - \mu)^2$
-2	$\frac{1}{8}$	4
-1	$\frac{3}{8}$	1
0	$\frac{1}{8}$	0
1	$\frac{1}{8}$	1
2	$\frac{2}{8}$	4

$$\mu = -2(\tfrac{1}{8}) + (-1)\tfrac{3}{8} + 0(\tfrac{1}{8}) + 1(\tfrac{1}{8}) + 2(\tfrac{2}{8})$$
$$= 0$$
$$\text{Variance of } 2X = 4(\tfrac{1}{8}) + 1(\tfrac{3}{8}) + 0(\tfrac{1}{8}) + 1(\tfrac{1}{8}) + 4(\tfrac{2}{8})$$
$$= 2 = 4(\tfrac{1}{2})$$
$$= 4(\text{variance of } X)$$

EXERCISES 10.4, page 410

1. .8944   **3.** .4013   **5.** .2417   **7.** .6170   **9.** $z = 1.75$ **11.** $z = 0.75$ **13.** $\mu = 6$, $\sigma = 2$ **15.** $\mu = 9$, $\sigma = 1$ **17.** $-\frac{8}{3}$
19. $15\frac{1}{2}$ **21.** .9772   **23.** .6247   **25.** .0002   **27.** .9876   **29.** .0122

EXERCISES 10.5, page 419

1. $\frac{25}{216}$ **3.** $\frac{3}{64}$ **5.** $\frac{992}{3125}$ **7.** (a) .1974 (b) .7888 (c) .9878   **9.** .0062   **11.** .0062   **13.** .0013

CHAPTER 10: SUPPLEMENTARY EXERCISES, page 421

1. (a), (b)

k	$\Pr(X = k)$	$k \cdot \Pr(X = k)$	$(k - \mu)^2$	$(k - \mu)^2 \Pr(X = k)$
0	$\frac{8}{27}$	0	1	$\frac{8}{27}$
1	$\frac{12}{27}$	$\frac{12}{27}$	0	0
2	$\frac{6}{27}$	$\frac{12}{27}$	1	$\frac{6}{27}$
3	$\frac{1}{27}$	$\frac{3}{27}$	4	$\frac{4}{27}$
		$\mu = 1$		$\sigma^2 = \frac{2}{3}$

3. .2857   **5.** $\geq \frac{8}{9}$ **7.** 10.56%   **9.** 0.0122   **11.** $z = 102.5$

CHAPTER 11

EXERCISES 11.1, page 430

1.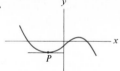

3.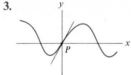

5.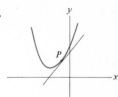

7. 1 **9.** -3 **11.** $\frac{1}{3}$ **13.** $-4, y - 4 = -4(x + 2)$ **15.** $\frac{8}{3}, y - \frac{16}{9} = \frac{8}{3}(x - \frac{4}{3})$ **17.** $y - 2.25 = 3(x - 1.5)$
19. $(\frac{5}{6}, \frac{25}{36})$ **21.** $(-\frac{1}{4}, \frac{1}{16})$ **23.** 12 **25.** $\frac{3}{4}$ **27.** $y + 1 = 3(x + 1)$

EXERCISES 11.2, page 439

1. 2 **3.** $8x^7$ **5.** $\frac{5}{2}x^{3/2}$ **7.** $\frac{1}{3}x^{-2/3}$ **9.** $-2x^{-3}$ **11.** $-\frac{1}{4}x^{-5/4}$ **13.** 0 **15.** $-3x^{-4}$ **17.** -192 **19.** $-\frac{1}{9}$ **21.** -1
23. $\frac{9}{2}$ **25.** 108 **27.** $\frac{1}{6}$ **29.** $25, -10$ **31.** $\frac{1}{32}, -\frac{5}{64}$ **33.** $16, \frac{8}{3}$ **35.** $48, y - 64 = 48(x - 4)$ **37.** $8x^7$ **39.** $\frac{3}{4}x^{-1/4}$
41. 0 **43.** $\frac{1}{5}x^{-4/5}$ **45.** $4, \frac{1}{3}$

EXERCISES 11.3, page 449

1. No limit **3.** 1 **5.** No limit **7.** -5 **9.** 5 **11.** No limit **13.** 288 **15.** 0 **17.** 3 **19.** -4 **21.** -8 **23.** $\frac{6}{7}$

25. No limit **27.** $-\frac{2}{11}$ **29.** 6 **31.** 3 **33.** $-\frac{2}{121}$ **35.** $\dfrac{-\sqrt{3}}{6}$ **37.** 0 **39.** 3 **41.** 0 **43.** 0 **45.** 2

EXERCISES 11.4, page 455

1. No **3.** Yes **5.** No **7.** No **9.** Yes **11.** No **13.** Continuous, differentiable
15. Continuous, not differentiable **17.** Continuous, not differentiable **19.** Not continuous, not differentiable

EXERCISES 11.5, page 461

1. $3x^2 + 2x$ **3.** $2x + 3$ **5.** $5x^4 - \dfrac{1}{x^2}$ **7.** $4x^3 + 3x^2 + 1$ **9.** $6x$ **11.** $3x^2 + 14x$ **13.** $-\dfrac{8}{x^3}$ **15.** $3 + \dfrac{1}{x^2}$

17. $x^2 - x$ **19.** $\dfrac{1}{x^6}$ **21.** $\dfrac{-1}{2\sqrt{x}}$ **23.** $30(3x + 1)^9$ **25.** $\dfrac{9x^2 + 1}{2\sqrt{3x^3 + x}}$ **27.** $6(4x - 1)(2x^2 - x + 4)^5$ **29.** $\frac{1}{3} - 3x^{-2}$

31. $10(1 - 5x)^{-2}$ **33.** $4x^3(1 - x^4)^{-2}$ **35.** $-2(x^2 + x)^{-3/2}(2x + 1)$

37. $\dfrac{3}{2}\left(\dfrac{\sqrt{x}}{2} + 1\right)^{1/2}\left(\dfrac{1}{4}x^{-1/2}\right)$ or $\dfrac{3}{8\sqrt{x}}\left(\dfrac{\sqrt{x}}{2} + 1\right)^{1/2}$ **39.** 4 **41.** 15 **43.** $f'(4) = 48, y = 48x - 191$
45. $f'(x) = 36x^3 + 18x^2 - 22x - 4 = 2(3x^2 + x - 2)(6x + 1)$ **47.** 6

EXERCISES 11.6, page 466

1. $10t(t^2 + 1)^4$ **3.** $(2t - 1)^{-1/2}$ **5.** $\frac{2}{3}(T^3 + 5T)^{-1/3}(3T^2 + 5)$ **7.** $6P - \frac{1}{2}$ **9.** $2a^2t + b^2$
11. $f'(x) = x - 7, f''(x) = 1$ **13.** $y' = \frac{1}{2}x^{-1/2}, y'' = -\frac{1}{4}x^{-3/2}$ **15.** $f'(r) = 2\pi(hr + 1), f''(r) = 2\pi h$
17. $g'(x) = -5, g''(x) = 0$ **19.** $f'(P) = 15(3P + 1)^4, f''(P) = 180(3P + 1)^3$ **21.** 20 **23.** 54 **25.** 34

27. $8k(2P - 1)^{-3}$ **29.** $f'(3) = -\frac{1}{2}, f''(3) = -\frac{1}{8}$ **31.** 20 **33.** (a) $f'''(x) = 60x^2 - 24x$ (b) $f'''(x) = \dfrac{15}{2\sqrt{x}}$

1. 13 3. 63 units/hr 5. 1 g/week 7. $-\frac{200}{27}$ units/day, $\frac{8}{5}$ units/day, increasing 9. (a) \$16 (b) extra cost = \$16.10
11. (a) \$9.01 (b) $\approx$ \$9.10 13. (a) 7 km/hr (b) 16.5 km
15. (a) 160 ft/sec (b) 96 ft/sec (c) -32 ft/sec^2 (d) $t = 10$ (e) -160 ft/sec 17. $t = 20$ 19. $(2 \pm \sqrt{2})/2$ 21. .12
23. 2.0025 25. It approximates the increase in the monthly payment per one percent increase in the interest rate.

CHAPTER 11: SUPPLEMENTARY EXERCISES, page 475

1. 8 3. 0 5. $18x$ 7. $1 + (2/x^2)$ 9. $7x^6 + 3x^2$ 11. $\dfrac{3}{\sqrt{x}}$ 13. $-\dfrac{3}{x^2}$ 15. $48x(3x^2 - 1)^7$ 17. $-\dfrac{5}{(5x - 1)^2}$

19. $\dfrac{x}{\sqrt{x^2 + 1}}$ 21. $-\dfrac{1}{4x^{5/4}}$ 23. 0 25. $10[x^5 - (x - 1)^5]^9[5x^4 - 5(x - 1)^4]$ 27. $\frac{3}{2}t^{-1/2} + \frac{3}{2}t^{-3/2}$ 29. $\dfrac{2(9t^2 - 1)}{(t - 3t^3)^2}$

31. $\frac{9}{4}x^{1/2} - 4x^{-1/3}$ 33. 28 35. 14, 3 37. $\frac{15}{2}$ 39. 33 41. $4x^3 - 4x$ 43. $-\frac{3}{2}(1 - 3P)^{-1/2}$ 45. 29
47. $300(5x + 1)^2$ 49. -2 51. $3x^{-1/2}$ 53. Slope -4; tangent $y = -4x + 6$ 55.

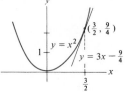

57. $y = 2$ 59. 96 ft/sec 61. 4 63. Does not exist 65. $-\frac{1}{50}$ 67. \$12.53

CHAPTER 12

1. (a), (e), (f) 3. (b), (c), (d)
5. Decreasing for $x < -2$, minimum point at $x = -2$, minimum value $= -2$, increasing for $x > -2$, concave up, y-intercept $(0, 0)$, x-intercepts $(0, 0)$ and $(-3.6, 0)$.
7. Decreasing for $x < 0$, minimum point at $x = 0$, increasing for $0 < x < 2$, maximum point at $x = 2$, decreasing for $x > 2$, concave up for $x < 1$, concave down for $x > 1$, inflection point at $(1, 3)$, y-intercept $(0, 2)$, x-intercept $(3.4, 0)$.
9. Decreasing for $x < 2$, min. at $(2, 3)$, increasing for $x > 2$, concave up for all x, no inflection points, defined for $x > 0$, the line $y = x$ is an asymptote, the y-axis is an asymptote.
11. Decreasing for $1 \le x < 3$, minimum point at $x = 3$, increasing for $x > 3$, maximum value $= 6$ (at $x = 1$), minimum value $= 2$ (at $x = 3$), inflection point at $x = 4$, concave up for $1 \le x < 4$, concave down for $x > 4$, the line $y = 4$ is an asymptote.
13. Slope increases for all x. 15. Slope decreases for $x < 3$, increases for $x > 3$. Minimum slope occurs at $x = 3$.
17. Oxygen content decreases until time a, at which time it reaches a minimum. After a, oxygen content steadily increases. The rate of increase increases until b, and then decreases. Time b is the time when oxygen content is increasing fastest.
19. y

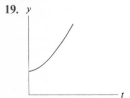

21. The parachutist's speed levels off to 15 ft/sec. 23. (a) Yes (b) Yes

25.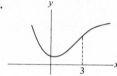

27.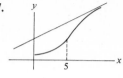

29.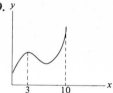

31. Relatively low

EXERCISES 12.2, page 494

1. (b), (c), (f)   **3.** (d), (e), (f)   **5.**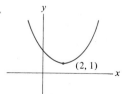

7.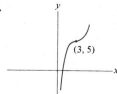

9.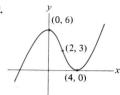

11. The second curve   **13.** The second curve

15.

17.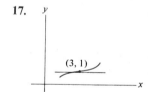

19.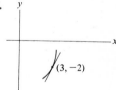

EXERCISES 12.3, page 503

1.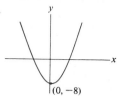

3.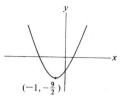

5.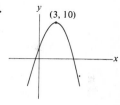

7.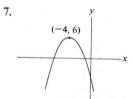

9.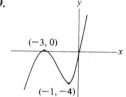

11.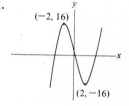

13.

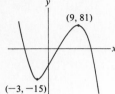

(9, 81)

(−3, −15)

15.

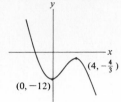

(4, −4/3)

(0, −12)

17.

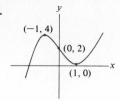

(−1, 4)

(0, 2)

(1, 0)

19.

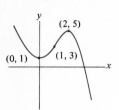

(2, 5)

(0, 1)   (1, 3)

21.

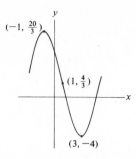

(−1, 20/3)

(1, 4/3)

(3, −4)

23.

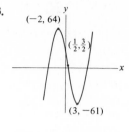

(−2, 64)

(1/2, 3/2)

(3, −61)

25. No, $f''(x) = 2a \neq 0$ **27.** (4, 3) min   **29.** (1, 5) max   **31.** (−.1, −3.05) min

EXERCISES 12.4, page 509

1. $\left(\dfrac{3 \pm \sqrt{5}}{2}, 0\right)$ **3.** $(-2, 0), (-\frac{1}{2}, 0)$ **5.** $(\frac{1}{2}, 0)$ **7.** The derivative $x^2 - 4x + 5$ has no zeros.

9.

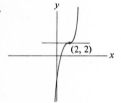

(2, 2)

11.

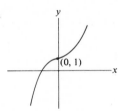

(0, 1)

13.

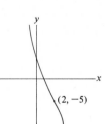

(2, −5)

15.

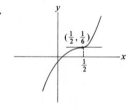

(1/2, 1/6)

1/2

17.

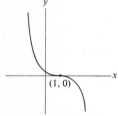

(1, 0)

19.

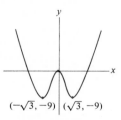

(−√3, −9)   (√3, −9)

21.

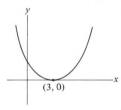

(3, 0)

23.

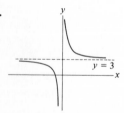

y = 3

25.

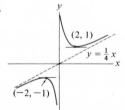

(2, 1)

$y = \frac{1}{4}x$

(−2, −1)

27.

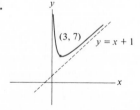

(3, 7)

y = x + 1

29.

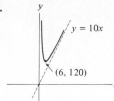

$y = 10x$
$(6, 120)$

31.

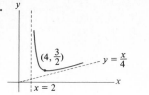

$\left(4, \frac{3}{2}\right)$
$y = \frac{x}{4}$
$x = 2$

33.

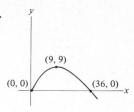

$(9, 9)$
$(0, 0)$
$(36, 0)$

EXERCISES 12.5, page 518

1. 36   **3.** $t = 4, f(4) = 8$ **5.** 20   **7.** 150 ft   **9.** 250   **11.** $x = 50, h = 50$ **13.** $x = 4$ in., $h = 2$ in.   **15.** $x = \dfrac{14}{4+\pi}$ ft

17. $x = 10$ ft, $h = 7.5$ ft   **19.** 20 cm × 20 cm × 20 cm   **21.** 5 ft × 5 ft × 6 ft

EXERCISES 12.6, page 525

1. 10 m × 10 m   **3.** $x = \dfrac{220}{\pi}$ yd   **5.** $\frac{30}{4}$ in.   **7.** 100 ft × 120 ft   **9.** $x = 6$ m, $h = 9$ m   **11.** 150   **13.** $1.10

15. $x = 40,000$ **17.** 400   **19.** $t = 20$ **21.** $2\sqrt{3} \times 6$

EXERCISES 12.7, page 537

1. $1   **3.** 32   **5.** 5   **7.** $x = 20$ units, $p = 133.33 **9.** 2 million tons, $156 per ton   **11.** (a) $1.00 (b) $1.15
13. (a) $x = 15 \cdot 10^5, p = 45. (b) No. Profit is maximized when price is increased to $50.

CHAPTER 12: SUPPLEMENTARY EXERCISES, page 540

1. Graph goes through (1, 2), increasing at $x = 1$.   **3.** Increasing and concave up at $x = 3$.
5. (10, 2) is a minimum point.   **7.** Graph goes through (5, −1), decreasing at $x = 5$.   **9.** (−2, 0) is a maximum point.

11.

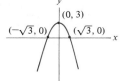

$(0, 3)$
$(-\sqrt{3}, 0)$ $(\sqrt{3}, 0)$

13.

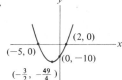

$(2, 0)$
$(-5, 0)$
$(0, -10)$
$\left(-\frac{3}{2}, -\frac{49}{4}\right)$

15.

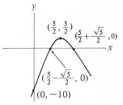

$\left(\frac{5}{2}, \frac{5}{2}\right)$ $\left(\frac{5}{2}+\frac{\sqrt{5}}{2}, 0\right)$
$\left(\frac{5}{2}-\frac{\sqrt{5}}{2}, 0\right)$
$(0, -10)$

17.

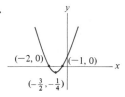

$(-2, 0)$ $(-1, 0)$
$\left(-\frac{3}{2}, -\frac{1}{4}\right)$

19.

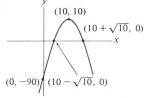

$(10, 10)$
$(10 + \sqrt{10}, 0)$
$(0, -90)$ $(10 - \sqrt{10}, 0)$

21.

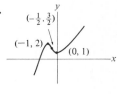

$\left(-\frac{1}{2}, \frac{3}{2}\right)$
$(-1, 2)$
$(0, 1)$

23.

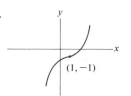

$(1, -1)$

25.

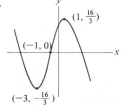

$\left(1, \frac{16}{3}\right)$
$(-1, 0)$
$\left(-3, -\frac{16}{3}\right)$

27.

$\left(-2, \frac{14}{3}\right)$

29.

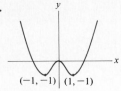

$(-1, -1)$ | $(1, -1)$

31. $f'(x) = 3x(x^2 + 2)^{1/2}, f'(0) = 0$

33. $f''(x) = -2x(1 + x^2)^{-2}$, $f''(x)$ is positive for $x < 0$ and negative for $x > 0$.
35. $1\frac{1}{2}$ ft wide $\times$ 3 ft long $\times$ 2 ft deep   **37.** $x = 25\sqrt{3}, y = 25\sqrt{6}$ **39.** 2000 cans

CHAPTER 13

EXERCISES 13.1, page 548

1. $(x + 1)(3x^2 + 5) + (x^3 + 5x + 2)(1)$ or $4x^3 + 3x^2 + 10x + 7$
3. $(3x^2 - x + 2)(4x) + (2x^2 - 1)(6x - 1)$ or $24x^3 - 6x^2 + 2x + 1$
5. $(x^4 + 1)(3) + (3x + 5)(4x^3)$ or $15x^4 + 20x^3 + 3$

7. $(x^2 - 3x + 1)\left(1 - \dfrac{2}{x^2}\right) + \left(x + \dfrac{2}{x}\right)(2x - 3)$ or $3x^2 - 6x + 3 - \dfrac{2}{x^2}$

9. $(2x - 7) \cdot 5(x - 1)^4 + (x - 1)^5(2)$ or $(x - 1)^4(12x - 37)$
11. $(x^2 + 3) \cdot 10(x^2 - 3)^9(2x) + (x^2 - 3)^{10}(2x)$ or $2x(x^2 - 3)^9(11x^2 + 27)$
13. $(2x + 9)^{-2}(-3) + (5 - 3x)(-2)(2x + 9)^{-3}(2)$ or $(6x - 47)(2x + 9)^{-3}$

15. $\sqrt{x} \cdot 3(3x^2 - 1)^2(6x) + (3x^2 - 1)^3 \cdot \dfrac{1}{2\sqrt{x}}$ or $(3x^2 - 1)^2(\frac{39}{2}x^{3/2} - \frac{1}{2}x^{-1/2})$

17. $(1 - 4x)\dfrac{x}{\sqrt{x^2 - 1}} + \sqrt{x^2 - 1}(-4)$ or $(x^2 - 1)^{-1/2}(-8x^2 + x + 4)$

19. $\dfrac{(x^2 + 1)(2x) - (x^2 - 1)(2x)}{(x^2 + 1)^2}$ or $\dfrac{4x}{(x^2 + 1)^2}$ **21.** $\dfrac{(-1)(10x + 2)}{(5x^2 + 2x + 5)^2}$ or $-\dfrac{10x + 2}{(5x^2 + 2x + 5)^2}$

23. $\dfrac{(x + 1)\dfrac{1}{2\sqrt{x}} - \sqrt{x}(1)}{(x + 1)^2}$ or $\dfrac{1 - x}{2\sqrt{x}(x + 1)^2}$ **25.** $\dfrac{(x^2 + 1)^2(1) - x \cdot 2(x^2 + 1)(2x)}{(x^2 + 1)^4}$ or $\dfrac{-3x^2 + 1}{(x^2 + 1)^3}$

27. $\dfrac{(x + 2)(1) - (x - 2)(1)}{(x + 2)^2}$ or $\dfrac{4}{(x + 2)^2}$ **29.** $\dfrac{\sqrt{4x + x^2}(-1) - (3 - x)\left(\dfrac{2 + x}{\sqrt{4x + x^2}}\right)}{4x + x^2}$ or $-\dfrac{5x + 6}{(4x + x^2)^{3/2}}$

31. $\dfrac{(3x^2 + 1)^2 \cdot 3(x^5 + 1)^2(5x^4) - (x^5 + 1)^3 \cdot 2(3x^2 + 1)(6x)}{(3x^2 + 1)^4}$ or $\dfrac{x(x^5 + 1)^2(33x^5 + 15x^3 - 12)}{(3x^2 + 1)^3}$

33. $v - 16 = 88(x - 3)$ **35.** 150; $AC(150) = 35 = C'(150)$ **37.** $2 \times 3 \times 1$
39. (a) $f(0) = 0, f(\frac{1}{2}) = \frac{1}{2}, f(1) = 1$ (c) Concave up for $0 < x < \frac{1}{2}$; concave down for $\frac{1}{2} < x < 1$; inflection point at $(\frac{1}{2}, \frac{1}{2})$ (d)

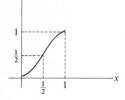

43. $f(x)$

EXERCISES 13.2, page 555

1. $\dfrac{1}{x^3 + 1}$ 3. $\sqrt{\dfrac{x^2}{x^2 - 1}}$ 5. x^4 7. $f(x) = \sqrt{x},\, g(x) = x^2 + 3x - 1$ 9. $f(x) = \sqrt{x},\, g(x) = \dfrac{x - 1}{x + 1}$

11. $f(x) = x^5,\, g(x) = x^3 + 9x - 2$ 13. $f(x) = \dfrac{x - 1}{x + 1},\, g(x) = x^2$ 15. $f(x) = \sqrt{x + 1},\, g(x) = x^{1/3}$

17. $f(x) = x^2 + 1,\, g(x) = x^3 + 1$ 19. $f(x) = x(x^2 + 1)^2,\, g(x) = x^2$ 21. $30x(x^2 + 1)^{14}$ 23. $2x^3(1 + x^4)^{-1/2}$

25. $3\left[\dfrac{2}{x} + 5(3x - 1)^4\right]^2\left[-\dfrac{2}{x^2} + 60(3x - 1)^3\right]$ 27. $8x(x^3 - 1)(x^2 + 1)^3 + 3x^2(x^2 + 1)^4$ or $x(x^2 + 1)^3(11x^3 + 3x - 8)$

29. $3(4 - x)^3(4 + x)^2 - 3(4 + x)^3(4 - x)^2$ or $-6x(16 - x^2)^2$

31. $\dfrac{1}{2}[2 + \sqrt{1 + x^2}]^{-1/2}\left[\dfrac{1}{2}(1 + x^2)^{-1/2}(2x)\right]$ or $\dfrac{x}{2}[2 + \sqrt{1 + x^2}]^{-1/2}(1 + x^2)^{-1/2}$

33. $\dfrac{\sqrt{1 + x^2} - x(\frac{1}{2})(1 + x^2)^{-1/2}(2x)}{1 + x^2}$ or $\dfrac{1}{(1 + x^2)^{3/2}}$

35. $\dfrac{(1 - \sqrt{2x}) \cdot \frac{1}{2}(2x)^{-1/2}(2) - (1 + \sqrt{2x}) \cdot (-\frac{1}{2})(2x)^{-1/2}(2)}{(1 - \sqrt{2x})^2}$ or $\dfrac{2(2x)^{-1/2}}{(1 - \sqrt{2x})^2}$

37.

39. $x^3 + 1$

EXERCISES 13.3, page 561

1. $-\dfrac{x}{y}$ 3. $\dfrac{1 + 6x}{5y^4}$ 5. $-\dfrac{y}{x}$ 7. $\dfrac{y}{2x}$ 9. $\dfrac{1 - y}{x + 1}$ 11. $\dfrac{2x}{2y - 1}$ 13. $\frac{1}{2}$ 15. $-\frac{2}{15}$ 17. $\frac{3}{5}$ 19. $-\dfrac{x^3}{y^3}\dfrac{dx}{dt}$

21. $\dfrac{2x - y}{x}\dfrac{dx}{dt}$ 23. $\dfrac{2x + 2y}{3y^2 - 2x}\dfrac{dx}{dt}$ 25. \$331,250 per month   27. $5\frac{1}{3}$ thousand books per year

29. (a) $\dfrac{dr}{dt} = \dfrac{1}{4\pi r^2}\dfrac{dV}{dt}$ (b) $\dfrac{5}{4\pi}\,(\approx .40)$ cm/sec

CHAPTER 13: SUPPLEMENTARY EXERCISES, page 564

1. $\sqrt{x}(3x^2) + (x^3 + 1) \cdot \dfrac{1}{2\sqrt{x}}$ or $(7x^3 + 1)$ 3. $\dfrac{(\sqrt{x} + 1) \cdot \frac{1}{2}x^{-1/2} - x^{1/2}(\frac{1}{2}x^{-1/2})}{(\sqrt{x} + 1)^2}$ or $\dfrac{1}{2\sqrt{x}(\sqrt{x} + 1)^2}$

5. $30x^2(x^2 + 5x + 9)(x^3 - 2)^9 + (x^3 - 2)^{10}(2x + 5)$

7. $\dfrac{(x^2 + 2)[2(x + 2) - 2(x - 2)] - [(x + 2)^2 - (x - 2)^2](2x)}{(x^2 + 2)^2}$ or $\dfrac{16 - 8x^2}{(x^2 + 2)^2}$

9. $2[x + 2 + (x + 2)^2][1 + 2(x + 2)]$ or $(x^2 + 5x + 6)(10 + 4x)$ 11. $[g(x)]^2 + g(x) + 3,\, 2g(x) + 1$

13. $[g(x)]^{3/2} + [g(x)]^{1/2},\, \frac{3}{2}[g(x)]^{1/2} + \frac{1}{2}]g(x)]^{1/2}g'(x)$ 15. $[3g(x) - 1]^5,\, 15[3g(x) - 1]^4g'(x)$ 17. $-15[g(x)]^{-4}g'(x)$

19. $-\frac{3}{4}[4 - g(x)]^{-1/4}g'(x)$ 21. $-[g(x) + 1]^{-3/2}g'(x)$ 23. $\dfrac{3x^2}{x^6 + 1}$ 25. $\dfrac{2x}{(x^2 + 1)^2 + 1}$ 27. $\frac{1}{2}\sqrt{1 - x}$ 29. $\frac{3}{2}$

31. 1.89 m^2/yr   33. -3 35. $\frac{3}{5}$ 37. $-\frac{1}{25}$ 39. 200 dishwashers per month

CHAPTER 14

EXERCISES 14.1, page 571

1. $2^{2x}, 3^{(1/2)x}, 3^{-2x}$ **3.** $2^{2x}, 3^{3x}, 2^{-3x}$ **5.** $2^{-4x}, 2^{9x}, 3^{-2x}$ **7.** $2^{(1/2)x}, 3^{(4/3)x}$ **9.** $2^{-2x}, 3^x$ **11.** $3^{2x}, 2^{6x}, 3^{-x}$
13. $2^x, 3^x, 3^x$ **15.** (a) 8 (b) $\frac{1}{8}$ (c) 5.66 (d) 17.15 (e) 1.15 (f) 1.87 (g) .18 (h) .07 **17.** 1 **19.** 2 **21.** -1 **23.** $\frac{1}{5}$
25. $\frac{5}{2}$ **27.** -1 **29.** 4 **31.** 2^h **33.** $2^h - 1$ **35.** $3^x + 1$ **37.** $3^{5x} + 1$

EXERCISES 14.2, page 576

1. 1.1612, 1.105, 1.10 **3.** 1.005, 1.002, 1 **5.** $10e^{10x}$ **7.** $4e^{4x}$ **9.** $e^{2x} + e^{5x}$ **11.** e^{1+x} **13.** e^{2x} **15.** 7.3891

17. .60653 **19.** $x = 4$ **21.** $x = 4, -2$ **23.** $xe^x + e^x$ **25.** $\dfrac{e^x}{(1 + e^x)^2}$ **27.** $20e^x(1 + 5e^x)^3$

EXERCISES 14.3, page 582

1. $-e^{-x}$ **3.** $5e^x$ **5.** $2te^{t^2}$ **7.** $\dfrac{e^x + e^{-x}}{2}$ **9.** $-2(e^{-2x} + 1)$ **11.** $3(e^x + e^{-x})^2(e^x - e^{-x})$ **13.** $-\frac{2}{3}e^{3-2x}$ **15.** $3e^{3t}$

17. $\left(3x^2 + 1 + \dfrac{1}{x^2}\right)e^{x^3 + x - (1/x)}$ **19.** $4(2x + 1 - e^{2x+1})^3(2 - 2e^{2x+1})$ **21.** $3x^2e^{x^2} + 2x^4e^{x^2}$ **23.** $3xe^{x^3} - x^{-2}e^{x^3}$

25. $-xe^{-x+2}$ **27.** $\left(-\dfrac{1}{x^2} + \dfrac{1}{x} + 3\right)e^x$ **29.** $\dfrac{2e^x}{(e^x + 1)^2}$ **31.** Max at $x = 1$ **33.** Min at $x = \frac{11}{2}$ **35.** Max at $x = 3$
37. Min at $x = 1$; max at $x = 3$ **39.** Max at $x = -6$; min at $x = -5$ **41.** $.02e^{-2e^{-.01x}}e^{-.01x}$ **43.** $y = Ce^{-4x}$
45. $y = e^{-.5x}$ **49.**

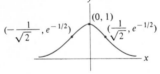

EXERCISES 14.4, page 588

1. -1 **3.** $-\ln 1.7$ **5.** $e^{2.2}$ **7.** 2 **9.** e **11.** 1 **13.** $\frac{1}{2}\ln 5$ **15.** $4 - e^{1/2}$ **17.** $\pm e^3$ **19.** $\dfrac{\ln(.5)}{-.00012}$ **21.** $\frac{3}{5}$ **23.** $\dfrac{e}{2}$

25. $3\ln\frac{9}{2}$ **27.** $5\ln 6$ **29.** $\frac{1}{5}\ln\frac{2}{5}$ **31.** $-\ln\frac{3}{2}$ **33.** $(-\ln 3, 3 - 3\ln 3)$, minimum **35.** $(\frac{1}{2}\ln\frac{3}{2}, \frac{1}{2})$, minimum
37.

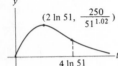

39. 109.947

EXERCISES 14.5, page 592

1. $\dfrac{1}{x}$ **3.** $\dfrac{1}{x + 5}$ **5.** $-\dfrac{\ln(x + 1)}{x^2} + \dfrac{1}{x(x + 1)}$ **7.** $\left(\dfrac{1}{x} + 1\right)e^{\ln x + x}$ **9.** $\dfrac{1}{x}$ **11.** $\dfrac{2\ln x}{x} + \dfrac{1}{x}$ **13.** $\dfrac{1}{x}$ **15.** $\dfrac{\ln x - 2}{(\ln x)^3}$

17. $2e^{2x}\ln x + \dfrac{e^{2x}}{x}$ **19.** $\dfrac{5e^{5x}}{e^{5x} + 1}$ **21.** $2(\ln 4)t$ **23.** $\dfrac{6\ln t - 3(\ln t)^2}{t^2}$ **25.** $y = 1$ **27.** $\left(e^2, \dfrac{2}{e}\right)$, yes

29.

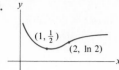

31. $\dfrac{1 + 3 \ln 10}{100}$

33. $P(x) = 300 \ln(x + 1) - 2x$ and $P'(149) = 0$. Since $P''(149) < 0$, the graph of $P(x)$ is concave down at $x = 149$. So $P(x)$ has a mximum point there.

35. $\dfrac{1}{e}$

CHAPTER 14: SUPPLEMENTARY EXERCISES, page 594

1. 81   **3.** $\frac{1}{25}$ **5.** 4   **7.** 9   **9.** e^{3x^2} **11.** e^{2x} **13.** $e^{11x} + 7e^x$ **15.** $x = 4$ **17.** $x = -5$ **19.** $70e^{7x}$ **21.** $e^{x^2} + 2x^2 e^{x^2}$
23. $e^x \cdot e^{e^x} = e^{x + e^x}$ **25.** $\dfrac{(2x - 1)(e^{3x} + 3) - 3e^{3x}(x^2 - x + 5)}{(e^{3x} + 3)^2}$ **27.** $y = Ce^{-t}$ **29.** $y = 2e^{1.5t}$

31.

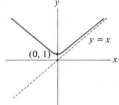

33.

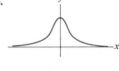

35. $5x$ **37.** 1   **39.** $\sqrt{5}$

41. $\frac{1}{2} \ln 5$ **43.** $e^{5/2}$ **45.** $2 + e^{8/5}$ **47.** $\dfrac{5}{5x - 7}$ **49.** $\dfrac{2 \ln x}{x}$ **51.** $\dfrac{6x^5 + 12x^3}{x^6 + 3x^4 + 1}$ **53.** $\dfrac{1}{x} e^{4x} + 4e^{4x} \ln x$ **55.** $\dfrac{1}{x \ln x}$
57. $\ln x$ **59.**

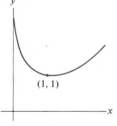

61.

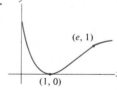

CHAPTER 15

EXERCISES 15.1, page 605

1. $P(t) = 400e^{.07t}$ **3.** (a) 5000 (b) $t = 6.9$ **5.** .017   **7.** (a) $P(t) = 4e^{.02t}$ (b) 6.7 billion (c) 1994   **9.** 71 min
11. (a) $P(t) = 30e^{-.08t}$ (b) 13.48 g   **13.** (a) $P(t) = 100e^{-.01t}$ (b) 74.082 g (c) 69 yr   **15.** (a) $-.13$ (b) 7.7105 g
17. 58.285%   **19.** 8990 yr   **21.** $f(t) = e^{-.081t}$ **23.** 184 yr   **25.** 20,022 yr   **27.** (a) $P(t) = 500e^{-.23t}$ (b) 10 months

EXERCISES 15.2, page 615

1. $1210   **3.** $10,000(1.02)^{12}$ **5.** $2316.37   **7.** $616.84   **9.** 11.45%   **11.** 39% interest compounded continuously
13. 8.9 yr   **15.** 1991   **17.** $446.26   **19.** 7.25%   **21.** $5488.10   **23.** 2.9%

EXERCISES 15.3, page 623

1. 20%, 4%   **3.** 30%, 30%   **5.** 60%, 300%   **7.** $-25\%, -10\%$ **9.** 12.5%   **11.** 5.8 yr   **13.** $p/(140 - p)$, elastic
15. $2p^2/(116 - p^2)$, inelastic   **17.** $p - 2$, elastic   **19.** (a) Inelastic (b) raised   **21.** (a) Inelastic (b) decrease
23. (a) 2 (b) yes

EXERCISES 15.4, page 635

1. (a) $f'(x) = 10e^{-2x} > 0$, $f(x)$ increasing; $f''(x) = -20e^{-2x} < 0$, $f(x)$ concave down

(b) As x becomes large, $e^{-2x} = \dfrac{1}{e^{2x}}$ approaches 0   (c)

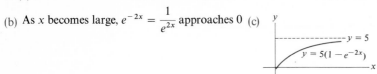

3. $y' = 2e^{-x} = 2 - (2 - 2e^{-x}) = 2 - y$
5. $y' = 30e^{-10x} = 30 - (30 - 30e^{-10x}) = 30 - 10y = 10(3 - y)$, $f(0) = 3(1 - 1) = 0$ **7.** 4.8 hr

CHAPTER 15: SUPPLEMENTARY EXERCISES, page 637

1. $29.92e^{-.2x}$ **3.** $5488.12   **5.** .058   **7.** $11.2e^{.024t}$ (b) 18.1 million (c) 1992
9. (a) $36,692.97 (b) the alternative investment is superior by $3859.   **11.** 400%   **13.** 3%, decrease   **15.** Increase
17. $100(1 - e^{-.083t})$

CHAPTER 16

EXERCISES 16.1, page 646

1. $\frac{1}{2}x^2 + C$ **3.** $\frac{1}{3}e^{3x} + C$ **5.** $3x + C$ **7.** $-\frac{1}{4}$ **9.** $\frac{2}{3}$ **11.** -2 **13.** $-\frac{5}{2}$ **15.** $\frac{1}{2}$ **17.** -1 **19.** 1   **21.** $\frac{1}{15}$
23. $\dfrac{x^3}{3} - \dfrac{x^2}{2} - x + C$ **25.** $4\sqrt{x} - 2x^{3/2} + C$ **27.** $4t + e^{-5t} + \dfrac{e^{2t}}{6} + C$ **29.** $\frac{2}{5}t^{5/2} + C$ **31.** C **33.** $\dfrac{x^2}{2} + 3$
35. $\frac{2}{3}x^{3/2} + x - \frac{28}{3}$ **37.** $2 \ln |x| + 2$ **39.** (a) $-16t^2 + 96t + 256$ (b) 8 sec (c) 400 ft
41. $P(t) = 60t + t^2 - \frac{1}{12}t^3$ **43.** $20 - 25e^{-.4t}\,°C$ **45.** $-95 + 1.3x + .03x^2 - .0006x^3$ **47.** $5875(e^{.016t} - 1)$

EXERCISES 16.2, page 654

1. 0   **3.** 5   **5.** 30   **7.** $\frac{4}{3}(1 - e^{-3})$ **9.** 14   **11.** $5(e - 1)$ **13.** $\ln 2$ **15.** $1\frac{7}{9}$ **17.** $\frac{1}{5}$ **19.** $3\frac{3}{4}$ **21.** $\frac{115}{6} + \ln \frac{7}{9}$
23. 323.2 km   **25.** 80.8   **27.** $1,185.75
29. The additional profit generated by raising the sales level from a to b units.   **31.** 10   **33.** $2(e^{1/2} - 1)$ **35.** $6\frac{3}{5}$
37. 15   **39.** 9   **43.** 10

EXERCISES 16.3, page 663

1. $\frac{64}{3}$ **3.** $\frac{52}{3}$ **5.** 20   **7.** 18   **9.** $\frac{32}{3}$ **11.** $\frac{32}{3}$ **13.** (a) $\frac{9}{2}$ (b) $\frac{19}{3}$ (c) $\frac{79}{6}$ **15.** $\frac{3}{2}$ **17.** $2 + 12 \ln(\frac{3}{2})$
19. 4.48 billion barrels saved

EXERCISES 16.4, page 675

1. .21875, .2421875, actual value .25   **3.** .23216, actual value .23254   **5.** 1.09286, actual value 1.09861   **7.** $20
9. $50   **11.** $200   **13.** $25   **15.** Intersection (100, 10), consumers' surplus = $100, producers' surplus = $250
17. $3236.68   **19.** $4.52P$ dollars   **21.** 3   **23.** 0   **25.** 8   **27.** 55 deg   **29.** ≈ 82 g
31. The sum is closely approximated by $\int_0^3 (3 - x)^2 \, dx$, and the value of this integral is 9.

33. The sum is closely approximated by $\int_0^1 (2x + x^3)\,dx$, and the value of this integral is $\frac{5}{4}$.

35. 100.08; they estimate the amount of oil consumed in 1976, 1977, 1978, and 1979.   **37.** $\frac{4}{3}\pi r^3$ **39.** $\dfrac{31\pi}{5}$

41. 8π **43.** $\dfrac{\pi}{2}\left(1 - \dfrac{1}{e^{2r}}\right)$

EXERCISES 16.5, page 685

1. $\frac{1}{6}(x^2 + 4)^6 + C$ **3.** $\frac{1}{4}(x^2 - 5x)^4 + C$ **5.** $e^{5x-3} + C$ **7.** $\ln|x^3 - 1| + C$ **9.** $\frac{2}{3}\sqrt{x^3 - 1} + C$ **11.** $-e^{(1/x)} + C$
13. $\frac{1}{3}\ln|x^3 + 3x + 2| + C$ **15.** $\frac{1}{11}(x^5 - 2x + 1)^{11} + C$ **17.** $\frac{3}{2}\ln|2x - 4| + C$ **19.** $-\frac{1}{2}e^{-x^2} + C$

21. $\frac{1}{5}xe^{5x} - \frac{1}{25}e^{5x} + C$ **23.** $\frac{x}{10}(2x + 1)^5 - \frac{1}{120}(2x + 1)^6 + C$ **25.** $\frac{2}{3}x(x + 1)^{3/2} - \frac{4}{15}(x + 1)^{5/2} + C$

27. $-3xe^{-x} - 3e^{-x} + C$ **29.** $x \ln x - x + C$ **31.** $\frac{1}{2}xe^{2x} - \frac{1}{4}e^{2x} + C$ **33.** $\frac{1}{2}e^{x^2} + C$
35. $\frac{2}{3}(2x - 1)(3x - 3)^{1/2} - \frac{8}{27}(3x - 3)^{3/2} + C$ **37.** $-\frac{1}{18}(6x^2 + 9x)^{-6} + C$ **39.** $2\sqrt{x}\ln x - 4\sqrt{x} + C$
41. $\ln|\ln 5x| + C$ **43.** $\$738.80$

EXERCISES 16.6, page 692

1. 0 **3.** No limit   **5.** $\frac{1}{4}$ **7.** 2 **9.** 5 **11.** 6 **13.** 2 **15.** $\frac{1}{2}$ **17.** 2 **19.** 1
21. Area under curve from 1 to b is $-4 + 4b^{1/4}$. This has no limit as $b \to \infty$.   **23.** $\frac{1}{2}$ **25.** Divergent   **27.** $\frac{1}{6}$
29. $\frac{2}{3}$ **31.** 1 **33.** $2e$ **35.** Divergent   **37.** $\frac{1}{2}$ **39.** 2 **41.** $\frac{1}{4}$ **43.** 2 **45.** $\frac{1}{3}$ **49.** $\$50,000$

EXERCISES 16.7, page 698

1. (a) $\frac{11}{32}$ (b) $\frac{7}{27}$ (c) $\frac{27}{32}$ (d) $\frac{27}{32}$ **3.** $\frac{3}{8}$ **5.** (a) $.18127$ (b) $.45085$ (c) $.13534$ (d) $.5$
7. (a) $f(x) = \frac{1}{3}, 0 \le x \le 3$ (b) $\frac{2}{3}$ (c) $\frac{1}{3}$ **9.** (a) $.25$ (b) 5000 acres   **11.** $.88$ **13.** (a) $.39347$ (b) $.36788$
15. $.22313$

CHAPTER 16: SUPPLEMENTARY EXERCISES, page 701

1. $-2e^{-x/2} + C$ **3.** $-\frac{1}{3}(x^3 - 3x + 2)^{-1} + C$ **5.** $\frac{1}{16}(2x + 3)^8 + C$ **7.** $\frac{1}{4}(e^x + 4)^4 + C$

9. $-2(2x + 1)e^{-x/2} - 8e^{-x/2} + C$ **11.** $\frac{1}{3}\ln|x^3 - 3x + 2| + C$ **13.** $\dfrac{x}{16}(2x + 3)^8 - \frac{1}{288}(2x + 3)^9 + C$

15. $\frac{1}{3}(x - 5)^3 - 7$ **17.** $\frac{3}{4}$ **19.** $\frac{1}{5} - \frac{1}{5}e^{-2}$ **21.** $\ln\left(\dfrac{e + e^{-1}}{2}\right)$ **23.** $\frac{76}{3}$ **25.** $\frac{125}{6}$ **27.** $\frac{15}{2}$ **29.** $\frac{40}{99}$, exact value: $.40547$

31. $\$433.33$ **33.** 15 **35.** $.02x^2 + 150x + 500$ dollars
39. (a) 68.5 quadrillion Btu (b) The improper integral represents the total reserves that will be used up during all years after $t = 0$ (1980). This should equal the reserves in 1980, assuming that gas is used until reserves are exhausted. In fact, $\int_0^\infty R(t)\,dt \approx 159$.
41. The total cubic centimeters of drug injected during the first 4 minutes   **43.** $\frac{1}{3}e^6$ **45.** $\frac{2}{25}$
49. (a) $e^{-1/3} \approx .72$ (b) $r(t) = e^{-t/72}$

CHAPTER 17

EXERCISES 17.1, page 711

1. $f(1, 0) = 1, f(0, 1) = 8, f(3, 2) = 25$ **3.** $f(0, 1) = 0, f(3, 12) = 18, f(a, b) = 3\sqrt{ab}$
5. $f(2, 3, 4) = -2, f(7, 46, 44) = \frac{7}{2}$
11. $\$50$. $\$50$ invested at 5% continuously compounded interest will yield $\$100$ in 13.8 years

13.

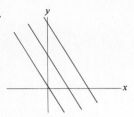

15. They correspond to the points having the same altitude above sea level.

EXERCISES 17.2, page 720

1. $5y, 5x$ **3.** $4xe^y, 2x^2e^y$ **5.** $-\dfrac{y^2}{x^2}, \dfrac{2y}{x}$ **7.** $4(2x - y + 5), -2(2x - y + 5)$ **9.** $(2xe^{3x} + 3x^2e^{3x})\ln y, x^2e^{3x}/y$

11. $\dfrac{2y}{(x + y)^2}, -\dfrac{2x}{(x + y)^2}$ **13.** $\dfrac{3\sqrt{K}}{2\sqrt{L}}$ **15.** $\dfrac{2xy}{z}, \dfrac{x^2}{z}, -\dfrac{1 + x^2y}{z^2}$ **17.** $ze^{yz}, xz^2e^{yz}, x(yz + 1)e^{yz}$ **19.** $1, 3$ **21.** -12

23. $\dfrac{\partial f}{\partial x} = 3x^2y + 2y^2, \dfrac{\partial^2 f}{\partial x^2} = 6xy, \dfrac{\partial f}{\partial y} = x^3 + 4xy, \dfrac{\partial^2 f}{\partial y^2} = 4x, \dfrac{\partial^2 f}{\partial y\,\partial x} = \dfrac{\partial^2 f}{\partial x\,\partial y} = 3x^2 + 4y$

25. (a) Marginal productivity of labor = 480; of capital = 40 (b) 240 fewer units produced
27. As the price of a bus ride increases, fewer people will ride the bus if the price of a train ticket remains constant. An increase in train ticket prices, coupled with constant bus fare, should cause more people to ride the bus.

29. $\dfrac{\partial f}{\partial r}, \dfrac{\partial f}{\partial m}, \dfrac{\partial f}{\partial s}$ **31.** $\dfrac{\partial V}{\partial P}(20,300) = -.06, \dfrac{\partial V}{\partial T} = .004$

33. $\dfrac{\partial^2 f}{\partial x^2} = -\tfrac{45}{4}x^{-5/4}y^{1/4}$. Marginal productivity of labor is decreasing.

EXERCISES 17.3, page 730

1. $(-2, 1)$ **3.** $(26, 11)$ **5.** $(1, -3), (-1, -3)$ **7.** $(\sqrt{5}, 1)(\sqrt{5}, -1); (-\sqrt{5}, 1); (-\sqrt{5}, -1)$ **9.** $(\tfrac{1}{3}, \tfrac{4}{3})$ **11.** $(0, 0)$ min
13. $(-1, -4)$ max   **15.** $(0, -1)$ min   **17.** $(-1, 2)$ max; $(1, 2)$ neither max nor min
19. $(\tfrac{1}{4}, 2)$ min; $(\tfrac{1}{4}, -2)$ neither max nor min   **21.** $(\tfrac{1}{2}, \tfrac{1}{6}, \tfrac{1}{2})$ **23.** 14 in. × 14 in. × 28 in.   **25.** $x = 120, y = 80$

EXERCISES 17.4, page 741

1. 58 at $x = 6, y = 2, \lambda = 12$ **3.** 13 at $x = 8, y = -3, \lambda = 13$ **5.** $x = \tfrac{1}{2}, y = 2$ **7.** $x = 2, y = 3, z = 1$

9. $x = 2, y = 2, z = 3$ **11.** $x = \dfrac{\sqrt{2}}{2}, y = \dfrac{\sqrt{2}}{2}$ **13.** $x = 4, y = 4, z = 2$ **15.** (a) $x = 81, y = 16$ (b) $\lambda = 3$

17. $x = 20, y = 60$

CHAPTER 17: SUPPLEMENTARY EXERCISES, page 744

1. $2, \tfrac{5}{6}, 0$ **3.** ≈ 20. Ten dollars increases to 20 dollars in 11.5 years.   **5.** $6x + y, x + 10y$ **7.** $\dfrac{1}{y}e^{x/y}, -\dfrac{x}{y^2}e^{x/y}$

9. $3x^2, -z^2, -2yz$ **11.** $6, 1$ **13.** $20x^3 - 12xy, 6y^2, -6x^2, -6x^2$
15. $-201, 5.5$. At the level $p = 25, t = 10,000$, an increase in price of \$1 will result in a loss in sales of approximately 201 calculators, and an increase in advertising of \$1 will result in the sales of approximately 5.5 additional calculators.
17. $(3, 2)$ **19.** $(0, 1), (-2, 1)$ **21.** Min at $(2, 3)$ **23.** Min at $(1, 4)$; neither max nor min at $(-1, 4)$
25. $20; x = 3, y = -1, \lambda = 8$ **27.** $x = \tfrac{1}{2}, y = \tfrac{3}{2}, z = 2$ **29.** $x = 10, y = 20$

Index